The Impact of Climatic Variations on Agriculture

VOLUME 2: ASSESSMENTS IN SEMI-ARID REGIONS

THE INTERNATIONAL INSTITUTE FOR APPLIED SYSTEMS ANALYSIS

is a nongovernmental research institution, bringing together scientists from around the world to work on problems of common concern. Situated in Laxenburg, Austria, IIASA was founded in October 1972 by the academies of science and equivalent organizations of twelve countries. Its founders gave IIASA a unique position outside national, disciplinary, and institutional boundaries so that it might take the broadest possible view in pursuing its objectives:

To promote international cooperation in solving problems from social, economic, technological, and environmental change

To create a network of institutions in the national member organization countries and elsewhere for joint scientific research

To develop and formalize systems analysis and the sciences contributing to it, and promote the use of analytical techniques needed to evaluate and address complex problems

To inform policy advisors and decision makers about the potential application of the Institute's work to such problems

The Institute now has national member organizations in the following countries:

Austria
The Austrian Academy of Sciences

Bulgaria
The National Committee for Applied Systems Analysis and Management

Canada
The Canadian Committee for IIASA

Czechoslovakia
The Committee for IIASA of the Czechoslovak Socialist Republic

Finland
The Finnish Committee for IIASA

France
The French Association for the Development of Systems Analysis

German Democratic Republic
The Academy of Sciences of the German Democratic Republic

Federal Republic of Germany
Association for the Advancement of IIASA

Hungary
The Hungarian Committee for Applied Systems Analysis

Italy
The National Research Council

Japan
The Japan Committee for IIASA

Netherlands
The Foundation IIASA–Netherlands

Poland
The Polish Academy of Sciences

Sweden
The Swedish Council for Planning and Coordination of Research

Union of Soviet Socialist Republics
The Academy of Sciences of the Union of Soviet Socialist Republics

United States of America
The American Academy of Arts and Sciences

The Impact of Climatic Variations on Agriculture

VOLUME 2: ASSESSMENTS IN SEMI-ARID REGIONS

Edited by

M. L. Parry

T. R. Carter

Department of Geography, The University of Birmingham, U.K. and IIASA, Laxenburg, Austria

and

N. T. Konijn

Department of Soil Science, Agricultural University, Wageningen, The Netherlands

KLUWER ACADEMIC PUBLISHERS
DORDRECHT / BOSTON / LONDON

INTERNATIONAL INSTITUTE FOR APPLIED SYSTEMS ANALYSIS
UNITED NATIONS ENVIRONMENT PROGRAMME

Library of Congress Cataloging in Publication Data

The Impact of climatic variations on agriculture.

On v. 1 t.p.: The International Institute for Applied Systems Analysis, United Nations Environment Program.
Includes index.
Contents: v. 1. Assessments in cool temperate and cold regions.
1. Crops and climate. 2. Climatic changes. 3. Meteorology, Agricultural. I. Parry, M. L. (Martin L.) II. Carter, Timothy (Timothy R.) III. Konijn, Nicolaas. IV. International Institute for Applied Systems Analysis. V. United Nations Environment Programme.
S600.7.C54I46 1988 630'.2'516 88-3146

ISBN 90-277-2719-8 (v. 2)
ISBN 90-277-2720-1 (v. 2 pbk.)

Published by Kluwer Academic Publishers,
P.O. Box 17, 3300 AA Dordrecht, The Netherlands.

Kluwer Academic Publishers incorporates
the publishing programmes of
D. Reidel, Martinus Nijhoff, Dr W. Junk and MTP Press.

Sold and distributed in the U.S.A. and Canada
by Kluwer Academic Publishers,
101 Philip Drive, Norwell, MA 02061, U.S.A.

In all other countries, sold and distributed
by Kluwer Academic Publishers,
P.O. Box 322, 3300 AH Dordrecht, The Netherlands.

Printed in The Netherlands

Preface

Three important studies were initiated in the 1970s to investigate the relationship between climatic variations and agriculture: by the National Defense University (1980) on *Crop Yields and Climate Change to the Year 2000*, by the U.S. Department of Transportation (1975) on *Impacts of Climatic Change on the Biosphere* and by the U.S. Department of Energy (1980) on *Environmental and Societal Consequences of a Possible CO_2-Induced Climatic Change* (the CIAP study). These were pioneering projects in a young field. Their emphasis was on measuring likely impacts of climatic variations rather than on evaluating possible responses, and they focused on first-order impacts (e.g., on crop yields) rather than on higher-order effects on society.

A logical next step was to look at higher-order effects and potential responses, as part of a more integrated approach to impact assessment. This was undertaken by the World Climate Impact Program (WCIP), which is directed by the United Nations Environment Program (UNEP). The WCIP is one of four aspects of the World Climate Program, which was initiated in 1979.

At a meeting in 1982, the Scientific Advisory Committee of WCIP accepted, in broad terms, a proposal from the International Institute for Applied Systems Analysis (IIASA) for an integrated climate impact assessment, with the proviso that the emphasis be on impacts in the agricultural sector. Martin Parry was asked to design and direct the project at IIASA. Funding was provided by UNEP, IIASA and the Austrian Government. The United Nations University provided additional funds for the case study in Ecuador.

This volume, and its companion volume, are the products of that project. Together they represent the work of 76 authors from 50 different scientific institutes in 17 countries. Yet they are not merely edited collections of chapters because each reports results from a *common set of impact assessments* that were designed, conducted and reported in a compatible manner by our collaborating scientists.

Discussion of the design of the assessments was initiated in September 1983 under the guidance of a steering committee, chaired by C.S. Holling, whose members were B. Bolin, W.C. Clark, M.L. Parry and P.E.O. Usher. Of particular interest were some of the ideas that had emerged that month at a meeting in Villach (Austria) on *The Sensitivity of Ecosystems and Society to Climatic Change.*

From October 1983 a core staff of three IIASA scientists (Martin Parry, Timothy Carter and Nicolaas Konijn) set about selecting case study regions (11 in all) with teams of 3 to 6 collaborating scientists in each case study. The teams came together at IIASA on two occasions, first to agree upon the details of the project design (to select impact models, develop climatic scenarios, etc.) and the second time, about six months later, to compare results and agree upon details of presentation. The "cool temperate and cold" teams (for Volume 1) met in April and October 1984, and the "semi-arid" teams (for Volume 2) met in October 1984 and April 1985. A mid-term review of the project was conducted by G.A. McKay following the October 1984 meetings.

The case study reports were submitted to IIASA over the next 18 months, and each received an international, interdisciplinary review. Individual sections of the case studies were also reviewed by specialists in appropriate fields. The reviewers for Volume 1, totaling 60 from 12 countries, were: S. Adalsteinsson, M. Andreasson, D.W. Anthony, L.M. Arthur, R.E. Bement, M. Black, M.R. Bryck, M.C. Bursa, M.G.R. Cannell, A.H.C. Carter, A.J.W. Catchpole, M. Cave, R.S. Chen, W.C. Clark, G.J.F. Copeman, M.D. Dennett, P. Driessen, S. Dunlop, J. Eadie, A. Eddy, W.R. Emanuel, K. Frohberg, H. Fukui, C.F. Green, D.S. Grundy, J. Hansen, M.N. Hough, T. Iwakiri, S.R. Johnson, R.W. Katz, W.W. Kellogg, H. van Keulen, C. van Kooten, S. Lebedeff, S. Le Duc, L. Lyles, A.J. McIntyre, H. Mansikkaniemi, T. Mela, D. Mitlin, R.E. Munn, P. Newbould, A.E.J. Ogilvie, M.S. Philip, L. Pölkki, C. Price, W.E. Riebsame, I. Saito, C.M. Sakamoto, G. Sirén, E. Siuruainen, P. Smithson, A. Smyshlaev, Y. Sugihara, J. Tarrant, J.H.M. Thornley, M. Torvella, C.C. Wallén, R.A. Warrick, and T.M.L. Wigley.

The reviewers for Volume 2 (altogether 53 from 16 countries) were: F.O. Agumba, R. Ananthakrishnan, G.F. Arkin, J.A.A. Berkhout, E. Berry, R. Bromley, I. Burton, P. Cooper, B. Currey, J.C. Dickinson, T.E. Downing, W.R. Emanuel, G. Fischer, P.A. Furley, B.E. Grandin, S. Gregory, R. Gwynne, M. Hamilton, G.H. Hargreaves, S. Hastenrath, R.L. Heathcote, C. Henfrey, B.G. Hunt, D. Jha, R.W. Katz, J.R. Kiniry, G. Knapp, V.E. Kousky, R. Lal, P. Legris, H.N. Le Houérou, M. Lipton, D.M. Liverman, J.G. Lockwood, L.O. Mearns, V.V.N. Murty, M. Nirenberg, L.J. Ogallo, J.P. Palutikof, P.W. Porter, D.A. Preston, G.W. Robertson, C. Sage, R.P. Sarker, P.L. Scandizzo, J. Shukla, K. Swindell, M. Thompson, W.I. Torry, J. Townsend, R.L. Vanderlip, F.J. Wang'ati, and A. Wood.

The reviews were then dispatched to authors, together with additional comments and general guidelines from the three volume editors, for final revisions of the case study reports.

Advice and encouragement were given by our scientific colleagues at IIASA (particularly Bill Clark, "Buzz" Holling, Jag Maini and Ted Munn) and by visitors to our project (particularly Bob Chen, Howard Ferguson, Michael Glantz, Bob Kates, Diana Liverman, Gordon McKay, Bob Watts and Dan Williams);

and Peter Usher, our program officer from UNEP, was there with the right kind of support when it was needed.

At Birmingham University and at IIASA Martin Parry and Tim Carter edited the reports into more coherent volumes. Dan Williams provided valuable editorial assistance and Marilyn Brandl, Lourdes Cornelio and Yasmin Taher keyboarded and corrected the text. Tim Grogan drew the figures. Bob Duis and his publications team at IIASA, in conjunction with David Larner at Reidel, did the final production work. Marilyn, Lourdes and Yasmin, secretaries to the Climate Project at IIASA, deserve our special gratitude for their efficiency and dedication.

In all, more than two hundred scientists and professional support staff have worked toward the publication of these two volumes. We thank them all and hope that they can take some satisfaction from the result.

Laxenburg, Austria
November, 1987

Martin L. Parry
Timothy R. Carter
Nicolaas T. Konijn

Contents

Part V. The Effects of Climatic Variations on Agriculture in Dry Tropical Regions of India

S. Gadgil, A.K.S. Huda, N.S. Jodha, R.P. Singh and S.M. Virmani

Part VI. The Effects of Climatic Variations on Agriculture in the Semi-arid Zone of the European USSR

A. The Stavropol Territory

A.A. Nikonov, L.N. Petrova, H.M. Stolyarova, V.Yu. Lebedev, S.O. Siptits, N.N. Milyutin and N.T. Konijn

PART I

The Impact of Climatic Variations on Agriculture: Introduction to the IIASA/UNEP Case Studies in Semi-Arid Regions

Contents, Part I

Contents, Part I

List of Contributors

PARRY, Dr. Martin L.
Atmospheric Impacts Research Group
Department of Geography
University of Birmingham
PO Box 363
Birmingham B15 2TT
United Kingdom

and

International Institute for
Applied Systems Analysis
Schlossplatz 1
A-2361 Laxenburg
Austria

CARTER, Mr. Timothy R.
Atmospheric Impacts Research Group
Department of Geography
University of Birmingham
PO Box 363
Birmingham B15 2TT
United Kingdom

and

International Institute for
Applied Systems Analysis
Schlossplatz 1
A-2361 Laxenburg
Austria

KONIJN, Dr. Nicolaas T.
Department of Soil Science
Agricultural University
PO Box 37
NL 6700 AA Wageningen
The Netherlands

LOCKWOOD, Dr. John G.
School of Geography
University of Leeds
Leeds LS2 9JT
United Kingdom

Abstracts of Part I

Section 1

This section outlines the aims and methods of the IIASA/UNEP project on the assessment of the effects of climatic variations on agriculture. It summarizes the results from case studies in semi-arid regions, details of which are presented elsewhere in this, the second of two volumes. The first volume reported the results of studies in cool temperate and cold regions in the northern hemisphere.

An overall goal of the project was to increase our understanding not only of *first-order effects* of climatic variations on agricultural productivity, but also of ***higher-order effects*** on local, regional and national economies. The first part of this section outlines some approaches for considering these interactions. Two broad types of experiment are described: *impact experiments* and *adjustment experiments*. The former provide estimates of the first-order effects of climatic variations on farming systems that are assumed to include no adjustments. The latter evaluate a number of adjustments available at the farm- or policy-level to offset or mitigate these effects. Experiments were conducted to examine the impacts of three common types of climatic scenario, each representing a short-term, drought-related climatic anomaly taken from the historical instrumental record.

The results of the case study experiments (in central and eastern Kenya, Northeast Brazil, the central Sierra of Ecuador, dry tropical regions of India, the Stavropol and Saratov regions of the southern European USSR, and the Australian wheatbelt) are summarized in the second part of the section. In spite of the large contrasts between the different regions, some general results can be noted.

Crop yields in all regions are very sensitive to seasonal rainfall amount, many exhibiting a positive linear or near-linear relationship over a typical range of climatic conditions. Moreover, this relationship appears to hold for crops grown under high as well as low levels of farm inputs (i.e., management and technology). Within-season variations in rainfall also contribute to crop yield variability, having a marked effect, in addition, on the quality of forage and hence on livestock nutrition.

Concurrent effects of climatic variations on fertilizer responsiveness, on soil moisture status and on longer-term soil nutrient status, which complicate the direct climate–yield relationship are examined, and experiments with adjustments to these (e.g., altered amounts and timing of irrigation or fertilizer application, fallowing practices, planting techniques, etc.) are reported that evaluate the potential for increasing or stabilizing crop productivity.

Higher-order effects of drought events are also examined, at the farm level (by means of questionnaire surveys in some subsistence economies), at the regional level (in terms of indicators such as gross production, farm income, regional GDP, employment and migration in some more developed regions) and at the national level (in cases where regional drought effects can be traced through to effects on the national economy). A range of adjustments to cope with drought effects are described, both at the farm level (including *ex ante* adjustments to optimize production and *ex post* adjustments to minimize losses) and at the public policy level (including food supply measures, ecological and infrastructural interventions and adjustments in land allocation).

Finally, some gaps in knowledge are identified and a number of priorities for future research are outlined.

Section 2

First-order impact models are the tools that are used throughout this volume to provide estimates of crop and livestock productivity under a range of climatic conditions (scenarios). This section provides a general introduction to the types of problems faced in modeling agricultural production in semi-arid regions, and summarizes the techniques employed in the case studies in this volume.

Apart from the obvious water limitation in semi-arid regions, models developed for less developed countries (in particular) must be able, in addition, to account for levels of regional productivity that represent only a small fraction of the potential, due to low levels of available technology. However, *in situ* model development is frequently hampered by inadequate or unreliable regional data bases. One solution is to introduce models from another region, but some local data are still required both for recalibrating and for running such models.

While the problems of modeling crop growth and yield are complex, the development of livestock production models poses a still more formidable challenge, for these often involve a two-stage modeling procedure: the climate–vegetation and vegetation–animal stages. Moreover, the quality of vegetation is as important as the quantity for livestock productivity, and both are closely related to climatic variations.

Section 3

In this section the climates of the study areas in Volume 2 are related to each other and to atmospheric circulation patterns within the tropics. The main features of tropical circulation patterns are briefly described. Tropical circulation patterns are particularly influenced by heat inputs from such sources as warm ocean surfaces acting through latent heat released in deep cumulus convection. Other tropical heat sources, such as high plateaus and equatorial rain forest, are also important. Strong atmosphere–ocean coupling is found in low latitudes, in contrast to the situation in temperate latitudes where it is weak. Tropical atmosphere–ocean coupling is observed in the "El Niño/Southern Oscillation phenomenon", which is a dominant cause of short-term variation in the tropics.

Rainfall variations and the incidence of drought in many tropical semi-arid regions are correlated with changes in sea-surface temperatures. Droughts can be caused to persist by increased surface albedo due to reduced vegetation, dry soils limiting local evaporation, and soil-derived dust increasing atmospheric stability. Climatic changes due to increasing atmospheric CO_2 content are unlikely to be of major importance in tropical semi-arid regions during the next 10–20 years, but will become so sometime in the next century. Reliable seasonal rainfall forecasts for tropical semi-arid regions may become available in the near future, and when such forecasts are available it will be necessary to consider carefully the sorts of responses in agriculture and other climate-sensitive activities that are likely to be most appropriate.

SECTION 1

The Assessment of Effects of Climatic Variations on Agriculture: A Summary of Results from Studies in Semi-Arid Regions

Martin Parry and Timothy Carter

1.1. Introduction

In 1982 and 1983 there occurred the most pronounced El Niño event ever recorded. In Northeast Brazil the total gross product of agriculture in 1983 was almost 16% below normal. In this drought-prone region 1983 was the fifth successive year of below-average rainfall, which contributed to the progressive worsening of living conditions of about 12 million *nordestinos*. Although the Brazilian government spent the equivalent of US $ 1 778 m in emergency programs over 1979–1983, assisting about 2.5 million people with employment in public works, there was widespread hardship and extensive migration both from farms to towns in the region and out of the region itself.

In the same year (1983) there were extensive floods in Ecuador and Peru, persistent droughts in southern and eastern Africa and in north-central India, and widespread forest fires in Australia. Wheat production in Australia in 1982–1983 was only 63% of the previous five-year average and would have been even lower if Western Australia had shared the drought conditions of the eastern states (Gibbs, 1984). In the US cornbelt abnormally low rainfall reduced average US maize (corn) yields by 29% (relative to 1982) and this, combined with reduced planting due to government policy, led to a fall in total US maize production of 50% (Parry *et al.*, 1985).

Only some of these short-term events have been connected with the El Niño/Southern Oscillation phenomenon, a periodic fluctuation in the intensity of the intertropical atmospheric and oceanic circulation (Southern Oscillation) that is usually coincident with an anomalous warming of the eastern tropical Pacific

Ocean (El Niño). In all cases, however, they can cause substantial decreases in productivity, large aggregate falls in production and, in extreme though not infrequent instances, starvation and death.

In the longer term it is likely that changes in climate will result from increasing concentrations of carbon dioxide (CO_2) and other radiatively active gases in the atmosphere. Present assessments indicate that increases in global mean annual temperatures of between 1.5 and 5.5°C are likely to occur as a result of increases in CO_2, probably between 2050 and 2100 (Bolin *et al.*, 1986). Such changes are likely to affect the frequency of short-term extreme events, for example, through changes in rainfall regime in some semi-arid areas.

Thus both short- and long-term variations of climate will probably continue to be important factors affecting agriculture in the future. The present volume is the second of two concerned with the estimation of such effects and with the evaluation of appropriate responses to them.

1.2. Aims

This volume considers, firstly, the range of effects that short-term variations of climate may have on agriculture in semi-arid regions of the world and, secondly, the range of adjustments available to mitigate or exploit these effects. The core of the volume is six case studies of effects on agriculture in different semi-arid regions. A companion volume reports results of similar case studies in cool temperate and cold regions (Parry *et al.*, 1988).

The two volumes are the outcome of a research project based at the International Institute for Applied Systems Analysis (IIASA) and funded jointly by IIASA and the United Nations Environment Programme (UNEP) as part of the World Climate Impact Programme (WCIP). The WCIP is one of four components of the World Climate Programme (WCP), which was initiated by the World Meteorological Organization (WMO) in 1979.

The purpose of the project was twofold:

(1) To investigate the effects of climatic change and variability* on agriculture.
(2) To evaluate alternative responses to these effects.

An important premise behind the study was that, while we are not yet in a position to forecast how the climate may change (either in the short or long term), we *can* estimate the potential consequences of each of a number of *possible* climatic changes. By considering the range of impacts from and adjustments to

*Throughout the volume reference is made to three forms of climatic perturbation: climatic variability, climatic change and climatic variations. Following Hare (1985), *climatic variability* describes the observed year-to-year differences in values of specific climatic variables *within* an averaging period (typically 30 years). *Climatic change* describes longer-term changes *between* averaging periods either in the mean values of climatic variables or in their variability. These two descriptors form subsets of the third collective term, *climatic variations*, which embraces short-, medium- and longer-term changes with time in values of climatic variables.

these possible events, we can improve both our techniques of impact analysis and our armory of potential responses. Thus, at some point in the future, when we *are* able to make reasonably accurate forecasts of climatic change, we shall also have acquired an ability both to assess their impact and to respond effectively to them.

The focus of this volume is almost exclusively on impacts from *short-term climatic variations*, especially droughts, whereas in Volume 1 there is greater emphasis on the effects of likely long-term CO_2-induced climatic change. There are two main reasons for this contrast in emphasis:

(1) In less developed countries, especially in semi-arid regions, the economic effects of climatic extremes are generally more severe than in cool temperate and cold regions, where highly developed agricultural systems are better able to withstand such events. The problem of short-term climatic variability, which is immediate and pressing in these regions, thus figures more prominently in the political arena than does the problem of long-term climatic change.

(2) At lower latitudes, where most of the semi-arid case studies in this volume are located, general circulation model (GCM)-based projections of future climate suggest that changes in air temperature brought about by increased atmospheric CO_2 concentrations are likely to be significantly smaller than those at middle and high latitudes. Moreover, the uncertainty attached to GCM estimates of changes in long-term regional precipitation (of obvious importance in semi-arid regions) is very much greater than that associated with changes in air temperature. These two features of GCM outputs were reasons for *not* including a GCM-based scenario of long-term climatic change in the present study.

1.3. Alternative Approaches to the Problem*

The past 15 years have seen encouraging developments in the method of climate impact assessment. With the benefit of hindsight it seems that our efforts up to the mid-1970s focused mainly on the (one-way) *impact* of climate on human activity. This has recently been replaced, and it is hoped improved upon, by a greater emphasis on the *interaction* between climate and human activity.

1.3.1. The emphasis on impacts

The *impact approach* is based upon the assumption of direct cause and effect where a climatic event (for example, a short-term variation of rainfall) operating on a given "exposure unit" (for example, a human activity) may have an impact

*This subsection is broadly similar to one in Volume 1 on cool temperate and cold regions. Readers familiar with it are advised to turn directly to Subsection 1.4.

or effect [*Figure 1.1*(*a*)]. This approach to impact assessment frequently sought relatively simple statistical relationships between climate and agriculture, often as regression equations between weather and yield, and giving little attention to an understanding of the nature of the relationships. Examples of such assessments are to be found in the National Defense University's study of possible effects of climatic changes on crop yields (NDU, 1980) and in the US Department of Transportation's Climate Impact Assessment Program (CIAP, 1975). Here the emphasis was on the search for relatively simple connections between a climatic variation on the one hand and a possible response on the other. In reality, of course, so many intervening factors operate that it is both misleading and quite impossible to treat these three study elements (climatic variation–exposure unit–impact) in isolation from their environmental and societal *milieu*, and very few studies have followed this method with success.

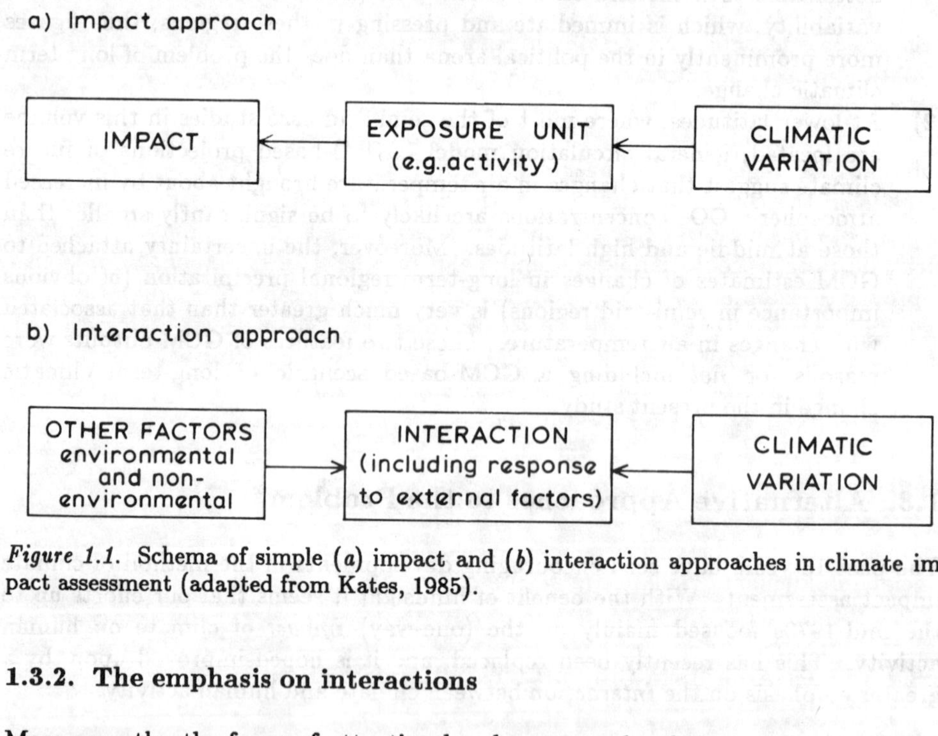

Figure 1.1. Schema of simple (*a*) impact and (*b*) interaction approaches in climate impact assessment (adapted from Kates, 1985).

1.3.2. The emphasis on interactions

More recently, the focus of attention has been to seek a better understanding of the interactions between climate and human activity by assuming that a climatic event is merely one of many processes (both societal and environmental in origin) that may affect the exposure unit [*Figure 1.1*(*b*)]. To illustrate, it may be argued that the magnitude of the effect of the 1930s drought on the Canadian prairies was substantially increased by the depression of farm prices and the desperate economic straits that had been reached by many prairie farmers before the beginning of the drought. In addition, widespread plowing of soils prone to wind erosion increased the amount of windblown dust that choked crops and reduced yields. Economics, weather and farming technology thus interacted to create a severe economic and social impact that was perhaps preconditioned by the Depression but triggered by drought.

The *interaction approach* was adopted by the project on *Drought and Man* of the International Federation of Institutes for Advanced Study (IFIAS) (Garcia, 1981). Here the emphasis was on comprehending the syndrome of political and economic as well as meteorological events that resulted in the severe effects of drought in the Sahel in the mid-1970s. An important conclusion of this and other studies is that, while meteorological drought may have triggered the Sahelian crisis, an underlying cause of the crisis was probably the decreasing resilience of Sahelian society to drought as a result of political, social and economic developments during the 1950s and 1960s.

1.3.3. Orders of interactions

It is possible to achieve a greater realism in simulating the connection between climate and human activity by considering the cascade of interactions that can occur from the first-order biophysical level, through second-order levels characterized by units of enterprise (farms, corporations, etc.) to third-order interactions at the regional and national levels (*Figure 1.2*). This approach was adopted in the companion to the present volume, which dealt with the possible effects both of long-term changes and short-term variations of climate on agricultural output in cool temperate and cold regions (Parry *et al.*, 1988). An appropriate example of the approach is the mixed hierarchy of simulation and regression models that was used in the Saskatchewan case study. Altered levels of spring wheat yield (estimated using a simulation model for several climatic scenarios) were used as inputs to farm simulation models, converted to production figures and aggregated by soil zone and farm size to give provincial production and commodity changes. These were used, in turn, as inputs to a regional input–output model to examine impacts both on the agricultural and the nonagricultural sectors. A third (employment) model was used to translate changes in output for various economic sectors into changes in sectoral and provincial employment (Williams *et al.*, 1988).

1.3.4. The search for a fully integrated approach

Additional complexity can be introduced by studying interactions of the same order, both *within* individual sectors (such as between different farming systems) and *between* different sectors (such as between the concurrent effects of a change in climate on agriculture, forestry, water resources, etc.), *and* the feedback effects operating between them (*Figure 1.3*). This form of fully integrated assessment has yet to be successfully implemented, largely because the full complement of systems models and our ability to connect them satisfactorily do not yet exist. However, some of the studies in this volume (such as that of the 1979 and 1983 droughts in Northeast Brazil) are partially integrated assessments in that they include investigations of impacts in several sectors and their links between them.

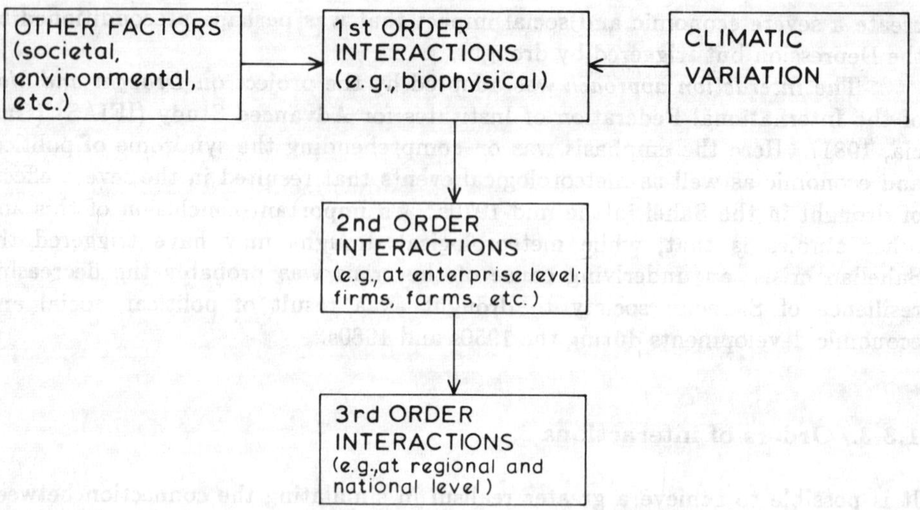

Figure 1.2. **Schema of interaction approach with ordered interactions.**

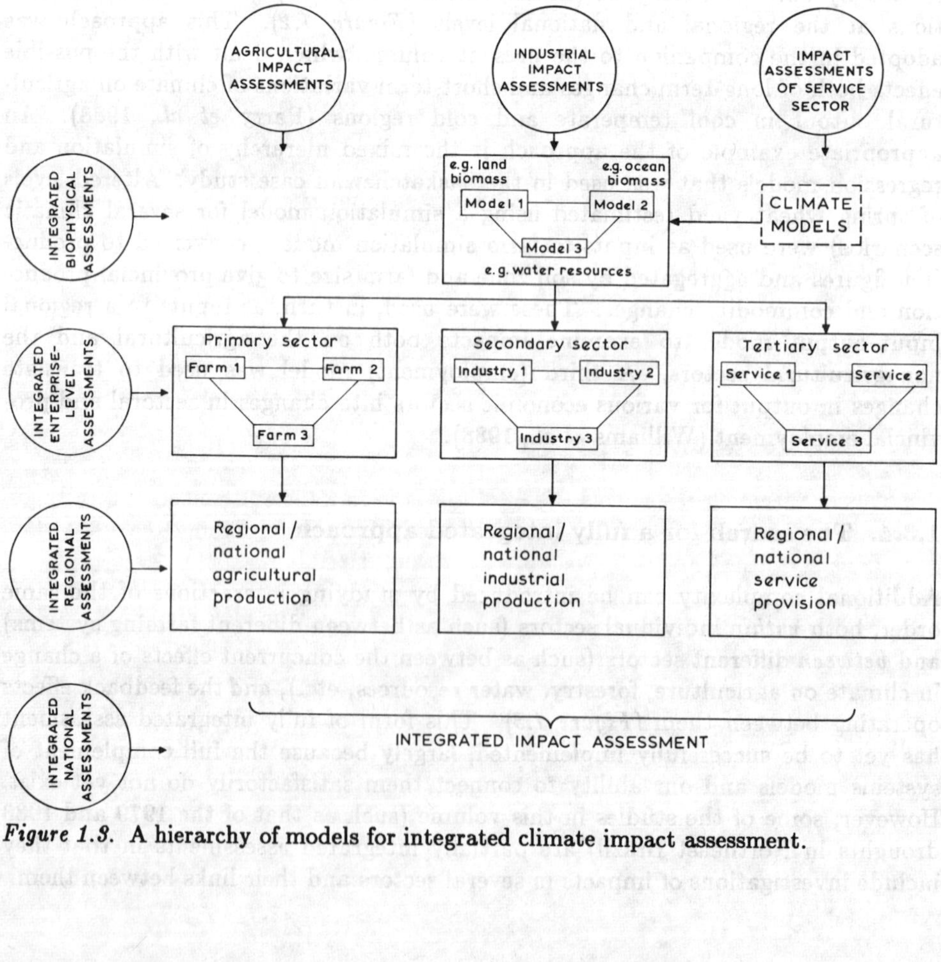

Figure 1.3. **A hierarchy of models for integrated climate impact assessment.**

1.4. An Outline of the IIASA/UNEP Approach

The approach adopted in the IIASA/UNEP project has three general characteristics. Firstly, it attempts a partially integrated assessment of impacts. Secondly, it adopts both a direct and an adjoint research method. Thirdly, it reports two types of experiment: impact and adjustment experiments.

1.4.1. A partially integrated approach

Three elements in the IIASA/UNEP project characterized its integrated approach:

Firstly, the project sought to assess the effect of climatic variations on agriculture by using a *hierarchy of models* that included the following:

- Models of climatic variation (based on analysis of the instrumental record, on outputs from global circulation models, or a combination of these).
- Biophysical models of first-order relationships, that is, those between certain climatic variables (e.g., temperature, precipitation, insolation, windspeed, etc) and biophysical supply or demand (e.g., biomass productivity, nutrition requirements, etc.). In particular these considered the effects on crop growth, livestock production, livestock health, etc.
- Economic models of second-order relationships, that is, those at the enterprise level (e.g., of farms, firms, agencies, institutions, etc.). These considered the effects (*inter alia*) of changes in farm-level production on farm profitability, farm purchases, etc.
- Economic models of higher-order relationships, for example, those between farm profitability and regional employment or gross domestic product.

Secondly, it was considered important to consider the effects of climatic variations and their interactions with other physical systems, distinguishing between:

- Those in which the effects of the climatic variation are transmitted through other physical systems (e.g., by pests and diseases; by changes in soil structure, soil nutrients, soil erosion, salinization, etc.).
- Those in which the effects of the climatic variation are themselves affected by other concurrent environmental trends (such as acid deposition, groundwater depletion, etc.).

Thirdly, the project considered two types of response to climate impacts:

- Adjustments at the enterprise level (which at the farm level might include changes of crops, increased irrigation, changes in fertilization, etc.).
- Policy responses at the regional, national and international level.

Figure 1.4 gives a schema of the project's study method, which incorporates the three elements outlined above.

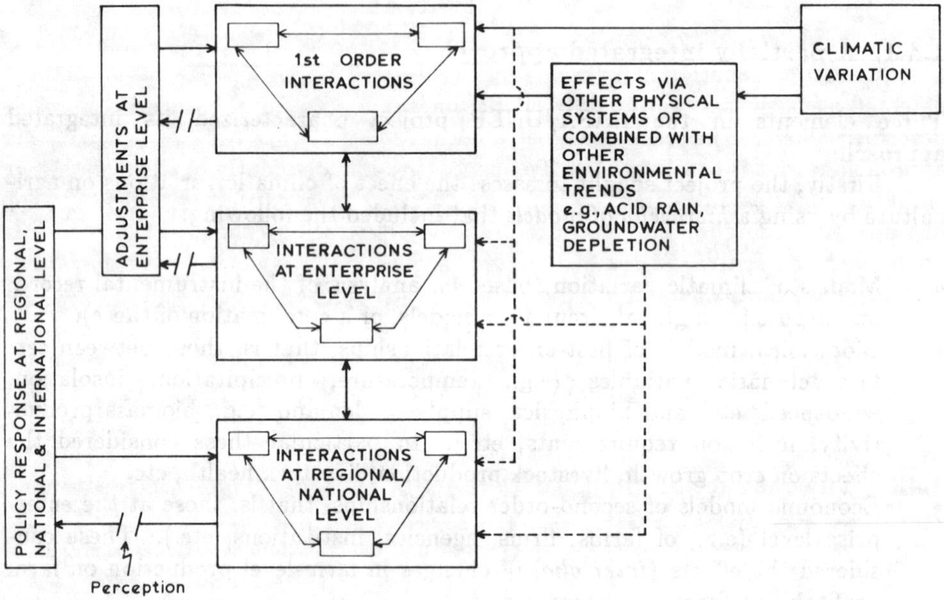

Figure 1.4. Schema of the IIASA/UNEP project's approach: an interactive approach to climate impact assessment with ordered interactions, interactions at each level, and some social and physical feedbacks.

1.4.2. The use of direct and adjoint methods

The scientific method most frequently adopted in climate impact assessment is the *direct method* in which, for example, the effects of a change in an input variable (such as a change in rainfall) are traced, in a number of steps, along several pathways (e.g., rainfall → crop biomass productivity → forage level → carrying capacity → livestock production → meat and milk supply, etc.) [*Figure 1.5(a)*]. The question is thus posed: For a given variation in climate, what is the effect on, for example, ecosystems, economy and society? An advantage of this approach is that assessments can be made even if the number of climatic scenarios is restricted (for example, in the case of projected possible future climates). The analysis is conducted on the basis of the character of the climatic changes rather than their likely impacts.

An alternative or *adjoint* method (Parry and Carter, 1984) focuses first on the sensitivity of the exposure unit and addresses the questions:

- To what aspects of climate is the exposure unit especially sensitive?
- What changes in these aspects are required to perturb the exposure unit significantly?

Climatic scenarios can then be characterized partly on the basis of these detected sensitivities and partly along lines adopted in the direct approach described above. The steps are illustrated in *Figure 1.5(b)*. An advantage of this approach is that it can help identify sensitivities independently of state-of-the-art climatic scenarios and can allow climatic changes to be expressed in a form that has direct meaning for the exposure unit. To illustrate, in the case study in this volume on the central Sierra of Ecuador, a questionnaire survey of farmers was first used to identify points of vulnerability in the local farming system (e.g., the pronounced effects of increased frost risk on loss of potato crops at high elevations). Possible variations of climate were then expressed, *inter alia*, as changes in frost risk in areas of potato cultivation (*see* Part IV).

1.4.3. The conduct of impact experiments and adjustment experiments

Two broad types of experiment were conducted (cf. *Figure 1.7*). Firstly, in a number of *impact experiments* each set of models in the hierarchy simulated a limited number of feedbacks within its own subsystems. The form of analysis was therefore essentially sequential, estimates of effects being based largely on assumed and essentially static sets of agronomic and economic responses.

Secondly, a number of *adjustment experiments* involved altering some of these assumptions to evaluate various options available to offset or mitigate the effects. These responses might occur at the enterprise level (for example, a farm-level decision to switch to a different crop) or through a change in government policy (such as a change in farm subsidies). Experiments involving use of different crops, varying amounts of fertilizer, etc., thus enabled a new set of impact estimates to be generated that could then be compared with the initial estimates, and thus help in evaluating appropriate policies of response to climatic variations.

In the studies reported here it has frequently proved useful to perform these experiments in a sequence, conducting impact experiments first on the (unrealistic) assumption that the economic and social systems will undergo no change and then conducting adjustment experiments that incorporate an increasingly complex pattern of assumed responses.

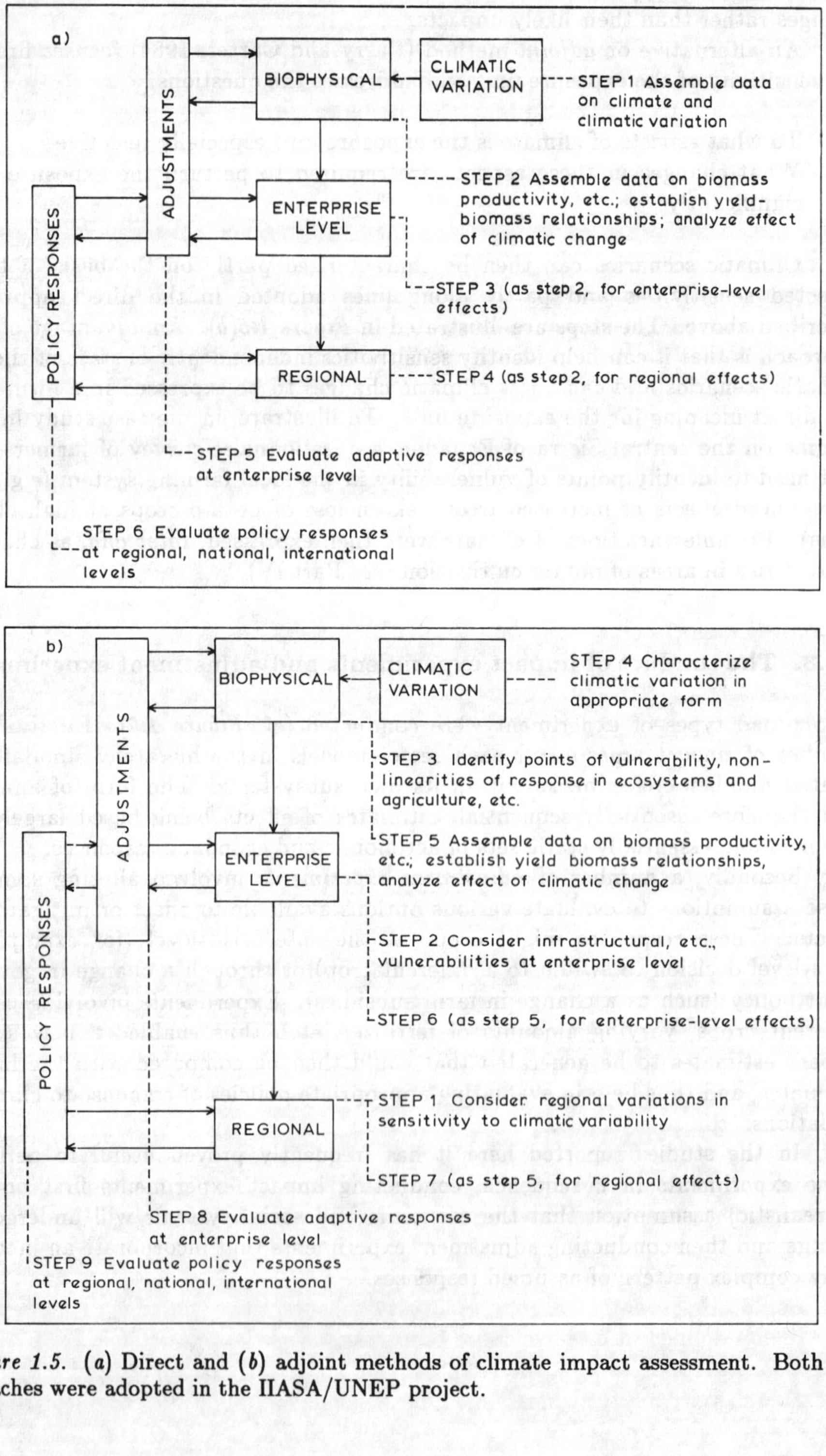

Figure 1.5. (*a*) Direct and (*b*) adjoint methods of climate impact assessment. Both approaches were adopted in the IIASA/UNEP project.

1.4.4. Some integrating concepts

Three broad concepts provided a common focus for the study: the evaluation of climatic changes as changes in risk, the mapping of impact areas and the emphasis on vulnerability of marginal systems to climatic change.

Climatic change as a change in risk

Government interest in impacts from climate is often a short-term concern (e.g., the effects of droughts, floods, cold spells, etc.). Long-term climatic changes, such as those that might result from increasing atmospheric CO_2, are frequently not perceived by governments to be of immediate importance. This suggests, therefore, that a useful way to express long-term climatic change for the policy maker is as a change in the *frequency* of short-term anomalies. For example, in the USSR case study in this volume, the effects of changes of climate are estimated from changes in the frequency of different weather types (*see* Part VI). One advantage of this type of approach is that the change can be expressed as a change in the *risk of impact* (Parry, 1985a). This would enable governments to devise programs that accommodated specified tolerable levels of risk, and adjust policies to match the new risk levels.

Delimiting impact areas

In order to conduct useful impact analyses, it is important to be specific about the exposure unit being studied and its location. One method of identifying areas that can be affected by climatic variations focuses on the shift of limits or margins representing boundaries between arbitrarily defined classes (Parry, 1985b). The classifications may be of land use, agroecological potential, vegetation, crop yields, etc., and the boundaries delimit zones on maps that may shift in response to a change in climate, thus defining "impact areas". This method can be combined with the change-in-risk approach to help identify agricultural areas that are particularly sensitive to climatic variations. Examples reported in this volume include spatial shifts of rainfall isohyets, of vegetation Life Zones, and of agricultural potential.

Vulnerability at the margin

Underlying the identification of spatial shifts of climatic resources is the idea that sensitivity to climatic variability may be more readily observed toward the margins of tolerance of an organism or activity, or at the margins of comparative advantage between two competing activities. We can define three types of marginality:

(1) *Spatial or geographical marginality*, which describes the edge of a specified region defined in biophysical or economic terms.
(2) *Economic marginality*, where returns on an activity barely exceed costs.
(3) *Social marginality*, where an underdeveloped population becomes isolated

from its indigenous resource base and is forced into marginal economies that contain fewer adaptive mechanisms for survival (Baird *et al.*, 1975).

These marginal areas or groups do not necessarily coincide on the ground. Marginal populations, for example, are not always found in areas that are poorly endowed with resources. But it is reasonable to suggest that, whatever the type of marginality, it is characterized by a special sensitivity to changes in resource availability (such as changes in climate, which can itself be regarded as a resource). The margins can be mapped and their shifts can be used to designate areas of impact from climatic change or climatic variability.

Marginality, risk and the spatial shift of these are, thus, three concepts that underlie the studies in this volume: The focus is on regions where rainfall is barely adequate for agriculture and where levels of agricultural risk are already high; close attention is given to changes in risk that may result from changes of climate; and there is emphasis on mapping these effects in order to identify areas of possible impacts.

1.4.5. The use of case studies

In order to develop and test the methodology in a manageable research context, the project focused on a number of case studies.

Four criteria determined the choice of case studies, namely, that they should:

(1) Be near the limits of agriculture with respect to available rainfall and thus, *prima facie*, sensitive to variations of rainfall.
(2) Include a variety of geographical locations, farming types, and political structures (i.e., centrally planned, market-oriented and subsistence economies).
(3) Have available for use a variety of crop–climate, livestock production and socioeconomic models already developed and validated for the region.
(4) Have scientists and policy advisers with sufficient interest and time to make a commitment to research on the project.

The case studies fall into two groups: those in cool temperate and cold regions (reported in Volume 1, Parry *et al.*, 1988) and those in semi-arid regions (reported in this volume).

Cool temperate and cold regions

Consideration of CO_2-induced climatic changes is a prominent feature of the studies in these regions, partly because CO_2-induced climatic warming is expected to be more pronounced at higher latitudes, and partly because agriculture in these more economically developed regions is less vulnerable to short-term climatic variability.

The regions chosen as case studies were: Saskatchewan (Canada); Iceland; Finland; the Leningrad, Central and Cherdyn regions of the northern USSR and northern Japan.

Semi-arid regions

Vulnerability to drought is symptomatic of some agricultural systems in these regions, where a high degree of sensitivity to present-day climatic variability is a major source of system disruption. Case studies were conducted in the central Sierra of Ecuador, Northeast Brazil, central and eastern Kenya, dry tropical India, the Stavropol and Saratov regions of the southern European USSR and the Australian wheatbelt (*Figure 1.6*).

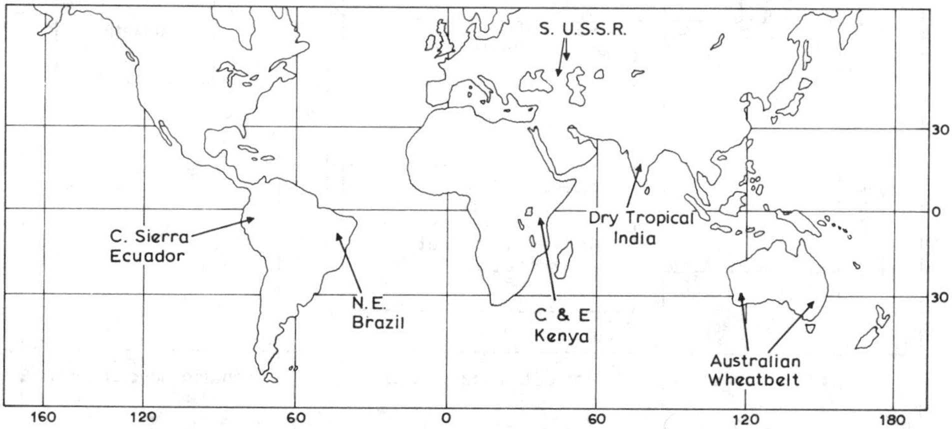

Figure 1.6. Location of the case study regions: central and eastern Kenya, Northeast Brazil, the central Sierra of Ecuador, dry tropical India, the Stavropol territory and Saratov region (USSR) and the Australian wheatbelt.

In each case study a team of between 4 and 7 scientists and policy advisers performed the experiments and submitted results. The teams met together at IIASA at the beginning of the project to establish common goals, agree upon climatic scenarios and upon methods of impact assessment, and to determine the intended outputs. They then returned to their home regions to complete the research, which was reported in draft form at a second meeting at IIASA. Omissions, overlaps and errors were discussed at that meeting, and results were submitted to IIASA for external reviewing and subsequent revision.

1.4.6. A common set of experiments

In order to establish a degree of comparability that would enable subsequent synthesis and the generalization of results, the case studies adopted similar climatic scenarios, similar model hierarchies and similar analytical approaches (*Figure 1.7*).

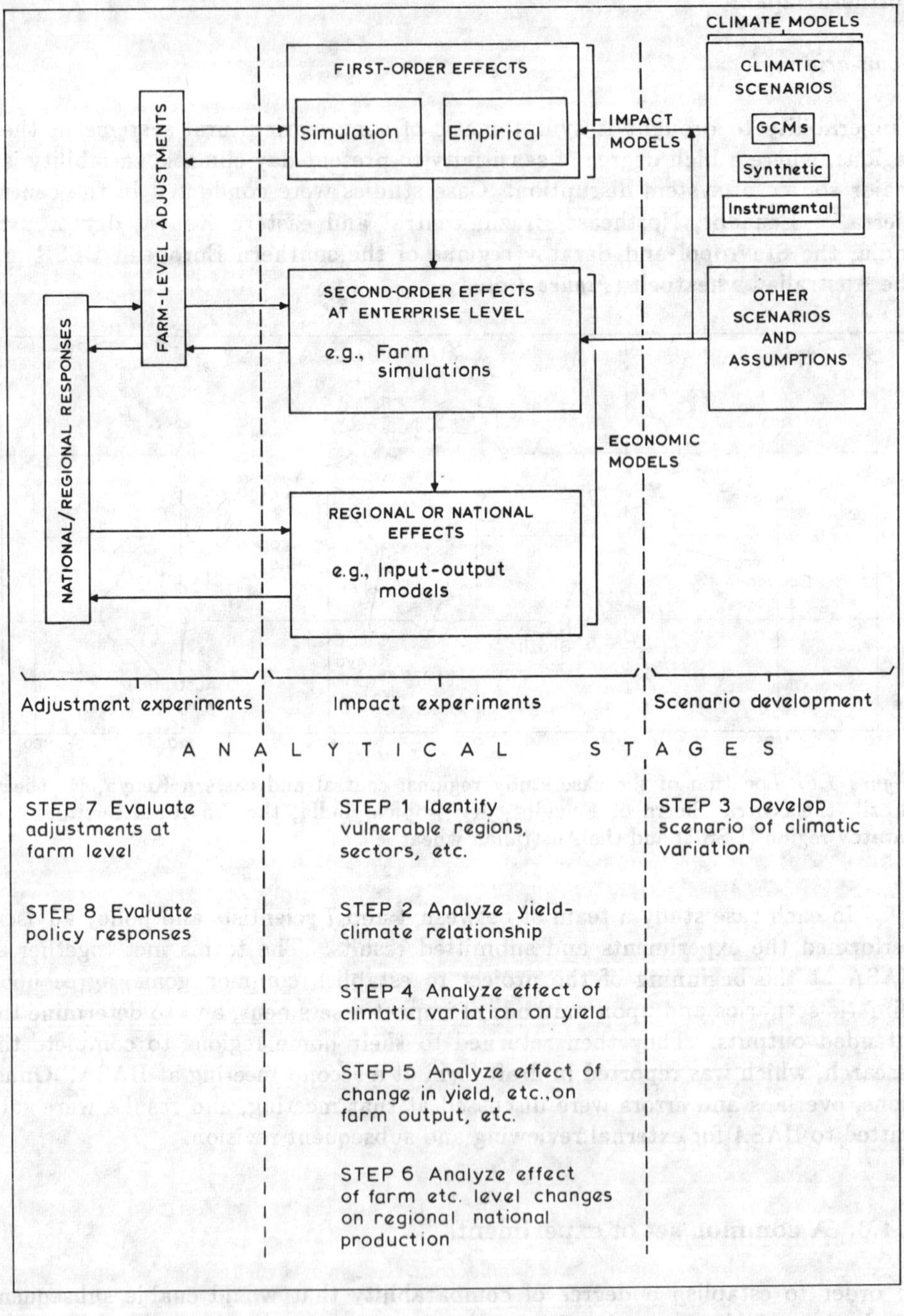

Figure 1.7. **Types of model used and stages of analysis adopted in the IIASA/UNEP project.**

Development of the climatic scenarios

Three broad types of climatic scenario were chosen as a common framework for case study experiments, each based on the historical instrumental climtic record:

(1) *A single, extreme drought event.* In most case studies a recent extreme event was selected so that observed effects could be compared with those estimated from the impact models. Extreme high rainfall events were also examined in some studies.

(2) *A drought event with a return period of 1 year in 10.* This was selected to represent events that are not the most extreme but have a frequency and magnitude that requires the adaptation of farming systems. The method of calculating this scenario is described below; 1-in-10 high rainfall events were also included in some case studies.

(3) *A "back-to-back" drought* characterized by two (or more) successive anomalously dry years. This was selected to allow investigation of the cumulative effects of consecutive extremes, which can be much more pronounced than those of single events.

The climatic scenarios chosen in each region are summarized in *Table 1.1.* Their selection is described in detail in the case studies themselves.

The 1-in-10 dry scenario was calculated by first ranking annual (or seasonal) precipitation totals for a long period of years from the lowest to the highest values. The frequency of occurrence of certain precipitation totals could then be cumulated and plotted on a cumulative probability graph (*see Figure 1.8*; for a further discussion of this technique, *see* Part I, Section 3). The cumulative probability, in this case, represents the probability that precipitation is *less than* a specified amount. The 1-in-10 drought, under this scheme, corresponds to the 10% cumulative probability i.e., there is a 10% (1-in-10) probability that precipitation is less than or equal to the amount specified.

Several methods were employed in the case studies to plot the cumulative probability graphs. A common procedure was to construct a best-fit curve through the (discrete) plotted points in order to read off the 10% probability rainfall value (as shown in *Figure 1.8*). Alternatively, normalized precipitation departures from the mean were organized into frequency classes, which were then cumulated and plotted as before (for example, *see* the Brazil case study, Part III). In both cases, either the 10-percentile precipitation value was used directly for impact experiments, or individual weather-years were identified in which a precipitation total similar to the 10-percentile value had been recorded. The advantage of the latter aproach is that *within-season* weather data for the chosen weather-years were often available for impact experiments in addition to the total precipitation.

A few other scenarios were selected that were of particular interest in specific case studies.

(1) *Weather-type frequency distributions.* Under this scheme, the climate of a region was classified according to the frequency of certain types of

4

Table 1.1. Types and characteristics of climatic scenarios selected in the IIASA/UNEP Project, Volume 2. Not all the years identified as precipitation-based scenarios were used in regional impact experiments.

		Climatic scenarios					
		Precipitation-based			*Other*		
Case study	*Baseline period*	*Extreme event*	*1-in-10 event*	*Consecutive drought*	*Long-term change*	*Synthetic*	*Miscellaneous*
C & E Kenya	1951– 1980^{m}	1984 (–)	1977(+) 1974(–)	1974– 1976			
NE Brazil	1915–80 (approx)	1984(+) 1983(–)	1967(+) 1952(–)	1979– 1980		P±,E±	
C. Sierra Ecuador	1964– 1983^{m}	1982/ 83(+) 1968(–)	1975(+) 1979/ 80(–)		T±	P±	$P_{35/65}$
Dry tropical India	1941– 1970^{m}	1956(+) 1972(–)	1971(–)	1971– 1972		T±,R±	$P_{30/70}$
S. USSR:							
Stavropol	1970–80	1976(–)			Wf		Wf
Saratov	1951–80	1975(–)			GISS 2 × CO_2	P±,T±	Wf
Australian wheatbelt	1951– 1980	1982/ 83(–)	Mapped	1963/64 –1964/65	P±		

P = Precipitation; T = Temperature; R = Solar radiation; E = Potential evapotranspiration; Wf = Weather-type frequency distribution; + = Above average; – = Below average; 35/65 = Percentiles; m = Median precipitation also used as baseline; GISS = Goddard Institute for Space Studies general circulation model experiment (*see* text for details).

weather-year. For example, in the USSR case study three types of weather-year were identified: years with "sufficient" water (for agriculture), years with "limited" water, and "dry" years. By assigning individual years to each category according to long-term historical records of annual weather conditions, the probability of occurrence of different weather types was evaluated. One advantage of this system is that particular types of impact can be associated with specific weather types, and assigned a probability. Moreover, a long-term change in climate can be usefully expressed as a change in probability of certain weather types, i.e., as a change in the frequency distribution (*see* Part VI).

(2) *Synthetic scenarios.* These comprise climatic data generated arbitrarily to simulate a climatic change, and are used to test the sensitivity of impact models. In the India case study, for example, intervals of 1° C for temperatures and 10% for solar radiation were used in adjusting climatic input data to a crop–weather model.

(3) *Long-term climatic change.* Although the issue of longer-term climatic change does not feature prominently in this volume, investigations have been conducted in some regions. For example, changes in the amount and distribution of annual precipitation between successive 30-year averaging periods have been described for the Australian wheatbelt (*see* Part VII).

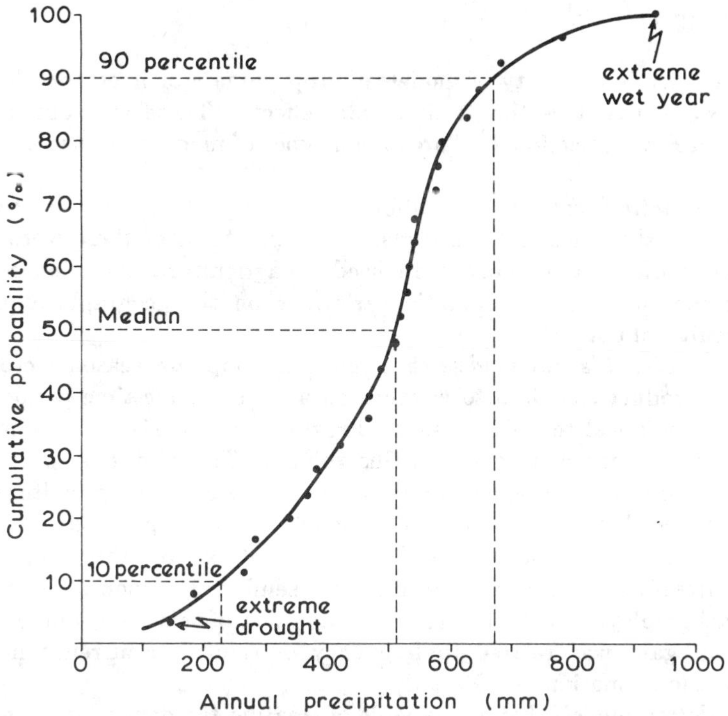

Figure 1.8. Example of a cumulative probability graph for annual precipitation over a 25-year period at a hypothetical station. The points are annual precipitation values through which the curve has been fitted by eye; 1-in-10 and median precipitation values can be read off the horizontal axis or represented by adjacent plotted points. Extreme precipitation values are also indicated (schematic).

Changes in mean annual temperature have been identified in the central Sierra of Ecuador that can be related to the incidence of frost (*see* Part IV). Finally, a scenario of future climatic change is used in the Saratov region (USSR) based on the Goddard Institute for Space Studies (GISS) general circulation model 2 × CO_2 experiment (Hansen *et al.*, 1983; 1984). This is compatible with case study experiments in cool temperate and cold regions in Volume 1 (Parry *et al.*, 1988).

In order to provide a reference against which to compare the above scenarios and their impacts, the period 1951–1980 was adopted as the standard climatological *baseline*. In practice, however, there were many cases where suitable data were not available for these years, and alternative periods had to be selected.

Also, on occasions when precipitation was the dominant criterion in scenario selection, it often proved useful to adopt the year that exhibited the *median* rainfall total as indicative of a "normal" or baseline rainfall year. Such instances are indicated in *Table 1.1*, together with most of the other climatic scenarios selected in this volume.

Linking the models

In each case study a variety of models of crop–climate or livestock–climate relationships were used to estimate first-order effects. These are summarized and compared in Part I, Section 2. Three main types of model were used:

(1) *Agroclimatic indices* that combine into a single term those meteorological variables that most influence plant growth. Most of these were moisture-based indices, and many were used in agricultural zonation schemes to examine the effect of climatic variations on the geographical pattern of agricultural potential.

(2) *Empirical–statistical models* that relate a sample of seasonal crop or livestock productivity data to weather data for the same time period and area using statistical techniques such as regression analysis. Only a few models of this type are represented in this volume. This is due, in part, to severe limitations in some regions on data (especially data on agricultural productivity) needed to develop such models, and in part to the restricted effectiveness of this approach in accounting for the complexities of climate–productivity relationships in semi-arid regions. The livestock model employed in the Kenya case study was of this type, where cattle liveweight gain was related to frequency of rainfall occurrence in a simple regression equation (*see* Part II).

(3) *Simulation models* that are capable of treating the dynamics of crop growth and the interrelationships of plant, soil and climatic factors. Since these are based on generally applicable functional relationships, they have the advantage of being adaptable to different production environments, providing there is a minimum data set available for local calibration purposes. Several models employed in experiments in this volume were originally developed in other regions and have been re-calibrated for conditions in case study regions. Another useful feature of these models is their capacity to simulate the effects of various management practices on crop growth and yield (i.e., in *adjustment experiments* that evaluate the potential for modifying practices such as planting, fertilizer application and irrigation).

Outputs from these models were used as input information for assessments of second- and higher-order effects. Some of these effects were evaluated using mathematical models. An example of these are estimates of crop yields generated by a crop production model for different climatic scenarios in the Stavropol region of the USSR (*see* Part VI). These were used as inputs for a crop area allocation model that evaluated the optimal allocation patterns for achieving specified production targets under a variety of agricultural technologies for each scenario. Most of the higher-order assessments reported in this volume are not model-based, however, but represent a range of *expert evaluations* of impact. Some of them are quantitative. For example, "output multipliers" were used in the Australian case study to translate drought effects on regional farm output and farm incomes into impacts on total regional output, income, employment and subsequently to evaluate effects on GNP (for details *see* Part VII). Others are semi-quantitative, where estimates of effects are judgmental, such as the estimates of decline in value of farmers' assets during drought years in the Indian study (*see* Part V). Finally, a large number of assessments are purely qualitative, reflecting information about typical farm-level effects of climatic variations (e.g., migration, off-farm employment, sales of livestock, etc.).

Analytical stages

Where possible the case studies adopted the adjoint approach, considering first the varying sensitivities of different regions, sectors and farming systems to different types of climatic variation and thus identifying, *a priori*, points of special vulnerability that should be the focus for study. Scenarios of climatic variation were then constructed using those climatic variables (or derived variables, such as potential evapotranspiration, onset of growing season, etc.), considered most appropriate for the particular impact study. Various orders of effect and response were then considered (*Figure 1.7*).

Table 1.2 is a checklist of the impact experiments conducted for each case study. It includes:

(1) Effects on derived climatic parameters (e.g., potential evapotranspiration).
(2) Effects on agricultural potential (e.g., shift of agroclimatic types, changes in biomass potential).
(3) Effects on crop yields.
(4) Effects on livestock.
(5) Higher-order effects on, for example, farm income, employment or rates of migration.

These were followed by adjustment experiments that considered the efficacy of various on-farm adjustments as responses to variations in climate, including:

(1) Changes of crop type or cultivar to suit the expected weather.
(2) Changes in the level of treatment applied, especially fertilizer applications and irrigation.
(3) Changes in the timing of farming operations to suit the weather (e.g., the timing of soil preparation, planting, fertilizing and irrigating.

More general policy responses at the regional and national level were also considered. These included:

(1) Changes in regional land-use allocations.
(2) Changes in regional levels of resource input.

Table 1.2. Checklist of experiments on the effects of climatic variations on agricultural production. Experiments in Part A assume *ca.* 1980 technology. Experiments in Part B assume adjusted technology. In each case study three or more climatic scenarios were considered (for details *see* text).

Type of experiment	*Case study regions*					
A. Impact experiments	*C & E Kenya*	*NE Brazil*	*C. Sierra Ecuador*	*Dry tropical India*	*S. USSR*	*Australian wheatbelt*
			Effects on derived parameters of climate			
Drought indices	X				X	X
Growth period	X			X		
Soil moisture		X		X	X	
Coherent rainfall zones				X		
Runoff				X		
Frost frequency			X			
			Effects on potential			
Biomes			X			
Cropping limits			X			
Agroclimatic zones	X			X		
Biomass potential	X					
Area affected by drought		X				
			Effects on crop yields			
Barley			X		X	
Beans		X	X			
Cotton		X				
Maize	X	X	X		X	
Manioc (Cassava)		X				
Millet					X	
Peas		X			X	
Potatoes			X			
Rice		X				
Sorghum				X		
Spring wheat					X	
Sugarcane		X				
Winter rye					X	
Winter wheat					X	X
			Effects on livestock			
Forage level	X					
Milk production	X					
Meat production	X					
Stock deaths	X					
Stock sales	X			X		
Livestock numbers		X				X
Total production	X					X
			Higher-order effects			
Farm income	X					X
Distribution of farm production between subsistence and sales				X		
Sale of assets	X			X		
Indebtedness				X		

Table 1.2 (*continued*)

Type of experiment	*Case study regions*					
A. Impact experiments (continued)	*C & E Kenya*	*NE Brazil*	*C. Sierra Ecuador*	*Dry tropical India*	*S. USSR*	*Australian wheatbelt*
			Higher-order effects (continued)			
Dependence on off-farm income	X			X		
Nutrition/diet	X					
Withdrawal of children from school				X		
Employment	X					X
Migration	X	X		X		
Production costs					X	
Regional/national production		X			X	X
GNP		X				X
B. Adjustment experiments						
			Changes in crop management			
Planting date	X		X	X		
Planting density			X	X		
Cultivar type				X		
Amount of irrigation				X	X	
Timing of irrigation				X	X	
Fertilizer application					X	
Fallowing			X		X	
Labor and mechanization					X	
			Changes in land allocation			
To stabilize returns					X	
To maximize average returns					X	
To maximize minimum returns					X	

1.5. Major Findings of the Present Study

1.5.1. First-order effects on agriculture: short-term

Effects on seasonal crop yield

When water is scarce, most plants close their stomata to avoid drying out. This closure, however, not only reduces transpiration but also the intake of carbon dioxide. Under these conditions little or no photosynthesis is possible, and growth is very limited. Put simply, this is why crop yields display a marked positive relationship with rainfall in most of the case studies reported in this volume. This relationship varies, of course, from region to region and crop to crop. In terms of impact, the most critical year-to-year differences are those that (as a result of differences between good years, average years and moderate or severe drought years) straddle the boundary between viable and nonviable production. For example, in the tea–coffee and maize zones in central and eastern Kenya, maize yields vary by almost one order of magnitude between good rainfall years (defined as occurring with a 10% probability) and severe drought years

(2% probability), an absolute difference of about 2000 kg/ha (*Figure 1.9*). Yet in the drier "marginal cotton" zone, which is marginal for maize cropping, even the good rainfall years give maize yields of only 1000 kg/ha. In all areas, however, the relationship between rainfall and maize yield is broadly linear, at least over the types of rainfall year considered in this study.

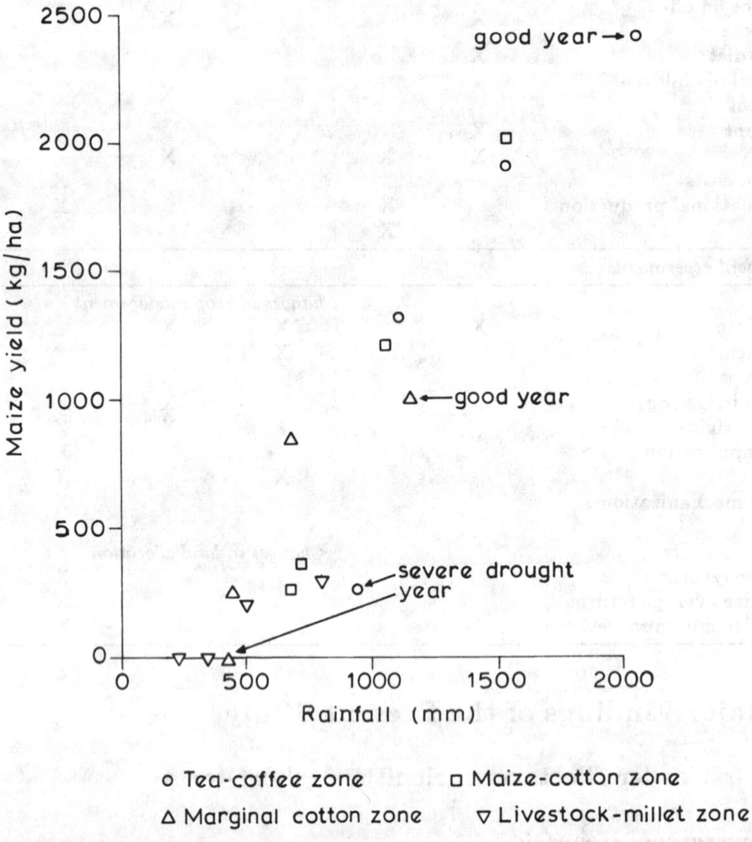

Figure 1.9. Estimated maize yield response to rainfall in different agroclimatic zones in central and eastern Kenya. Source: Part II.

A broadly similar set of relationships exists between rainfall and livestock forage. For example, the levels of production of forage grasses for livestock vary between severe drought years and good years in central and eastern Kenya over a range of 9000 to 19000 kg/ha of dry matter in the tea and coffee zone and 500 to 6000 kg/ha in the marginal cotton zone (*Figure 1.10*).

The data in *Figure 1.9* are qualitative assessments of *actual* farm yields of maize. A similar, broadly linear relationship exists between seasonal rainfall and *potential* maize yields (*Figure 1.11*), estimated using a crop simulation model that assumes optimal nutrients and crop management. However, these potential

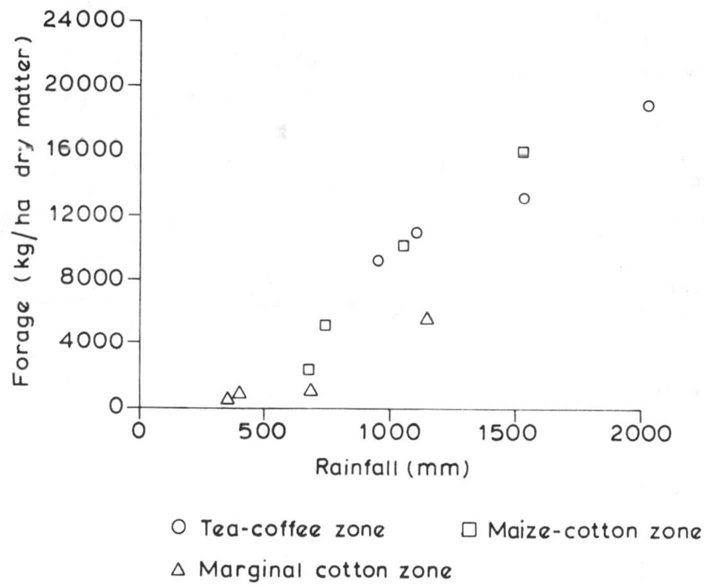

Figure 1.10. Estimated livestock forage response to rainfall in different agroclimatic zones in central and eastern Kenya. Source: Part II.

yields are almost one order of magnitude greater than those actually obtained, a result that is largely attributable to the low level of farm inputs (e.g., fertilizers, machinery, crop cultivars, etc.).

Season-to-season variability and yields

Because yield levels are often closely related to rainfall receipt, there often occur marked regional variations in interannual yield variability. For example, the coefficient of variation of winter wheat yields averaged over the Stavropol territory of the southern European USSR (*see* Part VI) is about 32%. Yet it varies from about 15–17% in the moderately humid zones of the territory to 52–63% in more arid zones.

Such figures, however, can often disguise an important aspect of the regional character of yield variations. This is that regional differences in maximum yields are relatively small, yield levels in good years being broadly similar in both moderately humid and less humid zones. In the Stavropol territory, for example, maximum yields of winter wheat differ from zone to zone within a narrow range of 13–26% (*see* Part VI). In contrast, regional differences in minimum yield levels are often very great; indeed, it is frequently only in the anomalously dry years that the differences in yields between humid and dry zones really become apparent. Again, taking the Stavropol example, yield levels in the driest years are 9 to 15 times less in the dry zones than those in the more humid zones.

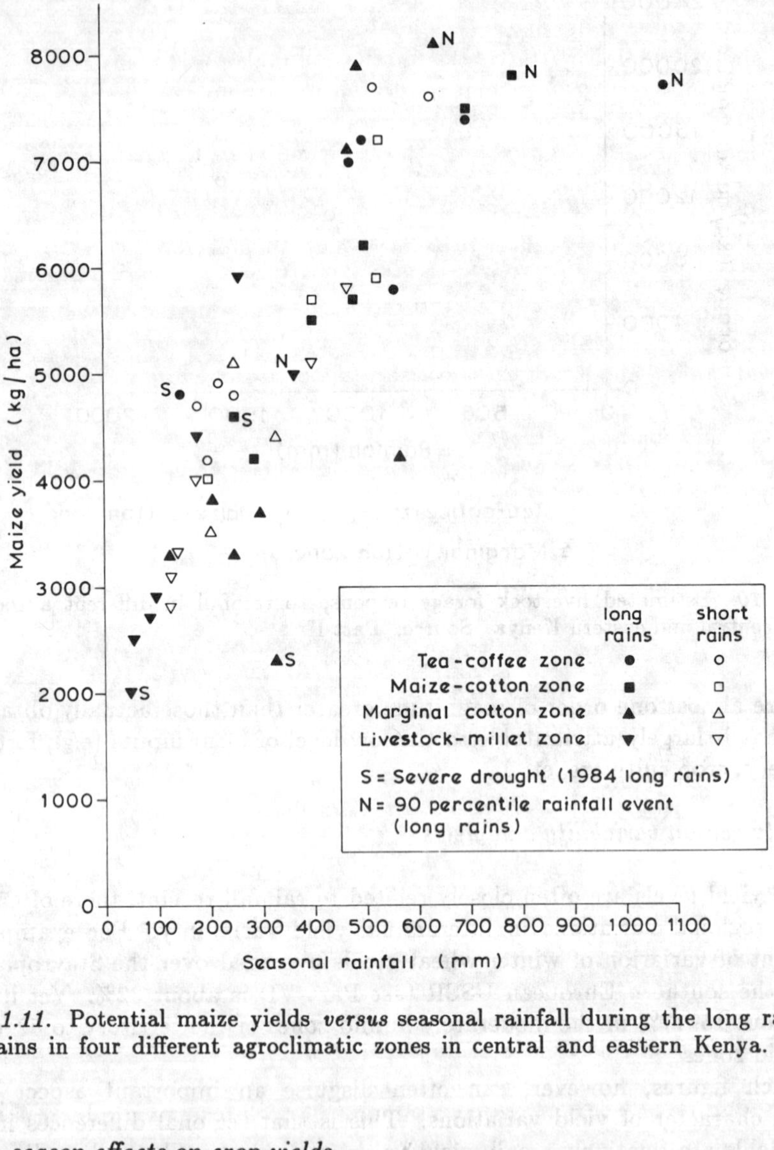

Figure 1.11. Potential maize yields *versus* seasonal rainfall during the long rains and short rains in four different agroclimatic zones in central and eastern Kenya. Source: Part II.

Within-season effects on crop yields

While there is a clear relationship between crop yields and seasonal rainfall totals in semi-arid regions, a significant component of season-to-season yield variability is attributable to *within-season* variations in rainfall. An illustration of this is given in *Table 1.3.* Here sorghum yields in Karnataka State (India) are compared for pairs of years displaying:

(1) A similar rainfall total and a similar rainfall pattern (i.e., a low "disparity index" between them – represented by the years 1965 and 1966). Sorghum yields in these years are very similar.
(2) A similar rainfall pattern but different annual totals (e.g., 1966 and 1969). Yields in the year with higher rainfall are nearly double those in the low rainfall year, in spite of the similar rainfall distribution.
(3) A similar total but different pattern of rainfall (e.g., 1969 and 1977). Here, the rainfall distribution in 1977 is clearly unfavorable for sorghum, limiting yields to levels comparable with those achieved with much lower annual rainfall totals but a different pattern (e.g., 1965 or 1966).

Of course, it is information such as this (accumulated by experience over the longer term), combined with their perception of rainfall reliability, that influence farmers' choices of the most suitable planting times for different crops.

Table 1.3. Comparison of the effects of rainfall total (cm) and pattern (disparity index) on sorghum yields between selected pairs of years for a subregion of Karnataka State, India. Source: Part V.

Paired years	*Annual rainfall* (cm)	*Disparity index*	*Sorghum yield* (t/ha)
1965	59.4	0.17	0.55
1966	59.5		0.51
1966	59.5	0.08	0.51
1969	76.9		0.97
1969	76.9	0.30	0.97
1977	76.5[a]		0.53

[a]Estimated from *Figure 2.8*, Part V.

Effects on yield quality

While much of the emphasis of first-order impact experiments in this volume is on estimating the effects of climatic variations on levels of plant production (especially crop yields), the quality of the plant product is also of great importance, and is itself closely related to weather conditions. In developed countries, such as Australia, where wheat production is largely export-oriented, variations in grain quality may effect its marketable value for processing industries, such as flour manufacture, and are closely monitored. In subsistence economies, the effects of variable crop quality are more directly manifested in human as well as animal nutrition. Experiments in Kenya, for example, have indicated that the nutritive value of forage for cattle is related to rainfall, though not as closely to rainfall amount (which is important for forage quantity) as to rainfall pattern and duration. An index of this latter rainfall characteristic has thus been used as a predictor of cattle liveweight gain in a simple livestock regression model (*see* Part II).

Effects on yield responsiveness to fertilizers

Aside from water, the level of soil nutrients is perhaps the most important factor limiting crop yields in semi-arid regions (both in developed and in less developed countries). Results from the case studies demonstrate clearly that yield responsiveness to nutrients is strongly influenced by climate, particularly rainfall. For example, a questionnaire survey of farmers in Chimborazo Province, Ecuador, provided information on the yields of barley, potatoes and broad beans obtained with and without applications of organic fertilizers during a wet year and a dry year (*Table 1.4*). While these data are clearly subjective, they point to substantial absolute benefits to be gained in applying fertilizers in wet years, though the gains in yield in dry years, while smaller than in wet years, are larger in relative terms. Also shown in *Table 1.4* are comparable model-based estimates of spring wheat yields in the Saratov region, USSR, again both with optimum levels of fertilizing and without fertilizing, during a wet year and a dry year. The findings are very similar to those in Ecuador.

Given these results, it is no surprise that the marginal returns on fertilizer applications are likely to be very low in semi-arid regions where, by definition, drought frequency is high. This helps to explain the low input levels of fertilizer application that are common in such regions, even in market-oriented farming regions such as the Australian wheatbelt. Aspects of the cost-effectiveness of fertilizer applications at regional level under a centrally planned agricultural system are discussed in Subsection 1.5.6.

Table 1.4. Comparison of estimated crop yields obtained with and without fertilizer applications in wet and dry years. Results are from Chimborazo Province, Ecuador, and the Saratov region, USSR.

		Crop yields (t/ha)			
		Chimborazo Province, Ecuador[a]			*Saratov region, USSR*[b]
Weather-year	*Fertilizer application*	*Barley*	*Potatoes*	*Broad beans*	*Spring wheat*
Wet	With	0.83	4.22	0.68	4.1
	Without	0.43	0.74	0.43	3.1
Dry	With	0.31	1.17	0.34	0.2
	Without	0.15	0.30	0.19	0.1

[a] Wet year is 1981–82; dry year is 1983–84. Based on survey responses by farmers. Source: *Tables 3.5–3.7*, Part IV.
[b] Wet year is 1978; dry year is 1975. Based on model simulation. Source: *Table 8.1*, Part VII.

Carry-over effects

One reason for using the back-to-back drought event as a climatic scenario in this volume was to study *carry-over* effects in consecutive seasons with reduced moisture supply. In some regions such effects on crop yields were found to be of minimal significance (for example, in the dry tropical regions of India, where the soil profile is nearly always dry at the beginning of the growing season for sorghum). However, in regions where soil moisture is normally replenished before the beginning of the growing season, but where such replenishment does not occur because of drought, the effect on yields may be substantial. This cumulative effect of rainfall deficiency is particularly severe under continuous cropping systems, but it can be offset to some extent by fallowing to conserve moisture. For example, yields of winter wheat obtained following fallow in a very dry year in the Ipatov region of the Stavropol territory, USSR, are almost twice those obtained following a previous winter wheat crop (*see* Part VI).

Nonlinear effects on agricultural production

Where output is the product of more than one system, the effects of climatic variability can be markedly nonlinear. An example is the sensitivity of milk production in Kenya; while the quantity of forage varies broadly linearly with rainfall (*see Figure 1.10*), milk production may be negligible, even in years that are only moderately dry and when forage quantity is a little below average (*Figure 1.12*). One reason for this abrupt decline in milk output is the concurrent effect of rainfall on forage quality, but another factor of importance is the stocking density of a region. Thus, with an annual rainfall total of about 1100 mm in an average year in the maize–cotton zone of the region, milk output is estimated to be about 3–5 kg/day, but with a similar annual rainfall total in a wet year in the drier marginal cotton zone (where stocking density is higher) production of only 0–1 kg/day would be expected. In both zones, however, there is a step-like relationship between milk production and rainfall receipt (*Figure 1.12*).

1.5.2. First-order effects on agriculture: long-term

Long-term effects on mean yield

The effects on mean crop yield of three types of long-term climatic change have been investigated in this volume: (a) synthetic scenarios involving systematic arbitrary changes in average climate, (b) estimates of future CO_2-induced climatic change and (c) changes in the frequency of different weather-types. The results are summarized below:

Synthetic scenarios have been used in three studies to test the sensitivity of crop yields to the types of changes in climatic variables that might occur over the long term. Some of the results are summarized in *Figure 1.13*, which shows the relationship of mean yields of various crops to anomalies of three climatic variables: air temperature, solar radiation and precipitation. It is assumed in each graph that all other climatic variables are fixed at their long-term average values. The following points emerge:

(1) Yields of rainy season sorghum in India and spring wheat in the Saratov region (USSR) exhibit a negative relationship to growing season

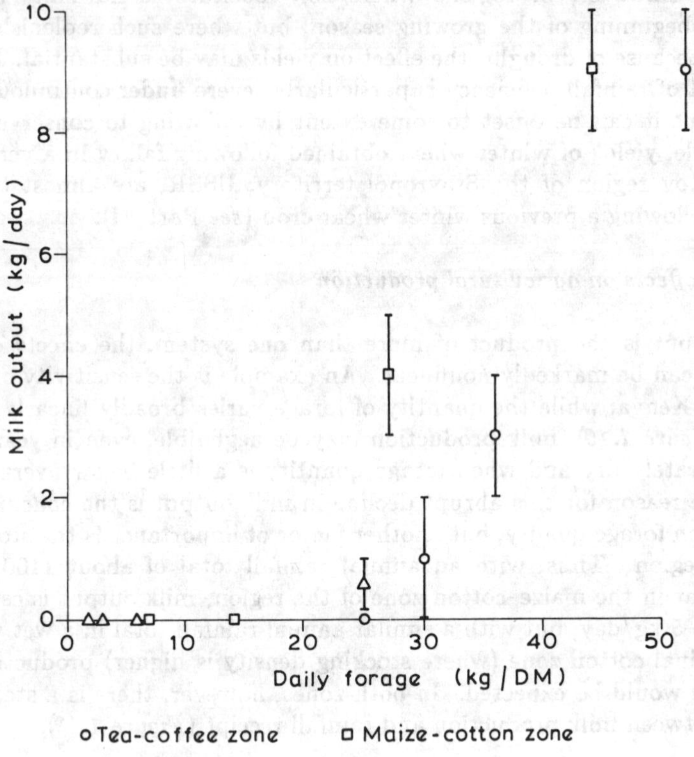

Figure 1.12. Estimated response in daily milk output to variations in daily forage production in different agroclimatic regions in eastern and central Kenya. Symbols as for *Figures 1.9* and *1.10*. Source: Part II.

temperature [*Figure 1.13(a)*]. In both cases this is because higher temperatures are associated with enhanced evapotranspiration, and hence increased water deficiency results in reduced yields. In contrast, yields of post-rainy season sorghum are positively related to temperature. In this case high temperatures, by decreasing the duration of the growing period, would leave crops exposed to drought stress for a shorter period, thus enhancing yields (for details, *see* Parts V and VI).

(2) Yields of sorghum are also sensitive to changes in solar radiation [*Figure 1.13(b)*]. In the rainy season, which is characteristically cloudy, above-average radiation leads to enhanced yields. However, in the post-rainy season the overriding effect of increased radiation is an increase in evapotranspiration losses, which is detrimental to crop yields (*see* Part V).

(3) The sensitivity of cowpea yields to changes in precipitation in Northeast Brazil (*see* Part III) is greater at a dry location (Solenade) than at a more humid location (Icó) [*Figure 1.13(c)*]. These results, and those for spring

wheat from the Saratov region, USSR (*see* Part VI), also illustrate the broadly linear, positive relationship between mean precipitation and crop yield, a finding similar to those illustrated for maize yield and forage production in central and eastern Kenya (cf. *Figures 1.9–1.11*).

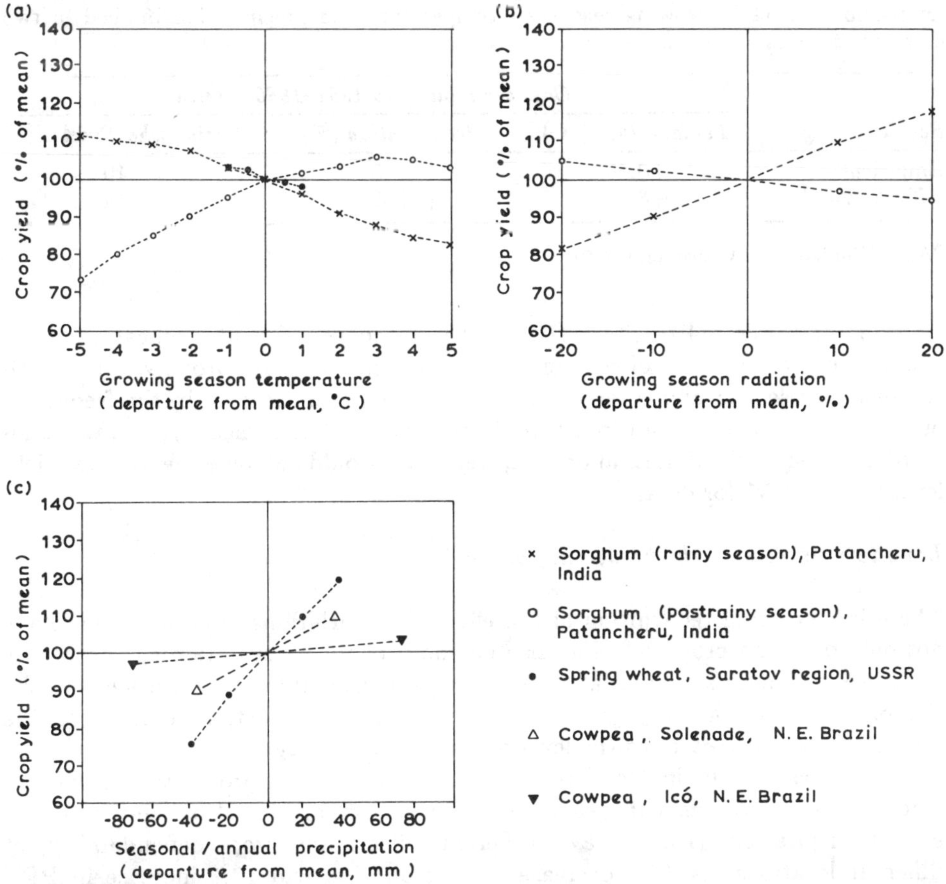

Figure 1.13. Modeled crop yield responses to (*a*) air temperature, (*b*) solar radiation and (*c*) precipitation, with other climatic variables assumed fixed at mean values. Each plot represents estimates by a particular model for a specific geographical region. Plotted points are individual model runs and dashed lines are interpolated by eye. Based on information in Parts III, V and VI.

Experiments to estimate the effects of *projected climatic changes* were conducted in only one case study area – the Saratov region (USSR). The response of spring wheat yields to two climatic scenarios was examined: the first based on an empirical technique relating regional temperatures and precipitation to mean annual northern hemisphere temperatures; the second, a projection of CO_2-induced climatic change (the GISS 2 × CO_2 scenario). The results are summarized in *Table 1.5.* Although both scenarios imply a regional warming, yields are more responsive to changes in precipitation than to changes in temperature, exhibiting the same positive relationship to precipitation that was shown in *Figure 1.13(c)*.

Table 1.5. Response of spring wheat yields at Ershov (Saratov region, USSR) to the long-term changes in growing season air temperature and precipitation implied in two climatic scenaros. Source: Part VI.

	Changes relative to 1951–1980 baseline		
Scenario	*Temperature* (°C)	*Precipitation* (%)	*Spring wheat yield* (%)
Empirical[a]	+0.45	–5	–10
GISS 2 × CO_2	+3.3	+22	+14

[a]After Vinnikov and Groisman (1979).

In the Stavropol territory (USSR) the responses of winter wheat yields to changes in the frequency of certain types of weather-year are broadly in line with responses presented above for the Saratov region. An increase in the frequency of dry or of very dry conditions would decrease long-term mean yields, while an increased frequency of normal or cool, wet years would enhance the average yield level (*see* Part VI for details).

Long-term effects on the probability of yields

There is increasing recognition of the effect that variations of climate may have not only on mean crop yield but also on the probability of obtaining acceptable yield levels. These are especially important in marginal farming areas where the risk of crop loss is already great. In some of the experiments reported here, the level of risk is referred to as the level of *crop dependability*.

The case study in Northeast Brazil provides an appropriate illustration. The effects of various changes in growing season precipitation and potential evapotranspiration (PET) have different effects on crop dependability at different locations. A 10% decrease in precipitation and 10% increase in PET reduces the probability of acceptable yields from 85% to 78% at Jaicos (in the more favored area of the drought polygon) and from 42% to 32% at Petrolina (in a more drought-prone area).

These estimations of risk assume that, in each scenario, the best planting season has been chosen. In reality, planting decisions are based on a limited knowledge of climatic risk and, as a result, the estimations of real crop dependability are substantially lower. In both cases, however, it is evident that long-term climatic variations have an important effect on risk of crop loss.

Concurrent effects on soil fertility

Farmers have traditionally employed a wide range of soil management techniques to optimize crop productivity under variable short-term climatic conditions. Examples to be found in this volume include the use of fertilizers, crop rotations and fallowing practices. Such techniques are certainly very effective in enhancing productivity over the short and medium term. However, their longer-term effects on soil nutrient status and fertility are often difficult to detect and thus less easy to evaluate.

Some estimates are presented in this volume for the Stavropol territory (USSR) based on simulations using a crop production model (*see* Part VI). The model computed the dynamics of humus and phosphorus content in the plowed layer for various levels of mineral fertilizer application in five agroclimatic zones of the territory under present-day (baseline) climatic conditions. The simulation period varied between 20 and 80 years, depending on the duration of crop rotations in each zone. Results indicate that in drier zones the rate of mineralization of organic matter exceeds the rate of decomposition and the humus content falls over the long term, whereas in more humid zones decomposition is more rapid and the humus content rises. In contrast, soil phosphorus content increases over the long run in all zones *except* for the most humid, where crop yields are highest and phosphorus recovery is greatest.

Results such as these clearly demonstrate how, given the concurrent changes in soil nutrient status that would occur over the longer term, the effects on crop yields of long-term climatic change may be quite different from those of year-to-year climatic variations of a similar magnitude.

1.5.3. Effects on the spatial pattern of agricultural potential

Spatial shifts of crop potential

Seasonal shifts. Some case studies in this volume consider the effect of seasonal variations in climate on the geographical location of agroclimatic zones. The zonation schemes are themselves a means of identifying broadly coherent regions of similar agricultural potential. The most common procedure is to combine several climatic variables into a single agroclimatic index using the longest run of data available. The zones thus characterized represent *averaged* agroclimatic potential.

Yet season-to-season variations in rainfall may create marked *seasonal* shifts in the spatial pattern of crop potential. The extent of these shifts is demonstrated in the case studies in Ecuador and Kenya. In Chimborazo Province (Ecuador) agroclimatic suitability zones vary greatly between the 1-in-10 wet year and the 1-in-10 dry year. Under the wet scenario (such as might occur in a mild El Niño year) the total area suitable for potatoes covers more than half of the region. Under the dry scenario it covers only about a fifth (*see* Part IV). In Kenya, where seasonal agroclimatic zone types were assigned to four stations based on conditions during four types of scenario-year (average, good, moderate drought and severe drought), the zones were found to be different for each station under each scenario year (*see Table 1.6*). Potential productivity varied by an order of magnitude between good and severe drought years for three of the locations.

Table 1.6. Agroclimatic zones (0, I, II, etc.[a]) and expected maize yields for four locations in Kenya under four climatic scenarios (kg/ha/yr). Source: Part II.

Scenario	*Kerugoya*		*Embu*		*Machakos*		*Makindu*	
Good year	0	>2375[b]	I	2375	I	2375	IV	850
Average year	I	2375	III	1275	IV	850	V	250
Moderate drought	III	1275	V	250	V	2250	VI	0
Severe drought	IV	850	V	<250	V	<250	VII	0

[a] Zones are: I = tea–coffee zone, II = marginal coffee zone, III = maize–cotton zone, IV = marginal cotton zone, V = livestock–millet zone, VI and VII = ranching zone.
[b] Subject to flooding and temperature effects.

It is thus evident that effective assessments of agroclimatic potential, especially in regions of large climatic variability such as semi-arid regions, require recognition of the importance of season-to-season differences in rainfall. In this sense the precise location of boundaries between zones of different potential is sometimes less important than the extent to which this boundary shifts from year to year with variations in seasonal rainfall. Farmers in the marginal zone covered by these boundary shifts may find themselves located in one zone in an average year, another in a good rainfall year and, possibly, yet another in a moderate drought year. The optimal cropping patterns for each zone may be quite distinct, yet farmers in the marginal zone need to strike a compromise between them. These suboptimal strategies at the interface between zones of different agroclimatic potential have received relatively scant attention to date.

Long-term shifts. Since many zonation schemes are based upon long-term mean values of the observed climate, any *change* in mean climate would be manifested in a long-term spatial shift of agroclimatic zones. The extent of such shifts was a focus of investigations in the companion to this volume, which was largely concerned with possible long-term changes of climate (Parry *et al.*, 1988). However, in the present volume the case study of Ecuador considers the effect of a 10% increase and a 10% decrease in long-term mean annual precipitation on the long-term location of boundaries between major ecological zones (*see* Part IV). The shifts of some of these boundaries under these scenarios is substantial, implying significant alteration of the spatial pattern of agroecological potential.

Spatial shifts of agricultural risk

Long-term changes in mean climate can be expected to alter the spatial patterns of agricultural risk. This is implied in the changes in crop dependability that have been calculated for specific stations under different climatic scenarios in Northeast Brazil (*see* Subsection 1.5.2, above) and in India (*see* Part V). These could be mapped as shifts of isopleths of crop productivity, although this has not been attempted in the case studies reported here. Such isopleth shifts would reflect, in a sense, the changes in the spatial pattern of probabilities of specific climatic events. Examples of those that have been mapped are shifts of isopleths of frost risk in Chimborazo Province, Ecuador (*see* Part IV, ***Figure 5.3***) and shifts of critical period 1-year-in-10 rainfalls (mapped for the Australian wheat-belt – *see* Part VII, ***Figures 2.13–2.16***).

1.5.4. Higher-order effects on agriculture

Differential effects within agriculture

The effects of drought can vary markedly according to farmers' attitudes toward risk, which in turn can be affected by their asset base.

Differences due to varying attitudes toward risk. Large landowners, small owners and nonowners frequently adopt different strategies to cope with risk. In Northeast Brazil, for example, the large owners opt generally for cash-generating activities such as cattle raising and transfer the more vulnerable activities, especially crop production, to renters and partners. Small owners and nonowners are engaged mainly in self-provision, which involves more risk. This is one reason why, over the drought period 1979–1982 in Northeast Brazil, the numbers of cattle actually increased, while large numbers of small owners, nonowners and their families were faced with severe personal hardship (*see* Part III).

Differences due to farmers' asset base. Since small owners and nonowners are largely concerned with subsistence production and also, in general, have a low level of access to irrigation and other drought-mitigating aspects of production, it is largely the crops in small-owner production that show the greatest effects of drought. In Northeast Brazil, for example, subsistence crops, such as beans, manioc and maize, showed substantial aggregate production falls in the 1980 drought, while some cash crops, such as sugarcane, were less affected (*see* Part III). In Kenya a similar pattern is apparent, with widespread losses of cattle amongst small producers in contrast to more limited effects on large-scale commercial agriculture (*see* Part II).

Effects at the household level

In spite of the variations mentioned above, it is possible to make some generalizations about the effects of climatic variations on the resource position of farmers.

Changes in household food consumption. In general, the proportion of production kept for subsistence tends to increase in drought years. In Northeast Brazil, for example, it increased by about 10 percentage points in the 1980 drought over the proportion in a normal rainfall year. Moreover, there is substantially greater elasticity in consumption in some foods than in others so that, for example, subsistence consumption of eggs, dairy products and meat decreases in drought years in favor of beans and oil seeds (*see* Part III).

Changes in patterns of household production. Farm households tend to respond to changes in consumption patterns by increasing the area under staple food production. This is evident for maize in Kenya during the 1984 drought and for beans and manioc in Northeast Brazil over 1979–1983. Where drought is a recurrent problem (or at least perceived as such by farmers) this may partly explain the long-term (*ca.* 1970–1987) increase in area under food crops in Northeast Brazil, which has offset an overall decline in productivity (due to more persistent cropping and the use of less suitable land).

In cases of extreme food shortage, consumption of seed grain may well prejudice the recovery of households when rains return. Up to a third of farmers in some parts of Kenya reported consumption of seed grain during the drought of 1984 (*see* Part II).

Changes in asset level of households. Attempts to sustain household income during drought periods include sale of assets, increases in borrowing, postponement of dues, etc. This varies greatly according to the asset position of households. Up to a quarter of households in northwest India reported sales of assets in drought years, and more than a half of households in the drier parts of central and eastern Kenya reported having sold, consumed or lost cattle during 1984 (*see* Parts V and II, respectively).

Effects on household income. Where small farmers are the main suppliers of food crops, there frequently occurs a sharp decline in food prices just after the harvest. This is due to the lack of capital and storage capacity that would enable small farmers to wait for the best moment to sell their produce. After a spate of post-harvest selling, prices begin to rise sharply, especially in drought years. This was evident in Northeast Brazil during the 1979 drought, when prices of beans and rice in the state of Ceará rose 90 to 100% above the rate of inflation. Rises of this scale are typical of price responses in local rural markets, but are often not reflected in consumer prices in the large cities, largely because of storage by middlemen (*see* Part III).

Effects on household purchases. Since a high proportion of the typical small-owner household budget is spent on food (e.g., about 45% during the 1979–1980 drought in Northeast Brazil), household expenditure on nonfood items tends to contract substantially in drought periods; and because these households are normally major consumers of industrial wage goods, such as processed food, clothing, footwear and medicine, the reduction in their income can have a major effect on the domestic market. Thus, while gross agricultural production in Northeast Brazil fell by about 24% during 1983, the change in real consumption of clothing fell by 33% (*see* Part III).

Effects on regional output

Only one case study in this volume has reported estimates of aggregate effects on output at the regional level, in this instance based on regional input–output analysis. These indicate, in Australia, that agricultural-to-regional output multipliers are of the order of 1.7 to 2. Thus, a reduction of AUS $ 140 m in rural sector output in the Darling Downs area of Queensland during the 1982–1983 drought led to a reduction of total regional output of AUS $ 240 m (*see* Part VII). These output multipliers are not as high as some (such as for government services about 3), but are important because of the relative instability of the agricultural sector.

Effects on regional income

Regional income multipliers are generally higher than output multipliers, for example, averaging about 3 in Australia. Consequently, climatic variations can induce substantial impacts on regional incomes and especially on employment. The case study on Australia in this volume includes an appropriate illustration of this multiplier effect, drawn from the work of Powell (1978): in a region where agriculture contributes 30% of regional output, if drought caused agricultural output to fall by 10% and thus regional output to fall by 3% (with an associated direct income decline of, say, 4%), the effect of an income multiplier of 3 would be a fall in regional income of more than 10% (*see* Part VII).

Wide-ranging social and environmental effects

In addition to the economic measures of effects of climatic variations in semi-arid areas outlined above, a wide range of other effects are considered in these case studies. They include:

- Effects on nutritional status.
- Effects on mortality.
- Effects on labor supply *via* migration.
- Effects on the structure of land ownership.
- Effects on the natural environment *via* associated long-term environmental degradation.

These effects are frequently specific to the political, social and environmental context of drought, and it would be misleading to generalize about them at this juncture. Details are given in each case study.

1.5.5. Potential farm-level adjustments to climatic variations

Many of the effects summarized above are part of a syndrome of changes in economic, technological and management conditions that occur in response to variations of climate. Some represent conscious or unconscious adjustments that are designed to mitigate effects of climatic variations. In considering these adjustments, it is useful to distinguish between two types: potential adjustments at the farm level and potential policy responses at the regional or national level. The former are considered here, the latter in Subsection 1.5.6.

The farm-level adjustments are of two broad types: (a) long-term adaptations to existing climatic variability, which reflect the accumulated efforts by generations of farmers to tune their farming systems as closely as possible to a fluctuating and uncertain agricultural resource base; and (b) short-term adjustments that are implemented on an occasional basis in response to the occurrence of drought or expectation of its occurrence.

Long-term adaptations to existing climatic variability

Traditional measures to cope with existing climatic variability generally involve some measure of *diversification*, as an attempt to spread intraseasonal risk over a number of seasons. Three types of diversification are common:

(1) *Crop-based diversification*, involving subplotting of cropped land, intercropping and planting of different crop combinations to combat intraseasonal climatic variations. Crop and livestock-based mixed farming is an additional form of adjustment.
(2) *Spatial diversification*, involving the spatial scatter of sub-farm plots either to cope with the spatial variability of climate by spreading the risk for single activities, or to exploit spatial climatic diversity (for example, at different altitudes in upland regions) by diversifying to a range of activities.
(3) *Occupational and income diversification*, involving the spreading of sources of income by sharing agricultural risk with others (e.g., through forms of mixed farm tenancies) or by seeking additional on-farm and off-farm income.

Farm-level diversity is greater in areas where there is a greater long-term variability of rainfall. However, the traditional use of diversification as an adaptation to climatic variability is not always compatible with a growing emphasis on specialized production of a few crops involving high levels of input.

Short-term adjustments to climatic variations

In addition to long-term adaptations built into farming systems, there are a number of short-term adjustments employed to mitigate the effects of different climatic situations.

Ex ante adjustments to climate. These include changes in the timing of crop management activities and adjustments in the allocation of resources in response to the climatic (particularly the moisture) situation at the beginning of the growing season. Most common are:

(1) Variation in the total cropped area, a practice that usually involves reducing the cropped area in expectation of a seasonal moisture deficiency.
(2) Variation in the location of cropping, including decisions on the site and orientation of cultivation. For example, at altitudes towards the upper limit of potato cultivation in the central Sierra of Ecuador, the most

favorable locations for planting are those that are least susceptible to frost (i.e., on east-facing slopes, both to catch the morning sun and to avoid the nocturnal pooling of cold air that might occur on flat ground – *see* Part IV).

(3) Variation in the extent of intercropping, with the aim of maximizing production of a diverse array of crops. Trials in central and eastern Kenya, for example, showed that the yield of intercropped fields was greater than that of mono-cropped fields in normal to good rainfall seasons, but was only 30–60% of the yield in mono-cropped fields during seasons with drought (*see* Part II).

(4) Variation in plant density to make the best use of available moisture. To illustrate: in a very dry year in the central Sierra of Ecuador, model estimates of yields of barley at a low plant density (25 plants per square meter) are up to four times greater than yields of a high-density crop (65 plants per square meter), depending on sowing date. Conversely, during a very wet year, yields are considerably greater at high than at low plant density (*see* Part IV). One disadvantage of low-density cropping under dry conditions, however, is the greater opportunity afforded for weeds to compete for water and light.

(5) Variation in the use of crops with low/high moisture needs, which is an effective strategy in regions where alternative crops are available. For instance, survey results from villages in dry tropical India indicate that larger areas of sorghum and castor may be planted in expectation of a dry season, while groundnuts, minor pulses, rainfed wheat and cotton are preferred for wetter conditions (*see* Part V).

(6) Variation in the use of short- *versus* long-duration crops, which can shorten the growth period to exploit scarce water resources or prolong growth to increase yields under favorable moisture conditions.

(7) Variation in the date of planting, which can have a critical effect on final crop yields. Results cited in the Kenya case study indicate that late planting can often lead to drought stress towards the end of the growing season. A planting delay of only 10 days following the rainy season may reduce maize yields by more than 20% (*see* Part II).

(8) Use of multiple planting times, which can spread the risk of climatic effects on crops through the season. At high altitudes in the central Sierra of Ecuador, for example, these staggered plantings (known as *suertes*) are used to limit the effects of early frosts (*see* Part IV).

(9) Variation in the fallowing before planting, which, in the absence of a crop (or weed) cover, conserves soil water by reducing losses through evapotranspiration. Results from Ecuador indicate that barley yields can be enhanced by increasing the *duration* of fallowing from one to two years (*see* Part IV); and experiments in the Stavropol region, USSR, testify to the significant gains in regional crop production that can be obtained by increasing the *extent* of fallowing, especially in the driest zones of the territory (*see* Part VI).

(10) Variation in the amount of tillage and mulching to limit erosion. The minimal working of soil is a characteristic response to wind erosion in some

semi-arid areas, and the use of mulch over the surface further protects soil, seeds and fertilizer against erosion by wind and water.

(11) Variation in the amount of fertilizer application, to enhance crop yields and thus compensate for adverse weather. However, as discussed in Subsection 1.5.1, the responsiveness of yields to fertilizer declines with low rainfall and, if very dry conditions are anticipated during the growing season, the large amount of fertilizer required to obtain a reasonable level of yield may be uneconomic.

(12) Variation in the amount and scheduling of irrigation to make the best use of limited water. To illustrate, model experiments indicate that early season irrigation of spring wheat in the Saratov region, USSR, is most effective in dry years when soil moisture reserves are low. With more moist early season conditions, the same amount of irrigation is better delayed until the anthesis stage of growth (*see* Part VI), a result that concurs with model-based estimates of irrigated sorghum yields for a location near Hyderabad in dry tropical India (*see* Part V).

As with diversification, increasing land shortage and a growing emphasis on mono-cropping in some regions has tended to reduce the opportunity for some of these types of response.

Ex post adjustments to minimize loss: (a) Adjustments in farm practice. These actions, taken after the onset of drought effects, vary considerably from region to region. However, some general adjustments can be identified, such as:

(1) Harvest of premature crops and/or crop by-products, which are common techniques for salvaging some return from crop cultivation.

(2) Abandonment of post-sowing operations, which are perceived as being ineffective in producing a harvestable grain yield, or less important than other salvage measures, such as collecting fodder, fuel, etc.

(3) Thinning and increased weeding of crops, which reduces moisture loss and can provide substitute fodder for livestock.

(4) Use of trees or bush for fodder and fuel, which serve as alternative sources of food and energy in the short term, but may contribute to environmental degradation over the longer term unless replanting is initiated following a drought episode.

(5) Variation in cropped area following drought, a practice often used to compensate for losses incurred during a dry season. As demonstrated in the Kenya study, however, this can be an expensive procedure, especially if a farmer's supply of seed for planting is limited (due to household consumption) or is of poor quality (*see* Part II).

As mentioned above, increasing emphasis on high-input mono-cropping has tended to reduce the scope for such adjustments.

Ex post adjustments to minimize loss: (b) Adjustments in household management. These include a wide range of economic and social responses, which again show considerable regional variation. The case studies in this volume refer to:

(1) Altered food consumption, as evidenced by enforced changes of diet or by malnutrition. For example, in central and eastern Kenya during the 1984 drought, imported yellow maize replaced the preferred local white maize in most households because the local crops had failed. Wild foods were a vital component of the diet in many of the drier regions. While increases in malnutrition were reported, a rapid recovery seems to have occurred by early 1985 (*see* Part II).

(2) Financial measures, ranging from the use of savings to the sale of vital assets. To illustrate: measures that were employed by rural households during the droughts in the Jodhpur (1963–64) and Sholapur (1972–73) districts of India included the postponement of private dues, the withdrawal of children from school, increased borrowing, and the sale of assets, both productive (e.g., livestock) and nonproductive (e.g., buildings). The effects on small landowners were more drastic than on large landowners, however, reflecting the greater ability of the latter to obtain short-term credit (*see* Part V).

(3) Migration, which is a common temporary response of individuals in a household, but may be more permanent and widespread in some cases. For example, the Northeast is the major region of out-migration in Brazil, the most drought-prone states being the main origins of migrants, and their major destinations being the southeast, center-west and north of the country (*see* Part III). In central and eastern Kenya, the destinations of temporary migrants from drought-affected regions include the areas with high potential for cash crops (such as tea, coffee or irrigated rice) or urban areas, a popular destination of young, educated, usually male members of a household (*see* Part II).

(4) Employment on public relief works, which ensures that the population affected by drought receives at least a minimum income. During the 1983 drought in Northeast Brazil, more than 2.5 million workers were hired by the government in an emergency program of construction of roads, dams, wells, etc. Workers were paid about half the legal minimum wage for a four-hour day, leaving them time for important farming activities, such as land preparation for the next year (*see* Part III).

As other adjustments have tended to lose their efficacy, these more serious responses, which frequently have a lasting effect on the farming household, have become more pronounced.

1.5.6. Potential policy responses

Existing types of public intervention

Four broad types of policy response to drought are illustrated in this volume. Some of these have their roots in early forms of crisis management and therefore retain a measure of crisis management practice, such as a persistence of *ad hoc* measures and inadequate attention to the causes of the crises (for example, the intrinsic vulnerability of certain farming systems and social systems).

Food, credit or income supply measures. These include provisions:

- To store food or fodder and increase its availability during and after drought.
- To improve the interregional transfer of food by improving transport networks and establishing appropriate trade arrangements.
- To increase supplies of food and fodder either by direct distribution or by subsidy of food and fodder costs.
- To maintain farm incomes by subsidizing agricultural interest rates and costs of farm inputs.
- To generate alternative employment or income by public works programs during and after drought, or by the long-term development of employment in the industrial and service sectors.

The provision of water. Large-scale provisions for the intertemporal or interregional transfer of water was particularly characteristic of early policies of drought relief, on the assumption that the overriding problem was one of water shortage and that the primary solution lay in large public irrigation works that both increased protection against future drought and provided employment in public works. Large-scale protective irrigation works of this kind formed, for example, the mainstays of drought policy in Brazil and India from *ca.* 1880 up to about 1950 (for details, *see* Parts III and V). More recently the emphasis has turned to smaller-scale developments, such as government encouragement of the widespread construction of farm ponds for water storage in Northeast Brazil over 1979–1986.

Changes in farm management infrastructure. These include the promotion of soil and soil-moisture conservation measures (such as contour bunding and occasional deep plowing) and the reorganization of farm structure and rural infrastructure (such as improvements to systems of transport, marketing and credit). Such measures were frequently typical of government policies in the 1950s and 1960s. Success was limited because the new technologies in agriculture (especially high-yielding crop varieties) were not then available.

Crop oriented measures. These include the development and dissemination of new technology, particularly with respect to high-yielding varieties of crops; to the management of weeds, disease and insect pests; and to crop management, such as timing of planting, spraying and harvesting. In most regions these measures were not widely implemented until the mid 1970s, with the growing availability of high-yielding cultivars. Although this approach has, overall, increased yields greatly, it can be criticized for a heavy emphasis on crop production, discouraging the diversified, flexible and low-risk production systems that often characterize traditional agriculture and provide it with some inherent resilience to drought.

Agroclimatological research has played an increasingly important role in this package of measures. For example, several case studies have demonstrated the value of climatological research in identifying regions that are roughly similar in terms of the character of rainfall receipt; they may therefore be treated as zones of broadly homogeneous cropping potential that deserve roughly similar farming strategies. In each of these zones it should be possible to model the effect of different levels of irrigation on crop yield under different climatic conditions, and thus identify appropriate levels and timings of irrigation that would optimize regional crop production. These measures form part of a more integrated framework of policy responses that are outlined further below.

Integrating the roles of information, technology and public intervention

In general, there has been a move away from *ad hoc* measures of drought relief, toward a more systematic effort to increase resilience to drought. An important part of this trend, highlighted by these case studies, is the need for an increased complementarity between improved information (especially on climatic variability), the new technology (particularly drought-resistant crop types) and various forms of public intervention. Four promising lines of intervention deserve to be emphasized.

Improved information on agroclimatic potential. There are many aspects of this intervention that merit consideration:

(1) Improved data collection, including a greater range of data of different types (e.g., on climate, soils, erosion, agricultural productivity, fertilizer use, etc.), a more dense network of observations, and more accurate and more frequent observations. The acquisition of adequate data is a basic prerequisite for the development, testing and validation of diagnostic models that can be used to simulate agricultural productivity.

(2) Improved data analysis, which would be valuable in those areas where adequate long-term data on climate, soils and agriculture already exist. Computer-trained analysts, along with appropriate hardware and software, are needed in greater numbers to take full advantage of available data.

(3) Agroclimatic "screening" of environments, which involves the analysis of long-term data to evaluate the agroclimatic potential of a region. For example, information could be compiled for a region on the amount and distribution of rainfall, or the potential for runoff, its collection and recycling, etc. Such information should be updated regularly through monitoring programs to improve the practical understanding of climatic variability and other environmental constraints on agriculture.

(4) Identification of broadly homogeneous regions, which can serve as a basis for agricultural planning. Techniques, such as principal components analysis, can be used to identify areas of dependable cropping or to demarcate areas of high or low potential for different crops or cropping systems. These techniques were employed in the India case study in this volume – *see* Part V.

(5) More extensive regional assessments of agricultural potential. Studies of more crops over larger areas and for different farming systems are needed, together with a better understanding of the effects of climate, soils, technology and management on crop yields, by means of experimental field trials and the development of diagnostic crop simulation models.

(6) Improved matching of agricultural activities to weather-types to optimize agricultural production. For example, the use of climatological data on the magnitude, frequency and duration of critical weather events can be used to evaluate the nature and timing of management responses, such as planting, weeding, fertilizer application, harvesting, etc. Such information would enable the identification of the strategies most appropriate to particular types of weather. A development of this is a concept introduced in the Kenya case study of "response farming" (i.e., practices that are varied in response to, for example, the date of onset of rains). Here, farming activities are adjusted on the basis of a statistically derived climatological forecast of seasonal conditions (*see* Part II).

(7) Development of seasonal weather forecasts, which provide a firm basis for prognostic decision making in agriculture. Although accurate forecasts are not yet available, it would be instructive to develop methods of assessing how accurate they will *need* to be as operational tools.

(8) Dissemination of available information. Firstly, more mechanisms are required for integrating the findings of scientific research into the actual planning process in drought-prone areas. Policy makers and planners need to be made aware of the value of new approaches, particularly by demonstrations and through well-conducted pilot projects. Secondly, extension services need to be developed further to provide advice for farmers on production methods and to disseminate information.

An improved use of new agricultural technology. Two lines of approach appear most promising:

(1) To increase production potential in good rainfall years and thus cushion the impact of scarcity in drought periods. For example, experiments with a crop growth simulation model for sorghum in the India case study showed that yields in average and good rainfall years estimated for high management (180 000 plants/ha and a hybrid variety) were about three times greater than those estimated for low management (50 000 plants/ha and a local variety), and nearly ten times greater than actual recorded farm yields (*see* Part V). Similarly, fertilizer applications in wet years can markedly enhance crop yields (*see Table 1.4*, above).

(2) To reduce losses in poor rainfall years, by use of crop types and management techniques that mitigate drought effects, e.g., the use of irrigation and fallowing to increase and to stabilize yields.

A focus on integrated regional development. In some case studies in this volume it was natural that the problem of drought should be viewed, in part, as a problem of underdevelopment. This view implies that, to be successful, government interventions need to be concerned with large-scale economic and social transformations. The focus here is not only on promoting access to water and improving post-drought rehabilitation, but on raising living standards, thus reducing the vulnerability of farming households to loss of income resulting from drought. An example of this is Brazil's *Projeto Nordeste*, which aims, over 15 years from 1985 to 2000, and at a cost of US$12 billion, to increase resilience to drought in the Northeast of the country by increasing land ownership by small rural producers; increasing public provision for education, sanitation and health in rural areas; and promoting access to water, credit and technical assistance (for details, *see* Part III).

The development of "rational" drought policies. A strictly rational response by farmers and governments to drought requires information on the probability of occurrence and duration of droughts, and on the relative costs and benefits of alternative responses. The "public-good" nature of this information justifies government intervention in its acquisition and provision. However, a strong case may be made (cf. the Australian case study in this volume) for attempting to restrict *ad hoc* government standby assistance, which, at least in developed economies, can add to the uncertainty facing farmers, and probably results in less effective farmer decisions leading to lower output than would otherwise be the case. Governments can facilitate rational individual responses by adopting a consistently applied and widely known set of policies and by encouraging an efficient capital market to enable farmers to balance the variability of their income against their (less variable) household and business expenses (*see* Part VII).

Adjustments in regional land-use allocation

In addition to the more general responses summarized above, a number of specific adjustments were tested in some case studies in this volume, in particular those concerned with land-use allocations. Adjustments of this kind would be more easily implemented in centrally planned economies, but could also be implemented elsewhere.

Changes of land use to stabilize production. Here the aim is to allocate regional land uses in such a way that the risk of heavy losses due to climatic variations is reduced. For example, two types of strategy have been examined in experiments in the Marks district of the Saratov region, USSR, for a range of crop responses in the region and on the basis of known frequencies of different weather-types (*see* Part VI):

(1) *Maximizing the minimum return*: a strategy that is oriented toward the best possible return under the worst possible conditions. In the Saratov region this strategy would lead to the selection of a crop allocation pattern that provided the optimal return on production (by value) in the driest weather-year, whilst satisfying other production constraints (e.g., on labor,

costs, prices, etc.). The result would be a marked shift from spring cereals to more drought-resistant winter cereals (particularly wheat).

(2) *Minimizing the variability of return*: a strategy that aims at obtaining returns with the lowest possible variability. To illustrate: in the Saratov region the same strategy as in (1) above would be preferred, probably because of the high frequency of drought in that area (*see* Part VI).

Changes of land use to optimize production. A less risk-averse strategy than the previous examples involves optimizing crop allocation to provide the greatest average return over the long-term, i.e., *maximizing the (statistically) average return.* With this objective, under variable climatic conditions, it may be advantageous to plan toward exploiting occasional favorable weather-years, at the expense of some losses in the most unfavorable years. For example, an optimized allocation oriented toward years with limited drought was the compromise solution fulfilling this objective in the Marks district, providing mean returns of 7.69 million roubles, compared with 7.44 million under a dry-year strategy and 6.72 million under a sufficient-year strategy (*see* Part VI, Section 7). Further experiments in the Stavropol region, USSR, have investigated the benefit of using fallowing in the crop rotation (which conserves soil moisture) to obtain a similar goal of increased mean production under different climatic scenarios (*see* Part VI, Section 3).

Adjustments in resource use

In addition to land-use changes, other policy-related decisions can be used to optimize agricultural production. In centrally planned economies (unlike market economies, where activities tend to be price-driven) resource use (e.g., fertilizer application, mechanization, labor, etc.) is strictly regulated at regional and at state levels. The Stavropol study offers a useful example of how changes in resource use (labeled "technologies" in the study), in association with changes in fallow allocation, can be used either to stabilize or to increase total production under different climatic scenarios, and at what cost. Under the present-day climate, production could be increased by one-sixth for an additional expenditure of 21% and an increase in fallow allocation; but if the climate were to change to a regime with a doubled frequency of drought, the cost of achieving the same production increase would be 39% higher than the present level (*see* Part VI, Section 3).

1.6. Conclusions and Recommendations

The results summarized below have been drawn from six regional case studies in Kenya, Brazil, Ecuador, India, the USSR and Australia. For details concerning the accuracy, meaning and context of the results, reference should be made to the case studies themselves.

1.6.1. Possible effects of climatic variations

First-order effects on agriculture: short-term

Variations in climate can have a short-term effect on agricultural productivity in semi-arid areas through:

(1) Variations in seasonal crop yields, which exhibit a positive, broadly linear relationship to seasonal precipitation over a range of technology levels.
(2) Season-to-season yield variability, which tends to be greater in drier zones where yield levels in the driest years may be considerably lower than in more humid zones.
(3) Within-season yield variability, which results from the relationship between variations in the distribution of seasonal rainfall and crop growth.
(4) Variations in yield quality, which is related as much to rainfall distribution as to rainfall amount and is an important factor governing livestock nutrition.
(5) Variations of yield responsiveness to fertilizers, which is markedly higher under moist than under dry conditions.
(6) Carry-over effects from a previous season, for example, of soil moisture, which may offset or exacerbate the effect of anomalous weather on crop yield.
(7) Combined impacts on more than one system that may induce a nonlinear response on agricultural output.

First-order effects on agriculture: long-term

Variations in climate can have a long-term effect on agricultural productivity in semi-arid areas through:

(1) The effects of possible long-term climatic changes on mean yields that are positively related to average seasonal precipitation, but which display conflicting responses to changes in air temperature and solar radiation.
(2) The effects of possible long-term climatic changes on the probability of a given yield. Quite large changes in level of crop dependability may result from quite small changes of mean climate owing to the marked nonlinear relationship between changes in mean climate and the frequency of anomalous weather-years.
(3) Concurrent effects on soil fertility, which may alter in response to certain agricultural practices continued over the long term (either in the presence or in the absence of climatic change) and affect the mean level of crop yield.

Effects on the spatial pattern of agricultural potential

In addition to the effect of variations in climate on yield at a given location, there are important differences of response between various plants and animals in various regions. This can result in substantial spatial shifts of agricultural potential, particularly:

(1) Short-term shifts of crop potential resulting from season-to-season variations, which can lead to major movements of the boundaries of agroclimatic zones.
(2) Long-term shifts of crop potential that would result from possible changes in mean climate, leading to the contraction or extension both specifically of physiological limits to certain crops and, more generally, of the area over which certain crops are at present grown in semi-arid regions.
(3) Spatial shifts of agricultural risk that would also result from possible changes in mean climate, leading to changes in the choice of crops grown by farmers.

Higher-order effects

Although the economic and social effects of drought vary greatly according to variations in farmers' asset base and their attitudes to risk, some effects were found to occur quite generally in the case studies in this volume, namely, effects on:

- Household food consumption and production.
- Household purchases.
- Household assets.
- Regional output.
- Regional income.
- Nutritional status, mortality, etc.
- Structure of land ownership.
- Rate of long-term environmental degradation.

Some potential adjustments at the farm level

A number of adjustments to climatic variations were considered in the case studies, particularly:

(1) *Long-term adaptations, viz:*
 - Crop-based diversification.
 - Spatial diversification.
 - Occupational and income diversification.
(2) *Short-term ex ante adjustments, viz:*
 - Variations in cropped area and location of cropping.
 - Variations in extent of intercropping.
 - Variations in use of crops with different moisture requirements and different growth periods.
 - Variations in date of planting, numbers of plantings and planting density.

- Variations in amount of tillage, mulching, fallowing and fertilizing.
- Variations in amount and scheduling of irrigation.

Increasing land shortage and a growing emphasis on mono-cropping are reducing the opportunity for some of these types of adjustment.

(3) *Short-term ex poste adjustments, viz:*
- Adjustments in farm practice, such as early harvesting, abandonment of post-sowing operations, increased weeding of crops, etc.
- Adjustments in household management, such as altered food consumption, financial measures ranging from the use of savings to the sale of assets, irrigation, and changes in employment.

Some potential policy responses at the regional and national levels

Existing types of public intervention, often *ad hoc* in form, include:

(1) Food, credit and income supply measures, by storing food and fodder, by substituting factors of production, or by generating alternative income.
(2) Large-scale provision for the storage or transfer of water.
(3) Changes in land management (e.g., soil moisture conservation) and reorganization of agricultural infrastructure (e.g., of marketing and credit systems).
(4) Introduction of drought-resistant crops, and drought-offsetting actions (e.g., improved timing of planting, spraying and harvesting).

There is a need for increased complementarity among information, technology and public intervention through:

(1) Improved information on agroclimatic potential, i.e., a greater range of data, more computer-based analysis, more agroclimatic screening of environments to match agricultural activities to regional weather-types and improved weather forecasting.
(2) Improved use of new agricultural technologies (such as high-yielding varieties) to increase production potential in good years and reduce losses in poor years.
(3) A focus on integrated regional development to reduce overall vulnerability to drought by increasing public provision for welfare and production.
(4) Development of a consistently applied and widely known set of drought policies to reduce the uncertainty that stems from *ad hoc* public intervention.
(5) Adjustments in regional land-use allocations to stabilize or optimize production.
(6) Adjustments in resource use (e.g., of expenditures on fertilizers, irrigation, etc.) to stabilize or maximize production.

1.6.2. Recommended research approaches

Four general research approaches have emerged from this study as valuable, complementary ways of assessing the possible effects of climatic variations:

(1) An *integrated approach* to climate impact assessment is needed in order to comprehend the interconnections between effects of climatic variations on various sectors of society and economy (e.g., on agriculture, energy, industry, etc.).
(2) An *iterative approach*, usually within a model framework, is useful, both for evaluating the effects of climatic variations and for evaluating a range of adaptive responses.
(3) An *adjoint approach* to the study of effects of climatic variations can be as valuable as the *direct approach.* The former focuses on the intrinsic sensitivities of economy and society to climate and then asks, "What type or degree of variation in climate would cause a significant perturbation of them?" The latter asks the question, "For a given climatic variation, what is the impact on (for example) ecosystems, economy and society?" Both approaches are valid, indeed complementary, lines of inquiry. An advantage of the former is that it can assist in the identification of *thresholds* or *distinct nonlinearities in response* to climatic change, helping us recognize what kinds of climatic change are critically important and what are not.
(4) The *spatial analysis* of effects is an essential part of climate impact assessment. There are numerous examples given in this volume of the value of specifying changes in climate as spatial shifts of climatic types, or vegetation zones, or levels of agricultural potential, etc.

Within each of these broad approaches a number of more specific tasks can be recommended. These are considered below.

1.6.3. Research priorities

This study has highlighted a number of priorities for future research. In summarizing these we shall refer to *Figure 1.14*, which depicts the approach adopted in the IIASA/UNEP project and identifies points that warrant further attention. These research needs are summarized below, together with examples (in italics) specifically related to effects of climatic variations on agriculture in semi-arid areas:

(1) More appropriate data on climatic variability and change:
- Greater spatial and temporal resolution and range of data on variability – *e.g., to enable screening of environments on the basis of rainfall variability and enable the identification of broadly homogeneous agroclimatic zones as a basis for agricultural planning.*
- More information on likely rates of change – *e.g., to enable consideration of effects on the frequency of drought events.*

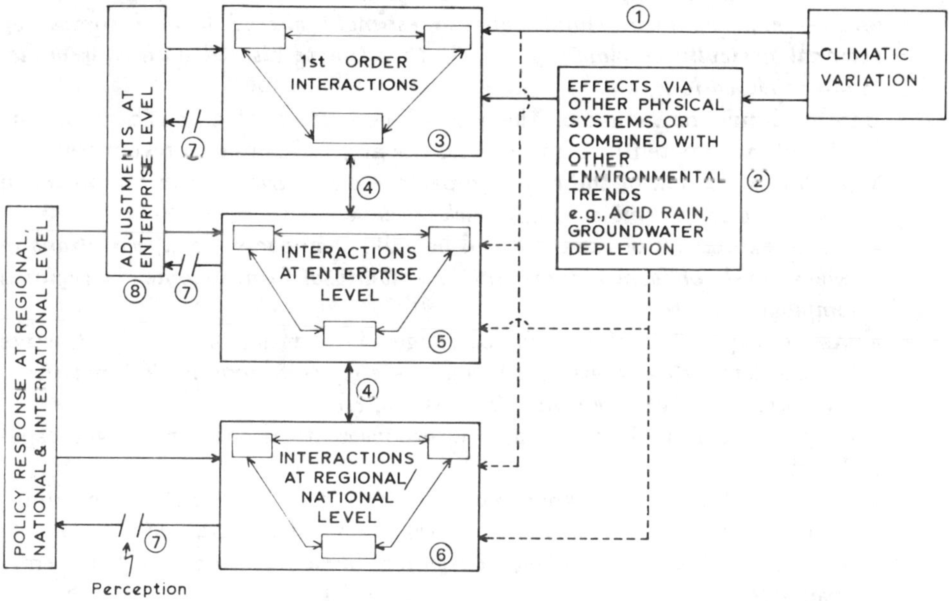

Figure 1.14. Research priorities in the assessment of effects of climatic variations: The schema of the IIASA/UNEP project's approach, with priority tasks identified by numbers that are described in the text.

- Improved seasonal weather forecasts – *e.g., to optimize the planting time, irrigation and fertilization of crops.*

(2) Improved understanding of the relationship between variations of climate and variations in other physical systems; specifically, climate and:
- Soil nutrients, soil erosion and salinization – *e.g., to enable consideration of effects of precipitation changes on nutrient mineralization rates.*
- Pest and diseases, and their vectors – *e.g., to enable analysis of changes in crop and livestock losses.*

(3) Improved understanding of the relationship between climatic variations and first-order impacts in agriculture; specifically:
- Climatic variations and production potential – *e.g., effects on indices of agricultural potential.*
- Climatic variations and agricultural yields – *e.g., by empirical or simulation modeling of crop-climate and livestock-climate relationships.*
- Direct impacts of climate on input requirements – *e.g., variations in fertilizer responsiveness.*

(4) Greater precision in tracing the higher-order effects of climatic variations in the secondary and tertiary sectors; specifically:
- By using hierarchies of linked climate, impact, and economic models to trace the pathways of impacts – *e.g., to trace effects of altered climate on crop yields, on gross production, on farm profitability, on regional competitiveness, etc.*

(5–6) Greater precision in simulating impacts at the enterprise level and regional level – *e.g., by farm simulations to estimate altered farm incomes, by regional agricultural planning models to estimate altered optimal regional land-use allocations, etc.*

(7) Improved understanding of the role of perception of climatic variations (their actual or potential impacts, and the array of potential responses):
- As an expression of the direct impacts – *e.g., variations in the mean, in frequency of extreme events, in risk levels, etc.*
- As an expression of the costs and benefits of impacts – *e.g., variations in mean levels of farm income, in the stability of farm income, in regional employment, etc.*
- As an expression of the choice range of potential or actual adjustive mechanisms; their costs and benefits – *e.g., variations in diet/nutrition, in education of children, in sale of assets, etc.*

(8) Improved understanding of adaptive responses at the enterprise level, especially of:
- Farm-level decision making, particularly in the face of risk, uncertainty, and constraints on choice – *e.g., the range of on-farm adjustments of management and technology (irrigation, diversification, etc.) currently available.*
- The resilience of economic systems – *e.g., the relative resilience of mono-cropping and of various degrees of diversification to climatic variations.*

(9) Strategy formulation at the regional, national and international levels, especially with respect to:
- The range of possible policy responses – *e.g., government price support, crop insurance schemes, food supply measures, etc.*
- The process of policy formulation – *e.g., centralized planning of production targets and land allocation, strategies of regional development, public measures to support the development, adoption and dissemination of new technology, etc.*
- The priority areas for resource allocation – *e.g., irrigation, plant breeding, agricultural extension services, fertilizer trials, etc.*
- The constraints on choice (with respect to other resource needs and problems) – *e.g., competing land uses, limits on expenditure, subsistence objectives, etc.*

1.6.4. The structure of this volume

The remainder of Part I of this volume considers, first, the choice of impact models and, second, the development of climatic scenarios for the case study experiments. The remaining six parts of the volume contain the results of the case studies.

References

Baird, A., O'Keefe, P., Westgate, K. and Wisner, B. (1975). Towards an Explanation and Reduction of Disaster Proneness. *Occasional Paper 11, University of Bradford, Disaster Research Unit.*

Bolin, B., Döös, B.R., Jäger, J. and Warrick, R.A. (1986). *The Greenhouse Effect, Climatic Change and Ecosystems*, SCOPE 29, Wiley, Chichester, 512 pp.

Climate Impact Assessment Program (CIAP) (1975). *Impacts of Climatic Change on the Biosphere*, Monograph 5, US Department of Transportation, Washington, D.C.

Garcia, R. (1981). *Drought and Man: The 1972 Case History. Vol. 1: Nature Pleads Not Guilty*, Pergamon, New York.

Gibbs, W.J. (1984). The great Australian drought: 1982–83. *Disasters*, **8**(2), 89–104.

Hansen, J., Lacis, A., Rind, D., Russell, G., Stone, P., Fung, I., Ruedy, R. and Lerner, J. (1984). Climate sensitivity: analysis of feedback mechanisms. In J. Hansen and T. Takahashi (Eds.), *Climate Processes and Climate Sensitivity*, Maurice Ewing Series, **5**, pp. 131–63, American Geophysical Union, Washington, D.C.

Hansen, J., Russell, G., Rind, D., Stone, P., Lacis, A., Lebedeff, S., Ruedy, R. and Travis, L. (1983). Efficient three-dimensional global models for climate studies: Models I and II. *Monthly Weather Review*, 111, 609–662.

Hare, F.K. (1985). Climatic variability and change. Ch. 2 in R.W. Kates, J.H. Ausubel and M. Berberian (Eds.), *Climate Impact Assessment: Studies of the Interacton of Climate and Society*, SCOPE 27, Wiley, Chichester, pp. 37–68.

Kates, R.W. (1985). The interaction of climate and society. Ch. 1 in R.W. Kates, J.H. Ausubel and M. Berberian (Eds.), *Climate Impact Assessment: Studies of the Interaction of Climate and Society*, SCOPE 27, Wiley, Chichester, pp. 3–36.

National Defense University (NDU) (1980). *Crop Yields and Climatic Change to the Year 2000*, Vol. 1, Washington, D.C., Fort Lesley, J. McNair.

Parry, M.L. (1985a). Estimating the sensitivity of natural ecosystems and agriculture to climatic change – guest editorial. *Climatic Change*, **7**, 1–3.

Parry, M.L. (1985b). Assessing the impact of climatic change on marginal areas. Ch. 14 in R.W. Kates, J.H. Ausubel and M. Berberian (Eds.), *Climate Impact Assessment: Studies of the Interaction of Climate and Society*, SCOPE 27, Wiley, Chichester, pp. 351–367.

Parry, M.L. and Carter, T.R. (Eds.) (1984). Assessing the Impact of Climatic Change in Cold Regions, *Summary Report SR-84-1*. International Institute for Applied Systems Analysis, Laxenburg, Austria, 42 pp.

Parry, M.L., Carter, T.R. and Konijn, N.T. (1985). Climate change. How vulnerable is agriculture? *Environment*, **27**, 4–5, 43.

Parry, M.L., Carter, T.R. and Konijn, N.T. (Eds.) (1988). *The Impact of Climatic Variations on Agriculture. Volume 1, Assessments in Cool Temperate and Cold Regions*, Reidel, Dordrecht, The Netherlands.

Powell, R.A. (1978). Stagnating economic growth: its effect on agriculture and regional implications. *Paper presented to the Fourth Advanced Studies Institute in Regional Science*, Siegen, Federal Republic of Germany.

Vinnikov, K.Ya. and Groisman, P.Ya. (1979). An empirical model of present-day climatic changes. *Meteorologia i Gidrologia*, **3**, 25–28 (in Russian). English translation available in: *Soviet Meteorol. Hydrol.*, **3**.

Williams, G.D.V., Fautley, R.A., Jones, K.H., Steward, R.B. and Wheaton, E.E. (1988). Estimating effects of climatic change on agriculture in Saskatchewan, Canada. In M.L. Parry, T.R. Carter and N.T. Konijn (Eds.), *The Impact of Climatic Variations on Agriculture. Volume 1, Assessments in Cool Temperate and Cold Regions*, Reidel, Dordrecht, The Netherlands.

SECTION 2

The Choice of First-Order Impact Models for Semi-Arid Regions

Timothy Carter and Nicolaas Konijn

2.1. Introduction

The variability of rainfall in space and time is such that, even in climates that are classified as humid, crop yield losses due to limited water availability can occur. In these circumstances the effect of a dry spell can often be minimized through the use of supplemental irrigation. However, in semi-arid regions not only is rainfall limited, but supplementary water resources are often absent, too, and irrigation of agricultural crops is possible only in restricted areas. Thus, agricultural production in these areas is based mainly on rainfed agriculture.

A rainfed production environment is very sensitive to relatively small variations in the climate. It requires from the farmer the capacity for rapid adjustment and adaptation to prevailing conditions. For example, the carrying capacity of the land for livestock may fluctuate markedly from year to year, and grasslands that are suitable for livestock one year may deteriorate quickly under adverse conditions the next. Difficulties associated with overgrazing, including the possibility of irreversible changes in the production environment through soil erosion, are an ever-present danger unless checked at an early stage. Attempts to model agricultural production in semi-arid regions therefore require an understanding of the production environment, the range of plant and animal responses and the tools available to the farmer to reduce the negative effects of water stress on crop and livestock productivity.

In this section, we provide a general introduction to the types of models that have been used in this volume for estimating the first-order effects of climatic variations on agricultural production in the semi-arid case study regions. We begin by describing some characteristics of agricultural production in semi-arid areas, including the physical constraints on agriculture, types of crop, cultivation practices, livestock production and other limiting factors. We next examine some methods of estimating the effects of climatic variations on agricultural production. The development, testing and use of agroclimatic models in climate impact assessment are discussed in depth in the companion volume (Carter *et al.*, 1988). Here we concentrate on aspects of particular importance for applications in semi-arid regions and consider, in addition to agroclimatic models, the use of livestock models and agroclimatic zonation schemes. These are illustrated with examples from the regional case studies and summarized in a model checklist.

2.2. Climatic Variations and Agricultural Production in Semi-Arid Regions

2.2.1. Characteristics of the physical production environment

Before characterizing particular features of semi-arid regions, it is useful to identify the geographical location and extent of these areas. We have adopted a climate-based classification of dry regions both to identify the distribution of semi-arid zones and to illustrate, in the most general terms, the average global moisture regime. Other possible classification criteria are discussed in Part I, Section 3. Subsequently, we have related these climatic zones to characteristic soil types.

Average moisture regime

Figure 2.1 shows the extent of dry regions based on the Thornthwaite (1948) moisture index and refined by grouping areas with a similar seasonal distribution of precipitation and similar temperatures for the coldest and warmest months (after Meigs, 1953). Three degrees of aridity are delimited: extremely arid, arid and semi-arid. The correspondence of these zones with the case study regions investigated in this volume can also be noted.

One of the characteristic features of semi-arid regions not illustrated in *Figure 2.1* is the high interannual variability both of the total and the distribution of precipitation. This, of course, has a critical bearing on year-to-year agricultural productivity, and forms a central theme of this section. A global-scale map of precipitation variability is presented in Section 3 (*Figure 3.8*).

Dominant soils

It is natural to expect the large-scale climatic zones (such as those depicted in *Figure 2.1*) to be mirrored closely by the spatial distribution of major soil classes. *Figure 2.2* provides a global picture of the distribution of soils over the three dry climatic zones identified above. It is based on the US Comprehensive Soil Classification System – 7th Approximation (Soil Survey Staff, 1960; 1976). Only the soils at the highest level of taxonomy (broadest category) are presented. This is a morphometric system where soils are classified in accordance with their

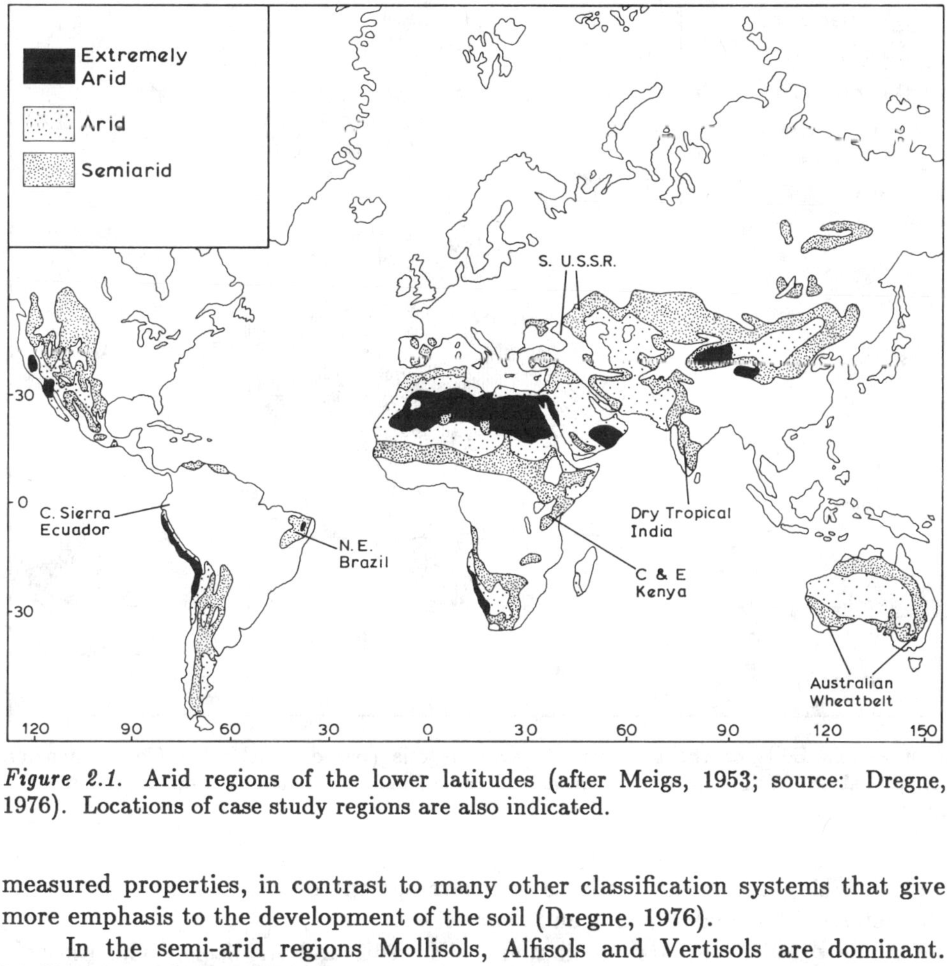

Figure 2.1. Arid regions of the lower latitudes (after Meigs, 1953; source: Dregne, 1976). Locations of case study regions are also indicated.

measured properties, in contrast to many other classification systems that give more emphasis to the development of the soil (Dregne, 1976).

In the semi-arid regions Mollisols, Alfisols and Vertisols are dominant. Mollisols are characterized by a dark top soil, usually the result of an accumulation of organic matter. In the older genetic classification systems they were called chernozem and chestnut soils; the first located in the wetter semi-arid regions, the second found under a slightly drier regime.

Alfisols are also found in the wetter semi-arid regions, these soils having a well-developed horizon with accumulated clay, which has been eluviated from the overlying horizons. In some regions of South America and Africa, Alfisols grade into Ultisols, although the latter are not included in *Figure 2.2.*

Vertisols are characterized by a high clay content and a typical clay mineral composition. The clay minerals, usually montmorillonite, will expand or shrink, depending on the moisture conditions. If there is a pronounced dry season during which the soil dries out and cracks are formed, soil aggregates of the topsoil will fill up the cracks. With rain the soil swells and becomes mixed, often creating a dark horizon some 40 cm or more thick. At the surface, this swelling

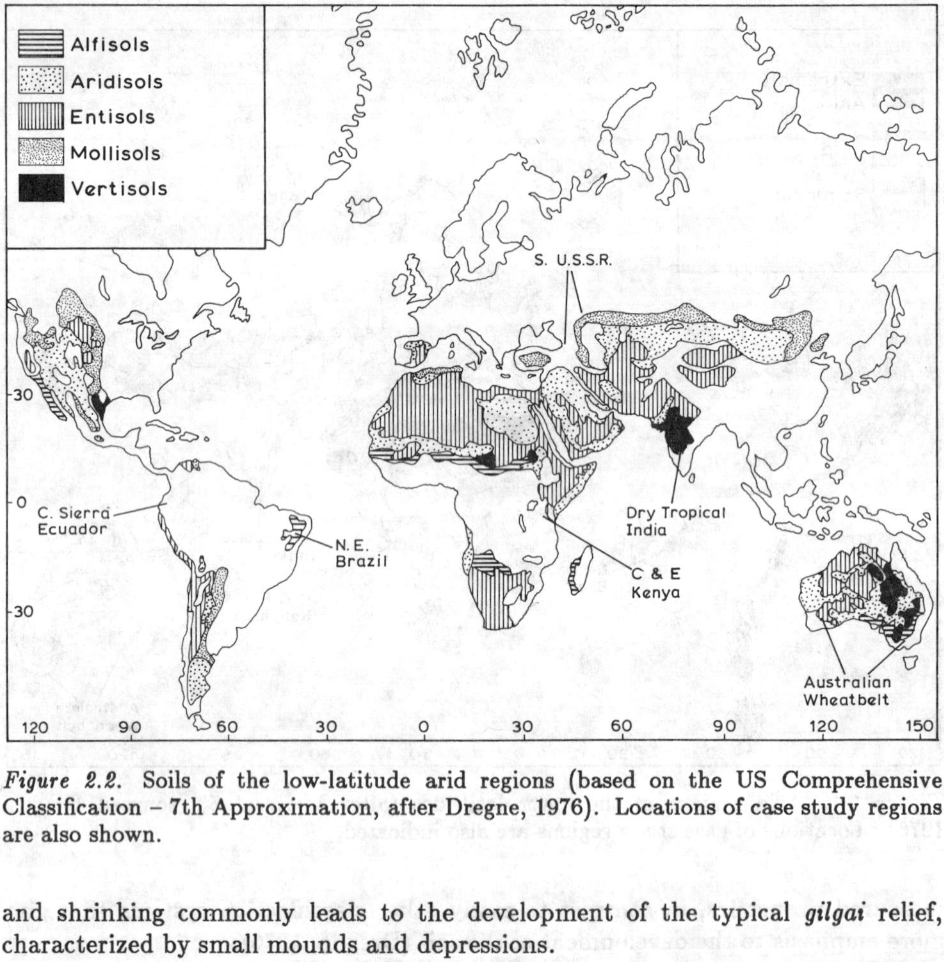

Figure 2.2. Soils of the low-latitude arid regions (based on the US Comprehensive Classification – 7th Approximation, after Dregne, 1976). Locations of case study regions are also shown.

and shrinking commonly leads to the development of the typical *gilgai* relief, characterized by small mounds and depressions.

The driest areas are dominated by Entisols and Aridisols. The Entisols show very little or no soil development. This might be due partly to the fact that little or no transport of material by means of water in the soil can take place. Often, however, the Entisols are reworked old material, recently deposited, such as blown sand. The Aridisols differ from Entisols in that they exhibit distinctions in their horizons that are characteristic of soil development in arid regions (e.g., soluble salts and carbonates). In cases where pronounced soil development can be observed, this may imply that the present soil is a relict of a former humid climate.

The case study regions are also shown in *Figure 2.2*, but the map only provides the most general information about regional soil types. More detail is presented, by region, in *Table 2.1.* It is clear from this table (itself highly generalized) that most of the study regions exhibit a wide range of soil characteristics, some of which do not fall conveniently into an "arid region" soil class. Since soil properties, together with the amount and distribution of precipitation, largely determine to what extent rainfed agriculture is possible, such a diversity

Table 2.1. Major soil types and general physical characteristics of the case study regions. Dominant soils are in solid capitals. Sources: FAO/UNESCO (various dates); Dregne (1976).

Region and part of study	*Soil types*	*General characteristics*
Central and eastern Kenya (Part II)	ALFISOLS INCEPTISOLS OXISOLS Vertisols lithic subgroups	Moist to intermediate savannah vegetation Complex soils, many of volcanic origin especially Inceptisols (suborder: andept); Alfisols in drier regions. Gently undulating topography.
Northeast Brazil (Part III)	ALFISOLS Entisols Oxisols Vertisols	Widespread *caatinga* vegetation with cacti and thorny woodland. Dominant Alfisols are moderately well drained and quite fertile. Topography is rolling and hilly.
Central Sierra of Ecuador (Part IV)	INCEPTISOLS (andepts) Mollisols Entisols lithic subgroups	*Paramo* vegetation ranging from low forest to tussock grassland. Highly varied topography with climate ranging from sub-humid to semi-arid. Widespread volcanic soils (suborder of Inceptisols: andept) grade into prairie soils (Mollisols) in dry regions.
Dry tropical India (Part V)	VERTISOLS Alfisols Aridisols Entisols	Mainly degraded savannah or tropical dry forest with small trees and shrubs. Vertisols are dominant soils, derived from basalt. A *gilgai* relief is common.
Saratov and Stavropol regions, S. USSR (Part VI)	MOLLISOLS	Steppe, ranging from sub-humid silvo-steppe to dry caspian steppe. Dominant soils are Mollisols: more fertile chernozems in humid areas, less fertile chestnut soils in drier areas.
Australian wheatbelt (Part VII)	ARIDISOLS ALFISOLS VERTISOLS Entisols	Vegetation ranges from sub-humid temperate woodland to semi-arid shrub. Soils are varied, but all are characteristic of arid regions.

of soil types needs to be adequately accounted for in agroclimatic models developed for these regions.

2.2.2. Physiological and morphological plant properties

Photosynthesis pathways

Practically all plants of agricultural importance are green and therefore employ photosynthesis to fix carbon dioxide and synthesize carbohydrates. According to the efficiency of photosynthesis, so-called C-3 and C-4 plants can be distinguished, after the first stable carbon compound formed in these plants. The agricultural crops of the C-4 plant type [e.g., sorghum (*Sorghum bicolor*) and pearl millet (*Pennisatum typhoideum*)] are generally better equipped to grow

Table 2.2. Average properties of four groups of crops. Footnotes relate crops to regions studied in this volume. Source: FAO (1978).

Properties	*Crop adaptability groups* *I*	*II*	*III*	*IV*
Photosynthetic pathway	C-3	C-3	C-4	C-4
Rate of photosynthesis at light saturation and optimum temperature (mg CO_2 dm^{-2} h^{-1})	20–30	40–50	70–100	70–100
Optimum temperature for maximum photosynthesis (°C)	15–20	25–30	30–35	20–30
Radiation intensity at maximum photosynthesis (W m^{-2})	140–420	210–560	>700	>700
Major crops	Barley[c] Wheat[e,f,g] Potato[c] Phaseolus bean (Ttrh)	Rice Soybean Cotton[b] Sweet potato Cassava Phaseolus bean (Tr)[b]	Maize (Tr) Pearl millet Sorghum (Tr)[d] Sugarcane	Maize (Ttrh)[a,c] Sorghum (Ttrh)

Tr = Tropical cultivars.
Ttrh = Temperate and tropical high altitude cultivars.
[a]Central and eastern Kenya (Part II).
[b]Northeast Brazil (Part III).
[c]Central Sierra of Ecuador (Part IV).
[d]Dry tropical India (Part V).
[e]Stavropol region, S. USSR (winter sown – Part VI).
[f]Saratov region, S. USSR (spring sown – Part VI).
[g]Australian wheatbelt (winter sown – Part VII).

under prolonged warm and dry conditions than those of the C-3 type [e.g., wheat (*Triticum aestivum*) and barley (*Hordeum vulgare*)]. *Table 2.2* presents a comparison between typical C-3 and C-4 plant characteristics.

Two properties of C-4 plants explain their ability to grow relatively well under water stress conditions. Firstly, when water stress occurs, the stomata in plants respond by closing, thus limiting transpiration. This leads to an increase in the plant's temperature, but since C-4 plants are more heat-tolerant than C-3 plants, they grow comparatively better. A second advantage of C-4 plants is their ability to maintain photosynthesis at a high level despite relatively low carbon dioxide concentrations. Carbon dioxide is absorbed through stomata, but this means that the reduction in stomatal aperture caused by water stress affects C-4 plants less than C-3 plants.

A third group of plants, the so-called Crassulacean Acid Metabolism (CAM) type, can also be distinguished that are well adapted to growth in arid conditions. They include pineapple (*Ananas cosmosus*) and sisal (*Agave sisalana*), but are only of minor agricultural significance in the case study areas examined in this volume.

Water availability is not the only climatic factor that determines the suitability of crops in different regions; temperature and radiation conditions are also important. Given the varied latitudinal and topographic locations of the case study regions, these factors help to explain the diversity of crop cultivars studied in the different semi-arid areas (*Table 2.2*).

Genetic properties and plant adaptation

Agricultural crops in semi-arid regions are required not only to survive under drought conditions, but also to yield a useable product. This entails careful selection and breeding of crops that are both adaptive to variations in environmental conditions and resilient to adverse circumstances, such as drought. Seed dormancy properties serve as a good example of this. At any time, as long as moisture and temperature conditions are adequate, non-dormant seeds can germinate. Yet in arid conditions this may involve a high degree of risk, for germination may occur at a time when subsequent water availability is insufficient for crop growth. Therefore, dormancy properties can be of great value for ensuring that germination takes place at the best possible time for subsequent crop development and yield.

In addition, given the short growing period that is commonly associated with semi-arid conditions, it is often important to select cultivars with a short growing cycle. For example, grain cultivars have been bred that develop rapidly up to the heading stage, in order to lengthen the grain filling period and enhance crop yield.

Crop plants in arid regions are adapted to the dry environment in various other respects as well. An increase in water use efficiency is achieved by some species through leaf rolling, a phenomenon that can be observed in several grasses under conditions of water stress. Sorghum has the ability to regenerate once a part of its above-ground foliage has died off, through activation of the dormant tillers that can be found at the bottom of the stem. If conditions are favorable these buds might shoot at a later growing stage and add to the production.

Most plants respond to dry conditions with a change in the shoot/root ratio. By increasing root density and rooting depth, plants can extract water from the soil more efficiently and to a greater depth (Hurd, 1968). Alternatively, certain plants develop a shallow, but horizontally extensive root system, which is a logical adaptation to situations where the water table is very deep or absent, and where there is limited rainfall, which only wets the soil superficially. Moreover, it has also been observed that certain plants have the ability to prevent all others in their vicinity from competing for water. It has been hypothesized that they produce germination inhibitors that act not only on other species, but also on their own species (Went, 1948). Only following their death may other plants replace them.

There are numerous examples of adaptations in plant morphology. For instance, under high radiation conditions plants with a waxy surface reflect radiation more efficiently, reducing the danger of heat stress. A typical example is sorghum, for which the configuration and position of the stomata also reflect drought resistance: the stomata might be located only on the shaded side of the leaf, and are small and numerous rather than large and few. The latter aspect is not necessarily determined genetically; many plants develop this property, even within a single life cycle, and it can be considered as an automatic hardening response mechanism.

2.2.3. Cultivation practices

Plant cultivation practices in dry regions are designed to conserve water and to reduce water losses through evaporation and transpiration. One of the oldest practices is that of dry farming. Fallow periods in the crop rotation are introduced to increase the soil water storage. During these periods the soil is kept weedless in order to reduce transpiration, and often the creation of a loose topsoil layer helps to reduce evaporation. Weed control is essential in arid regions. Weeds compete for soil moisture with the cultivated plant, and since low plant densities do better under dry conditions, improved light conditions for weeds enables them more actively to compete with the crop.

Reductions of evaporation and transpiration can be achieved by various other methods. Shelterbelts, whether natural (e.g., trees) or artificial (e.g., fences) will reduce wind speed and consequently the drying power of the air is decreased. Such an effect can be quantified by means of evaporation measures that incorporate a wind speed term (e.g., the Penman equation, Penman, 1948). Of course, this is not the only advantage of shelterbelts; they also offer protection from wind erosion and, in the colder semi-arid regions such as the southern European USSR (*see* Part VI), they might act as snow fences in the winter months, helping to conserve soil moisture.

More modern methods, although certainly not economic under all circumstances, include the reduction of plant transpiration by the use of chemical sprays that act as stomatal inhibitors (Waggoner and Bravdo, 1967). Beside their direct anti-transpirant properties, these chemicals can also increase plant reflectivity, thus indirectly decreasing transpiration through reduced plant temperature. The use of plastic covers (polyethylene) in horticulture helps to reduce soil evaporation, but the high investment required is rarely justifiable on economic grounds. In fact, this effect can be achieved, to a limited extent, by the old practice of creating a mulch layer (e.g., mixing the topsoil with residual plant matter).

It is not always easy to assess the wider consequences of certain cultural practices; it requires considerable knowledge of the processes involved to anticipate the full effects. For example, mulch layer formation might help to conserve water, but at the same time it affects the soil temperature. This might be beneficial in warm dry regions for, during germination, temperatures near the soil surface may otherwise rise to damagingly high levels. Some positive effects of shelterbelts have been mentioned above, but a decrease in wind speed can also mean reduced replenishment of the carbon dioxide concentration within a crop, with likely adverse consequences for the assimilation of carbon dioxide.

Conservation of water can also be achieved by improving its infiltration into the soil. Rainfall often occurs as storms with a high intensity that exceeds the infiltration rate, leading to rapid runoff and unnecessary water losses. Cultivation practices, such as soil terracing or contour plowing, will decrease the rate of runoff, thus increasing infiltration. Mulch layer formation or tillage practices, such as chiseling, may also be effective in increasing the water intake by soil. In some special cases deep plowing might also be helpful by bringing to the surface a soil layer with a better infiltration capacity.

The amount of rainfall and the rainfall distribution will dictate whether runoff reduction is an advantage for crop growth. If rainfall is very low it might be better to create runoff, such as in the practice of water harvesting. By various means rainfall is forced into runoff (e.g., by increasing the density of the topsoil) and guided to locations where water is most needed.

2.2.4. Livestock production

In spite of the emphasis in this volume on the responses of field crops to climatic variations, the role of livestock production in most of the study regions should not be understated. In the less developed agricultural regions, livestock fulfill a vital role in risk-reduction strategies for small farmers, some of which are directly beneficial for crop production (including the provision of valuable natural fertilizer for agricultural crops, draft power for tillage and income through sales for the purchase of cropping inputs). Livestock production also accounts for a sizeable proportion of agricultural output (by value) in developed countries, such as the USSR and Australia.

In semi-arid regions, as for crop production, it is water supply that is the single major limiting factor for livestock well-being. In dry years, when both the quantity and (often) the quality of normal grazing land are reduced, unless alternative sources of grazing or supplementary feed are available, animal nutrition is inevitably reduced. Lactation is the most nutritionally demanding aspect of livestock production, and a reduction in milk production is a common early indicator of decreased nutrition. Similarly, in the case of meat production, good nutritive *quality* of feed, as well as sufficient quantity, appear to be necessary conditions for efficient liveweight gain. Both conditions are closely related to rainfall receipt, with results from semi-arid areas of Kenya indicating that whereas the quantity of feed is related to *total* rainfall, its quality is more closely related to the *distribution* (i.e., duration) of rainfall (*see* Part II, Section 5). Indeed, experimental evidence from the semi-arid Sahel points to a positive relationship between mean annual rainfall (MAR) and the biomass production of rangelands, but a negative relationship between MAR and the protein content of rangeland vegetation (Breman and de Wit, 1983).

Farmers in subsistence economies have several options for maintaining stock levels during a drought situation. Crop residues and communal grazing land can be used as supplementary sources of feed, and those farmers with the resources can purchase feed. However, if these strategies fail, the farmer's final recourse for survival is to sell the stock. In this situation, animals are often in poor condition, prices are low and animal mortality is inevitable. With a subsequent return to adequate grazing conditions, however, a rise in prices often presents a significant obstacle to stock replacement following a drought. Climatic variations may thus be regarded as only one (perhaps triggering) mechanism contributing to variations in livestock production. Economic, social and political factors are also of fundamental importance.

2.2.5. The role of technology

In the foregoing subsections numerous aspects of the physical environment have been described that control agricultural production in semi-arid regions. It is clear that under rainfed conditions farmers only have limited means to combat the effects of drought. However, farmers in developed countries (DCs) are in a much better position to cope with drought events than their counterparts in less developed countries (LDCs). We can illustrate this with reference to four aspects of crop production.

(1) *Land preparation* for crops is often impeded in semi-arid regions because of problematic soil conditions. For example, the topsoil of heavy soils, such as Vertisols, is usually dry and hard at the time of plowing, and in many LDCs the implements available are often inadequate for proper cultivation.

(2) Even if land preparation is successful and planting is timely, these benefits are likely to be negated if a *crop's growth cycle* is not effectively matched to the length of growing period. The best-adapted and highest-yielding cultivars are often hybrids that farmers cannot produce themselves, and these are usually prohibitively expensive for farmers in LDCs to purchase.

(3) *Weeds and pests* pose a continual threat to crop production. Under conditions of limited water, low sowing densities of seed tend to produce higher crop yields. However, at the same time this practice provides a better opportunity for weeds to grow, making weed control indispensable. The labor requirement for weed control is high, but this is usually a cheaper option in LDCs than the use of expensive herbicides. Likewise, pesticide use is often restricted in LDCs on economic grounds.

(4) Alongside water deficiency, *nutrient deficiency* is probably the major limiting factor affecting crop production (Breman and de Wit, 1983). This is especially true in situations where precipitation receipt is above average, and where plants are unable to take full advantage of the available water because nutrients are limiting. The application of fertilizers is the obvious solution for improving production, but use of fertilizers is largely price-driven and hence is limited in LDCs.

2.3. Modeling the Effects of Climatic Variations on Agricultural Production

Three broad approaches are described in the case studies presented in this volume for characterizing regional agroclimates and for quantifying the effects of climatic variations on agricultural production:

(1) Agricultural and ecological zonation schemes.
(2) Agroclimatic model experiments.
(3) Livestock production modeling.

2.3.1. Agricultural and ecological zonation

A useful framework for assessing the climatic suitability of areas for crop cultivation or livestock production is provided by agricultural and/or ecological zonation schemes. Zones are mapped, usually at the national or regional level, relating the suitability of crop and forage plants to local conditions of climate, soils and topography.

Three main uses of zonation schemes can be identified that are of value in climate impact analysis:

(1) For identifying coherent agroclimatic subregions.
(2) For delineating seasonal or annual geographical shifts in crop potential.
(3) For demarcating possible longer-term shifts in agroecological potential.

We will illustrate each of these with examples from the case studies; *Table 2.3* summarizes some of the methods employed.

Coherent agroclimatic subregions

The highly variable temporal and spatial precipitation conditions that are characteristic of semi-arid areas make it very difficult to conduct quantitative assessments of agricultural sensitivity to climate over large geographical areas. Zonation schemes provide a means of identifying broadly coherent agroclimatic subregions, within which specific climate impact experiments can be assumed to be of general applicability.

Most zonation schemes were designed for agricultural planning purposes, indicating which areas in a region are climatically suitable for different crops. The most common procedure is to combine several climatic variables into an *agroclimatic index*, which is calculated over a network of locations and then mapped as a series of isolines. This approach is used to map agroclimatic zones in Kenya (*see* Part II), employing a ratio of mean annual precipitation to mean annual potential evaporation, and to map agricultural zones in the Stavropol and Saratov regions of the southern European USSR (*see* Part VI), based on an index of aridity (the hydrothermal coefficient). In both of the above examples, the zones serve as the basic regional units for describing the climate, soils and

Table 2.3. Some methods of agricultural and ecological zonation.

Zonation	*Region of use & part of study*	*Data requirements*	*Method*	*Purpose of use*
Agroclimatic zones	Central and eastern Kenya (Part II)	Precipitation; potential evaporation; altitude	Compute index; map	Characterization of long-term and seasonal agricultural potential
Agroclimatic suitability zones	Chimborazo Province, Ecuador (Part IV)	Precipitation; temperature; critical altitude	Identify critical isolines; map	Delimitation of crop suitability for barley and potatoes
Holdridge Life Zones	Chimborazo Province, Ecuador (Part IV)	Temperature; precipitation; altitude; soil types	Derive biotemperature and isohyets; map	Location of potential vegetation zones
Coherent agroclimatic regions	Karnataka State, India (Part V)	Precipitation; crop zones	P.C.A.[a]; N.N.A.[b]; correlation; map	Identification of cropping zones with coherent rainfall patterns
Agricultural zones	Stavropol & Saratov regions, S. USSR (Part VI)	Precipitation; temperature	Compute hydrothermal coefficient; map	Delimitation of zones of aridity
Wheatbelt	Australia (Part VII)	Precipitation	Identify critical isohyets	Location of present-day wheatbelt

[a]Principal components analysis.
[b]Nearest neighbor analysis.

agricultural practices and, subsequently, for conducting climate impact experiments. The indices are simple to calculate, and data requirements are not demanding.

Another promising approach to agroclimatic zonation is illustrated for the southern Indian state of Karnataka (*see* Part V). The statistical techniques of principal components analysis and nearest neighbor analysis are used to delimit agricultural zones within Karnataka State, made up of clusters of Taluks (agricultural districts) with similar cropping patterns. Subregions of the agricultural zones that are coherent with respect to annual rainfall pattern and totals are then identified on the basis of cross-correlations of rainfall data between meteorological stations. It is clear that this technique is capable of defining coherent subregions that are sufficiently homogeneous, both in terms of cropping pattern and climate, to provide a basis for further study of crop responses to climatic variations. One drawback of the approach, however, is its sophistication: apart from problems of acquiring data from a sufficiently dense network of stations, the statistical techniques are not easy to apply, the computation requires access to appropriate computer software and the results are sometimes difficult to interpret.

Seasonal shifts in crop potential

Agroclimatic zones can also be mapped that represent the potential geographical limits to cultivation of specific crops according to seasonal climatic conditions. The limit to cultivation can be specified on the basis of known crop requirements for heat, light and moisture. One of the uses of such a zonation scheme in climate impact assessment is that the impact of seasonal variations in climate can be expressed as a shift in geographical location of the agroclimatic zones.

For example, agroclimatic suitability zones have been mapped for barley and potatoes in Chimborazo Province, Ecuador (*see* Part IV). Because of its topography, the region exhibits marked contrasts in climatic conditions between locations at different altitudes. The two crops have different tolerances to moisture and temperature conditions, but both are susceptible to frost damage (usually at high altitude), and drought (predominantly at lower altitudes).

Zones of suitability for barley alone, for barley and potatoes, and for potatoes alone have been mapped according to annual climatic conditions. The effects of anomalously wet and anomalously dry years on potential crop suitability have then been studied as spatial (and altitudinal) shifts of the mapped zones (*see* Part IV). Similar experiments have been conducted to study the seasonal shifts of agroclimatic zones in Kenya (described above; *see* Part II).

A common criticism of studies of seasonal shifts in crop potential is that such shifts do not bear comparison with reality. Certainly, in the absence of accurate seasonal weather forecasts, it is unrealistic to expect farmers to be able to match crop cultivation patterns to the pattern of seasonal climate. However, the approach does give an impression of the potential mismatch between climatic resources and cropping practices during anomalous climatic episodes, and the potential adjustments that would be required if farmers *were* to be afforded prior notice of such episodes at some point in the future.

Until that time, it is reasonable to assume that present-day cropping patterns are more closely associated with long-term average climatic conditions (*see* below).

Long-term shifts of agroecological potential

Many of the zonation schemes already described employ agroclimatic indices that are calculated on the basis of long-term mean values of the observed climate. The zones delimited are thus representative of "average" conditions for plant growth, although they do not account for the interannual variability of climate. The notion that present-day cropping patterns or vegetation types are a manifestation of long-term agroecological adaptation to the environment, implies that any *change* in mean climate would lead, over the long term, to a change in agroecological potential and hence to geographical shifts in cultivation and natural vegetation patterns.

The method of analyzing long-term shifts is analogous to that described above for seasonal shifts: only the time period of averaging is different. For example, mapping of potential vegetation zones according to the Holdridge Life Zone classification has been carried out in Chimborazo Province, Ecuador (*see*

Part IV). Maps were constructed on the basis of long-term average (baseline) climatic conditions, and assuming both a 10% increase and a 10% decrease in long-term mean annual precipitation. Shifts in the location of key Life Zones between these different climatic scenarios were identified, which further could be related to likely shifts in agricultural crop potential in the region.

Similarly, the present-day dry limit of wheat cultivation appears to follow certain critical isohyets in different regions of Australia (*see* Part VII). Long-term changes in precipitation amount and distribution have been recorded during the past century, and shifts of isohyets have been mapped for several regions. This raises the interesting question of how much such changes might have affected (or could in the future affect) the location of the wheat limit. In all likelihood, however, this boundary is controlled as much by economic as by climatic factors.

2.3.2. Agroclimatic models

Problems of evaluating the sensitivity of crop productivity to climatic variations are occasionally tackled using suitability assessments similar to those outlined in the previous subsection. However, the nature and scale of these studies generally precludes detailed evaluation. A more discriminating method of analysis at regional and local levels, especially for the purpose of estimating specific impacts of climate on crop yield, utilizes agroclimatic models, in particular, crop–weather models.

Agroclimatic models are the tools with which we can attempt to quantify the effects of climate on crop productivity. We will provide here only a brief summary of the types of agroclimatic models, their calibration and testing and some of their weaknesses. These issues are treated in depth by the authors in the companion to this volume (Carter *et al.*, 1988). We will follow this with a short discussion of the special difficulties faced in modeling crop–climate interactions in semi-arid regions, paying particular attention to the models employed throughout this volume.

Types of models

Two general techniques for examining the response of agricultural crops to climatic variations can be identified, involving the use of (a) agroclimatic indices or (b) crop–climate models.

Agroclimatic indices were introduced above (*see* Subsection 2.3.1), and can be described as derived variables that provide a *relative* measure of crop suitability or of crop productivity. An agroclimatic index has the advantage of being simple to calculate and, since it constitutes a single term to which crop growth and development are found to be particularly responsive, relatively simple to apply. Examples range from simple aridity indices (such as the hydrothermal coefficient employed in he USSR case study – *see* Part VI) to simulation schemes involving multiple calculations of an index (such as the yield index for cowpea utilized in the Brazil study – *see* Part III).

Crop–climate models are mathematical devices that can be used over a range of time scales (and hence, for a variety of climatic scenarios) to relate crop variables (such as development stage, leaf area or grain yield) to meteorological variables (e.g., seasonal precipitation, air temperature or solar radiation). Two broad classes of model can be distinguished:

(1) *Empirical–statistical* models relate a sample of crop yield data to a sample of weather data for the same area and time period using statistical techniques, such as regression analysis. This approach does not easily lead to a causal explanation of crop–climate relationships, but can provide quite effective estimates of crop yield, especially in regions where crop growth and yield are governed by a *single* major weather factor (for example, rainfall and crop yields in Northeast Brazil, *see* Part III).
(2) *Simulation* models are developed on the basis of an understanding of the dynamic relationships between the basic processes of crop growth (such as photosynthesis, respiration or translocation of assimilates) and environmental and management factors (such as climate, soils and cultivation practices). Most of the models used for impact experiments in this volume are of this type, which is probably a reflection of the complexity of crop–climate relationships in semi-arid regions and the need for explanatory models that can account for large variations in environmental conditions occurring over short time periods during the growing season.

In view of their widespread use in this volume, and of their potential usefulness in future studies, the main focus of the following discussion is on crop–climate models.

Data requirements and model performance

The data requirements for developing and running empirical–statistical models are generally less demanding (i.e., there are fewer "explanatory" variables, and these are commonly averaged over monthly, seasonal or even annual time periods) than for simulation models, and herein lies an unfortunate irony. For while simple empirical–statistical techniques have been successfully applied to many cool temperate and cold regions in the relatively data-rich developed world (as evidenced by the experiments in Volume 1; Parry *et al.*, 1988), these have not proved very effective in some semi-arid regions. Here, the complexity of crop–climate relationships is such that explanatory simulation models often offer the only effective predictive capability, yet the demanding input requirements of such models are not usually matched (especially in LDCs) by adequate regional data bases.

An essential prerequisite for the effective use of crop–climate models is that they should have been adequately tested and validated. *Sensitivity tests* can provide useful information about the stability of a model and its effective range of operation. These commonly involve an analysis of model responses to systematic adjustments in input variables and model parameters. Moreover, in order to appraise its performance, it is necessary to *validate* a model's estimates against observed data. However, this poses considerable problems in many semi-arid regions where recorded measurements of crop growth and yield, if they exist at all, are often scarce or unreliable. In this context, a simulation model that has been developed and validated for another region can sometimes be used in data-scarce regions, providing there is a minimum data set (of at least one growing season) with which to calibrate the model (for example, this was the case with the barley simulation model used in the Ecuador study – *see* Part IV). Empirical–statistical models, in contrast, are developed using climatic and crop data that are *specific* to a region and are thus not easily transportable.

Model constraints in semi-arid regions

Climatic constraints. In our discussion of agroclimate models in Volume 1, several climatic variables were identified as being of importance in influencing crop cultivation and production in cool temperate and cold regions (Carter *et al.*, 1988). Temperature and radiation conditions, in particular, were seen to exert marked constraints on the types of crops that can be grown, and on their development and productivity. So, for the purposes of simulating crop–weather relationships in these regions, those factors (especially temperature) form vital components in the development of agroclimatic models.

In contrast, the temperature and radiation regimes in semi-arid regions usually provide an abundant supply of heat and light energy for cultivated crops. In Subsection 2.2.2 we discussed how, given the selection of suitable crops, these conditions can promote rapid rates of plant photosynthesis and crop growth, offering much greater potential for annual productivity than that achievable in cooler climates. However, the dominant constraint on crop productivity in semi-arid regions is, by implication, moisture availability.

The introduction of a moisture component to crop–weather models, especially where that is the overriding constraint on growth, immediately switches the emphasis of the modeling approach; for, while in the simulation of temperature and radiation effects on plant growth it is often sufficient to consider only the boundary-layer meteorological effects on plant foliage and crop development (given some assumptions about nutrients and water supply), modeling the availability of moisture inevitably involves additional considerations of soil–plant–atmosphere interactions. These considerations, unlike the effects of temperature and radiation, require a *continuous accounting* procedure in order to assess the soil moisture status before planting as well as during the growth period. Furthermore, the response of crops to moisture may be magnified by the active role that water plays in influencing nutrient release (whether nutrients are derived from natural or anthropogenic sources). Modeling these interrelationships, and others not touched upon, poses real difficulties in these regions, not least because of the highly variable temporal and spatial precipitation conditions characteristic of semi-arid areas.

Technological constraints. Another feature of agriculture in less developed regions, already touched upon in Subsection 2.2.5, is the relatively low level of mechanization and technology. One consequence of this situation is that levels of crop productivity are significantly lower than those potentially achievable with optimum crop management for given environmental conditions. All too often, however, crop–weather models (particularly simulation models) are developed on the basis of experimental crop trials under controlled conditions of management and technology. Thus, model estimates often reflect crop productivity under a climatic limitation *alone*, with simulated yields commonly well in excess of observed farm-level yields. Estimates can be improved if the effect of nutrients and of management characteristics (e.g., planting time, sowing density, etc.) can be included in the model, but these considerations require quantitative information on farm practices that is often difficult to obtain. Moreover, some human aspects of crop production systems in subsistence economies are very difficult to simulate. For example, labor shortages may impose severe restrictions on the efficiency of crop management, leading to delays in sowing or harvesting, to poorly timed nutrient applications or pest control measures and to reduced attention to weed control.

Model applications in semi-arid regions.

In spite of the many difficulties described above, when properly tested and validated, agroclimatic models can provide two types of information:

- Estimates of crop yields under a range of climatic conditions.
- Insights into a range of crop management adjustments that are available for improving levels of productivity under a variable climate.

Some aspects of these are discussed below.

Crop yield estimates. Depending on the modeling objectives, various methods of quantifying crop yields are presented in the different case studies and can be listed as follows:

(1) *Relative measures of yield* or potential productivity are common outputs from agroclimatic indices (described above), which can sometimes be related to absolute yields through statistical procedures. For example, the agroclimatic index used in the Kenya case study (*see Table 2.3*) has been related to maize yields under different climatic scenarios, both qualitatively (to observed farm-level yields) and quantitatively (to model-simulated potential yields) (*see* Part II).

(2) *Potential yield* is a measure provided by many simulation models that have been run assuming that cultivation practices, soil nutrients and water availability are optimal for crop productivity. Since the model equations are commonly developed on the basis of carefully controlled field experiments, model estimates of this type can often be verified against experimental data. Moreover, by comparing these estimates with observed farm-level

yield under the same climatic conditions, the gap (attributable to shortcomings in management, technology and resources) between actual and potential yields can easily be quantified. Examples include model estimates of sorghum yields at high and low management in dry tropical India (*see* Part V) and simulations of potential and actual spring wheat yields in the Saratov region, USSR (*see* Part VI).

(3) An important focus of investigations in this volume is on the impacts of a range of seasonal climatic conditions. The climatic conditions are commonly described as frequency distributions of weather-types, and climatic scenarios have been selected on the basis of climatic events with particular probabilities of occurrence. Similarly, there are advantages in expressing model estimates of crop yields over a period of years in terms of ***yield probabilities.*** In this way, it is possible to assess not only the mean level of yield at a particular location, but also the ***crop certainty*** (i.e., the probability of exceeding a specified yield level in any one season). Crop yields are represented as probabilities in experiments in Northeast Brazil (for relative yields of cowpea – *see* Part III) and in dry tropical India (for simulated sorghum yields – *see* Part V).

(4) As well as providing estimates of crop yield based on climatic observation data, simulation models also have the capacity (given accurate weather data) to provide *seasonal forecasts of yield.* At present, however, reasonably accurate weather forecasts are only available for periods of a few days ahead. Nevertheless, some useful information about likely end-of-season crop yields can be obtained in advance using crop–weather simulation models, simply from a knowledge of spring soil water storage and early-season weather conditions, and by substituting proxy data (e.g., long-term average weather conditions) to represent late-season conditions. Experiments of this kind are reported for the Saratov region, USSR (*see* Part VI).

Crop management adjustments. Perhaps the most useful feature of crop–weather models (especially simulation models) is their capacity to simulate the effects of various management practices on crop growth and yield. Adjustments in management practices offer considerable potential for ameliorating the detrimental effects of climate, and figure prominently in model experiments in this volume. Some examples are listed below:

(1) *Adjustments of sowing dates* will affect the match between the timing of crop development and the availability of environmental resources. Experiments have been conducted to evaluate the effects of altered sowing dates on yields of maize in Kenya (*see* Part II), of barley in Ecuador (*see* Part IV), of sorghum in India (*see* Part V) and of spring wheat in the Saratov region, USSR (*see* Part VI).

(2) *Adjustments of sowing density* can influence the water use of a crop, but also affect the likely level of competition for water and light from weeds. These effects were examined for barley in Ecuador (*see* Part IV) and for sorghum in India (*see* Part V).

(3) *Fallowing practices* can lead to more efficient retention of moisture in soils and, hence, to improved crop yields. These effects are evaluated in the Stavropol region, USSR, for winter wheat cropping (*see* Part VI) and in Chimborazo Province, Ecuador, for barley cultivation (*see* Part IV).

(4) *Substitution of crop cultivars* for better-adapted or higher-yielding varieties can also be simulated, providing the physiological properties of the introduced cultivars are suitably parameterized in the model. A comparison of simulated yields from a traditional variety of sorghum with those from a hybrid are presented in the India case study (*see* Part V).

(5) *Adjustments of nutrient applications* can be enormously beneficial for crop yields, especially when timed to coincide with suitable weather conditions. Investigations of nutrient effects on crop yield are reported for both the Stavropol and the Saratov regions of the USSR (*see* Part VI).

(6) *Scheduling of irrigation* is a critical activity in regions where irrigation water is available, but in limited supply. Simulation models can provide information about the preferred timing and amount of irrigation required to achieve a desired level of yield. Experiments of this kind are presented for sorghum yields in India (*see* Part V) and for spring wheat yields in the Saratov region, USSR (*see* Part VI).

A model checklist. The major characteristics of the different models that are employed in the case studies in this volume are summarized in *Figure 2.3.* The columns on the left-hand side of this checklist comprise descriptive elements of each model: model type and region of application, intended outputs, input data requirements (including climate, soils, management, other factors such as day length, and derived variables such as potential evapotranspiration) and the model's spatial resolution.

The right-hand side of the checklist contains subjective assessments of the models by the authors (on the basis of the information presented or cited in this volume). Firstly, the demonstrated validity of each model is examined, including specific assessments of internal model validation and of external validation of model estimates against independent observations (both for the study area itself and for other regions), and a general assessment. Secondly, the effectiveness of the models for use in climate impact experiments is assessed, with categories evaluating the availability and quantity of data required as model inputs, the demonstrated range of model applicability (for example, to extreme climatic conditions) and the extent to which a model can be used to investigate farm-level adjustments (such as irrigation scheduling, altered planting dates, varied sowing density, etc).

2.3.3. Livestock models

The problems of quantifying the effects of climatic variations on animal health and production are probably more formidable than those faced in estimating crop productivity. In part this is because, in addition to the direct effects of climate on animal well-being, the pathway of impacts is dominantly a two-stage

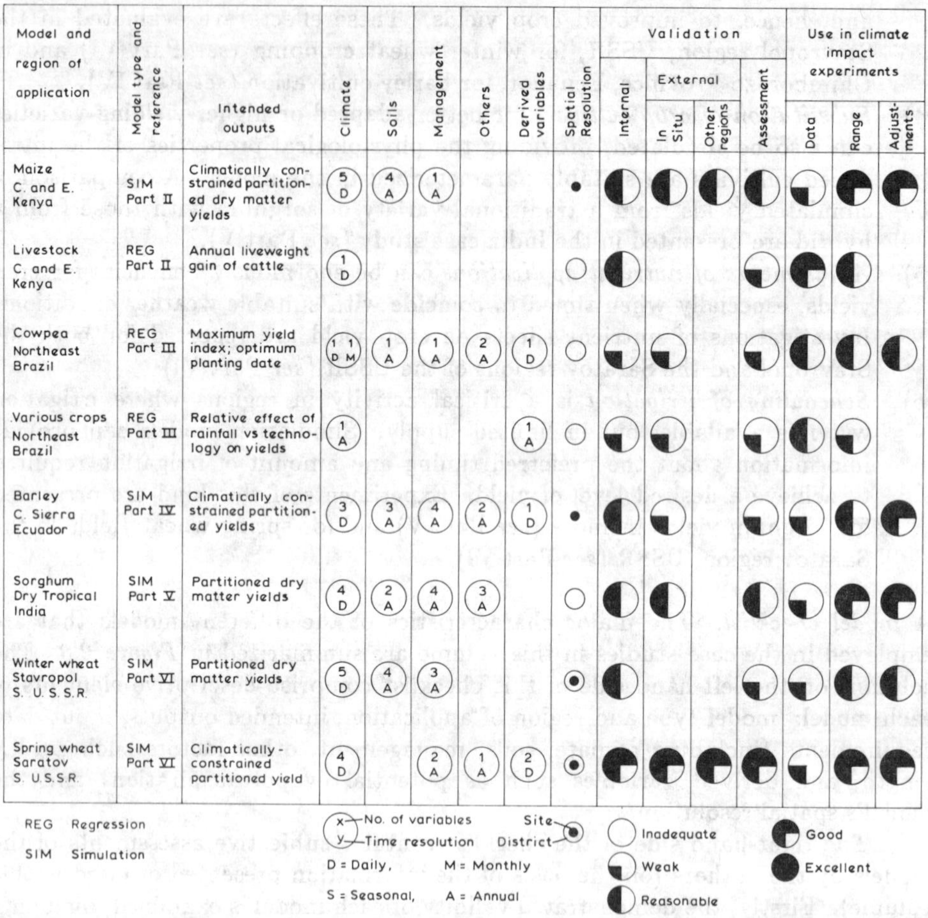

Figure 2.3. A checklist for comparing the characteristics and the usefulness of models employed in this volume.

process: the climate–vegetation and vegetation–animal stages, each stage characterized by a range of cause and effect relationships.

A further problem concerns the type of livestock output affected by climate, for while crop production is usually measured in terms of production per unit area of a single plant component (e.g., grain, root or leaf), livestock production can be measured in terms of a range of output parameters, including: growth (liveweight gain), carcass quality, milk, hides/wools, and reproduction (e.g., conception rates, calving interval, age at first service, etc).

Notwithstanding the various difficulties outlined above, significant progress has been made in modeling livestock management systems in some regions. Several models are available that simulate forage productivity as a function of climate. An example is the empirical–statistical pasture yield model for Iceland, presented in the companion to this volume (Bergthórsson *et al.*, 1988), which was used to estimate, for given grazing pressure and levels of supplementary feed, potential milk production under a range of climatic conditions.

Similarly, models of animal production have been developed that relate carcass weight or liveweight gain to climate, utilizing climatic data as input variables [for example, the empirical relationship developed for the Makindu regon of Kenya between potential liveweight gain of cattle and number of 10-day periods per year recording more than 20 mm of rainfall (*see* Part II, Section 5; this model is also included in the checklist presented in *Figure 2.3*), or the temperature/lamb carcass weight model developed for Iceland (Bergthórsson *et al.*, 1988)].

The linkage of forage models to animal production models is a logical next step, as it permits the inclusion of feedback mechanisms, such as the influence of stocking rate and grazing pressure on forage supply. Such a model framework has been developed for cattle production in two regions of Tanzania (Sullivan, 1983). A further advantage of this type of simulation approach is that it facilitates the evaluation of various options for managing scarce forage resources, such as adjustments in herd numbers and age structure, or supplemental feeding. However, the major disadvantages of such schemes are the demanding data requirements, both for calibrating and for testing models, and the geographically limited area of their application.

Finally, given the mixed nature of subsistence farming systems in many regions, a desirable goal for future studies of climate and agriculture involves the *integration* of stock and crop models within an interactive framework. This more comprehensive treatment of first-order farm-level effects of climatic variations would thus provide a firmer basis for evaluating the role (emphasized in Subsection 2.2.4) of economic, social and political factors on livestock (and crop) production.

2.4. Conclusions

In this section we have presented a brief overview of the factors that play a role in the agricultural production environment in semi-arid regions. It has been pointed out that the modeling of agricultural production is usually more complicated in less developed countries than in developed countries. Two major reasons explain this:

(1) Agricultural productivity is often very low in LDCs, usually reflecting limited technological resources. These factors have to be accounted for in relating model estimates to reality.
(2) While additional information on factors such as technology might compensate for the more complex modeling requirements, unfortunately data availability in these regions is usually limiting as well.

It is against this background that the performance of first-order impact models should be judged. The quality of empirical–statistical models and agroclimatic indices depends, to a large extent, on the agricultural and climatic data available locally. Simulation models, based on generally applicable functional relationships, are more easily adaptable to different production environments. However, no simulation model presented in this volume incorporates all of the relevant production-determining processes. Some features of the different models are summarized and evaluated in a model checklist.

Livestock modeling is even more complicated, for it usually involves a two-stage approach: firstly, forage production needs to be modeled as a function of climate; and secondly, livestock products, such as meat, milk, wool or hide, are related to forage (i.e., food intake). Estimates of forage production are commonly limited to the quantity produced, but the quality, in many instances, is equally important.

In conclusion, first-order impact models are of limited application in semi-arid regions. Nevertheless, bearing in mind their limitations, they can provide useful information both on the *impacts* of climatic variations on agricultural production, and on possible farming *responses* to counter the negative effects of climate, particularly drought.

Acknowledgment

We are grateful to Dr. Harry Potter for some helpful comments on livestock modeling.

References

Bergthórsson, P., Björnsson, H., Dýrmundsson, O., Gudmundsson, B., Helgadóttir, A. and Jónmundsson, J.V. (1988). The effect of climatic variations on agriculture in Iceland. In M.L. Parry, T.R. Carter and N.T. Konijn (Eds.), *The Impact of Climatic Variations on Agriculture. Volume 1, Assessments in Cool Temperate and Cold Regions*, Reidel, Dordrecht, The Netherlands.

Breman, H. and de Wit, C.T. (1983). Rangeland productivity and exploitation in the Sahel. *Science*, **221**, 1341–1347.

Carter, T.R., Konijn, N.T. and Watts, R.G. (1988). The choice of first-order impact models. In M.L. Parry, T.R. Carter and N.T. Konijn (Eds.), *The Impact of Climatic Variations on Agriculture. Volume 1, Assessments in Cool Temperate and Cold Regions*, Reidel, Dordrecht, The Netherlands.

Dregne, H.E. (1976). Soils of arid regions. *Developments in Soil Science*, **6**. Elsevier, Amsterdam, 237 pp.

FAO (1978). Report on the Agro-Ecological Zones Project. Volume 1, Methodology and Results for Africa. *World Soil Resources Report*, **48**, Food and Agriculture Organization, Rome, Italy.

FAO/UNESCO (various dates). *Soil Map of the World, 1:5 000 000.* Volume IV. South America (1971); Volume V. Europe (1981); Volume VI. Africa (1977a); Volume VII. South Asia (1977b); Volume X. Australasia (1978). UNESCO, Paris.

Hurd, E.A. (1968). Growth of roots of seven varieties of spring wheat at high and low moisture levels. *Agron. J.*, **60**, 201–205.

Meigs, P. (1953). World distribution of arid and semi-arid homoclimates. In UNESCO, Reviews of Research on Arid Zone Hydrology, *Arid Zone Res.*, **I**, 203–210.

Penman, H.L. (1948). Natural evaporation from open water, bare soil and grass. *Proc. Roy. Soc., Lond. Ser. A*, **193**, 120–145.

Soil Survey Staff (1960). *Soil Classification – A Comprehensive System – 7th Approximation.* Soil Conservation Service, US Department of Agriculture, Washington, D.C., 265 pp.

Soil Survey Staff (1976). *Soil Taxonomy.* Soil Conservation Service, US Department of Agriculture, Washington, D.C.

Sullivan, G.M. (1983). Modeling livestock management systems under variable climatic conditions in East Africa. In W.K. Lauenroth, G.V. Skogerboe and M. Flug (Eds.), Analysis of Ecological Systems: State-of-the-Art in Ecological Modelling, *Developments in Ecological Modelling*, **5**, Elsevier, Amsterdam.

Thornthwaite, C.W. (1948). An approach toward a rational classification of climate. *Geogr. Rev.*, **38**, 55–94.

Waggoner, P.E. and Bravdo, B.A. (1967). Stomata and the hydrologic cycle. *Proc. Nat. Acad. Sci. U.S.*, **57**, 1096–1102.

Went, F.W. (1948). Ecology of desert plants. I. Observations on germination in the Joshua Tree National Monument, California. *Ecology*, **29**, 242–253.

SECTION 3

Climate and Climatic Variability in Semi-Arid Regions at Low Latitudes

John G. Lockwood

3.1. Introduction

The case study areas considered in this volume are all located within semi-arid climatic regions. The causes of aridity and drought are complex and are not necessarily the same in each case study area. The study areas considered represent a selection of the semi-arid environments found predominantly at low latitudes, often with widely differing causes. The detailed climates of each case study area are described later in separate parts of the volume. In this section the climates of the study areas are related to each other and to atmospheric circulation patterns.

Horizontal temperature gradients in the tropical atmosphere are small so that heating of the lower atmosphere causes rising motion. In contrast, in middle and high latitudes, horizontal temperature gradients are often large, and the horizontal advection of cold air can balance heating. Tropical circulation patterns are particularly influenced by heat inputs from such sources as warm ocean surfaces acting through latent heat released in deep cumulus convection. Other heat sources, such as high plateaus and equatorial rain forest, are also important. The heat sources show marked latitudinal and longitudinal variations, and also a marked tendency to vary on both annual and, in the case of the oceans, on non-annual scales. Thus tropical circulation patterns are organized in a complex manner in both north–south and east–west directions and also show a marked tendency to vary on both annual and greater-than-annual cycles. One consequence is that rainfall patterns show both annual and greater-than-annual variations and also marked teleconnections with distant locations.

The climatic system is extremely heterogeneous, containing subsystems that are all interactive to some extent. Examples of subsystems in this context are the atmosphere, surface layers of the ocean, soils, forests, etc. One important property distinguishing each of the various subsystems is its "time scale", that is, the time it would take for dissipative processes acting alone in the absence of continued forcing to remove departures from equilibrium. A short time scale indicates that the system (or subsystem) has the inherent capability of responding quickly to any perturbations from its equilibria. In particular the response time of the atmosphere is relatively fast (a few days), while that of the deep oceans and ice sheets is very slow (a thousand years or more). In general, subsystems with fast response times tend to adjust quasi-statically to the changing boundary conditions imposed by the slow-response subsystems. Thus subsystems with slow response times tend to "carry along" the subsystems with fast response times and to form the memory of the climatic system. In the tropics, the surface layers of the ocean, which have a memory much longer than that of the atmosphere, are of central importance to long-term climatic change.

In the tropics warm ocean surfaces can influence atmospheric circulation patterns via surface evaporation and deep cumulus convection. If the winds should stop blowing altogether, then a horizontal redistribution of heat will cause horizontal density gradients in the ocean to weaken considerably. In low latitudes this can happen within a matter of weeks or months because the speed of the waves that effect the oceanic adjustment to a change in wind conditions increases with decreasing latitude. It is for this reason that the intense Somali and equatorial undercurrents in the Indian Ocean, where the monsoons reverse direction, can be generated seasonally, whereas it would take on the order of a decade to generate the North Atlantic Gulf Stream from a state of rest. This strong atmosphere/ocean coupling in the tropics is important both for explaining climatic changes and also because it gives rise to the possibility of making reliable seasonal weather forecasts using coupled atmosphere–ocean numerical models.

3.2. Tropical Circulation Patterns

The two major causes of the global wind circulation are inequalities in radiation distribution over the earth's surface and the earth's rotation. Global radiation distribution drives the global circulation, whereas the earth's rotation determines its shape.

3.2.1. Meridional circulation

A schematic representation of the mean meridional circulation in the northern hemisphere during winter is shown in *Figure 3.1*. The same general representation applies to the southern hemisphere. A simple direct cell circulation, known as the Hadley cell, is clearly seen south of 30° N. Eastward angular momentum is transported from equatorial latitudes by nearly horizontal eddies, 1 000 km or more across, moving in the upper troposphere and lower stratosphere. Such a transport, together with the dynamics of the middle-latitude atmosphere, leads to an accumulation of eastward momentum between 30° and 40° latitude, where

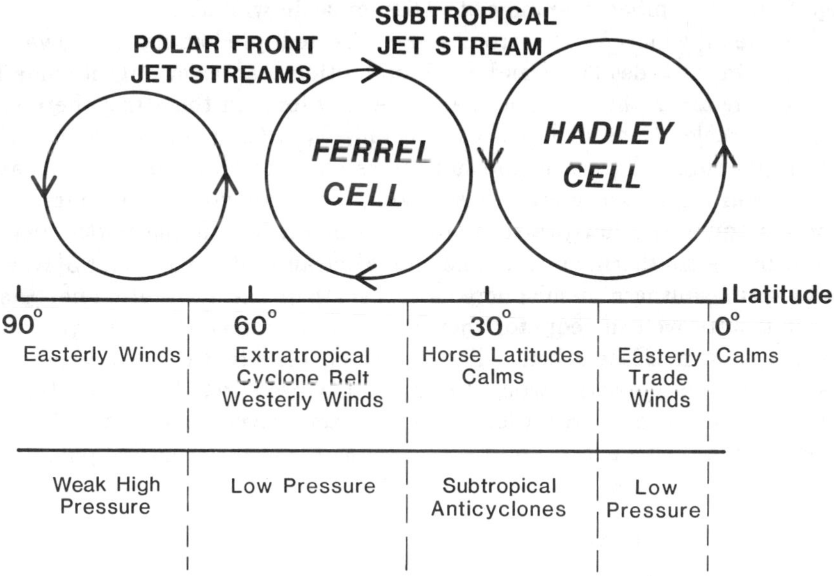

Figure 3.1. A schematic representation of the mean meridional circulation of the northern hemisphere.

a strong meandering current of air, known as the subtropical westerly jet stream, develops. The cores of the subtropical westerly jet stream in both hemispheres and in both seasons occur at an altitude of about 12 km. These subtropical jet streams mark the boundaries of the tropical world, and the air subsiding from them forms the subtropical high-pressure belt. More momentum than is necessary to maintain the subtropical jet streams against dissipation through internal friction is transported to these zones of strong winds. The excess is transported downward to maintain the eastward-flowing surface winds of the middle latitudes against ground friction. Since the eastward angular momentum is generated by the earth's rotation, which is constant, seasonal and long-term changes in the positions of both the westerly subtropical jet streams and the subtropical high-pressure belts will be small.

The two belts of subtropical high pressure at about 30° N and S contain several quasi-permanent anticyclonic cells separated from each other by cols. In the northern hemisphere, the most notable ones are the two oceanic highs, in the Pacific and the Atlantic respectively, and the North African high, which fails to show up at sea level, but emerges clearly at the 3 km level. In the southern hemisphere major anticyclones are found over the Pacific, Atlantic and Indian Oceans. Subtropical anticyclones show great permanence, for they tend to be located at fixed positions on the globe and undergo only slight seasonal variations, amounting on average to about 5° of latitude. The high-pressure belt is nearest to the equator in winter, but there is slight asymmetry with regard to the geographical equator, for the southern ridge is situated about 5° latitude closer to it in the mean than the northern one. At the subtropical ridge lines (the axes of the anticyclones), pressure is practically equal in both hemispheres, varying on average from 1015 mbar in summer to 1020 mbar in winter.

The mean winds in the tropics are directed predominantly toward the equator within latitudes 30° N and S. Such motions can exist only if some force accelerates the air in that direction, and the only force in the atmosphere known to produce accelerations of large-scale windfields is the pressure force. Therefore, high pressure will exist in the subtropics and low pressure in the equatorial zone, the resulting pressure gradients driving the mean winds. The average latitude of the equatorial low-pressure trough is about 5° S in the northern winter and 15° N in the southern winter. The annual mean latitude is about 5° N, so the meteorological southern hemisphere is larger than the northern one. As the mean winds approach the equator they absorb large amounts of energy from the oceans in the form of latent heat. The release of this latent heat by condensation and rainfall in the equatorial trough produces low pressure. Therefore, the tropical wind systems can be regarded as largely self-driven and, if this is so, the equatorial trough is an effect, not a cause, of the tropical circulation patterns.

3.2.2. The trade winds

Between the equatorial low-pressure trough and the subtropical high-pressure belt lie two belts of tropical easterlies, each with an equatorial meridional flow. These wind systems are known as the northeast trade winds in the northern hemisphere and the southeast trade winds in the southern. The extreme steadiness of the trades reflects the permanence of the subtropical anticyclones, which are inclined to be the most intense in winter, making the trade winds strongest in winter and weakest in summer. All tropical oceans, except the northern Indian Ocean, have extensive areas of trade winds. Each trade wind region contains definite centers of high resultant wind speed that reach 6–8 ms^{-1}. In contrast, wind speeds are low in the subtropics and in the equatorial trough zone. An important feature of the trade winds is a low-level inversion known as the trade-wind inversion. Subsidence is most marked at the eastern ends of the subtropical anticyclonic cells, that is to say, along the desert cold-water coasts of the western edges of America and Africa, for it is here that the trade-wind inversion is at its lowest. Normally as the trade winds approach the equator, the trade inversion increases in altitude and conditions become less arid. Over the oceans the intense tropical radiation evaporates water, which is carried aloft by convective thermals and eventually distributed throughout the layer below the trade inversion. The result is that the layer below the inversion becomes more moist as the trade winds near the equator and the continual convection in the cool layer force the trade inversion to rise in height.

3.2.3. The equatorial trough and the ITCZ

Trade winds from the northern and southern hemisphere meet in the equatorial trough, which is a shallow trough of low pressure, generally situated on or near the equator. Its position is clear-cut in January, apart from the central South Pacific, and in July the trough can be located with ease over Africa, the Atlantic and the Pacific to about 150° E. The equatorial trough shows a marked tendency to meander with longitude in January from 17° S to 8° N, in July from 2° N to 27° N. This is largely derived from wide oscillations in the monsoon regions, that is, the whole of southern Asia plus North Africa in the northern summer, and southern Africa and Australia in the southern summer. The excursions into the southern hemisphere are smaller than those into the northern, because the more constant westerly circulation of the southern hemisphere middle latitudes constrains the equatorial trough to near the equator. Over southern Asia and the Indian Ocean mean seasonal positions are around 15° N and 5° S, a difference of 20° latitude. In contrast the trough is quasi-stationary over the Pacific and Atlantic, where seasonal displacement is restricted to 5° latitude or less. Winds in the equatorial trough are generally calm or light easterly. Over some areas, such as the Indian Ocean and Southeast Asia, the structure of the trough is very complex and surface westerlies are found.

The Intertropical Convergence Zone (ITCZ) can be found within the equatorial trough zone. Defined as a surface discontinuity separating the trade winds of the two hemispheres, it can be identified readily on climatic charts, but it is not easily found on daily weather maps. There may be virtually no weather activity or clouds along it, or there may be two separate active convergence zones at different latitudes with good weather in between. In practice it is found that the rising limb of the Hadley cell is restricted to massive convective clouds within tropical synoptic systems, which may be associated with the ITCZ.

3.2.4. The Walker circulation and the El Niño/Southern Oscillation phenomenon

Dietrich and Kalle (1957) have mapped the difference between sea-surface temperature (SST) and its average global value along the part of each latitude circle situated over the oceans (*Figure 3.2*). Cold-water regions are defined as ocean areas with sea-surface temperatures below the global average. By far the most extensive cold-water area is the South Pacific cold water, for it stretches westward from the coast of South America by about 85° longitude; in contrast the South Atlantic cold water continues westward from the coast of Africa by only about 40° longitude.

Bjerknes (1966, 1969) considers that when the Pacific cold ocean water along the equator is well developed, the air above will be too cold to take part in the ascending motion of Hadley cell circulation. Instead, the equatorial air flows westward between the Hadley cell circulation of the two hemispheres to the warm West Pacific where, having been heated and supplied with moisture from the warmer waters, the equatorial air can take part in large-scale, moist adiabatic ascent. This thermally driven circulation (*Figure 3.3*) between an equatorial heat centre – the "maritime continent" of Indonesia – and a cooling area in the eastern Pacific is often known as a Walker circulation. As shown in *Figure 3.3*, the equatorial Atlantic is analogous to the equatorial Pacific in that the

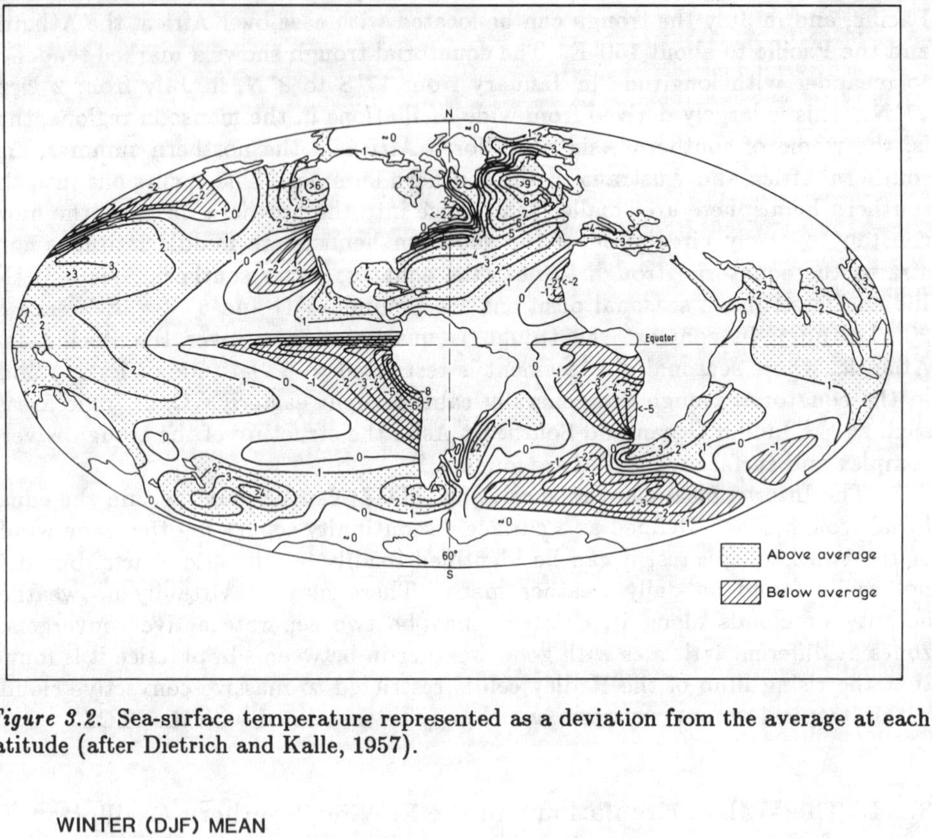

Figure 3.2. Sea-surface temperature represented as a deviation from the average at each latitude (after Dietrich and Kalle, 1957).

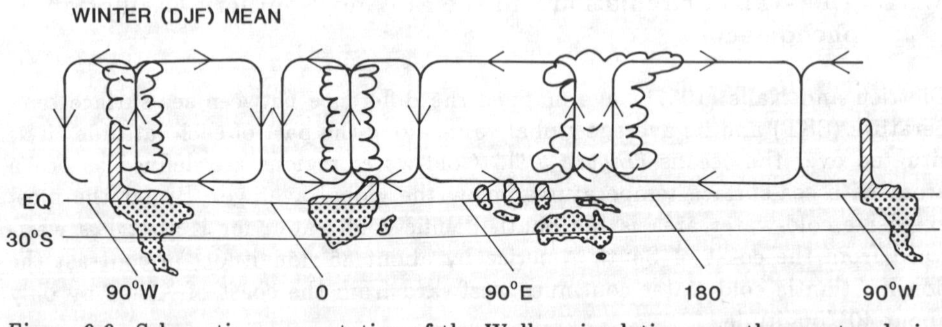

Figure 3.3. Schematic representation of the Walker circulation near the equator during the northern winter.

warmest part is in the west, at the coast of Brazil, but west–east contrasts of water temperature are much smaller than in the Pacific. However, in January a thermally driven Walker circulation may operate from the Gulf of Guinea to the Andes, with the axis of the circulation near the mouth of the Amazon.

The strength of the Walker circulation varies with the surface water temperature of the Pacific Ocean and in turn forms part of a wider tropical fluctuation known as the Southern Oscillation. According to Berlage (1966):

> The southern oscillation is a fluctuation of the intensity of the intertropical general atmospheric and hydrospheric circulation. This fluctuation, dominated by an exchange of air between the southeast Pacific subtropical high and the Indonesian equatorial low, is generated spontaneously. The period varies between roughly 1 and 5 years and amounts to 30 months on the average.

Associated with the Southern Oscillation is the El Niño phenomenon, when the eastern tropical Pacific Ocean occasionally becomes much warmer than average for periods of a year or more. Historically, El Niño is a phenomenon that manifested itself off Peru at Christmas. Thus, fishermen of the port of Paita gave it the name of *Corriente del Niño* – in English, "Current of the (Christ) Child". This designation was used by many past authors, although today it is a tacit convention to use the term "phenomenon" instead of "current" since it involves a transitory irregularity in the ocean–atmosphere system.

Data sets that describe El Niño/Southern Oscillation events are most reliable for the period since World War II. Philander (1983) comments that of the nine events that have occurred since 1950, seven evolved in a remarkably similar manner. These events (*Figure 3.4*) started as modest anomalies off the coast of Ecuador and Peru during the northern spring and then expanded westward until the entire tropical Pacific Ocean was affected six months later. The minor event of 1963 and the major 1982 event evolved differently because in the latter case unusually warm surface waters first appeared in the central Pacific Ocean. The coast of South America was affected subsequently, so that the phase of the atypical phenomena was very different from the phase of the more common events.

According to Kousky *et al.* (1984) the usual El Niño/Southern Oscillation event begins with a strengthening of the southeast trades in the Pacific and an accumulation of warm water in the western equatorial Pacific. The trades then weaken, usually during the southern summer, and the warm water is transported eastward across the equatorial Pacific. Thus, the normal east–west SST gradient is weakened, and the normal Walker circulation is greatly altered. Due to the zonally symmetric distribution of anomalously warm water in the equatorial Pacific, a stronger-than-normal Hadley circulation is generated and, as a consequence, upper-level subtropical jets are stronger and more persistent than normal in both hemispheres. The stronger Hadley circulation favors an intensification of the trade winds, which tends to restore the SSTs to their pre-El Niño state. Thus, the Southern Oscillation generally proceeds to its mature stage within 6–12 months after initiation, which usually occurs during April to June. Then, the ocean–atmosphere system takes an additional six months to return to its normal state.

3.2.5. The Asian monsoon circulation

Originally the term "monsoon" was applied to the surface winds of southern Asia, which reverse between winter and summer. The characteristics of the

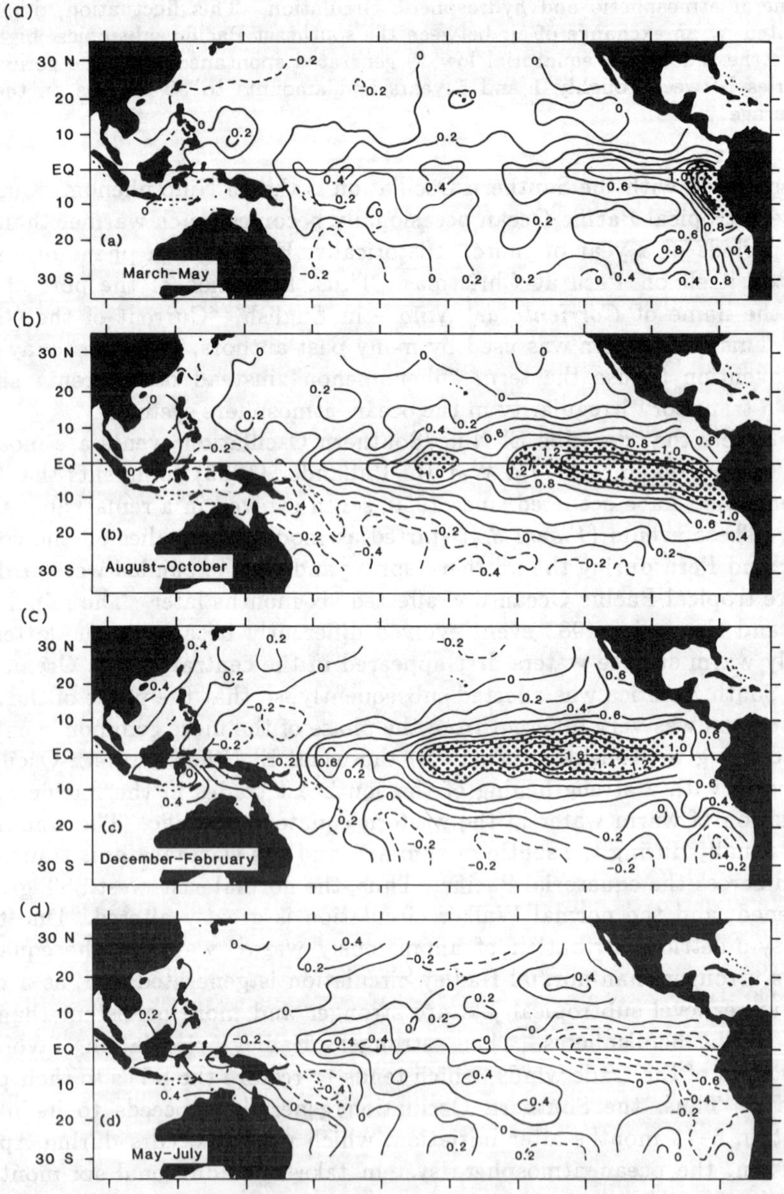

Figure 3.4. Sea-surface temperature anomalies (in °C) during a typical El Niño/Southern Oscillation event obtained by averaging data for the events between 1950 and 1973. (*a*) March, April and May after the onset; (*b*) the following August, September and October; (*c*) the following December, January and February; (*d*) May, June and July, more than a year after the onset (after Philander, 1983; Rasmusson and Carpenter, 1982). Reprinted with permission from *Nature*, © 1983, Macmillan Journals Ltd.

monsoon climate are to be found mainly in the Indian subcontinent, where over much of the region the annual changes may conveniently be divided into the northeast (dry) and southwest (wet) monsoon seasons.

When upper winds are taken into account, it is found that the Asian monsoon is a fairly complex system. During the northern winter season (*Figure 3.5*), the subtropical westerly jet stream lies over southern Asia, with its core located at about 12 km altitude. It divides in the region of the Tibetan Plateau, with one branch flowing to the north of the plateau, and the other to the south. The two branches merge to the east of the plateau and form an immense upper convergence zone over China. In May and June the subtropical jet stream over northern India slowly weakens and disintegrates, causing the main westerly flow to move north into central Asia. While this is occurring, an easterly jet stream, mainly at about 14 km, builds up over the equatorial Indian Ocean and expands westward into Africa. The formation of the equatorial easterly jet stream is connected with the formation of an upper-level high-pressure system over Tibet. In October the reverse process occurs; the equatorial easterly jet stream and the Tibetan high disintegrate, while the subtropical westerly jet stream reforms over northern India. The driving force for the Indian summer monsoon is high-level heating over the Tibetan Plateau together with latent heat release in tropical storms over India. It is a good example of tropical circulation patterns being modified by a powerful heat source. The southwest monsoon is of particular importance to India because it provides much of the rainfall required for agriculture (*see* Part V, this volume).

The Himalayan–Tibetan plateau is of importance because it appears to accelerate the onset of the Asian monsoon and to increase its ultimate intensity. Satellite data indicate that the central and southeastern parts of Tibet remain free of snow throughout most of the year. Hence, the plateau must heat rapidly during the northern spring. This direct warming of the middle troposphere creates an upper-level anticyclone, which is readily observed on synoptic charts, upper-level divergence and low-level convergence. Thus, suitable conditions are produced for the Asian monsoon in the northern spring. Latent heat released in intense tropical storms over India keeps the system functioning during the northern summer. Since a complex feedback system produces the southwest monsoon, failures in the system are common and produce extensive breaks in the monsoon rains when the whole system shows signs of collapse. Variations in winter snow cover over Tibet will influence the start and intensity of the southwest monsoon. General cooling over Southern Asia at the end of the northern summer causes its collapse.

According to Miller and Keshavamurthy (1968), the southwest monsoon current in the lower 5 km near India consists of two main branches: the Bay of Bengal branch, influencing the weather over the northeast part of India and Burma; and the Arabian Sea branch, dominating the weather over the west, central and northwest parts of India. The low-level flow across the equator during the southwest monsoon is not evenly distributed between the latitudes 40° E and 80° E, as was previously thought, but has been found by Findlater (1969, 1972) to take the form of low-level high-speed southerly currents, which are

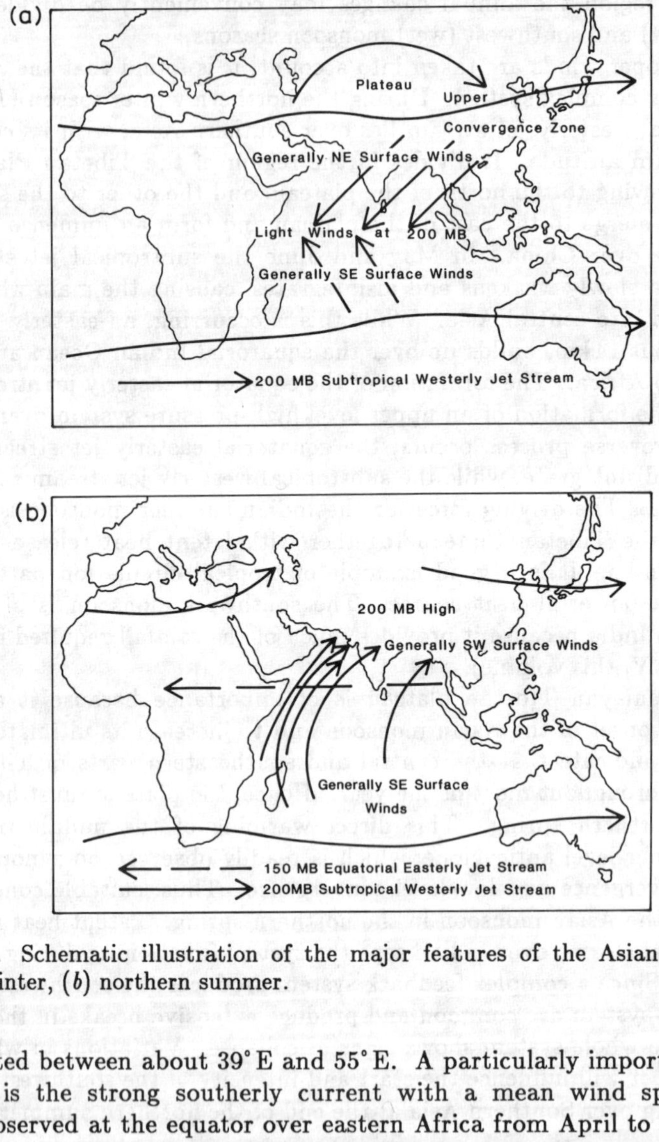

Figure 3.5. Schematic illustration of the major features of the Asian monsoon: (*a*) northern winter, (*b*) northern summer.

concentrated between about 39° E and 55° E. A particularly important feature of this flow is the strong southerly current with a mean wind speed of about 14 ms^{-1} observed at the equator over eastern Africa from April to October. The strongest flow occurs near the 1–5 km level, but it often increases to more than 25 ms^{-1} and occasionally to more than 45 ms^{-1} at heights between 1–2 and 2–5 km. According to Findlater, this high-speed current flows intermittently during the southwest monsoon from the vicinity of Mauritius through Madagascar, Kenya, eastern Ethiopia, Somalia, and then across the Indian Ocean toward India. The low-level southerly current off the coast of East Africa is generated by the dynamics of flow patterns of this nature and has fundamental consequences for the climate of Kenya (*see* Part II).

3.2.6. Deserts

Deserts exist on earth for either one or a combination of the following reasons:

Downward leg of the Hadley cell

Deserts occur in the subtropical latitudes where widespread, persistent subsidence occurs because they correspond to the downward leg of the Hadley cell below the subtropical jet streams. These subsidence zones move by latitude with the seasons. The latitudes with maximum seasonal variability are, of course, those which are semi-desertic, while those with minimum variability have subsidence throughout the year and are therefore perpetual deserts. Superimposed on this meridional circulation is the zonal Walker circulation, which complicates the dynamics of the tropical and subtropical atmosphere but has to be taken into account in any discussion of the variability of climate over the subdesertic regions of the world.

Lee of mountains

At several locations around the world, such as western North America, western Southern Argentina and parts of inner Asia, dryness is caused by mountain barriers that make air descend on the lee side of the mountains.

In the tropics moisture available for rain formation is trapped in a shallow layer below the trade-wind inversion. The depth of this moist layer varies, but it is usually between 1 and 2 km. If a mountain barrier projects through the moist layer, it will interrupt the surface flow, and the surface moist layer will not penetrate behind the mountain range. Even if the mountains do not completely block the moist layer, the reduction in moisture advection to the lee of the range can still be substantial. For example, under tropical conditions an airstream with a surface moist layer 2 km deep, and which has its lowest kilometer blocked by mountains, will lose between 60 and 70% of its precipitable water content. Thus, many tropical plateaus are dry, while nearby coastal lowlands have high rainfall.

Absence of water vapor and/or of rain-inducing disturbances

Desertic conditions may also be caused either by the absence of water vapor in the air or by the absence of rain-inducing disturbances or both. The best examples of these circumstances are the downwind sides of cold oceanic currents, and inner continental regions that are remote from sources of water vapor. Precipitation (P) within a continental area is partly formed of moisture advected (P_A) into the region by the winds and partly of moisture that has evaporated locally. The surface evaporative loss has three main components:

(1) Water evaporated from wet vegetation canopies during and after precipitation (E_I).
(2) Water evaporated from the soil surface (E_S).
(3) Water transpired through vegetation (E_T).

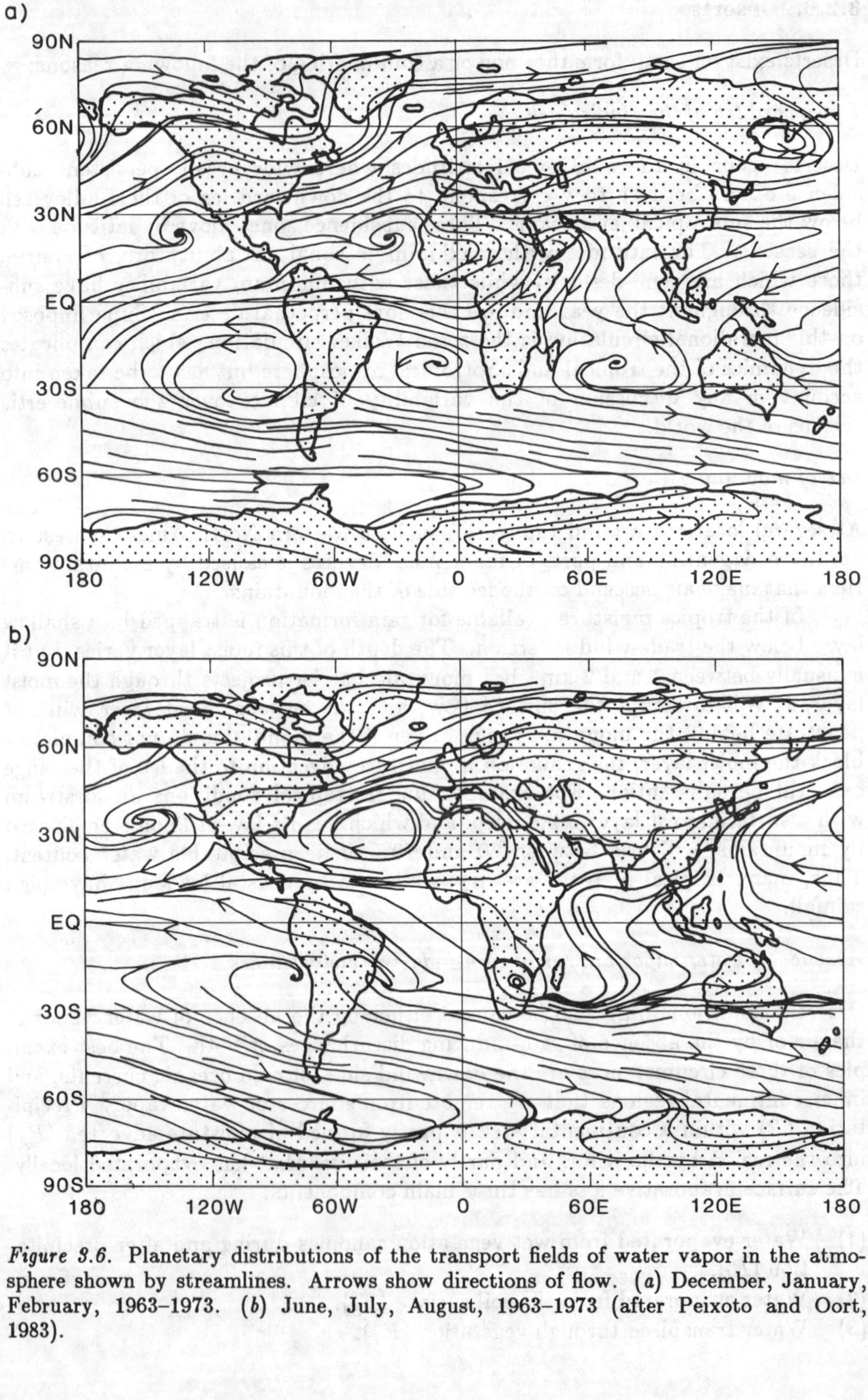

Figure 3.6. Planetary distributions of the transport fields of water vapor in the atmosphere shown by streamlines. Arrows show directions of flow. (*a*) December, January, February, 1963–1973. (*b*) June, July, August, 1963–1973 (after Peixoto and Oort, 1983).

It is therefore possible to add two further components to the precipitation: local precipitation (P_L) formed mostly of E_I, and regional precipitation (P_R) formed mostly of E_T and E_S. Thus, P becomes:

$$P = P_A + P_L + P_R \tag{3.1}$$

In maritime climates with a vigorous atmospheric circulation, most of the precipitation consists of P_A. In many tropical climates with a weak atmospheric circulation, precipitation can have a large component of P_L and P_R. Significant amounts of water vapor are transported inland by a process of continual recycling between the atmosphere and the land surface, a process in which vegetation has a key role.

Figure 3.6 shows the vector transport fields of water vapor in the atmosphere. In a hypothetical steady state, the streamlines indicate the prevailing paths of water vapor after its release from the various source regions at the earth's surface. The source regions are mainly the oceans. There is little water vapor advection into East Africa, the southern USSR, much of Australia, and India in the northern winter. Water vapor is mainly transported by the low-level atmospheric flow, the patterns of which are clearly apparent in *Figure 3.6*. These should be compared with *Figure 3.7*, which shows annual rainfall over the tropics and which indicates the distribution of the arid and semi-arid zones. The southern part of the USSR is just on the northern edge of the map.

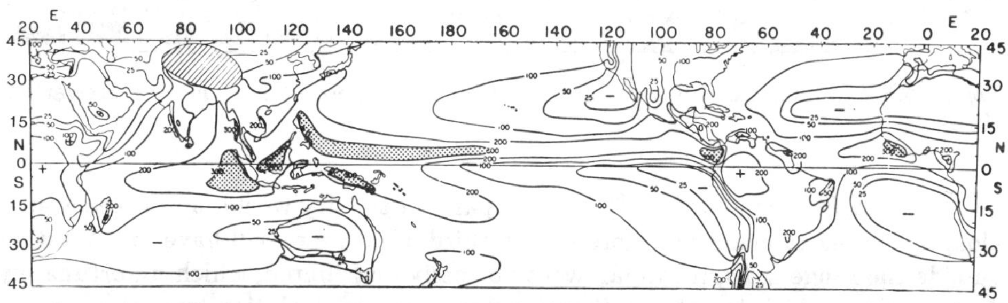

Figure 3.7. Mean annual precipitation in the low latitudes (cm); stippled areas above 300 cm y^{-1}. (After Riehl, 1979.)

3.3. Rainfall Variability

It is crucial for farmers to know how "reliable" different totals of annual precipitation are in a region. This is especially true for marginal regions where the average annual amount barely suffices for the wide range of activities undertaken.

A frequently used measure of rainfall variability is the coefficient of variation, defined as:

$$V_\sigma = \frac{\sigma}{\bar{P}} \times 100 \tag{3.2}$$

where σ = standard deviation of precipitation, and $\bar{P}$ = mean annual precipitation.

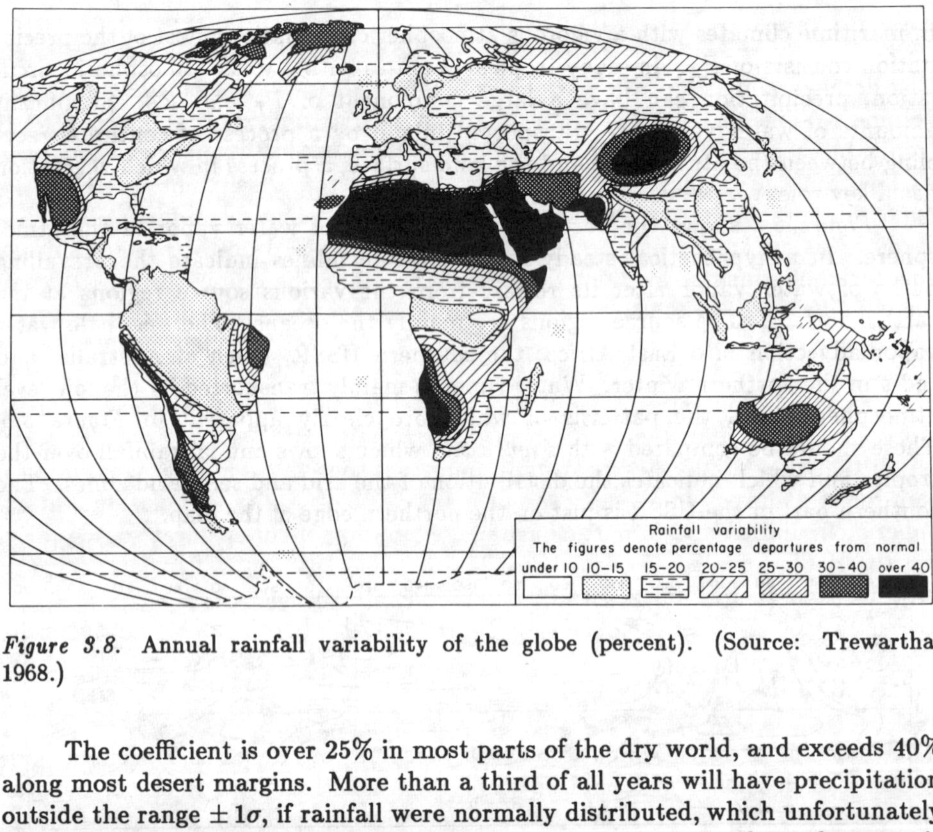

Figure 3.8. Annual rainfall variability of the globe (percent). (Source: Trewartha, 1968.)

The coefficient is over 25% in most parts of the dry world, and exceeds 40% along most desert margins. More than a third of all years will have precipitation outside the range $\pm 1\sigma$, if rainfall were normally distributed, which unfortunately for the arid world it is not. In *Figure 3.8* the average deviation from the mean is plotted, expressed as a percentage of this mean. The map is largely schematic, but does give a crude picture of rainfall variability over the globe. It suggests that the highest variability within any latitude is where the mean annual rainfall is lowest. Regions with variability of less than 15% are restricted in the main to the Zaire River Basin, the Amazon Basin, and parts of Indonesia in the tropics. From there, variability increases toward the subtropical deserts and also Northeast Brazil, where variability is very high compared to mean rainfall.

In most parts of the arid world rainfall is not normally distributed, and Wallén (1968) has made use of the interannual variability

$$V_I = 100\,\frac{\sum(P_{n-1} - P_n)}{\bar{P}(N-1)} \tag{3.3}$$

(where n is an individual year in a series of N years), which is the percentage ratio of the mean interannual precipitation–difference (taken regardless of sign) to the mean annual precipitation. For the Middle East the ratio exceeds 50% along the desert margin, and is generally in the range 25–35% in dry farming areas.

3.4. The Case Study Regions

The detailed climates of the various case study areas are described later in this volume. While all the areas are in relatively low latitudes, not all are in strictly tropical locations. The Kenyan, Indian, Ecuadorean and Brazilian studies are all well within the tropics, but the studies in both the USSR and Australia are partly in marginal tropical–temperate areas. The Kenyan and Brazilian case study areas are interesting because they are both located on the eastern sides of equatorial continents, yet the dynamics of the atmosphere in each case are completely different.

The causes of aridity have been discussed and are reflected in the climates of the case study areas. In some cases the exact causes of the aridity are complex and not completely clear. For example, the semi-arid areas in both Northeast Brazil and Kenya are in may ways anomalous, and are both interesting in that they are located near major moisture sources, such as oceans, tropical rain forest or lakes. In all cases the primary cause of aridity is the general atmospheric circulation and in particular the distribution of surface heat sources driving the Hadley–Walker–monsoon circulations. The aridity is often intensified by local features such as relief, lack of vegetation and dry soils.

3.4.1. Kenya

The most impressive climatic anomaly in all of Africa is probably the widespread deficiency of rainfall in tropical East Africa. The difference between this area and Northeast Brazil is that the atmospheric circulation patterns are dominated by the monsoon circulations over southern Asia as described earlier. In general there is an association between the ITCZ and rainfall, on the one hand, and strong monsoonal winds and dry conditions, on the other. The movement of the ITCZ in general causes March, April, October and November to be the wettest months. The aridity is partly due to tropical circulation patterns and also to a lack of moisture advection inland because of the relief. The climate of the study area in central and eastern Kenya is described in Part II.

3.4.2. Northeast Brazil

The detailed climate of this region is described in Part III. There it is noted that Northeast Brazil, located between 1° and 18° S and between 35° and 47° W (that is, just east of the Amazon rain forest) is an anomalous region on the tropical continents because it presents a semi-arid climate, which is due to the high variability of the spatial and temporal distribution of rainfall. The prime causes of the relative aridity are the thermally driven Hadley–Walker cells described earlier. In particular the Amazon River Basin acts as a local heat source with ascent taking place in deep cumulus clouds, while the corresponding descent takes place over the subtropical Atlantic Ocean, including Northeast Brazil. The aridity may be intensified by relief and other factors described later.

More than one major mechanism produces rainfall in Northeast Brazil; therefore, the time of monthly maximum rainfall varies between regions. In northern areas the meridional displacement and strength of the ITCZ are important, while further south cold fronts or their remains from the high latitudes of the southern hemisphere produce rainfall.

3.4.3. Ecuador

In this case the rainfall distribution is related to relief – an example of a high-altitude tropical climate – the valleys being dry and the mountains humid. Since temperature falls with altitude, crop production may be reduced not only by lack of rainfall but also by low temperatures. The detailed climate of the study area (in the central Sierra of Ecuador) is described in Part IV.

3.4.4. India

The Indian monsoon circulation has been described briefly earlier, and the climate of the dry tropical Indian case study area is considered later in Part V. Nearly half of the Indian subcontinent is arid or semi-arid, and by far the larger part of this dry area is located in northeast India and southwest Pakistan. Semi-arid climates extend across central India to the southeast corner of the country. The aridity is caused by a combination of the topographic features and the monsoon circulation patterns.

3.4.5. Southern USSR

This is an example of a semi-arid environment in a continental interior. The aridity is due to distance from the sea, and the area lies on the margins of the tropics. The climate is described in detail in Part VI. In many ways the two study areas (the Stavropol and Saratov regions of the southern European USSR) are related more closely to the middle latitudes than to the tropics. In particular they are areas where climatic change in the future may be dominated by increasing atmospheric carbon dioxide rather than by changes in the Southern Oscillation, as appears to be the case in much of the tropics.

3.4.6. Australia

The climate of the Australian wheatbelt is described in Part VII. It is of interest in that it covers a variety of semi-arid climatic types across the southern tropical high-pressure ridge and is partly on the borders of the tropics. The aridity is related to the subsiding limb of the Hadley cell over Australia.

3.5. Drought

One of man's worst natural enemies in semi-arid lands is drought. Its beginning is subtle, its progress is insidious and its effect can be devastating (Hounam *et al.*, 1975). Since rainfall is still the main source of fresh water for agricultural, domestic and industrial use, drought can have an impact ranging from slight personal inconvenience to endangered nationhood.

3.5.1. Definitions

A study of drought requires an objective definition, but, according to Hounam *et al.*, no universally acceptable definition has so far been developed. It can be assumed that the basic cause of drought is inadequate precipitation, but the exact meaning of the word "inadequate" needs careful definition. The distribution of average annual precipitation does not by itself give an indication of drought incidence or intensity, although obviously it has a marked control over normal land use. The use of such a distribution may lead to areas of low average rainfall being identified as drought areas. Aridity is usually defined in terms of low average rainfall and, ignoring the possibility of climatic change, is a permanent climatic feature of a region. Hounam *et al.* suggest that drought is a temporary feature in the sense that, considered in the context of variability, it is experienced only when rainfall deviates appreciably below normal. Aridity is, by definition, restricted to regions of low rainfall and, usually, of high temperature, whereas drought is possible in virtually any rainfall or temperature regime. Activities in arid zones are often poorly geared to cope with the permanence of aridity, but a drought situation results in at least some interruption of normal activities in all zones. Man's use (or overuse) of the rainfall resource in semi-arid zones has produced recurrent drought.

Drought should be regarded as a hydrologically extreme state in exactly the same manner that flooding is regarded as a state of extreme river flow. Dry spells with large soil moisture deficits are normal at many sites in the dry season. Such dry spells should not be referred to as droughts unless they reach an unusual intensity or are abnormally long for the particular sites. Thus, drought should be defined in terms of prolonged and abnormal moisture deficiency.

One major reason for the lack of a universally agreed definition of drought is that concepts of drought are related to particular water uses. For instance, agriculture is chiefly concerned with adequate wet season rainfall to offset the evapotranspiration that occurs in that season and gives rise to large soil moisture deficits. On the other hand, water supply interests lie much more with rain that provides runoff for reservoirs and percolation to recharge aquifers. On a world scale, ideas on agricultural drought vary, depending on the exact type of agricultural activity and the normal local climate.

Agricultural drought may be expressed in terms of the degree to which growing plants have been adversely affected by an abnormal soil moisture deficiency. The deficiency may result either from an unusually small moisture supply or unusually large moisture demand, and will depend on both the crop and the rainfall distribution during the growing season. While a deficiency in the water supply for livestock may be regarded as a facet of agricultural drought, it is really a different problem – because it does not depend directly on soil moisture.

3.5.2. Statistical frequency distributions

To understand the availability of water, it is necessary to examine the statistical properties of frequency distributions of rainfall. Gibbs (1981) states that, unfortunately, the frequency distribution of totals of most daily rainfalls, many monthly rainfalls and some annual rainfalls are far from normal (Gaussian), and the use of arithmetic means and standard deviations is quite inappropriate for many purposes including the study of drought. Gibbs (1981) remarks that complications arise when frequency distributions of rainfall amounts contain a number of zero values, as is the case with all hourly, most daily, many monthly and some annual totals.

A useful device, employed in most of the case study regions, is to use a diagram in which the ordinate is percentage cumulative frequency and the abscissa the magnitude of the element. A curve plotted on such a diagram shows the percentage of occasions a particular value is exceeded. If the sample of observations is sufficiently large, the curve may be used to deduce the probability of occurrence. Gibbs (1981) suggested that the square roots of hourly, daily and monthly rainfall tend to be normally distributed, and has used a square root scale for the abscissa. The scale of the ordinate is such that, if the magnitude of the elements represented on the abscissa is normally distributed, the curve will be a straight line.

In practice most plots approximate straight lines and indicate the range of probability of rainfall. After examining a study of the history of drought in Australia, Gibbs and Maher (1967) found good agreement between drought and the occurrence of annual rainfall in the first decile range. This suggests that in marginal lands, such as Australia, man uses available water so that he experiences a deficiency (drought) one year in ten – a return period adopted as one of the climatic scenarios used in the case studies in this volume.

3.6. Causes of Climatic Variability

Arid and semi-arid lands, where interannual rainfall variability is inherently extreme, dominate large parts of the tropics, and severe droughts are a common occurrence in nearly every region. Important features of these droughts are their markedly synchronous occurrence over large and often remotely separated areas; the extreme spatial coherence of rainfall anomalies and their persistence over long time periods; and the sometimes apparently normal location of the intertropical convergence zone during drought years.

Among the main explanations of these persistent rainfall anomalies are the Southern Oscillation and associated changes in sea-surface temperature (i.e., El Niño), changes in vegetation cover and albedo, changes in soil moisture, atmospheric dust and interactions with the middle-latitude atmosphere.

3.6.1. The Southern Oscillation

In his search for precursors to the Indian monsoon rainfall, Walker (1923) uncovered a tendency, particularly south of the equator, for a variation in pressure between the eastern and western Pacific over a period of years, an alternation he called the Southern Oscillation. Bjerknes (1966, 1969, 1972) showed that this alternation in surface pressure was related to variations in rainfall and sea-surface temperature (SST) in the equatorial eastern Pacific and, in particular, to the anomalously warm SST off the coast of Peru known as El Niño (*see* Subsection 3.2.4, above). Using 24 years of zonal mean sea-surface temperature, Chiu and Newell (1983) have shown that the annual and semi-annual variations account for more than 90% of the variance outside the tropics. In the tropics they found a high persistence and substantial low-frequency variations. A broad region in the tropical Pacific has less than 80% of the total variance explained by the seasonal cycle, and between 7.5° N and 2.5° N the seasonal cycle explains only about 40% of the total variance. Chiu and Newell also found a very strong persistence in the SST of the tropical and subtropical Pacific from 22.5° N to 7.5° S. Wright (1979) and Fleer (1981a and b) have also noted a strong persistence in rainfall in the central equatorial Pacific.

Walker (1923) first pointed out that the Southern Oscillation phenomenon is a dominant feature in global weather. Eigenvector analysis of global sea-level pressure substantiated Walker's claim (Kidson, 1975). A number of studies have shown strong relationships between tropical weather patterns in particular, and global patterns in general, and El Niño events and the Southern Oscillation. *Figure 3.9* shows the variation since 1958 of SST in the region 0–10° S, 180–90° W, zonally averaged tropospheric temperature in the tropics between 30° N and 30° S, average summer monsoon (June–September) rainfall for India, and a normalized Southern Oscillation index, where normalization was achieved by dividing the departure from average of the mean monthly pressure difference between Tahiti and Darwin by the standard deviation of the values of that

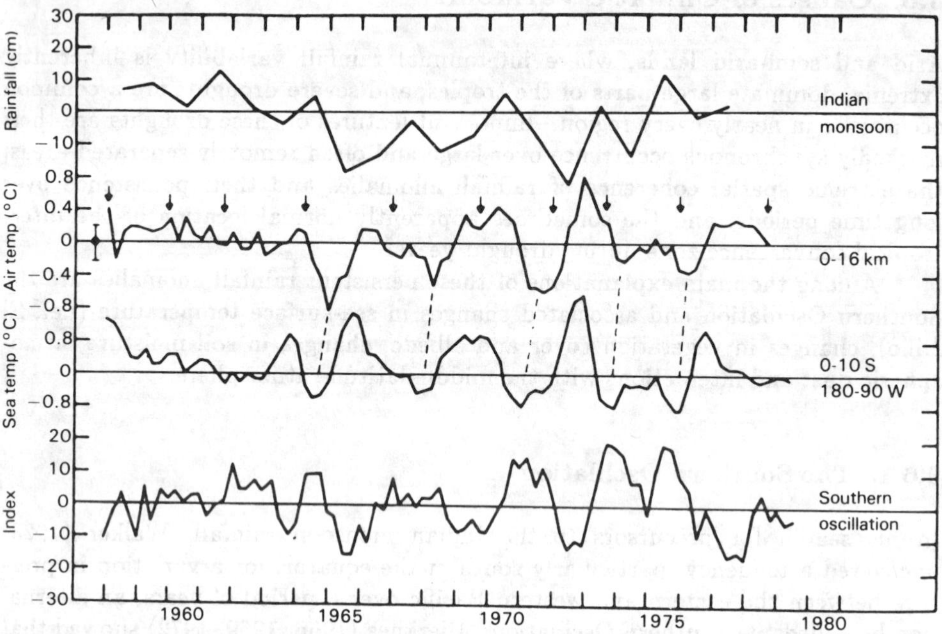

Figure 3.9. Comparison of time variations (from bottom to top) in Southern Oscillation index (normalized pressure difference between Tahiti and Darwin), sea-surface temperature, zonally averaged tropospheric temperature in the tropics, and Indian summer monsoon rainfall (after Angell, 1981).

month. The data are plotted by season, with the annual variation eliminated through the use of deviations from long-term seasonal means.

Figure 3.9 shows that between 1958 and 1979 there has been an almost exact out-of-phase relation between SST and Southern Oscillation index, i.e., warm SST has been associated with relatively high pressure at Darwin and/or low pressure at Tahiti. Angell (1981) found that the inverse relation between SST and Southern Oscillation index is significant at the 99.9% level. *Figure 3.9* also shows a close relationship between SST and zonally averaged tropospheric (0–16 km) temperature in the tropics.

The curve of Indian summer monsoon rainfall appears related to the other curves of *Figure 3.9.* The relation can be expressed as either relatively small rainfall amounts slightly preceding warm SST, or as relatively large rainfall amounts slightly following warm tropospheric temperature in the tropics. According to Angell (1981), there have been 20 years in which SST was at least 0.8° C above average during the El Niño season (DJF); and during each of these years the summer monsoon rainfall in India two seasons earlier was below average. In 17 of the 20 years between 1958 and 1977, above-average monsoon rainfall has followed (by about two seasons) warm tropospheric temperatures in the tropics, and below-average rainfall has followed cool temperatures.

Fleer (1981a) has undertaken a global survey of rainfall teleconnections. He computed more than 1000 cross-spectra using several reference stations.

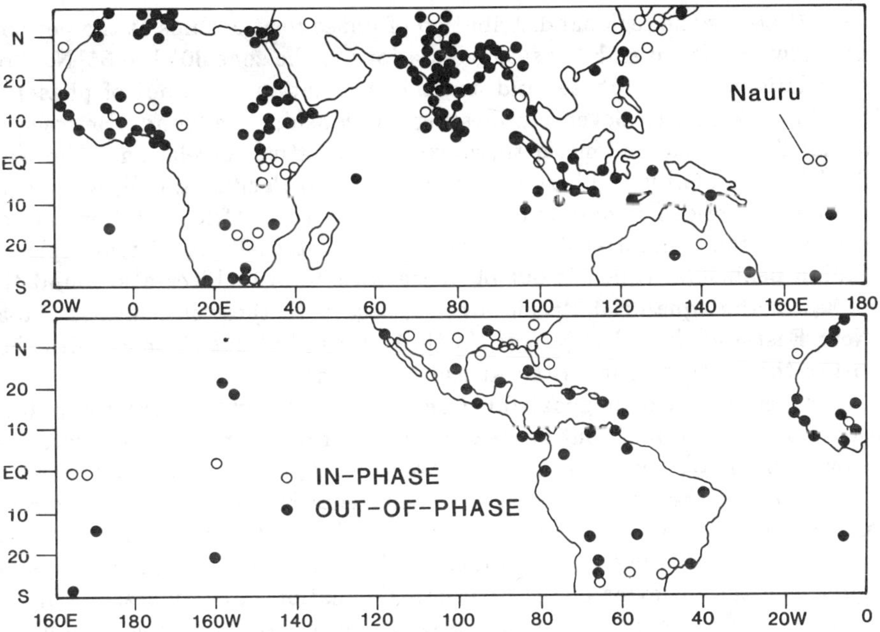

Figure 3.10. Regional distribution of coherent relationships between rainfall anomalies: reference station Nauru (after Fleer, 1981a and b).

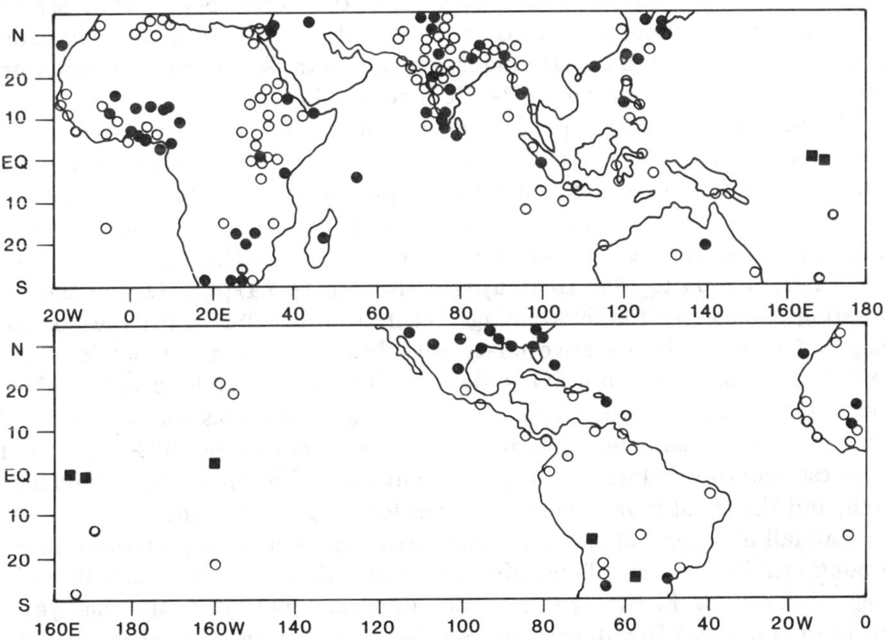

Figure 3.11. Regional distribution of the phase relation between seasonal rainfall anomalies and the Southern Oscillation index. Open symbols: in phase. Solid symbols: out of phase (after Fleer 1981a).

Figure 3.10 shows the regional distribution of phase relationships at the period of 5 years between Nauru and 289 selected stations in the zone 30° S to 35° N. Positive correlations (i.e., in phase) and negative correlations (i.e., out of phase) are indicated in this figure. Several interesting relationships are found. Between the equatorial Pacific and Indonesia there exists a negative correlation. The equatorial African rainfall is in phase with the equatorial Pacific rainfall, while equatorial South American rainfall and the majority of West African stations are out of phase with the equatorial Pacific. There is an internally consistent region on the Indian peninsula, which is out of phase with Nauru. Fleer also found that anomalies in the equatorial Pacific are in phase with those in southern Japan, the Near East and the subtropics of North America, but out of phase with those in most of the northern part of the African continent.

Using empirical orthogonal function analysis, Wright (1975) computed a Southern Oscillation index out of mean seasonal atmospheric pressure data from Capetown, Bombay, Djakarta, Darwin, Adelaide, Apia, Honolulu and Santiago. For every season the time series of the coefficients of the first eigenvector was normalized so that the mean value was zero and the standard deviation unity. The time series consisting of these dimensionless numbers are homogeneous and interpret well the behavior of the Southern Oscillation in every season. A high index corresponds to high atmospheric pressure in the southeast Pacific subtropical high and to low pressure in the Malayan–Indonesian low. Coherence analysis by Fleer (1981a) between this Southern Oscillation index and seasonal rainfall shows that the highest share of fluctuation is in the period range of approximately 5 years. A high index (*Figure 3.11*) that corresponds to a strongly developed Walker circulation is connected with high Indonesian, low South Indian, high Middle and North Indian, high South American, low equatorial Pacific and high North and South Pacific precipitation.

Parker (1983) has attempted to satisfy the need for a tabulation of a Southern Oscillation index that is simple enough to be updated readily by any user. He considers that the requirement for a simple but reliable index appears to be met best by the index recommended by Chen (1982) for diagnostic studies: Tahiti minus Darwin mean sea-level pressure. Values of this index are listed in *Table 3.1* from 1935 to 1985 (March) [*see also Figure 3.12(d)*]. One of the most interesting changes in Parker's listings occurs during 1982, when the Southern Oscillation entered the negative phase with below-normal mean sea-level pressure at Tahiti, and above-normal at Darwin. This corresponds to warmth in the usually cold areas of the equatorial Pacific. The lowest seasonal values for the series for northern summer, autumn and winter occurred in 1982–1983, as did the lowest monthly values for August, September, November, and January to March, and the equal-lowest monthly values for June and October.

Rainfall in several of the case study areas shows a strong relationship with the Southern Oscillation. Thus, Kousky *et al.* (1984) consider that there is a strong tendency for El Niño to occur simultaneously with or within one year of drought in Northeast Brazil, with the Southern Oscillation index showing a similar relationship. Of the six highest peaks in their Southern Oscillation index, five are associated with either simultaneous drought in Northeast Brazil or drought the following year. Since these five droughts (1915, 1919, 1932, 1942, 1958) are

Table 3.1. Monthly values (mb) of the difference of mean sea-level pressure between Tahiti (18°S, 150°W) and Darwin (12°S, 131°E) (after Parker, 1983). Mr. D.E. Parker is thanked for providing unpublished values for 1983, 1984 and 1985. (The table is reproduced with the permission of the Controller of Her Majesty's Stationery Office.)

Year	*Jan.*	*Feb.*	*Mar.*	*Apr.*	*May*	*June*	*July*	*Aug.*	*Sept.*	*Oct.*	*Nov.*	*Dec.*
1935	5.6	3.4	6.0	2.5	0.6	0.7	0.7	1.8	3.2	3.9	3.3	2.5
1936	3.8	4.5	4.1	4.9	2.0	0.8	1.4	0.1	2.0	2.7	0.6	3.4
1937	6.2	3.3	4.9	2.4	1.4	1.4	-0.1	2.0	2.1	2.3	2.4	4.6
1938	5.8	5.1	3.1	2.6	3.1	3.2	3.6	3.5	3.4	4.8	3.0	6.0
1939	7.8	6.0	5.9	3.5	1.5	1.0	2.2	1.6	0.8	0.5	1.7	1.8
1940	4.4	3.7	2.0	1.2	-0.2	-1.2	-1.4	-1.2	-0.9	-0.1	1.9	-2.3
1941	2.4	1.3	2.0	1.0	0.8	-0.6	-2.2	-1.3	1.0	-0.4	1.5	1.8
1942	1.7	3.8	2.9	1.7	2.3	2.2	0.8	2.3	3.8	4.3	2.3	6.2
1943	6.4	6.8	4.7	4.0	2.0	0.2	1.4	2.9	3.3	4.4	3.5	1.8
1944	2.7	5.4	5.0	1.7	1.5	0.7	-0.4	2.2	2.8	1.5	1.9	4.3
1945	5.5	5.9	6.4	1.5	1.6	2.2	1.5	3.5	3.8	3.3	2.4	4.8
1946	3.9	5.5	3.6	1.2	0.2	0.0	-0.6	1.0	-0.3	0.9	2.7	2.4
1947	3.4	3.7	6.1	1.8	-0.1	1.5	2.4	2.8	4.3	2.6	4.3	4.5
1948	3.8	4.0	3.2	2.7	2.1	0.6	1.1	1.0	1.1	3.9	3.6	2.4
1949	2.9	5.0	5.0	2.5	0.9	-0.3	0.7	1.0	2.7	3.8	2.0	5.0
1950	5.5	8.2	7.2	4.3	2.6	4.5	4.2	3.6	3.5	5.7	4.8	8.0
1951	7.2	5.9	3.0	1.4	0.0	1.0	-1.1	0.8	0.4	0.9	1.6	2.1
1952	2.7	3.1	4.2	1.5	2.6	2.1	1.7	1.3	2.0	3.4	3.0	1.2
1953	4.9	3.3	2.9	2.3	-1.9	0.9	0.8	-1.0	0.1	2.8	2.5	2.6
1954	5.6	3.7	3.7	3.1	2.1	0.9	1.5	3.2	2.7	3.2	3.3	6.0
1955	3.4	7.7	4.4	1.7	3.1	3.1	3.7	3.9	4.8	5.5	5.3	5.3
1956	6.8	7.2	5.7	3.6	3.8	2.7	2.8	3.4	2.5	6.0	3.3	5.4
1957	5.5	4.1	3.6	2.4	0.0	1.0	1.2	0.3	0.7	2.8	1.2	2.9
1958	0.9	3.3	3.7	2.6	0.3	1.2	1.5	2.9	1.8	2.8	2.2	2.3
1959	2.7	1.7	5.6	3.0	2.1	0.5	0.3	1.0	2.4	3.6	4.7	5.1
1960	4.6	4.4	5.1	3.4	2.2	0.9	1.7	2.7	3.6	2.9	4.0	4.9
1961	3.9	5.9	-0.1	3.5	1.8	0.9	0.9	1.6	2.5	2.1	4.0	6.2
1962	8.0	5.4	3.5	2.3	1.2	2.0	-1.0	2.4	3.2	4.5	3.6	3.8
1963	6.0	3.3	5.1	3.1	1.9	-0.2	0.6	1.2	1.3	0.5	1.5	1.2
1964	3.7	4.3	5.3	4.0	2.5	2.1	1.8	4.0	4.7	5.1	3.3	2.9
1965	3.6	5.1	4.6	1.0	1.5	-0.3	-2.5	0.0	0.0	1.1	0.3	3.8
1966	1.9	3.8	1.5	1.6	0.5	1.3	0.9	2.4	2.0	2.5	3.0	2.8
1967	7.5	7.3	5.5	1.9	1.2	1.9	1.1	2.6	3.3	2.8	2.2	2.4
1968	5.3	6.6	3.4	2.0	3.5	2.7	2.0	1.7	1.9	2.6	2.4	3.8
1969	1.6	3.2	4.0	1.3	0.8	1.1	-0.1	1.0	0.6	1.0	2.9	4.2
1970	2.3	2.3	4.3	1.8	1.9	2.5	0.1	2.3	4.5	4.6	5.9	6.9
1971	5.0	7.9	7.5	5.1	2.8	1.5	1.2	4.0	5.0	5.8	4.0	3.9
1972	5.2	6.2	4.4	1.7	-0.3	-0.3	-1.9	0.8	-0.1	1.1	2.4	1.1
1973	3.8	1.7	4.1	2.1	2.0	2.7	1.9	3.6	4.6	4.5	7.7	6.8
1974	8.8	8.0	7.7	3.7	3.0	1.5	2.8	2.7	4.4	4.3	2.7	3.4
1975	3.3	5.7	6.1	4.1	2.4	3.1	4.2	4.9	6.1	5.8	5.0	7.2
1976	6.9	7.3	6.4	2.5	1.9	1.3	-1.0	-0.2	0.2	3.4	4.4	2.8
1977	3.6	6.5	2.2	1.2	0.3	-1.0	-1.5	-0.2	0.8	0.8	0.7	1.5
1978	3.8	-0.7	2.9	1.4	3.7	1.9	1.8	2.0	2.5	2.0	2.6	3.3
1979	3.6	6.0	3.4	1.7	2.1	1.9	-0.3	0.9	2.6	2.5	2.2	2.1
1980	5.1	4.8	2.4	0.8	1.2	0.7	0.7	1.9	1.5	2.7	2.4	3.3
1981	5.0	3.9	0.9	1.7	2.7	3.0	2.3	2.5	3.4	2.0	3.3	4.4
1982	6.4	4.7	4.3	2.1	0.6	-1.2	-2.0	-2.0	-1.1	-0.4	-1.8	-0.7
1983	-2.0	-2.5	-1.2	0.3	2.3	0.8	-0.2	1.5	4.0	3.6	2.8	3.5
1984	4.7	5.8	2.8	2.6	1.6	0.1	1.1	2.0	2.7	2.1	3.5	3.2
1985	3.6	6.5	4.6									

the severest during the period, it is evident that monitoring sea-level pressure anomalies in the Australasia–Indian Ocean region may be useful in forecasting severe drought in Northeast Brazil.

Kousky *et al.* (1984) report that the rainfall pattern over Brazil during the 1982-1983 El Niño event was similar to that for the warm events of 1972 and 1976. *Table 3.2* gives the rainfall anomalies for the three events at selected stations in the Amazon Basin and Northeast Brazil. Negative rainfall anomalies during the period January–May were recorded in the Amazon Basin and the Northeast during all three events, with the greatest deficiencies occurring during the 1982–1983 event. Kousky *et al.* (1984) consider that the increased strength of the trades in the Atlantic during the Southern Oscillation events creates negative SST anomalies which, as shown by Hastenrath and Heller (1977), Markham and McLain (1977) and Moura and Shukla (1981), is an important characteristic of drought years in Northeast Brazil. In particular Moura and Shukla (1981) suggest that a possible mechanism for the occurrence of severe droughts over Northeast Brazil is the establishment of a thermally direct local circulation, which has its ascending branch at about 10° N and its descending branch over Northeast Brazil and the adjoining oceanic region. The driving force for this anomalous circulation is provided by warming due to enhanced moist convection associated with warmer sea-surface temperature anomalies over the northern tropical Atlantic, and cooling associated with colder sea-surface temperature anomalies in the southern tropical Atlantic. The combined effects of thermally forced subsidence and reduced evaporation and moisture flux convergence produces severe drought conditions over Northeast Brazil.

Table 3.2. January–May rainfall anomalies (mm) for selected stations in the Amazon Basin and Northeast parts of Brazil, during the last three El Niño/Southern Oscillation events. The five-month normal rainfall (mm) is given in the last column on the right (after Kousky *et al.*, 1984).

	January–May			
Station	*1972*	*1976*	*1983*	*Mean*
Sao Gabriel do Cachoeira 0°8'S, 67°5'W	-6	-93	-407	1394
Manaus 3°7'S, 60°1'W	-78	-19[a]	-416	1334
Belem 1°28'S, 48°29'W	-224	-168	-387	1808
Barra do Corda 5°30'S, 45°16'W	-92	-42	-133	796
Conceicao do Araguaia 8°16'S, 49°17'W	-189	-234	-178	978
Fortaleza 3°43'S, 38°33'W	-458	+102	-456	1146
Quixeramobim 5°12'S, 39°18'W	-151	-39	-291	563

[a] February 1976 departure at Manaus is absent.

A very extensive study of SST anomalies and drought for Australia has been made by Streten (1981, 1983). He has shown that for the period 1950–1969 major droughts covering extensive areas of Australia can be identified with cold SST around the continent and that transitions from dry to wet years and vice versa can be seen in the SST anomalies. Voice and Hunt (1984) have used a global circulation model to explore observed relationships between cold SST anomalies in the Australian region and drought over the Australian continent. Two experiments were performed: one involving a large cold sea-surface temperature anomaly located north of Australia but extending well into the Pacific and Indian Oceans, together with a smaller cold anomaly in the tropical Atlantic; the other added a warm anomaly in the tropical eastern Pacific to the previous anomalies. The overall results were very similar in the two experiments. Drought-like conditions were produced over parts of Australia, southern Africa, South America and North America, but the dominant response was in the Australian region. The authors found that the impact of the SST perturbations on the model was, to a large extent, controlled by the cold anomaly in the central and western Pacific areas. The analysis of the model experiments suggests that different precipitation, or drought, distributions over Australia would result from different initial conditions or from varying a given SST anomaly pattern relative to a selected synoptic distribution.

An association between Indian summer monsoon rainfall and the Southern Oscillation has already been noted. This relationship has been further explored by Mooley and Parthasarathy (1983). In *Table 3.3* they give details of strong and moderate El Niño years, along with all-India rainfall and the seasonal rainfall departures from the mean, expressed as percentages. The average departure of the all-India rainfall during these 22 El Niño years is −8.8%. There are 17 years (i.e., 77% of the total number of El Niño years) of below-normal rainfall, out of which eight years show drought situations. Of the five El Niño years with above-normal monsoon rainfall, one year shows excess rainfall. Some of the strong El Niño years – namely, 1877, 1899, 1918 and 1972 – show high monsoon rainfall deficiencies. After further statistical analysis Mooley and Parthasarathy (1983) came to the following conclusions for moderate to strong El Niño occurrences:

(1) During the 22 El Niño years during the period 1871 to 1978, the monsoon rainfall over most parts of the country was below normal.
(2) Association between El Niño events and deficient rainfall or drought conditions over India is statistically significant.
(3) The worst years for the Indian summer monsoon are generally El Niño years.
(4) Deficient rainfall or drought conditions over 19 meteorological subdivisions show significant association with the El Niño phenomenon. The association is significant at the 1% level for northwestern India, the northern west coast and the central peninsula.

These findings agree with those of Fleer (1981b) discussed earlier.

Table 3.3. Details of El Niño years and all-India summer monsoon rainfall. Normal seasonal rainfall = 85.31 cm; standard deviation (σ) = 8.29 cm (after Mooley and Parthasarathy, 1983).

Series	*Year*	*El Niño intensity: M = moderate S = strong*	*All-India seasonal rainfall* Amount (cm)	*% departure from normal*
1	1871	M	84.57	–0.9
2	1877	S	60.37	–29.2
3	1880	M	81.68	–4.3
4	1884	S	92.91	+8.9
5	1887	M	89.65	+5.1
6	1891	S	78.90	–7.5
7	1896	M	82.41	–3.4
8	1899	S	62.81	–26.4
9	1902	M	79.11	–7.3
10	1905	M	71.51	–16.2
11	1911	S	73.29	–14.1
12	1914	M	89.89	+5.4
13	1918	S	64.81	–24.0
14	1925	S	80.27	–5.9
15	1929	M	81.93	–4.0
16	1939	M	78.86	–7.6
17	1941	S	72.90	–14.6
18	1953	M	91.97	+7.8
19	1957	S	78.43	–8.1
20	1965	M	70.68	–17.2
21	1972	S	65.32	–23.4
22	1976	M	85.47	+0.2
Averages			78.08	–8.8

3.6.2. Albedo, soil moisture and atmospheric dust

While the Southern Oscillation may be a prime cause of drought in many tropical semi-arid areas. low rainfalls may be enhanced and caused to persist by a number of local factors. Northern summer radiation budgets of the surface–atmosphere system are positive almost everywhere to the north of 10° S with a maximum in the cloudless zones of the northern areas of the subtropical oceans. Negative radiation balances are observed over the desert regions of North Africa and the Middle East. There are several reasons for these negative balances during summer over the northern deserts. Because the surface is dry there is no evaporation, and surface temperatures become high with an associated large infrared radiation loss. The lack of clouds and of water vapor in the air allows unusually large amounts of radiation to escape from the surface to space. Lastly, the albedo of desert surfaces tends to be high, probably between 30 and 40%, as compared with vegetated surfaces. Because of this negative radiation imbalance, energy must be imported into the desert region to fill the deficit. The extra energy is supplied by the Hadley cell circulation. Thus, the

Hadley cell circulation creates the subtropical deserts, but the deserts in turn can intensify the Hadley cell.

Albedo effects

The biogeophysical feedback mechanism that accounts for the persistence of drought in the Sahel was first proposed by Charney (1975). The effect of drought, reducing soil moisture and thus evaporation and cloud cover, and increasing surface albedo as plant cover is destroyed, is generally to increase ground and near-surface air temperature while reducing both the surface radiation balance and also that of the local surface–atmosphere system. Charney's theory (1975) postulates that overgrazing and general loss of vegetation in tropical areas increases the albedo and decreases the radiation balance, making it less positive or more negative. In order to maintain thermal equilibrium and in the absence of advection, the air must descend, compress and warm adiabatically, thus decreasing the relative humidity and precipitation. Even though Charney applied his theory to deserts with negative radiation balance at the top of the atmosphere, the theory is also valid for deserts with a positive radiation balance. The main assumption is that adiabatic warming (descending motion), rather than advection (horizontal motion), is the main compensation for the decrease in the radiation balance.

Sud and Fennessy (1982) have recently used a general circulation model to investigate surface albedo changes. In tropical areas albedo normally increases with vegetation loss. In the first integration, called the control run, the surface albedo was normally prescribed, whereas in the second integration, called the anomaly run, the surface albedo was modified due to loss of vegetation in four regions: the Sahel in Africa, the western Great Plains in the United States, the Thar Desert border in the Indian subcontinent and Northeast Brazil in South America. An analysis of two simulations for July shows that, in the Sahel and the Thar Desert border regions, the current results suggest reduced precipitation with increased surface albedo in accordance with Charney (1975). The semi-arid Northeast Brazil, which had a winter circulation, also conformed to Charney's (1975) hypothesis. There was a general lack of albedo impact on precipitation in the Great Plains. This experiment provides support for Charney's hypothesis (1975) regarding the influence of surface albedo on mean monthly climatology in the subtropical desert margin regions.

Courel *et al.* (1984) have investigated, using satellite measurements, the albedo history of the western Sahara and Sahel. Some of their results for four strips 2° wide in latitude from 14° to 22° N, and extending from longitude 16° to 0° W are shown in *Figure 3.12(b)*. The albedo of the Saharan zone, curves 1 and 2, is relatively stable throughout the period. For the period 1967–1972, when the Sahel drought was intensifying [*Figure 3.12(a)*], Sahel albedo (curves 3 and 4) increased according to theory, and then showed a decrease up to 1975 as rainfall conditions improved. In contrast, Courel *et al.* (1984) found that although drought conditions still persisted in the Sahel, the surface albedo declined significantly between 1973 and 1978, in contradiction, they claim, to the expectations of the Charney albedo feedback mechanism.

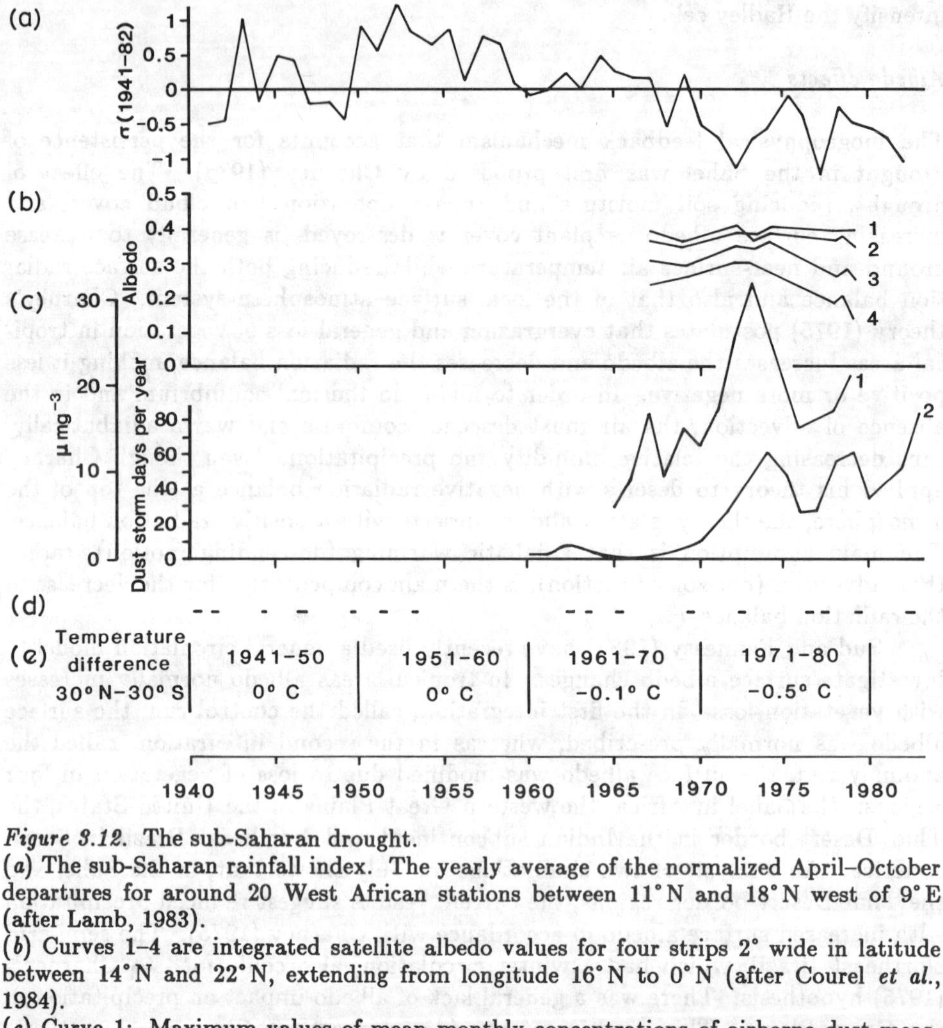

Figure 3.12. The sub-Saharan drought.
(*a*) The sub-Saharan rainfall index. The yearly average of the normalized April–October departures for around 20 West African stations between 11° N and 18° N west of 9° E (after Lamb, 1983).
(*b*) Curves 1–4 are integrated satellite albedo values for four strips 2° wide in latitude between 14° N and 22° N, extending from longitude 16° N to 0° N (after Courel *et al.*, 1984).
(*c*) Curve 1: Maximum values of mean monthly concentrations of airborne dust measured at Barbados (after Newell and Kidson, 1984). Curve 2: Annual dust storm frequencies at Nouakchott (18° 07' N, 15° 56' W) (after Middleton, 1985).
(*d*) Periods when the monthly values (mb) of the difference of mean sea-level pressure between Tahiti (18° S, 150° W) and Darwin (12° S, 131° E) were negative. This is a measure of the Southern Oscillation. Values taken from Parker (1983).
(*e*) The difference (in °C) between the zonally averaged decadal sea-surface temperature anomalies at 30° N and 30° S (after Folland *et al.*, 1984).

Soil moisture effects

Figure 3.13 illustrates the influence of a vegetation cover on evaporative loss. The exact nature of the evaporative loss will depend on the weather input, and the nature of the soil and vegetation. The vegetation here is assumed to be of

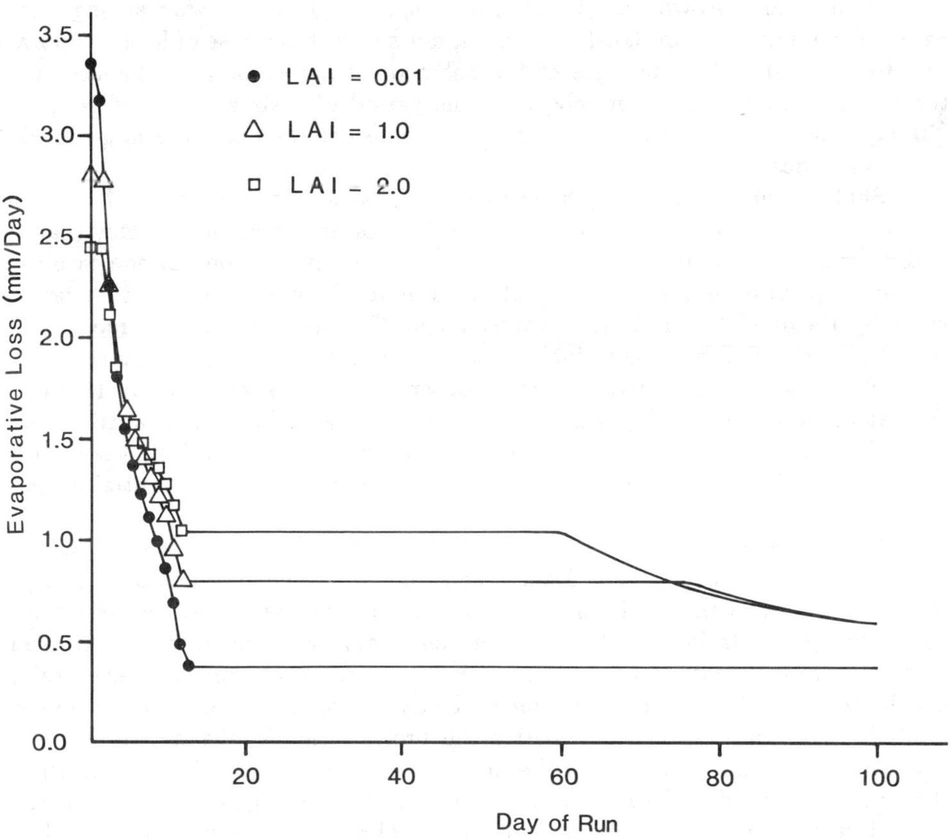

Figure 3.13. Predicted evaporative loss (soil evaporation plus transpiration) for various densities of vegetation and repeated input of 24 hourly data for a hot sunny day. The soil is assumed to be initially at field capacity and the vegetation to be coniferous trees 10 m high. The meteorological input is from the well-mixed atmosphere above the trees, and the radiation and energy balance of the surface adjusts as the soil loses moisture. LAI is leaf area index, and 0.01 refers to bare soil.

the coniferous forest type and the weather sequence to be repeated hourly data for a hot dry day. The soil is assumed to be initially at field capacity. The interesting features are that the initial evaporative loss from a bare soil is very high, but rapidly falls to low values after about a week as the surface layers of the soil dry. The introduction of a sparse vegetation cover reduces the initial evaporative loss, but also keeps evaporation at relatively moderate values for a period of time since transpiration reduces soil moisture at depth. Vegetation with a leaf area index around 4 may allow an evaporative loss of around 1.5 mm/day for nearly 2 months. The important point is that, with sparse vegetation covers, evaporation will fall to low values in warm climates unless there are relatively frequent rainstorms. This may be important if a large component of the rainfall is produced by local evaporation.

Walker and Rowntree (1977) have made a good case for saying that Sahelian rainfall feeds on Sahelian rainfall, not so much because of local evaporation from moist soil, as because of the behavior of baroclinic disturbances over the moist surface. Unfortunately, the time period of their experiment is small (20 days), so they may have just observed a transient response to a large shock, such as a widespread and rapid flood.

Shukla and Mintz (1982) have used a global atmospheric general circulation model to investigate the influence of soil moisture on climate. They placed two different constraints on the land-surface evapotranspiration: in one case the evapotranspiration is always set equal to the potential evapotranspiration calculated by the model (this is the evapotranspiration when the soil is moist and completely covered by vegetation); in the other case no evapotranspiration is allowed to take place. Starting from an observed atmospheric state on 15 June, the calculations for the dry-soil and wet-soil cases were carried forward for 60 days. The results discussed here are the time-averaged fields for July, the month when the northern hemisphere extratropics have the maximum potential evapotranspiration.

In the wet-soil case, the precipitation over Europe and over most of Asia is about 4 mm/day and does not differ much from the calculated potential evapotranspiration. In contrast in the dry-soil case, Europe and most of Asia have almost no precipitation. Only over Southeast Asia and India, in the dry-soil case, is there transport of water vapor from the ocean that produces heavy rain, and in this case the precipitation most closely resembles the observed summer rainfall. Over most of North America the precipitation in the wet-soil case is between 3 and 6 mm/day, but in the dry-soil case the precipitation over most of the continent is reduced to about 1 mm/day or less. Over South America, the rainfall near the equator in the wet-soil case is about 6 mm/day, and in the dry-soil case the rainfall is almost as large, all of it being water transported from the ocean. Along the coasts of Central America and Northeast Brazil the land precipitation decreases and the ocean precipitation increases in going from the wet-soil to the dry-soil case. Across Africa, at about 10° N, the precipitation in the wet-soil case is about 4 mm/day larger than the local evapotranspiration, but north and south of the rain band the precipitation is about 2 to 3 mm/day smaller than the evapotranspiration. In the dry-soil case, there is a band of rain of 3 to 4 mm/day at about 14° N, and this precipitation is about the same as the amount by which the precipitation exceeded the evapotranspiration in the wet-soil case.

Shukla and Mintz (1982) state that if their calculations are indeed applicable to nature, the implication for forecasting extratropical summer rainfall is clear. In about the month of May the continental rainfall changes from the large-scale upglide condensation type to the cumulus convection type. If, after this change takes place, there is a large amount of moisture stored in the soil, the summer months that follow can have a large or small amount of rainfall, depending on the circulation conditions. If the soil is dry, so that there is little or no evapotranspiration to keep the atmospheric planetary boundary layer moist, the remaining summer months will have little rainfall. Similar considerations could apply in the tropics in relation to wet and dry seasons. Shukla and Mintz consider that surface evapotranspiration, which requires moisture in the soil, is a necessary (though not sufficient) condition for extratropical summer precipitation. Such considerations could be important in the southern USSR and southern Australia. For further details of these models, *see* Mintz (1984).

Effects of atmospheric dust

Deserts and semi-arid regions are a major source of dust, and *Figure 3.12(c)* illustrates that the severe drought in the Sahel has been instrumental in producing an increased output of soil-derived aerosols into the atmosphere. Between 100 and 400 million tons of dust are blown over the Atlantic in a normal year and reach Barbados and Miami regularly (Prospero *et al.*, 1981). Carlson (1979) has evaluated the climatic impact of the Saharan dust layer over the Atlantic and found that aerosol concentrations that are frequent over the eastern Atlantic could enhance the heating rates in the 1000–500 mbar layer by as much as 1°C/day. This may provide a way of explaining the persistence of drought, because as the drought begins more dust is generated, heats the atmospheric layers and thereby increases the meridional temperature gradient, moving the ITCZ equatorward, and thus aiding the continuation of the drought.

The influence of the surface wind speed is shown by the following calculations by Lettau (1959) and Deacon (1969). Assuming a geostrophic wind speed of 25 ms^{-1} at latitude 43°, Deacon showed that over scrub with a roughness length of 100 cm, the wind speed at 10 m above the ground would be 7 ms^{-1}; but if the scrub were cleared and replaced by bare soil with a roughness length of 0.1 cm, the 10 m wind speed would increase to 13 ms^{-1}. He comments that in a semi-arid climate a wind speed of 13 ms^{-1} is sufficient to cause soil to start blowing away.

The concept is popular that violent, dust- and gas-producing volcanic eruptions are important climatologically because they often distribute a dust veil into the stratosphere that effectively screens incoming solar radiation, thus cooling the earth's surface. The warming of world climate during the 1920s and 1930s can be attributed in part to the absence of such volcanic eruptions at that time; and the cooling that culminated in the 1960s can be attributed to a renewal of eruptive activity that included the major eruption of Agung in 1963. A more recent major eruption, that of El Chichon in Mexico in the spring of 1982, was expected to result in a cooling in the years following. This cooling, however, is not apparent in the temperature record, a fact that may be explained by the contrary warming effect of the El Niño event of 1982–1983, in which enormous quantities of heat were transferred to the atmosphere from the warmer-than-normal tropical Pacific Ocean.

3.7. Future Climate

Over the last 30,000 years there is ample evidence of major changes in the hydrology of the semi-arid regions. These are reviewed in many papers, e.g., Street-Perrott and Harrison (1984) and Lockwood (1985). At various times during this period the semi-arid regions have been both considerably wetter (8 Kyr B.P.) and drier (18 Kyr B.P.) than at present. These changes are probably connected with the Milankovitch mechanism of earth orbital changes, changes in the distribution of ice sheets, and sea-surface temperatures resulting from the melting of the temperate-latitude ice sheets. These changes are therefore unlikely to be repeated in the near future since they are long-term climatic variations.

Comment has already been made that there is a strong link between sea-surface temperature and drought in many tropical areas. For example, colder-than-average tropical North Atlantic sea-surface temperatures are associated with drought in the Sahel. Folland *et al.* (1984) have investigated worldwide marine temperature fluctuations, 1856–1981. Their results show a worldwide temperature fluctuation of about 0.6° C, with the coldest period being centered around 1905–1910 and the warmest occurring in the 1940s. The major climatic fluctuation in the first half of the present century occurred simultaneously in both hemispheres but with greater amplitude in the northern hemisphere. The mid-latitudes of the two hemispheres now appear to be fluctuating out of phase, as they were to some extent before 1900. In ***Figure 3.12(e)*** are shown some values for the zonally averaged sea-surface temperature anomaly differences between 30° N and 30° S. It is seen that since the decade 1961–1970, the northern oceans have been cooling relative to the southern. This could have consequences for long-term rainfall trends in such areas as Northeast Brazil and the Sahel.

The carbon dioxide content of the atmosphere is increasing at a rate that has ranged over the past 20 years between about 0.5 and 1.5 ppm yr^{-1}, with an average of about 1.00 ppm yr^{-1}. A recent summary has concluded that, if present trends continue, the combined concentrations of atmospheric CO_2 and other trace gases would be radiatively equivalent to a doubling of CO_2 from pre-industrial levels as early as the 2030s (WMO, 1986). The likely effects of a doubling of atmospheric CO_2 on future climate are not yet well understood, but the most advanced experiments with general circulation models (GCMs) of the climatic system show increases in global mean equilibrium surface temperature of between 1.5 and 4.5° C (WMO, 1986). Case studies in the companion to this volume indicate that climatic changes of this order would cause major changes in agricultural potential in cool temperate and cold areas (Parry *et al.*, 1988). While surface warming in the tropics is likely to be less than at higher latitudes, there may be significant reductions of soil moisture in a zonal belt centred around 42° latitude (Manabe and Wetherald, 1980). Monsoonal precipitation may also increase significantly. In the immediate future the tropical temperature changes due to increasing atmospheric CO_2 are hardly significant for plant growth. In the longer term, however, the predicted decreases in subtropical soil moisture and the increased monsoonal rainfall may be of great importance. Manabe *et al.* (1981), again using GCMs, have predicted an overall increase in aridity during the summer season in the middle- and high-latitude continents with increasing CO_2 and this may have consequences for the southern USSR. Marked decreases in soil moisture are also predicted for parts of Australia.

At present there is no clear evidence of any climatic changes within the tropics arising as the direct result of increases in atmospheric CO_2 content during this century. A number of mechanisms have been discussed that could lead to changes in rainfall in semi-arid regions, and during the next 10 to 20 years these could be more important than CO_2-induced changes. Nevertheless, Schneider (1984) comments that the advent of a reasonably clear CO_2 global temperature signal this decade is a real possibility.

The climate anomalies associated with El Niño/Southern Oscillation events are highly persistent and nearly global in extent. The persistent nature of these events derives from the strong coupling between the tropical atmosphere and the tropical ocean. Thus, drought in Australia, Indonesia, India, West Africa and Northeast Brazil as well as excessive rainfall in the central and eastern Pacific, Peru, Ecuador and southern Brazil are all related to the Southern Oscillation. In the next 20 to 30 years rainfall changes in the tropics are likely to be dominated by the Southern Oscillation with some modification due to other causes described previously. Sometime next century, changes due to increasing atmospheric CO_2 will start to become of major importance in the tropics. Since the causes of the Southern Oscillation are not fully understood at this time, it is not possible to give long-term forecasts of its occurrence. But this may become possible in the near future, and when such forecasts are available it will be necessary to consider carefully the sorts of responses in agriculture and other climate-sensitive activities that are likely to be most appropriate. The case studies of effects of climatic variations that are the focus of this volume thus have two goals: to improve the adaptation of agriculture to climatic variability *now*, and to establish the basis for appropriate adjustment of agriculture to possible climatic variations *in the future*.

References

Angell, J.K. (1981). Comparisons of variations in atmospheric quantities with sea surface temperature variations in the equatorial eastern Pacific. *Monthly Weather Rev.*, **109**, 230–243.

Berlage, H.P. (1966). The Southern Oscillation and world weather. *Mededeelingen en Verhandelingen*, **88**. Korunklijk Nederlands Meteorologisch Instituut, The Hague.

Bjerknes, J. (1966). A possible response of the atmospheric Hadley circulation to equatorial anomalies of ocean temperature. *Tellus*, **18**, 820–829.

Bjerknes, J. (1969). Atmospheric teleconnections from the equatorial Pacific. *Monthly Weather Rev.*, **97**, 163–172.

Bjerknes, J. (1972). Large-scale atmospheric response to the 1964–65 Pacific equatorial warming. *J. of Physical Oceanography*, **2**, 212–217.

Carlson, T.N. (1979). Atmospheric turbidity in Saharan dust outbreaks as determined by analysis of satellite brightness data. *Monthly Weather Rev.*, **107**, 322–325.

Charney, J.G. (1975). Dynamics of deserts and drought in the Sahel. *Q. J. of the Royal Meteorol. Society*, **101**, 193–202.

Chen, W.Y. (1982). Assessment of Southern Oscillation sea-level pressure indices. *Monthly Weather Rev.*, **110**, 800–807.

Chiu, L.S. and Newell, R.E. (1983). Variations of zonal mean sea-surface temperature and large-scale air-sea interaction. *Q. J. of the Royal Meteorol. Society*, **109**, 153-168.

Courel, M.F., Kandel, R.S. and Rasool, S.I. (1984). Surface albedo and the Sahel drought. *Nature*, **307**, 528–531.

Deacon, E.L. (1969). Physical processes near the surface of the earth. In H.Flohn (ed.), *General Climatology*. World Survey of Climatology, Volume 2, Elsevier, Amsterdam, pp. 39–104.

Dietrich, G. and Kalle, K. (1957). *Allgemeine Meereskunde*. Gebrüder Borntraeger, Berlin.

Findlater, J. (1969). A major low-level air current near the Indian Ocean during the northern summer. *Q. J. of the Royal Meteorol. Society*, **95**, 362–380.

Findlater, J. (1972). Aerial explorations of the low-level cross-equatorial current over eastern Africa. *Q. J. of the Royal Meteorol. Society*, **98**, 274–289.

Fleer, H. (1981a). Large-scale tropical rainfall anomalies. *Bonner Meteorologische Abhandlungen*, **26**. Bonn.

Fleer, H.E. (1981b). Teleconnections of rainfall anomalies in tropics and sub-tropics. In J. Lighthill and R.P. Pearce (eds.), *Monsoon Dynamics*, Cambridge University Press, Cambridge, pp. 5–8.

Folland, C.K., Parker, D.E. and Kates, F.E. (1984). Worldwide marine temperature fluctuations 1856–1981. *Nature*, **310**, 670–673.

Gibbs, W.J. (1981). The nature of drought and strategies to reduce its effects. In *Proceedings of the Technical Conference on Climate – Asia and Western Pacific*. World Meteorological Organization, Geneva.

Gibbs, W.J. and Maher, J.V. (1967). Rainfall Deciles as Drought Indicators. *Bulletin 48*. Australian Bureau of Meteorology.

Hastenrath, S. and Heller, L. (1977). Dynamics of climatic hazards in northeast Brazil. *Q. J. of the Royal Meteorol. Society*, **103**, 77–92.

Hounam, C.E., Burgos, J.J., Kalik, M.S., Palmer, W.C., and Rodda, J. (1975). *Drought and Agriculture*. Technical Note 38. World Meteorological Organization, Geneva.

Kidson, J.W. (1975). Tropical eigenvector analysis and the Southern Oscillation. *Monthly Weather Rev.*, **103**, 187–196.

Kousky, V.E., Kagano, M.T. and Cavalcanti, F.A. (1984). A review of the Southern Oscillation: oceanic–atmospheric circulation changes and related rainfall anomalies. *Tellus*, **36A**, 490–504.

Lamb, P.J. (1983). Sub-Saharan rainfall update for 1982: continued drought. *J. of Climatology*, **3**, 419–422.

Lettau, H.H. (1959). Wind profile, surface and geostrophic drag coefficients in the atmospheric surface layer. *Advances in Geophysics*, **6**, 241–257.

Lockwood, J.G. (1985). *World Climatic Systems*. Edward Arnold, London.

Manabe, S. and Wetherald, R.T. (1980). On the distribution of climatic change resulting from an increase in CO_2 content of the atmosphere. *J. of Atmospheric Sciences*, **37**, 99–118.

Manabe, S., Wetherald, R.T. and Stouffer, R.J. (1981). Summer dryness due to an increase of atmospheric CO_2 concentration. *Climatic Change*, **3**, 347–386.

Markham, C.G. and McLain, D.R. (1977). Sea surface temperature related to rain in Ceará, Northeast Brazil. *Nature*, **265**, 320–323.

Middleton, N.J. (1985). Effect of drought on dust production in the Sahel. *Nature*, **316**, 431–434.

Miller, F.R. and Keshavamurthy, R.N. (1968). *Structure of an Arabian Sea Summer Monsoon System*. East–West Center, Honolulu.

Mintz, Y. (1984). The sensitivity of numerically simulated climates to land-surface boundary conditions. In J.T.Houghton (ed.) *The Global Climate*, Cambridge University Press, Cambridge, pp. 79–105.

Mooley, D.A. and Parthasarathy, B. (1983). Indian summer monsoon and El Niño. *Pure and Applied Geophysics*, **121**, 339–352.

Moura, A.D. and Shukla, J. (1981). On the dynamics of drought in north-east Brazil: observations, theory and numerical experiments with a general circulation model. *J. of Atmospheric Science*, **38**, 2653–2675.

Newell, R.E. and Kidson, J.W. (1984). African mean wind changes between Sahelian wet and dry periods. *J. of Climatology*, **4**, 27–33.

Parker, D.E. (1983). Documentation of a Southern Oscillation index. *Meteorol. Magazine*, **112**, 184–188.

Parry, M.L., Carter, T.R. and Konijn, N.T. (eds.) (1988). *The Impact of Climatic Variations on Agriculture. Volume 1, Assessments in Cool Temperate and Cold Regions*, Reidel, Dordrecht, The Netherlands.

Peixoto, J.P. and Oort, A.H. (1983). The atmospheric branch of the hydrological cycle and climate. In A. Street-Perrott, M.Beran and R.Ratcliffe (eds.) *Variations in the Global Water Budget.* Reidel, Dordrecht.

Philander, S.G.H. (1983). El Niño Southern Oscillation phenomena. *Nature*, **302**, 295–301.

Prospero, J.M., Glacuum, R.A. and Nees, R.T. (1981). Atmospheric transport of soil dust from Africa to South America. *Nature*, **289**, 570.

Rasmusson, E.M. and Carpenter, T.H. (1982). Variations in tropical sea surface temperature and surface wind fields associated with the Southern Oscillation/El Niño. *Monthly Weather Rev.*, **110**, 354–384.

Riehl, H. (1979). *Climate and Weather in the Tropics.* Academic Press, London.

Schneider, S.H. (1984). On the empirical verification of model-predicted CO_2-induced climatic effects. In J.E. Hansen and T. Takahashi (eds.) *Climate Processes and Climatic Sensitivity*, Maurice Ewing Series, **5**, pp. 187–201, American Geophysical Union, Washington, D.C.

Shukla, J. and Mintz, Y. (1982). Influence of land-surface evapotranspiration on the earth's climate. *Science*, **215**, 1498-1500.

Street-Perrott, A. and Harrison, S.P. (1984). Temporal variations in lake levels since 30,000 yr. B.P. – an index of the global hydrological cycle. In J.E. Hansen and T. Takahashi (eds.) *Climate Processes and Climate Sensitivity*, Maurice Ewing Series, **5**, pp. 118–129, American Geophysical Union, Washington, D.C.

Streten, N.A. (1981). Southern hemisphere sea surface temperature variability and apparent associations with Australian rainfall. *J. of Geophysical Research*, **86**, 485-497.

Streten, N.A. (1983). Extreme distributions of Australian annual rainfall in relation to sea surface temperature. *J. of Climatology*, **3**, 143-153.

Sud, Y.C. and Fennessy, M. (1982). A study of the influence of surface albedo on July circulation in semi-arid regions using the GLAS GCM. *J. of Climatology*, **2**, 105-125.

Trewartha, G.T. (1968). *An Introduction to Climate*, 4th edition, McGraw-Hill, New York, 408 pp.

Voice, M.E. and Hunt, B.G. (1984). A study of the dynamics of drought initiation using a global general circulation model. *J. of Geophysical Research*, **89**, 9504-9520.

Walker, G.T. (1923). Correlations in seasonal variations of weather VIII: a preliminary study of world weather (World Weather I). *Memoirs of the India Meteorological Department*, **24**, 75-131.

Walker, J. and Rowntree, P.R. (1977). The effect of soil moisture on circulation and rainfall in a tropical model. *Q. J. of the Royal Meteorol. Society*, **103**, 29-46.

Wallén, C.C. (1968). Agroclimatological studies in the Levant. In *Agroclimatological Methods*, Proceedings of the Reading Symposium, UNESCO, Paris, pp. 225-233.

WMO (1986). *Report of the International Conference on the Assessment of the Role of Carbon Dioxide and of Other Greenhouse Gases in Climate Variations and Associated Impacts*, Villach, Austria, 9–15 October 1985. WMO-No. 661, World Meteorological Organization, Geneva, 78 pp.

Wright, P.B. (1975). An index of the Southern Oscillation. *CRU RP4*. Climatic Research Unit, University of East Anglia, England.

Wright, P.B. (1979). Persistence of rainfall anomalies in the central Pacific. *Nature*, **277**, 371-374.

PART II

The Effects of Climatic Variations on Agriculture in Central and Eastern Kenya

Contents, Part II

List of Contributors

AKONG'A, Dr. J.
Institute of African Studies
University of Nairobi
PO Box 30197
Nairobi
Kenya

DOWNING, Mr. T.E.
CENTED
Clark University
Worcester, MA 01610
USA

KONIJN, Dr. N.T.
Department of Soil Science and Geology
Agricultural University
PO Box 37
NL 6700 AA Wageningen
The Netherlands

MUNGAI, Mr. D.N.
Department of Geography
University of Nairobi
PO Box 30197
Nairobi
Kenya

MUTURI, Mr. H.R.
National Environment Programme
PO Box 67839
Nairobi
Kenya

POTTER, Dr. H.L.
Kenya Agricultural Research Institute
PO Box 57811
Nairobi
Kenya

Abstract of Part II

Climatic variability contributes to severe problems for the people of the semi-arid parts of the Africa through drought-related food shortages. The 1984 drought that afflicted the Sahelian region and eastern and southern Africa focused attention on this, and in Kenya the National Environment Secretariat initiated a project to study this problem in an area of central and eastern Kenya. This case study presents a detailed climate impact assessment for three districts – Kirinyaga, Embu and Machakos – which include a range of conditions from a humid tea-growing agroclimate to semi-arid grazing lands. The study area's climate is described and the nature of drought discussed. Kenya's past and present policies toward development of the semi-arid (traditionally drought-prone) areas and drought preparedness are reviewed. Various agroclimatic and socioeconomic aspects and some linkages among them are explored, utilizing several models. The policy implications of the results are then discussed.

The main climatological basis for the analyses is the 1951–1980 base period. Conditions are usually considered for the two rainy seasons or for the 12 months beginning in October, and results are presented according to the five agroclimatic zones in the study area. Five climatic scenarios are defined, based on cumulative probabilities of occurrence at four reference stations:

(1) Good year (or season): rainfall amount exceeded in 10% of the base period years.
(2) Average year (or season): median rainfall.
(3) Moderate drought: rainfall amount exceeded in 90% of the base period years.
(4) Severe drought: rainfall amount exceeded in 98% of the base period years.
(5) Back-to-back drought: successive years or seasons with rainfall amount that would be exceeded in 90% of the base period years.

In addition various seasons from the historical records are used to illustrate the effects in a range from good seasons to severe droughts. These are the March–May and October–December rainy seasons in 1974, 1975, 1976 and 1977, and the March–May season of 1984.

The impact of climatic variability on agricultural production is estimated, using a simple agroclimatic zonation in conjunction with the ratio between total rainfall and potential evaporation on an annual or seasonal basis. A dynamic crop production model using meteorological data for 10-day periods and some soils data are also utilized for examining impacts on maize yields. Livestock are quite important in the study area, and relationships between climate, grass, meat and milk production are employed in assessing climatic impacts on the livestock enterprise. There follows a review of the coping strategies to lessen drought impacts; strategies available to and used by the smallholder farmers, who constitute a major part of Kenya's population. General trends in smallholders' vulnerability to drought are documented.

The case study emphasizes the need to consider the whole chain of events from a meteorological drought through to human consequences and responses. Integrated assessments that consider not only the climate and the crops and livestock, but also further parts of the chain, are needed in order to identify linkages that are critical and amenable to policy interventions, and to identify knowledge gaps, such as the lack of time series of crop yield data. An example of a critical linkage might be the use of meteorological data at an early stage in a rainy season to make assessments of crop prospects to aid farmers in deciding whether or not to fertilize, thin the crop, or intercrop. Consideration of the whole chain or system can provide useful insight with respect to policies relating to such matters as agricultural extension, soil conservation, marketing, credit and prices, livestock development, and food security. Recalling the 1984 drought and previous drought episodes emphasizes the importance of formulating future policies to reduce individual and national vulnerability to drought on the basis of integrated assessments that recognize the role of climatic variability in the development of the country.

SECTION 1

Introduction to the Kenya Case Study

1.1. Climatic Adversity and Public Response

1.1.1. Climatic variability and drought in Kenya

Recent widespread droughts in southern Africa, the Sahelian region and eastern Africa have sparked speculation as to whether Africa is experiencing a pronounced climatic change. There is a distinct possibility that the global climate will undergo significant warming in the decade ahead due to increases in atmospheric carbon dioxide and trace gases, such as chlorofluorocarbons, resulting in some highly productive agricultural areas becoming drier, with more frequent incidences of drought, and culminating in economic marginalization (Parry, 1985). There is as yet no clear picture of the expected changes in the climates of the tropics (*see* Part I, Section 3 of this volume). Certainly, climatic variability is a dominant feature of the climate of Africa, and drought is a common phenomenon. The historical record, whether oral or written, contains abundant references to delayed or curtailed wet seasons and prolonged periods of no rains at all.

Climatic variability is the departure from normal or the difference in magnitude between climatic episodes. As such, it includes good and bad seasons. In East Africa, two generalizations are warranted: climatic variability is largely dominated by rainfall; and bad seasons, characterized by drought, have far greater impact on social and economic systems than good seasons. Therefore, this case study focuses on drought, although it includes discussion of strategies for maximizing production in good seasons as a mechanism for coping with drought.

Reduced agricultural production may be caused by poor weather, crop disease, ecological degradation, failure of national input supply and marketing

Table 1.1. Drought years in Kenya.

Drought years	*Region*	*Impacts*
1896–1898	Kikuyu and Kamba lands	Failure of three consecutive rains; rinderpest, locusts; thousands died (est. 25–75% of population); smallpox, war, famine, railway construction added to food shortages.
1907–1911	Lake Victoria area, Machakos & Kitui	Minor food scarcities in various localities.
1913–1918	Ethiopia, Kenya & Kamba lands	Coincided with Sahelian drought.
1925	Kerio valley, coast	Local food shortages.
1933–1934	Coast, Kikuyu lands	Drought, crop losses, locusts; large livestock losses (50%) in Baringo District; soil erosion and overstocking are recognized as problems.
1936	–	Minor drought and locusts.
1942–1944	Kenya, Uganda	Drought, locusts, smallpox and military demands caused food shortages; deaths under 200 people.
1949	Kikuyu lands, coast	Rains failed; good harvests in Nyanza and Rift Valley Provinces.
1952–1955	Kitui & other districts	Drought reduced exports by 20–25%.
1960–1961	Widespread: Maasai-land, S. Kenya, Machakos, Kitui, Rift Valley, northern districts	Drought followed by floods caused famine requiring US aid. Cattle mortality among Maasai 70–80%. Widespread crop failure. Ten million pounds spent on food relief in 1961.
1974–1976	Kajiado, Kitui, Machakos, lower Meru, Tana R., Turkana	Government projects (food reserve, drought-resistant crops, livestock improvement schemes) averted worse impacts. Maasai cattle losses as high as 80%.
1981	Eastern Province	Known as "famine with cash in my pocket". Large imports required, partly due to depletion of maize reserves by early exports.
1983	Coastal hinterland, Kitui, Machakos, Meru, Kakamega, Nyanza	Poor long and short rains; high prices in local markets; some water shortages. Cattle and human migration.
1984	Central, Rift Valley, Eastern & NE Provinces	Severe drought in long rains over all of Kenya except Western and Coast Provinces. Required large food imports. Short rains normal.

Sources: Berry *et al.* (1980); Downing (1982); Odingo (1982); Ojany and Ogendo (1973); Wisner (1977).

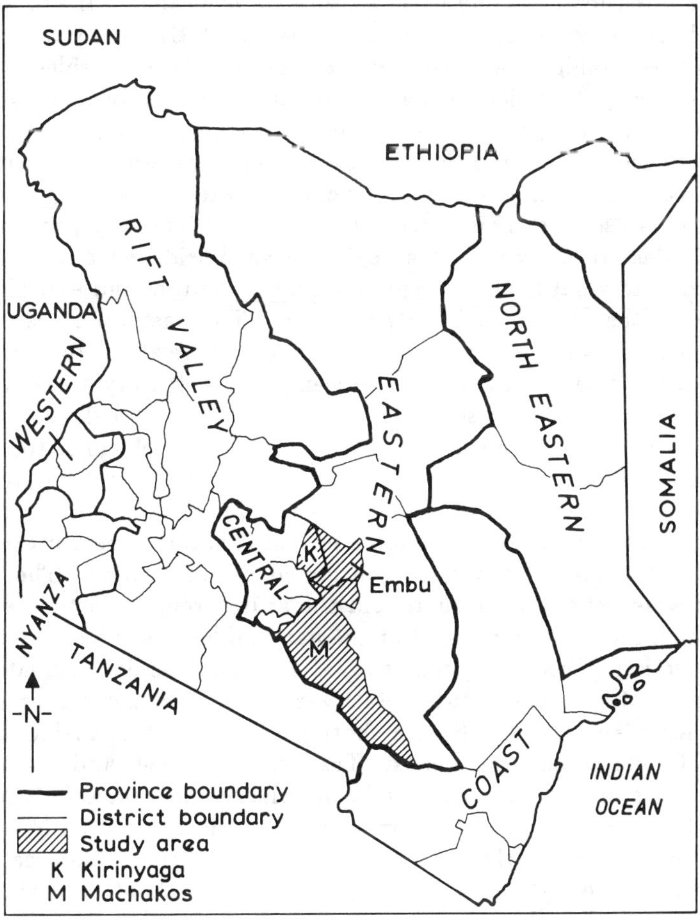

Figure 1.1. Location map.

systems, and poor farm management. Overall food availability may be further affected by market forces, resource limitations, storage problems, import and distribution constraints, and political and economic marginality. Famine, or deficient food consumption by individuals, may result from the above, as well as from poverty and the inability to procure available food. Underlying processes that disrupt social networks and contribute to population growth and underdevelopment are also factors in the chain of events from shortfalls in expected agricultural production to famine. In the interaction between natural and social forces, it is never certain which explanations are the most valid. However, it is clear that weather is a fundamental variable and is often identified as a cause, or at least a trigger, of drought and famine.

The effect of climatic variability on agricultural production is extremely important to Kenya's economy. About one-third of the GDP (1981) is from agriculture (Central Bureau of Statistics, 1982a). Cereals (maize, wheat and others) accounted for 13%, and livestock 20%, of the value of the total marketed agricultural production in 1981 (Central Bureau of Statistics, 1982a). The effects are not only economic: over 80% of Kenya's population resides in the rural areas and is largely dependent on agriculture and livestock for its livelihood.

There can be few people in Kenya who have been untouched by the effects of drought (*Table 1.1*). Traditionally, two areas have received the most attention in Kenya concerning the impacts of climatic variability on agriculture, livestock and human society (*Figure 1.1*). The Eastern Plateau, approximately coincident with Eastern Province, is a region of semi-arid, low-altitude land (formerly called "marginal lands"), beginning east of Nairobi and extending around Mt. Kenya on the eastern and northern sides. The pastoral areas of northern and southern Kenya have long been studied in relation to their climatic resources. More recently, attention has been given to the dry areas around Lake Victoria, the new settlement schemes in the Rift Valley and the coastal agricultural lands, where too much water is often as serious a problem as too little.

The 1984 drought has provided an immediate background to this case study. The drought actually began with localized crop failures in Northeast and Rift Provinces in Kenya. It was notable in that it affected the central highlands, an area that has not usually been considered drought-prone. The succeeding short rains were generally adequate and the acute drought, fortunately, did not persist. The impact of the drought at the national level was a severe shortfall in cereal production (*Table 1.2*), requiring at first commercial and later donated food imports. Marketed livestock totals were over 50% above average. Subsequently, since 1986 there has been a shortage of livestock available to the abattoirs and of milk to the creameries. Tea prices in 1984, and coffee prices in 1985–1986, were higher than usual and somewhat compensated for the increased foreign exchange required for food imports. The timely decision by the government to purchase food on the international market averted widespread famine in the country. The burden of the drought at the household level was the most severe for those with little cash to purchase food, or in relatively remote areas (in the drier zones of the country) where food was not always readily available in the local market.

1.1.2. The history of public response to problems of semi-arid areas

Through the first three quarters of this century, six eras may be distinguished in the history of the development of the semi-arid (traditionally drought-prone) areas in Kenya (Ngutter, 1979). These are outlined below, followed by an assessment of recent government policies.

Financing East African investments

After the railway from Mombasa to the Kenya highlands was completed in 1900, the British government began to encourage Europeans to settle in Kenya to

Table 1.2. Marketed agricultural production in Kenya, 1980–1984.

Commodity	*1980*	*1981*	*1982*	*1983*	*1984*
Cereals:					
Quantum index[a]	58.7	85.5	100.0	105.8	83.4
K. pounds, million[b]	35.3	48.2	59.7	81.4	71.4
Livestock and products:					
Quantum index	80.6	82.0	100.0	100.3	168.5
K. pounds, million	353.1	386.9	448.9	555.5	788.8
Permanent crops:[c]					
K. pounds, million	204.5	195.6	232.9	316.6	551.8

[a]Quantum index is quantity produced relative to 1982.
[b]US$1.00 = K. pounds 0.81 (1985).
[c]Includes coffee, tea, sisal and others.
Source: Central Bureau of Statistics (1985).

develop surplus and export production to make the railway (and other investments) self-financing. The Crown Lands Act (first passed in 1902 and strengthened in 1915) alienated land for European settlement, which reached a peak in 1908. African lands were generally left in the lower zones, and development of the semi-arid areas was ignored.

Dual policy

In 1923 the British government issued a declaration that African interests should be paramount in all matters affecting Kenya, including land. This ended the alienation of African lands and led to a dual policy of agricultural development on the European and African lands. Nominal beginnings to develop the African areas included establishment of two agricultural schools, local land-use regulatory authorities and District Betterment Funds, development of infrastructures, and identification of overstocking and overgrazing problems. These efforts were largely on paper and interrupted by World War II. The major focus, again, was on the more humid zones. The semi-arid areas supplied meat during the war, but received little attention in this era.

African land development

The displacement of Africans from the highlands, human and livestock population growth, and neglect of the semi-arid areas gave rise to serious land degradation. An African Settlement Board was set up in 1945 to resettle Africans on less used lands, thus relieving the problems of overstocking. The resettlement approach soon gave way to development alternatives.

The first major policy statement for the semi-arid areas was the Ten Year Development Plan for 1946–1955. The African Land Development Organization (ALDEV) was established to finance, survey and plan projects in the African lands. It largely focused on the semi-arid areas. Projects included settlement schemes, grazing control and management schemes, and afforestation.

The Ten Year Plan was followed by the Swynnerton Plan, 1955–1960, to promote agricultural intensification through land consolidation and registration, cash cropping, improved livestock management, ranching schemes and infrastructural development (including extension services). For the first time the semi-arid areas were specifically recognized as requiring special consideration. They were thought to be capable of food self-sufficiency, as well as being a source of livestock and livestock products. Projects in the semi-arid areas included controlled stocking rates, livestock marketing, water development and tsetse eradication.

Land transfer and resettlement

With the emergence of Kenya as an independent country, the formal distinction between European and African lands was abolished and many of the former European lands were transferred to African owners and offered to African settlers. The Million Acre Scheme in the 1960s was the most famous program, but many farms were bought by corporations or cooperatives and are still being subdivided and settled. The resettlement programs largely ignored the special characteristics of the semi-arid areas. Indeed, many of the settlement schemes are in dry areas with individual holdings that are too small to be viable even in average years.

Rangeland management

During the resettlement era, increasing concern began to emerge about the need to develop coherent policies for the rangelands – including the drier semi-arid lands and the arid zones. The Range Development Program in the mid-1960s included ranch planning, research and training. Institutions were developed to implement policies on the rangelands. Group title to rangelands was recognized.

Dryland farming

The gap between the programs for the humid areas of the European highlands and the rangelands become apparent in the 1970s, coincidental with a major expansion of the agricultural frontier into the semi-arid areas. The two five-year development plans, 1974/78 and 1979/83, set major policies for the semi-arid areas. They included development of drought-resistant crop varieties, improved cropping systems, land rehabilitation and infrastructural development. Institutional developments included expansion of research stations; establishment of regional development authorities, the Kenya Rangeland Ecological Monitoring Unit and the Arid and Semi-Arid Lands Unit; and implementation of a number of systematic surveys and pre-investment studies.

1.1.3. Current policy

In the two decades since independence the government has carried forward programs at several levels to reduce the drought hazard. These have been succinctly outlined in the Sessional Paper on food policy (Government of Kenya, 1981). The guiding objectives of Kenya's food policy set forth therein are to:

(1) Maintain a position of broad self-sufficiency in the main foodstuffs in order to feed the nation without using scarce foreign exchange on food imports.
(2) Achieve a calculated degree of security of food supply for each area of the country.
(3) Ensure that these foodstuffs are distributed in such a manner that every member of the population has a nutritionally adequate diet.

Specifically referring to national food security, the Sessional Paper (pp. 19–20) sets out a number of methods for achieving the above objectives:

(a) Increase food production in all areas of the country.
(b) Emphasize drought-resistant crops, such as sorghum and millet, in the dryland areas.
(c) Establish a food commodity monitoring and reporting system.
(d) Improve monitoring and forecasting of weather conditions in the main agricultural zones, and disseminate information more widely on expected weather trends.
(e) Regulate food exports to maintain domestic supplies and import food necessary to meet nutritional requirements.
(f) Accumulate a multi-commodity strategic food reserve from domestic surpluses and grain supplied on concessional terms to be used during periods of crop failure or other emergency situations.

The recent five-year development plan, for 1984–1988, continues to recognize the special needs for development of the semi-arid areas, particularly under the policy of decentralized planning, the "District Focus for Rural Development" (Government of Kenya, 1983). A third of the projected increases in agricultural production in Kenya is expected to come from new acreage, largely in the semi-arid areas. The specific policies for agricultural improvement (p. 197) are:

- Small farm focus.
- More intensive resource use through improved crop and livestock husbandry practices.
- Technology improvement through an increase in research and extension effort.
- Market incentives that not only provide fair prices, but also insure prompt payment for deliveries of produce to quasi-governmental bodies.
- Increased emphasis on arid and semi-arid lands by providing improved services, including output-increasing technology.

- Improved management and administrative procedures to insure that resources are used efficiently.

Significant progress has been made in recent years in meeting many of these objectives. Within the longest time frame are research projects to develop suitable crops and farming practices for the dry areas and better livestock development approaches. These include maize breeding, experiments with xerophytic (adapted to dry conditions) trees and grasses, livestock disease control, and agricultural tool development. In several of the drier districts the government has implemented integrated development projects to reinforce the District Focus for Rural Development – the decentralization of government services and decision making. These projects have attempted to strengthen the rural infrastructure, develop the productive sectors of the economy and train local leadership. At the national level the government maintains a systematic monitoring network for weather, food production and sales, and health. These data are used to indicate when imports may be required to augment the existing stores. A strategic reserve of 360 000 metric tons of cereals is targeted to be kept by the National Cereals and Produce Board (NCPB). This represents about one-third of the NCPB storage capacity (Government of Kenya, 1981).

There are as yet many constraints to achieving satisfactory food security at the national and local levels. The need for accurate, timely information and responsible decisions often conflicts with bureaucratic budgeting, protocol and politics. In addition it is difficult to implement national policies at the farm level and to promote improved agricultural management. As the following sections indicate, food security at the household level is not solely a matter of agricultural production (or of food policy), but must include economic development and the roles of markets and off-farm incomes.

1.2. The Kenya Case Study

As the realities of the drought in 1984 became apparent, the National Environment Secretariat (NES) formulated a project to analyze the background conditions for drought, and document the effects of the 1984 drought in a six-district area of central and eastern Kenya. NES is a department of the Ministry of Environment and Natural Resources, located in Nairobi. It is responsible for coordinating policy on environment and monitoring the state of the environment in Kenya. The study area includes parts of the central highlands and semi-arid Eastern Plateau. The districts are Kiambu, Murang'a, Kirinyaga, Embu, Machakos and Kitui (*Figure 1.1*).

This section introduces the Kenya case study. Subsequent sections illustrate for three districts – Kirinyaga, Embu and Machakos – the application of a systematic climate impact assessment in Kenya. The assessment has three notable features:

(1) The entire chain of drought hazard is assessed in the scope of one project. As with other case studies in this volume, the analysis integrates the

climatology, biophysical climate impacts, social responses and policy implications. Part of the integration is to connect the separate methods of each component in the analysis.

(2) A consistent set of climatic scenarios is used as input into each section on climate impacts, responses and policy. The use of scenarios forms the common thread between the sections. The scenarios, based on the historical climatic record for 1951–1980, are simulations of future climatic episodes. They are then used to model the impacts on agriculture, livestock and smallholder socioeconomic conditions, and to assess the policy implications of these impacts. The specifics of the scenarios are elaborated in Section 2. Five scenarios are defined:

   (a) Good season.
   (b) Average season.
   (c) Moderate drought.
   (d) Severe drought.
   (e) Multiple seasons of moderate or worse drought.

   The use of scenarios facilitates the quantification of the impacts of climatic variability and identification of thresholds for socioeconomic responses. For most impacts the effect of multiple-season drought is additive. For example, there is little additional impact of successive droughts on crop productivity. The importance of multiple seasons of drought are most clear in livestock impacts (Section 5) and socioeconomic vulnerability to drought (Section 6).

(3) The focus of the case study on integration is intended for the nonspecialist reader responsible for drought policy. The emphasis is on the practical implications of a large body of research work on drought in Kenya, rather than reviewing issues of more academic concern. The individual sections and the case study as a whole have been reviewed by competent scientists familiar with Kenya and the relevant disciplines.

Table 1.3. Median rainfall and coefficient of variation at four reference stations, 1951–1980 (in mm).

Period	*Kerugoya*	*Embu*	*Machakos*	*Makindu*
Mar–May (long) rains				
Median	702	527	397	175
CV	28	32	40	52
Oct–Dec (short) rains				
Median	461	354	344	261
CV	57	54	67	58
Oct–Sept (AY)				
Median	1525	1955	905	553
CV	25	27	36	41

Notes: CV is coefficient of variation, in %. AY is agricultural year.

Throughout the case study four reference rainfall stations have been used to provide comparisons between the climate and climate impacts papers (*Table 1.3* and *Figures 1.2* and *1.3*). The stations were chosen to represent specific agroclimatic zones in the Kenya Soil Survey's agroclimatic classification. Data for 36 rainfall stations are used in Sections 2 and 3 to provide spatial detail in the study area. Additional stations are used in Sections 3, 4, 5 and 6 to provide comparisons and where data are lacking for the four reference stations. Machakos is on the border between zones III and IV, and is replaced by Katumani in the livestock section (5) and conclusions (Section 7), because Katumani is more representative of the livestock economy in zone IV. For other purposes Machakos was chosen to portray the climatic situation of zone IV because it has a longer record, which allows computation of drought frequencies and trends.

Agroclimatic zone		*Reference rainfall station*
I/II	Tea/coffee	Kerugoya
III	Maize/cotton	Embu
IV	Marginal cotton	Machakos and Katumani
V	Livestock/millet	Makindu

This study effectively began at a meeting hosted by the United Nations Environment Programme (UNEP) on behalf of the International Institute for Applied Systems Analysis (IIASA). The participants discussed the scope of the Kenya case study and its relevance both to Kenya and to the international community. There is no one institution in Kenya responsible for climate impact assessment and no ready source of funds to draw upon for research assistants, data collection and analysis, purchase of relevant documents, photocopying and manuscript preparation. Gradually, funding was obtained, and qualified authors with sufficient time to contribute to the analysis and writing were enlisted.

Section 2 addresses the climate of the three districts, particularly with reference to the nature of meteorological drought, using available instrumental records of rainfall. Five specific topics are addressed:

(1) Incidence of drought in the study area.
(2) Trends in average climatic parameters.
(3) Trends in climatic variability.
(4) Spatial correlations between the drought records.
(5) Prospects for drought forecasting.

The climatic scenarios, based on the probability of a good or bad season, and selected to show the impact of weather on agriculture and household economies, are described. Historical seasons are mapped to display a range of climatic patterns.

Maize has been chosen to illustrate the impact of climatic variability on agricultural production. Although sorghum and millet are more suitable crops in

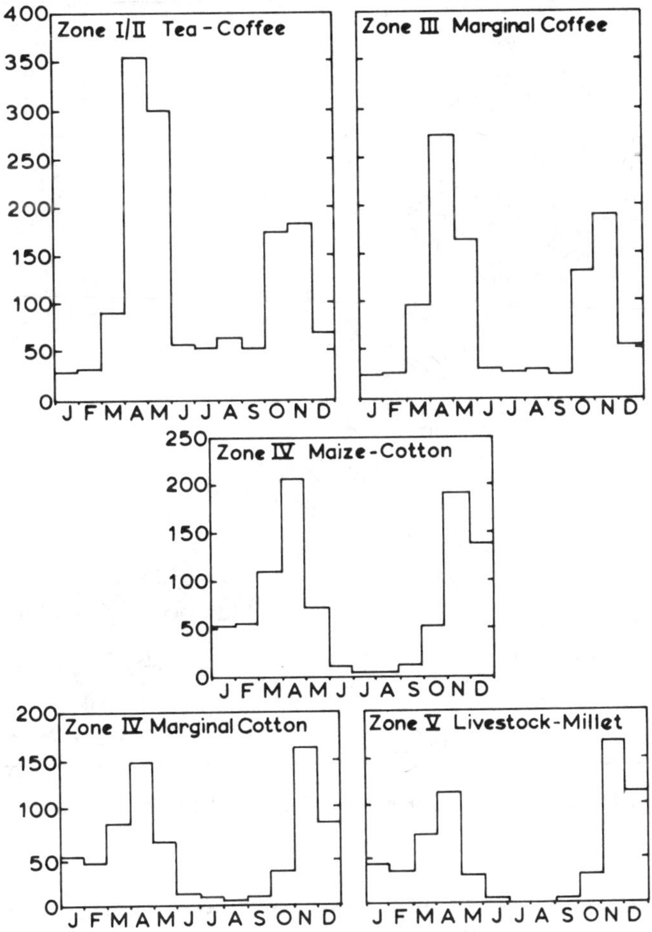

Figure 1.2. Monthly rainfall patterns at selected stations. For station locations, *see Figure 1.3.*

the lower zones (IV and V), they are not as widely grown as maize, nor are they as important in the local diet. Unless specifically stated, "maize" refers to the predominant local variety. In the upper zones this is local maize and hybrids; and in the lower zones, local maize and Katumani Composite B, a drought-resistant variety.

Three different models were used to interpret the climatic record as it would affect agriculture (maize) and livestock. The first model, presented in Section 3, is very simple, can be computed by hand and illustrates a method that could be applied on a real-time basis with minimal resources. The second model (Section 4) follows the FAO water budget procedures, utilizing 10-day averaged meteorological data. It is an example of a more complex model that could be

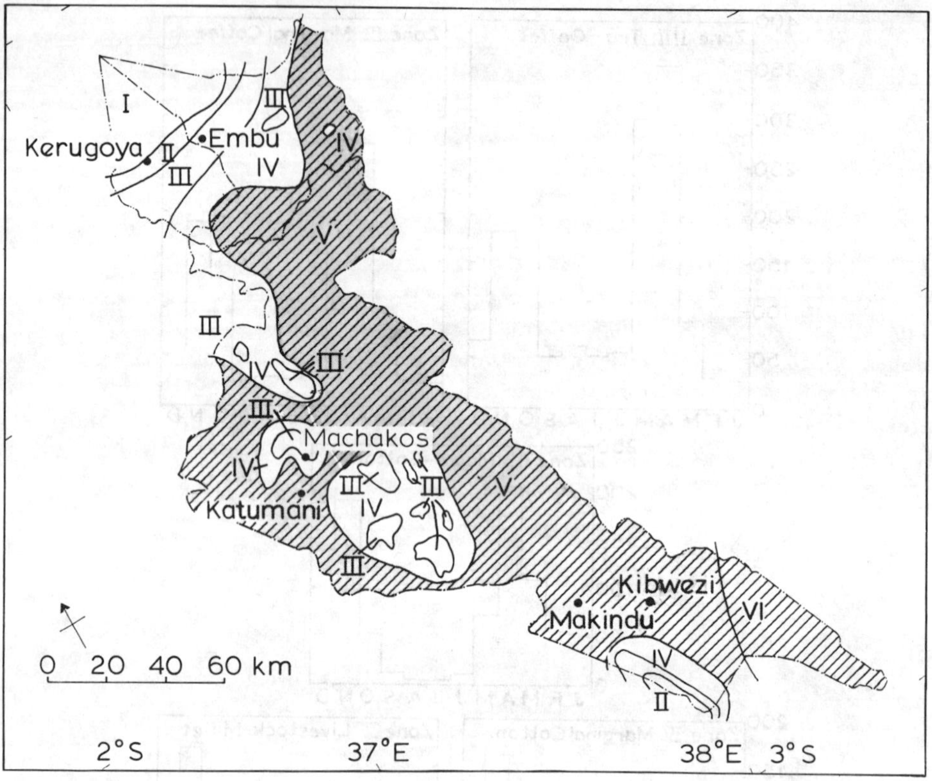

Figure 1.3. Agroclimatic zones in Kirinyaga, Embu and Machakos Districts (Sombroek *et al.*, 1982).

used to verify simpler methods. Both crop models are constrained by the lack of time series of yield data to calibrate the models and validate their results.

The third model (Section 5) charts the relationship between rainfall and livestock production, utilizing several years of experimental and field data. Dairy production is emphasized for the higher altitudes and beef production for the lower zones. Section 5 uses some relevant methods to determine the impact of climate on integrated crop and livestock small-farm systems that should be explored further.

The analysis of climatic impacts is traced one step further, into the realm of smallholder socioeconomics, in Section 6. The analysis discusses vulnerability to climatic variability in terms of long-term trends (or adaptations), and short-term adjustments that may be employed in a drought year. Adaptations and adjustments are further characterized by the nature of their effects on food availability to households (food entitlements) – on-farm production, market exchanges or savings, donations and relief – and the elements of a household objective function. The socioeconomic nature of vulnerability to drought is a real concern to policy makers. The usual sources of food and sources during a drought have implications for long-term development policy and short-term government interventions.

The last section distills the implications of the several analyses for both national and district planners. It includes a summary of the case study, using the same climatic scenarios computed for an entire agricultural year (October–September) for ease of comparison and to portray the livestock situation more realistically.

1.3. Description of the Study Area

The three districts included in this paper form a continuum from the upper slopes of Mt. Kenya (a national park and national forest) through the humid tea and coffee zones, down to the drier cotton zones, and in the southern and eastern areas of Machakos, the semi-arid and arid livestock–millet and ranching zones (*Figure 1.3*). There is a general downward altitude gradient from north to south, with scattered hills in Embu and Machakos (*Figure 1.4*).

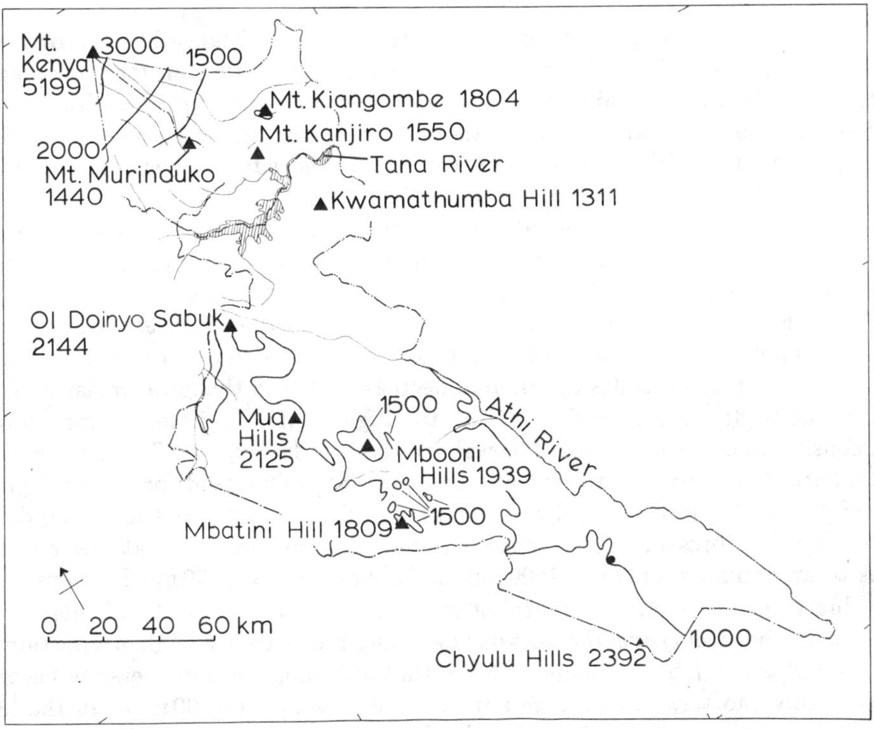

Figure 1.4. Topography and relief of the study area; all altitudes shown in meters.

Altitude is the principal determinant of rainfall and evapotranspiration and, hence, of agricultural zonation. Economic development also generally correlates with altitude, largely due to the historical success of tea and coffee in the upper zones and lack of comparable crops (except in a few irrigated areas) in the lower zones.

The northern study area is part of the Tana River watershed, while the southern area is included in the Athi River basin. Nairobi is the nearest large city (about one million people) and the economic centre of the region, although Embu (about 20 000 people) and Machakos (about 100 000 people) are important service towns.

Kirinyaga District is a traditional Kikuyu homeland, while Machakos is largely settled by the Akamba. Embu District is predominately occupied by the Embu in the upper zones and the Mbeere in the southeast. Both groups are settled agriculturalists with livestock as an important element of their farming systems. The Kikuyu have a longer tradition in agriculture than the Akamba, who have in the past kept large herds of livestock, but in Machakos are now agriculturalists.

1.3.1. Climatic patterns

The climatic factor of greatest economic and social significance in East Africa is rainfall (Griffiths, 1972). Much of East Africa has too little rain for settled agriculture. In the more humid areas, the rain is concentrated into definite seasons that must be utilized to the full by the agriculturalist and the pastoralist. The average annual rainfall is mapped in *Figure 1.5*, and summary statistics for the four reference stations are presented in *Table 1.3*.

Rainfall totals are regionally very different. The total annual average ranges between 500 and 2300 mm. It is largely influenced by altitude and aspect – exposure to moisture-laden winds. On the southeastern slopes of Mt. Kenya, in Kirinyaga and Embu Districts, due to the effects of the southeast winds which bring rainfall, a maximum annual average of over 2200 mm is reached at about 2500 m. At higher altitudes the rainfall decreases, due to the lower moisture content of the trade wind inversion. Given the high rainfall and steep slopes, forest and conservation areas (national parks) are the best land uses. The lower herbaceous part of the tropical alpine zone has often been proposed for seasonal grazing of livestock for farmers living below the forests, due to the demands of increasing land pressure. The average annual rainfall decreases at lower elevations to an annual average of 1600 mm at an elevation of 1600 m. Low evaporation due to cloud cover and morning mists, and storage capacity of more than 200 mm in the deeply weathered volcanic soils, mean that a high proportion of the rainfall is available to plants. Below that altitude, the rainy seasons become considerably shorter, and average annual rainfall is under 1000 mm. In the hot, dry, lower zones near the Tana River, the average annual rainfall is less than 650 mm.

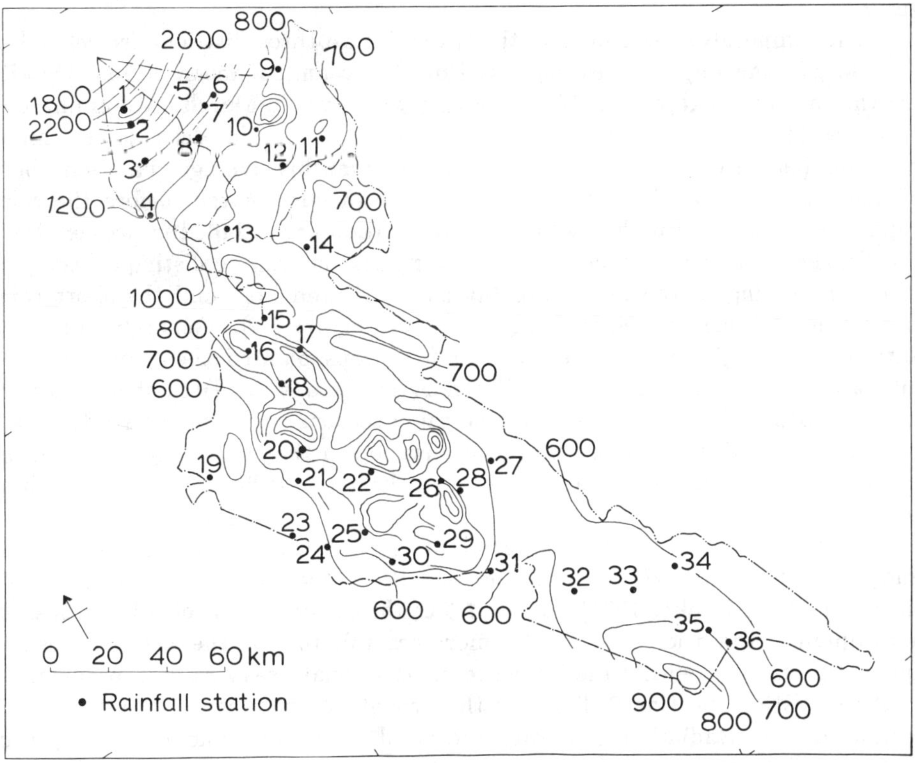

Figure 1.5. Average annual rainfall (in mm). (1) Upper Kamweti Forest Station, (2) Castle Forest Station, (3) Kerugoya Water Development, (4) Sagana Fish Culture Farm, (5) Irangi Forest Station, (6) Runyenjes DO, (7) Kevote Catholic Mission, (8) Embu DO, (9) Embu, Ishiara, (10) Siakago Girl's Secondary School, (11) Kiambere Market, (12) Kiritiri Chief's Camp, (13) Karaba Chief's Camp, (14) Masinga Chief's Camp, (15) Doinyo Sabuk, Ngoliba Estate, (16) Doinyo Sabuk, Kianzabe, (17) Kabaa Catholic Mission, (18) Kangundo DO, (19) Athi River Station, (20) Machakos DO, (21) Katumani Experimental Research Station, (22) Mbooni Chief's Office, (23) Konza Railway Station, (24) Kiu Primary School, (25) Kilome DO, (26) Makueni Unoa Hill, (27) Makueni Athi Camp, (28) Makueni Kampi Ya Mawe, (29) Machakos, Matiliku, (30) Kasikeu Secondary School, (31) Simba Railway Station, (32) Makindu Meteorological Station, (33) Kibwezi, Dwa Plantation, (34) Masongaleni Estate, (35) Darajani Railway Station, (36) Mtito Andei, Tsavo National Park. DO = District Office.

The center of Machakos District mainly consists of hills and small plateaus rising to 1800–2100 m, surrounded by a large plateau, at about 1700 m in the west and sloping down to 700 m in the southeast. This undulating plateau is broken by isolated mountains, such as Oldoinyo Sabuk in the northwest, by the volcanic outflow of the Yatta Plateau in the east, and by a range of volcanic ash, the Chyulu Hills, in the southeast. Precipitation follows the pattern of altitude, with the wettest hills receiving an annual average of 1000 mm. In contrast, the lower areas to the southeast receive only 600–700 mm on average. However, aspect is also important, with the southern and eastern slopes generally being wettest.

Air temperature and evaporation rates in the three districts are related to altitude. In the upper zones, e.g., at Embu, the annual mean is 19.5 °C, with monthly means ranging from 17.2 °C in July to 21.9 °C in March. Minimum temperatures in the upper zones may be below 10 °C. In the drier zones annual mean temperatures range from 20–24 °C, with monthly averages between about 15° and 30 °C. July is usually the coolest month, and March, before the rains begin, the warmest. Further details on evaporation are included in Section 3.

In the area under study, two main rainy seasons can be distinguished – the long rains falling in the months of March, April and May and the short rains that fall in October, November and December (*Figure 1.2*). Generally, the agricultural year begins in October, with a few crops planted then not harvested until after the March–May rains. The harvest seasons are December–February and July–September. The length of the rainy seasons varies in the study area – the use of "short" and "long" is a generalization in that the "long rains" in the southern parts of the study area often are shorter and lighter than the "short rains".

Spatial patterns of climatic variability have been well documented in the study area (Braun, 1977; Dennett *et al.*, 1982; Jaetzold and Schmidt, 1983; Musembi, 1984; Ogallo, 1981). *Table 1.3* confirms the results of other researchers: rainfall in the drier zones is also more variable than in the wetter zones. In some respects this reflects the influence of occasional heavy rainfall occurrences on the statistics. Section 2 discusses the drought climatology of the study area in more detail. Radiation, humidity and wind, essential elements for crop production models, are considered in Section 4.

1.3.2. Farming systems

Table 1.4 summarizes some of the key aspects of farming systems in the study area for each of the agroclimatic zones (*Figure 1.3*). Data on population, farm production, and economic development are discussed below, with references provided in the notes to the table. Several themes are recurrent in the development of these areas: high levels of population growth, which require increased investment in the rural infrastructure and lead to smaller landholdings; poor profitability of many cash crops due to deficiencies in marketing, infrastructures and management; slow rate of increase in agricultural productivity, associated with the continued dominance of maize in unsuitable agroclimatic zones; capitalization of the household economy with the increased importance of off-farm income in meeting essential needs, and the differentiation of households based on access to the cash economy; and an individualization of social structures with less emphasis on communal control of resources. These themes, discussed more fully in Section 6, form the background for a description of the agricultural and economic situation within each zone.

Agroclimatic zone I, where tea and coffee are the major crops, has the highest population density and the smallest average size of landholding.

Population growth is about 4.1% per year, and population densities are over 460 people per square kilometer. Maize productivity on average is highest in this zone (above 2375 kg/ha for farmers using only a few inputs), but limited in the upper areas due to temperature. Maize acreage on small farms is typically 40% of the planted area. Improved seeds are used by almost all farmers. The use of fertilizers and insecticides is more common than in other zones, but not heavily applied to maize. Most households keep a few livestock – one or two grade cows and several goats or sheep. Income from coffee and tea has spurred development in the zone, leading to higher levels of education and off-farm employment. Rural roads are well developed for transport of tea to factories. The availability of credit and inputs through the coffee and tea cooperatives and local supply stores are important aspects of the economic development of the area.

The marginal coffee zone (II) has agricultural and economic patterns similar to zone I, with the notable absence of tea as a cash crop. The zone is densely settled (436 people per sq. km), although farms are slightly larger than in zone I. The population is growing at 5.1% per year, indicating a continuing ability of the zone to provide economic opportunities. Typically, a fifth of the farm is under permanent crops (mainly coffee). Coffee production gives the smallholder access to credit through the cooperative societies. However, coffee is not usually very profitable for smallholders in this zone, in spite of the boom in coffee prices in 1985–1986. Maize yields are lower than in zone I (under 1500 kg/ha for most producers), but sufficient to supply the average household with its staple food. The average number of animals per household is twice that of zone I, but dependent on local circumstances, such as the availability of grazing land. Educational facilities, rural infrastructure and employment opportunities are adequate, but expansion is a continual need due to population growth. The rainfall station at Kerugoya was chosen to represent zones I and II, which have similar climatic patterns.

The maize–cotton zone (III) appears to be the most vulnerable to the social and economic disruptions caused by climatic variability. Although absolute densities are not high (about 230 people per sq. km), land pressure is severe, given the lower productivity of the land; and the population is still growing, at a rate of 2.2% per year. Average household holdings are five to six hectares, of which less than half might be planted. Maize is still the dominant food crop (with average yields slightly lower than in zone II), and along with cotton, tobacco and livestock, the major source of cash from on-farm activities. There are very few permanent crops, although cultivation of citrus fruits is increasing. Improved maize varieties are common. The absence of a reliable cash crop, but fairly good access to off-farm employment, puts this zone on the boundary between capitalist and subsistence economies. The Embu rainfall station is representative of the upper areas of this zone.

The marginal cotton zone (IV) is characterized by a number of crops suitable to the dry areas – millet, sorghum, cow peas and pigeon peas – although maize is still a common crop. Cultivation of maize in this zone is risky (averaging less than 1000 kg/ha for producers with low inputs), but persists as it requires less labor than sorghum and its taste is preferred. The lack of a reliable cash crop in this zone reduces farm production to a subsistence level, although

Table 1.4. Farming systems of Kirinyaga, Embu and Machakos districts.

	Agroclimatic Zone				
	I	*II*	*III*	*IV*	*V*
Zone description	Tea–coffee	Marginal coffee	Maize–cotton	Marginal cotton	Livestock millet
Population and land					
Annual average rainfall (mm)	> 1400	1000–1400	900–1000	800–900	600–800
Population density per sq km[a]	463	436	233	112	45
Population growth rate (%/yr)[a]	4.1	5.1	2.2	5.4	3.8
Average household size[a]	5.8	5.7	7.2	5.4	5.5
Total land area per household (ha)[a]	1.4	1.5	3.2	5.9	13.9
Household farm resources					
Adult labor per household (average)[b]	2.3	2.6	2.8	2.5	3.3
Average size of landholding (ha)[b]	2.2	2.8	5.6	5.2	4.5
% of holding in:					
Permanent crops	26	21	1	1	8
Annual crops	37	52	40	57	28
Pasture, fodder fallow[b]	31	21	52	43	60
Area planted per household (ha)					
Maize	0.7	1.1	1.2	1.7	1.9
Nonfood crops	0.6	0.5	0.2	0.5	0.5
Total planted[b]	1.7	2.4	2.2	4.1	3.8
% of crop planted with improved seed					
Maize – pure stand	95	90	50	70	90
Maize – interplanted[b]	95	100	40	65	35
Disposal of maize					
NCPB	18%	0%	9%	28%	12%
Local market	20	30	21	21	32
Own consumption[c]	61	70	70	51	57
Household consumption of on-farm produce					
Maize (kg/yr)	1088	1022	975	1021	1281
Beans (kg/yr)	245	265	340	230	376
Cal/yr/person[b]	1764	2483	1534	2411	1282
Average no. of livestock per household[d]	6.7	14.6	11.7	14.3	31.7
Average annual yields for production levels I and II (kg/ha)[e]					
Maize	2375–4000	1425–2425	1275–2250	850–1500	–
Beans	900–1750	775–1500	700–1100	300–600	400–700
Potatoes	7250–14 000	–	–	–	–

Table 1.4 continued.

	Agroclimatic Zone				
	I	*II*	*III*	*IV*	*V*
Zone description	Tea–coffee	Marginal coffee	Maize–cotton	Marginal cotton	Livestock millet
Average annual yields for production levels I and II (kg/ha)[e]					
Pigeon peas	–	–	450–900	400–700	–
Brown millet	–	–	700–1500	450–1100	275–850
Sorghum	–	1800–2900	1900–2650	1400–2500	–
Coffee	725–1225	350–550	–	–	–
Tea	5375–6625	–	–	–	–
Pyrethrum	450–800	–	–	–	–
Cotton	–	–	350–900	300–600	–

[a]Central Bureau of Statistics (1981c). Includes households with only 1 or 2 people. Population by zone allocated according to the subdivision.

[b]Jaetzold and Schmidt (1983). Small farm surveys conducted in 1977. There were about 90 respondents in zone I, 60 in zone IV, and 30 each in zones II, III and V. Thus, these data should be treated as illustrative of specific communities in each zone, rather than representative of the zone itself.

[c]*Ibid.* NCPB is the National Cereals and Produce Board, a quasi-governmental body.

[d]*Ibid.* Includes cattle, sheep and goats. Numbers are not reliable due to the small sample.

[e]*Ibid.* Estimates are from Table 15, based on available yield data for each zone. Production level I includes the majority (perhaps 75–80%) of Kenya's smallholders who use few inputs and simple technologies. Production level II should be attainable by 60% of the farmers in the next 15 years.

household reproduction requires cash for school fees, clothes, food supplements, transportation and other needs. Livestock is an important aspect of the farming system, but household herd sizes are not much larger than in zones II and III due to the scarcity of grazing lands. Population densities are somewhat over 100 people per sq. km, but the population is growing at 5.4% per year due in part to immigration from wetter areas. Economic development has lagged behind that of the wetter areas, although there is access to opportunities in the neighboring zones (II and III). Machakos (District Office), represents a typical climatic pattern in this zone, although it is at the upper boundary of the zone.

The livestock–millet zone (V) is very sparsely settled (about 50 people per sq. km) with relatively ample common grazing land available. However, the population growth rate is high (3.8% per year) for an area of inherent low productivity. Food crops are limited to millet, sorghum, green grams, cow peas and some varieties of beans. Some maize, using improved varieties, is cultivated, but production is highly unreliable. Controlling runoff with terraces or bunds is a necessary practice to improve production. The mainstay of the economy is livestock, although the area is permanently settled, and herd migrations are limited by the availability of water, labor and security. The rural infrastructure (schools, access roads, water supply points, health centers, technical advisers, etc.) is considerably less well developed than in the other zones. Many households in this zone have permanently employed family members or depend on casual wage labor to supplement their farm production. Makindu is representative of the climate of zone V.

Zone VI is suitable for ranching, and nomadic pastoralism characterizes the southeast corner of the study area, which is part of the Tsavo East National Park. It is not included in the following analyses.

SECTION 2

Drought Climatology and Development of the Climatic Scenarios

2.1. Introduction

Drought, however it may be measured, is a deficit in available water for the specific purposes for which it is required (Palutikof *et al.*, 1982). It can be defined from three points of view: meteorological, hydrological and agricultural. The first involves an arbitrary criterion of a shortfall in expected moisture (rainfall) resulting from a change in the region's rain-producing mechanisms, some of which may be linked to shifts in the atmospheric circulation. The second definition is concerned with low streamflows and groundwater levels and, hence, inadequate water supply for irrigation, domestic and industrial uses. Agricultural drought considers supplies of water for crops and grasslands (Hounan *et al.*, 1975).

This section explores the drought climatology of Kirinyaga, Embu and Machakos Districts, defining drought on the basis of seasonal rainfall. Succeeding sections extend the analysis by incorporating more climatic variables: Section 3 computes a moisture availability index using rainfall and potential evaporation; the crop production model (Section 4) requires data on potential evaporation, radiation, humidity, temperature, wind, runoff, soil moisture-holding capacity and plant characteristics; and Section 5 utilizes the number of rainy periods, rather than total rainfall, in describing the drier zones of the study area. The within-season distribution of rainfall is included in both Sections 4 and 5.

Subsection 2.2 relates the climate of Kirinyaga, Embu and Machakos to the regional circulation patterns. Subsequent attention is given to a number of crucial questions:

(1) What is the incidence of drought in the study area?
(2) Is average rainfall increasing or decreasing in the three districts (within the time scale of the next decade)?

(3) Is rainfall variability changing, resulting in greater or smaller probabilities of meteorological drought?
(4) Is drought in one area connected to drought in another area? Are drought episodes national, regional or local?
(5) What are the prospects for drought prediction and drought monitoring?

The climatic scenarios to be used later in the case study are defined, and historic examples from the 1970s and 1984 are given in Subsection 2.4. Finally, conclusions regarding the climate of the study area are presented.

Decadal (10-day period), monthly and annual rainfall data were obtained from the Kenya Meteorological Department. There are about 200 rainfall stations in the three districts, of which 37 have more than 30 years of records, and 14 more than 50 years. In a few places records extend back to the late 1890s for rainfall. Stations with long records may give a biased impression of the climate of the area, as they are located in relatively favorable areas where settlements have persisted. The network of full meteorological stations (5 in the study area) has only been developed in the last 20 years. Data on evaporation, wind, temperature, humidity and radiation have thus been collected at relatively few stations and for only the last 10 to 15 years.

2.2. Regional Circulation Patterns and the Climate of Kirinyaga, Embu and Machakos

The dominant controls on the weather and climate of East Africa (Griffiths, 1972) are: (a) regional circulation patterns and the intertropical convergence zone (ITCZ; *see Figure 2.1*); (b) latitude, which affects the timing of rainfall minima and maxima; (c) topography and aspect, which influence the intensity of the ITCZ and the amount of rainfall; and (d) inland lakes, e.g., Lake Victoria, which provide local sources of moisture.

2.2.1. Regional circulation and ITCZ

The ITCZ is the area of convergence between the northeast and southeast trade winds. It is not a sharply defined discontinuity of air masses and is usually identified by the coincident cloud cover and thunderstorms with heavy rainfall. The ITCZ lies about 5° S in the northern winter and about 15° N in the southern winter. The annual mean position is 5° N. The ITCZ is responsible for much of the annual rainfall of East Africa. The progression of the intertropical convergence zone, due to differential heating coupled with unequal distribution of land and water between the two hemispheres, causes a monsoonal type of flow. The northeast monsoon blows from December to February, with a maximum development in January; and the southeast monsoon blows from May to September, with a maximum in July. There is a general association of the ITCZ with rain, and of the strong monsoonal winds with dry conditions.

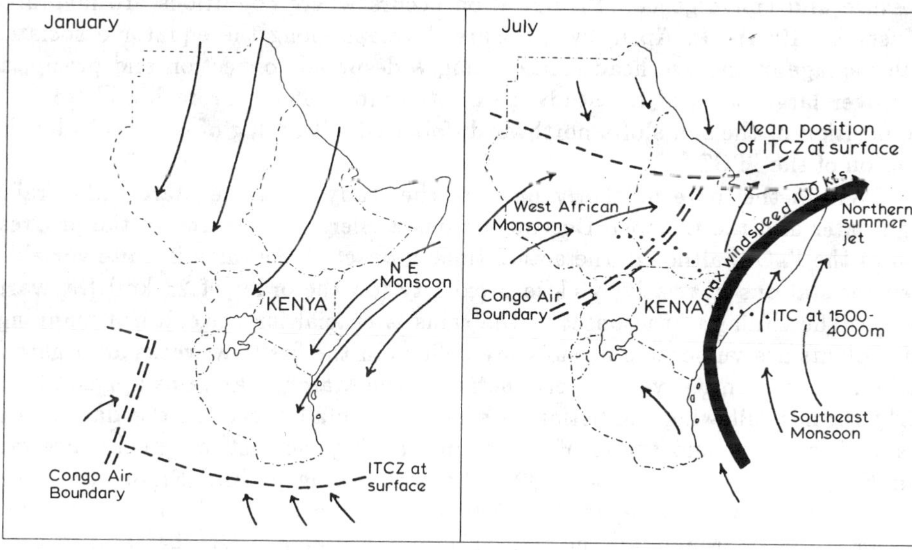

Figure 2.1. Regional circulation patterns in January and July (Sources: Ardunuy, 1979; Downing, 1982; Findlater, 1977; Griffiths, 1972; Nieuwolt, 1977; Trewartha, 1961).

The dryness of the monsoons is due to several factors (Farmer, 1981):

(1) Both the NE and SE monsoons are divergent, reducing the possibility of instability due to convergence.
(2) The NE monsoon is predominantly continental, and is dry by the time it reaches East Africa.
(3) The SE monsoon is shallow, reinforced by inversions related to the subtropical high pressure found over southern Africa in the southern winter. The highlands also reduce the effective depth of the monsoons with some of the high mountains protruding through the monsoonal air into the drier upper easterlies.
(4) The wind flow is parallel to the coast and highland ridges in all but the transition seasons. Thus, it is not effective in transporting moisture inland.
(5) Upper air layers have a marked decline in moisture content.
(6) Intensified divergence is caused by the north–south alignment of low pressure at the 300 mb level over Lake Victoria, which inhibits advection of moisture into Kenya from the Indian Ocean or Central Africa.

2.2.2. Latitude and seasons

The movement of the ITCZ over the area disrupts the monsoonal wind flow and provides the basis for increased rainfall activity. This results in March–April and October–November being the wettest months. In the October–December rainy season the ITCZ is located south of the equator and is a zone of low pressure and convergence. Precipitation occurs where conditions are favorable (Musembi, 1984). In April low pressure develops along the equator associated with passage of the overhead sun, causing widespread convection and precipitation over large parts of the study area (Musembi, 1984). From March to May the position of the sun shifts northward, followed with a lag of 4–5 weeks by the position of the ITCZ.

Within the more northerly parts of the study area, the March–May rains begin later and the October–December rains earlier. This is due to the progression of the ITCZ, although the actual time of onset of the rains is quite variable, even for stations in relatively close proximity (on the order of 25 km) (Stewart, 1983). The timing of the onset of the rains is crucial for agricultural planning, since plants are vulnerable to moisture deficits in the first few weeks after germination. In eastern Kenya, the termination of the March–May rains appears to be fairly regular, allowing prediction of seasonal rainfall based on the date of the onset of the rains and the rainfall amounts during the first few weeks (Stewart and Faught, 1984; Kashasha, 1982). At both Kerugoya and Embu, the rainy seasons begin, on average, in mid-October and the last week of March; while further south, at Makindu, the rains begin at the end of October and in early March (Jaetzold and Schmidt, 1983; Musembi, 1984; Alusa, 1978).

The March–May season is generally called the "long rains", and the October–December season the "short rains", although season length and reliability vary across the study area. The length of the October–December rainy season becomes longer as one goes toward the southern parts of Kenya. This season is sometimes referred to as the "long rains" in the lower parts of the study area (southern Machakos District), where they are more reliable (i.e., last longer with higher totals) than the March–May rains. In this case study, "long rains" refers to the March–May season and "short rains" to the October–December season.

2.2.3. Topography and aspect

Topography is particularly important in determining average rainfall in the study area (*see Figures 1.4* and *1.5*) (Brown and Cochemé, 1973; Porter, 1976). Average annual rainfall in the Nairobi area has been correlated with altitude ($r = 0.65$) (Nieuwolt, 1980). Aspect, because it affects exposure to the moisture-bearing winds, is also important (Brown and Cochemé, 1973).

2.2.4. Sources of moisture

The primary source of moisture for rainfall in the three districts is the Indian Ocean. Lake Victoria and the Congo Basin are important moisture sources in western Kenya, and occasionally pulses in the regional circulation will bring this air mass into central and eastern Kenya. Much of the study area is dry and local moisture sources, such as forests, grasslands, cropped areas and reservoirs, are probably minor compared to moisture advected from the ocean.

2.3. Drought Incidence, Trends and Prediction

2.3.1. Drought definition and incidence

Drought frequencies and magnitudes are basic data required by agricultural planners and officials responsible for monitoring food supply and demand. The frequency of drought in the three-district area can be interpreted from the maps of rainfall reliability, which estimate the amount of rainfall likely to be exceeded in 6 out of 10 years (*Figures 2.2* and *2.3*). The use of 60% reliability is not intended to indicate an estimate of farmers' adaptation to the local rainfall regime. The maps should not be confused with measures of rainfall variability (such as the coefficient of variability). They indicate the probability of sufficient seasonal moisture for successful crop production. In 6 out of 10 growing seasons (i.e., rainy seasons) large portions of the area receive less than 300 mm – a minimum amount considered essential for dryland maize agriculture (Muhammed, 1986). Around Mt. Kenya the March–May rains are more reliable than the October–December rains. The reverse is true in southern Machakos and in the Machakos Hills.

Calculation of drought frequencies based on a simple drought index shows the relative incidence of rainfall anomalies (Ogallo and Nassib, 1984). The index (*DI*) was computed, as the negative standardized scores for seasonal rainfall, for the four reference stations (Kerugoya, Embu, Machakos and Makindu):

$$DI = \frac{P - \bar{X}}{S} \tag{2.1}$$

where P = seasonal precipitation, $\bar{X}$ = long-term average for that season, and S = seasonal standard deviation from P.

The DI is normalized with mean of 0 and standard deviation of 1; if DI is positive it is set to 0. The long rains seasonal rainfall at Kerugoya, Embu and Machakos is not highly skewed (*Table 2.1*), and use of the mean and standard deviations is appropriate. For the short rains and at Makindu, the seasonal rainfall is skewed, and means are 25–70 mm more than the seasonal medians. Because of this skewness, the drought situation for the short rains and dry zones will tend to be slightly overemphasized. Using departures from the mean or median and some measure of variability, such as the standard deviation, assumes that productive and social systems are adapted to the local average conditions, and that the impact of meteorological drought on agricultural and social systems corresponds to the deviation from "normal" conditions. This is acceptable only as a first approximation of the incidence of meteorological drought. Typically, productive systems are risk-averse with adaptations oriented toward poor years (Porter, 1979), although this situation may be changing in central and eastern Kenya (Section 6). More refined analyses should be carried out to document drought frequencies in relation to specific needs for water, as in the following subsections.

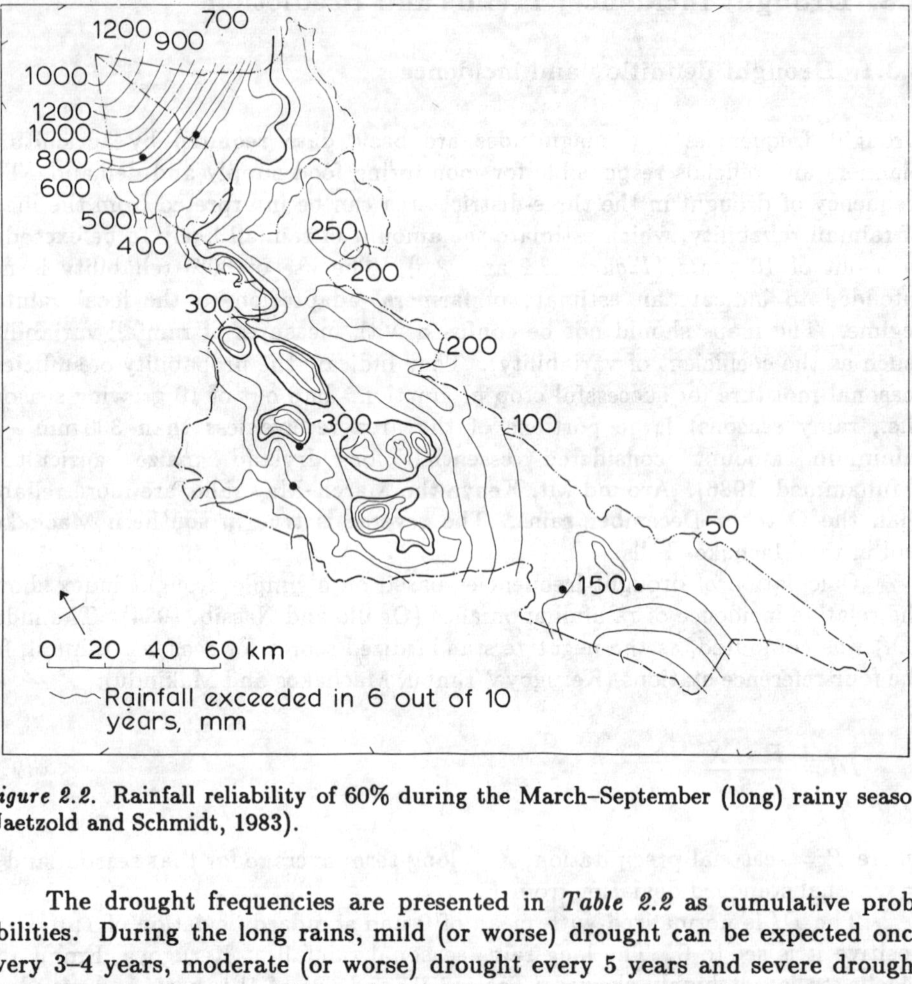

Figure 2.2. Rainfall reliability of 60% during the March–September (long) rainy season (Jaetzold and Schmidt, 1983).

The drought frequencies are presented in *Table 2.2* as cumulative probabilities. During the long rains, mild (or worse) drought can be expected once every 3–4 years, moderate (or worse) drought every 5 years and severe drought once every 12–20 years. The probabilities are much the same for the four reference stations. For the short rains, drought is more common: mild (or worse) drought averages once every 1.8–2.4 years, moderate (or worse) drought once every 2.5–4 years, and severe drought once every 5.5–9 years. Makindu appears to be more prone to mild or moderate drought than the other stations. It should be noted that these averages are statements of probability of drought (in a particular season) occurring in any one year.

2.3.2. Secular climatic trends

One issue of relevance to the capability of forecasting drought is the possibility, due to carbon dioxide or other causes, that the climatic parameters (e.g., means and variability) are changing over time. If the climatic parameters are unstable, then the historical statistical record may not be useful in weather prediction.

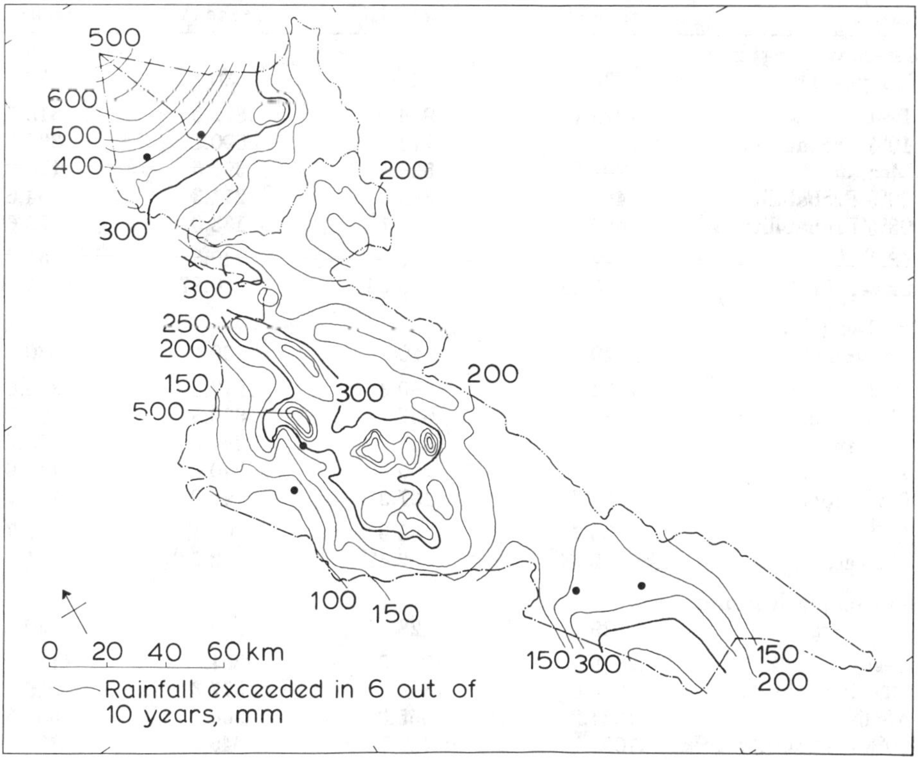

Figure 2.3. Rainfall reliability of 60% during the October–February (short) rainy season (Jaetzold and Schmidt, 1983).

Linear least-squares regression lines were fitted to the entire period of record at the four reference stations for seasonal rainfall, omitting missing values, and for 11-year moving averages with missing values interpolated from neighboring stations adjusted for altitude (*Table 2.3* and *Figures 2.4* and *2.5*). The use of 11-year periods is to smooth the rainfall series. It is an arbitrary period; it is not connected to sunspot cycles or any physical rationale. For the long rains, three of the stations have negative trends, but only the 11-year moving average trends for Machakos and Makindu are statistically significant. For the short rains, only Makindu has a negative trend. The three positive trends in 11-year moving averages are statistically significant. There does not appear to be a widespread general trend in the smoothed series. In every case, the range of annual rainfall is much greater than the difference between one average mean and the next. Thus, there is no evidence on the time scale of the next decade or so of significant changes in the climatic parameters (*see also* Farmer, 1981). However, longer-term climatic changes have occurred in East Africa, and the climate of the next century may not be like the climate of this century (Downing, 1982; Farmer and Wigley, 1985; Nicholson, 1981).

Table 2.1. Seasonal and agricultural year rainfall statistics and probabilities[a] (in mm).

Season	*Kerugoya* (*zone* I/II)	*Embu* (*zone* III)	*Machakos* (*zone* IV)	*Makindu* (*zone* V)
Mar–May (long) rains				
No. years	29	28	30	30
Best	1228.1	936.9	872.0	510.7
10% Probability	1029.4	774.9	590.1	327.5
Median	701.6	526.5	396.5	175.2
90% Probability	480.8	349.9	211.3	94.6
98% Probability	450.1	230.7	133.0	23.6
CV[b]	27.7	31.9	40.3	51.9
Skewness	0.52	0.38	0.87	1.00[c]
Oct–Dec (short) rains				
No. years	29	29	30	30
Best	1584.8	1280.3	1577.0	816.6
10% Probability	730.3	657.1	670.5	603.0
Median	461.2	353.8	344.0	260.5
90% Probability	226.8	200.5	240.0	148.9
98% Probability	202.7	137.9	198.0	102.7
CV[b]	56.7	53.9	67.4	57.7
Skewness	2.34[c]	2.42[c]	2.77[c]	1.24[c]
Oct–Sep (agricultural year)				
No. years	29	28	30	30
Best	2818.4	1994.5	2261.0	1113.0
10% Probability	2043.0	1516.0	1377.3	1016.3
Median	1524.9	1055.3	904.7	552.5
90% Probability	1096.7	717.7	645.0	330.5
98% Probability	937.3	676.3	581.0	208.9
CV[b]	24.9	29.6	36.2	38.4
Skewness	1.33[c]	0.65	1.85[c]	1.07[c]

[a]Probabilities are for cumulative distribution for exceeding stated value.
[b]CV is coefficient of variation.
[c]Significant at 1% probability level.

Related to trends in rainfall, is the question of whether or not cycles occur in the rainfall series. There has been considerable speculation and research into the existence and causes of cycles in rainfall in eastern and southern Africa (Morth, 1973; Ogallo, 1979; Rodhe and Virji, 1976; Tyson *et al.*, 1975). In Kenya several serious droughts have occurred at the end of the decade, often followed by very good years. Autocorrelations were tested for the four reference stations in the study area, for the seasonal rainfall drought index (*Table 2.4*). For the long rains, the strongest cycles common to all of the stations are at a lag of 2 (negative), 9 (negative) and 10–11 (positive) years. There is less evidence of cycles in the short rains, with the strongest autocorrelations at lags of 7 (negative) and 9 (negative) years. These data may be useful in understanding the mechanisms behind drought events. However, the correlations explain only a small percentage – less than 10% – of the variability in the drought index, which

Table 2.2. Seasonal drought probabilities using a normalized drought index.[a]

Season	*Kerugoya*	*Embu*	*Machakos*	*Makindu*
Mar–May (long) rains				
No. years	49	76	91	81
Mild drought	0.31	0.36	0.25	0.43
Moderate drought	0.22	0.22	0.20	0.23
Severe drought	0.06	0.08	0.05	0.06
Oct–Dec (short) rains				
No. years	50	76	90	80
Mild drought	0.46	0.41	0.49	0.56
Moderate drought	0.32	0.29	0.26	0.40
Severe drought	0.14	0.18	0.11	0.11

[a]Drought is defined on a cumulative basis, such that "mild drought" incorporates all years where the drought index (DI) ≤ 0.2. Thus, the probabilities do not total to unity. Definitions of drought index: no drought = DI > -0.2, mild drought = DI ≤ -0.2, moderate drought = DI ≤ -0.5, severe drought = DI ≤ -0.8.

Table 2.3. Linear least-squares regression of seasonal rainfall, and 11-year moving averages.

	Correlation (r) with year			
Season	*Kerugoya*	*Embu*	*Machakos*	*Makindu*
Mar–May (long) rains				
Rainfall	−0.11	0.03	−0.16	−0.09
11-year moving average	−0.20	0.24	−0.48[a]	−0.38[a]
Oct–Dec (short) rains				
Rainfall	0.20	0.10	0.08	−0.06
11-year moving average	0.53[a]	0.40[a]	0.33[a]	−0.22

[a]T-test significant at 5% probability.

is too little for them to have any practical value for predicting drought occurrences in any given year.

Trends in rainfall variability were analyzed using 11-year running periods for the length of record for the four reference stations. A measure of interannual variability (*IAV*) was calculated as:

$$IAV = \left[\left(\frac{\sum_{i=2+m}^{n+m} (|X_i - X_{i-1}|)\, q}{n-1} \right) \left(\frac{\sum_{i=2+m}^{n+m} X_i}{n} \right)^{-1} \left(\sum_{i}^{m} q \right)^{-1} \right] \times 100 \qquad (2.2)$$

where $\sum$ is the sum, i is the year, m is a counter for each year of the running period, n is the averaging period (11 years), X is the seasonal rainfall, and q is a weight applied to each value: if m = 1 to 6 $q = m$; if m = 7 to 11 $q = 12-m$. The advantage of the *IAV* is that it is dimensionless, i.e., it is not specific to the

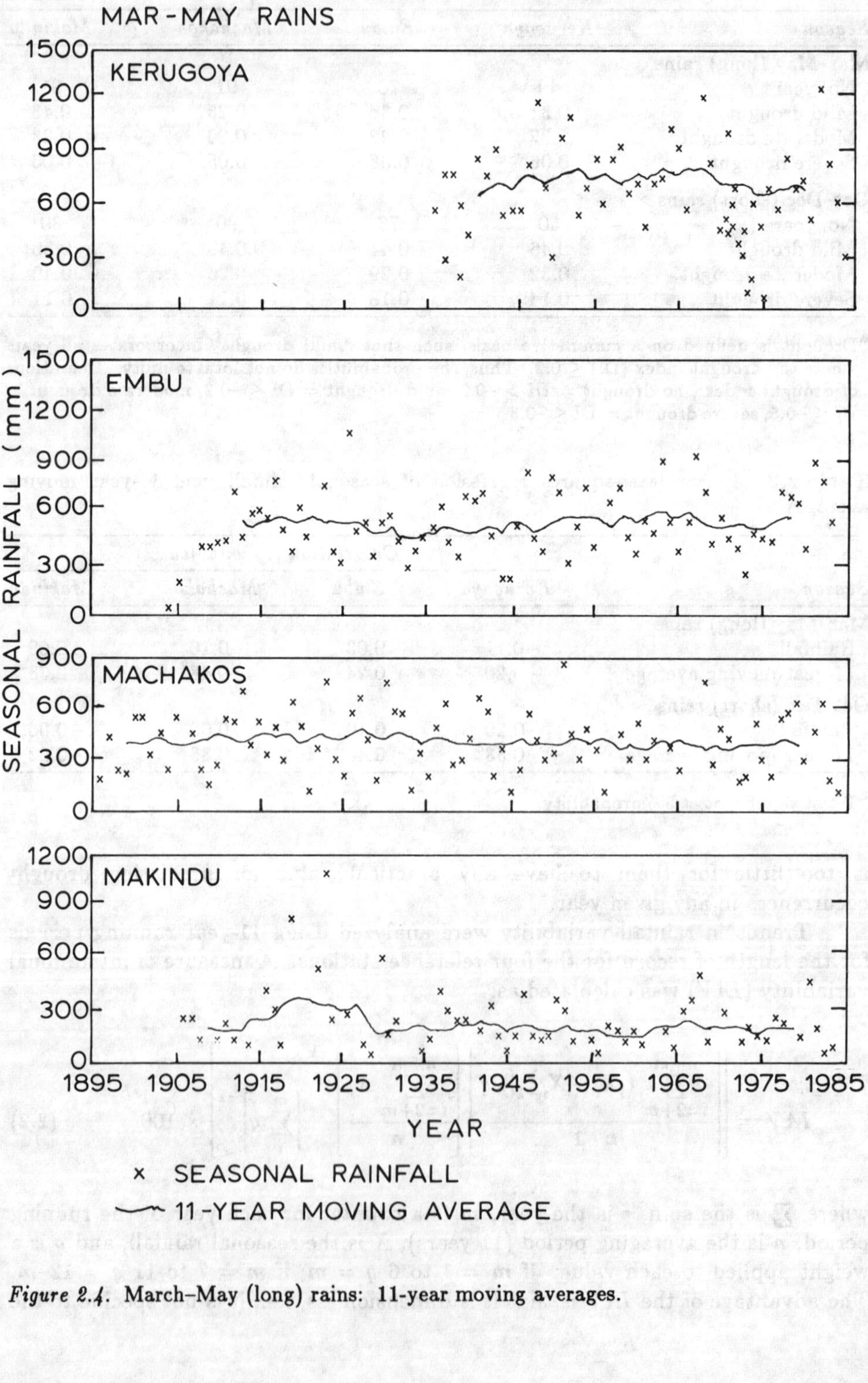

Figure 2.4. March-May (long) rains: 11-year moving averages.

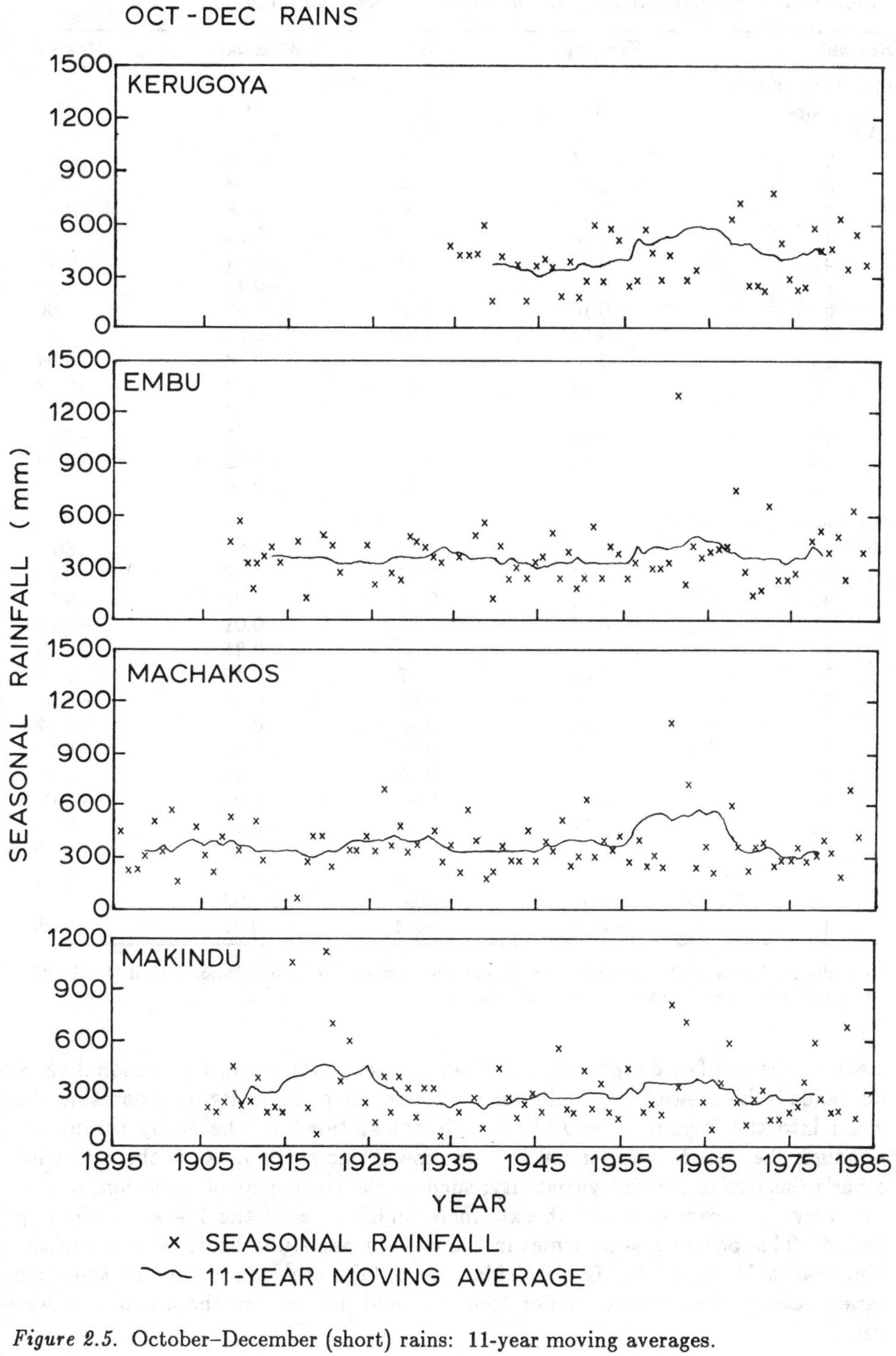

Figure 2.5. October–December (short) rains: 11-year moving averages.

Table 2.4. Autocorrelations[a] of drought index for seasonal rainfall.

Season	*Kerugoya*	*Embu*	*Machakos*	*Makindu*
Mar–May rains				
No. years	50	75	91	81
$\lambda = 1$	−0.12	0.14	−0.01	0.14
2	−0.24	−0.24	−0.25	−0.21
3	−0.12	−0.15	−0.04	−0.11
4	0.21	0.06	0.08	0.04
5	0.03	0.08	−0.15	−0.08
6	0.02	0.00	−0.11	−0.16
7	−0.18	−0.08	−0.11	−0.12
8	−0.10	−0.02	0.09	0.06
9	−0.04	−0.16	−0.15	−0.11
10	0.11	0.15	0.14	−0.04
11	0.24	0.12	0.18	0.33
12	−0.06	−0.15	0.16	0.00
13	0.14	−0.05	0.03	−0.01
14	0.03	0.04	−0.05	−0.03
15	0.05	0.09	0.19	0.09
Oct–Dec rains				
No. years	49	75	90	80
$\lambda = 1$	0.05	−0.03	−0.17	−0.09
2	−0.12	0.02	−0.03	−0.12
3	−0.04	0.03	−0.01	−0.05
4	−0.07	−0.00	0.21	0.01
5	−0.08	−0.17	−0.19	−0.00
6	0.10	−0.05	0.07	−0.07
7	0.01	−0.15	−0.11	−0.12
8	−0.08	0.05	0.04	0.04
9	−0.21	−0.25	−0.12	0.02
10	−0.06	0.01	−0.04	−0.09
11	−0.20	0.19	−0.01	0.10
12	−0.01	−0.09	−0.16	−0.10
13	0.10	−0.04	−0.14	−0.22
14	−0.12	−0.00	−0.03	−0.16
15	−0.06	−0.02	0.04	0.16

[a] λ is the lag between the current year (t) and the year with which it is correlated ($t+\lambda$). The drought index is as computed in Subsection 2.3.

mean of the rainfall distribution. As can be seen from *Table 2.1*, seasonal rainfall tends to be skewed, especially for the short rain season. By incorporating the mean into the formula, the *IAV* at different stations can be easily compared, although the critical comparison is over time at one station. As is the case with other measures of rainfall variability, such as the coefficient of variation, one or two very wet years can lead to extremely high values of the 11-year averaging period. This occurs several times in the data in our study area. The use of the binomial weights in the *IAV* calculations reduces this distortion in the time series. Conclusions about secular trends should not rely on the anomalous wet years.

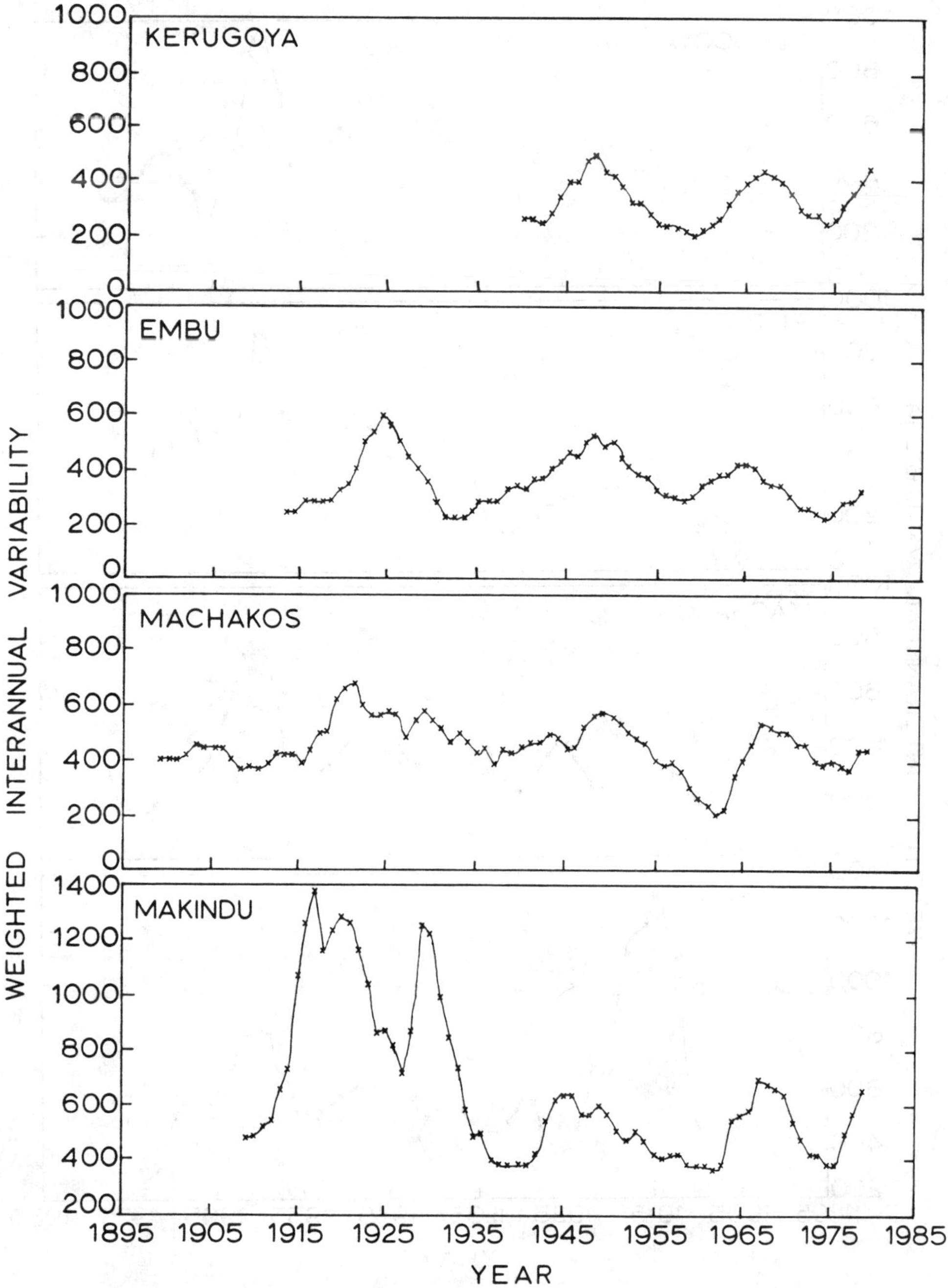

Figure 2.6. Interannual variability for 11-year running periods, March–May (long) rains.

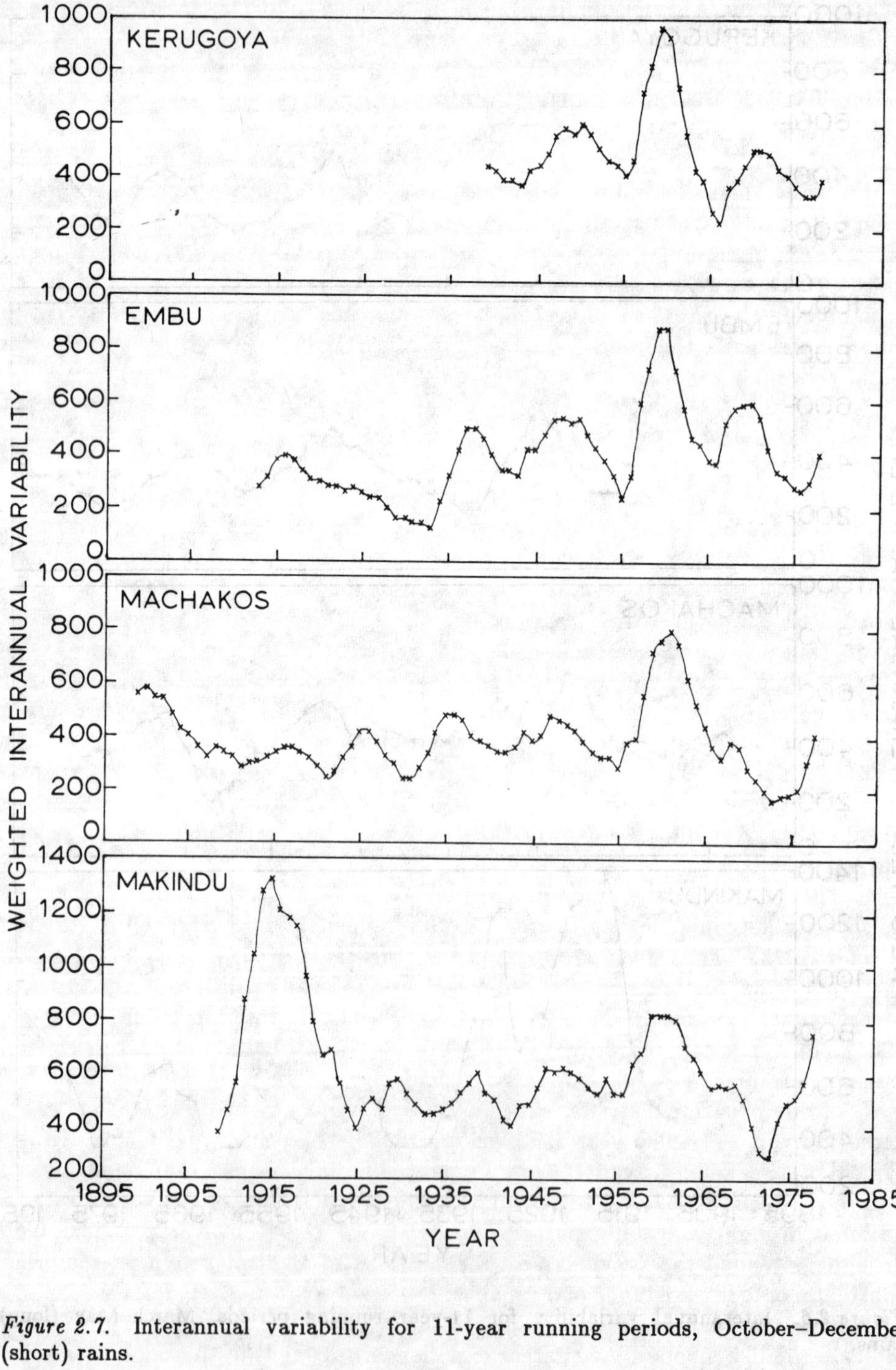

Figure 2.7. Interannual variability for 11-year running periods, October–December (short) rains.

The results for the March–May (long) and October–December (short) rains are presented in *Figures 2.6* and *2.7*. The *IAV* statistic for the March–May rains is lower and more stable for Kerugoya and Embu than for the other stations, with Makindu showing the highest values and largest fluctuations. The late 1930s to early 1940s and 1950s to early 1960s were periods of low interannual variability. The 1920s, late 1940s and early 1970s were periods of high variability. The values during the late 1970s and the 1980s are about average for the time series. In Kerugoya, Embu and Makindu there was recent increased variability, while Machakos shows somewhat lower variabilities at present than in the recent past. For the October–December rains, all of the stations show recent increases in interannual variability. The lowest variability was around 1975 and in the 1920s to 1930s, while the late 1950s to mid 1960s were periods of high variability. The short rains are more variable than the long rains at Kerugoya and Embu, while the reverse is true for Machakos. The two seasons appear equally variable at Makindu.

2.3.3. Spatial patterns of drought

Another topic of continuing interest to policy makers (including farmers) concerns the probability of drought occurring in several areas at the same time. This topic is relevant at several scales. Within our study area the probability of drought occurring in several zones at once is of interest to the district official, as it may affect streamflows, migration and requirement for food assistance. The Kenyan planner in the National Cereals and Produce Board (NCPB) may want to know the correlation of drought among the traditional surplus-producing areas. If drought is likely to affect several productive areas at once, he may decide to increase the nation's strategic food reserve. The World Food Program representative may want to know, if there is a drought in Kenya, what is the likelihood of drought elsewhere in Eastern Africa, thus influencing where he can buy surplus food.

The geographical extent of a drought episode is generally correlated with its severity, and inversely correlated with its frequency of occurrence (Mbithi and Wisner, 1972). Thus, *local* droughts in the study area occur every year. Even in the bumper harvest year of 1981–82 there were several localities, on the order of several square kilometers in extent, that experienced total crop failure (Murage, 1985). Such a local crop failure is likely due to a combination of drought, pests, disease, late planting and poor management. A *regional* drought occurs on the scale of one or a few districts, with an average frequency in Kenya of one every three to five years. The year 1983 was a drought year in Machakos and Kitui – typical of a regional drought. A *national* drought affects more than one province and requires concerted and extraordinary efforts by the government. In Kenya a national drought may be expected once every 10 years on average. The 1984 drought extended the previous year's conditions to Central Province, requiring considerable food assistance and imports.

These estimates of the frequency of the geographic extent of drought do not indicate the likelihood of given areas experiencing drought in the same year. For

Table 2.5. Correlations of seasonal drought index between reference stations, 1934–1983.[a]

Season	*Kerugoya*	*Embu*	*Machakos*	*Makindu*
Oct–Dec				
Kerugoya	–			
Embu	0.83[b]	–		
Machakos	0.25	0.40[b]	–	
Makindu	0.37[b]	0.43[b]	0.55[b]	–
Mar–May				
Kerugoya	–			
Embu	0.44[b]	–		
Machakos	0.20	0.41[b]	–	
Makindu	0.35[b]	0.21	0.27	–

[a]Drought index is as defined in Subsection 2.3.1. N = 49 years: 1976–77 is missing.
[b]T-test significant at 5% probability.

the eastern African region, serious drought in Kenya appears to be most highly correlated with drought in Uganda and Tanzania, and very poorly correlated with drought in Sudan, Ethiopia or Somalia.

Within Kenya the regional coincidence of drought has been examined cartographically (Ministry of Water Development, 1979) and for rainfall in general by principal component analysis (Farmer, 1981; Mungai, 1984). Detailed studies of compensating zones, i.e., areas with inverse correlations of drought, have not been carried out.

Using the drought index described above, cross-correlations were computed among the four reference stations (*Table 2.5*). For a region of almost 16 000 square kilometers, the drought indices show reasonable correlations. Neighboring stations show the highest correlations, although surprisingly in both seasons there is a significant correlation between Kerugoya and Makindu. All of the correlations are positive, indicating a likelihood of drought coincidental through the three districts. Drought incidence and magnitude are regional in nature, affecting large areas at the same time. Thus, while monthly tropical rainfall between stations may not be well correlated (Jackson, 1981), drought (measured by seasonal drought indices) appears to have a more consistent pattern. Based on the R^2, some 15–25% of the variability in the drought index series throughout the study area can be related to these regional correlations. This is not a strong enough correlation to warrant predicting drought in the entire area for a particular season based on data from only a few stations.

2.3.4. Drought forecasting

What are the prospects for forecasting rainfall, or drought, over the short, medium or long term? The ability to forecast rainfall for various levels of certainty and time periods would enable the scheduling of agricultural activities that could take advantage of good rain or mitigate the impacts of a drought.

The synoptic climatology of the tropics is still poorly understood. The synoptic stations are few, the rain-producing mechanisms are numerous, the rain itself is variable. The 1984 short rains in Kenya illustrate the difficulties. These started in mid-September with the not-uncommon "grass" rains that precede the October–December rains. After the severe drought, these rains were very welcome and many farmers started planting. However the synoptic conditions did not resemble those normally associated with the October–December rains (a strong ITCZ), and the Meteorological Department advised people that the grass rains might not last until the short rains. Nature was merciful in this case, and the grass rains were followed by the short rains after a dry spell of only about a week. They continued to be heavy and long. Given the difficulties of short-term forecasting, the prospects for reliable seasonal forecasts are dim, certainly not yet within the range of accuracy required for scheduling the importation and distribution of inputs, or for anticipating a surplus to be purchased by the NCPB.

The most promising area for forecasting rainfall appears to be to make better use of the available data and the existing rainfall monitoring network. There are about 2000 active rainfall stations in the country, of which about 800 send timely reports every month and about 50 report every week. The remainder send in their recording cards after a month or more delay. Currently, the Meteorological Department releases a weekly weather summary and monthly maps of rainfall, its seasonal total and its seasonal departure from normal. Coupled with the expected rainfall for the rest of the season, this can be very valuable information. Researchers have shown the potential for estimating total seasonal rainfall, given the date of onset of the rains and the rainfall amounts over the first few weeks. For several stations in Machakos and Kitui Districts, patterns emerge quite early in the season, justifying such responses as changing the levels of inputs, or planting different crops (Kashasha, 1982; Stewart and Hash, 1982; Stewart and Faught, 1984). These methods have not yet been developed to the point of attempting an on-farm test with extension agents, but the methods deserve further attention.

Around the world laymen and scientists over the past several years have attributed all sorts of anomalous weather events to the effect of El Niño/Southern Oscillation (ENSO) episodes (Glantz, 1984). The 1982–1983 ENSO episode was associated with floods in the eastern Pacific, droughts in Australia, Indonesia and Africa, and changes in the Indian monsoons (*see* Part I, Section 3 of this volume). An understanding of the causes and effects of large-scale sea-surface temperature and pressure changes could be used to forecast seasons and regions of extraordinary weather. Although every ENSO episode is somewhat different, preliminary investigations have established some possible linkages between the eastern African climate and such global anomalies as the ENSO (Degefu, 1985; Farmer, 1987; Njau, 1987; Ogallo, 1987).

Table 2.6. Definition of climatic scenarios at reference stations (mm precipitation).[a]

Season/Scenario	*Kerugoya*	*Embu*	*Machakos*	*Makindu*
Mar–May (long) rains				
Good year	1029	775	590	328
Average year	702	527	397	175
Moderate drought	481	350	211	95
Severe drought	450	231	133	24
Oct–Dec (short) rains				
Good year	730	657	671	603
Average year	461	354	344	261
Moderate drought	227	201	240	149
Severe drought	203	138	198	103
Oct–Sep (agricultural year)				
Good year	2043	1516	1377	1016
Average year	1525	1055	905	552
Moderate drought	1097	718	645	331
Severe drought	937	676	581	209

Scenario	*Definition*
Good year	Exceeded in 10% of the years
Average year	Median
Moderate drought	Exceeded in 90% of the years
Severe drought	Exceeded in 98% of the years
Back-to-back drought	Successive years (seasons) with rainfall exceeded in 90% of the above

[a]Rainfall is for the 1951–1980 period, as in *Table 2.1.* The cumulative probabilities are taken from the actual data, interpolated to give the 10%, 50%, 90% and 98% scenarios.

2.4. Development of the Climatic Scenarios

For further analysis of the importance of climatic variability on agricultural production, data for the four reference stations, selected in the introductory section to represent different agroclimatic zones, were examined in greater detail. Scenarios of seasonal rainfall were chosen for each station based on cumulative probabilities for the 1951–1980 period (*Table 2.6*). Comparison with data for the years of record was also made. These scenarios give a probabilistic estimate of climatic events, which can then be related to changes in agricultural production. The cumulative probability reflects the rank order of each event, and is independent of the mean. The median is the 50% cumulative probability, i.e., half of the years are above the median. The 10% and 90% events can be expected once in 10 years, and the 2% and 98% events, expected once in 50 years, are extreme cases. The scenarios are only defined at these four stations. Scenarios with spatial detail for the entire study area would require computing the cumulative probabilities and impact models at all of the 36 stations. This was beyond the scope of the present study.

As an alternative, historical episodes that are still remembered by rural residents and planners, and are thus concrete illustrations of drought and its impacts, are presented, expressed as the percentage of seasonal rainfall compared

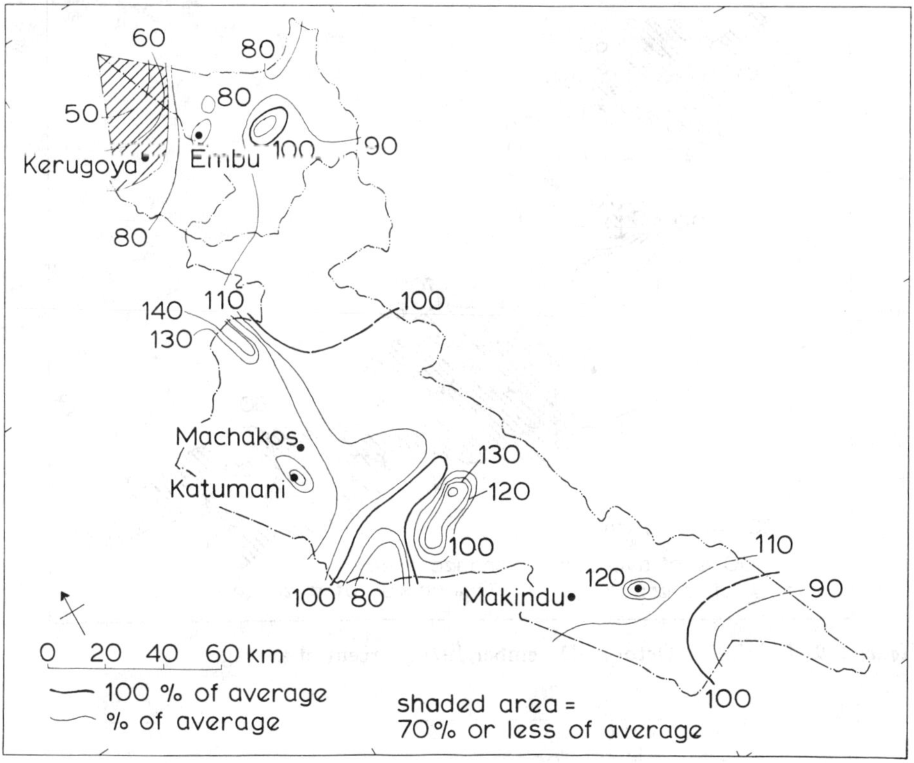

Figure 2.8. Rainfall in March–May 1974, percent of average.

to normal, in *Figures 2.8* to *2.16*. The areas receiving less than 70% of normal rainfall are shaded to highlight the spatial patterns of drought. The use of 70% is for convenience, although it is considered a drought threshold in India (Jodha, 1985). Eight seasons in five years were chosen to illustrate the variability of climate in the three districts:

(1) 1974 – Beginning of the mid-1970s' drought episode with the short rains; the worst drought in decades for many stations, particularly in lowlands and eastern side of Mt. Kenya.
(2) 1975 – Widespread drought with large areas below 60% of average; somewhat wetter near the hills of central Machakos.
(3) 1976 – Widespread drought in the long rains; higher rainfall on the upper slopes of Mt. Kenya. Short rains season begins recovery from drought, particularly in southern and central (at higher elevations) Machakos.
(4) 1977 – One of the best years on record, with no areas recorded below 90% of average rainfall for the year.
(5) 1984 – One of the worst droughts in the long rains in this century, widespread in central and eastern Kenya.

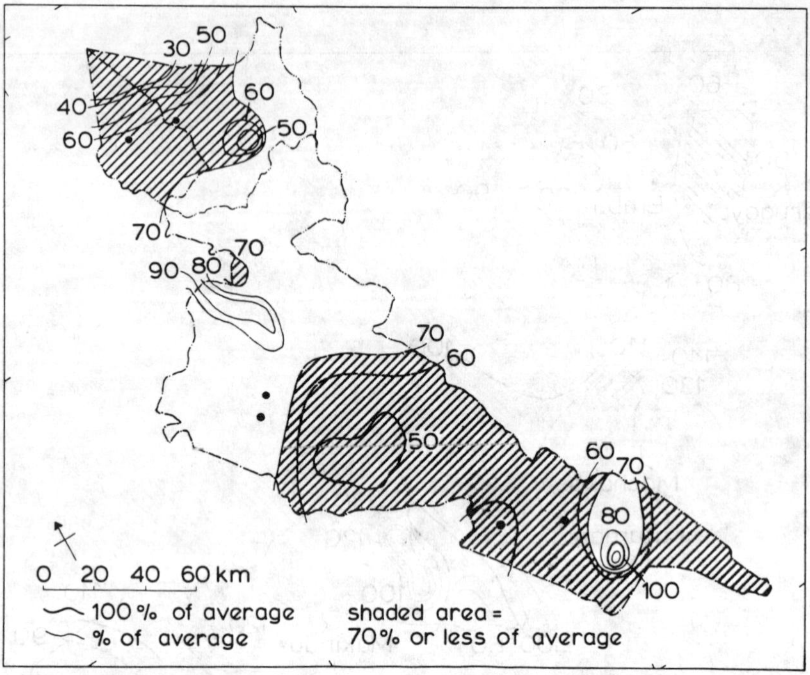

Figure 2.9. **Rainfall in October–December 1974, percent of average.**

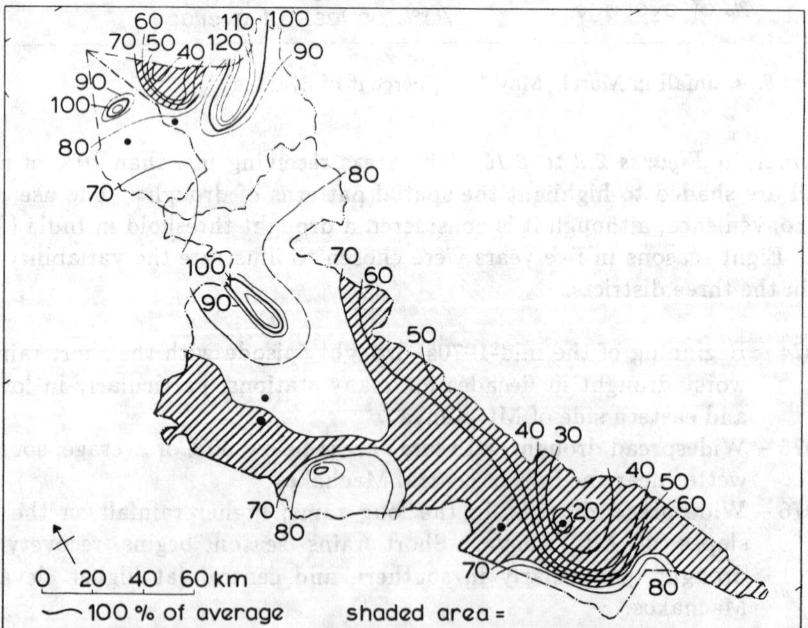

Figure 2.10. **Rainfall in March–May 1975, percent of average.**

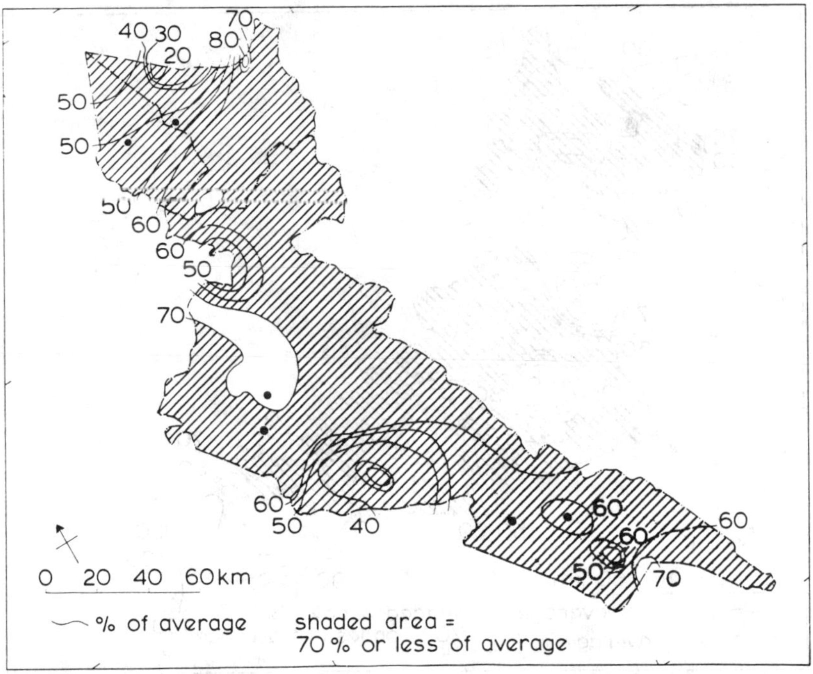

Figure 2.11. **Rainfall in October–December 1975, percent of average.**

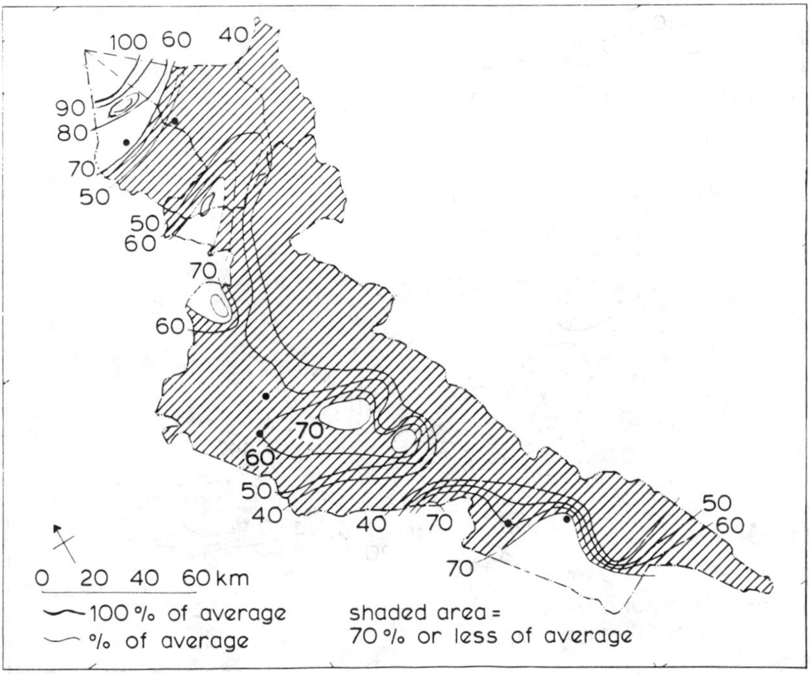

Figure 2.12. **Rainfall in March–May 1976, percent of average.**

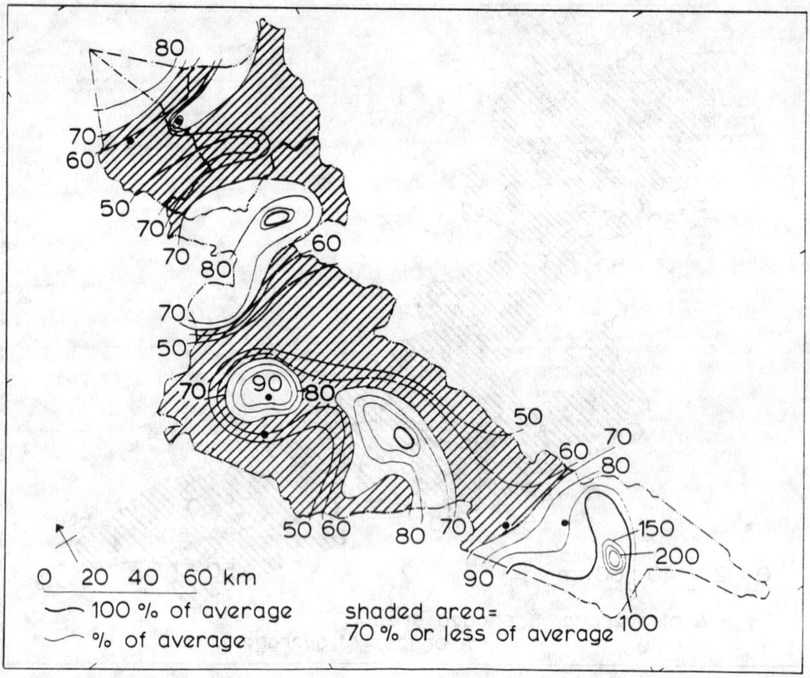

Figure 2.13. **Rainfall in October–December 1976, percent of average.**

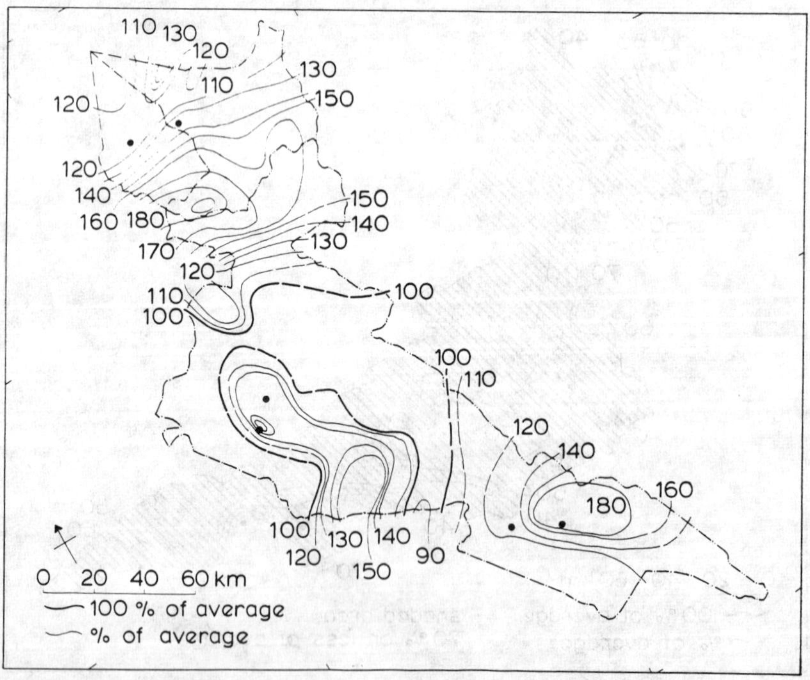

Figure 2.14. **Rainfall in March–May 1977, percent of average.**

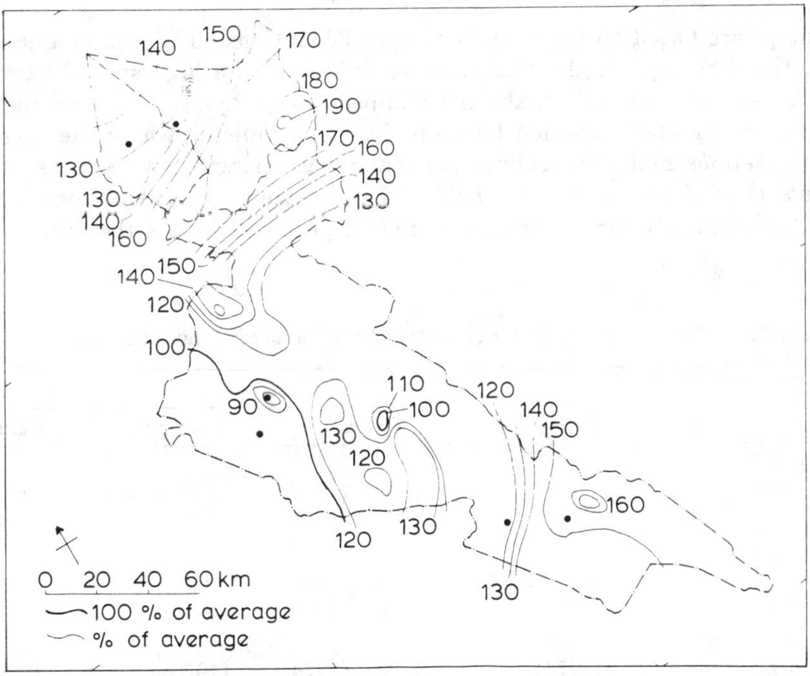

Figure 2.15. **Rainfall in October–December 1977, percent of average.**

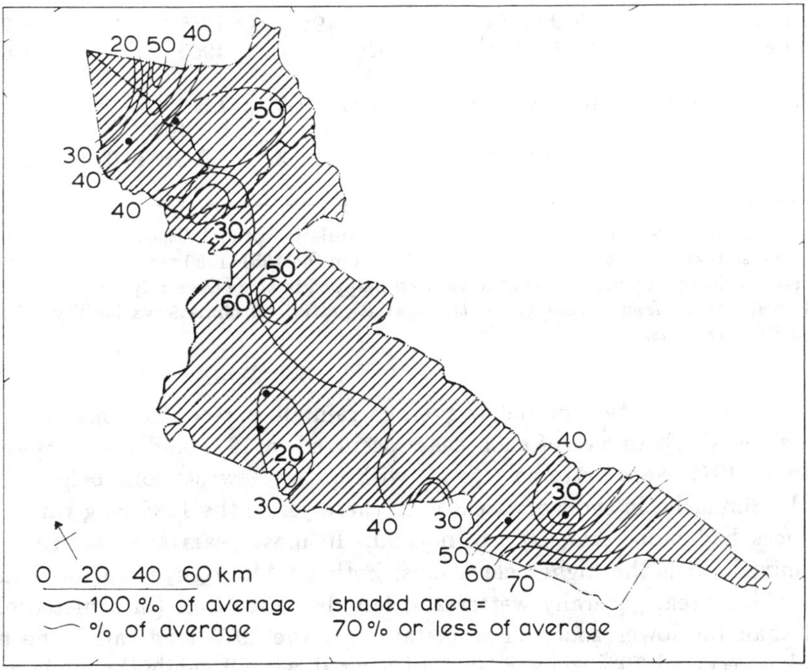

Figure 2.16. **Rainfall in March–May 1984, percent of average.**

The maps are based on the records of some 36 stations. The lack of a dense network, the differing lengths of station records, and the high spatial variability, even for seasonal rainfall, make the mapped patterns only a general picture of the seasonal rainfall. Dashed lines on the maps indicate where the scarcity of rainfall stations make the isolines particularly uncertain. Several aspects of the region's rainfall are apparent. *Table 2.7* notes which historical years approximate the scenarios derived from cumulative probabilities for the four reference stations.

Table 2.7. Historical examples for climatic scenarios at reference stations.[a]

	Station			
Scenario	*Kerugoya*	*Embu*	*Machakos*	*Makindu*
Good year				
Mar–May	1970	1982 (1977)	1978 (1977)	1967 (1977)
Oct–Dec	1968	1972	1963	1978
Average year				
Mar–May	1959 (1975)	1966 (1974)	1967	1964 (1974)
Oct–Dec	1978	1964 (1976)	1971 (1976)	1969
Moderate drought				
Mar–May	(1974)	1965 (1974)	(1976)	(1984)
Oct–Dec	1970 (1976)	1962	1960 (1977)	1973 (1974) (1975)
Severe drought				
Mar–May	1972 (1984)	1973 (1984)	1956 (1984)	1972
Oct–Dec	1971 (1975)	1970	1966	1966
Back-to-back drought	LR-SR 1969	LR-SR 1952	LR 1965–SR 1966	(SR 1974–LR 1976)
	(LR-SR 1974)	LR-SR 1973	LR-SR 1973	(LR 1983–LR 1984)

[a]The climatic scenarios are defined according to cumulative probabilities. The first year given is the closest example of the scenario probability from the 1951–1980 record. Parentheses indicate reasonable approximations that have been mapped in this case study. The maps are provided to illustrate, from recent years, the spatial pattern of climatic variability. SR: short rains; LR: long rains.

Widespread above- or below-average rainfall is recorded only in the very extreme cases. Examples of such cases occurred in 1977 and 1984. Both rainy seasons in 1977 were extremely good, with only a few stations below 100% of normal. Similarly, in the worst season in these years, the 1984 long rains, only a few places had more than 50% of normal. In most years the pattern is much more mixed, with the higher elevations, both on Mt. Kenya and on hills elsewhere in the area, generally wetter and less drought-prone (in a meteorological sense) than the lower zones. For instance, in the 1975 long rains, the hills of Machakos received 70% to over 100% of normal rainfall, while the lowlands near Kibwezi experienced notable drought. The upper slopes of Mount Kenya have the lowest coefficients of variability, as expected, given their altitude and exposure to the moisture-bearing southeasterly winds, and cannot be considered drought-prone, given the current land use.

In 1974 a very wet long rains season was followed by the onset of a severe drought. The 1976 short rains and 1977 long rains illustrate how quickly such droughts can end. Multiple-season drought occurred in 1974–1976. In the lower zones of the study area there is little moisture stored between seasons. In some cases multiple-season moisture is critical for agriculture, as noted in the effect of fallow (Porter, 1983) and for two-season crops, such as pigeon peas. The impact of long-duration, multi-season drought is more critical to livestock production (Section 5) and to smallholder economics and coping strategies (Section 6).

2.5. Conclusions

The principal features of the climate of central and eastern Kenya, particularly as it relates to drought, can be summarized as follows:

- The dominant control over the rainfall of the study area is the movement of the ITCZ.
- Rainfall amounts, rainfall variability, and potential evaporation, are correlated with altitude, although aspect is locally important.
- Rainfall, in both extremely good and bad seasons, is more spatially uniform than in the average years, which suggests the importance of region-wide synoptic conditions in influencing the weather in these anomalous extreme years.
- Drought is a recurring phenomenon in the climatic record, and drought episodes in any part of the study area tend to be positively correlated with those in other parts.
- Given that there are no strong trends in either the average rainfall or its variability over the past few decades, there is no reason to believe that the climate of the next few decades will be markedly different in its fundamental parameters than that of the current historical record.
- Drought does not appear to occur in well-defined cycles that could be reliably used to predict future episodes.
- The most promising avenue for future climatic research may be that of monitoring seasonal rainfall, and predicting its total amount early in the season in order to help in decision making with respect to agricultural investment and government responses.

SECTION 3

The Effects on Agroclimatic Zones and Agricultural Production

3.1. Introduction

All climatic elements, whether singly or in combination, affect crop and animal production. The climatic elements of greatest significance to agricultural production in Kenya are rainfall, temperature and evapotranspiration. The balance between rainfall and evapotranspiration is particularly important. Characterization of an area based on climatic parameters relevant to agriculture (i.e., agroclimatic) is a useful practical exercise. The resulting agroclimatic zones indicate which areas are climatically suitable for different crops. Additional information on the length and intensity of the rainy and dry periods and on soil characteristics is necessary to designate agroclimatic zones.

There have been several attempts to characterize agricultural site potential in Kenya, based on the ecological factors of climate, soils and vegetation. The first maps of ecological zones based on the natural vegetation were interpreted in terms of site potential for agriculture (Edwards, 1940 and 1956). Pratt and Gwynne's (1977) work on Kenya was similarly based on vegetation, but included climatic terms in the keys. Pratt *et al.* (1966) calculated moisture indices according to the Thornthwaite (1948) system. The moisture indices were used in conjunction with vegetation to define and delimit ecoclimatic zones in Kenya. Woodhead (1970), on the other hand, calculated and mapped an available water index (AWI) for many stations in Kenya and equated the AWI with the ecological zones of Pratt *et al.* (1966). Pant and Rwamdusya (1971) and Obasi and Kiangi (1977) calculated water balance components, which they used to arrive at the distribution of climatic zones in Kenya. Sombroek *et al.* (1982), at the Kenya Soil Survey (KSS), used the r/Eo ratio (where r = average annual rainfall and Eo = average annual potential evaporation) and mean annual temperature to arrive at moisture availability and temperature zones for Kenya. The r/Eo ratio is similar to Woodhead's (1970) AWI. The agroclimatic zones of

Sombroek *et al.* were further developed by Jaetzold and Schmidt (1983) by using a water balance crop production model to define and map the agroecological zones that appear in their *Farm Management Handbook of Kenya.* The agroecological zones are overlain on a map of soil types and characteristics. The work of Jaetzold and Schmidt provides additional information for the assessment of potential agricultural production in Kenya.

A major disadvantage of most of the (agro) climatic zonation systems outlined in the preceding paragraph is that they are based on average climatic data. This means, for instance, that the agroclimatic zone boundaries as depicted in the KSS and other similar works are the means of a large number of years. For agricultural planning strategies, it is necessary to examine the shift from year to year (or season to season) of the agroclimatic zone boundaries. This section describes a simple method for examining the seasonal shifts in agroclimatic zones for the three districts of this case study.

3.2. Agroclimatic Zones of Central and Eastern Kenya

The two previous sections discussed the agroclimatic zones and climate of central and eastern Kenya. The rationale for and derivation of the agroclimatic zones will now be considered in more detail.

The most important controls of the weather and climate in the case study area are: (a) the ITCZ; (b) altitude; and (c) aspect. Rainfall increases toward the higher elevations, while potential evaporation decreases. The major characteristics of climate that affect plant growth are: (1) the balance between rainfall and evaporation, and (2) temperature. With regard to rainfall, the length and intensity of the rainy and dry seasons and their variation from year to year are of particular importance. In the three districts the average annual rainfall ranges from 500 mm to 2300 mm; the average potential evaporation, from less than 1200 mm to 2500 mm; and average annual temperatures, from 10 °C to 30 °C. More details on the spatial patterns of evaporation are found in the *National Atlas of Kenya* and in the work of Woodhead (1968). Clearly, there are many different types of rainfall distributions and agroclimatic zones within the study area.

Since 1973, the Kenya Soil Survey has used the r/Eo ratio (average annual rainfall/average annual potential evaporation from pan A measurements) when delimiting the agroclimatic zones for their 1:100 000 reconnaissance soil surveys. The same method with some modifications has been used to delimit moisture availability zones for the national map of the agroclimatic zones. The purpose of the agroclimatic zone map, according to Sombroeck *et al.* (1982; p. 43), is "to provide a tool for assessing which areas are climatically suitable for various land use alternatives with particular emphasis on the suitability for crops or crop varieties". The agroclimatic map has been overlain on a soils map, and temperature zones were defined to indicate agroecological zones, although the three maps are not otherwise related.

Table 3.1. Moisture availability indices and potential production in tons dry matter (DM). Source: Sombroek *et al.* (1982).

	Agroclimatic zone						
Characteristic	*I*	*II*	*III*	*IV*	*V*	*VI*	*VII*
Primary crop	Tea/ coffee	Marginal coffee	Maize/ cotton	Marginal cotton	Livestock/ millet	Ranching	
r/Eo ratio (%)	>80	65–80	50–65	40–50	25–40	15–25	<15
Climatic designation	Humid	Sub-humid	Semi-humid	Semi-humid/ semi-arid	Semi-arid	Arid	Very arid
Average no. growing days/year	365	290–365	235–290	180–235	110–180	75–110	<75
Potential annual plant biomass (tons DM/ha)	>30	20–30	12–20	7–12	3–7	1–3	<1

The ratio of average annual rainfall to average annual potential evaporation, r/Eo, is a first guide for evaluating the agricultural potential of an area. Because the object of the agroclimatic zonation is to provide broad zones of land potential, round figures for the moisture availability zones were used. The r/Eo ratio was expressed as a percentage (*Table 3.1*). The boundaries between zones are arbitrary, but based on the general suitability for crops. The KSS agroclimatic zones are widely used in Kenya (being regularly incorporated into District Development Plans) and are easy to compute.

The agroclimatic zones occurring in the study area, defined on the basis of the r/Eo ratio, are mapped in *Figure 1.3* in Section 1. Reflecting the influence of altitude, zone I, the wettest with the lowest potential evaporation, is found on the upper slopes of Mt. Kenya. The majority of Machakos is in zone V, with low and unreliable rainfall and high potential evaporation. The intermediate zones are found toward Mt. Kenya and on isolated hills. To the south is zone VI, a small portion of the study area devoted to a national park. The driest zone in Kenya (VII) is not found in the study area.

Agroclimatic zone I is the zone of tea and coffee, which are major crops in Kirinyaga and Embu districts. In this zone, low temperatures, low evaporation and high precipitation are recorded. Agroclimatic zone II is the marginal coffee zone. Here temperature and evaporation values increase, while precipitation decreases slightly. Agroclimatic zone III is the maize–cotton zone. Cotton is the major cash crop, while maize is used mainly as a subsistence crop, but also for sale when there is a surplus. Temperatures and evaporation increase in this zone and rainfall decreases. Agroclimatic zone IV is the marginal cotton zone. The climate patterns in zone IV are similar to zone III, but drier and more variable. Agroclimatic zone V is the livestock–millet zone. In this zone the climate is very variable and dryland crop production is not reliable.

In addition to climatic factors, soils play a crucial role in determining the water balance for agriculture. Systematic studies of water-holding capacities have not been done in the three districts. The soils in zones I and II are mountain soils, developed on olivene basalts and well-drained with medium to heavy texture, and soils on volcanic foot ridges, also well-drained with a heavy texture. The soils in zones III and IV are diverse, but the most common are those formed on Tertiary basic igneous rocks (well-drained, heavy) and on undifferentiated basement system rocks (well-drained, medium to heavy). The soils of zone V are very often shallow and are formed on various levels of uplands (on quartzites and basement system rocks) and erosional plains. The textures vary, but generally they have a higher content of sand than the volcanic soils and therefore a lower water-holding capacity. In places soils seal in the top layer during the rains, thus increasing runoff unless plowed. A hardpan under the soil exists in some localities which decreases infiltration. Given the lack of data and variability of the soils, their contribution to the agricultural impacts of climatic variability have been oversimplified – a matter that should be addressed in the near future. In the absence of better data, an agroclimatic zone map based on rainfall and evaporation could be overlain on a map of estimated soil moisture-holding capacity. The development of more systematic adjustments could be helped by relating water budget models to the simpler r/Eo ratios.

3.3. Agroclimatic Zones for Particular Seasons and Agricultural Years

The agroclimatic zones discussed above are based on climatic averages, but investigation of the impacts for particular climatic scenarios requires consideration of the conditions of specific years and seasons. However, some additional calculations with respect to potential evaporation are needed first.

3.3.1. Seasonal potential evaporation

Annual average potential evaporation, Eo, for 36 stations was estimated using the empirical relationship developed by Woodhead (1970), based on the altitude of the station:

$$Eo = 2422 - 0.358h$$

where Eo = evaporation (pan A) in mm, and h = altitudes in meters.

The Eo values calculated using this formula are annual averages. Since the rainfall pattern in the study area is bimodal, average Eo needed to be determined for two seasons, i.e., the March–May and October–December rains. Five meteorological stations that are in or near the study area and measure and report monthly evaporation were selected, corresponding to the five agroclimatic zones. The average monthly evaporation was expressed as a percentage of the annual average (*Table 3.2*). Seasonal Eo coefficients were calculated for the two seasons by summing the values for the three months of each season (*Table 3.3.*).

Table 3.2. Monthly percentages of average annual evaporation, *Eo*.[a]

Month	*Kimakia*	*Muguga*	*Dagoretti*	*Embu*	*Makindu*
Jan	10.9	11.4	11.4	9.7	8.3
Feb	10.3	10.6	10.6	10.0	8.8
Mar	10.9	10.9	10.7	9.7	9.6
Apr	9.5	7.0	8.2	8.3	8.2
May	8.1	6.1	7.2	7.0	7.5
Jun	6.0	5.5	5.8	6.1	7.3
Jul	4.1	5.1	5.0	5.7	7.0
Aug	4.6	5.8	5.6	6.8	8.0
Sep	7.7	8.6	8.2	9.0	9.4
Oct	9.7	10.4	9.5	10.7	10.2
Nov	8.5	8.2	7.9	7.8	8.4
Dec	9.7	9.9	9.9	9.0	7.4

[a]Period of record varies, from about 5 to 10 years. Evaporation is from pan A measurements. Source: East African Meteorological Department (1975).

Table 3.3. Seasonal percentages of average annual evaporation, *Eo*.[a]

Month	*Kimakia*	*Muguga*	*Dagoretti*	*Embu*	*Makindu*
Mar–May	28.5	24.6	26.1	25	25.3
Oct–Dec	27.9	28.5	27.3	27.5	26.0

[a]See note to *Table 3.2*.

Table 3.4. Comparison of agroclimatic indices at Makindu derived from estimates of *Eo* for specific years, and for average conditions.

Measurement	*1979*	*1980*	*1981*	*1982*
Rainfall, mm	777	402	661	938
Annual *Eo* (P), mm[a]	2026	2148	1874	2249
Eo departure from average, mm[b]	−38	84	−190	185
r/*Eo* (P)[c]	0.38	0.19	0.35	0.42
r/*Eo* (W)[d]	0.38	0.20	0.32	0.45
% Error in ratios	0	+5	−9	+7

[a]*Eo* estimated by Penman method by Meteorological Department, using a crop factor of 0.75 to convert to *Eo*.
[b]Average *Eo* = 2064 mm, estimated using Woodhead's formula based on the altitude for Makindu.
[c]Rainfall/*Eo*, using Penman estimate of *Eo* for each year.
[d]Rainfall/*Eo*, using Woodhead estimate of average *Eo* (see note b).

For 36 rainfall stations in the study area, average seasonal *Eo* values were computed by taking the seasonal percentage (for the appropriate zone) of the estimated annual *Eo* based on Woodhead's formula. This method provides a

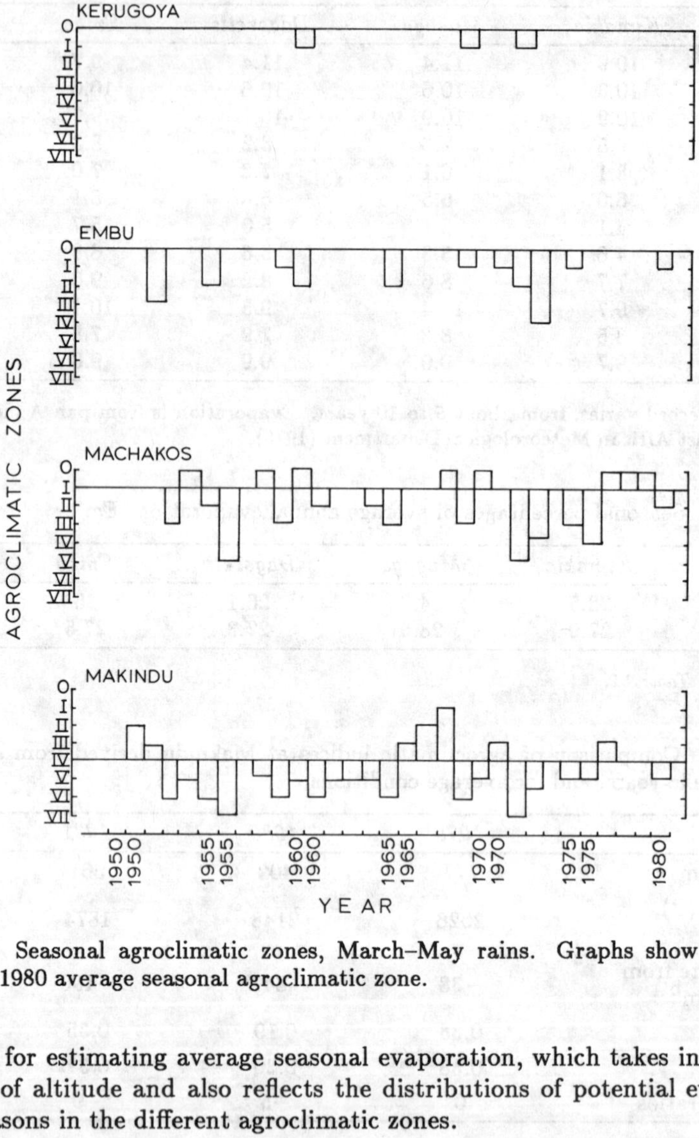

Figure 3.1. Seasonal agroclimatic zones, March–May rains. Graphs show departures from 1950–1980 average seasonal agroclimatic zone.

procedure for estimating average seasonal evaporation, which takes into account the effect of altitude and also reflects the distributions of potential evaporation among seasons in the different agroclimatic zones.

A test of the adequacy of using average annual *Eo* values based on altitude was performed for Makindu by comparing the *r*/*Eo* ratio derived from annual evaporation data for individual years with that computed using average annual *Eo* estimates from Woodhead's formula (*Table 3.4*). Annual *Eo* computed for particular years by the Penman method departed from the average based on Woodhead's formula by −190 mm to +185 mm. The values of the errors indicated by differences between the two methods of computing *r*/*Eo* ranged from 0 to 9%. In comparisons of several years of data for Katumani by Porter (1983),

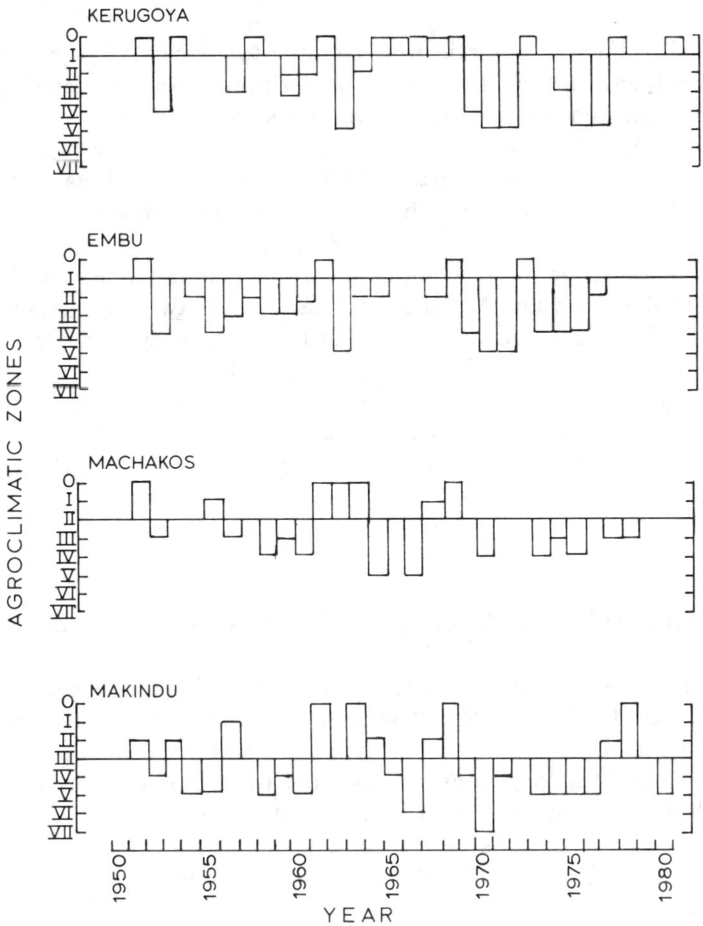

Figure 3.2. Seasonal agroclimatic zones, October–December rains. Graphs show departures from 1950–1980 average seasonal agroclimatic zone.

Penman estimates and pan A measurements were found to be highly correlated, so it was deemed appropriate to use the Penman method here (*Table 3.4*).

In other studies, regression equations developed at the Central Bureau of Statistics (Kirimi, 1986) predicted monthly evaporation with errors on the order of 50 mm, based on altitude, number of rain days and distance from the sea (not an important variable at the scale of the study area). Clearly, rainfall is the dominant variable affecting moisture availability; and in any event, potential evaporation is more variable in the dry months than in the rainy seasons. Refinement of the available evaporation data or computational procedures would probably not significantly alter the pattern of agroclimatic zones, even for individual years or seasons.

3.3.2. Seasonal zones

Before applying the agroclimatic zone concept to individual years and seasons, one further adjustment was thought to be desirable. An additional category, 0, was added to the zones (I to VII) to indicate seasons or years where r/Eo was greater than 1.00. Computing r/Eo using only the rainy season months of the year shifts the zonation to a wetter classification, so zone 0 was introduced to differentiate the very wet seasons from those that otherwise would be classified as zone I.

Agroclimatic zones were computed for the four reference stations (Kerugoya, Embu, Machakos and Makindu) for each of the two seasons for each year, 1951–1980 (*Figures 3.1* and *3.2*). For the March to May rains, Kerugoya and Embu average zone 0, with infrequent departures to drier conditions. Machakos falls on average into zone I, but with a more variable pattern. Makindu averages zone IV, with a few instances of wetter seasons, but frequent departures to zones VI and VII. The October to December rains show somewhat the reverse patterns, with Kerugoya and Embu being drier and more variable than they are for March–May, and Machakos and Makindu wetter and less variable.

3.4. Spatial Shifts in Agroclimatic Zones in Selected Seasons

Shifts in agroclimatic zones based on r/Eo for individual seasons were mapped for the years 1974–1977 and the 1984 long rains. The maps are based on data for 36 stations in the study area. For each year, two maps were prepared, covering the March to May long rains and October to December short rains (*Figures 3.3* through *3.11*). The percentage of the study area classified by each of these seasonal agroclimatic zones is tabulated in *Table 3.5* for all the selected seasons. In the maps and table, areas for which a seasonal zone = 0 result was obtained are included with zone I.

The 1974 October–December (short) rains was one of the worst seasons mapped. Most of the study area fell under seasonal agroclimatic zones III, IV and V. The three districts had a shortfall in food production, resulting in famine, mostly in the lower parts of the study area. The 1977 March–May (long) rains was the best season mapped. Kirinyaga and Embu districts fell under seasonal agroclimatic zone I (or 0), while Machakos District fell under zones I (or 0), II and III. In that year high yields were recorded throughout the study area. The good (bad) seasons seem to be more uniformly wet (dry), while the average seasons (such as in 1975) present more complex patterns, with pockets of extreme rainfall and drought.

3.5. Climatic Scenarios and Potential Productivity

The seasonal and annual agroclimatic zones for each reference station were computed for the climatic scenarios defined in Section 2 (*Table 3.6*). The results suggest that drought is not a serious problem in the March–May rains at Kerugoya,

Table 3.5. Percentage of study area[a] included in seasonal agroclimatic zones for selected seasons.

	Agroclimatic zone						
Season	*I*	*II*	*III*	*IV*	*V*	*VI*	*VII*
Mar–May 1974	27.0	18.2	34.8	15.6	4.3	0	0
Oct–Dec 1974	2.3	2.5	8.1	20.5	66.4	0	0
Mar–May 1975	18.7	8.4	11.5	23.1	22.9	9.0	6.8
Oct–Dec 1975	1.6	2.4	7.2	30.6	52.0	6.0	0.1
Mar–May 1976	5.7	5.4	8.1	15.0	60.4	3.4	1.9
Oct–Dec 1976	4.6	5.8	28.7	52.2	8.7	0	0
Mar–May 1977	66.6	10.2	22.5	0.3	0.4	0	0
Oct–Dec 1977	48.8	21.7	29.5	0	0	0	0
Mar–May 1984	2.7	0.9	2.6	9.0	20.8	32.6	31.4

[a]Total area of the three districts is 18 034 sq km.

Table 3.6. Agroclimatic zone (ACZ) based on average climate, and ACZ and *r/Eo* for seasons and agricultural years for climatic scenarios.[a]

Season/ Scenario	*Kerugoya*		*Embu*		*Machakos*		*Makindu*	
	ACZ	*r/Eo*	*ACZ*	*r/Eo*	*ACZ*	*r/Eo*	*ACZ*	*r/Eo*
Mar–May (long) rains								
Eo	485		491		458		520	
Good season	0	2.12	0	1.58	0	1.29	III	0.63
Average season	0	1.45	0	1.07	I	0.87	V	0.34
Moderate drought	I	0.99	II	0.71	IV	0.46	VI	0.18
Severe drought	I	0.93	IV	0.47	V	0.29	VII	0.05
Oct–Dec (short) rains								
Eo	516		515		502		535	
Good season	0	1.41	0	1.28	0	1.34	0	1.13
Average season	I	0.89	II	0.69	II	0.69	IV	0.49
Moderate drought	IV	0.44	V	0.39	IV	0.48	V	0.28
Severe drought	V	0.39	V	0.27	V	0.39	VI	0.19
Oct–Sep (agricultural year)								
Eo	1857		1888		1833		2064	
Good year	0	1.10	I	0.80	II	0.75	IV	0.49
Average year	I	0.82	III	0.56	IV	0.49	V	0.28
Moderate drought	III	0.59	V	0.38	V	0.35	VI	0.16
Severe drought	IV	0.50	V	0.36	V	0.31	VII	0.10
Average climate	I/II		III		IV		V	

[a]Based on rainfall scenarios defined in *Table 2.6*, using the 1951–1980 record. For comparison, the zones based on average climate, as shown in *Table 2.1*, are also listed here.

as rainfall is equivalent to more than 90% of the potential evaporation. For both seasons at Machakos and Makindu severe drought results in crop failure, with rainfall less than 40% of potential evaporation at Machakos and 20% at Makindu.

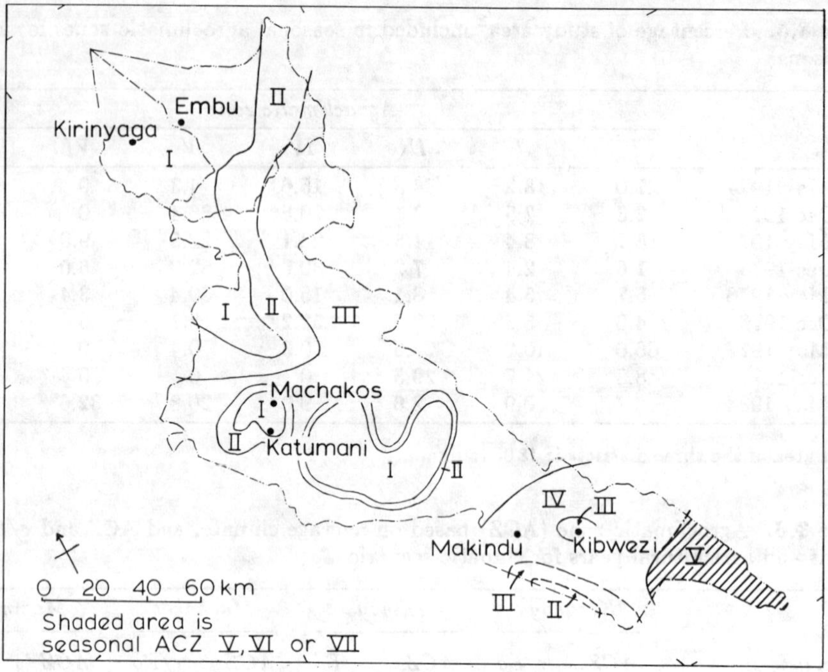

Figure 3.3. **Seasonal agroclimatic zones in March–May 1974.**

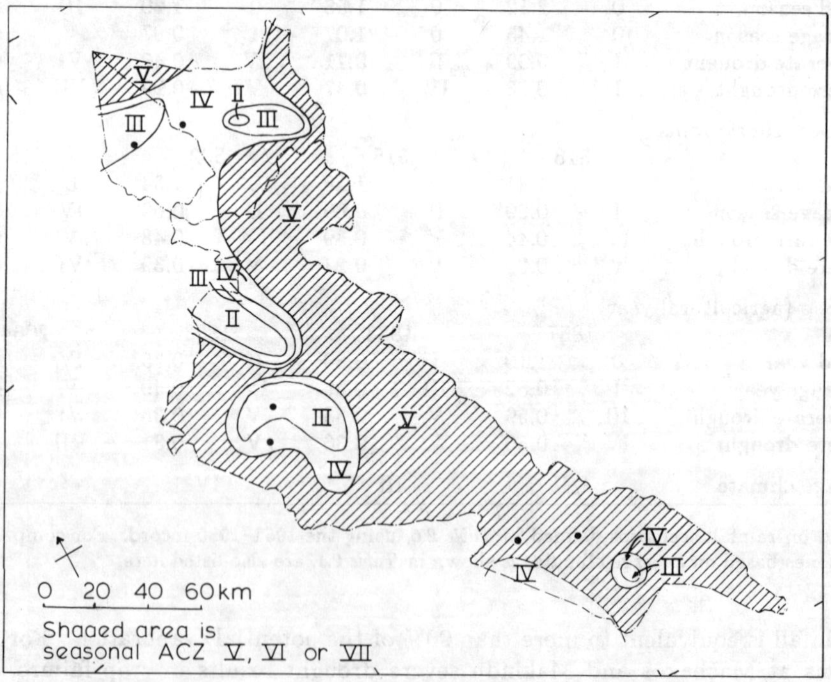

Figure 3.4. **Seasonal agroclimatic zones in October–December 1974.**

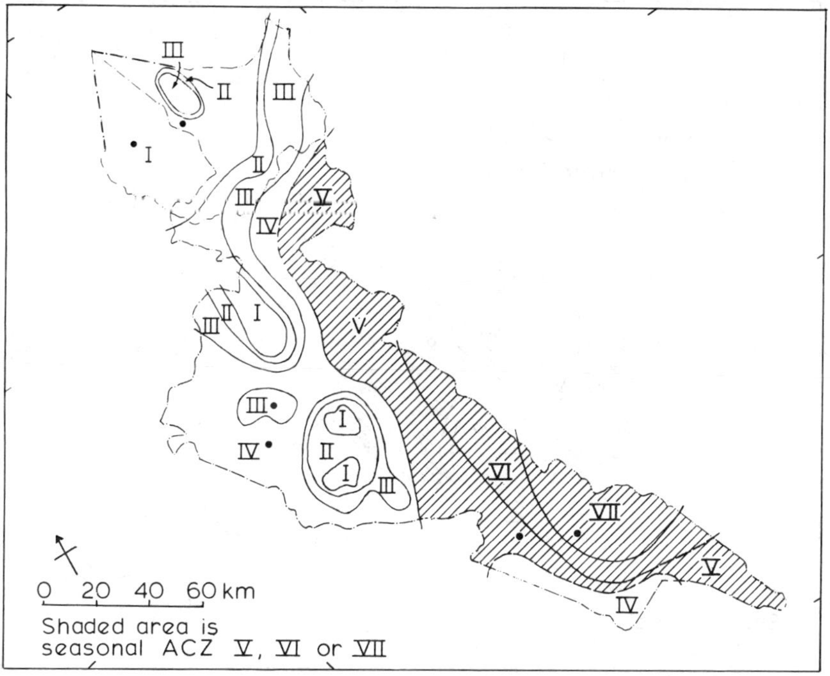

Figure 3.5. **Seasonal agroclimatic zones in March–May 1975.**

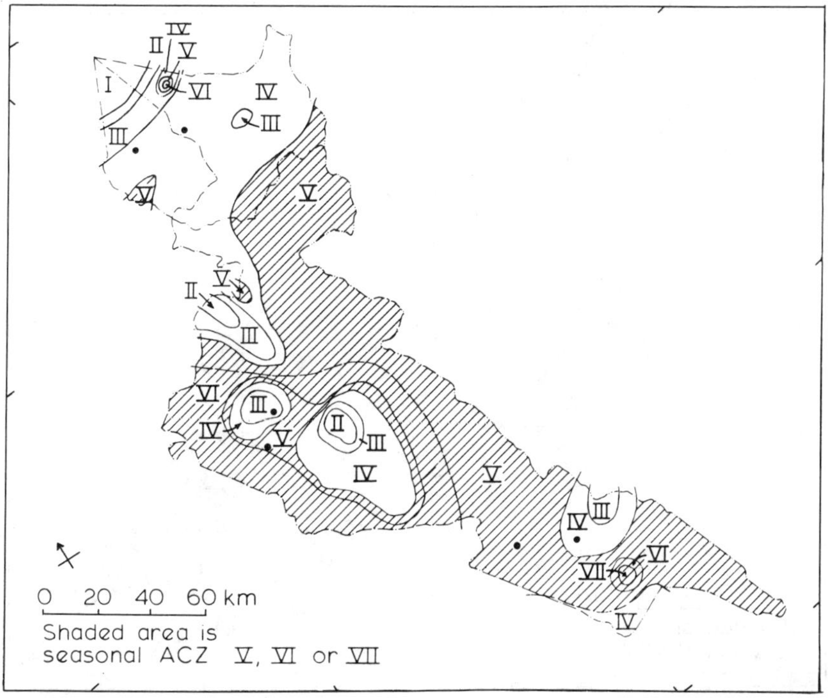

Figure 3.6. **Seasonal agroclimatic zones in October–December 1975.**

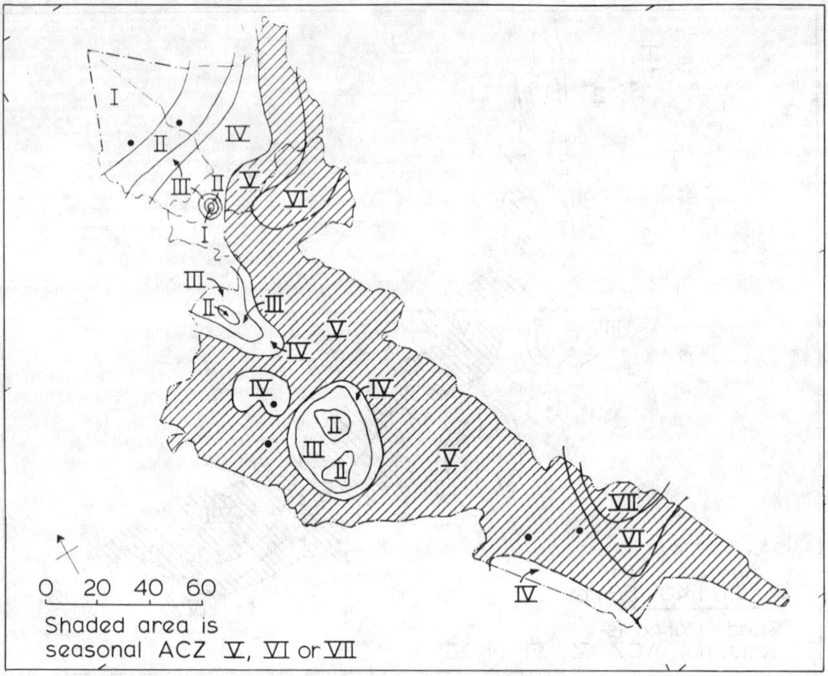

Figure 3.7. **Seasonal agroclimatic zones in March–May 1976.**

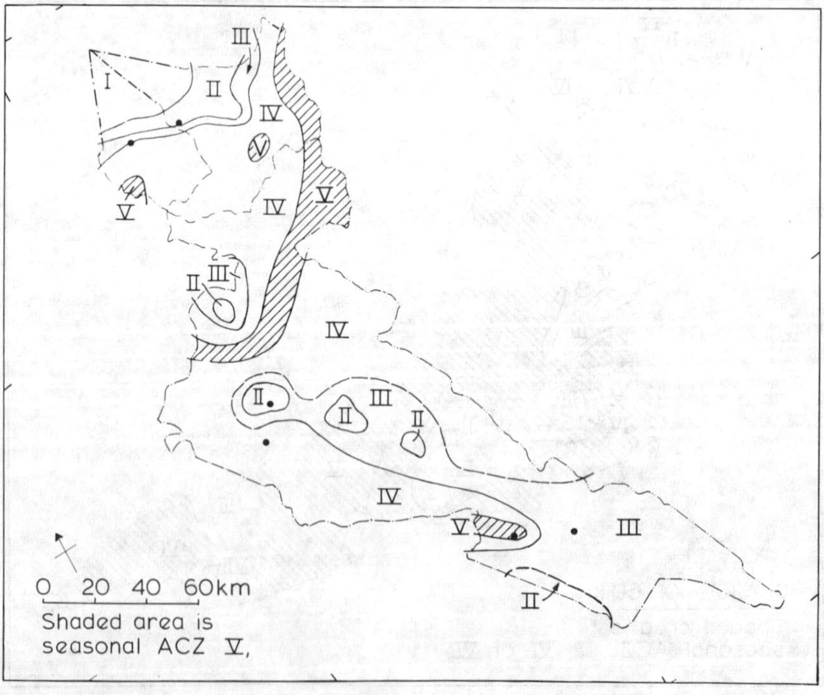

Figure 3.8. **Seasonal agroclimatic zones in October–December 1976.**

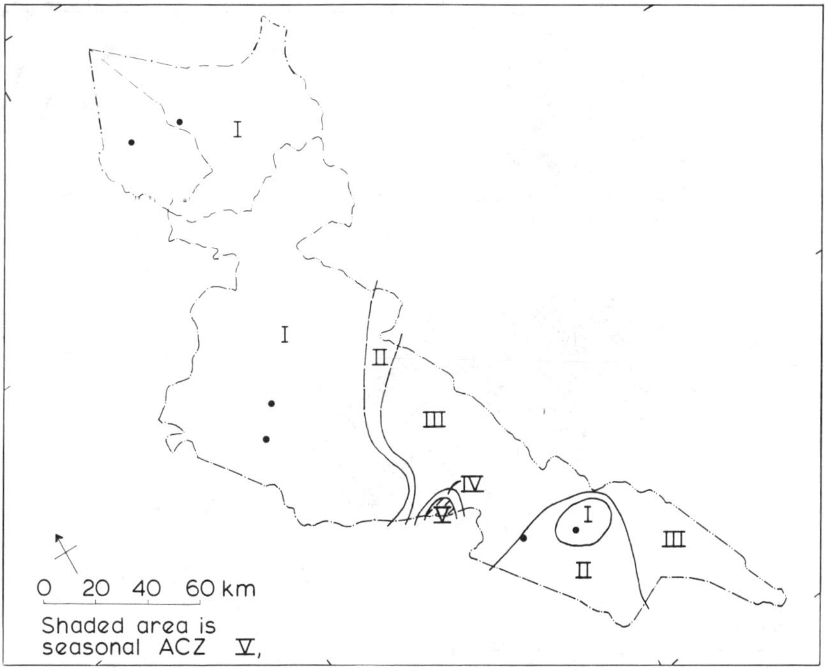

Figure 3.9. Seasonal agroclimatic zones in March–May 1977.

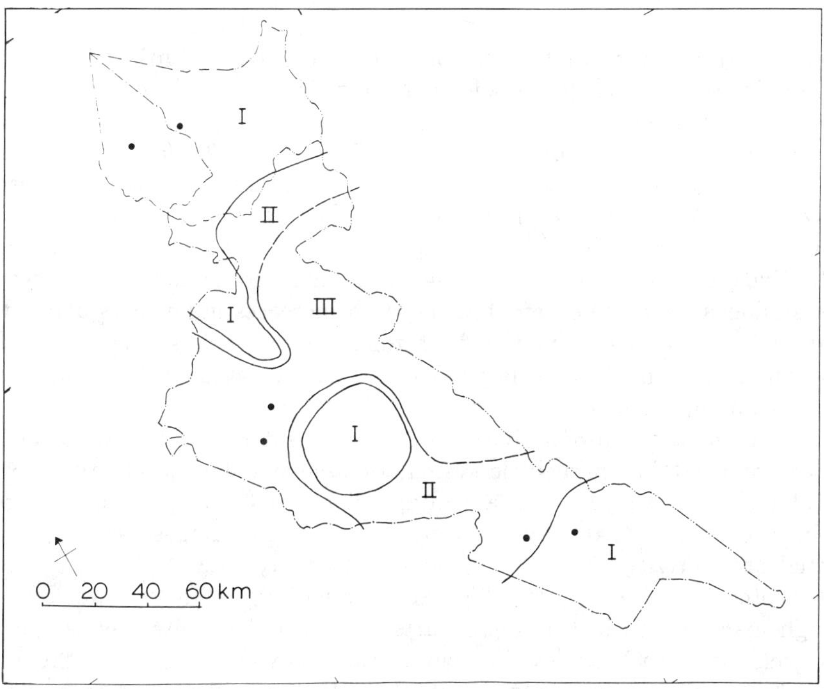

Figure 3.10. Seasonal agroclimatic zones in October–December 1977.

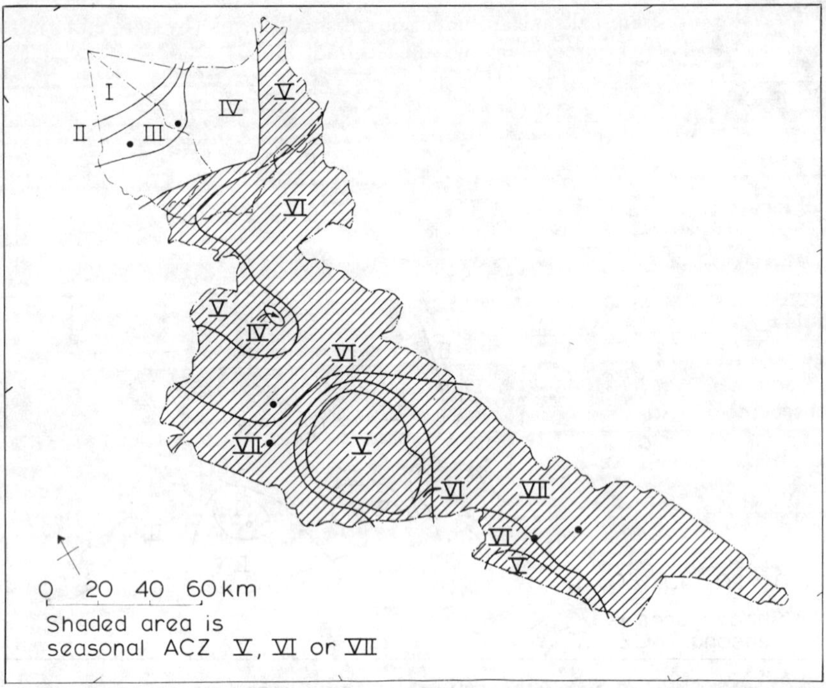

Figure 3.11. Seasonal agroclimatic zones in March–May 1984.

The major factors that determine the productive potential of an area are agroclimatic, soil and crop management conditions. Based on the r/Eo ratio, Sombroek *et al.* (1982) estimated the potential annual production (biomass) of the moisture availability zones in Kenya (*Table 3.1*). *Table 3.7* provides estimates of average annual maize yield, average maize area planted per household, and average potential maize production per household by agroclimatic zone, on the basis of data from the *Farm Management Handbook* (Jaetzold and Schmidt, 1983). Except in zone V, the average potential production per household is rather similar among the different zones, with increases in hectares planted compensating for lower yields in the drier zones. Nevertheless, the lower-altitude zones will have a more variable production, as indicated by the changing seasonal zones mapped above.

The estimates of productivity should be considered at best as expert opinion, because there has been little systematic assessment of productivity or agricultural yields in Kenya. The advantage of characterizing the study area by agroclimatic zone is that it enables one to make use of these sketchy data on potential productivity. They help give a preliminary spatial picture of the food supply potential. Clearly, the highest productivity is in the wettest zones, although maize yields in the upper parts of zone I are limited by temperature. Maize yields in zone V are too low, on average, to warrant growing the crop, or at least investing in many inputs. Actual production will depend on many other

Table 3.7. Potential annual maize production on small farms (by agroclimatic zone as indicated in *Table 3.1*). Source: Jaetzold and Schmidt (1983).

	Agroclimatic zone					
	I	*I/II*	*III*	*III/IV*	*IV*	*V*
Average maize yield on small farms[a] (kg/ha)	2375	1000	1275	1050	850	250[b]
Average maize area (ha) planted per household[c]	0.7	0.9	1.2	1.5	1.7	1.9
Average potential maize production[d] (kg/household)	1663	1710	1530	1575	1445	475

[a]Lowest production level, as farming is commonly practiced by 75% of smallholders, as defined in source.
[b]No yield estimates available; 250 kg/ha is a reasonable guess, by the case study authors.
[c]Based on a limited small-farm survey conducted in 1977, cited in the source.
[d]Does not include differences between area planted and area harvested. These figures are simply products of multiplications of the first two rows, and are not actual measurements either of yields or of household average production.

Table 3.8. Agroclimatic zone (based on agricultural years) and expected maize yields for climatic scenarios (kg/ha/yr), based on average annual maize yields given in *Table 3.7.*

	Kerugoya		*Embu*		*Machakos*		*Makindu*	
Scenario	*ACZ*	*Yield*	*ACZ*	*Yield*	*ACZ*	*Yield*	*ACZ*	*Yield*
Good year	0	>2375[a]	I	2375	I	2375	IV	850
Average year	I	2375	III	1275	IV	850	V	250
Moderate drought	III	1275	V	250	V	250	VI	0
Severe drought	IV	850	V	<250	V	<250	VII	0

[a]Subject to flooding and temperature effects.

variables, including other climatic factors, such as those incorporated in the model in Section 4.

The potential production data are derived from average climatic data; using information reflecting the variations from season to season can give estimates that deviate significantly from these values. Utilizing the data on yields, shifts in potential agricultural production can be analyzed in relation to shifts in the agroclimatic zones. This was done for the four reference stations and the agricultural year climatic scenarios described in Section 2. The resulting estimates of potential productivity for the climatic scenarios for good, average, moderate drought and severe drought are presented in *Table 3.8.* The differences in potential productivity for the same area in different years are remarkable – an order of magnitude difference between good years and severe drought years at Embu, Machakos and Makindu. At Kerugoya the estimated yield for good years is over 250% of that in severe droughts, although even in the latter conditions some maize will be harvested.

There are several obstacles to be overcome before the lines of analysis begun in this section can be followed up with more detailed research and practical applications. Since the seasonal zonation is qualitatively different from the annual agroclimatic index, the data on average annual crop yields by agroclimatic zones cannot be converted to seasonal yields. Also, time series of measured yield data are not available in Kenya to validate the estimation of yields for different years based on climatic data. The crop data that are available are averages developed for the purposes of land suitability mapping rather than assessment of seasonal fluctuations. Applications involving computation of aggregate potential production for each season and the entire study area would require data on the amount of land planted and harvested with various crops or devoted to grasslands. This figure can vary from season to season, and in some cases may be more important than the expected yields. Without data on area planted or harvested, it is not possible to give more than an indication of the magnitude of the expected change in production from normal. Data on area planted are not currently available at a scale that allows comparison with spatial shifts in the agroclimatic zones.

3.6. Conclusion

A simple agroclimatic index developed for determining static zones can be readily adapted to illustrate climatic variability over time and space. The resulting zones are easier to relate to potential agricultural production than actual rainfall figures. A simple monitoring system, utilizing these zones, estimates of expected yields, and data on area planted, could give a first indication of when climatic conditions may lead to food shortages.

SECTION 4

The Effects on Maize Yields

4.1. Introduction

In this section we simulate the effects of various climatic scenarios on maize yields using a dynamic crop production model, applied in conjunction with data that describe the production environment of the districts of Kirinyaga, Embu and Machakos in Kenya. An analysis of the climate characteristics is presented below. These data, together with soil characteristics, are essential input for the production model. The model is described briefly and the potential maize yields generated for different climatic scenarios are reported.

4.2. The Production Environment

4.2.1. The climate

The dynamic crop production model estimates crop yields based on the following climate characteristics:

- Air temperature
- Windspeed
- Relative humidity
- Global radiation
- Rainfall

Information for these variables are shown for the climatological station of Machakos Dam (*Figure 4.1*). The plotted averages are decadal values, i.e., values for 10-day periods, and are based on observations for the period 1963–1982. Each of the average decadal values of the variables is accompanied by its 25 and 75 percentile values as a measure of the variability.

The annual amplitude of the air temperature is small, with a range of about 4 °C between the warmest and coolest months. The coolest period of the

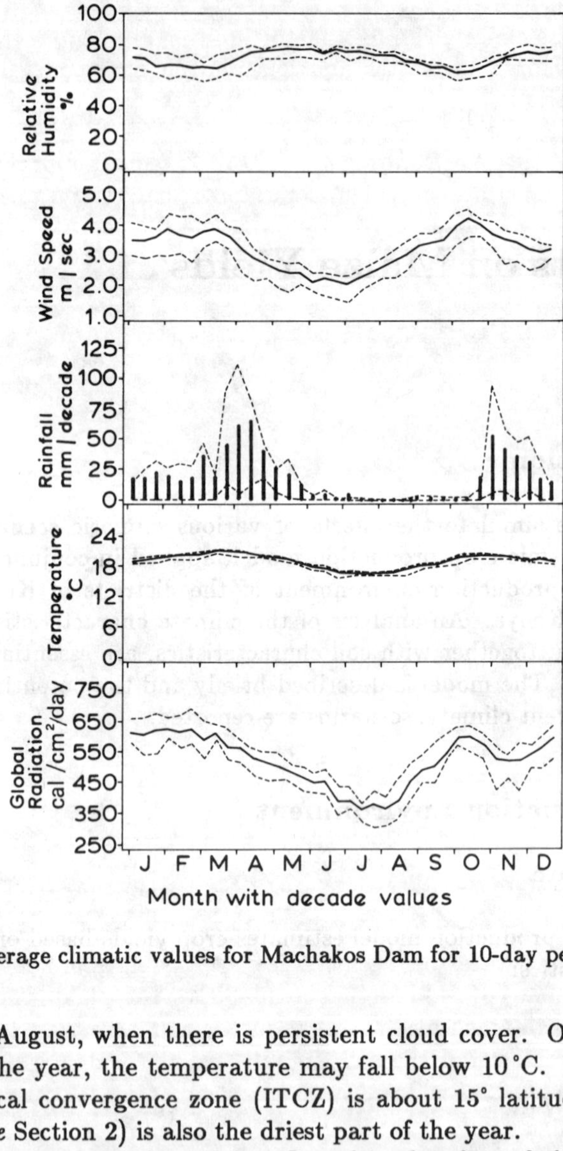

Figure 4.1. Average climatic values for Machakos Dam for 10-day periods (decades).

year is June–August, when there is persistent cloud cover. On a few nights, at that time of the year, the temperature may fall below 10 °C. This period, when the intertropical convergence zone (ITCZ) is about 15° latitude to the north of the region (*see* Section 2) is also the driest part of the year.

The presence of cloud cover is reflected in the plots of global radiation. At the end of July, average radiation is about 1470 Joule $cm^{-2} day^{-1}$, roughly 60% of the level of January–February. The radiation reaches its highest values just before and at the beginning of the rains; this is especially apparent when the short rains start (in October).

The seasonal rainfall exhibits a typically bimodal pattern. Machakos Dam receives relatively more rain during the period of the long rains (March–April). Farther north, the dominance of the long rains is more accentuated, while to the south it is the short rains that bring relatively more rain (*see also Figure 4.7*).

4.2.2. The soils

Soil fertility can be expected to be a factor limiting crop productivity, because of the generally low organic matter and low phosphorous content. However, the only soil characteristic taken into account in this study was water-holding capacity. The soil water properties employed were obtained from the Kenya Soil Survey and from Van de Weg and Mbuvi (1975). *Figure 4.2* presents the relation between the soil moisture content and the force that retains the soil water (soil moisture tension) for a sandy clay/clay soil. The soil texture is the major factor determining the soil moisture retention curve.

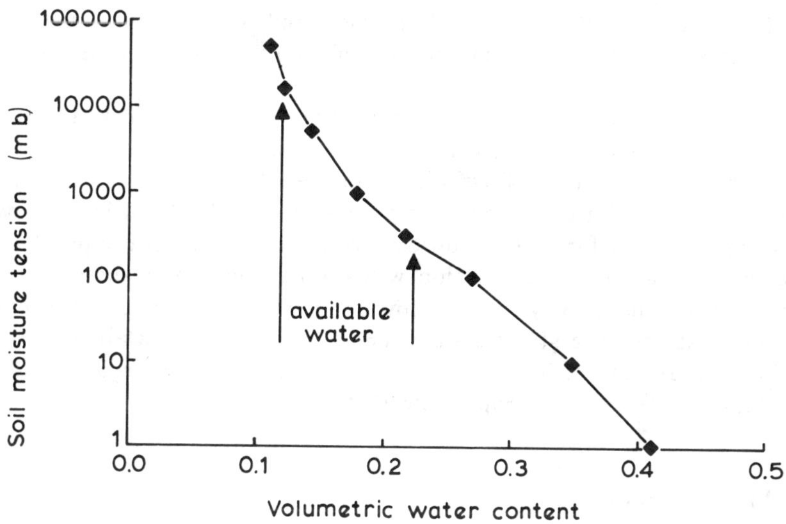

Figure 4.2. Soil moisture retention curve.

The amount of water between water suction values of about 330 mbar (pF 2.5) and 15 000 mbar (pF 4.2) is that available to the plant. The plant cannot take up the water retained by the soil at suctions greater than 15 000 mbar (permanent wilting point), and the soil cannot hold the water against the gravitational force below pF 2.5 (field capacity).

4.2.3. The crop

The most popular food crop in Kenya and the study area is maize (*Zea mays*). Food crops, such as sorghum (*Sorghum bicolor*) and millet (*Pennisetum typhoides*), are less in demand and are only grown where the unreliable rainfall makes maize production impossible.

The maize variety that is used in this study is the hybrid Katumani Composite B. It was released in 1967 by the Katumani Station and thrives well at altitudes between 1000 and 1700 meters above sea level. Compared with the traditional varieties, the hybrid Katumani Composite B is much better adapted to dry conditions. Its growing cycle is much shorter, only about 120 days, and it can therefore mature before the intense drought of the dry season begins, although Njoroge (1984) notes that there is evidence that this variety still does not flower early enough to escape drought in certain areas of Eastern Province.

The crop properties for maize have been described by Nadar (1984) and Mwenda (1983). According to Mwenda (1983), maize needs at least 200 mm of rain during the growing season. However, it depends very much on the soil type and the initial soil moisture content whether that amount of rainfall can produce a reasonable yield.

Research in Kenya and elsewhere suggests that possible effects of temperature on crop production might need to be considered. For example, Law and Cooper (1976) explained a yield decline associated with late planting at Kitale not by water stress, but as a result of low soil temperatures during the growing season.

Temperatures below about 10°C may affect the growth of C-4 plants such as maize (C-4 plants are discussed further in Subsection 4.3.2 below), because of the blocking of the transport of assimilates formed during the day. This problem arises particularly under conditions of high daytime light intensities. If low night temperatures continue for several consecutive nights, injury of the plant's chloroplasts caused by starch accumulation will reduce the effectiveness of the photosynthesis mechanism (Taylor and Rowley, 1971; West, 1973). However, the altitude of most the case study area where maize is grown is about 500 to 700 meters lower than that of locations where this problems has occurred, and even at night temperatures rarely drop below 10 °C.

4.2.4. Management

Most of the smallholders in the study area do not use equipment for mechanical cultivation. The low level of technology makes each of the cultivation practices required for successful crop production extremely labor-intensive.

When the rains are of a long enough duration, planting is ideally done sometime after the rainy season starts, in order to allow time to kill the germinated weeds. However, the rainy period is usually short, and planting therefore has to begin soon after its onset. Little capital is available for herbicides, and all the manual labor is usually required for planting. Therefore, weed competition is a major factor limiting production.

4.3. The Crop Production Model

4.3.1. The structure of the model

The crop production model estimates water-limited yields (Y_{wl}) on the basis of characteristics that describe the climate, site, soil, and crop together with some aspects of management (Konijn, 1984). Water-limited yields are determined by the soil water, and the radiative and temperature regime. The effects of weed and pest control are not considered, and plant nutrients are assumed to be available in abundance.

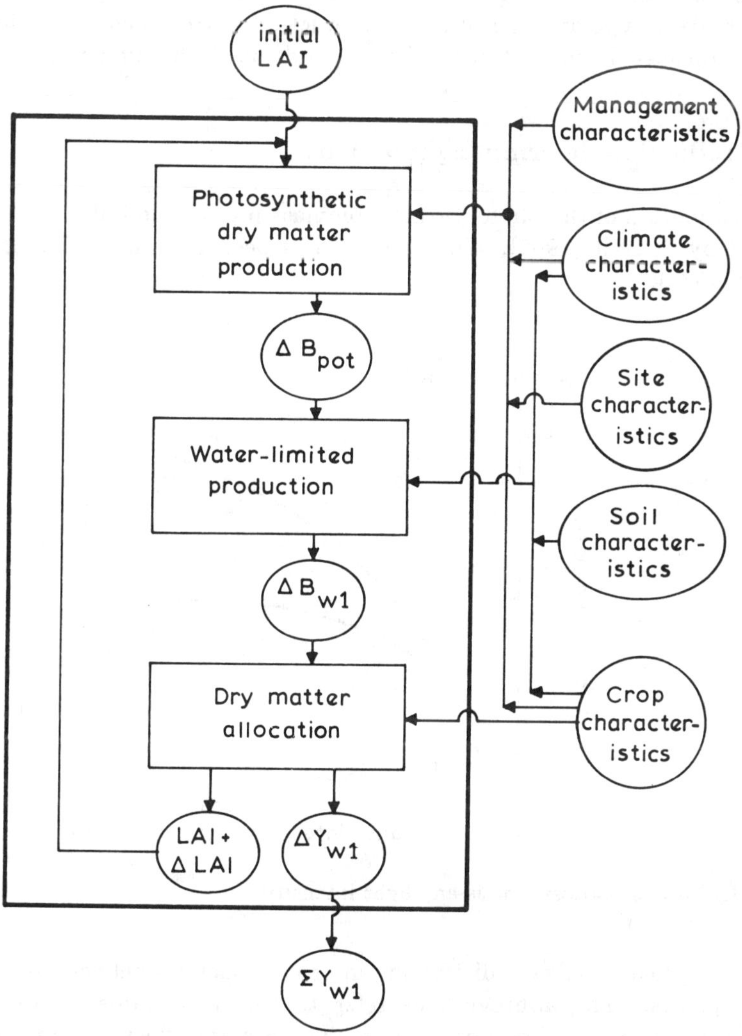

Figure 4.3. The structure of the crop production model (Konijn, 1984). LAI = Leaf Area Index.

Figure 4.3 depicts the operations of the model. First the potential biomass production (B_{pot}) is estimated as a function of leaf area based on the radiative and temperature regime. Next the effect of the soil water regimes is estimated, through computation of the water-limited biomass production (B_{wl}). Based on the crop properties, the biomass is allocated to the various plant organs (leaf, stem, root and grain) and converted into plant-specific dry matter. From this result we obtain the final product, the water-limited yield (Y_{wl}). The characteristics required at the various stages of the estimations are indicated in *Figure 4.3.* The estimations are carried out with daily time steps.

The leaf area index (LAI) develops over time from its initial value, depending on the plant type and the physical production environment. The initial leaf area is a crop specific characteristic that depends on the seeding rate.

4.3.2. Production-determining functions

The determination of the photosynthetic biomass production follows an approach developed by de Wit (1965). This method has been revised by Goudriaan and van Laar (1978).

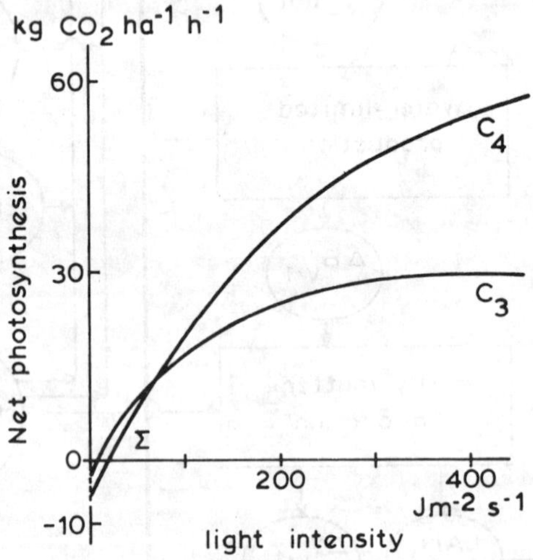

Figure 4.4. Rate of photosynthesis and light intensity.

The approach makes a distinction in crop properties between so-called C-3 and C-4 plants. The abbreviations refer to the first stable organic product created in the process of photosynthesis. These plant types respond differently to light intensity, as is illustrated in *Figure 4.4.* Under the present carbon dioxide concentration in the atmosphere, the C-4 plant is more productive at higher light intensities. The crop selected for this study, maize, belongs to the C-4 plant type. The temperature regime is another of the factors determining the dry matter production. For example, under conditions of low temperature, production by the C-4 plants will not significantly differ from that of C-3 types. As mentioned previously, night temperatures below 10 °C have been found to cause damage to the photosynthesis mechanism, because they prevent the C-4 plants from transporting the assimilates produced in the daytime period. This might be important at the higher elevations in the Kenyan highlands. In the case study area, such low temperatures can only be expected near Mt. Kenya, which is at present not used for agricultural purposes.

Constraints on the transpiration of a plant due, for example, to a lack of available water, reduce plant production, because no transpiration takes place if the stomata are closed. Furthermore, when the stomata are closed, the uptake of carbon dioxide, which is essential in photosynthesis, stagnates. The changes in water availability during the growing season are estimated by means of a water balance, which may be expressed as follows:

$$S_{t+\Delta t} = S_t + P_{\Delta t} - R_{\Delta t} + I_{\Delta t} - ET_{\Delta t} \tag{4.1}$$

The soil moisture content at the beginning of a time step (S_t) may increase by rainfall (P) or irrigation (I). The soil moisture content will decrease as a result of evaporation and transpiration (ET), and runoff (R). Capillary soil water rise has been ignored, because in the area under study the groundwater level is too deep to contribute water to the rooting zone. Under these arid conditions, deep percolation can also be ignored.

Although the rainfall data are required on a decadal base, i.e., for 10-day periods, the model generates daily rainfall values internally, based on frequency distributions and numbers of days with rainfall that are derived from local rainfall observations. A proportion of the rainfall will become runoff depending on the infiltration rate, which in turn depends on the actual condition at the soil surface and the soil moisture content. The relation between the rainfall and the overland flow is estimated using the curve number equation of the US Soil Conservation Service (1972). Curve numbers express the conditions at the soil surface. The values, which change with the kind of land use (crop) and the infiltration rate, have been assembled from experiments that covered a wide range of soil surface conditions.

To estimate the actual evapotranspiration, potential evapotranspiration was calculated using the equation of Penman (1948), and assuming the reflection of a green surface (Monteith, 1973). Of course, evapotranspiration often does not reach the potential level because the crop cannot fulfill the atmospheric demand for water, or because insufficient water is available in the soil. An equation expresses the sensitivity of the crop to the evaporative demand of the atmosphere and the available water in the soil (FAO, 1977).

If the actual transpiration is lower than the potential, a reduction in the dry matter production can be expected. The reduction can be expressed as follows (de Wit, 1958):

$$\frac{\Delta B_{wl}}{\Delta B_{pot}} = \frac{ET_{wl}}{ET_{pot}} \tag{4.2}$$

where

ΔB_{pot} = increase in potential biomass

ΔB_{wl} = increase in water-limited biomass
ET_{wl} = actual evapotranspiration
ET_{pot} = potential evapotranspiration

Once the water-stressed biomass (B_{wl}) has been estimated, it is allocated over the various plant organs. The allocation is determined by the growth stage of the crop. Unless modified according to temperature conditions and/or day-length, the allocation follows a fixed pattern for specific crops or varieties of crop.

During growth, plants need to respire a part of the produced assimilates in order to keep their metabolisms functioning. The maintenance respiration is temperature-dependent, and at 25 °C amounts to about 1.5% of the total accumulated living dry matter.

Another process, growth respiration, converts the biomass into plant material. Crops with a high protein content, such as legumes, produce less material than, for example, coarse grains, such as maize. Penning de Vries (1975) determined the growth respiration for different plant organic compounds. The effect of the maintenance and growth respiration can be expressed as follows:

$$\Delta Y_n = CV\,(\Delta Y_g - mr \cdot Y) \tag{4.3}$$

where

ΔY_n = gain in net dry matter
ΔY_g = gain in gross dry matter
Y = standing dry matter
CV = conversion factor
mr = maintenance respiration

The part of the dry matter that is allocated to the leaves will extend the leaf area, which increases the possible CO_2 assimilation for the next time-step calculation. The calculations produce results similar to those depicted in *Figure 4.5*, which are from a model run performed for maize. The dynamics of the allocation to the various plant organs is clearly demonstrated. In this example, by harvest time the harvest index, which expresses the grain as a proportion of the above-the-ground dry matter, was about 0.4.

4.3.3. Data requirements and data processing

The crop production model calculates yields with daily time steps. Although daily climatic data would serve well, it was decided to utilize the climatic characteristics by 10-day intervals, as they are more frequently available on that basis,

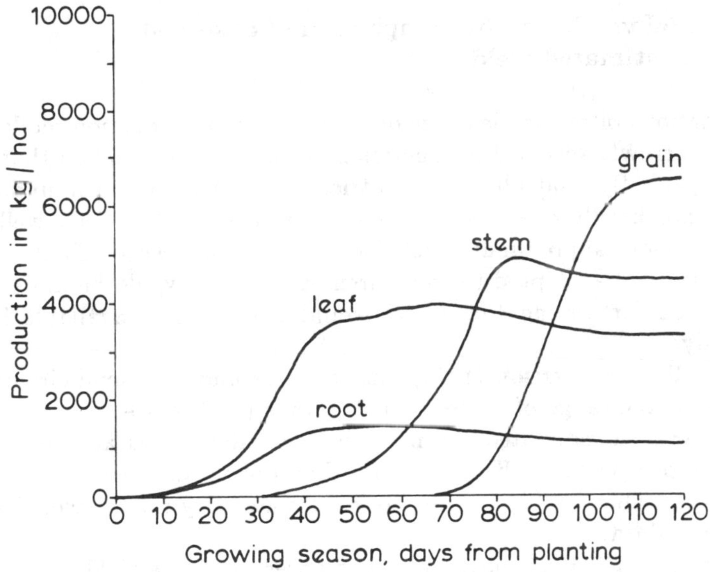

Figure 4.5. Example of production for maize according to the crop production model.

often as standard meteorological measurements. Such data still allow the model to respond to interseasonal fluctuations in the climate.

As has been pointed out, the crop production model can only function when the appropriate climatological parameters (temperature, windspeed, relative humidity, global radiation and rainfall) are all available. As noted in Subsection 4.2.1 above, there are three climatological stations in the study area where at present all these parameters are measured: Embu Mwea, Machakos Dam and Kampi ya Mawe. The first two stations have records that go back to 1963. Machakos Dam, in particular, has a good series of data. Embu Mwea has an independent rainfall record, and Kampi ya Mawe lacks windspeed measurements.

However, there is a quite dense network of stations with rainfall measurements. Since the other parameters can be shown to apply to rather large areas and play only a minor role in crop production, then by combining data from the three mentioned stations with rainfall data from other stations, a more detailed study of the crop production patterns in different years can be carried out. Thus, the model was run for the four reference stations discussed in Section 2 – Kerugoya, Embu, Machakos and Makindu.

In order to be able to select the 10% and 90% seasons out of the series of rainfall years for the four reference stations (Subsection 4.4), cumulative probability curves have been calculated.

With respect to the soil characteristics, the soil moisture retention curve has been derived from information provided by the Kenya Soil Survey (Subsection 4.2.2).

4.3.4. Model validation by comparison of observed and estimated yields

The estimations of maize yields made with the crop production model are based on rather narrowly specified production situations. It is assumed that weeds and pests are perfectly controlled and nutrients are available in abundance. Such situations are hardly ever encountered in the real world. Only well-controlled field experiments approximate such ideal conditions. Although the model to a large extent can be adapted to local circumstances, it would be unlikely that the data required for testing the model would match those available in the area under study.

According to Harrison (1970), maize yields under optimal circumstances in Kenya fall in the range of 10 to 12 metric tons per hectare. With the crop production model we estimated the maximum possible yields at about 9.5 tons of dry matter per hectare. If we assume that the yields mentioned by Harrison would contain about 15% moisture, our estimated dry matter would correspond to Harrison's data.

Bakhtri *et al.* (1984) have published observed maize yields for five consecutive growing periods – namely, the long and short rain periods of 1981 and 1982, and the long rain growing period of 1983. These data were collected in the central part of the Machakos district from seven farmers. The crop production model was run for the same growing seasons, using the rainfall data for Machakos D.O. and Makindu. It was assumed that in all cases the initial soil moisture content was at a favorable level, although in practice this would be possible only after a period of water conservation with no preceding crop. The water-holding properties of the soil are as indicated by ***Figure 4.2***.

Although the observed yields are much smaller than the estimated yields, for both rainfall stations there is a good correlation between the observed and estimated values (***Figure 4.6***). The Machakos D.O. data are somewhat better related, probably because the climatological station there is located closer to the farmer's fields where the yield data were obtained. The large, but not unexpected gap between the levels of the yields is partly due to the low level of technology among the smallholders in the case study area. The assumptions made regarding initial soil moisture content and the water-holding capacity of the soil may have contributed to the difference as well. Njeru (1984) mentions that observed yields in the area are about 15% of the maximum possible, and this seems to be in line with the ratios between the observed data published by Bakhtri *et al.* (1984) and those estimated by the crop model here. Also, as mentioned above, the maximum yields from this model seem in agreement with the maxima quoted by Harrison (1970).

4.4. Results

The crop production model described in Subsection 4.3 was not developed specifically for the Kenya semi-arid case study. Nevertheless, it proved to be applicable, mainly because the required data were available locally. The quality

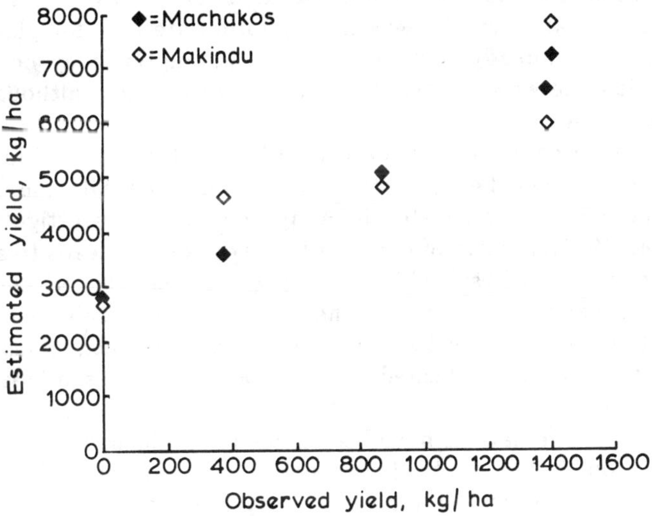

Figure 4.6. Relation between observed and estimated maize yield.

of the results depends mainly on the general applicability of the relationships used for determining production. Most of these relationships are based on sound principles that make them generally applicable (de Wit, 1958 and 1965). But, the parameterization, and checks on the parameters employed, may limit their use, because the necessary local experimental data are not always available.

As has been demonstrated (Subsection 4.3.4), observed and estimated yields can differ greatly because of inadequate parameterization and because of the effects of factors other than the radiative, temperature and soil water regimes. In spite of the large differences between measured and observed yields, water stress seems to be a major factor in explaining yield differences between consecutive growing seasons. However, more work needs to be done on validation, preferably with the help of field experiments specifically designed to obtain data for production situations that can be handled by the crop production model. How smallholders deal with the constraints under which they operate (Section 6) would need to be taken into account.

In view of the fact that rainfall is unequally distributed through the year, the planting times chosen by farmers are closely related to the times of onset of the rainfall seasons. The start and end of the rainfall season is determined by the movement of the intertropical convergence zone (ITCZ) across the region. Accurate determinations of the beginning and end of the season, or prediction of the amount of rainfall, are not possible at present. A convective rainstorm can be confused with a start of the rainy season and lead to an untimely planting of the crop resulting in serious yield losses.

Comparing the length of the crop growth cycle to the average length of the growing season indicates that late planting will frequently lead to drought stress at the end of the growing season. Thus, timely planting is very important.

However, traditional methods of planting are very time-consuming: it takes about 85 man-hours to plant a hectare (van Heemst *et al.*, 1981). In our calculations we assume that when the seasonally appropriate time for planting arrives, land preparation is already finished; but this assumption may not be justified. Likewise, inadequate weed control limits actual production, although this is not considered in the model.

Yields have been determined according to the 10-day period in which planting took place, so that the effect of planting time on yields can be examined. This effect has been demonstrated in Kenya by various investigators (Dowker, 1964; Goldson, 1963); a delay of planting by 10 days easily leads to a yield loss of 20%. The question arises: which planting time should be chosen for the scenarios? For validation purposes, the planting time resulting in the highest water-limited yield was selected, since the most common practice was not known. The optimal time indicated by this procedure was usually just before the rainy season starts.

The scenario simulations have been made for initial soil moisture tension of 500 mbar suction, i.e., a relatively moist condition. High initial soil moisture tensions would lead to much lower yields. Dry farming is a technique that aims at conservation of soil water; the soil is kept fallow for a period long enough to accumulate the required amount of water. The amount of water that may be stored depends on characteristics of the soil. The storage capacity used for these calculations is for a deep soil with a clay texture, having a storage capacity of 0.95 mm water per centimeter of soil (*Figure 4.2*). This is a typical soil for the highlands (Kerugoya and Embu), while Machakos and Makindu have much shallower soils with lower storage capacities.

4.4.1. Comparing the estimated yields with the r/Eo results

A comparison was made between the yields estimated with the crop production model and the agroclimatic index, r/Eo (r = rainfall, Eo = potential evaporation). The determination of r/Eo was described in detail in Section 3. The estimations cover the selected historical episodes of 1974, 1975, 1976, 1977 and 1984, introduced in Subsection 2.4, and the values have been plotted against each other in *Figure 4.7*. A clear relation between the two sets of results exists, which is not unexpected since rainfall and evapotranspiration play an important role in both methods.

Of course, the two sets of results are not in complete agreement. Differences result from the fact that the crop production model takes the within-season fluctuations of the various meteorological parameters into account, while the r/Eo index works with the total seasonal rainfall.

We may conclude that the r/Eo index is a useful, though somewhat crude, method for providing easily calculated estimates for characterizing the degree of drought. Which method is best depends entirely on the degree of sophistication required.

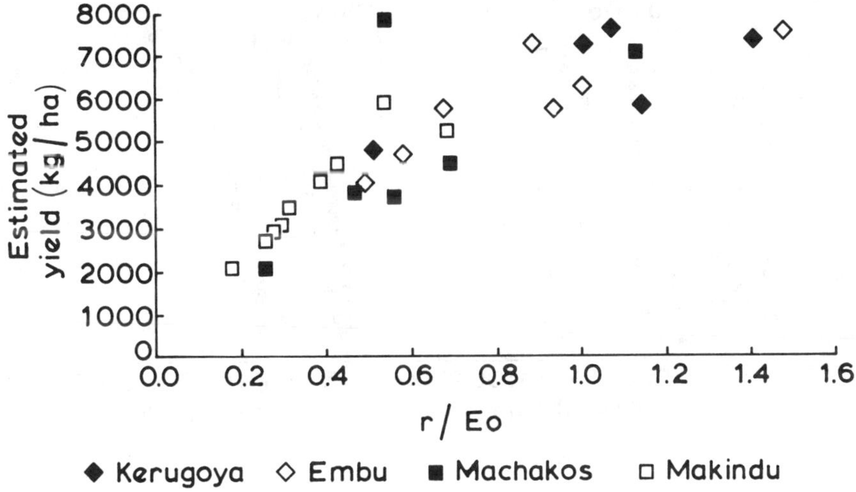

Figure 4.7. Relation between estimated yields by crop production model and the r/Eo ratio.

4.4.2. Results of the scenarios for the four reference stations

Data for the four stations representing different agroclimatic zones (Kerugoya, Embu, Machakos and Makindu) were used in analyzing the different climatic scenarios.

The 10% and 90% scenarios

The rainfall values selected for the calculations for these scenarios were obtained from cumulative probability analyses (curves are shown in *Figure 4.8*). Note that, compared with other sections in this study, the convention for selecting percentiles is *reversed* – i.e., the 10% scenario indicates a moderate drought while the 90% scenario denotes a "good" rainfall season. The yields as estimated by the model, and the estimated yields reduced to the levels that can be expected at the present level of technology in the area (*Figure 4.9*), correspond clearly to the amount of rainfall. However, the total seasonal rainfall (shown at top of each of the yield estimations in *Figure 4.9*) only partially explains the differences in estimated yields between the 10% and 90% scenarios. The rainfall distribution within the season also has an important influence on these model estimates.

The driest location (Makindu) shows the largest relative variability in yields. The differences between the 10% and 90% scenario results are pronounced for both seasons. The yields for the wetter stations, Kerugoya and Embu D.O., fluctuate much less. The yields related to the 10% seasonal rainfall value become lower, going from north to south in the case study area. The picture for the yields related to the 90% rainfall is more irregular.

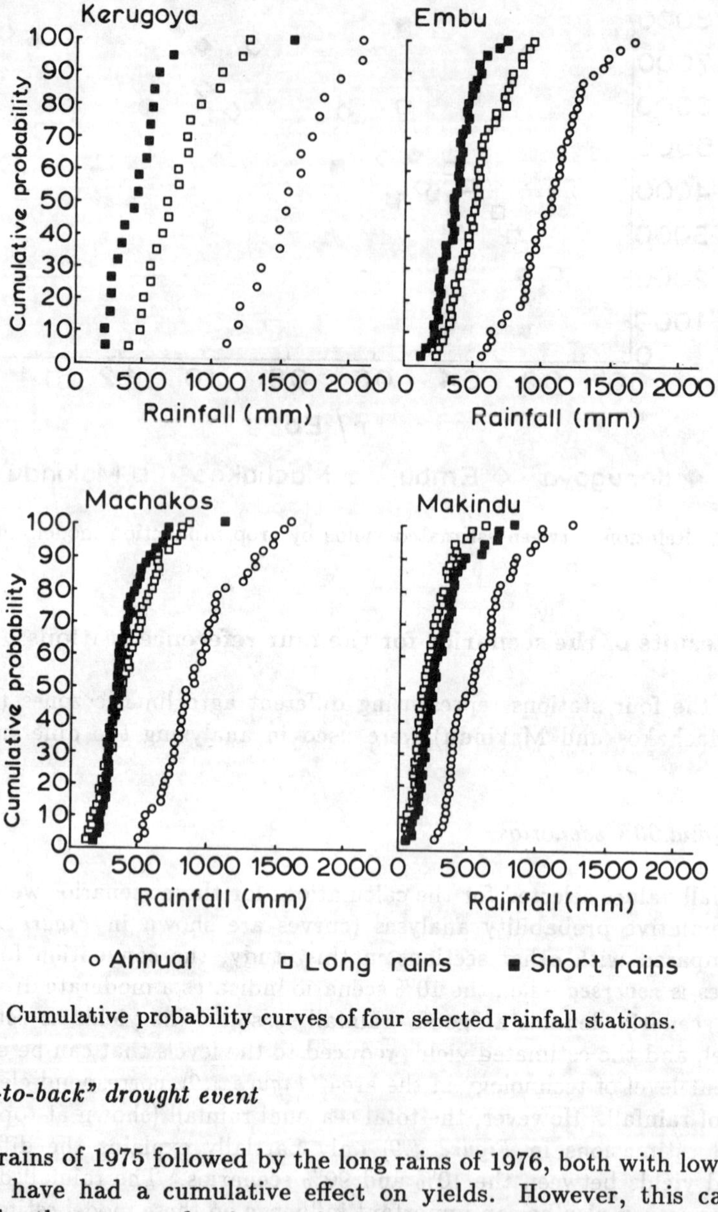

Figure 4.8. Cumulative probability curves of four selected rainfall stations.

The "back-to-back" drought event

The short rains of 1975 followed by the long rains of 1976, both with low rainfall, could well have had a cumulative effect on yields. However, this cannot be estimated by the crop production model because the initial soil moisture tension is always assumed to be 500 mbars.

It is difficult to give a clear quantitative picture of the effect of different initial soil moisture tensions, for a change in the initial soil moisture content leads to a change in the optimal planting time. If the initial soil moisture tension is higher, the optimal planting time is later, and yields would fall much faster with any delay of planting after the optimum time. A back-to-back drought event might have an additional impact on the smallholder during the second season of

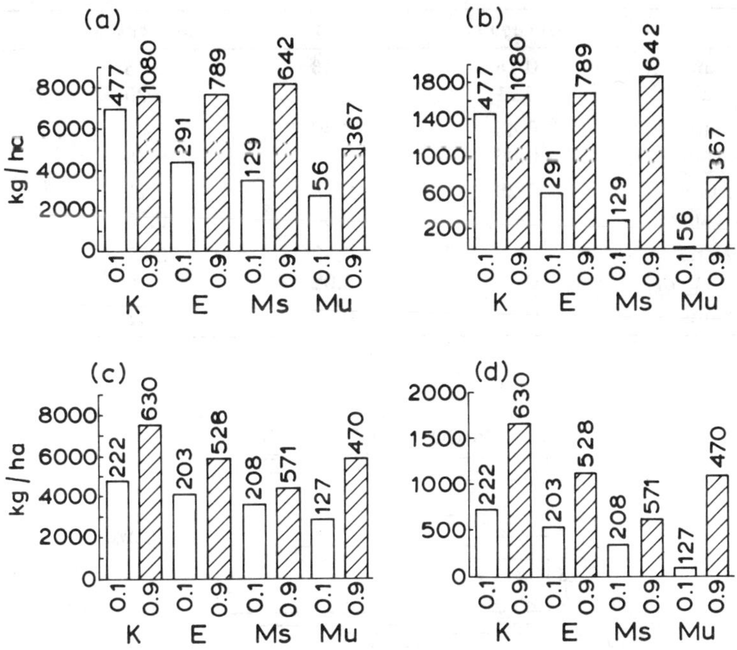

Figure 4.9. Maize yields (kg/ha) for the 10% and 90% seasonal rainfall years. (*a*) and (*c*) Model results for yield with restrictions from climatic variables only. (*b*) and (*d*) Yields reduced to expected levels given low level of technology. K = Kerugoya; E = Embu; Ms = Machakos; Mu = Makindu. Note: Figures above bars are seasonal rainfall total (mm): Long rains [(*a*) and (*b*)], short rains [(*c*) and (*d*)].

the drought, if the first season had already resulted in such low yields that little seed is left for the next growing season. This is a common situation in areas with dominantly subsistence farming.

The selected historical seasons

The growing seasons of the years 1974, 1975, 1976, 1977 and 1984 have been described in Subsection 2.4. The cumulative probability of seasonal rainfall (*Figure 4.7*) is shown in *Table 4.1* for each of these cases. The spatial variability of those probabilities is quite evident. No direct conclusion about the severity of the moisture stress can be drawn from these values, since the level of rainfall is quite different for the four stations. For example, the amount of rainfall during the long rains in Kerugoya is much higher than during the short rains. Consequently, the estimated yield for the long rains in 1974 is much higher than for the short rains, although the cumulative probability for the long rains in 1974 has the lowest value. The short rains bring relatively more rain in the south than the long rains, but the total amount during the short rains is quite low (e.g., Makindu).

Table 4.1. Cumulative probability for selected historical seasons.

Season	*Kerugoya*	*Embu*	*Machakos*	*Makindu*
1974 long rains	0.15	0.43	0.75	0.63
1974 short rains	0.21	0.27	0.39	0.20
1975 long rains	0.45	0.40	0.29	0.39
1975 short rains	0.05	0.12	0.15	0.24
1976 long rains	0.25	0.28	0.21	0.31
1976 short rains	–	0.50	0.55	0.41
1977 long rains	–	0.80	0.67	0.75
1977 short rains	0.74	0.92	–	0.80
1984 long rains	0.04	0.07	0.48	0.03

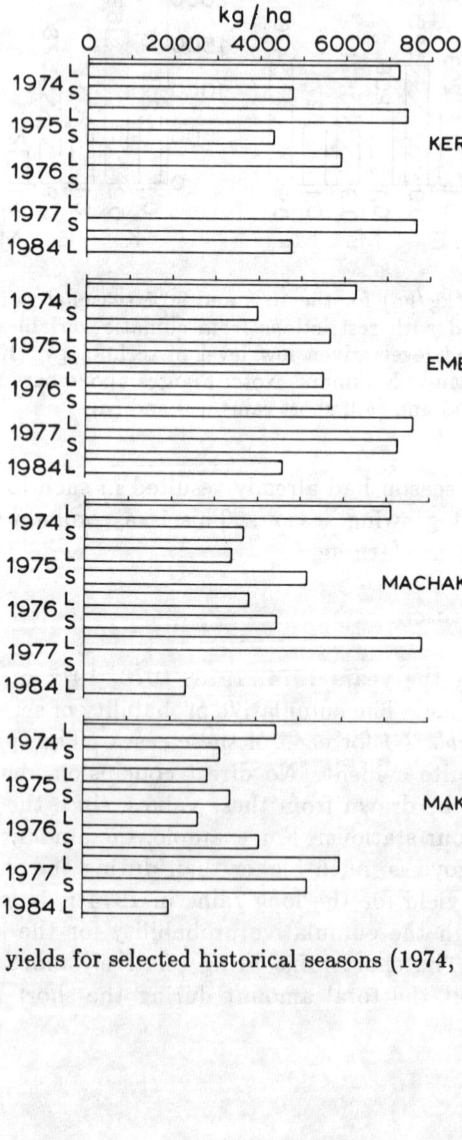

Figure 4.10. Maize yields for selected historical seasons (1974, 1975, 1976, 1977, 1984).

The yield is to a large extent determined by the total amount of rainfall for the season. However, the distribution of the rainfall through the growing season has a significant effect; rainstorms of great intensity lead to water loss by overland flow, whereas the same amount spread over a longer period allows more infiltration. Consequently, different yields are often obtained with the same amount of seasonal rainfall.

The maize yields estimated by the model for the selected seasons are presented in *Figure 4.10.* The socioeconomic implications of these and other seasons are considered in some detail in Section 6. *Figure 4.10* illustrates that crop yields not only fluctuate from season to season, but they may also exhibit large spatial variations. While the results support, on the whole, the characterization of the selected scenario years presented in Section 2, the four locations do not show perfect uniformity in their responses. Even during the 1984 long rains season, with exceptional and widespread drought that must have affected production considerably, the conditions were not unprecedented for this century at the two more northerly stations, Kerugoya and Embu.

Although the model results support the qualitative descriptions of the selected scenario seasons in Section 2, a complete spatial picture cannot be given on the basis of the four reference stations. To obtain such a picture would require a more comprehensive study involving information on the extent and location of the area under maize and the spatial pattern of rainfall and other climatic and edaphic elements.

4.5. Conclusions

The application of a dynamic crop production model to data from the case study area showed a very large gap between the measured and the estimated yields. Although the validation that was carried out had some serious limitations, since parameterization was possible only to a limited extent, the model nevertheless provides a clear indication that water availability is quite important and in certain years can lead to a complete crop loss. The gap between the measured and estimated yields appears to be due mainly to the lack of mechanical equipment to carry out timely weed control and the limited availability of fertilizers. Both can be attributed to the lack of capital. These and other factors affecting the vulnerability of smallholders to drought are considered in further detail in Section 6.

SECTION 5

The Effects on Livestock Production

5.1. Background

The major climatic, edaphic and present social and agricultural characteristics of the study area have been described in Section 1. With regard to animal production, the area can be divided roughly into four enterprise zones, which closely follow the agroclimatic potential zonation depicted in *Figure 1.3* (Jaetzold and Schmidt, 1983; Pratt *et al.* 1966; Sombroek *et al.* 1982). For each of the four different enterprise zones, the sample sites in *Figure 1.3* will be considered in relation to the climatic scenarios selected in Section 2.

For this analysis relating to livestock production, agroclimatic zone I, which is primarily forest reserve, was not included. *Agroclimatic zone II* is of only limited extent within the study area, located on the southeastern slopes of Mt. Kenya, between 1600 and 2000 m altitude. This area includes some of the best coffee-growing conditions in the country, so that great emphasis is placed on arable cropping, particularly with coffee, for cash generation. Farm size is generally small, between 1 and 6 ha. The most commonly found livestock enterprise is dairying, with an average of 3.9 livestock units per holding, on about 1 ha of grazing/forage (Jaetzold and Schmidt, 1983; Nkanata, 1985). The animals are typically Ayrshire or Friesian grade animals, that is, animals that have been derived from the original Zebu stock, improved by cross-breeding with exotic animals. Government policy has been to encourage this upgrading process by providing a heavily subsidized artificial insemination service, one insemination costing the farmer *ca.* US$0.10. Increasingly, as farm sizes are reduced through subdivision, the animals are kept on a more or less zero-grazing or stall-feeding system based on Napier grass (*Pennisetum purpureum*) as the main forage source. Kerugoya is the chosen sample site for this agroclimatic zone.

Agroclimatic zone III is represented mainly in the areas of between 1500 and 1800 m altitude, with an annual average rainfall of over 900 mm. As in agroclimatic zone II, the pressure on land is very high and farm size is typically between 2 and 5 ha. The cropping pattern and livestock systems are similar to those of zone II, but the somewhat lower average annual rainfall, together with greater variability of rainfall amount and seasonal distribution, result in greater climate-related risks for the farmers in zone III. Embu is the chosen sample site for this agroclimatic zone.

Agroclimatic zone IV represents the marginal areas situated between the traditional cropping (wetter) zones and the drier zones that have not been considered suitable for crop production in the past. The pressure on land in this zone is generally not quite so high as in the wetter zone III, so farm size is somewhat larger, between 4 and 12 ha, although the pressure is increasing as the higher potential areas fill up. The holdings in this zone typically have up to a maximum of 3 ha of arable crops, normally maize intercropped with a legume, such as pigeon peas or beans. This upper limit of 3 ha appears to be the maximum area that can reasonably be prepared, planted and weeded using presently available animal- or hand-powered cultivation resources within the short growing season. No appreciable opportunities exist at present for cash cropping, due to the lack of a crop suited to the rather variable rainfall. Low crop yields, or even crop failures, are common as the effects of poor distribution and low total rainfall are confounded with the use of cropping technologies originating in, and more suited to, the wetter agroclimatic zones. Cattle are kept for subsistence production of milk and for draught power for arable cropping. The cattle are predominantly local Zebu type, the number of grade-type animals being limited not only by lack of suitable breeding stock and absence of artificial insemination services in the area, but also by farmers' recognition of the greater general hardiness of the local stock in the prevailing conditions. Small stock are common, both hair sheep and goats, owing not only to their ability to obtain their dietary requirements even in dry conditions, but also to their convenient role as a cash-producing resource (Pollard, 1981; Rukandema *et al.*, 1981). Although cattle also have some role as a source of realizable cash, the usually much greater value of the cow means that sale of cattle are restricted, as much as possible, to major occasions, such as marriages. The role of livestock in this zone is particularly important as the generally low, variable yields from the arable cropping result in little opportunity for generating income from the crop enterprise. The sample site chosen for this zone is Katumani, 10 km south of Machakos Town, due to the availability of relevant field data.

Agroclimatic zone V covers the greater part of the study area. The low, highly variable rainfall has precluded successful arable cropping to any significant level so that livestock production has been, and is likely to remain, the most important type of land-use. The traditional land-use system in this area was nomadic pastoralism. A combination of land pressure from an increasing human and livestock population, together with a desire by the government to encourage better land management through land title registration, now limits mobility as a strategy to reduce the effect of shortages of feed and water (Potter, 1981). Livestock in this zone were traditionally used to supply the subsistence requirements of the pastoralists, with milk being the single most important product. In spite of the central role of milk in the production system in this agroclimatic zone, climatic effects on levels of milk output and also on the breeding process have not yet been studied for a period sufficient for adequate evaluation of any possible production model. In the present study the effects of climatic change on livestock production in agroclimatic zone V will therefore be related to the growth of animals destined to be slaughtered for beef, as research data are available from a study involving more than 10 years of field data (KARI, 1985). This approach also appears appropriate when it is realized that beef production is developing as a major source of livelihood on the larger holdings as the cash economy is expanded. Considerable immigration into the area by settlers unable to obtain land in higher rainfall areas has occurred in recent years. Sales by land speculators, or sub-division of cooperatively owned units into individual holdings, has led to the emergence of considerable numbers of small (<10 ha) holdings in agroclimatic zone V. The difficulties mentioned above for the mixed farms in zone IV are magnified for the smallholders in the drier zone V. Makindu is the sample site for agroclimatic zone V.

5.2. Site Details

Throughout the study area, the climatic component most influencing herbage growth, and therefore potential for animal production, is rainfall. Because of the moderate altitude, temperatures are not normally limiting. Problems have been encountered in the analysis of the rainfall data for the area. The most commonly used descriptor for rainfall is the mean, perhaps qualified by a standard error. Particularly where the rainfall is low, the distribution of annual totals may be skewed, rather than normal, so the mean and standard deviation may not be the most appropriate statistics to use (Norton-Griffiths, 1977).

For the purposes of this section, it was necessary to have a probability function for rainfall – in this case, annual rainfall. The cumulative probability curve (Te Chow, 1964) has therefore been used for expressing the characteristics of the selected scenarios. The rainfall parameters chosen to characterize the sample sites are the median and percentiles derived from the cumulative probability curves for the period of instrumental record. The median is much less affected than the mean by the occasional anomalous year with very heavy rainfall.

A further point regarding the rainfall records in the area relates to the spatial distribution of the rainfall gauges. Although the network of nearly 40 rain gauges within the study area represents a level of coverage that is relatively good for a developing country, care is required in interpretation of the instrumental records, in the light of the representativeness of the distribution of the gauges. An examination of the sites indicates that those with the longest periods of record are those nearest to centers of permanent settlement. These settlements may well have been founded because of specially favorable conditions – for example, permanent water supply or availability of fuel wood – which may have been the result of locally favorable climatic conditions. Fuelwood is found most commonly in the more hilly areas, which may cause topographic effects on rainfall. It may therefore be inferred that the available instrumental records will exhibit an upward bias for rainfall, compared to the "true" value, especially in the drier parts of the study area in zone V.

Some of the significant agricultural characteristics of the locations chosen to represent the agroclimatic zones are presented below. These characteristics are based on the most recently available research data (Jaetzold and Schmidt, 1983; Pollard, 1981; Rukandema *et al.*, 1981; and Van der Valk, 1985). Rainfall station numbers refer to the list included in Section 1, *Figure 1.5*.

(1) Kerugoya: Rainfall station No. 3. Agroclimatic zone II. Median rainfall (1951–1980) 1525 mm. Smallholder mixed cropping area. Farm size 2.8 ha (1 ha forage and grazing); 3.9 livestock units of 400 kg grade animal.

(2) Embu: Rainfall station No. 8. Agroclimatic zone III. Median rainfall (1951–1980) 1055 mm. Smallholder mixed cropping area. Farm size 3 ha (1/3 farm size forage and grazing); 2.5 livestock units of 400 kg grade animal.

(3) Katumani: Rainfall station No. 21. Agroclimatic zone IV. Median rainfall (1956–1982) 660 mm. Marginal smallholder mixed cropping area. Farm size 5 ha (1.5 ha forage and grazing); 8 livestock units of 300 kg Zebu animal.

(4) Makindu: Rainfall station No. 32. Agroclimatic zone V. Median rainfall (1951–1980) 550 mm. Pastoral area. Farm size 100 ha (all natural grazing/browse); stocking rate 3 ha per 300 kg Zebu animal.

The agroclimatic zone allocation takes into account the diminishing effectiveness of rainfall as mean temperature increases with reduction in altitude from northwest to southeast within the study area.

5.3. The Climatic Scenarios

Figure 5.1 indicates the pattern of annual rainfall cumulative probability for the four sites for the years 1951–1980. The curves have been smoothed by the least-squares method.

The scenarios considered were as follows:

(a) "Good" year rainfall = that exceeded in only 10% of the years.
(b) "Average" year rainfall = the median as derived from the cumulative probability curve.
(c) "Poor year" rainfall = that exceeded in 90% of the years.
(d) "Disaster" year = the lowest rainfall year during the instrumental record, and corresponds to that exceeded in about 98% of the years.
(e) "Back-to-back" scenario = the worst two consecutive years in the record period.

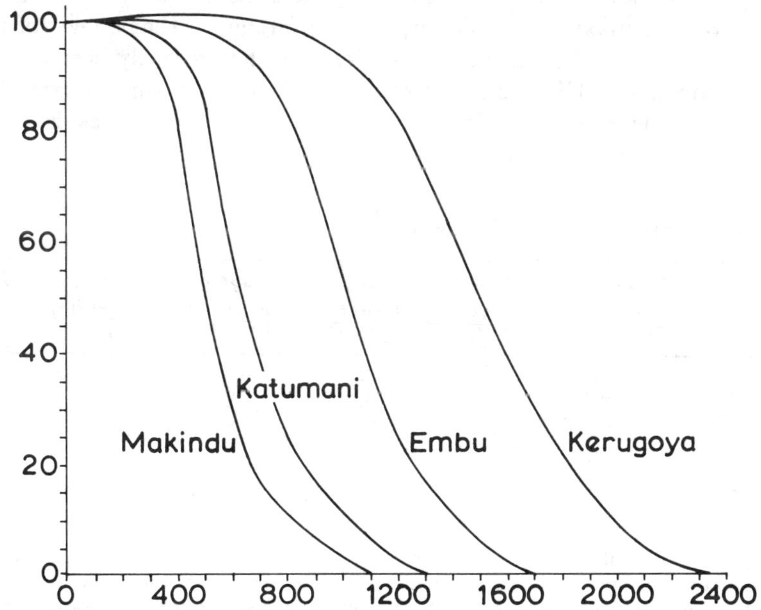

Figure 5.1. Cumulative probability curves of rainfall for reference stations.

5.4. Results

The results are presented in terms of the production of Napier grass forage and its conversion to animal product (milk) based on the available data for the crop in Kenya (Jaetzold and Schmidt, 1983; Karanja, 1985; KARI, 1983; KARI, 1984; Potter and Anindo, 1985; and Wouters, 1984). For the purposes of the present study, calculations are based on the total annual forage production, although it is recognized that in practice the variation in the supply of forage from year to year may be lower than that indicated by the growing conditions, because of the effect of the cutting system used.

5.4.1. Kerugoya – agroclimatic zone II

Table 5.1 indicates the results estimated for Kerugoya for the five scenarios. The farm parameters used are those obtained by the most recent survey in the area, with the production resources and methods at the "low" level, as in over 75% of existing farms (Jaetzold and Schmidt, 1983). Local feeding trials indicate that for the grade cow a daily intake of 14 kg of Napier forage dry matter is attainable, supporting a milk yield of 10–11 liters of milk per day, without recourse to concentrates (KARI, 1983; Potter and Anindo, 1985). This milk output is considerably higher than the 5–7 liters per day recorded on farms typical of this agroclimatic zone (Stotz, 1983). Even in the average scenario (median) year, with the present level of stocking the forage available is only about 80% of the 14 kg required for maximum milk output from forage, so that a level of milk output of only 4–5 kg can be expected. Napier grass has recently been estimated to provide as much as 80% of the diet of the stock on such small farms (Wouters, 1984), so it is clear that the feed availability for existing stock is imposing a severe restriction on milk output.

Table 5.1. Scenario results for Kerugoya.

Scenario	*Annual rainfall* (mm)	*Farm forage* (kg DM/yr)[a]	*Daily forage* (kg DM)	*Forage per livestock unit* (kg/day)	*% of potential intake*	*Estimated milk output* (kg/day)
Good	2040	19 000	52	13	95	8–10
Average	1525	13 000	36	9	65	2–4
Poor	1100	11 000	30	8	55	0–2
Disaster[b]	940	9 000	25	6	45	
Back-to-back	1300/940	c	–	–	–	–

[a]DM = dry matter.
[b]Napier grass would still survive in this scenario, but the quantity of forage available would be sufficient to provide for only the bare maintenance requirements of the stock.
[c]Although little milk production would be expected during this scenario, stock deaths would only be expected where no other feed resources (e.g., crop residues, off-farm grazing) were available.

The effects of a reduction in rainfall may therefore be predicted. Initially a reduction in forage supply will lead to a lowering of the level of milk output, to the point at which the animal will no longer be provided with more than its maintenance requirements, at which stage the animal will dry up. This would be expected to occur under the poor scenario condition. Although it is theoretically possible for the farmer to ensure continued production by purchasing feeds, the resources are unlikely to be available. The persistence of the Napier grass will not be threatened, even in the disaster scenario; but given the present stocking rates, only the bare maintenance requirements for the stock may be available. Complete failure of the food crops is unlikely, so even though the crop yields will be low, the subsistence food requirements for the farm family will still be met from within the farm. In addition, the payment system for the cash crops – for example, coffee – results in much of the income being received during the year following the delivery of the crop. This serves to cushion the effects of a bad year, so that the consequences of a single bad year are mitigated.

With the back-to-back drought scenario, again the possibility of complete food crop failure will be slight, so that farm family subsistence needs should still be met. Little, if any, cash income will be likely in the second year, as the cash crop will have provided little for sale in the first year. However, as the survival of the stock and forage should not be under threat, the ability of the farmer to make a rapid recovery after the drought is high. In this zone (II), because of the already well-developed cash economy based on cash cropping and possible milk sales, there is a reasonable provision of credit, both formally through the banking system and informally through local traders. Thus, even in the drought years food can be purchased if required.

5.4.2. Embu – agroclimatic zone III

The results of the good and average scenarios, as presented in *Table 5.2*, indicate that there are very close parallels between Embu and Kerugoya. The parallels do not extend into the drier scenarios. As the rainfall drops below the poor scenario level, the survival of Napier grass becomes doubtful, so that the major source of forage will no longer exist, resulting in stock deaths. This was observed in 1984, the worst drought year since records began.

Table 5.2. Scenario results for Embu.

Scenario	*Annual rainfall* (mm)	*Farm forage* (kg DM/yr)	*Daily forage* (kg DM)	*Forage per livestock unit* (kg/day)	*% of potential intake*	*Estimated milk output* (kg/day)
Good	1520	16 000	44	18	125	8–10
Average	1055	10 000	27	11	78	3–5
Poor	720	5 000	14	6	39	
Disaster[a]	675	2 500	7	3	20	
Back-to-back	750/675	b	–	–	–	–

[a] In this scenario Napier grass does not survive, and it is therefore likely that stock will face death by starvation in the absence of other feeds (e.g., crop residues, off-farm grazing), which would also be limited by the poor season. This actually happened in 1984, when 30–40% of all grade cattle in Eastern and Central Provinces are estimated to have died in the worst drought for 50 years (Nkanata, 1985).

[b] Such a scenario occurred in the years 1959/1960. At that time the land pressure was not as high as at present, and most of the stock were the local Zebu type with lower demands than for grade animals. A recurrence of such conditions now would be expected to result in 70–80% mortality of all grade stock on the affected farms. Considerable though the economic loss would be to the farmer (as the animal represents an asset valued at US$300–600, compared with an annual farm cash surplus of US$800–1000), even at the rainfall level in the poor scenario years there will not be a complete failure of the arable crops on the farm. Maize yields will be extremely low, but the bean crop will still give a reasonable yield, so the farmer will still be able to provide most, if not all, of his household's food requirements. Under the disaster and back-to-back scenarios, the comments regarding the role of delayed payment for the cash crop and the role of credit made for Kerugoya would also apply to Embu. As stock losses will have occurred due to the lack of Napier forage, especially in the back-to-back scenario conditions, the role of credit in assisting the replacement of stock and replanting of forage would be vital.

5.4.3. Katumani – agroclimatic zone IV

The climatic scenarios for this zone are considered for a 5 ha farm unit with 1.5 ha of on-farm grazing, which is almost invariably indigenous grassland. Although Napier grass and other forages have been tried in this zone in trials over the last 30 years, no suitable species has been found. A few species appear to show promise in plant introduction nurseries; but the establishment, grazing management and persistence studies required for thorough evaluation have not yet been carried out. Such studies will be essential to ensure that investment of limited resources in the use of specific forage or pasture materials is fully justified for the farmer. The combination of erratic rainfall and the farmer's concentration of his limited available resources on food cropping severely constrains the possibility of establishing a forage crop. Surveys have confirmed that at present off-farm grazing in communal areas provides almost all of the feed for the stock in this zone (Pollard, 1981; Rukandema *et al.*, 1981). The availability of such off-farm feed is being continually reduced through settlement of immigrants squeezed out of the wetter zones (Potter, 1982). In dry years, even at present, what remains of these off-farm resources are very heavily used, so that animal production is severely curtailed and stock deaths become common.

Table 5.3 shows the estimated consequences of the climatic scenarios in the Katumani situation. In the bimodal rainfall zone at the altitude concerned (*ca.* 1500 m), an annual total of 500 mm rainfall may be considered the lower limit for the successful growth of at least one maize crop in a year. Cropping technologies for growing maize in even drier conditions have not yet been developed to the point where they are suitable for the actual level of resources likely to be available to the farmer. In addition, other crops, such as sorghum, which would tolerate dry conditions better than maize, are unacceptable to the farmer at present, often for very practical reasons. In the case of sorghum, the availability of education to virtually all children has reduced the ability of the farmer to reduce bird damage, so that losses would reach unacceptably high levels.

Table 5.3. Scenario results for Katumani.

Scenario	*Annual rainfall* (mm)	*Farm forage* (kg DM/yr)	*Daily forage* (kg DM)	*Forage per livestock unit* (kg/day)	*% of potential intake*	*Estimated milk output* (kg/day)
Good[a]	1150	9000	25	3.1	40	0–1
Average[a]	680	2350	6.4	0.8	11	
Poor	450	900	2.5	0.3	4	
Disaster	440	750	2.1	0.25	3.4	
Back-to-back	450/480	b	–	–	–	–

[a]In these scenarios little or no milk output could be expected from on-farm grazing or forage supply. The actual level of milk output would be directly related to the availability of crop residues and off-farm grazing or roughage collection, these being highly site-specific.

[b]In this scenario the lack of on-farm grazing, as well as the lack of crop residues in seasons of crop failure, would place extreme demands on the limited off-farm feed sources. In the poor seasons of this scenario, this common feed supply would be inadequate so that mortality rates of 70–80% of stock may be anticipated.

In a moderate drought (poor scenario), the arable crops will fail completely with not even subsistence levels being achieved. Owing to the lack of forage, both on-farm and in the communal areas, very little milk output for human use will be produced. In this case the farmer would be forced to sell off stock so that food could be purchased. As the drought is likely to be extensive, the farmer is at a disadvantage in selling stock. Not only are there likely to be many animals for sale at the same time, causing prices to decline, but the animals will be in poor condition, giving low grades of meat and thus further lowering prices.

If a back-to-back drought occurs, the reduction in stock numbers through sales or deaths occurring in the first year would result in the farmer not having stock to sell in the second year, when crop failure and milk shortage recur. No relief would be available to the farmer from cash crop sales as no suitable crop has been so far identified for these marginal areas. With no other resources to provide food, the farmer would face starvation unless famine relief or some off-farm income were available. The relatively frequent occurrence of droughts in these marginal areas limits the willingness of possible credit sources to provide assistance, so that in this zone the effects of drought are exceptionally severe.

5.4.4. Makindu – agroclimatic zone V

The ever-increasing immigration into agroclimatic zone V, due to land pressure elsewhere, has created great difficulties not only in the study area but also in other zone V regions in Kenya. In less than one year in three is there enough annual rainfall (500 mm) for the production of a single maize crop in a year in this zone. In addition, high ambient temperatures and, hence, high evaporation rates at the lower altitudes of zone V mean that less of the rainfall is actually available for crop growth. Higher variability of rainfall as well as lower rainfall totals than in zone IV also add to the difficulties of arable cropping for the smallholders. Although such areas may be expected to show greater success with crops, such as sorghum or millet, which are more able than maize to mature on little rainfall, maize remains the preferred crop. Undoubtedly, much of the reason for this is social pressure relating to to maize as a "modern" crop. Such factors as shortage of labor for bird-scaring in small grain crops and the greater processing and cooking time requirements are among the more practical constraints limiting the use of alternatives to maize. Until technologies are available for the reliable production of food crops in such poor rainfall areas, frequent crop failures appear to be inevitable.

Past usage of the agroclimatic zone V areas suggests that livestock may be suited to the prevailing conditions better than crops. However, there has been pressure to increase the output of animal product per unit area, to provide for the subsistence requirements of a growing human population as well as to generate potentially saleable surpluses. To date, the evaluation of high-quality forages for use in this zone has not progressed beyond the plant introduction nursery stage, so that livestock have only the indigenous vegetation to rely on for feed. Lactation is the most nutritionally demanding aspect of livestock production (Williamson and Payne, 1965), and lack of high-quality fodder limits the potential milk production of animals in this zone to the 1–2 liters per day recorded (Allan, 1965; Semenye, 1981).

The past requirement for large numbers of such low-producing animals to supply the subsistence needs for the household becomes less practicable in a settled pastoral system, as the feed and water supplies vary in a highly seasonal manner, and extensive grazing areas are no longer available to provide a buffer against the effects of this variation. A change to a system of purchase of human food, financed by the sale of large and small stock for meat, has therefore developed in this zone (White and Meadows, 1981), in some areas where the holdings are of suitable size (>100 ha).

The Makindu site is therefore considered in relation to the production of liveweight, and therefore meat, as affected by rainfall. Field experiments carried out at the Kenya Agricultural Research Institute over the last 20 years have indicated that animal performance may be more closely related to the *duration* of the rainfall seasons rather than to the overall *quantity* of rain. This is because dietary intake quantity and efficiency of feed conversion into liveweight have been shown to be directly correlated with the nutritive value of the feed (Potter, 1985). Nutritive value or feed quality is related to the presence or, more accurately, duration of rainfall, whereas the quantity of feed is related to the total amount of rain.

Table 5.4. Scenario results for Makindu.

Scenario	*No. of growth periods*[a]	*Liveweight per head*	*Liveweight gain* (kg/year/head)
Good	12	170	57
Average	9	137	46
Poor	6	104	35
Disaster	5	93	31
Back-to-back[b]	5/6	93	31 + (?)

[a]Growth period = 10 days with more than 20 mm rainfall.
[b]Performance in the second year will depend on the quantity of forage available at the end of the first year, as little new growth will be produced in the second year. During the second year, effective diet selection will become progressively less possible, and stock deaths may be expected as the animals become unable to obtain even maintenance requirements.

Table 5.4 indicates the effect of the climatic scenarios on potential liveweight gains of Zebu-type cattle where the rainfall is considered in terms of periods of rainfall in the year. A growth period is considered to be 10 days in which more than 20 mm of rain falls. A regression equation derived from more than 10 years of relevant field data is used to estimate the liveweight gain (KARI, 1985):

$$Y = 11.04a + 37.53 \qquad (5.1)$$

where a is the number of growth periods in the year, 37.53 is a constant representing the growth of the animal even in a completely dry year (made possible by selection within the available dry sward of the more nutritive fraction), and Y is the annual liveweight gain in kg per animal. The gain per animal appears to be relatively independent of stocking rate over quite a large range of stocking rates, as it depends on the quality of vegetation rather than quantity.

However, the amount of herbage will be important for the determination of the efficiency of diet selection.

In the present case, this means that the impact of a back-to-back drought would depend on the status of the sward at the onset of the two years concerned. The amount of standing forage would depend on the past rainfall and stocking rate. For a particular producer, the stocking rate would be set for the individual circumstances according to the risk level adopted. The higher the stocking rate, the greater the production in terms of meat per unit area in the good years, but the greater the likelihood of herbage shortage in dry year (Potter, 1981). Overall resource availability level would be a highly holding-specific characteristic. This makes it rather difficult to make general statements for production per holding in this zone. As an indication of the production possibilities under practical field conditions, results from the KARI field station in agroclimatic zone V indicate that with a stocking rate of 2–3 ha/head for cross-bred steers weighing an average of 300 kg, over a 10-year period the average annual liveweight gain was approximately 130 kg/head, representing about 55 kg meat per hectare per year in an average rainfall of 560 mm. During the 10-year period, the annual liveweight gain varied from 90 to 165 kg per head associated with annual rainfalls of 250 and 820 mm (KARI, 1985).

It has become clear that the meat production system has considerable ability to deal with the vagaries of the climate in zone V, so the producers in this zone can make a living even in the difficult conditions. This flexibility is increased when small stock are present. A major problem with the beef production system is that it can provide a livelihood for only a low human population density, especially with the present low value of meat (<US$1.00 per kg), so that smallholder units of about 10 ha or less are not able to adopt it. With the difficulties noted above for other possible production systems on smallholdings in the drier areas, the future for such holdings, definitely in the short term and probably in the medium to long term, looks bleak.

5.5. Conclusions

Of the four enterprise systems considered in this study, it is the Katumani situation (zone IV) that appears most sensitive to climatic change. Cropping technologies that would ensure success in food production even in the drier years, and are appropriate to the level of resources likely to be available, have not been developed. The small size of the holdings severely limits the ability of livestock to provide for the subsistence requirements of the occupiers, let alone generating cash income. With the continuation of migration into this zone from the wetter areas, and the increase in land pressure due to a high rate of local population increase within agroclimatic zone IV, it is clear that this zone is going to experience major difficulties in future. The meat production system discussed for agroclimatic zone V is not practicable in zone IV to any major extent, due to the small size of holding.

Although the meat production system discussed for zone V appears to show considerable ability to cope with the vagaries of the climate, it is only practicable on the larger holdings. A sustainable production of either crops or livestock on the emerging smallholdings in agroclimatic zone V appears most unlikely in the near future, given the land and other resources available to the farmer and the technology that has so far been developed from research studies. Although local research, and transfer of technology from other countries, may increase the potential for production in zone V, there seems rather little likelihood that the necessary resources will become available to the smallholder. In the medium to longer term, the migration of population into the drier areas will have to be reversed to prevent massive future famine, as the land will be unable to support the people on it. Alternative employment opportunities will be required to allow the human population to be reduced to the level supportable by a sustainable production system, such as that described for beef. This issue will be discussed further in Section 7.

In agroclimatic zone III, subsistence production is possible in almost all years, with the carry-over effects of the cash-cropping payment system providing a flexibility with regard to dealing with the inevitable dry years. Further subdivision of holdings due to population pressure will, however, lead to the problems characteristic of the drier zones.

SECTION 6

Smallholder Vulnerability and Response to Drought

6.1. Introduction

Previous sections have outlined the drought climatology of the Kirinyaga–Embu–Machakos area and the results of several agroclimatic models that indicate agricultural and livestock potentials in different climatic scenarios. Even using very simple models, it is clear that climate and weather are the primary variables in determining potential productivity (and its temporal and spatial patterns) in the area. The next step is to trace the importance of variability in agricultural and livestock production to smallholder agriculturalists. Over 80% of the population of Kenya resides in the rural areas, and the majority are smallholder farmers. Development policy in Kenya has long recognized the importance of drought in planning agricultural and livestock development. It must also incorporate consideration of the smallholders' vulnerability to drought and their means of coping with recurrent droughts of various magnitudes and durations.

In general, most smallholder farmers can be characterized as "satisfyzers" (Rempel and Lobdell, 1985) – balancing labor and capital investment in agriculture with a number of other activities and satisfied with a reasonable level of productivity, rather than trying to maximize output in all, or any distribution of, seasons. Climatic variability includes extremes of good and bad years. A complete description of smallholder agriculture must be related to expectations in both good and bad years. Thus, maize is widely grown in places where sorghum is more suitable. Livestock are kept as insurance, even in areas of acute shortages of grazing land. As discussed in Subsection 6.3 below, one strategy for coping with drought is to take maximum advantage of good years, and save the surplus (food or cash) for times of need. The primary concern in this section is with the impact of, and responses to, the bad years.

Impacts of climatic variability, drought in particular, are difficult to trace in a socioeconomic environment that includes many other variables. Subsection 6.2 provides a framework for identifying the linkages between drought and household economics, by presenting a typology of food entitlement and household characteristics. Subsections 6.3 to 6.7 utilize this typology to describe smallholder vulnerability to drought and drought coping strategies. Two scenarios (10% and 2% probability of drought) defined in previous sections are used here. As in the case of the agroclimatic models, only a few relevant variables (which largely determine the patterns of impacts and responses and are amenable to policy decisions and interventions) have been chosen for discussion. The methodology is admittedly descriptive. A fuller treatment of the relationship between climatic variability and household economics would require elaboration of household budget models, for which the present data base is inadequate. A more quantified approach would require computerization and analysis of the Central Bureau of Statistics (CBS) household budget survey of 1981–1982 and the CBS/NES drought survey. Both of these data sets are near the stage of systematic analysis. The present analysis, therefore, does not indicate the percentage of population likely to experience certain impacts or undertake various coping strategies (except in rough categories shown in *Table 6.3*). An indication of the distribution of the effect of food shortages may be inferred from the distribution of wealth. Within a community, economic differentiation among rural smallholders is occurring according to access to the cash economy, rather than through differing levels of agricultural productivity. Between agroecological zones the primary differentiation is the ability to raise productive cash crops, rather than inherent differences in productivity or food self-sufficiency.

Typically, a smallholder in Kenya is defined as a rural landowner with less than 22 hectares (Central Bureau of Statistics, 1985). In the drier areas, this may be increased somewhat so that sedentary agropastoralists are included. The discussion focuses on smallholders and generally excludes the landless and urban dwellers, due to the lack of relevant data, although the impact of climatic variability is certainly significant among these groups of people. There are few pastoralists in the study area (*see* Campbell, 1985, for an assessment of drought responses in the neighboring pastoral areas). Large landowners in this area tend to be managers of capitalist farms and ranches and are thus quite distinct in their vulnerability and response to climatic variability.

An important observation of this section is that household well-being and responses to drought are primarily determined by nonclimatic factors. This is supported by national nutritional surveys. Nutritional status is an indicator of vulnerability to drought and the impact of changes in sources of food supply. Where nutrition in "normal" times is poor, due to any number of conditions, the population is likely to be even more vulnerable to drought: there is little margin for food shortages. Young children are most sensitive to nutritional deficiencies and various measures are commonly used. The percentage of children below a standard height for age (HA) index indicates those who are stunted, reflecting a long-term nutritional deficiency. Wasting, which may be a short-term phenomena, is measured by low weights for age (WA) or weights for height (WH), if age is unknown. In the most recent nutritional survey (1982), children

Table 6.1. Nutritional status and selected health indicators in Central and Eastern Provinces, 1982. Source: Central Bureau of Statistics (1984).

	Province	
Indicator	*Central*	*Eastern*[a]
% Stunted – < 90% height for age [b]	20	23
% Wasted – < 80% weight for height[b]	3	3
Mortality, 0–5 yrs old, % per thousand	85	128
% Mothers with no education	30	46
% Households without piped water	77	88
% Households without sewage facilities	2	38

[a]Excludes Isiolo and Marsabit Districts.
[b]For children 3–60 months old.

in Eastern Province had a slightly higher frequency of stunting than those in Central Province (*Table 6.1*). More importantly, the related health indicators for Eastern Province are all well above those for Central Province. This is consistent with the generally harsher physical environment in the lower altitude zones. Analysis of these data by agroecological zones indicated high rates of chronic malnourishment east of the Rift Valley in the coffee and lower cotton zones. Thus, vulnerability to changes in food entitlements (i.e., availability to the household) cannot be explained by a simple agroecological zonation. The food and nutrition section of the 1979–1983 Development Plan identified six groups at risk of protein energy malnutrition (Government of Kenya, 1979; *see also* Government of Kenya, 1983):

(1)–(3) Smallholders
- with low incomes
- without required inputs (including sufficient land)
- involved in cash cropping, especially seasonal crops, for some areas

(4) Urban poor
(5) Pastoralists
(6) Preschool children and pregnant and lactating women.

6.2. Household Food Entitlement and Vulnerability to Drought

This discussion of the socioeconomic impacts of climatic fluctuations follows the paradigm of human ecology, which characterizes households as actors in their environment, adjusting and adapting to its changes (for example, for East Africa *see* Berry *et al.*, 1971; Mascarenhas, 1973 and Porter, 1979). The framework presented below draws upon the work of Sen (1981), an economist who uses the term "food entitlements" to describe a household's various sources of food and its means of acquisition; Wisner and Mbithi (Wisner, 1977; Wisner and Mbithi, 1972; Mbithi, 1972; Mbithi and Wisner, 1974) who documented drought coping strategies in eastern Kenya; and Rempel and Lobdell (1985), who provide a framework for discussing household objectives and income.

The basic unit of analysis in this paper, and of social organization in Kenya, is the household. It is a dynamic grouping of individuals, usually related, and is defined by the Central Bureau of Statistics (1981a) as living in the same compound, eating from the same sources and answerable to the same head. This is the definition used in the CBS/NES survey reported in this chapter. The household is dynamic in its composition, due to growth, births and deaths, and migration, and in its objectives and decision-making. Although the paper discusses household vulnerability to drought in general terms, without operationalizing the model described below, the definition of the household is important. In particular it limits the household members to those living in the same compound. Thus, many of the households are headed by women whose husbands are employed elsewhere. The household composition can be directly related to available labor for farming and food requirements. The definition of a household is always arbitrary and problematical; it may be misconstrued to indicate a social stasis that ignores the importance of the network of social relations that households maintain and utilize in farming and other economic activity. Although there is a trend toward increasing individualization, the kinship network should not be ignored (*see* O'Leary, 1984, for examples of its importance in Kitui).

6.2.1. A household objectives model

A formal model of household objectives proposed by Rempel and Lobdell (1985) is useful in characterizing household vulnerability to drought. In their model households are satisfyzing units that seek to achieve a "desired level of household income" (Y') (*ibid.*, p. 8):

$$Y' = (C + R) \cdot A + S \qquad (6.1)$$

where

C = the minimum level of consumption that is socially acceptable, per adult equivalent member
R = the minimum expenditure on maintaining social relationships in the community, per adult-equivalent member
A = the number of adult-equivalent members
S = the desired level of savings (cash needs) and surplus (investment).

Household income (Y), which households strive to balance with desired income (Y'), is composed of (*ibid.*, p. 11–12):

$$Y = \sum_j (L_j \cdot W_j) + \sum_i (T_i) \qquad (6.2)$$

where

L_j = the labor units engaged in the jth activity, per adult-equivalent member
W_j = the average return to labor in the jth activity, per adult-equivalent member
T_i = other transfers from or to the household, including taxes and donations, changes in stocks and savings, and the net change in income due to changes in the household composition when members leave the household, for the ith purpose.

The objectives and socioeconomic environment of rural households in Kenya have undergone dramatic changes in the two decades since independence. In an elementary sense, smallholders have two major objectives (Franzel, 1983):

(1) To provide their families with sufficient food for consumption, storage and seed [C in equation (6.1)].
(2) To provide cash for essential purchases such as school fees, medical care and self-help projects, and supplementary foods and consumer items [R and S in equation (6.1)]. Household agricultural production is typical of self-provisioning or subsistence economies, with fairly low levels of inputs. However, household reproduction is dependent on the cash (capitalist) economy – for education, raised standards of living evidenced by consumer items, and investment in agriculture (Kodamaya, 1984).

Households have four resources to allocate to achieve their objectives: land, labor, capital and managerial expertise (partly related to labor). Of these, labor is the most mobile, and therefore more critical in the short-term. It is well documented that in most parts of central and eastern Kenya labor is a critical constraint to improving agricultural production, due to the seasonal peaks for plowing, planting and weeding, the social division of agricultural labor, and the emphasis on education (Franzel, 1983; Hunt, 1978; Llano-Alcober *et al.*, 1983 and Rukandema *et al.*, 1981 and 1983). As elaborated in *Table 6.2*, household allocations can be aimed at changing productivity (W), changing the range of productive activities (j) or changing the nature of the household's transfers (T).

These characteristics of household economics are classified according to the household's food entitlement: from subsistence production, market exchanges or donations. The result leads to a typology of household vulnerability to drought according to their ability to command food entitlements, and to specific household characteristics. As elaborated below, households that are least vulnerable to drought have access to a diversity of nonagricultural market exchanges. They

Table 6.2. Household economics, food entitlements and drought responses.[a]

	Income				Expenditure		
Food entitlements	*Agricultural productivity* (Wa)	*Nonagricultural productivity* (Wa')	*Diversity* (j)	*Transfers* (T)	*Consumption* (C)	*Social relationships* (R)	*Surplus* (S)
Subsistence production	Soil conservation Irrigation Drought resistant crops Draught animals Planting time Intercrop Fallow system Inputs Weed Thin crop stand Water conservation Improved livestock Fodder production	Trading/shops	Number of crops Disperse livestock Relay planting Area planted/cultivated Wild foods Mixed livestock herds Multiple/scattered plots/farms	Improved storage	Food from farm produce Wild foods	Farm produce used in feasts, special occasions and hospitality	Stores of seed and food from farm Livestock
Market exchanges	Cash crops Livestock sales	Education and training Cooperatives	Wage labor Permanent employment Local or urban businesses	Loans/borrowing Insurance Sell assets	Purchased foods Essential purchases Consumer purchases	Food sales	Savings/investment
Donations			Kin and friendship network Government/NGO contacts	Famine relief Donations/gifts	Donated food	Remittances/gifts to others	Taxes Donations

[a]Symbols in heading are as in equations (6.1) and (6.2). Sources: Food entitlements adapted from Sen (1981); household economics adapted from Rempell and Lobdell (1985); drought responses adapted from Wisner (1977), Akong'a (1982) and interviews by the authors.

will experience the effect of drought through the local and national economy as it affects food supply and prices in the local markets. Climatic variability has a direct influence on households that rely heavily on their own production to meet their subsistence needs. For those households in this category with small holdings, poor soils, and insufficient labor and capital, the effect of drought may be acute.

6.2.2. Household coping strategies and climatic adversity

Households employ a variety of strategies aimed at trying to make (Y) approach (Y'). Some of the strategies employed in response to their entire environment may also be effective in drought years. These strategies have been termed adaptations, i.e., long-term processes built into the socioeconomic system that affect their sensitivity to weather variables. They are "routine practices which do not vary much from season to season..." (Wisner, 1977, p. 79).

By contrast, during a drought episode households balance their resources and objectives by employing adjustments, which are short-term coping strategies: "elements in a resource-use system with...drought mediating, escaping or coping properties, which are only involved when there is a moisture crisis. They constitute and are generally perceived as departures from 'normal' farming or ranching practise" (*ibid.*). There is no hard rule as to what constitutes adaptations or adjustments, for it will differ from one individual season or region to another. An adjustment perceived as being reliable whenever needed can also be classified as an adaptation. Thus, although the distinction between adaptations and adjustments is useful, in this section the more general term "coping strategies" is employed to refer to both adaptations and adjustments.

Table 6.2 indicates a range of potential adaptations and adjustments to drought, categorized by food entitlements, and matched against elements in the household's income (Y) and desired expenditure (Y'). It represents specific strategies that households may employ to try to balance Y and Y' during drought years.

(1) Strategies that improve subsistence agriculture (for household consumption) generally involve knowing or making better use of water and land resources. It is the food source most directly sensitive to weather, and includes production grown on land owned, rented or used by the household, employing family or hired labor.

(2) The next most directly vulnerable food source is food obtained through exchange for one's own production in a noncash barter arrangement. In Kenya this is a minor source of food for most households (Fisher, 1977). Strategies to diversify the economic base of the household beyond subsistence agriculture involve market or monetary exchanges. This may be confined to the local market, where cow peas might be sold and beans bought, or in regional or even international markets, e.g., where income from tea sales is used to buy the household's food. Market price fluctuations, particularly for unprocessed foodstuffs, are very important in the

local markets. Income from cash crops, traded outside the local economy, and off-farm employment or businesses, enable a household to purchase food in a volatile local market. However, some of these sources of income may be vulnerable to drought as well.

(3) Remittances and donations from relatives, friends and agencies are a third source of food or cash. The level of donations received depends upon the donors' perception of the drought, as well as other factors, such as the strength of kinship ties, integration of the local area into the cash economy, effectiveness of the area's political leaders in attracting assistance, and the infrastructure for trade and distribution of food relief.

These three spheres of food entitlements are differentially vulnerable to climatic fluctuations and natural resource and economic trends. Taken together, for a particular household, they characterize the household's vulnerability to drought, or its food entitlement map (Sen, 1981). Referring to the model above, if the realized income Y is less than the desired expenditures Y', then consumption C and savings and surplus S are affected. Similarly, the drought impact on the household can be viewed as depending on shifts in food entitlements. A failure of all entitlements will lead to the most serious nutritional, social and economic impacts, as the household must bear the consequences of the drought. More common in Kenya are partial changes in each entitlement, with socioeconomic impacts, such as loss of cash reserves or capital, and migration.

6.3. Drought Vulnerability and Coping Strategies

Table 6.3 rates the most common or most promising drought coping strategies in central and eastern Kenya. The ratings are subjective assessments by the Kenya case study authors, guided where possible by data from recent rural surveys, the National Environment Secretariat's drought responses survey (NES, 1985a and 1985b) and other available sources. Prevalence refers to the percentge of households in the area that employ the strategy. Effectiveness represents the ability of the strategy to achieve the household's objectives in an average year and two levels of drought years, as defined in Sections 1 and 2. A moderate drought is one occurring once in 10 years (90% probability of moisture conditions being better), and a severe drought has a frequency of once in about 50 years (98% probability), roughly typified by the 1984 long rains drought. The ratings for recovery indicate whether the strategy has a positive or negative effect on recovery from drought, i.e., when the rains come in the next good season. The last columns indicate the constraints on the household's ability to implement the strategy. Labor includes household members only. Capital is cash for tools, equipment, inputs and wage labor. Technology includes new kinds of tools or equipment, draught animals, new cropping systems, improved breeds of plants, etc. Education/skills/information denotes a lack of awareness about the strategy, a lack of knowledge or skills to implement it, or a need for certain information before or during the season. Land includes space as well as quality of soils and their suitability for various uses.

Table 6.3. Characteristics of selected coping strategies.[a]

Response/ coping strategy	Prevalence	Effectiveness: Normal	Effectiveness: Drought Moderate	Effectiveness: Drought Severe	Recovery	Constraints: Labor	Constraints: Capital	Constraints: Technology	Constraints: Education/ information	Constraints: Land
Subsistence production										
Soil conservation	M-H	M	M	L	0	−	−	−	+	−
Water conservation	L	M	H	L	0	−	−	−	+	−
Irrigation	L	H	H	H?	+	−	+	+	+	+
Multiple farms	L	M	M	L	0	−	+	−	−	+
Inter/relay cropping	H	H	H-L	L	+	−	−	−	+	−
Dry planting	M	H	H	L	0	−	−	+	−	−
Mixed livestock herds	M-H	H	H	M	+	+	−	−	−	+
Dispersed grazing	H	H	H	M	+	+	−	−	−	+
Fodder production[b]	M	H	H	M	+	+	−	+	+	+
Drought-resistant crops	H	M	H	M	+	−	−	+	+	−
Monetary activity										
Local wage labor	M-H	H	H	H	+	+	−	−	+	−
Migrant wage labor	M	H	H	H	−	+	−	−	+	−
Permanent employment	M	H	H	H	+	+	−	−	+	−
Local business	L	M	M	L	0	+	+	−	+	−
Cash crop[c]	M	H	M	L	+	+	+	+	+	+
Sell capital assets	M	H	M	L	−	−	−	−	−	+
Livestock sales	H	H	M-H	M-L	−	−	−	−	+	−
Remittances/donations										
Relatives/friends	M-H	M	H	H	+	−	−	−	−	−
Government and others[d]	M-H	?	H	H	?	−	−	−	+	−
Loans/credit	L	H	H	H	+	−	+	−	+	−

Key:

Effectiveness
H = High
M-H = Mod. high
M = Moderate
L = Low

Prevalence
>50%
30–50
15–30
0–15

Constraints
− = negative, i.e., impedes recovery, is a constraint
+ = positive, i.e., aids recovery, is not a constraint
0 = neutral, i.e., no effect on recovery
? = uncertain or variable

[a]Consensus agreement by case study authors based on available data. Ratings are intended to be qualitative and relative as no systematic survey data are available. In many cases, these are hypotheses to be verified. [b]Very common in the upper zones, almost nonexistent in the lower zones. [c]Does not include food crops. Very common in upper zones, rare in lower zones. [d]High in the lower zones, low in the upper zones.

The NES survey of smallholders' responses to the 1984 drought recorded major foods consumed during the drought, and thus the extent of shifts in food entitlements (*Table 6.4*). In the drier areas about three-fourths of the households reported a shift from their customary staple food to some other food. In the wetter zones, the figure was about 30%. For most of the households the shift was from white maize to yellow (imported) maize.

Table 6.4. CBS/NES survey of smallholder response to 1984 drought: preliminary results. (Data collected in January 1985 by the Central Bureau of Statistics on behalf of the National Environment Secretariat).

Household (HH) *survey item*	*Agroclimatic zone*				
	I	*II*	*III*	*IV*	*V*
Number of HHs	40	31	81	34	44
Average HH size	6.5	5.5	5.8	7.0	6.0
Percentage of HHs in which:					
Member moved in 1984	13	13	26	29	25
Nuclear family member is permanently employed	30	16	22	9	18[a]
Major food changed in 1984	30	29	72	79	86
Famine relief received in 1984 from govt. or NGO	8	42	40	56	66
Larger maize area planted in 1984 short rains than in 1983 short rains	8	0	17	6	7
Had enough maize seed for planting in 1984 short rains:[b]					
Yes	13	13	26	35	7
No	53	45	53	62	48
Had no cattle at end of 1984	20	16	35	18	11
Sold, consumed or lost cattle	33	48	42	62	59
Average number of cattle at end of 1984 (absolute value)	2.5	2.3	4.2	3.8	4.1
Received cash remittances from relatives or friends:	23	19	30	68	32
Usual source of cash[c]	35	25	41	46	42
Received during drought	30	25	37	49	40
Increased or stayed same during drought	38	33	55	69	49

[a]Half of this figure were government employees in a road work camp.
[b]Remainder either did not respond or did not plant maize.
[c]In normal years as well as during droughts.

An increase in total household food supply might overcome famine in central and eastern Kenya while leaving the population largely undernourished. Malnutrition can be caused not only by food shortages, but also by the low nutrient value of the food ingested, poor methods of food preparation or food taboos barring some members of the family from eating certain food (Onchere, 1976). For example, it has been observed that there is widespread malnutrition in children in the Mwea rice scheme in lower Kirinyaga and Embu districts as a result of a lack of variety in the diet. This is because "it is difficult for those with rice plots to plant vegetables because water use is restricted, and...also livestock keeping in the rice villages is prohibited" (Scotney and Nyambura, 1979, p. 6). The disposition of the cash payments for the rice only to the men is also a factor (Chambers and Moris, 1973). Neither of these causes is related to climate.

The discussion that follows reviews, for each category of food entitlement, the long-term trends in drought vulnerability, and analyzes specific adjustments and adaptations that households may employ.

6.4. Vulnerability and Coping Strategies in Subsistence Production

Small-farm agricultural production in the study area has generally been increasing as more land is brought under cultivation and farming practices improve. However, there are several trends that affect the vulnerability of household farm production to climatic episodes.

Table 6.5. Population growth in Kirinyaga, Embu and Machakos Districts.[a] Sources: Central Bureau of Statistics (1971 and 1981c).

	1969			*1979*			*1969/79 growth*	
Zone	*Pop*	*Area*	*Density*	*Pop*	*Area*	*Density*	*%*	*Annual %*
I	132 065	424	311	179 252	387	463	49	4.1
II	77 011	292	264	108 143	248	436	65	5.1
III	235 727	1259	187	280 495	1206	233	25	2.2
IV	271 893	4141	66	382 373	3413	112	70	5.4
V	367 375	11 918	31	474 264	10 542	45	45	3.8

[a] Area is in sq. km. and density in people per sq. km. Data exclude townships. Area differences between censuses are due to changes in location boundaries, inclusion of gazetted forests and parks and measurement errors. Data were computed from the census reports by sublocation. Sublocations on boundaries between agroclimatic zones were apportioned to each zone by the area in the zone and the average population density of that zone.

6.4.1. Land pressure

One of the most notable trends in the rural landscape of Kenya is the high rate of population growth and resulting subdivision of farms. *Table 6.5* indicates that population densities have increased the most in zones II and IV, and the least in zone III. The increase between 1969 and 1979 in zones I and V was in the 45 to 50% range. Zones I and II have probably reached the maximum density that can be reasonably supported by the current levels of agricultural production. With a population still growing, zone III has been marked by outmigration as the agricultural frontier has expanded into zones IV and V. Subdivision of land takes more land out of production as houses, roads and fences are constructed. In a densely populated district in western Kenya (Kisii), 14% of the district is taken up by hedges, roads, structures and market centers. A further 41% is grazing land, fallow or in woodlots (Epp *et al.*, 1983).

The effect of population growth is partially offset by an expansion of the cultivated area. However, this is often onto marginal lands and areas that would otherwise have remained rangeland, further exposing a large number of families to the risk of reduced production due to the loss of this grazing land. This migration into the drier areas from more productive zones has been documented by KREMU (Agatsiva and Mwendwa, 1982) and described by O'Leary (1984) for Kitui. The situation is made morc precarious in that no successful smallholder farming technology has yet been identified in the arid and semi-arid regions of Kenya. In addition, the agricultural expansion further depletes the remaining indigenous forests, valuable for their genetic resources, usable materials, and for their role in protecting much needed watersheds (National Environment Secretariat, 1984).

6.4.2. Ecological degradation

Rates of soil erosion have increased in the last few decades as the agricultural area expands onto less stable soils and steeper slopes, and as soil conservation structures deteriorate. Data from zones I, II and III indicate soil loss on annual food crops is well above the rate of soil formation. Maize fields are particularly vulnerable, with losses over 50 metric tons per hectare measured in the rainy season in Kiambu and Murang'a – areas typical of agroclimatic zones I, II and III (Lewis *et al.*, 1983). Plots with low vegetative cover in Machakos (in agroclimatic zones III and IV) recorded 20–30 metric tons of soil loss per hectare per year (Styczen, 1983). Erosion from rangelands is lower, on the order of 4 to 15 m tons/hectare/year (Dunne and Wahome, 1981). Equally serious is gulley erosion – often triggered by runoff from roads, drainage ditches or urban areas (Thomas, 1974).

Much emphasis has been placed in recent years on soil conservation utilizing a variety of methods. Progress has been remarkable: some 9000 km per year of bench terraces have been constructed in Machakos in the last few years (Norton-Griffiths, 1986). But the magnitude and extent of the problem overshadow the progress to date. In zones II through V soil erosion may be particularly important, in that the loss of the humic layer decreases the soil's water-holding capacity. In areas where rainy days are scattered, with dry intervals between, this would increase the number of stress days for the crop.

Terracing is the major water and soil conservation practice in the study area. In Machakos it is carried out by women through participation in *miethya* (singular: *mwethya*) groups, traditional reciprocal work groups whose membership cuts across clan boundaries. Each member in turn assists in terracing, cultivation, planting, weeding, harvesting, and construction of houses, schools, dams, cattle dips and access roads. Some of the groups also participate in commercial enterprises, such as making bricks, keeping poultry and goats, and the management of group farms. Through the *miethya*, women, who, with their children, are the most vulnerable during periods of food scarcity, find social and psychological support in that they are not suffering alone. Through communal participation, including exchanging information, singing, dancing and eating, the women transform heavy tasks into occasions for entertainment. They use traditional sanctions, such as fines or being barred from access to a public utility, to enforce participation. They are an important base for rural development in eastern Kenya and their role as media for diffusing innovation should be recognized (Neunhauser *et al.*, 1983).

One method of improving soil fertility for higher productivity is to use chemical fertilizers. However, only a few people use fertilizers in the study area. Likewise, only a small proportion of the farmers use the inexpensive and readily available dung manure, which is much better for the arid and semi-arid lands in that it contains humus. In a study of lower Embu, Rukandema *et al.* (1983) reported that only 14% of the farmers in their sample from Kathera and 6% from Kamarandi applied natural manure. As a long-term strategy of agricultural development, livestock should be more fully integrated into the crop production system.

6.4.3. Land tenure

Historically, landholdings were fragmented, often in several agroecological zones, and clan elders allocated usufruct. Since colonial times most of the land has been consolidated into adjoining parcels, and private ownership is now recognized. In the early 1970s several studies noted that land consolidation increased farming risk, and adjudication has led to an active land market and gradual accumulation by the local elite (Brokensha and Njeru, 1977; Haugerud, 1981; Mwaniki, 1982). The pace of accumulation appears to have slowed, owing to more stringent requirements for approval of land transfers. However, the trend is irreversible and cumulative. For the smallholder who sells a small parcel during a drought, vulnerability increases accordingly. Population growth, a slow degradation of the resource base, and a largely stagnant technology contribute to the process of marginalization whereby a household may be able to sustain itself, but cannot reproduce its livelihood in the next generation. The result is a growing landless population – some remain in the rural areas, others migrate to urban centers in search of employment.

Another trend in land tenure since independence has been the purchase and subdivision of the former white highlands (Abrams, 1979; Odingo, 1971). The opportunity to acquire land, or for those already owning farms, additional holdings, probably increased subsistence production on the former estates. However, many of the new settlers were unfamiliar with the ecology of their new farmlands and appropriate farming practices. Thus, agricultural production and the migrant households were more vulnerable to drought.

6.4.4. Pests and wars

Two causes of widespread famine in the past appear to have lessened. Locusts and interethnic wars were a great handicap to successful farming activities. Locusts used to destroy any green cover, often spreading over very large regions (Odingo, 1982). Plagues of this sort seem to be well controlled by the increased monitoring and response activities of the government and such agencies as the Desert Locust Control Organization. The experience in central and eastern Kenya in 1984 with army worms, which were effectively overcome and thus prevented from causing widespread crop failure, supports this conclusion.

Interethnic wars were especially lamentable because they were man-made and thus avoidable. Not only were food crops and stores destroyed when found, but also, because of insecurity, people could not work in the fields. The Mbeere in Embu, for example, lived in perpetual fear of the Embu, Maasai and Chuka, and spent much of their time in hiding. The available food was also taken to the war front by women who should have been engaged in cultivation at the time. When the Mbeere and the Embu were equally affected by famine, the Embu would seize livestock from the Mbeere to replenish their own herds.

6.4.5. Labor

Another trend is in the composition of farm labor. Parents have recognized the value of education and have been building self-help primary and secondary schools (Hunt, 1978). The absence of child labor, particularly to work as herders and to scare away animals near harvest time, has contributed to the change from cultivation of millet and sorghum to maize, and to reduced access to distant common grazing lands. Among adult men, few who have been educated beyond primary school appear to be very interested in agriculture – the return to labor is low and the work hard. The exodus of adults to towns and cities deprives the rural household of labor and, where the migrant is the head of the household, managerial experience and financial investment, traditionally a male responsibility.

6.4.6. Agricultural technology

Intercropping is a common practice, the objective being to maximize the harvests of a diverse array of crops. In arid and semi-arid areas, farmers intercrop maize, beans, pigeon peas, cow peas, green grams and other crops on the same plot of land. The number of crops interplanted depends on the region and the individual farmer's preferences for particular crops. In normal or good years, intercropping increases yields through nutrient exchanges and reduces risk, as planting and harvesting are spread out over time and advantage is taken of the different crops' responses to the seasonal rainfall distribution. It also reduces the peak labor requirements and soil erosion. Recent trials at Katumani in Machakos compared the yield of intercropped and monocropped fields. In normal to good seasons, with rainfall over 325 mm, the intercropped fields yielded more per hectare. In a mild drought season, the intercropped yields were slightly lower than the comparable area yields from monocropped fields. With rainfall less than 285 mm, the intercropped fields produced 30 to 60% of the yield levels for monocropped fields (Stewart, n.d.). Further experiments should be carried out to identify the crops that would yield maximally under drought conditions if interplanted. Trials should be conducted under field conditions similar to those of the farmers. Research station plots are usually plowed by tractors, kept weed-free, and sprayed for pests. Hence, responses are generally much higher than could be expected on farmers' fields, and it is unlikely that the actual response functions have the same shape as the experimental ones (Franzel, 1983).

Table 6.6. Adjustments to drought in Eastern Kenya.[a] Source: Wisner (1977, p. 122).

Type of adjustment	*Practice*	*Percentage practicing*
Affect or know rainfall source	Pray	93
Increase moisture	Plant before first rain	80
	Plant dry	71
	Plant with first rain	59
Reduce moisture need	Plant drought-resistant crops	72
	Weed more	64
Diversify	Seek wage employment	60
Distribute or share loss	Ask help from kinsmen	66
Bear loss	Buy food	96
	Sell cattle	53

[a]Responses from northern Kitui, lower Meru and lower Embu, typical of the drier areas in the Kirinyaga–Embu–Machakos study area. Multiple responses were recorded, e.g., a farmer may plant some of his land early and some late.

Scheduling for optimum soil moisture involves planting with, or even before, the first rains (Section 4). Responses in *Table 6.6* indicate that more than 50% of the respondents in eastern Kenya plant with the first rains as a strategy for maximizing the use of available moisture. Several socioeconomic factors constrain early planting. In parts of the region, slash-and-burn shifting cultivation is still practiced. This has to be done when the rains have already come in order to avoid overgrowth of weeds. Many of those who cultivate with the plow start preparing the fields after the rains have started. Not all the families can afford to buy plows or have draught animals. Those who hire plowmen or depend on friends and relatives have to wait their turn, further delaying planting. In some areas the soils are too heavy to plow until the rains have begun. The consequences of late planting may be severe: in one study a 3–4% reduction in yield was reported for every day planting was delayed after the onset of the rains (Dowker, 1964). The results from Section 4 confirm the necessity of planting early.

Famine conditions may exacerbate planting problems:

> ...for during the period of famine it is not uncommon to find peasants consuming all the stock that they had set aside for seed. The side effect of this is that not only does it result in late planting but also in using seeds that may not be particularly suitable for the low-rainfall areas (Owako, 1969, p. 215).

Reducing moisture waste may involve such practices as more weeding and prolonged fallow (*Table 6.6*). Weeding requires considerable labor and special tools and is often not done thoroughly or on time (Llano-Alcober *et al.*, 1983). Few farmers use draught animals in weeding, which would alleviate the labor requirements.

The crop water and nutrient requirements may be changed by planting drought-resistant crops or planting many varieties of crops during the same season. Evidence shows that the people in the semi-arid areas of eastern Kenya have traditionally planted drought-resistant or early maturing crops such as bulrush millet, sorghum, cow peas, pigeon peas and certain beans. To assist farmers, the government has focused its attention on an early-maturing maize variety, Katumani (Composite B). Unfortunately, Katumani maize has not significantly reduced the incidence of frequent food shortages in this region. Mbithi (1972) estimated that a family of five adults, growing Katumani maize, requires five acres of land to be self-sufficient in food production. However, the average plot devoted to maize was 1.2 acres, 40% of which was allocated to Katumani maize. Given the adoption rate of Katumani, the land available for maize production, and the marginal increase in yield between Katumani and local maize, Katumani has not made a great impact on the food situation. Mbithi's results are similar to those of Kolkena (n.d.), who found that although the average cultivated area for Katumani maize in Machakos District was 0.72 ha, 50% of the households had planted 0.61 ha or less. The proportion of the population using Katumani maize seeds in Onchere's sample from Machakos is shown in *Table 6.7*.

Table 6.7. Percentage of population using Katumani maize seed in Machakos, 1975. Source: Onchere (1976, p. 51).

	Location				
Planted	*Kangundo*	*Tala*	*Kinyui*	*Katheka*	*Kathama*
Katumani	0	10	12	0	17
Katumani × 2	26	25	15	30	24
Total	26	35	27	30	41

Note: Katumani × 2 is the residue of what was once Katumani Composite B maize seed.

Factors that militate against the widespread acceptance and efficacy of Katumani maize seed include:

(1) Distribution problems: Katumani maize seed is produced near Kitale in the highlands, where only one maize crop in a calendar year can be grown, owing to the wet, cold climatic conditions there. Thus, when the seed is required in eastern Kenya, it may not be available. Moreover, Kitale is geographically distant (over 450 km) from eastern Kenya.

(2) Competition with other crops: In places such as Mbeere, Tharaka, Kitui and Machakos, Katumani maize competes with the available drought-resistant crops and with livestock production. In Machakos and Kitui, Katumani and the local maize varieties have almost succeeded in replacing traditional drought-resistant crops (e.g., sorghum and millet), a phenomenon that has increased the vulnerability of the population to crop failure and famine. This is part of a long-term shift toward maize cultivation, due to taste, social preferences and shortage of labor to scare away birds from sorghum and millet fields at harvest time.

(3) Cost: Few people can afford to buy Katumani maize seed each year. Franzel (1983) found that, in Kirinyaga, those who planted hybrid maize seed had high incomes and probably did not depend entirely on farming for their subsistence foods. Asked to say why they did not plant Katumani maize seed, respondents in Machakos indicated that lack of money to buy seeds is not the primary impediment to the acquisition of Katumani maize seeds (*Table 6.8*), but is still an important constraint. In all five of the areas surveyed, the majority of the respondents still prefer the local variety. The implication is that those who plant the hybrid variety are selective. They could be higher-income earners, as Franzel observed, but may include those who have other socioeconomic characteristics, such as having been out of the local area for one reason or another for a significant length of time.

(4) Local selection: Katumani is a composite, meant to be grown in succeeding seasons from seeds harvested from the first crop. However, farmers usually save the largest cobs for seeds and may plant Katumani where it can be cross-pollinated from local maize plants. Thus, there has been a drift toward slower-maturing plants, and later generations may not be much different from the local varieties.

Table 6.8. Reasons for not growing Katumani maize in Machakos, 1975. Source: Adapted from Onchere (1976, p. 51).

	Location				
Responses	*Kangundo*	*Tala*	*Kinyui*	*Katheka*	*Kathama*
Prefer local seed	52	35	60	60	42
Lacked money	7	5	4	–	10
Katumani yields less	7	–	–	–	3
Katumani not found	7	25	9	10	4
Sample number	70	65	73	70	59

6.4.7. Rainfall prediction and utilization

Knowing or affecting the rainfall source, such as through the performance of traditional rites or participation in prayer with the hope of making it rain, is an important adjustment that is based on belief and tradition. In Mbeere Division of Embu District, for example, "rainmakers would say that there were some ill-omened women who caused rainfall shortages if they planted first" (Njeru, 1979, p. 144). To this day, it is widely believed that, if women who have reached their menopause plant first, then there will be adequate rainfall and good harvests. This belief is based on the understanding that women of such age are unlikely to commit sexual sins and are pure in body and at heart. People in villages therefore wait for such women to plant, a phenomenon that is likely to delay planting if the women cannot plant early for one reason or another. Other traditions exist to predict when it will rain and how long the rains will be. This skill is often associated with witchcraft, but may rely on biological or physical indicators as well.

Prayer for rain, on the other hand, is a common feature of modern Kenya as demonstrated by the most recent widespread national food shortages in 1979–1980 and 1984–1985 when the government instructed religious groups in the country to arrange for their own or interdenominational prayers for rain. When the rains started falling in some parts of the country, prayers of thanksgiving were arranged.

Various types of monitoring systems, using statistics to predict the seasonal rainfall, may be important in the future (Stewart and Hash, 1982; Stewart and Faught, 1984). The approach suggested by Stewart and his colleagues proposes to monitor the first few weeks of the rainy season on farmers' fields or for neighboring localities. The rainfall probabilities for the rest of the season can then be determined and used to adjust the crop practices accordingly. Additional inputs and hired labor for weeding may be justified if the season will be wet. In a mild drought season, thinning the crop and early weeding by family members may be warranted. Coupling modern and traditional weather prediction might increase local reliability and acceptability of a monitoring system.

Changing locations increases water resources available to the household and involves such procedures as controlled grazing, sending livestock to kin or nonkin, planting a larger area or in scattered fields in different regions, as a means of spreading the risk of loss.

6.4.8. Wild foods

For many people in zone V of the study area, wild foods were a mainstay of the diet in the 1984 drought. Baobab, fig and tamarind fruits, local wild vegetables, and a type of wild groundnut were collected (usually by children) and consumed. The collection of honey for food and for sale is traditionally an important industry in the region, with the potential of increased production if market outlets can be found and processing plants established. Wild foods are available only in areas where there is plentiful common land. Wild animals were also hunted and consumed in the rural areas near game parks and other sparsely settled lands. Conversely, those near the wild lands are subject to greater damage by the wild animals and the diseases they transmit. One of the field enumerators reported the situation in one area of zone V:

> There was *ngaatu*. These are wild seeds more or less like groundnuts.... Women would rise very early in the morning after their husbands went to look for casual labor, leaving behind the most important word, *ndunkambikwatye*, i.e., "don't rely on me". The women would go looking for alternatives. They would dig a very big area – one would think there had been a tractor plowing – and then fill a basket with these wild groundnuts. These they would take home and start pealing the outside coat. Then they gave them to the kids to eat. And that day would have passed (NES, 1985b).

6.4.9. Drought scenarios and effective coping strategies in subsistence production

Generally, the agricultural practices adopted in response to drought are highly effective in moderate (10% probability) droughts, but of very limited effectiveness during severe (2% probability) droughts (*Table 6.3*). During a moderate drought, soil and water conservation techniques may be crucial for crop production. Even during the 1984 drought, some farms with well-terraced fields produced reasonable yields; but, in general, soil and water conservation are not effective during a severe drought. In addition, they may be counterproductive in the more humid zones during wet seasons if they contribute to landslides.

Labor to construct permanent conservation structures and labor and materials for temporary approaches (mulching and ridging) are the major requirements for more widespread adoption. Fertilizer is wasted in drought seasons, when moisture is the limiting factor, but should be part of a strategy to make better use of the wet seasons. In most areas of central and eastern Kenya commercial fertilizer is readily available, but farmers often need credit to purchase it. In drought seasons intercropping may be effective only if it benefits from an unusual distribution of the rainfall (Stewart, n.d.).

Several strategies for coping with drought may have negative consequences in normal or good seasons. Katumani Composite B maize yields less than local maize in good years. Terraces may contribute to waterlogging. Multiple distant farms may reduce the labor available on each farm, and thus hinder adoption of improved agricultural practices.

6.5. Vulnerability in Monetary Exchanges

In his study of Machakos District, Onchere (1976) found that nonfarm sources of household income contributed up to about 50% of the total household income.

Mbithi (1972, p. 10) underscores the role of nonfarm coping strategies:

> Our studies on on-farm occupation show that the development of these areas must rest on the principle of money-earning diversification ... thus we found that those normally called "farmers" are the ones involved in cattle trade, beer making, crafts, shopkeeping, witchcraft healing, pottery, weaving etc.

Table 6.9. Household maize consumption, sales and purchases 1978–1979 (000 tons). Source: Central Bureau of Statistics (1982b).

		Central Province				Eastern Province			
Cycle	*Dates*[a]	*Cons*	*Sales*	*Purc*	*P/C*[b]	*Cons*	*Sales*	*Purc*	*P/C*
14	16.5–12.6	12.5	10.8	4.2	0.34	14.2	2.8	1.9	0.13
15	13.6–10.7	10.3	21.6	4.5	0.44	12.7	5.0	4.2	0.33
16	11.7–7.8	10.8	14.0	9.5	0.88	13.3	4.3	4.0	0.30
17	8.8–4.9	9.6	8.2	3.2	0.33	15.3	6.5	5.6	0.37
18	5.9–2.10	11.3	9.9	2.8	0.25	15.3	6.0	6.6	0.43
19	3.10–30.10	11.7	13.0	3.5	0.30	16.2	5.3	8.0	0.49
20	31.10–27.11	5.9	7.4	1.0	0.17	18.1	4.8	10.9	0.60
21	28.11–25.12	13.7	17.6	5.0	0.36	12.0	6.2	8.5	0.71
22	26.12–22.1	21.0	15.4	5.1	0.24	25.1	5.6	13.2	0.53
23	23.1–19.2	14.2	8.8	2.7	0.19	16.3	2.0	4.6	0.28
24	20.2–19.3	15.7	16.6	3.7	0.24	14.7	0.8	1.9	0.13
25	20.3–16.4	15.4	11.5	2.7	0.18	15.3	1.1	1.8	0.12
26	17.4–15.5	14.5	11.8	4.0	0.28	15.3	4.1	2.2	0.14
Annual		166.7	166.5	51.8	0.31	204.0	54.4	73.3	0.36

[a]Based on an integrated rural survey (IRS4) begun in May 1978.
[b]*P/C* = ratio of purchases to consumption.

The dominant market exchanges are cash transactions. Data at the provincial level indicate that about one-third of the annual maize consumption comes from purchases (*Table 6.9*). This ratio reaches over 50% during the short rains in Eastern Province (typical of Machakos) before the harvest. Central Province, representative of upper Kirinyaga and Embu, is more self-sufficient in maize production. These aggregate statistics mask notable variations for certain localities and certain households. But they indicate family reliance on market exchanges for a large proportion of their basic household needs. In Mwala (Machakos District), 92% of the families purchase some food (Rukandema *et al.*, 1981).

6.5.1. Maize markets

There are two maize markets in Kenya (Schmidt, 1979). The formal market is controlled by several statutes and the district administration. By law, all maize transported across district boundaries must be sold to the National Cereals and Produce Board (NCPB), which then sells it to licensed traders for distribution in deficit areas and during the months of peak demand. The informal market includes local traders, who buy and sell local produce in open markets, and itinerant traders who import produce from distant areas without proper licenses. The effect of the controlled market has been:

(1) To give the NCPB, a quasi-governmental organization, ready access to surplus production, which can be exported or stored, and more reliable data on the national production, which can be used in planning imports.
(2) To impose severe restrictions on the interdistrict flow of maize, often causing shortages in some areas (although they may not persist for long).
(3) To increase the local consumer cost of maize, due to inefficiencies and excess profits.
(4) To make the local prices of maize more volatile as price differentials between surplus and deficit areas cannot be easily reduced.
(5) To suppress national maize production levels.

In a survey of 30 local markets in the study area, just after the 1984 drought, white maize prices (averaged for several markets) increased between 43% and 272%, depending on the zone (*Table 6.10*). Bean prices also increased dramatically, by as much as 290% in zone IV. These prices are based on reports given by women vendors and shopkeepers in January 1985 for 1984 by three-month intervals, and thus are somewhat speculative. In only a few markets were there any months when maize or beans were not available for extended periods, although they were sometimes scarce and expensive. Yellow maize became widely available around September. It was sold at somewhat lower prices, being considered an inferior food. The very high prices in October reflect a demand for seed as much as for food. Access to a cash income is critical in recovering from drought, to purchase seed for the next season. Many people planted yellow maize, but it did not germinate.

6.5.2. Employment

Data on household cash income in the study area are not readily available, and probably not reliable. Data from a survey of 229 households divided among the five agroclimatic zones of the study area indicate that few of the households have permanently employed members of the nuclear family (either living with the household or not) (*Table 6.4*). The highest percentage (30%) is in zone I and lowest in zones IV and V. Permanent employment is the only cash income that is fairly insensitive to climatic fluctuations, at least in the short term. Long droughts, over more than one season, may force the closure of some agroindustries. Several firms making grain products (bakeries, millers, etc.) were closed temporarily during the 1984 drought.

Table 6.10. Average local market prices in 1984 (in Kshs./kg) and percentage increase over the year.[a] Source: NES (1985a); data collected in January 1985 – *see* text for further details.

Crop/ quarter	*Agroclimatic zone* I	II	III	IV	V
White maize:					
Jan–Mar	3.2	3.7	3.0	1.8	3.4
Apr–Jun	4.4	4.3	4.5	3.2	4.1
Jul–Sep	5.0	5.2	6.1	4.0	5.2
Oct–Dec	4.6	5.8	7.1	6.7	6.0
Jan–Dec incl. (%)	44	57	137	272	76
Beans:					
Jan–Mar	9.2	8.8	7.6	3.0	8.9
Apr–Jun	11.8	10.8	10.4	3.7	10.6
Jul–Sep	14.7	14.7	13.2	9.0	13.7
Oct–Dec	16.0	16.3	17.6	11.7	16.6
Jan–Dec incl. (%)	74	85	132	290	87

[a]US$1.00 = Kshs. 16.5 in 1984.

6.5.3. Migration

During periods of food crisis, some men and women leave home to go to neighboring districts or areas to look for casual labor so they can send money or food to their families. For example, the Mbeere have traditionally survived by going to the upper parts of Embu District. In more recent times such temporary migrations have tended to restrict themselves within district boundaries. People from the lower parts of Embu and Kirinyaga Districts would move to the high potential areas to provide casual labor on tea and coffee farms, especially during the harvest season when more labor is required. In the lower parts, the Mwea rice scheme is a significant destination of casual labor migrants during the planting and harvest seasons. In the western parts of Machakos District, coffee picking demands more labor than family members can provide. For example, Mbithi and Wisner (1974, p. 94) observed that "short term migration to the nearest upland area to search for wage employment was very common in the marginal zones of Meru, Embu and Kitui. Mbeere travelled to the lower slopes of Mount Kenya on the Embu side and to Chuka".

Apart from rural-to-rural migration aimed at providing casual labor on farms or as herders, there are rural-to-urban migrants, whose destinations are Kenya's major towns: Nairobi and Mombasa. Those who go to urban areas are relatively young, well-educated and predominately male, a phenomenon that removes from the rural areas many able-bodied, innovative people. Districts of origin often suffer from this exodus of skill and labor. However, if the migrants remit some money to members of the family at home, then their absence from home is an advantage.

6.5.4. Cash crops

In zones I and II income from coffee and tea is fairly reliable, and supplied many households in 1984 with cash to buy food. Tea farmers, in particular, experienced a boom year as world prices rose sharply, following drought in India. Drought has a delayed effect on coffee and tea income, since full payment is received long after the harvest. These tree crops recover fairly quickly after a drought, and good harvests were realized after the short rains in 1984–1985. In the lower zones cotton, sisal and tobacco are the chief cash (nonfood) crops, although food crops are sold as well.

Cash crops have been promoted more widely than food crops in Kenya. The foreign exchange from coffee and tea is a major element of the national economy. While some argue that this land could be used to feed people (Rempel, 1985), households growing a large proportion of cash crops can probably buy food in times of scarcity, as long as it is available at reasonable prices. In some cases there is a sexual division of agricultural work and responsibilities for securing the household's food. Where the woman must feed the family and the man is paid for the cash crop, the money may not be spent on food for the entire household, subject to the husband's desires when paid. In such cases, on-farm food production is essential.

Charcoal burning increases dramatically during a drought, as the poor are left with few alternatives for cash. It could be an important rural industry on a sustainable yield basis with improved methods.

6.5.5. Livestock

Livestock, the traditional source of cash in the drier zones, suffers drastically in a drought. In his study, Kolkena (n.d.) found that 74% of his sample of households in Machakos owned goats; 47% owned cows and 40% had bulls and oxen. In 1984 livestock losses in the study area were estimated to be between 50% and 60%. Recent data from Kitui, a drier district to the east of Machakos, may be used to supplement the available data for the lower zones of the study area. Akong'a (1982) found that out of 345 heads of households in Kitui District, 75% owned some livestock. Of those who owned livestock, 68% had sold some between January 1980 and the time the survey started in September of the same year (a drought year). A question in Akong'a's survey required respondents to say what they did with the last harvest. The data indicate that, in the drier areas, for a family to obtain cash income, it must have some livestock. If livestock, which can be sold, are not available, then the little food harvested during the drought will be insufficient when stored food from previous seasons is depleted (*see also* Pollard, 1981). Respondents were asked to name their sources of food during the period of famine. By far the most common source was purchased food.

The Kenya Meat Commission is responsible for interdistrict livestock purchases, but is not able to handle the entire supply during a drought. Also, their prices are often lower than in the local markets. In a drought the local demand for meat is reduced as there are alternative needs for cash. Thus, the markets quickly become saturated and prices plummet. Problems in transporting and slaughtering animals outside the impact area are further constraints to an effective emergency livestock market (Airey, 1985). Mobile abattoirs, once operated by the colonial government, could be a strategy to overcome these difficulties. In one area of zone V, the price of a goat in 1984 in the local market was only enough to purchase a 2 kg packet of yellow maize. The traditional reluctance to sell livestock during drought is probably due to these poor market arrangements and high costs of replacing the animals after the drought. Natural increase is slow, and herd owners may not have cash reserves for purchasing livestock. Successive droughts, even if separated by several years, are particularly crippling to the livestock owner who has not been able to rebuild his herds.

6.5.6. Drought scenarios and monetary exchanges

Selling capital assets, unless there is investment in productive enterprises, is a coping strategy of last resort, and is generally of limited effectiveness beyond the immediate need. Local businesses, e.g., small shops along the road, are not usually very productive and during droughts are dependent on the local residents' ability to purchase supplementary foods and non-essential consumer items. Monetary activities other than selling capital assets may be highly effective in normal years and in both moderate and severe droughts.

The nature of the drought – in particular, its length and geographic coverage – may influence the effectiveness of wage labor as a coping strategy. If the drought affects the regional or national economy enough to force the closing of firms and reduce spending on private or public projects, the resulting unemployment will have a significant impact on household vulnerability to drought. This was not a widespread phenomenon in Kenya in 1984, but should not be discounted on the basis of the experience of one episode – a severe, but short-lived drought.

6.6. Vulnerability in Entitlement to Donations and Famine Relief

The third source of food, and usually one of last resort in modern economies, consists of donations and famine relief from government, religious groups, nongovernmental organizations, and private citizens. Eastern Province has been a traditional recipient of food assistance in Kenya, although Machakos and Embu have received relatively less than the other drier districts of the province. Central Province received on the order of 1/100th the food assistance provided to Eastern Province in the years 1973–1983 (Kliest, 1985). The drought of 1984 was exceptional in that both provinces required assistance.

The government famine relief efforts in 1984 were coordinated by an *ad hoc* interministerial committee in the Office of the President, with a parallel committee of NGOs coordinating the private relief efforts. Government relief foods were distributed by the district administration through the local chiefs. Each chief identified eligible recipients in his location and distributed the food he received accordingly. He was assisted by the local elders and leaders.

Preliminary figures from the NES smallholder drought response survey indicate that in zones IV and V over 50% of the respondents received food assistance from the government, religious groups or NGOs (*Table 6.4*). These figures are for recipients of any food, regardless of amount. In zones II and III the figure was about 40%, and under 10% in zone I. Although famine relief was more widely available than in previous famines, the data for zones III, IV and V indicate that screening of recipients was fairly effective. The survey did not reveal anyone who should have received assistance but did not. However, the majority of the foods arrived in the rural markets several months after local supplies were virtually exhausted. On a regional basis, the drought-affected areas east of the Rift Valley were targeted for food relief and distribution of the imported yellow maize. The white maize surplus areas were well supplied with the preferred white maize.

There did not seem to be large transfers of food from relatives to their drought-stricken families. The controls on food shipments and transport costs are constraints, and cash is more easily sent.

In some respects the 1984 drought was exceptional. The government had ample foreign exchange reserves to import food – the first few months of supplies were all commercially obtained. Although bilateral and World Food Program shipments took a long time to arrive, there were good reserves in the country from the previous two years. The transport and distribution network was adequate. And western and coastal Kenya received normal rainfall and produced average crops. However, these advantageous conditions that helped reduce the impact of the 1984 drought might not prevail in a future drought episode. The reliability of donations as a household food source to avert disaster in future droughts should not be assumed.

Remittances from relatives and friends (self-help donations) and loans from cooperatives or local money lenders are common in all years. The flow of remittances may decrease in severe, prolonged droughts if the giver is also affected by the drought. Likewise rural credit organizations may curtail loans if the prospects of recovery appear to be small. On the other hand, government, nongovernmental organizations and religious groups usually increase their donations during severe drought, as in 1984, but may provide little, if any, relief in moderate droughts, as in 1983 in Machakos.

6.7. Bearing the Consequences

When all sources of food are insufficient to meet the household's needs, the household may use up its savings or sell its capital investments, such as livestock or land, in order to purchase food. More serious social consequences may include abandonment of the farm (at least temporarily). Nutritional and health effects may begin early in a drought as families ration their food or change their diet. Morbidity and mortality are thus the ultimate measures of drought impacts.

The NES survey did not reveal many households that had abandoned their farms during the 1984 drought. A quarter of the households reported a household member had left the compound due to the drought. But only in a few cases did entire households leave (perhaps 1–2% of the sample, although it is difficult to estimate), and there could equally well have been other social reasons (divorce, old age and sickness, change in employment).

There was certainly an increase in malnutrition during the drought. Reports from health centers and nutrition monitoring projects indicate at least a several-fold increase in reported malnutrition. But the baseline data are uncertain. The NES survey measured mid-arm circumferences (MAC) for children under five years of age. MAC is a composite measure of stunting and wasting and has a considerable margin of error (Central Bureau of Statistics, 1977). Nevertheless, it is a useful preliminary indicator of malnutrition. In central and eastern Kenya, the data from the NES survey do not indicate long-term, lasting nutritional impacts from the 1984 drought (*Table 6.11*). The values are not appreciably different from those measured in 1977 by the Central Bureau of Statistics (1977). Assuming that there was increased malnutrition and that the sample reported here was representative, recovery, by January 1985, seems to have been fairly rapid.

Table 6.11. Mid-arm circumference (MAC) by agroclimatic zone.[a] Source: NES (1985b).

	Agroclimatic zone		
MAC	*I–II*	*III–IV*	*V–VI*
< 11 cm	0 (0)	6 (2)	0 (0)
11–11.9	2 (2)	7 (3)	4 (3)
12–12.9	1 (1)	21 (8)	9 (7)
13–13.9	14 (11)	46 (17)	31 (24)
14–14.9	36 (30)	72 (27)	35 (27)
> 15	69 (57)	117 (43)	51 (39)
Total	122	269	130

[a]Numbers in parentheses are percentages of column total.

6.8. Policy Implications of Smallholder Vulnerability to Drought

The main thesis of this paper is that vulnerability to drought in central and eastern Kenya is a result of social and economic trends, rather than the specific characteristics of the climate of the area. Further, while the range of coping strategies practiced by rural smallholders appears to have become smaller in the last few decades, the most effective and widely adopted coping strategies are adaptations, i.e., fundamental long-term changes in the household's economy, which diversify the household's sources of income. This fundamental trend toward increasing reliance on off-farm, monetary income has almost certainly lessened the household's vulnerability to drought. At the same time, it shifts the burden of drought to the regional and national economy – most immediately in requiring a sufficient national strategic reserve to ensure that rural markets have adequate supplies at reasonable prices. In a longer-term perspective, the most effective national drought policy will be to maintain a healthy economy, particularly one in which rural dwellers can freely participate. The government's policy to encourage rural growth centers is helpful in this regard.

There is a large number of households that does not have permanently employed family members and relies on casual labor for their major cash needs. These households are more vulnerable to drought, as the availability of casual work, particularly agricultural, may change during a drought. For these households the most effective policy is likely to be emergency work programs that maintain income to the household to purchase food. Food for Work program also fulfill this objective in an indirect manner, but are more difficult to organize and are more costly. As long as the market is supplied with food, it is simpler to ensure households are able to purchase their food needs.

Emergency food relief can then be targeted to those households that are less integrated into the monetary economy and are more dependent on their own production. These households are not necessarily confined to certain agroclimatic zones. Rather they reflect the characteristics of the household (perhaps elderly or headed by widows), their available land and capital resources, and other coincidental misfortunes (previous droughts, sickness, etc.).

Understanding the climatic resources of the area is important for regional planning and for household operations. The long-term expected climate is important for planning where agricultural investment and innovation will be most fruitful. This is the primary responsibility of the government, through its research organizations, extension agents and projects to develop the agricultural infrastructure. It may not be practical to expect each district to be self-sufficient in food. In the drier areas, increasing agricultural productivity may require relatively more investment than in the wetter zones, and the limit of productivity in the foreseeable future is not very high. Thus, economic development in the drier zones based on agricultural production is limited, and nonagricultural alternatives will be necessary.

For the household, the most important application of climatic knowledge relates to the time scale of a season. While there exists a large number of cultivation techniques that can be applied in a poor year to make good use of the available water, the most promising application of weather monitoring and prediction is to determine when it will be an average or good season, so that additional inputs and hired labor may be worthwhile. Greater access to and use of draught animals may also be critical. Surplus production could then be sold or stored for the inevitable drought. An improved practical understanding of climatic variability, and the matching of its monitoring and prediction to the needs of smallholder farming systems and decision making, may well be the key to intensifying agricultural production (and reducing the consequences of drought) in the drier zones of central and eastern Kenya.

SECTION 7

Implications for Policy

7.1. Policy Background

The overall economic strategy adopted by the Kenya Government since independence in 1963 has been based on an open market system in which profit to individuals or organizations has been an accepted driving force. In the agricultural sector, the involvement of government at both production and processing levels has been rather limited, although involvement in marketing and distribution has been considerable. During the 1960s and 1970s agricultural production as a whole expanded at an overall rate of about 3.5% per annum (Government of Kenya, 1981). However, with an annual population growth rate of over 4% per annum, even this remarkable growth in output has not kept pace with the food needs of the population.

Subsection 7.2 summarizes the implications of the previous papers in this case study. Subsection 7.3 examines various aspects of government policy to determine where changes may be possible not only to increase overall output, but also to reduce the effects of the vagaries of the weather on agricultural production and farmers' living standards. Conclusions are given in Subsection 7.4.

7.2. Summary of the Kenya Case Study

The previous sections of this case study have considered the climate, agricultural production and smallholder socioeconomics of the study area – Kirinyaga, Embu and Machakos Districts in Central and Eastern Kenya. Five scenarios were developed, based on the cumulative probability of seasonal rainfall for the 1951–1980 record:

- A good season with a 10% probability of occurrence.
- An average (median) season.

Table 7.1. Summary of climatic impact assessment in the tea–coffee zone, I/II (Kerugoya).

	Agroclimatology			*Production*		
Scenario	*Normal rainfall* (mm)	*% of normal*	*ACZ*	*Forage* (DM t/ha)	*Maize* (kg/ha)	*Smallholder economics*
Good year	2040	134	I	19	2400	Good to very good yield from food and cash crops. Off-farm and cash crop income available for investment, education and household purchases. Adequate feed for grade cows.
Average year	1525	100	II	13	1900	Good crop yields for food and cash crops. Income available as above. Grade cows require use of off-farm feed and crop residues to supplement on-farm forage.
Moderate drought	1100	72	III	11	1300	Fair crop yields. Proceeds from previous cash crops main source of farm income. Grade cows severely affected by feed shortage resulting in little milk production and calf losses.
Severe drought	940	62	IV/V	9	250	Poor crop yields, but total crop failure unlikely. Proceeds from previous year's cash crop vital to farm income. Grade cows severely affected as even off-farm grazing and crop residues in limited supply. Very little milk produced; calf mortality approaching 100%
Back-to-back drought	1300 /940	–	–	–	–	Poor crop yields in both years. Little cash crop income in second year so credit and off-farm income required to purchase farm needs, food and subsequent reinvestment. Stock deaths expected as feed availability shrinks in second year.

- A moderate drought season with a 10% probability.
- A severe drought with a 2% probability of occurrence.
- Back-to-back drought (successive seasons of drought with a 10% or less probability).

Table 7.2. Summary of climatic impact assessment in the maize–cotton zone, III (Embu).

	Agroclimatology			*Production*		
Scenario	*Normal rainfall* (mm)	*% of normal*	*ACZ*	*Forage* (DM t/ha)	*Maize* (kg/ha)	*Smallholder economics*
Good year	1520	144	II	16	2000	Good to very good yield from food and cash crops. Off-farm and cash crop income available for farm requirements. Adequate feed for grade cows.
Average year	1055	100	III	10	1200	Fair to good crop yields. Forage for grade cows will require supplementation from off-farm or crop residue sources.
Moderate drought	720	68	IV	5	350	Cash and food crop harvests poor. Previous year's cash proceeds and off-farm income important for farm purchases. Survival of Napier grass marginal; feed supply sub-maintenance with possible stock deaths.
Severe drought	675	64	V	2.5	250	Very little crop or forage growth. Proceeds from previous year's cash crop needed for food and other purchases. Napier grass will die out; stock losses may reach 70–80% in absence of other feed from off-farm resources.
Back-to-back drought	750/ 675	–	–	–	–	Little or no crop yield or forage production in first year, followed by similar second year, resulting in no cash crop income and no food available on farm. Credit and off-farm income vital for provision of food and reinvestment in forage, livestock and seeds after drought.

The implications of each scenario, as developed in the preceding sections, are presented in *Tables 7.1* through *7.4* for the four major ecological zones in the study area, utilizing data from the four representative stations. Katumani has been used instead of Machakos to represent zone IV, as it is more representative of the livestock situation in the zone. It is somewhat drier than the Machakos station, which is at the upper limit of the zone. The scenarios have been applied to an entire year, rather than individual seasons, to make them more compatible between zones, and because the estimates of yields (maize and grasslands) are available on an annual basis. The annual rainfall and percentage of normal are climatic measures of the scenario. The agroclimatic zone (ACZ) is based on the ratio of rainfall to evaporation, as described in Section 3. Maize production is estimated from data in Jaetzold and Schmidt (1983). These are very imprecise data, and should be interpreted in a relative sense rather than as an indication of absolute production (*see* comments made in Sections 3 and 4). The livestock production figures are from Section 5, and the description of smallholder socioeconomics from Sections 5 and 6.

In the tea–coffee zone (agroclimatic zones I and II), as typified by Kerugoya, good and average years and moderate drought years range between annual zonations of I and III. Dry matter production and maize yields in these scenarios will be adequate to meet household subsistence needs, and in the better years will provide a surplus for sale or storage. Income from agricultural produce in these years can be used for development, education and household necessities and luxuries. The situation changes markedly in a serious drought. Given the usually small areas of the holdings, grade livestock may not survive on on-farm fodder production and food will have to be purchased, unless previous harvests were stored. The serious drought will not be critical unless it is prolonged, although the few household with little savings or off-farm income may need special assistance.

The maize–sunflower zone (III), as typified by Embu, is drier than zones I and II, and the annual agroclimatic zone (ACZ) varies between II and V. Thus, in a good year, agricultural productivity is sufficient to provide surplus for investment, education and luxuries. Depending on the size of holding and access to off-farm grazing, average years may also provide sufficient surpluses. A moderate drought requires households to utilize stored production or off-farm income to meet subsistence needs. In a severe drought livestock sales will be required for households with no off-farm income to meet their food needs. Food assistance may be needed by many households.

The marginal cotton zone (IV), as typified by Katumani, is on the margin between successful maize and livestock (on-farm) production. There is very limited cash cropping in this zone, owing to the variable climate and lack of a suitable crop (including the marketing infrastructure). In a good year the area is productive (zone II/III), and maize yields are above subsistence needs. Livestock production will depend on farm size and off-farm grazing. With adequate resources, surplus milk may be sold in the good year. The average scenario indicates little yield from maize and very low milk production. If there is insufficient fodder or off-farm grazing is unavailable, livestock will have to be sold. In the moderate and severe drought, crops will fail and livestock mortality will be high unless extensive off-farm resources are available. Famine relief will be necessary for households without off-farm income, particularly in the event of a prolonged drought. In the back-to-back drought, livestock will have been sold or slaughtered by the second year. Assistance will have to be provided for recovery from the drought (seed and stock), as few households will have cash income for purchasing farm inputs.

Table 7.3. Summary of climatic impact assessement in the marginal cotton zone, IV (Machakos and Katumani).[a]

	Agroclimatology			*Production*		
Scenario	*Normal rainfall* (mm)	*% of normal*	*ACZ*	*Forage* (DM t/ha)	*Maize* (kg/ha)	*Smallholder economics*
Good year	1150	169	II/III	6	1000	Food crops give fair to good yield with some surplus. Very limited cash cropping. Milk output from crossbreeds depends on off-farm resources and crop residues, on-farm resources adequate for little more than maintenance.
Average year	680	100	IV	1.6	850	Poor crop yields, insufficient even for subsistence. Milk output extremely low, depending on off-farm feed resources as crop residues very limited. Reliance on stock sales and off-farm income for food purchases.
Moderate drought	450	66	V	.6	250	Crops yield little. Stock deaths high due to lack of feed resources both on and off-farm. Without off-farm income, famine relief necessary.
Severe drought	440	65	V/VI	.5	0	Crop failure. Stock deaths may reach 70-80%. Extensive famine relief required.
Back-to-back drought	450/ 480	–	–	–	–	Situation in year 1 as for severe drought, but in year 2 stock sales will not be practicable as stock left will be very limited. Extensive famine relief will be required, followed by credit or grants for restocking and planting after drought.

[a]Data are for Katumani, which represents the average conditions in the zone for livestock and most smallholders more accurately than Machakos.

The situation in the livestock–millet zone (V), as typified by Makindu, is different from the wetter areas. Livestock and drought-resistant crops are the agricultural mainstays. Even in a good year (when the area is classified as zone IV), there will be little harvest from maize. In moderate and severe droughts (classified as zone VI and VII), households must sell livestock or have off-farm income to purchase their food requirements. Famine relief is probably required

Table 7.4. Summary of climatic impact assessment in the livestock–millet zone, V (Makindu).

	Agroclimatology			*Production*		
Scenario	*Normal rainfall* (mm)	*% of normal*	*ACZ*	*Forage* (DM t/ha)	*Maize* (kg/ha)	*Smallholder economics*
Good year	750	150	IV	25	250	Maize crop minimal, more drought-resistant crops, such as sorghum, yield better. Livestock produce little milk; main income from sale of meat. Maximum meat production from natural grazing.
Average year	500	100	V	20	200	Crops comments as above. Reasonable meat production offers some cash income.
Moderate drought	350	70	VI	15	0	Almost complete crop failure. Forage quality good only for short period, meat production limited by lack of grazing of adequate nutrititve value. Without off-farm income, stock sales required; but market saturation gives low returns for stock in poor condition. Famine relief probably required.
Severe drought	220	44	VII	<10	0	Complete crop failure. Very little meat production due to lack of forage. Previous range condition critical to stock survival. Stock deaths widespread. Famine relief or off-farm income extensively required; limited possibility for stock sales, as above.
Back-to-back drought	300/ 250	–	–	–	–	First year as for severe drought scenario. Second year effects depend on previous range condition, but stock deaths widespread. Famine relief required, followed by assistance in restocking and seed purchase after drought.

for many households in the moderate drought, and will be a widespread need in severe and prolonged droughts. Long-term storage of food (over a year) was formerly prevalent in this zone. Since the 1984 drought, people may revive this practice. Improvement in the livestock marketing system in drought years is necessary to enable households to sell stock and purchase food.

To summarize the foregoing, drought in the high potential zones (I and II) is relatively less frequent, rarely as severe and has less impact at the household level than drought in the lower zones. The presence of permanent cash crops and a higher level of development mean there are fewer households that will need emergency assistance during drought episodes. This is not to say there are not significant economic impacts of drought in the coffee and tea zones. That has been clearly shown in the foregoing sections. However, the drought impacts differ in character from those in the semi-arid zones, III through V. In these lower zones, the reliance on an annual crop for food and cash, and the vagaries of livestock production and marketing, accentuate the vulnerability to drought. The following discussion examines the history of government policy in the semi-arid areas of Kenya, as background to the recommendations given in Subsection 7.4.

7.3. Implications of the Case Study Results for Future Policy

7.3.1. Production technology improvement

Much of the sustained growth in agricultural output during the past two decades in Kenya has resulted from a rather unique set of circumstances, which are unlikely to recur:

(1) Rapid expansion of the cultivated area, especially in the high potential areas, has followed the allocation and subdivision of farms previously occupied by European settlers.
(2) The enthusiasm of the immediate post-independence period encouraged investment in the agricultural sector.
(3) Several research-based improvements in production technology became available which provided very rapid "quantum leaps" in level of output. (These included hybrid seeds and improved breeds of livestock).

In the future, improvement in levels of production may be expected to be less spectacular, with the emphasis on reducing production costs by methods appropriate to the small farm size and lack of cheap agrochemicals and farm machinery. Efforts will be required to close the gap between research trial yields and those on farms and to increase security of yields without necessarily raising the maximum yield potential.

Further evaluation of the efficacy of "Response Farming" is of particular relevance here as it is very suited to the situation in the semi-arid zones of the study area. With this technique, examination of past seasonal weather patterns may be used to indicate criteria during the early part of the growing season that give a useful prediction of entire season. Results so far obtained from sites within the study area suggest that the date of onset of the rains and the early season rainfall within the first three or four weeks may be useful predictors of whether the season is likely to be good, bad or average (Stewart and Faught, 1984). Further development of this approach offers the prospect of greater stability of crop yields, especially in a poor season, together with a more efficient use of scarce resources. For example, possible alternative strategies in a maize–beans system could be:

(1) Poor season – No fertilizer top-dressing, no intercropping and reduction in plant population by thinning.
(2) Average season – No fertilizer top-dressing, intercropping permitted and standard plant population.
(3) Good season – Fertilizer top-dressing recommended, intercropping permitted and plant population kept at standard level.

The possible benefits from such an approach would also be enhanced if improvements in understanding of regional weather patterns enabled more accurate seasonal weather forecasting to be done, as the approach can readily incorporate such forecast information. Research results to date suggest that Response Farming methods may have considerable promise, but much more work is required to evaluate their on-farm practicability.

Improvements in extension services to provide advice on production methods in Kenya are currently being assisted by the World Bank under the "Training and Visit" system. In this system particular aspects of production of the commodity in question are dealt with in detail at an appropriate point in the growing season using the farmers' own fields or animal units. Such efforts, however, will only be successful if the farmer can be encouraged to produce the product for a guaranteed market. Development of the Training and Visit techniques to help livestock enterprises is at a very early stage, with no transferable experience being available from other countries.

7.3.2. Marketing

As mentioned above, the area of the agricultural sector in which government involvement has been considerable has been in marketing and distribution. Agricultural production over and above subsistence needs has been distributed through one of three main channels, depending on the product concerned. Without effective marketing systems, price and availability fluctuations from year to year will remain very large. In years of good harvests, prices fall and storage becomes a problem. In poor years, however, due to the lack of suitable on-farm storage and the need for cash, the farmer may have to sell his crop immediately after harvest and then purchase the same, or similar, food several months later at two or three times the price.

Government marketing agencies

Marketing of the major food products has been by agencies through which the government can directly influence levels of production by controlling prices not only of sales by farmers, but also of costs to distributors or processors. The major statutory bodies concerned at present are:

(1) Maize and wheat – National Cereals and Produce Board (NCPB).
(2) Milk – Kenya Cooperative Creameries (KCC).
(3) Meat – Kenya Meat Commission (KMC).

These bodies have powers derived from statutes that can be invoked to control not only prices, but also the movement of these products through the country. In the case of the first two bodies, the purchase of their products is a legal monopoly, for the NCPB on a national basis, and for KCC within its operating zone (basically the high-potential farming areas).

The establishment of these agencies was aimed at ensuring continued, adequate national food supply (Government of Kenya, 1981). Difficulties have arisen in the operation of these agencies in the past few years, and their operations are undergoing careful examination at present. In the case of NCPB, limits to storage facilities available, cumbersome payment systems and frequent uncertainties over the stability of the realization price of the product between the beginning of the season and harvest have been disincentives to production. For the two livestock agencies, the major problem has been in delay in payments. In a situation where inputs are required continually, rather than seasonally as with crops, such delays have serious effects on the level of production.

Quasi-governmental marketing agencies

For the agricultural products destined for export, a number of agencies have been established that are in theory under government control but in practice are relatively independent. Coffee, tea, pyrethrum and horticultural crops fall into the category of crops handled by these agencies. Expansion of output in these products has been more rapid than for the food crops, due to relatively attractive prices being offered to the farmers as a result of high world demand and international marketing agreements. In recent years, however, the market for these products has become more volatile, as the growth in world demand has slackened and new competing producing countries have entered the market. In general, however, these quasi-governmental marketing agencies have worked much more successfully than have the government marketing agencies.

The system of payment used by the quasi-governmental agencies involves a basic initial sum to the farmer upon delivery of the product, followed by a later payment of the balance on the basis of the price realized internationally. The initial payment largely covers the requirements by the farmer for his immediate recurrent needs, while the payment of the balance somewhat later provides for major investments and may also be a much-needed source of income during a poor year following a good season.

In the current Kenya Development Policy, emphasis on expansion of cash crop production is being reduced, with efforts being directed more toward expansion of food crops to increase national food security (Government of Kenya, 1983).

Private marketing agencies

For certain agricultural products in Kenya, direct arrangements are made between producer and distributor or processor. Barley, for Kenya Breweries, oil seed crops for East African Industries, sugar cane for the various sugar mills and seed for the seed companies are examples of this category of product. Under such arrangements, the farmers have been assured of a guaranteed market at an agreed price, usually paid promptly following delivery of the crop. The requirements for high quality control of the product have led to the development of effective private extension services to provide the necessary advice on production techniques, together with an input resource package supplied on credit against the expected value of the crop. Although there are possible dangers of exploitation of the farmers by relatively sophisticated commercial interests, experience to date suggests that the interests of both parties are met reasonably satisfactorily. In general, this type of market system has proved acceptable to the farmers concerned. However, in some areas where there is little production of food crops and the cash income does not go toward household food consumption, malnutrition rates are well above the national average. One of the factors contributing to this is that payments for the cash crop usually go to the men, with the result that the income may be devoted to activities such as drinking, rather than to food purchases or investment for further farm development.

The role of cooperatives

Large numbers of local marketing cooperatives have been established with government encouragement, in an attempt to provide farmers with easier access to inputs and markets than would have been possible if they operated on an individual basis. With very few exceptions, agricultural cooperatives have proved to be a drain rather than a benefit to the farmer, due to inefficient management and, frequently, corruption involving misuse of funds.

Future marketing policy options

The government food policy document (Government of Kenya, 1981) suggests an increase in the scope of the private marketing sector available to the farmer. This may be seen as a realization by the government that flexibility in marketing may be a source of encouragement to the farmer to increase investment of resources in agricultural production. The continuation of the government agencies alongside the private markets would enable the government to ensure, with state agencies acting as buyer of last resort, the maintenance of strategic reserves of the major food items.

7.3.3. Livestock development policy

The value of livestock to the small farmer has been discussed in some detail in Sections 5 and 6. The establishment in 1979 of a Ministry of Livestock

Development marked the acceptance by the government of the importance of stock in the national economy. The National Livestock Development Policy document (Government of Kenya, 1980b) summarized the strategy to be adopted for the expansion of the livestock industry. Rationalization of marketing was identified as vital, particularly in the semi-arid areas where easier sale of stock at reasonable prices would allow for reduced stocking levels not only by provision of greater income from fewer, higher-grade animals, but also by reducing the absolute dependence on livestock for subsistence food needs. The proceeds from the stock sales may be used to procure a more varied diet. There have already been studies indicating that improvement in the local marketing situation may contribute to rapid changes in animal management systems with benefits to herd condition, grazing resources and farm income (White and Meadows, 1981). Expansion of breeding centers for both small stock and cattle has already commenced (Government of Kenya, 1980a) and will provide for more rapid recovery from stock deaths in droughts, provided a suitable credit system can be established.

7.3.4. Credit

The National Food Policy document (Government of Kenya, 1981) acknowledged the vital role of credit in the development and stabilization of the agricultural industry. The document does not, however, discuss in any detail the institutional framework required, other than stressing the need for some form of decentralization, and the preference for credit to be payable in kind rather than in cash.

The provision of credit financing for seasonal or long-term use to the small-farm sector is not attractive at present to the commercial banking sector, due to the lack of collateral. Although the farm land title may be available for collateral on an individual basis, the social costs of dispossession in case of nonpayment not only throw a heavy burden on the state to provide for the affected family in future, but also generate considerable animosity towards the banking sector. The provision of credit to the small-farm sector is clearly associated with greater risk than for large-scale farming and will therefore require considerable involvement in underwriting possible losses.

Past experience in Kenya suggests that the methodology of the provision of credit will require careful study. Credit schemes developed in the past have run into great difficulties because of poor repayment. Losses have occurred even where the product – for example, maize – is in theory only marketable through official channels, so that recovery of the loan out of the sale proceeds should have been possible. Extension of grace periods before repayment comes due, and even writing off of bad debts within these government schemes, have encouraged the feeling that the finance can be considered a grant rather than a loan that has to be repaid. The problems already noted for the marketing sector have further hampered the efficient use of such government credit schemes. However, where the provision of credit has been closely associated with extension advice, provision of inputs and marketing facilities, the use of credit and repayment records have been much better. This has been noted, for example, in the Machakos Integrated Development Program supported by the European Common Market Development Fund, and with the Dairy Development Project, supported by the Dutch Government. The relative success of such schemes confirms the general willingness of farmers to increase investment of resources, whether their own or through credit, provided that there is confidence in the availability of inputs and there is prompt payment at a reasonable price for the output.

7.3.5. Pricing

There are considerable political benefits available to any government that is able to maintain low food prices to the consumer in the urban areas. There is evidence that in Kenya consumer food prices have been kept low by means of low producer prices payable to the farmers, who have therefore been disinclined to raise output (Government of Kenya, 1980a). The effects of this have been confounded with the high costs associated with the maintenance of marketing agencies that absorb considerable fractions of the consumer price, leaving a small residue for payment to farmers. In the case of milk, the farmer has received only about 50% of the price paid by the consumer, with the remaining 50% being "lost" in the marketing and distribution costs of the Kenya Cooperative Creameries (KCC) operations.

In a basically free market economy, such as that in Kenya, direct government subsidy of food items to the consumer would be unexpected, as a matter of policy, apart from the strain placed on the national economy. However, evaluation of the possible role of food subsidies in the maintenance of output by the farmers, while at the same time ensuring an adequate local supply of relatively cheap food to the consumer, is likely to be given close attention in future. Although subsidization by the government may be a very expensive procedure, a thorough examination of the social and monetary costs of emergency food importation if the local food supply is inadequate may make subsidization of local production preferable. In 1984, for example, drought conditions in much of the country were such that rainfed agricultural production failed, whatever the level of technology used. Maintenance of a larger strategic reserve of food, if necessary from production supported through farmer subsidies, might have been advantageous in such a situation. An early announcement of increased producer prices for the 1984 short rains crop resulted in a very rapid recovery from a potentially more serious situation, so that by the middle of 1985 the strategic reserve in fact exceeded the recommended six-month requirement.

7.3.6. Food security

Present government policy is to hold a total of approximately six months' national requirements for the main food crop, maize, as a strategic reserve (Government of Kenya, 1981). In cases of depletion of this reserve, immediate importation is permitted for replenishment before the locally planted crop is harvested. Thus, in 1984, when it became apparent that widespread crop failure could be expected, due to the very poor rainfall, the effect on this strategic reserve was anticipated. Arrangements were therefore set in hand for the urgent importation of maize. The initial importations were arranged on a strictly commercial basis on the world markets, through the use of Kenya Government controlled funds, whether those funds were of Kenyan or other origin. This allowed for rapid action without the possibility of political or other *quid pro quoo* with long-term consequences, which might have been the case if imports had been arranged by requests for food aid from possible donors. Considerable publicity was given to these imports, and this had undoubted benefits in maintaining confidence in the government in a potentially difficult time, and almost completely avoided "panic" buying by the urban consumers. During the later stages of the drought-free distribution of food as famine relief was arranged by the government, with the assistance of private charities and donors. But the emphasis was on the distribution of food for purchase to reduce the dangers of long-term dependence on food aid, which has been found to reduce initiative for recovery following the end of drought in some other countries. The response to the drought in 1984 by the Kenya Government has clearly indicated the value of rapid, decisive, well-publicized action in alleviating the effects of a major climatic perturbation on food supply and in maintaining the confidence of the population.

7.3.7. Particular problems of the semi-arid areas

Throughout the Kenya case study, the particular sensitivity of the semi-arid areas to the effects of climatic variability have been apparent. The rapidly increasing migration into the zone by people with little relevant experience of the conditions, together with the high natural population increase (over 4% per annum) of the resident population, pose serious problems. Development of appropriate production schemes involving not only production technologies, but also credit and marketing arrangements, are urgently required to provide for both food security in these areas and for the ability to recover rapidly from the effects of the inescapable agricultural droughts.

The present government appreciates the problems of such areas, as evidenced by the provision of research and extension facilities. Integration of these technical services with an effective credit and marketing system will provide an effective environment for the expansion of output to cater to the growing population. In support of these purely agricultural requirements, the development of local industries, probably based on agricultural processing, in the first instance, could provide for employment in the semi-arid areas, to reduce direct dependence on agriculture for subsistence and income.

7.4. Conclusion

The case study of the agricultural effects of climatic variability and its socioeconomic impacts in Kenya has drawn several implications for policy, as noted above. These are tentative, due to the lack of key pieces of data, such as reliable crop yield measurements. The primary feature of this case study is the attempt to link analyses from different disciplines into an integrated assessment. Only by assessing the entire chain of events that link meteorological drought to human consequences and responses can effective interventions be devised to reduce vulnerability to drought or ameliorate its impacts. Integrated assessments make two major contributions. One is identifying those linkages in the drought hazard chain that are critical in determining the nature of the drought (both in time and in the magnitude of impacts) and that are particularly amenable to policy interventions. The second is noting where data are most unreliable and most needed. Extensive surveys and elaborate monitoring systems cannot be supported by the government unless their results contribute to policy decisions on a timely basis. The demand for data, and a consideration of who will use it and how it will be included in the next level of analysis, should guide its collection.

That climatic variability is a real and recurrent aspect of development in Africa cannot be denied. The experience of the Government of Kenya in 1984 illustrates many ways in which the effects of drought can be reduced. Future policies to reduce individual and national vulnerability to drought and prepare for its inevitable recurrence should be formulated by drawing upon integrated assessments of the role of climatic variability in the development of the country.

Acknowledgments

Many organizations and people have contributed, at least indirectly, to this study. Each of them is hereby acknowledged and thanked. The authors do not view the study of the relationships between climatic variability, agricultural production and socioeconomic impacts as being completed, and look forward to their colleagues' comments and further collaboration. In particular, they wish to thank the Ford Foundation and the International Institute for Applied Systems Analysis (IIASA) for their financial support. Finally, while these acknowledgments are many and sincere, the views expressed in this study are the authors' and do not represent the official positions of the authors' host institutions, the Ford Foundation or IIASA.

References

Abrams, P.D. (1979). *Kenya's Land Resettlement Story.* Challenge Publishers, Nairobi.

Agatsiva, J.L. and Mwendwa, H. (1982). *Land Use Mapping of Kenya Using Remote Sensing Techniques.* Kenya Rangeland Ecological Monitoring Unit, Nairobi.

Airey, J. (1985). Livestock marketing improvement measures. *Proceedings of the Animal Production Society of Kenya.* Nairobi.

Akong'a, J. (1982). *Famine, Famine Relief and Public Policy in Kitui District.* Institute for Development Studies, Nairobi.

Allan, W. (1965). *The African Husbandman.* Oliver & Boyd, London.

Alusa, A.L. (1978). A note on the onset of the rains of East Africa. *Research Report No. 3/78.* Meteorological Department, Nairobi.

Ardunuy, P. (1979). On the observed diurnal oscillation of the Somali Jet. *Monthly Weather Rev.* **107(12)**, 1694–1700.

Bahktri, M.N., Gavotti, S. and Kimemia, J.K. (1984). On-farm research at Katumani: pre-extension trials experience (with special reference to semi-arid areas of Eastern Province of Kenya). *E. Afr. Agric. For. J.* **44**, 437–443.

Berry, L. *et al.* (1971). *Human Adjustment to Agricultural Drought in Tanzania: Pilot Investigations.* University of Dar-es-Salaam, Bureau of Resource Assessment and Land Use Planning, Dar-es-Salaam.

Berry, L. *et al.* (1980). *Eastern Africa Country Profiles: Kenya.* Clark University, Worcester, MA.

Braun, H.M.H. (1977). The reliability of the rainy seasons in Machakos and Kitui Districts. *Miscellaneous Paper M12.* Kenya Soil Survey, Nairobi.

Brokensha, D. and Njeru, E.H.D. (1977). *Some Consequences for Land Adjudication in Mbeere Division, Embu.* Institute for Development Studies, Embu.

Brown, L.H. and Cochemé, J. (1973). A study of the agroclimatology of the highlands of Eastern Africa. *Technical Note No. 125.* World Meteorological Organization, Geneva.

Campbell, D. (1985). Response to drought among farmers and herders in southern Kajiado District, Kenya. *Human Ecology* **12(1)**, 35–64.

Central Bureau of Statistics (1971). *1969 Census of Population.* Government Printers, Nairobi.

Central Bureau of Statistics (1977). The rural Kenya nutrition survey. *Social Perspectives* **2(4)**.

Central Bureau of Statistics (1981a). *Household Budget Survey 1981/82, Enumerator's Reference Manual.* Ministry of Economic Planning and Development, Nairobi.

Central Bureau of Statistics (1981b). Infant mortality in Kenya: past and present differentials. *Social Perspectives* **6(1)**.

Central Bureau of Statistics (1981c). *Kenya Population Census, 1979, Volume 1.* Government Printers, Nairobi.

Central Bureau of Statistics (1981d). *The Integrated Rural Surveys 1976–79.* Ministry of Economic Planning and Development, Nairobi.

Central Bureau of Statistics (1982a). *Economic Survey, 1982.* Central Bureau of Statistics, Nairobi.

Central Bureau of Statistics (1982b). *Seasonal Variations in Food Crops: Evidence from IRS4.* Ministry of Economic Planning and Development, Nairobi.

Central Bureau of Statistics (1984). *Situation Analysis of Children and Women in Kenya: Section 4; The Wellbeing of Children.* United Nations Children's Fund, Nairobi.

Central Bureau of Statistics (1985). *Economic Survey, 1985.* Ministry of Finance and Planning, Nairobi.

Chambers, R. and Moris, J. (Eds) (1973). Mwea: an irrigated rice settlement in Central Kenya. *Afrika-Studien Nr. 83.* Weltforum-Verlag, Munich.

Degefu, W. (1985). *Some Aspects of Meteorological Drought in Ethiopia.* Ethiopia National Meteorological Services Agency, Addis Ababa.

Dennett, M.D., Rodgers, J.A. and Stern, R.D. (1982). Rainfall at Kampi-Ya-Mawe and Katumani, Kenya. Manuscript. *Tropical Agricultural Meteorology Group.* University of Reading, Reading.

Dowker, B.D. (1964). A note on the reduction in yield of Taboran maize by late planting. *E. Afr. Agric. For. J.* **30**, 33–34.

Downing, T.E. (1982). *Eastern Africa Regional Studies: Climate.* Clark University, Worcester, MA.

Dunne, T. and Wahome, E.K. (1981). *Soil Erosion Index Map.* Kenya Rangeland Ecological Monitoring Unit, Nairobi.

East African Meteorological Department (1975). *Climatological Statistics for East Africa, Part 1.* Eastern African Community, Nairobi.

Edwards, E.C. (1940). A vegetation map of Kenya with particular reference to grassland types. *J. of Ecology* **28**, 337–85.

Edwards, E.C. (1956). The ecological regions of Kenya: their classifications in relation to agricultural development. *J. Experimental Agricul.* **24**, 89–108.

Epp, H. *et al.* (1983). *Land Use in Kisii District.* Kenya Rangelands Ecological Monitoring Unit, Nairobi.

Farmer, G. (1981). *Regionalisation and Study of an Alleged Change in the Rainfall Climatology of East Africa.* PhD Dissertation. University of Sheffield, Sheffield, UK.

Farmer, G. (1987). A Rainfall Database for Eastern Africa with some kenyan Examples. *Paper presented at the First Technical Conference on Meteorological Research in Eastern and Southern Africa*, 6–9 January 1987. Kenya Meteorological Department, Nairobi.

Farmer, G. and Wigley, T.M.L. (1985). Climatic Trends for Tropical Africa. *A Research Report for the Overseas Development Administration.* Climatic Research Unit, Norwich, UK.

Findlater, H. (1977). Observational aspects of the low level cross-equatorial jet stream of the western Indian Ocean. *J. of Pure and Applied Geophysics* **11(516)**, 251–62.

Fisher, P.M. (1977). *Risk Aversion by Small Farmers – Can It Be Measured? Results of a Survey of 69 Small Farmers, Kabare Location, Kirinyaga District, March 1977.* MA Thesis. University of Nairobi, Nairobi.

Food and Agriculture Organization (FAO) (1977). Crop water requirements. *FAO Irrigation and Drainage Paper No. 24.* FAO, Rome.

Franzel, S.C. (1983). *Planning an Adaptive Production Research Program for Small Farmers: A Case Study of Farming Systems Research in Kirinyaga District, Kenya.* PhD Dissertation. University of Michigan, Ann Arbor, MI.

Glantz, M.H. (1984). El Ninõ – should it take the blame for disasters? *Mazingira* (March).

Goldson, J.R. (1963). The effect of time of planting on maize yield. *E. Afr. Agric. For. J.* **20**, 160–163.

Goudriaan, J. and van Laar, H.H. (1978). Calculation of daily totals of the gross CO_2 assimilation of leaf canopies. *Neth. J. Agric. Sci.* **26**, 373–382.

Government of Kenya (1979). *Development Plan 1979–1983.* Government Printers, Nairobi.

Government of Kenya (1980a). *Report of the Ad Hoc Committee on Maize and Pasture Research.* Ministry of Agriculture, Nairobi.

Government of Kenya (1980b). *National Livestock Development Policy.* Ministry of Livestock Development, Nairobi.

Government of Kenya (1981). *Sessional Paper No. 4 of 1981 on National Food Policy.* Government Printers, Nairobi.

Government of Kenya (1983). *National Five Year Development Plan for 1984–1988.* Ministry of Finance and Planning, Nairobi.

Greener, R. (1986). Food Studies Group, Central Bureau of Statistics, Nairobi. Personal communication.

Griffiths, J.F. (1972). *Climates of Africa.* Elsevier, New York.

Harrison, M.N. (1970). Maize improvement in East Africa. In C.L.A. Leaky (ed.), *Crop Improvement in East Africa.* Farnham Royal.

Haugerud, A. (1981). *Development and Household Economy in Two Ecozones of Embu District.* Institute for Development Studies, Nairobi.

Heemst, H.D.J. van, Merkelijn, J.J. and van Keulen, H. (1981). Labour Requirements in Various Agricultural Systems. *Quart. J. of Int. Agr.* **20**, 178–201.

Hounan, D.E. *et al.* (1975). Drought and agriculture. *Technical Note 138.* World Meteorological Organization, Geneva.

Hunt, D. (1978). Chayanov's model of peasant household resource allocation and its relevance to Mbeere Division, Eastern Kenya. *J. Development Studies* **15**, 59–86.

Jackson, I.J. (1981). *Local Differences in the Patterns of Variability of Tropical Rainfall: Some Characteristics and Implications.* Unpublished manuscript. University of New England, Armidale, NSW, Australia.

Jaetzold, R. and Schmidt, H. (1983). *Farm Management Handbook of Kenya: Natural Conditions and Farm Management Information.* Ministry of Agriculture, Nairobi.

Jodha, N.S. (1985). International Institute of Applied Systems Analysis, Laxenburg, Austria. Personal communication.

KARI (1983, 1984, 1985). *Annual Record of Research*, Animal Production Research Department, Kenya Agricultural Research Institute, Muguga, Kenya.

Karanja, G. (1985). Senior Pasture Agronomist, Embu Agricultural Research Station. Personal communication.

Kashasha, D.A.R. (1982). *A Study of Effective Rainfall for Crop Production.* MSc. Thesis. Department of Meteorology, University of Nairobi, Nairobi.

Kenya Department of Meteorology (1984). *Climatological Statistics.* Department of Meteorology, Nairobi.

Kirimi, J. (1986). Central Bureau of Statistics, Nairobi. Personal communication.

Kliest, T. (1985). *Regional and Seasonal Food Problems in Kenya.* Africa Studies Centre, Leiden.

Kodamaya, S. (1984). *Implication of Non-Farm Occupations for Farm Production among Smallholders in Murang'a District, Kenya.* Institute for Development Studies, Nairobi.

Kolkena, T.F.M. (n.d.). *Report on a Socio-economic Survey in Two Rural Areas of Machakos District, Kenya.* A Joint Research of the Royal Tropical Institute and the Department of Tropical Hygiene, Amsterdam. Royal Tropical Institute and the Department of Tropical Hygiene, Amsterdam.

Konijn, N.T. (1984). *A Crop Production and Environment Model for Long-Term Consequences of Agricultural Production*, Working Paper WP-84-51. International Institute for Applied Systems Analysis, Laxenburg, Austria.

Law, R. and Cooper, P.J.M. (1976). The effect and importance of soil temperature in determining the early growth rate and final grain yields of maize in western Kenya. *E. Afr. Agric. For. J.* **41**, 189–200.

Lewis, L., Kamau, G. and Cheruiyot, R.C. (1983). *Land Degradation Monitoring Programme of the National Environment and Human Settlements Secretariat, Kiambu District.* Clark University Program for International Development, Worcester, MA.

Llano-Alcober, D., Cornelis, J., Medland, J., Mrama, G.C., and Prayag, S. (1983). Agricultural development and research priorities for a semi-arid area of Machakos District, Kenya. *ICRA Bulletin 10.* Agricultural University, Wageningen, Netherlands.

Mascarenhas, A.C. (1973). Studies in famines and food shortages. *J. Geographical Association of Tanzania*, No.**8**.

Mbithi, P.M. (1972). *Drought, Welfare and Rural Development: Two Progress Reports and Provisional Recommendations Arising from a Joint Study of Drought in Kenya.* Institute for Development Studies and Department of Geography, Nairobi.

Mbithi, P. and Wisner, B. (1972). *Drought and Famine in Kenya: Magnitude and Attempted Solutions.* Institute for Development Studies, Nairobi.

Mbithi, P.M. and Wisner, B.G. (1974). Drought in eastern Kenya: nutritional status and farm activity. In G.F. White (ed.), *Natural Hazards: Local, National, Global.* John Wiley, New York.

Ministry of Water Development (1979). *Kenya National Master Water Plan.* Ministry of Water Development, Nairobi.

Monteith, J.L. (1973). *Principles of Environmental Physics.* Edward Arnold, London.

Morth, H.T. (1973). *A Study of the Areal and Temporal Distributions of Rainfall Anomalies in East Africa.* East African Meteorological Department, Nairobi.

Muhammed, L. (1986). Katumani Dryland Research Station, Machakos. Personal Communication.

Mungai, D.N. (1984). *Analysis of Some Seasonal Rainfall Characteristics in the Lake Victoria Region of Kenya.* MA Thesis. University of Nairobi, Nairobi.

Murage, F. (1985). National Cereals and Produce Board. Personal communication.

Musembi, D.K. (1984). *Identification of Crop-Growing Seasons of Semi-Arid Kenya by Analysis of the Soil Moisture Patterns.* MSc. Thesis. Texas A&M University, Austin, TX.

Mwaniki, N. (1982). *Social and Economic Impacts of Land Reform in Mbeere.* Institute for Development Studies, Nairobi.

Mwenda, E.W. (1983). Medium maturing maize breeding programme in Central Kenya. In J.C. Mohucs and W.H. Tahir (eds.), *More Food from Better Technology.* Food and Agriculture Organization, Rome.

Nadar, H.M. (1984). Matching rainfall patterns and maize growth stages under Machakos area conditions. *E. Afr. Agric. For. J.* **44**, 52–56.

National Environment Secretariat (1984). Endangered resources for development. *Proceedings of Workshop on the Status and Options for Management of Plant Communities in Kenya.* Environmental Training and Management in Africa, Nairobi.

National Environment Secretariat (1985a). *CBS/NES Survey of Drought Responses, Preliminary Results on Income Sources.* National Environment Secretariat, Nairobi.

National Environment Secretariat (1985b). *CBS/NES Survey of Drought Responses, Preliminary Findings.* National Environment Secretariat, Nairobi.

Neunhauser, P., Bayreuther, H., Engel, A., Friesenegger, M., Magelassa, A., Neves, A.M., Renneke, V. and Salzer, W. (1983). *Appropriate Land Use Systems for Smallholder Farms: A Survey of Ecological and Socio-Economic Conditions in Machakos district, Kenya.* Institute of Socio-Economics of Agriculture Development. Technical University of Berlin, Berlin.

Ngutter, L.G.K. (1979). Government policy in semi-arid lands: its evolution. *Paper presented at the Workshop on Semi-Arid Areas in Kenya*, 23–27 July. Institute for Development Studies, Nairobi.

Njau, T. (1987). Seasonal variability of Kenya rainfall and its teleconnections. *Paper presented at the First Technical Conference on Meteorological Research in Eastern and Southern Africa*, 6–9 January. Kenya Meteorological Department, Nairobi.

Nicholson, S.E. (1981). The historical climatology of Africa. In T.M.L. Wigley, M.J. Ingram and G. Farmer (eds.), *Climate and History.* Cambridge University Press, Cambridge, UK.

Nieuwolt, S. (1977). *Tropical Climatology.* John Wiley, New York.

Nieuwolt, S. (1980). The interpolation of rainfall in the Nairobi area. *Research Report 8/80.* Meteorological Department, Nairobi.

Njeru, E.H.N. (1979). *Social Structure and Environment: Two Case Studies from Eastern Kenya.* MA Thesis. University of Nairobi, Nairobi.

Njeru, N. (1984). Some physiological properties of different maize varieties at NARS Kitale. In J.C. Mohucs and W.H. Tahir (eds.), *More Food from Better Technology.* Food and Agriculture Organization, Rome.

Njoroge, K. (1984). Maize breeding at Katumani: the first 25 years (Abstract). *E. Afr. Agric. For. J.* **44**, 287.

Nkanata, J. (1985). *Workshop on Provision of Livestock Production Research Information for Use by Extension Workers in the Training and Visit Programme*, February. Kenya Agricultural Research Institute, Muguga.

Norton-Griffiths, M. (1977). *Aspects of Climate in Kajiado District.* Kenya Wildlife Management Project, Ken/71/526, Project Working Document No. 13. Food and Agriculture Organization, Rome.

Norton-Griffiths, M. (1986). Ecosystems, Ltd., Nairobi. Personal communication.

Obasi, G.O.P. and Kiangi, P.M.R. (1977). *Monthly Distribution of Water Balance Components in Kenya.* Kenya Meteorological Department, Nairobi.

Odingo, R. (1971). *The Kenya Highlands.* East African Publishing House, Nairobi.

Odingo, R. (1982). A study of the causes, consequences and policy recommendations on drought in Kenya. In R. Garcia and J. Escudero (eds.), *Drought and Man: The 1972 Case Study.* Pergamon, New York.

Ogallo, L.J. (1979). Rainfall Variability in Africa. *Mon. Wea. Rev.* **107**, 1133–1139.

Ogallo, L.J. (1981). Reliability of rainfall in East Africa. *Research Report No. 5/81.* Institute for Meteorological Training and Research, Nairobi.

Ogallo, L.J. (1987). Kenya Meteorological Department, Nairobi. Personal communication.

Ogallo, L.J. and Nassib, I.R. (1984). Drought patterns and famines in East Africa during 1922–83. *Paper presented at the Second WMO Symposium on Meteorological*

Aspects of Tropical Droughts, Forteleza, Brazil, 24–28 September. World Meteorological Organization, Geneva.

Ojany, F. and Ogendo, R. (1973). *Kenya: A Study in Physical and Human Geography*. Longman, Nairobi.

O'Leary, M. (1984). *The Kitui Akamba*. Heinemann, Nairobi.

Onchere, S.R. (1976). *Structure and Performance of Agricultural Production and Input Markets in the Northern Division of Machakos District, Kenya*. MA Thesis. University of Nairobi, Nairobi.

Owako, F.N. (1969). *The Machakos Problem: A Study of Some Aspects of the Agrarian Problem of Machakos District of Kenya*. PhD Dissertation. University of London, London.

Palutikof, J.P., Farmer, G. and Wigley, T.M.L. (1982). Strategies for the amelioration of agricultural drought in Africa. In *Proceedings of the Technical Conference on Climate of Africa*, pp. 228–249. World Meteorological Organization, Geneva.

Pant, P.S. and Rwamdusya, E.M. (1971). *Climates of East Africa*. East African Meteorological Department, Nairobi.

Parry, M.L. (1985). *The Sensitivity of Natural Ecosystems and Agriculture to Climatic Change*. Research Report RR-85-001. International Institute for Applied Systems Analysis, Laxenburg, Austria. (Reprinted from *Climatic Change* **7**.)

Penman, H.L. (1948). Natural evaporation from open water, bare soil and grass. *Proc. Royal Society Ser. A* **193**, 120–145. London.

Penning de Vries, F.W.T. (1975). Use of assimilates in higher plants. In J.P. Cooper (ed.), *Photosynthesis and Productivity in Different Environments*, pp. 459–480. Cambridge University Press, Cambridge, UK.

Pollard, S.J. (1981). *Report on the Nzaui/Machakos Farming System Study*. Agricultural Economics Unit/Agricultural Machinery Testing Unit, Ministry of Agriculture, Nairobi.

Porter, P.W. (1976). Climate and agriculture in East Africa. In C.G. Knight and J.L. Newman (eds.), *Contemporary Africa, Geography and Change*. Prentice-Hall, Englewood Cliffs, NJ.

Porter, P.W. (1979). *Food and Development in the Semi-Arid Zone of East Africa*. Maxwell School of Citizenship and Public Affairs, Syracuse University, Syracuse, NY.

Porter, P.W. (1983). Problems of Agrometeorological Modeling in Kenya. In D.F. Cusack (ed.), *Agroclimate Information for Development: Reviving the Green Revolution*. Westview Press, Boulder, CO.

Potter, H.L. (1981). Strategies recommended for the settled pastoralist. *Proceedings of the Animal Production Society of Kenya* **XIII**, 18–27.

Potter, H.L. (1982). Livestock feed resources in semi-arid areas. *Combined Workshop on the Review of the FAO Dryland Farming and the USAID Dryland Cropping Projects*, November. Embu Institute of Agriculture (mimeo). Kenya Agricultural Research Institute, Nairobi.

Potter, H.L. (1985). *Aspects of Climate, Herbage Growth and Animal Production in a Semi-Arid Area of Kenya*. PhD Thesis. University of Nairobi, Nairobi.

Potter, H.L. and Anindo, D.M.O. (1985). Napier grass for feeding of livestock. *Proceedings of the Animal Production Society of Kenya* **XVII**.

Pratt, D.J. and Gwynne, M.D. (1977). *Rangeland Management and Ecology in East Africa*. Hodder and Stoughton, London.

Pratt, D.J. Greenway, P.J. and Gwynne, M.D. (1966). A classification of the East African rangeland with an appendix on terminology. *J. Applied Ecology* **3**, 369–382.

Rempel, H. (1985). *Food Security Options for Sub-Saharan Africa.* Institute for Development Studies, Nairobi.

Rempel, H. and Lobdell, R.A. (1985). *A Model of Labour Allocation Decision-Making in Peasant-Type Households.* Institute for Development Studies, Nairobi.

Rodhe, H. and Virgi, H. (1976). Trends and periodicities in East Africa rainfall data. *Mon. Wea. Rev.* **104**, 307–315.

Rukandema, M., Mavua, J.K. and Audi, P.O. (1981). *Report on Survey Results from Mwala Location (Machakos).* Dryland Farming Research and Development Project. Ministry of Agriculture, Machakos.

Rukandema, M., Muhammed, L. and Jeza, A. (1983). *The Farming Systems of Semi-Arid Lower Embu, Eastern Kenya.* Dryland Farming and Development Project. Ministry of Agriculture, Machakos.

Schmidt, G. (1979). Effectiveness of maize marketing controls in Kenya. In J.T. Mukui (ed.), *Price and Marketing in Kenya,* O.P. No. 32. Institute of Development Studies, Nairobi.

Scotney, J. and Nyambura, R. (1979). *Ministry of Housing and Social Services, Annual Reports, Kerugoya.* Kirinyaga Family Life Training Centre. Ministry of Housing and Social Services, Kerugoya.

Semenye, P.P. (1981). Practical application of range areas monitoring. *Proceedings of the Animal Production Society Kenya* **XIII**, 28–36.

Sen, A.K. (1981). *Poverty and Famine.* Oxford University Press, Oxford, UK.

Sombroek, W.G., Braun, H.M.H. and van der Pauw, B.J.A. (1982). *Exploratory Soil Map and Agroclimatic Zone Map of Kenya.* Kenya Soil Survey, Nairobi.

Stewart, J.I. (1983). Kenya Agricultural Research Institute, Muguya. Personal communication.

Stewart, J.I. (n.d.). *How Research Assists in Managing Crops to Maximize Water Availability, Yields and Returns.* USDA/USAID/GOK Dryland Crop Systems Research Project. Kenya Agricultural Research Institute, Muguga.

Stewart, J.I. and Faught, W.A. (1984). Response farming of maize and beans at Katumani. In W.A. Faught *et al.* (eds.), *Proceedings of the Symposium on Dryland Farming Research in Kenya. East Africa Agricultural and Forestry Journal.* **44**, 29–51.

Stewart, J.I. and Hash, T.C. (1982). Impact of weather analysis on agricultural production and planning decisions for semi-arid areas of Kenya. *J. Applied Meteorology* **27**, 447–494.

Stotz, D. (1983). *Production Techniques and Economics of Smallholder Livestock Production Systems in Kenya.* Ministry of Livestock Development, Nairobi.

Styczen, M. (1983). *Monitoring of Erosion on Rangelands – Results from 1982 Short Rains.* Dryland Agricultural Research Station, Katumani, Machakos.

Survey of Kenya (1970). *National Atlas of Kenya.* Government Printers, Nairobi.

Taylor, A.O. and Rowley, J.A. (1971). Plants under climatic stress. 1. Low temperature, high light effects on photosynthesis. *Plant Physiol.* **47**, 713–718.

Te Chow, V. (1964). Statistical and probability analysis of hydrological data. In V. Te Chow (ed.), *Handbook of Applied Hydrology.* McGraw-Hill, New York.

Thomas, D.B. (1974). *Air Photo Analysis of Trends in Soil Erosion and Land Use in Part of Machakos District, Kenya.* MSc Thesis. University of Reading, Reading, UK.

Thornthwaite, C.W. (1948). An approach toward a rational classification of climate. *Geographical Review* **38**, 55–94.

Trewartha, G. (1961). *The Earth's Problem Climates.* University of Wisconsin Press, Madison, WI.

Tyson, P.E., Dyer, T.G.J. and Mametse, M.N. (1975). Secular changes in South African rainfall. *Q. J. Royal Meteorological Society* **102**, 817–833.

US Soil Conservation Service (1972). *SCS National Engineering Handbook. Section 4: Hydrology.* Government Printing Office, Washington, DC.

Van der Valk, Y.S. (1985). *Annual Report for the 1983–1984 Year of Recording Farms.* Dutch Dairy Development Project, Ministry of Agriculture and Livestock Development. National Animal Husbandry Research Station, Naivasha, Kenya.

Weg, R.F. van de and Mbuvi, J.P. (1975). Soils of the Kindaruma area. *Reconnaissance Soil Survey Report No. R1*, Kenya Soil Survey, Nairobi.

West, S.H. (1973). Carbohydrate metabolism and photosynthesis of tropical grasses subjected to low temperatures. In Plant Response to Climatic Factors. *Proc. Uppsala Symp., 1970.* UNESCO.

White, J.M. and Meadows, S.J. (1981). *Evaluation of the Contribution of Group and Individual Ranches in Kajiado District, Kenya to Economic Development and Pastoral Production Strategies.* Ministry of Livestock Development, Nairobi.

Williamson, G. and Payne, W.J.A. (1965). *An Introduction to Animal Husbandry in the Tropics.* Longman, London.

Wisner, B.G. (1977). *The Human Ecology of Drought in Eastern Kenya.* PhD Dissertation. Clark University, Worcester, MA.

Wisner, B.G. and Mbithi, P.M. (1972). *Drought in Eastern Kenya: Comparative Observations of Nutritional Status and Farm Activity at 17 Sites.* Institute for Development Studies, Nairobi.

Wit, C.T. de (1958). Transpiration and crop yields, *Agr. Res. Rep. 646.* Pudoc, Wageningen, Netherlands.

Wit, C.T. de (1965). Photosynthesis of leaf canopies. *Agr. Res. Rep. 663.* Pudoc, Wageningen, Netherlands.

Woodhead, T. (1968). *Studies of Potential Evaporation in Kenya.* East Africa Agricultural and Forestry Research Organization, Nairobi.

Woodhead, T. (1970). A classification of the East African rangeland, II: The water balance as a guide to site potential. *J. Applied Ecology* **7**, 647–672.

Wouters, A. (1984). *Result of Napier Plots.* Dutch Dairy Development Project, Ministry of Agriculture and Livestock Development. National Animal Husbandry Research Station, Naivasha, Kenya.

PART III

The Effects of Climatic Variations on Agriculture in Northeast Brazil

Contents, Part III

List of Contributors

MAGALHÃES, Dr. Antonio R.
Secretaria de Planejamento
Esplanada dos Ministérios
70, andar – sala 787
70063 Brasilia – DF
Brazil

FILHO, Mr. Homero C.
DMQ/EMBRAPA
Ed. Venancio 2000, 7. andar
70333 Brasilia – DF
Brazil

GARAGORRY, Dr. Fernando L.
Pakistan Agricultural Research Council
Food and Agricultural Organization
PO Box 1476
Islamabad
Pakistan

GASQUES, Mr. José G.
Instituto de Planejamento
Econômico e Social (IPEA)
CPR/IPLAN
Ed. BNDE – 15. andar – sala 01
70076 Brasilia – DF
Brazil

MOLION, Dr. Luis Carlos B.
Instituto de Pesquisas Espaciais
INPE
Caixa Postal 515
12200 São Jose dos Campos – SP
Brazil

NETO, Mr. Malaquias da S.A.
DMQ/EMBRAPA
Ed. Venancio 2000, 7. andar
70333 Brasilia – DF
Brazil

NOBRE, Dr. Carlos A.
Instituto de Pesquisas Espaciais
INPE
Caixa Postal 515
12200 São Jose dos Campos – SP
Brazil

PORTO, Mr. Everaldo R.
DMQ/EMBRAPA
Ed. Venancio 2000, 7. andar
70333 Brasilia – DF
Brazil

REBOUÇAS, Dr. Osmundo E.
SEPLAN/CE
Centro Administrativo – Cambeba
60000 Fortaleza – CE
Brazil

Abstract of Part III

In Northeast Brazil (NEB), which includes 18% of the area and 29% of the population of Brazil, over half the area is semi-arid, afflicted by recurrent severe droughts and sometimes floods. About a third of NEB's population live in this semi-arid zone, mainly in conditions of poverty in which they are very vulnerable to the adverse effects of such climatic events. This problem has been a major concern of government agencies and others for over 100 years. The present case study reviews the physical basis of this phenomenon and the history of public response. It then employs several techniques to explore meteorological, biological, agronomic and economic aspects of the problem and linkages among them, and makes several recommendations as to appropriate policies of response.

The main climatological basis for the case study is a pair of time series: one for northern NEB for 1912 to 1985; the other, for southern NEB for 1911 to 1972. Climatic scenarios adopted for this case study, based on these series or comparable records for individual locations, include:

(1) An anomalously dry year, 1983.
(2) A "back-to-back" drought situation, the years 1979–1980.
(3) A tenth percentile rainfall year (10% of years would be drier).
(4) A ninetieth percentile rainfall year (10% of years would be wetter).
(5) Other anomalous years identified in the climatological or crop records.

Scenarios of types (3) and (4), and, in most cases, (5), are based on the agricultural year, a 12-month period that may or may not start in January.

A simulation model of the water–soil–plant type, originally designed for determining best planting times, was used to estimate the effects on yields of cowpea beans (a common NEB food crop) of several climatic situations, for six representative semi-arid municipalities. The climatic situations included combinations of 10% increases or decreases in rainfall and potential evapotranspiration. Estimates were also made for the various scenarios listed above. Results indicated yields more than 60% below normal in extreme drought years.

The trends over time, 1960 to 1983, were examined for important food and non-food crops for different NEB states; and the impacts of droughts, such as those of 1983 and 1979–1980, were considered. Food crop yields have been generally declining, with production increases obtained only by expanding the crop areas. Droughts cause further sharp reductions in yields, and in severe drought years, such as 1983, the crop area may also be reduced. Growing of food crops, particularly beans and manioc, is the economic activity most affected by drought, and the small farmers are the people most directly affected. Droughts result in food shortages and high food prices for the urban poor. They also depress the regional economy and reduce its contribution to national production. The reduction by nearly one quarter in NEB gross agricultural production due to the 1983 drought reduced the region's contribution to national corn, bean and cotton production by half and substantially cut demand for a whole range of goods and services in the region. Attempts to apply multiple regression analysis to quantifying the effects of rainfall and time trend on crop yields were not altogether successful, although some encouraging results were obtained.

This case study leads to several recommendations. Review of the meteorological background suggests that it may be feasible to obtain some warning of severe droughts two or three months in advance, and this should be investigated. From the simulation and empirical analyses, the importance of water is obvious; programs to develop and promote irrigation are needed, but they should be part of comprehensive development programs. The need is also emphasized for programs to enhance educational levels, employment opportunities and agricultural productivity to reduce the vulnerability of the population to drought impacts. Finally, there is a great need for integrated climate–economy–society research, to improve our understanding of climate processes and how they affect society, and to provide information to assist those charged with developing and implementing policies for coping with these impacts.

SECTION 1

Introduction: Drought as a Policy and Planning Issue in Northeast Brazil

The Nordeste or Northeast Brazil (NEB) has been a focus of major concern for governments, planners, social scientists and other experts over the last century, principally because of recurrent droughts and the poverty that affects the majority of the population in its semi-arid zone.

The 1983 drought was one of the most serious in this century. This was the fifth successive year of below-average rainfall, which greatly reduced food and livestock production and adversely affected the living conditions of about 12 million people. Droughts are a persistent, recurring problem in NEB, but the population has not yet learned how to cope with them, in spite of government efforts to increase drought resilience. The present need is to design new response strategies to reduce the scale of impact on the regional economy and society.

The purpose of this case study is to assess the agricultural and social impacts of droughts, and to suggest decision making strategies to increase the capacity of society to cope with this problem.

1.1. Regional Background

NEB comprises an area of approximately 1.5 million km^2, equivalent to about 18% of Brazil's total surface area (*see Figure 1.1* and *Table 1.1*). This area is larger than the combined area of France, Spain and Portugal. Its geographical situation lies between latitudes 4°–16° S and longitudes 33°–46° W. Its population is about 38.1 million (1985), and an increasing proportion is in urban areas: 51% of the population now, as against 42% in 1970, 34% in 1960, and 26% in 1950 (IBGE AEB, 1981). The labor force of 12.5 million people has a very low income level, 73% of them earning less than US$1400 annually. Half the population is illiterate, and infant mortality rate is 93 per 1000, as compared to 66 for

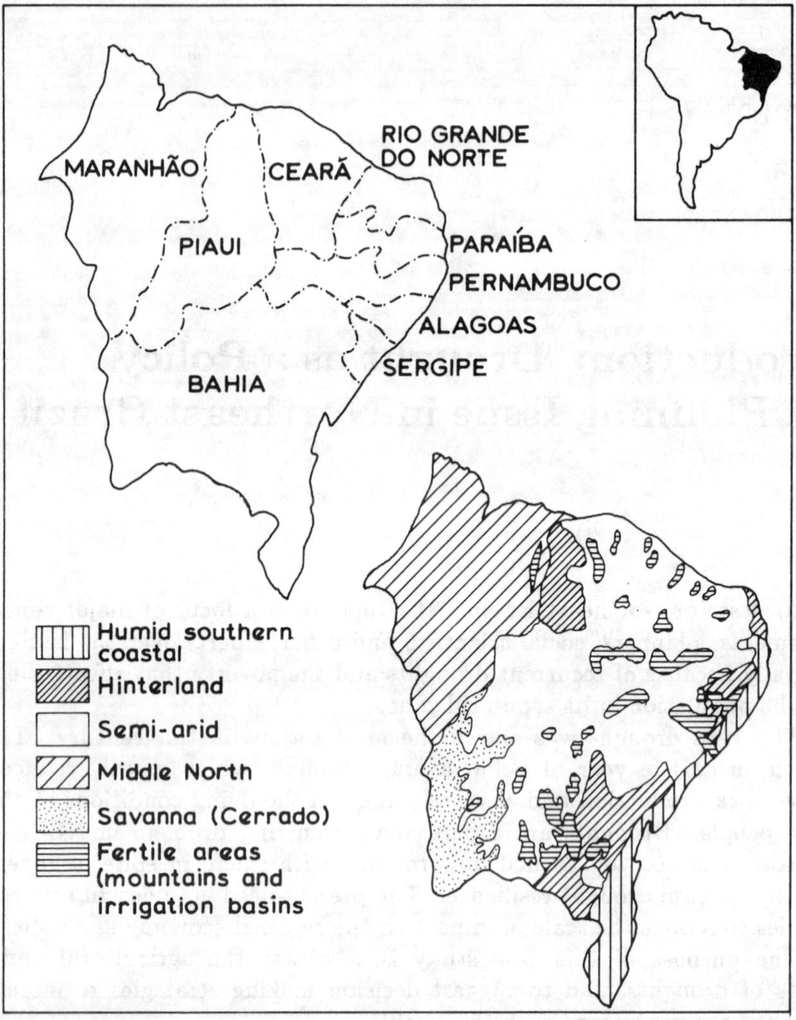

Figure 1.1. Northeast Brazil, with geographic zones. The boundary of the area follows that defined by SUDENE (Superintendência do Desenvolvimento do Nordeste).

Brazil as a whole. Life expectancy is around 53 years, while for the whole country it is 60 years. The region contains 29% of the Brazilian population, living in 9 of the 23 states of the country, but generates only 15% of the gross national product. In 1983, regional annual per capita income was US$828, compared with US$1639 for Brazil as a whole. In 1975 (a normal rainfall year), 27% of the regional GNP was generated by agriculture, 29% by manufacturing and 44% by services. In drought periods agriculture's share declines – for example, it was 20% in 1979, as compared with 30% for manufacturing and 50% for services.

Table 1.1. Descriptive regional (NEB) and national Brazilian statistics.

Indicator	*NEB*	*Brazil*	*NE/BR* (%)
1. Area (1000 km^2)[a]	1 548.7	8 511.9	18.2
2. Population, 1980 (1000)[a]	34 753.0[b]	119 099.0	29.2
3. Labor force, 1980 (% of total population)[c]	32.9	36.8	89.4
4. Total annual rate of population growth (1970–80)[c]	2.18	2.49	87.6
5. Natural annual rate of population growth(1970–80)[c]	2.73	2.19	124.7
6. Crude death rate, 1980 (per 1000)[c]	9.4	8.3	113.3
7. Literacy rate (%)[c]	50.7	72.5	69.9
8. Life expectancy, 1970–80 (years)[c]	52.6	60.0	87.7
9. Infant mortality rate, 1980 (per 1000 live births)[c]	92.6	65.5	141.4
10. Underemployment rate, 1980 (%)[c]	47.7	29.2	163.4
11. Labor force receiving up to 2 minimum wages per month (%)[c,e]	73.5	59.5	123.5
12. GNP, 1983 (1000 million Cruzeiros)[d,f]	17 736.5	121 055.4	14.6
13. Per capita GNP, 1983 (1000 Cruzeiros)[d,f]	477.1	944.5	50.5

[a]Source: IBGE AEB (1981).
[b]The NEB population in 1985 was estimated to be 38.1 million. See SUDENE PNUD Banco Mundial (1984).
[c]Source: SUDENE AQSN (1985).
[d]Source: IPEA IRN (1985).
[e]The value of the minimum wage is periodically determined by law and is equivalent to *ca.* US$50 per month.
[f]The average rate of exchange in 1983 was CR$576.2 = US$1.00.

Social and institutional structures are traditionally strongly resistant to structural changes. Political power is still based, to a large extent, on land ownership, although the urban industrial sector is increasingly important. Land ownership is quite concentrated (*Table 1.2*).

Table 1.2. NEB structure of land ownership (1980).

Size of farms (ha)	*Number of farms*	*Percentage of total number*	*Total area of farms*	*Percentage of total area*
< 5	1 366 184	55.75	2 474 359	2.76
5–10	291 968	11.91	2 013 603	2.25
10–20	243 718	9.94	3 312 690	3.70
20–50	263 603	10.76	8 120 839	9.07
50–100	129 723	5.29	8 847 285	9.88
100–500	124 924	5.10	24 991 972	27.91
500–1000	15 696	0.64	10 542 557	11.77
> 1000	10 299	0.42	29 250 199	32.66
No information	4 585	0.19		
Total	2 450 700	100.00	89 553 504	100.00

Sources: IBGE CAB (1980) and MINTER–SUDENE (1984, vol. 3).

1.2. Divisions of the Northeastern Region

There are four main zones in the Northeast that are distinct in terms of soil, climate and production systems (*Figure 1.2*):

(1) *Humid area/southern coastal zone* (*Zona da mata/Litoral Meridional*): this zone occupies 7% of NEB area and holds 32% of the population. Annual precipitation of up to 2000 mm/year is recorded in this region. Soils being rich, the basic crops are sugarcane, cocoa and coconut.

(2) *Hinterland* (*Agreste*): an intermediate zone, where annual precipitation is low (800 to 1000 mm) and there is a mixed spatial pattern of rich and poor soils. This zone occupies 11% of the regional area and contains 14% of its population. The agricultural economy is based on food production (beans, maize, fruits, vegetables), tobacco, cotton and cattle raising.

(3) *Interior and northern coastal zone* (*Sertão/Litoral Setentrional*): the typical vegetation in this zone has, as its basic element, the *caatinga*, a heterogeneous vegetation, very rich in xerophilous species, comprising stunted, thorny trees with small and few leaves that are shed in the absence of rainfall. It occupies 60% of NEB and contains 42% of the population. The climate is semi-arid (precipitation up to 800 mm/year, concentrated in three months); soils are poor, siliceous and stony. A large part (about two-thirds) of the subsoil is crystalline and thus almost impermeable. Rains occur in summer, precipitation is irregular and rivers are not permanent. The rural economy is mainly based on the production of cattle, cotton and subsistence crops.

(4) *Middle North* (*Meio Norte*): this area is delimited between the western bank of Paraíba river and the edge of the Amazon rain forest. It occupies 22% of NEB area and supports 12% of its population. Soils are generally poor, but there are spots of rich soil. Economic activities are based on the fruits of babassu palms, rice and cattle.

1.3. The Semi-Arid Zone

Coinciding with the interior and northern coastal zone [division (3) above], the semi-arid zone is 860 342 km^2 in size, representing more than half of the total area of NEB and supporting about one-third of the population. The main climatic characteristics are:

- Annual mean precipitation of 400–800 mm, with a coefficient of variability of up to 60%, concentrated predominantly in three months of the year.
- High air temperatures (annual average 23–27 °C).
- High insolation (average annual sunshine hours totaling approximately 2800).
- Low relative humidity.
- High potential evaporation (averaging about 2000 mm/year).

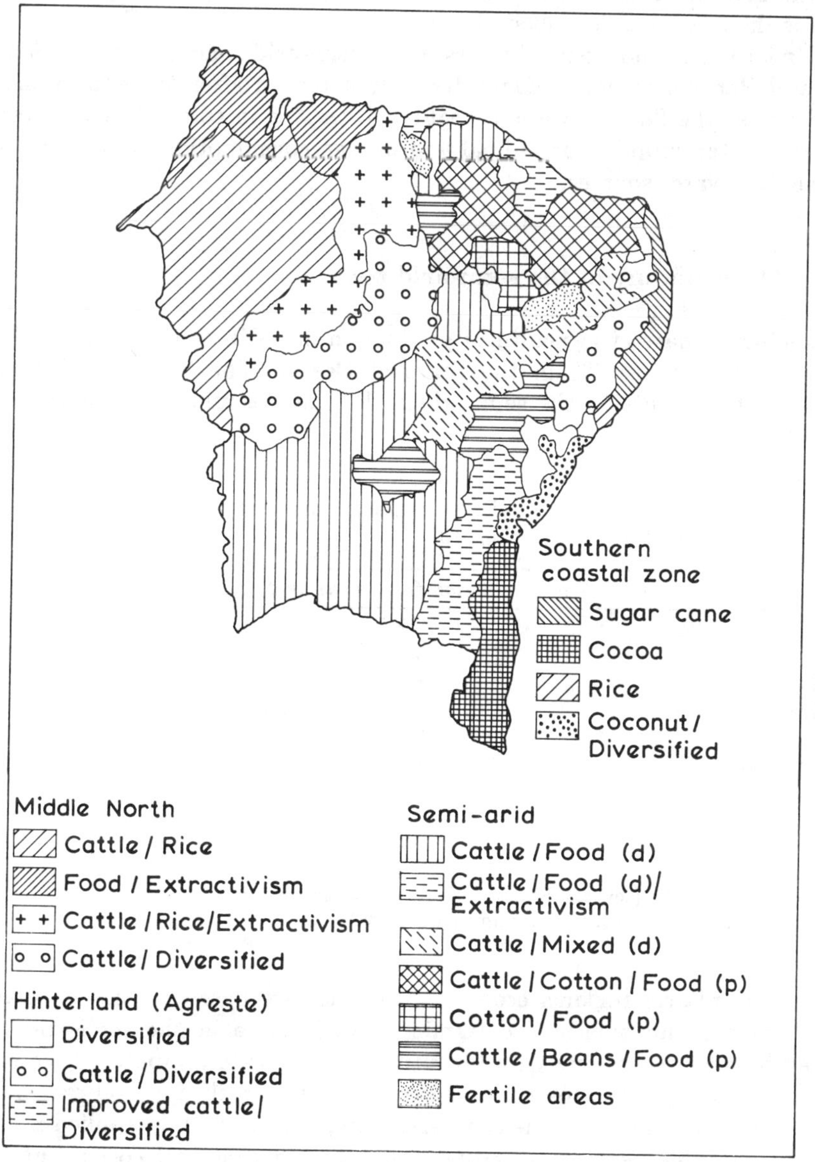

Figure 1.2. Production system types in Northeast Brazil: d = diversified food; p = food production. Source Jatibá (1985).

The explanation and prediction of the precipitation fluctuations of this region constitute one of the challenging issues of tropical meteorology.

Most soils are poor, crystalline (shallow lithosols with a low water intake rate) and almost impermeable, which restricts the processes of water absorption and infiltration. Other soils are sedimentary. The typical vegetation of the region is the caatinga, described above.

Under such conditions all rivers in the semi-arid zone, except the São Francisco and Paraíba rivers, become dry only a month after the end of the rainy period unless the flow of water is regulated, as it is in some rivers, by reservoir storage. Water supplies for animal and human consumption are derived mainly from surface water sources.

1.3.1. The economy of the semi-arid zone

The semi-arid share of agricultural production in NEB varies from 23.5% in the State of Bahia to 88.7% in the State of Ceará. The State of Maranhão is situated entirely outside the semi-arid region, in the middle north zone (*Table 1.3*).

Table 1.3. Percentage share of NEB's semi-arid zone in the value of state agricultural production, 1975.

State	*% of production from semi-arid zone*
Maranhão	–
Piauí	53.4
Ceará	88.7
Rio Grande de Norte	74.8
Paraíba	65.6
Pernambuco	16.8
Alagoas	26.3
Sergipe	58.2
Bahia	23.5

Source: IBGE CAB (several issues). Data organized by José Otamar de Carvalho (personal communication, 1985).

The semi-arid region's economic activities are based on the association of cattle, cotton, and subsistence farming. Agriculture in the semi-arid zone is marked by poor techniques with low utilization of modern inputs. Selected seeds are used basically for industrial and exports crops, rarely for food crops. Fertilizers are not used much in the semi-arid zone, but are mainly applied in production of high value per hectare crops, such as sugarcane and cocoa, outside this zone. In general, semi-arid zone productivity is low and technology is traditional.

There is a high concentration of land ownership: at the small-farm end of the scale 29% of the farms occupy only 1.6% of the area, while at the other extreme only 3.4% of the farms occupy 40% of the area. The Gini coefficient of land concentration – a measure of inequality that expresses the ratio between actual and maximum inequalities (Gastwirth, 1974), varying from 0 (no inequality) to 1 (maximum inequality) – is close to 0.8 and the Lorenz curves show a great area of inequality (*Figure 1.3*).

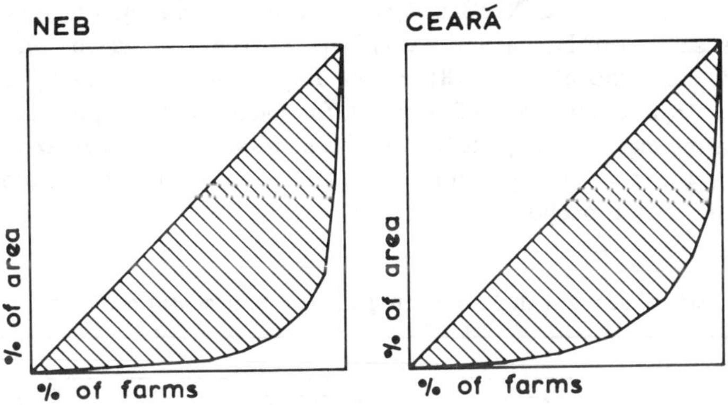

Figure 1.3. Lorenz curves of land area and farm size in NEB and the State of Ceará. Source: IBGE CAB (1980).

The definition of large or small landowner depends not only on the land size but also on the quality of natural resources for specific uses. In general, a farm having more than 1000 ha is large, between 100 ha and 1000 ha is medium, and below 100 ha is small. In NEB the mean farm area is 36.5 ha. The large landowners are mainly involved in cattle and cotton production. Small farmers, with or without land, usually cultivate subsistence crops, such as beans and corn. These subsistence crops account for 15% of the agricultural income in the semi-arid zone (de Lima Viana, 1982, p. 164), fiber and oleaginous plants (mainly cotton) produce 35% of the income; cattle 28%, and other products account for 24%.

Semi-arid agricultural production is undertaken predominantly by small farmers. They account for 76.2% of the value of rice production, 82% of beans, 94% of manioc (cassava), 85.4% of corn and 61.6% of cotton.

1.3.2. Droughts in the semi-arid zone

There is no established definition of drought in the Northeast, despite some attempts to produce one. This is because the characterization of drought depends both on the quantity of rainfall and on its time and space distribution. Sometimes a 30% decrease in precipitation, combined with fairly good rain distribution, can result in a "normal" year, i.e., average agricultural production and employment. On the other hand, there may be years with "normal" precipitation that is poorly distributed, which result in drought conditions.

For the purpose of this case study, the following working definitions will be used:

(1) An *extreme drought year* occurs when total precipitation falls 50% or more below normal.
(2) A *drought year* occurs when precipitation falls 25%, in relation to the 30-year average, with a concentration in two months or less.
(3) A *partial drought* is one involving only part of NEB.

In the history of the Northeast, there are remarks on drought events as early as the seventeenth century, when the Portuguese arrived in the region (Alves, 1982; Carvalho *et al.*, 1973) (*Table 1.4*). There have been, on the average, between 18 and 20 years of drought for each 100-year period. The most serious consequences of droughts are in the semi-arid zone, but extreme events (i.e., a fall in mean annual precipitation of 50% or more) also extend into the other areas of NEB, especially the Agreste.

Table 1.4. Years of drought in NEB during the last four centuries.

Century				
Seventeenth	*Eighteenth*	*Nineteenth*	*Twentieth*	
1603	1711	1804	1900*	1951
1614*	1721	1809	1902–03	1953
1692	1723–24*	1810	1907	1958*
	1736–37	1816–17	1915*	1970
	1744–46	1824–25*	1919*	1979
	1754	1827	1932–33	1980
	1760	1830–33	1936	1981
	1772	1845	1941–44	1983*
	1776–77	1877–79*		
	1784	1888–89		
	1790–94*	1891		
		1898		

* = Years of extreme events.
Source: Moura and Shukla (1981).

In the semi-arid zone there is a pronounced variation of drought occurrence between one area and another. The most affected states are Piauí, Ceará, Rio Grande do Norte, Paraíba and Pernambuco. In other states (Maranhão, Alagoas, Sergipe and Bahia), droughts are usually partial in terms of size of affected area, and have less impact on agriculture and on animal and human life (*Figure 1.4*).

In the beginning of the territorial occupation of the Northeast (sixteenth and seventeenth centuries), droughts were not an important problem because the population was concentrated in the more humid areas, and population density was low. The problem intensified with the population increase in the semi-arid zone.

The impact of drought on society and the economy in the semi-arid zone and in NEB as a whole has been very great. The population, especially the poor, has been directly affected by the lack of drinking water and the loss of work. The result was often hunger and thirst leading to starvation and death. Migration to the coast and to other regions of Brazil, especially the Amazon and the south, has been a common response. The impact on agricultural production used to be very high – first on subsistence crops, then cash crops and cattle

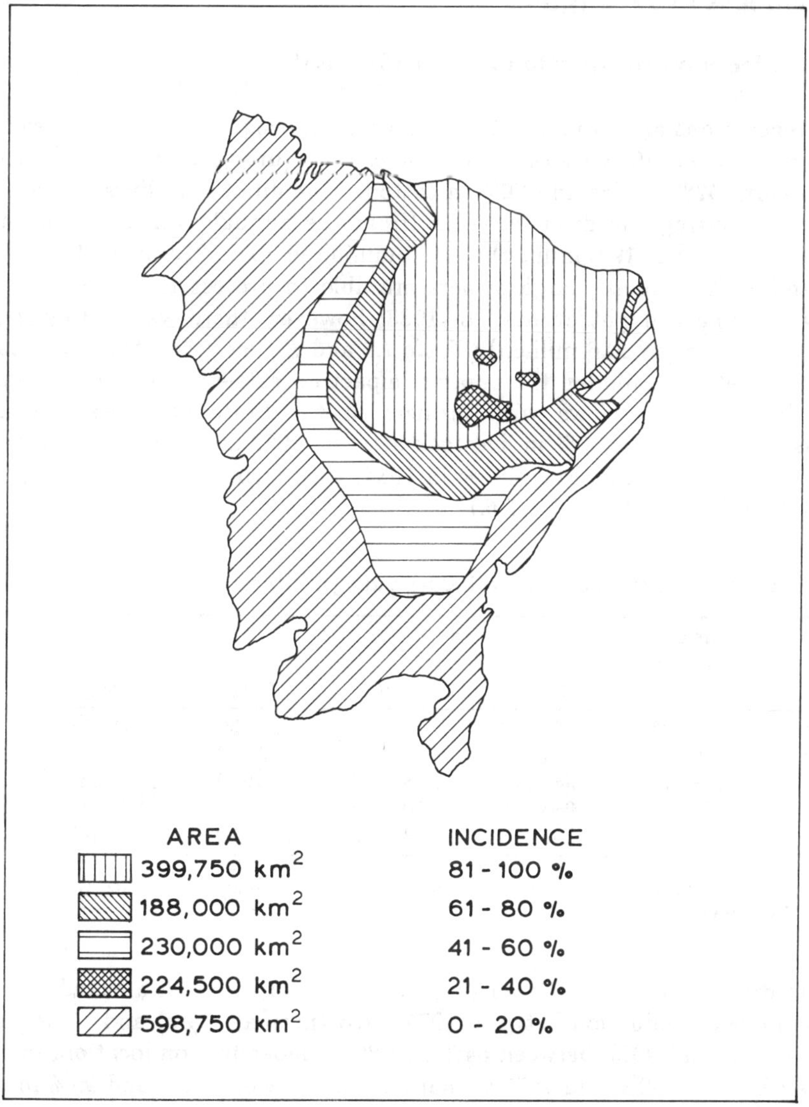

Figure 1.4. Incidence of drought in Northeast Brazil (% years).

raising. Though available data are not accurate, it has been estimated that approximately half of the population died in the drought-related famine of 1877–1879 (estimated total population in the affected area: 1.7 million; Carvalho *et al.*, 1973). Other sources indicate that some 100 000 to 200 000 people died, and many migrated to the Amazon Region (Furtado, 1964). Both livestock and agricultural production were almost totally destroyed.

1.4. Effects of Drought

1.4.1. Effects on production and productivity

Subsistence crops are more affected than cattle raising. The losses in production observed between 1978 (normal year) and 1980 (drought) were 72% in beans, 82% in corn, 52% in rice and 70% in cotton. The number of livestock continued to increase during the drought years (from the annual average in the normal 1968–1978 period of 19.6 million) to 20.5 million in 1979, 21.9 in 1980 and 22.1 in 1981 and 1982. Indeed, large owners are able to protect their activity, except during extreme drought events. In 1983, however, there was a decline in the number of cattle (to 20.6 million) (*Table 1.5* and *Figure 1.5*). Aggregate data for the Northeast have hidden the greater impact on cattle raising in the semi-arid part. Personal interviews with farmers in the State of Ceará indicated a decrease of as much as 80% in livestock between 1982 and 1983. (The interviews were conducted by A.R. Magalhães in 1984, with farmers of the municipality of Sobral, Ceará. *See also Table 1.6.*)

Table 1.5. NEB cattle (numbers in 1000s) and agricultural production (tons).

	Annual average 1976–78	*1979*	*1980*	*1981*	*1982*	*1983*
Cattle	19 603	20 512	21 876	22 136	22 113	20 605
Cotton	620 836	451 105	384 024	343 446	420 208	179 519
Rice	1 388 771	1 343 714	1 483 538	929 949	1 977 045	623 040
Beans	600 879	683 082	501 283	449 952	730 144	227 567
Corn	1 389 892	1 255 510	830 460	505 730	1 133 485	306 916

Sources: IBGE (various dates), data organized by Fundação Joaquim Nabuco (up to 1980) and IPEA (1981–1983).

In different places in Ceará (in general, one of the most affected states) the decline in rice production between 1975–1976 (normal years) and 1980 (drought year) was around 90%; between 54% and 89%, depending on location, in cowpea (beans); between 30% and 79% in manioc, and between 77% and 99% in cotton. Though rice consumes much water, it is cultivated, as a secondary crop, in some selected areas of Ceará in *vazantes* (margins of dams) and, to a small extent, in irrigated areas.

Sharp reductions also appear in productivity (production per unit area). In Ceará State, between 1975–1976 and 1980, there were reductions of 18% in rice (cultivated in some specific areas of the state where there is sufficient water), from 18% to 54% in beans, from 36% to 52% in manioc, and between 23% and 71% in cotton. This decline was due to both a decrease in the cultivated area and in yield per unit area (*Table 1.6*).

Large owners, small owners and nonowners adopt different strategies to cope with risk. Usually the large owners choose cash activities, such as cattle raising, and transfer the more vulnerable agricultural activities, especially food

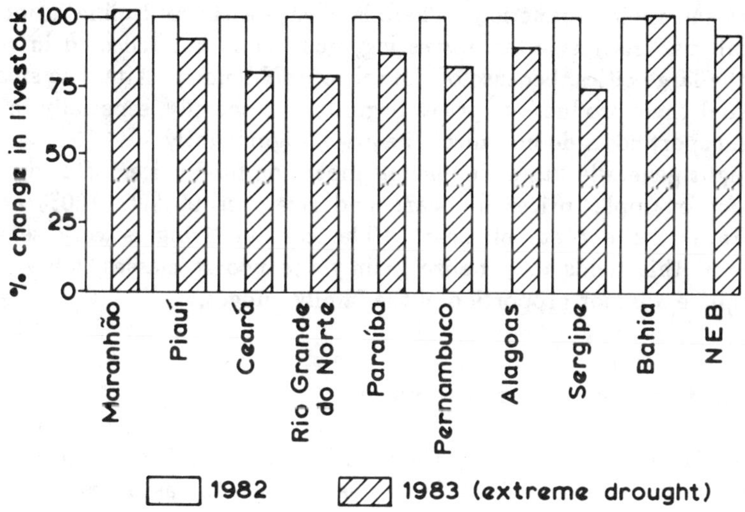

Figure 1.5. Livestock changes, 1982–1983 (%).

Table 1.6. Ceará State: cultivated area, production and productivity by main agricultural product.[a]

	Nondrought year: 1978			*Drought year: 1983*		
Main products	*Planted area* (ha)	*Production* (t)	*Yield* (t/ha)	*Planted area* (ha)	*Production* (t)	*Yield* (t/ha)
Tree cotton	1 200 000	237 600	0.198	675 202	47 264	0.070
Annual cotton	84 000	27 720	0.330	74 367	17 034	0.229
Rice	56 000	60 200	1.075	16 292	30 077	1.846[b]
Beans	400 000	120 000	0.300	167 291	24 811	0.148
Corn	480 000	259 200	0.540	146 092	175 310	0.120
Manioc (cassava)	175 000	1 575 000	9.0	82 974	442 088	5.0

[a]Source: IBGE.
[b]The increase in rice productivity in a drought year is probably due to the fact that production continued only in irrigated areas.

production, to renters and partners. Small owners and nonowners are engaged in a basic needs strategy, which involves more risk. Both, especially nonowners, use few inputs other than labor and land.

1.4.2. Effects on food prices

Prices of food crops, such as beans and corn, are usually determined by local market forces, while prices of raw materials (e.g., cotton) and cattle are influenced by the national and international markets. Moreover small farmers lack the capital and storage capacity that would enable them to wait for the best moment to sell their production. There is always a sharp decline in food prices in the local markets just after harvesting, but this is not reflected in consumer prices in the large cities, because of the role of middlemen. After the small farmers have sold their production, prices begin to rise sharply, especially in drought years. Transport costs do not affect the prices significantly.

Drought generally has a significant impact on food prices. During the 1979 drought, for example, prices of beans and rice rose 90% to 100% above the inflation rate in the State of Ceará. The cost of living is very sensitive to drought, and this tends to penalize mainly the poorer classes, for whom food items comprise a major proportion of the family budget.

1.4.3. Effects on the regional economy

One effect of drought is to reduce the contribution of agriculture to the regional income, while increasing activity in the services sector as a result of informal movements of people to urban centers. In fact, unplanned urbanization is a consequence of such migrations. *Table 1.7* shows the decline of the share of agriculture in terms of real output in years of drought (1970 and 1979), as compared with nondrought years (1965 and 1975). The industrial manufacturing sector is, as a whole, relatively unaffected by drought. The industrial product continued to grow both through drought years, such as 1970, and nondrought years. The industrial share of Northeast Brazil's total gross national product (GNP) increased steadily even in drought years.

Table 1.7. NEB sector share (%) of gross national product.

Sectors	*1965*	*1970*	*1975*	*1979*
Agriculture	27.9	20.0	26.9	20.0
Manufacturing	22.6	26.9	28.5	29.8
Services	49.5	53.1	44.6	50.2

Source: SUDENE PFB (1979).

There are some manufacturing sectors, such as food processing and textile industries, that are dependent on agricultural output. However, those sectors have been able to adapt to regional shortages of raw materials through diversification of supply sources from other regions and even from abroad.

1.4.4. Effects on migrations

During drought periods there are many waves of migration toward towns in the interior or capital cities of the region, and to other regions of Brazil. Such centers or regions may experience sharp increases in population during these periods, exacerbating slum proliferation and housing shortage problems.

Field research conducted by the Fundação Joaquim Nabuco during the droughts of 1979–1980 showed that 24.2% of workers affected by the droughts wanted to move. In the absence of drought, only 0.7% would have wanted to move. The 75.8% not wanting to move gave their reasons as:

- Government public works assured them a job (32.7%).
- Reluctance to leave their present homes (28.1%).
- No need to move (19.1%).
- Other motive (e.g., the situation for the poor is the same everywhere; no documents nor money to travel; accustomed to drought; next year will be better, etc.).

There are no available specific data on migration during drought years. General interregional migration data are presented in *Table 1.8.* The Northeast is the main region of emigration in Brazil. In the 1970s there was a net increase in this migration from the Northeast of more than 2 million people. The main origins of migration are the more drought-prone states: Ceará, Paraíba, Piauí, Rio Grande do Norte and Sergipe. The main destinations most recently (1980) have been in the southeast and center west and north regions. Earlier the south had been the most important in-migration region.

Table 1.8. Migration movement by regions, 1950–1980.

Regions	*1950*	*1960*	*1970*	*1980*
In-migration				
North	223 652	249 061	356 876	1 065 795
Northeast	984 029	1 400 215	1 666 646	2 293 902
Southeast	2 663 266	4 147 701	6 603 818	9 250 949
South	857 542	1 977 763	2 886 972	2 682 535
Center West	359 434	864 661	1 665 945	2 662 860
Out-migration				
North	148 655	106 319	251 449	375 415
Northeast	1 920 209	3 627 480	5 139 604	7 789 514
Southeast	2 668 524	1 858 944	5 963 629	6 360 003
South	395 634	802 694	1 521 434	2 674 852
Center West	73 297	143 964	304 141	756 257
Balance (in – out)				
North	74 997	82 742	105 427	600 380
Northeast	−936 180	−2 227 265	−3 472 958	−5 495 612
Southeast	5 258	248 757	640 189	2 890 946
South	461 908	1 175 069	1 365 538	7 683
Center West	286 137	720 697	1 361 804	1 906 603

Source: IBGE – Diretoria Técnica Departamento de Estudos de População (unpublished data).

The temporal pattern of migration can be quite complex. As would be expected, people start migrating during drought years when it becomes evident that there will be no rain. However, the *time* when this happens varies from state to state, according to the normal local timing of the rainy season. In Ceará State, the migration starts by March. Elsewhere it may be earlier or later. Those who return do so when the drought is over, but the perception of the end of a drought is also determined by "normal" weather and farmers' expectations of it.

1.4.5. Effects on society

The impacts of drought differ sharply among social classes in the northeastern semi-arid zone. The consequences of rainfall shortage and irregularity are particularly severe for wage earners, partners, dwellers and renters. The wage earner is hired by farmers for the execution of a specific task and is paid in money, usually on a daily basis. The partner is a nonowner. He produces on the owner's land, taking all risk, and paying the owner with a percentage of production, up to 50%. The renter pays in cash for the rent of the land. The dweller is a permanent worker. He lives on the farm and has the choice of working a certain number of days for the owner, or making a payment in cash or goods. Usually he is allowed to produce beans and maize for his family's subsistence. The squatter (*posseiro*) is a nonowner producer who does not pay a rent for the land use. The land where he produces may be public or private, with or without the owner's consent. In certain cases, a dweller may also be a wage earner or a partner, and a partner may also be a renter. Virtually all (99%) of the land in the semi-arid zone is directly operated by owners or squatters, in conjunction with partners (19%), dwellers (11%) and renters (1%). *See* de Lima Viana (1982, p. 119).

These consequences are also very serious for small farmers, even though they own some land. For owners of medium-sized and large farms, drought is a problem of production, while for the other classes it is a problem of subsistence (*Table 1.9*).

In fact, for wage earners, partners, dwellers and renters, a drought can preclude the possibility of working on the owner's land, partly because subsistence agriculture becomes too risky and expensive, and also because the owner wants all his resources devoted to protecting his livestock investment. For the small landowners, it is possible to use some financial savings or loans secured by their land. Only as conditions continue to deteriorate in a prolonged drought do they decide to migrate or to engage in the government's emergency front (i.e., public) works. According to field research by Fundação Joaquim Nabuco (1983), in the drought of 1980 the workers in the emergency front works were classified as:

nonowners	63.3%
wage earners	5.3%
partners	29.5%
dwellers	15.9%
squatters (posseiros)	3.4%
renters	1.1%
others	8.0%

Table 1.9. NEB: responses to drought by different groups.

Population group	*Response to drought*
Wage earners	Loss of employment; emigration; work in government financed public works
Tenant farmers	Loss of access to land; emigration; work in government financed public works
Small landowners	Loss of production and income; work in government emergency programs (public works); sale of land, normally to large owners
Medium and large farm owners	Loss of agricultural production and livestock; fall of income from agricultural activity; benefits from credit subsidies and direct investment, financed by the government, through the emergency program; acquisition of more land from small owners

For the medium and large farm owners, the first recourses are to utilize water resources from reservoirs, to adapt cattle raising to changed pasture conditions, to obtain subsidized bank credit and to use the now-abundant manpower for farm restoration. For them, the problem is basically one of production.

1.4.6. Effects on public spending

The Brazilian Government spent the equivalent of US$1777.9 million in emergency programs during the 1979–1983 drought period (*Table 1.10*). Some 2.5 million workers were assisted in 1983, and 88% of the total area of NEB was affected.

The emergency program was designed to meet three broad objectives:

(1) To create direct employment in public works, in order to assure that the poor population affected by drought received at least a minimum level of income. In 1983, more than 2.5 million workers were hired by the government and employed in the construction of roads, dams, wells, and other works. Each worker received a wage roughly equivalent to half a legal minimum wage, but was normally allowed to work four hours a day, leaving the rest of his time for other occupations, such as preparing his own land for the next year. Many of these public works were also located on private farms, provided that the private water capacity storage generated by the program was made available for the community as a whole.

(2) To attack the problem of malnutrition and hunger, through distribution of food baskets to the affected families.

(3) To provide water for human consumption, through the transportation of water in tank trucks over distances up to 200 km.

Table 1.10. NEB: percentage of total area affected by drought, annual growth rate of agricultural GNP, number of assisted people, and emergency spending during the 1979–1983 drought period.

Year	*Affected area of total NEB* (%)	*Growth rates of agricultural GNP – total NEB* (%)	*Assisted people*[a]	*Emergency expenditures* (US$ million[b])
1979	32	0.0	459 830	275.0
1980	84	−0.9	709 850	482.0
1981	87	−5.6	1 168 700	467.5
1982	84	+12.1	746 800	179.0
1983	88	−15.8	2 502 300[c]	374.4
Total	–	–	–	1777.9

Sources: Area: MINTER (Ministry of Interior), 1983. GNP rates: SUDENE (Superintendency of the Development of the Northeast)/CPR (Coodenadoria de Planejamento Regional). Divisão de Contas Regionais up to 1981, 1982 and 1983, estimated by ETENE. Assisted people and emergency expenditures: MINTER–Coodenadoria de Defesa Civil.

[a] Number of persons working in emergency public works and receiving wage from the Federal Government. Refers to the period from June of one year to May of the next year.

[b] December 1983 exchange rate.

[c] June–November 1983.

The emergency program was beneficial in that it precluded starvation and thousands of deaths, as well as massive out-migration. The aggregate salaries paid by the government kept the rural economy functioning, with positive effects on the regional economy as a whole. Although no estimate has been made of the number indirectly employed as a result of the program, it was certainly large.

Historically, the emergency programs have also contributed to the creation of infrastructure (dams, roads, etc.). Notwithstanding these positive effects, it was recognized that the programs were not sufficient to prevent deterioration of nutrition levels and increasing mortality rates. As relief programs, they did not help to solve structural problems, a prerequisite for reducing poverty and increasing resilience in coping with droughts.

During 1979 and 1980, the emphasis was on works to increase the future productive capacity on private holdings. From 1981 to 1983, the program was directed to small public works for collective use, such as roads; but it still allowed for the building of dams on private farms, if the owners allowed access to the water stored by these dams. The wages were paid directly by government officials.

Before they entered the emergency program, the workers' survival depended mainly on daily wages (39.9%) and, in the case of small farmers, the consumption of some production stocks (23.6%). The wages were very low, about 50% to 80% of the legal minimum wage. This minimum wage is equivalent to approximately US$50.00 per month and is normally not enough to assure a family's subsistence. Usually, however, there is more than one person contributing to the family's budget. Moreover, caution must be exercised in using the US dollar equivalent to measure the purchasing power of the minimum wage. In reality, this purchasing power may be two or three times larger because of the local price structure, and because the scarcity of US dollars tends to inflate the exchange rate.

It is interesting to note that, during the 1979–1980 drought, 80.5% of the total family budget of workers enlisted in the emergency program was spent either on food (45.6%) or to repay personal debts (34.9%) (Fundação Joaquim Nabuco, 1983).

1.5. Policies to Cope with Drought

Ever since the second half of the nineteenth century it has been the practice to adopt deliberate regional policies for Northeast Brazil, largely as a consequence of its susceptibility to impact from drought. Such policies have evolved according to the changing conception of the regional problem, as viewed by government, society and politicians in general.

The climatic problem of NEB, given its consequences for the population and the economy, has always been the most important factor to consider in the formulation and implementation of regional policies. As these policies have developed, there have been improvements in related planning procedures, resulting both from a better understanding of the region and the adoption of modern planning techniques.

1.5.1. Historical development of regional policy

Four periods in the evolution of regional policy in NEB may be distinguished (with some overlap between periods):

1. The engineering or hydraulic phase (1877 to ca. 1950)

In 1877 there was a severe drought, after three decades of near-normal years. Although the available data are inadequate to demonstrate it quantitatively, that drought was the most serious disaster recorded historically in NEB. Estimates of the number of deaths it caused range from 100 000 to more than 1 000 000 people (Furtado, 1980). That disaster prompted the then Emperor of Brazil to begin an official action aimed at increasing the resilience of the region with respect to future droughts. In 1877 the Emperor established a commission to study the problem and make suggestions.

The basic assumption at this time was that the regional problem was due to water shortage, and that the primary solution was to be found in the storage of water in normal years through the construction of reservoirs, thus providing employment in public works. In addition, building of a transport infrastructure was advocated, largely to facilitate out-migration during drought periods. The intended goals were largely attained, since the mortality rate was apparently sharply reduced in subsequent droughts, though there is a lack of specific data to prove this.

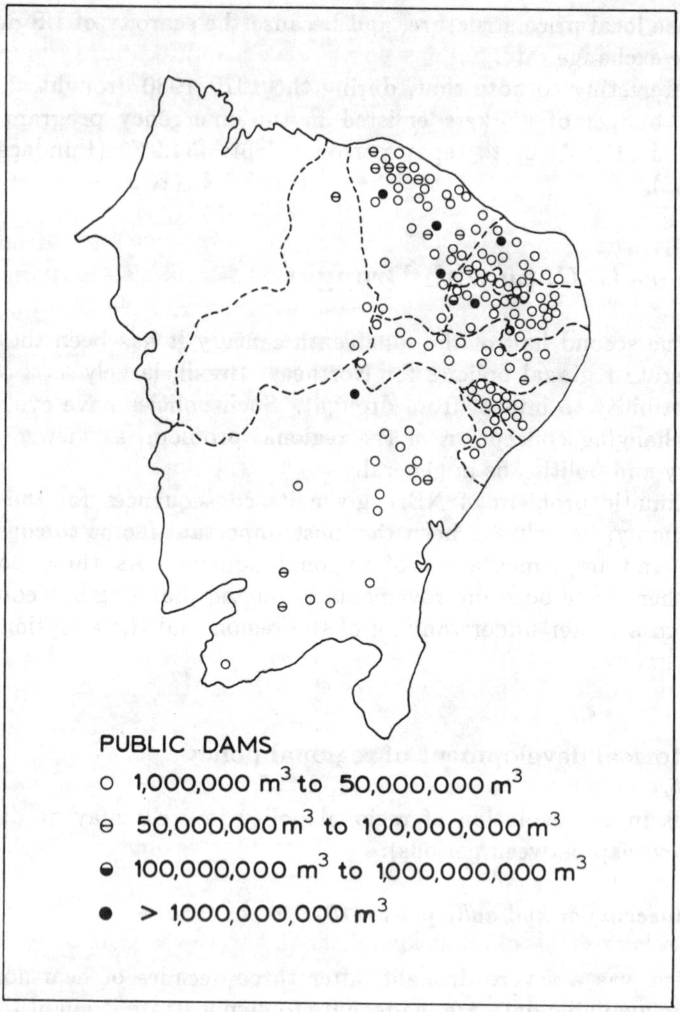

Figure 1.6. Location of storage dams in Northeast Brazil. Source: IBGE (1977).

During this period, institutions specializing in hydraulics works were created, and still exist today. This phase can be regarded as a public works planning operation rather than a regional or spatial planning initiative. By 1960, 987 dams had been built, with water storage capacity of 16.5 thousand million cubic meters. These data do not include small dams constructed by private farmers (*Figure 1.6*) (MINTER–SUDENE, 1985).

2. The ecological phase (ca. 1950 to ca. 1960)

In this phase the emphasis was on more appropriate adaptation to the climate of the semi-arid zone. It was recommended that drought-resistant (xerophilous) plants should be introduced to these regions, such as drought-tolerant cotton, fodder palm, etc.

The results were not very successful, even though in principle the recommendation was valid. Xerophilous plants have a limited utility because of constraints on their yield and production potential. The main studies on xerophilous plants have been done by Duque (1973).

3. The development phase (ca. 1954 to ca. 1975)

More recently, the assumption was that the problem of NEB was one of underdevelopment in all sectors and of dependence on the central–southern part of the country. The main recommendation was to reduce interregional disparities and to promote regional growth through industrialization. Agriculture, it was argued, should be developed to guarantee food supply in urban centers, keeping real wages low so as to stimulate industrial development.

In this phase, the Bank of the Northeast (Banco do Nordeste do Brazil) was founded, a federal agency in charge of long-term financing of regional investments. In addition, a federal authority was created (SUDENE) with the aim of promoting regional planning and coordination. Several development plans were prepared and implemented, and fiscal incentives were created and are still offered today to stimulate new investments.

In 1958, a report written by the Task Force for the Development of the Northeast (Grupo de Trabalho para o Desenvolvimento do Nordeste), headed by Celso Furtado (MINTER–SUDENE, 1967), included the following points:

(1) The inequality of income levels between NEB and the central–southern region was the main problem to be faced during that phase of national development.
(2) The growth rate of the NEB economy, in the previous decade, was substantially less than that of the central–southern region.
(3) In addition to the traditional causes of backwardness in the regional economy (relative lack of good land, inadequate rainfall, income concentration in the sugarcane economy, predominance of the subsistence sector in the semi-arid zone), there were new causes related to the process of industrialization of the country.
(4) As a production crisis, drought is particularly serious because the impact is concentrated on the most vulnerable part of the system – subsistence agriculture.

The report proposed a Plan of Action to:

- Intensify investment in the manufacturing sector.
- Transform the agricultural economy of the humid coastal zone with more emphasis on food production.
- Transform the economy of the semi-arid zone to make it more productive and resistant to drought.
- Occupy the new agricultural frontier on the fringes of NEB (largely in the State of Maranhão), to receive the excess semi-arid zone population.

The main development activity was the industrialization program conducted by SUDENE. Through this program, fiscal incentives from the federal government were granted to private enterprises throughout the whole country to induce them to establish new industrial capacity in NEB. There was some success. Field research conducted by SUDENE and the Banco do Nordeste do Brasil in 1978 in 763 industrial projects (Magalhães, 1983) concluded that:

(1) The program had turned the industrial manufacturing sector into the most dynamic economic activity in the region, replacing the previous role of the export sector and public expenditures.
(2) New manufacturing industry in the Northeast was efficient in comparison with other manufacturing industries in Brazil.
(3) The contribution to the creation of new employment had been low, in relation to the region's needs, because of the characteristics of modern industry and the need to compete in the national market. By 1978, 183 000 permanent jobs had been created. (The corresponding number, updated to 1986, is estimated to be about 400 000).

Industrialization was concentrated in three state capitals (Salvador in Bahia State, 48% of total investment; Recife in Pernambuco State, 17% of total investment; and Fortaleza in Ceará State, 4% of total investment). The main subsectors were: chemicals, 40%; textiles, 11%; metallurgy, 14% – each expressed as a percent of total fixed investment (Magalhães, 1983).

Despite various problems, the NEB economy has grown quite rapidly. The growth rate of regional GNP, which had been 6.4% a year on average over 1950–1960, rose to 8.3% p.a. over 1970–1980. This growth was led by the manufacturing sector, which averaged 8.1% p.a. over 1950–1960, and 10.4% p.a. over 1970–1980 (Magalhães, 1983). The per capita income of NEB, expressed as a percentage of that of Brazil as a whole, increased from 40.3% in 1960 to 43.7% in 1975 (Ramos Viana, 1981) and 45.7% in 1980 (IPEA IRN, 1985).

4. The socioeconomic approach (ca. 1975 to present)

The concept underlying this new approach is that backwardness in NEB is due not only to the lack of water or to insufficient economic growth, but is essentially a problem of the poverty of the great majority of the population. This idea stemmed from a reexamination of public policies and their effects on society, mainly in the rural areas of NEB. It is clear that populations are still vulnerable to droughts, have limited access to means of production (land, water, capital and technology), and exhibit an inability to accumulate wealth to resist the effects of future droughts and other hazards.

Recommendations regarding development to overcome these problems follow two different lines:

(1) To retain the current objectives of economic development, while endeavoring to create the maximum number of jobs and ensure that the benefits of this development are felt inside the region.

(2) To adopt policies that will benefit the vulnerable population directly, in terms of increasing their earnings. The common economic instruments, such as credit and fiscal incentives, are supposed to benefit indirectly the vulnerable population by increasing work opportunities, but in reality their effects on these people have been very small.

Together with the continuation of the use of fiscal incentives for the private sector to promote industrial development, new projects have been designed that go beyond the provision of basic needs and try to incorporate the rural population into the production process. The main projects are described in *Table 1.11*.

The previous phase (1954–1975) encouraged, in practice, the development of the manufacturing sector as a strategy to increase GNP, reduce regional inequalities and, as a consequence, create new jobs. The present phase (since 1975), though a continuation of the previous one, is more comprehensive. It assumes that industrialization and economic growth alone will not be able to create all the required jobs and spread the benefits to all the population. It is necessary to adopt a more comprehensive diagnosis, based not only on economic and climatic variables, but also on social variables, so as to establish goals aimed at eliminating poverty and increasing economic growth and resilience with respect to drought.

The recent evaluation of those programs and projects has led to their aggregation into the so-called Projeto Nordeste (Northeast Project), a very ambitious program that has three main goals: (i) to provide small farmers with access to means of production; (ii) to use appropriate technology; and (iii) to improve human resources, education and health (*Table 1.12*) (World Bank, 1983; MINTER–SUDENE, 1984; SUDENE, 1984).

It is recognized that emergency programs during droughts should be integrated with long-term programs. This integration may be beneficial through:

- Reduction in the competition for financial and human resources. During droughts, emergency programs in the past have used part of the resources assigned to long-term programs.
- Use of emergency programs to build works that will serve the objectives of the permanent programs.

Projeto Nordeste is not a project in itself, but a process of reexamining and redefining the whole development policy of NEB, based on a comprehensive social and economic diagnosis and linking general goals and specific lines of action. It consists of three main parts:

(1) Establishment of general goals and directions, and definition of priorities. This is done in the Development Plan of the Northeast and the National Development Plan.
(2) Adoption of specific strategies to make possible the implementation of the defined priorities.
(3) Definition of projects for the implementation of such strategies.

Table 1.11. NEB: recent development projects.

Program/year of creation	*Implementing agencies*	*Geographical area*	*Objectives*	*Beneficiaries*	*Results*
Polonordeste 1974	SUDENE States	47 sub-areas 850 000 km^2	Rural development in priority areas. Improve living conditions	Small producers	160 000 employments. 340 000 families attended with rural extension and 160 000 with rural credit. Some improvement in living conditions.
Irrigation 1971	DNOCS CODEVASF	Any irrigable place in NEB	Increase production and productivity. Reduce dependency in droughts	Small producers and small enterprises	24 000 employments. 104 enterprises. Families attended (7277) reported improvement in living conditions.
Sertanejo 1976	SUDENE DNOCS CODEVASF States	Semi-arid 850 000 km^2	Increase resilience of farms to drought	Small and middle producers	14 200 employments. Some improvement in living conditions.
Prohidro 1979	SUDENE Federal and state entities	Semi-arid	Increase water availability for human and agricultural use	Urban and rural population, farmers	Water supply: 400 000 families. Irrigation: 4000 families. 1800 public wells. 11 400 private wells and dams. Some improvement in living conditions.
Agroindustry 1974	BNB	NEB	Develop agroindustry. Create jobs. Interiorize development	Agroindustrial enterprises	14 000 direct jobs. 70 agroindustrial plants. Some improvement in living conditions.
Procanor 1979	SUDENE States	Sugarcane coastal zone	Improve living conditions of poor people in the coastal zone	Poor families	8000 families attended with subsidized credit social infrastructure etc. Some improvement in living conditions.
Fiscal incentives 1961	SUDENE	NEB	Promote industrialization	Private enterprises	400 000 jobs. 2200 industrial and agricultural private projects. Increase in the rate of growth of GNP. Urban concentration.
Emergency drought years 1979–1983	SUDENE States	Semi-arid	Reduce the direct impact of drought on the more vulnerable population	Rural and small towns – poor population	2.7 million employments. Reduced outmigration. Avoided starvation.

Sources: von Oven (1983); MINTER–SUDENE(1984); SUDENE BNB (1986).

Table 1.12. *Projeto Nordeste*: rural development strategy.

Projects	*Implementing agencies*	*Objectives*	*Strategy*
PAPP – Program for Support of the Small Producer	SUDENE, States, Federal entities	Eliminate rural poverty. Reach 2.0 million families in 15 years.	Implement integrated projects in priority areas. Promote access of each family to means of production, technology and the market.
Irrigation	Ministry of Irrigation, CODEVASF, DNOCS, DNOS, SUDENE, States	Increase food production and productivity. Irrigate 1.0 million ha by 1990.	Implement public and private irrigation projects.
Small non-agricultural businesses	SUDENE, BNB	Foster nonagricultural activities in rural areas and small towns. Promote entrepreneurial capacity.	Provide financial and technical assistance to small private and cooperative projects.
Education	Ministry of Education, SUDENE, States	Eliminate illiteracy in 15 years.	Organize all means and efforts to attack the main bottlenecks in the rural education system. Integrate with other programs.
Health	Ministry of Health, Ministry of Social Security, SUDENE, States	Improve health condition of the population. Eradicate endemic diseases.	Integrate health services. Improve institutions. Organize action from the city level to the local rural areas.
Sanitation	Ministry of Health, Ministry of Urban Development, SUDENE, States	Promote access to good-quality water for all rural population.	Design state programs and institutional improvement. Implement with the community taking co-responsibility. Subsidize financing.

The rural problem has been designated as a priority. The Rural Development Strategy aims at eliminating poverty, increasing agricultural production and productivity, increasing resilience with respect to drought and thus improving living standards. The intention is to enable each rural family to generate sufficient income to rise above subsistence levels.

The strategy will be implemented through six components (*see Table 1.12*):

(1) Support of small rural producers.
(2) Irrigation.
(3) Support of nonagricultural rural producers.
(4) Education in the rural area.
(5) Health in the rural area.
(6) Basic sanitation in the rural area.

The Program for Support of the Small Rural Producer (PAPP) consists of components designed to:

(1) Promote access to the ownership of land for all small rural producers. This component is a part of the National Plan of Agrarian Reform, approved in January 1986.
(2) Promote access to water at the farm level (water resources for human, agricultural and animal use).
(3) Promote access to rural credit (investment and working capital).
(4) Provide technical assistance and rural extension.
(5) Adapt agricultural research to meet the needs of small producers.
(6) Support commercialization of inputs and outputs.
(7) Support the development of rural communities.

This program will be based on 10 state subprograms and is intended to reach 2 million rural families in 15 years, i.e., by the year 2000. It replaces the other rural programs (POLONORDESTE, SERTANEJO, PROHIDRO, PROCANOR) and will be implemented progressively, beginning in priority areas of each state. The adaptation of the previous programs began in 1985. At that time three state programs were approved and their implementation was begun (in Sergipe, Bahia and Rio Grande do Norte). Programs in the other states began in 1986.

The coordination authority is SUDENE, with the support of several federal agencies, but the specific planning and implementation is carried out by the states. Financing is provided by the federal government, with budget funds and World Bank loans. The estimated cost is US$6000 per family, with a total of US$12 thousand million in 15 years.

The irrigation component of the rural development strategy (*Table 1.12*) is a continuation of the previous irrigation program in the region. There are now some 200 000 irrigated hectares in NEB, of which 60 000 hectares are in public projects and the rest in private. The new program is intended to irrigate 1 million hectares in the 5 years from 1986 to 1990.

The irrigation will be carried out through public (40%) and private projects (60%). It is expected that this program will increase agricultural production, especially food production, and will directly create 750 000 jobs. The estimated cost is US$4500 per hectare, giving a total of US$4.5 thousand million in the 1986–1990 period. The Ministry of Irrigation, created in 1986, implements the irrigation program, with the participation of DNOCS, CODEVASF, DNOS, the Ministry of Agriculture and SUDENE. At the time of writing (1987), the other rural development strategy programs are still being defined.

1.5.2. Challenges for policy-makers

After a century of public policies, results are still less than satisfactory, although, as stated earlier, drought is no longer such a disaster as it used to be in the last century, in terms of human deaths. The poor population is still vulnerable to drought and still suffers from it.

Technologies imported from developed countries are not appropriate for the semi-arid zone. It is essential to develop techniques that meet the specific requirements of local conditions, taking into account the people's needs. Such techniques should be designed to make small farms economically feasible. It is necessary to foster a better understanding of the interaction between soil, water, plants and climate at each location. It is also important to promote the rational planning of water use. Water must be provided in drought periods for the whole population, animals and agricultural use. These challenges have not yet been met.

In the last three decades there has been a growing concern for the social consequences of NEB underdevelopment. It is important to note that there are crucial political aspects to this question. The poverty in NEB has presented an unfavorable image of Brazil both at home and abroad and has served to intensify a widespread feeling that public policies for regional development must be improved. A comprehensive approach to this issue, based on the above assessment of 100 years of regional policies, offers some hope that it is possible to apply measures that are socially efficient and effective.

1.6. Objectives of the Present Case Study

The aims of this study are:

(1) To consider the climatology of droughts and their predictability in NEB.
(2) To present alternative methods for modeling crop impacts and socioeconomic impacts that occur as a result of anomalous climatic events.
(3) To use the results thus obtained to provide a basis for an improved understanding of the impact of droughts and for designing policies to increase the resilience of agriculture to drought.

1.7. Structure of the Case Study

Following this introduction, Section 2 analyzes the climatology of drought in NEB with special emphasis on the characteristic patterns of drought and frequency of occurrence of particular drought types, as well as possibilities for drought prediction. Scenarios of the most important drought types are also established. Because of the particular characteristics of these selected scenarios, however, they do not always provide a sufficient basis for illustrating the analysis of agricultural droughts, especially in simulation work. For this reason alternative additional scenarios are introduced later where appropriate. Section 3 reports on experiments with a simulation model to evaluate the impact of climatic variations on potential yields of several crop types. Section 4 studies the effects of climate on the production and productivity of the region. Finally, Section 5 presents conclusions and outlines some policy implications.

SECTION 2

The Climatology of Droughts and Drought Prediction

2.1. Introduction

This section reviews some aspects of the climate of NEB (Northeast Brazil, i.e., the "Nordeste"), focusing in Subsection 2.2 on rainfall, its interannual variability and the mechanisms that produce rain. Subsection 2.3 provides a short review of the state of the art of drought prediction for NEB. It also discusses the occurrence of drought and describes the coincidence of drought episodes and the El Niño/Southern Oscillation phenomenon. A set of agroclimatic scenarios of different types of drought episode that can be used in crop production experiments and computer analyses are identified in Subsection 2.4. A summary and conclusion are given in Subsection 2.5.

The climate of any region is determined, to a large extent, by the general circulation of the atmosphere. This circulation is driven by the differential heating of the rotating globe by solar radiation and by the asymmetric distribution of oceans and continents. Circulation patterns that are generated in the atmosphere redistribute heat, moisture and momentum around the globe. However, this redistribution is an imperfect process. Sometimes it acts to diminish variations in climatic elements within a region. At other times it increases regional differences in the same climatic elements, such as temperature and rainfall, that have a significant bearing on human activities. Rainfall is one of the most critical elements affecting agricultural activities in low latitude regions, such as NEB.

NEB is located between 4° and 16° S and 33° and 46° W, i.e., just to the east of the Amazonian rainforest, with an approximate area of 1.5 million km^2. It is considered an anomalous region in tropical continents because, in contrast to other regions in this latitudinal range, it has a semi-arid climate. This is due to the relatively low values of rainfall over most of the area. Although some eastern coastal areas annually receive 1600 mm or more, some interior valley areas have

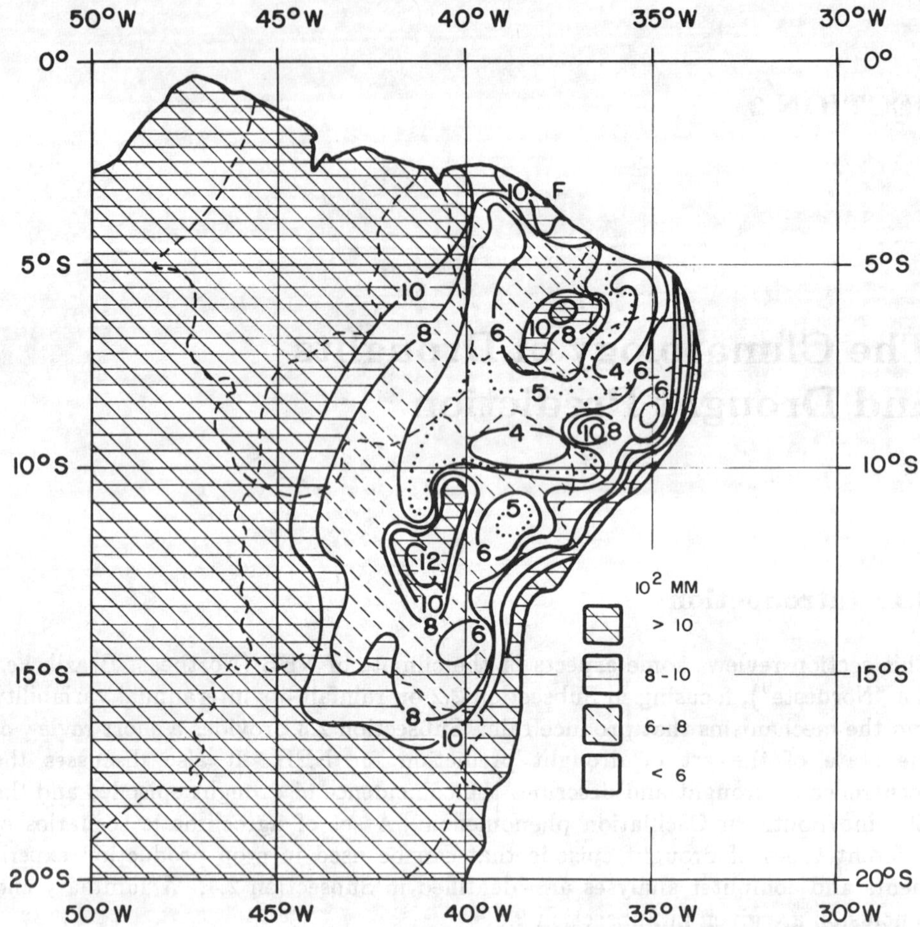

Figure 2.1. Annual precipitation in decimeters (1 dm = 100 mm) over NEB. Source: adapted from Kousky and Chu (1978).

an annual average of less than 400 mm (*Figure 2.1*). As is typical in other semi-arid regions of the world, e.g., the Sahel, northeastern Africa and parts of India, spatial and temporal variability of rainfall are very high. *Figure 2.2* shows precipitation indices (areal averages of precipitation departures from the long-term mean standardized by the standard deviation for each station) for northern and southern NEB. The areas represented by each graph are delimited in *Figure 2.3*. The year-to-year variations are outstanding. For example, during the rainy season of 1985 in northern NEB there was a positive deviation of about 200%, while in 1983 a negative deviation of nearly 100% was observed. The spatial variability is illustrated by the year 1970, when northern NEB experienced drought, while the southern part was slightly wetter than normal.

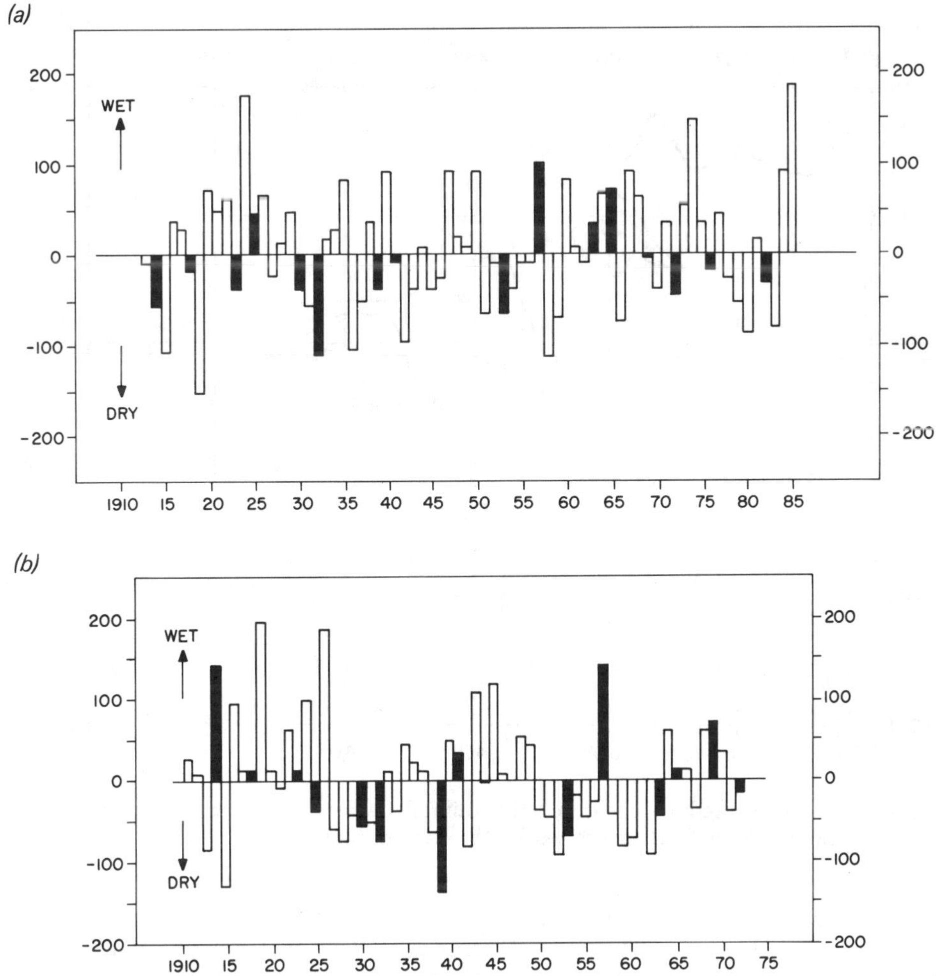

Figure 2.2. Rainfall departure indices for (*a*) northern NEB and (*b*) southern NEB. The indices are calculated as areal averages of seasonal (March–May) in (*a*) and annual (July–June) in (*b*) rainfall departures from the mean standardized by the standard deviation for 30 stations in northern NEB and 14 stations in southern NEB. Solid bars denote El Niño years. Northern and southern parts of NEB are shown in *Figure 2.3*. Sources: (*a*) Hastenrath (1986); and (*b*) Hastenrath *et al.* (1984).

The causes of this great interannual variability are not yet fully understood. However, an analysis of the rainfall distribution over NEB and neighboring areas suggests that the relative dryness may be caused primarily by mechanisms in the general circulation, more specifically by a thermally driven circulation cell of the Hadley–Walker type (*see* Part I, Section 3, of this volume), and therefore induced by climatic controls external to the region. Their mechanism can be explained in simple terms as follows.

Immediately to the west of NEB is the Amazon basin, which presents very cloudy conditions and high rainfall rates over a large area. Most of this

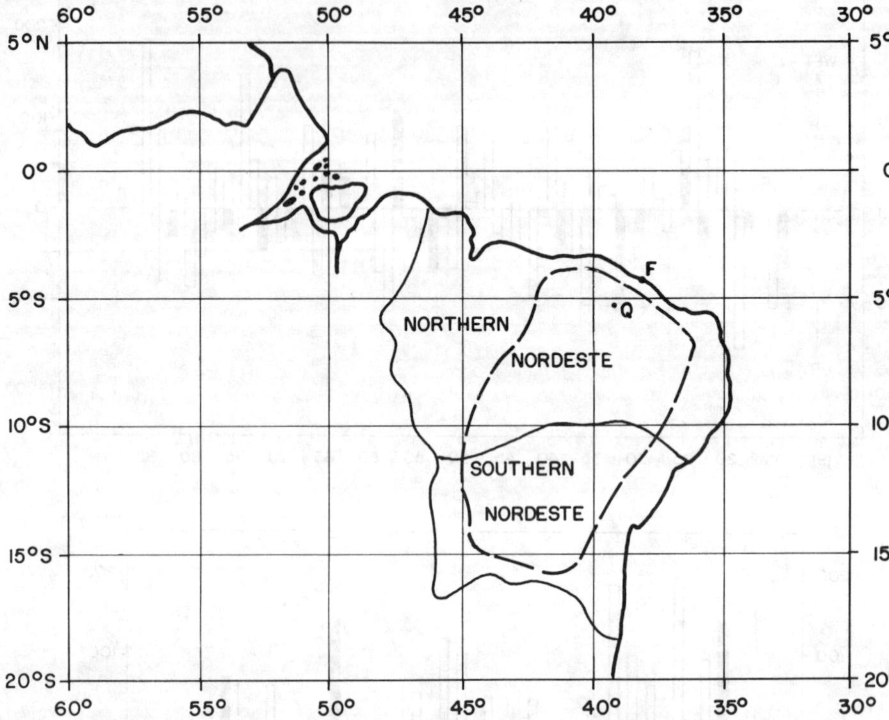

Figure 2.3. Areas used to construct the rainfall departure indices of *Figure 2.2* for northern and southern NEB (delimited by the dashed line).

precipitation comes from towering cumulus clouds associated with warm, moist upward convection currents called "thermals". When water vapor condenses in these clouds to form water droplets, latent heat of condensation is released, which warms the air. This warming causes the air to ascend. The air rising over Amazonia descends over the entire subtropical Atlantic Ocean, including NEB and possibly the western coast of Africa. The subsiding (sinking) air is not conducive to cloud formation and thus serves to inhibit rainfall. The effects are reflected in data on the average rainfall distribution (Jäger, 1976) and satellite cloud cover (Miller and Feddes, 1971). This is predominantly an east-to-west circulation (known as a Walker-type circulation).

Similarly, the north–south circulation cells, with ascending motion in the intertropical convergence zone (ITCZ) over the equatorial Atlantic Ocean – the region where the northern hemisphere and the southern hemisphere trade winds meet – and descending motion over the subtropical Atlantic of both hemispheres, called the Hadley circulation, also contribute to the aridity of areas such as NEB. The combined effect of these two types of circulation, and their interannual variability both in strength and geographical position, appear to be important factors behind the semi-aridity of NEB and the high interannual variability of its

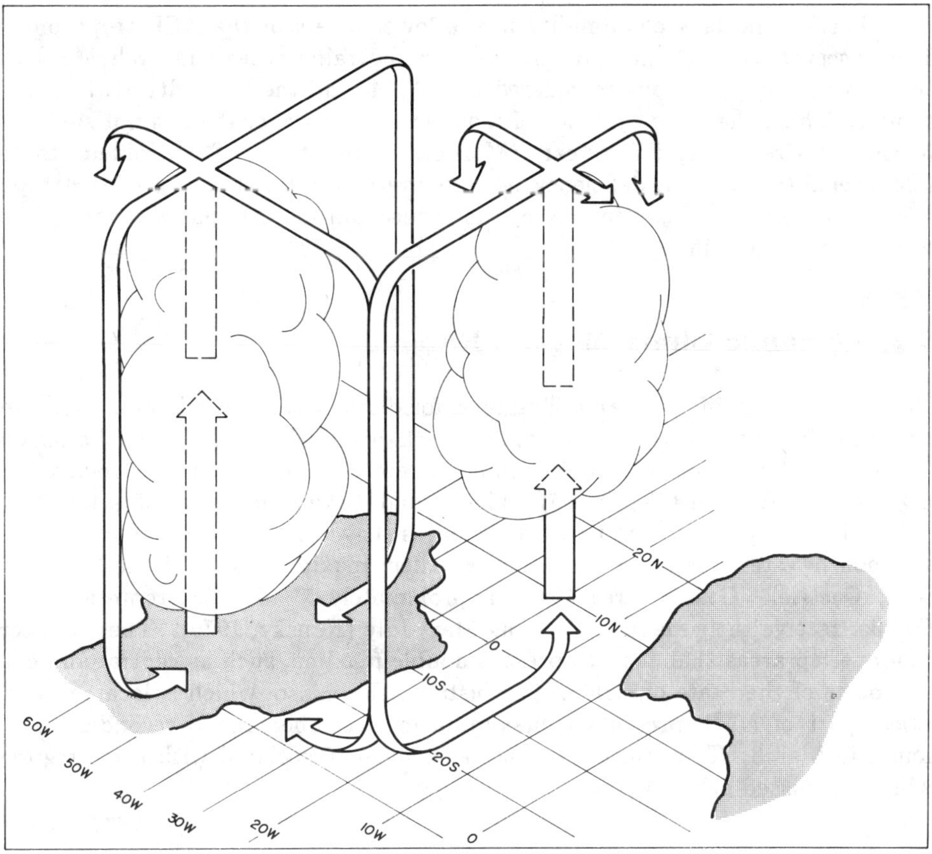

Figure 2.4. Schematic diagram of the Hadley–Walker circulation over South America and the South Atlantic sector. The cloudiness over the ocean represents convective activity in the intertropical convergence zone (ITCZ).

precipitation. *Figure 2.4* shows in schematic form the Hadley–Walker circulation over South America and the South Atlantic sector.

In addition, factors such as high surface albedo and topographic features may enhance the dryness associated with the general circulation patterns. Charney (1975) proposed an explanation for self-maintenance of a desert region (*see* Part I, Section 3, of this volume), which has been shown to be at least partially applicable to NEB (Gomes Filho, 1979). The principle is that with high surface albedo (due to lack of vegetation), more solar radiation is reflected from an arid region than from surrounding areas – in the case of NEB these are the Amazon rainforest and the Atlantic Ocean – and the atmospheric column over the region is relatively cooler. In order to compensate for this thermal contrast, air subsides, compresses and thus warms. This sinking motion serves to inhibit rainfall. In NEB this feedback mechanism may well contribute to the enhancement and prolongation of dry episodes, although the underlying control is the large-scale atmospheric circulation.

There is no lack of humidity in the lower layers of the NEB troposphere. The observed vertical moisture profiles for the rainy season (March–May) for two contrasting years, one considered wet (1974) and the other dry (1976), were similar. Thus, the mere presence of moisture in the troposphere is not sufficient to characterize wet and dry years (Marques *et al.*, 1983). The problem, therefore, seems to be not a deficiency of moisture in the lower layers of the troposphere, but a lack of continuous dynamic mechanisms capable of transforming water vapor into rain.

2.2. Dynamic Climatology of Rainfall

The existence of different rainfall regimes for distinct parts of NEB tends to support the idea that more than one major mechanism affects rainfall in the region. Rainfall reaches a monthly maximum in the northern portions of the region during March and April (*Figure 2.5*). The annual distribution of rainfall for Quixeramobim, a typical station for that region, is shown in the figure. The southern and southwestern parts of NEB receive their maximum rainfall in December (e.g., Caetite). Coastal areas from approximately 5° to 18° S, represented by Olinda, receive their maximum during May–July (Kousky, 1979). There are certain overlap areas that tend to present double maxima, such as central and coastal parts of the State of Bahia. For instance, Remanso, which is located in the driest part of NEB, presents a maximum in December and a secondary maximum in March. The rain producing mechanisms associated with these regimes will be discussed below.

2.2.1. Intertropical convergence zone (ITCZ) and sea-surface temperatures

It is likely that the most important factor in determining how deficient or abundant the rainy season will be in the drier part of NEB (the part with rainfall maximum in March in *Figure 2.5*) is the latitudinal position of the ITCZ over the equatorial Atlantic, more specifically, over the western Atlantic near the coast of South America. Normally the ITCZ migrates seasonally from its northernmost position at about 14° N in August–September to its southernmost displacement at about 2° S in March–April. In drought years the ITCZ usually does not cross the equator in its southward migration, and therefore does not reach NEB (*Figure 2.6*). Under those circumstances NEB lies south of the region of high rainfall and in a region of predominantly sinking, rainfall-inhibiting, motion. On the other hand, in wet years the ITCZ moves as far south as 5° S near the coast of NEB. Dynamically, the ITCZ is a band of low pressure and convergence of the airflow in the lower layers of the atmosphere near the surface. Convergence of moist air near the surface produces the conditions favorable for ascending motion, condensation of water vapor, cloudiness and high precipitation rates. It follows that, when the ITCZ is in the immediate vicinity of NEB, the large-

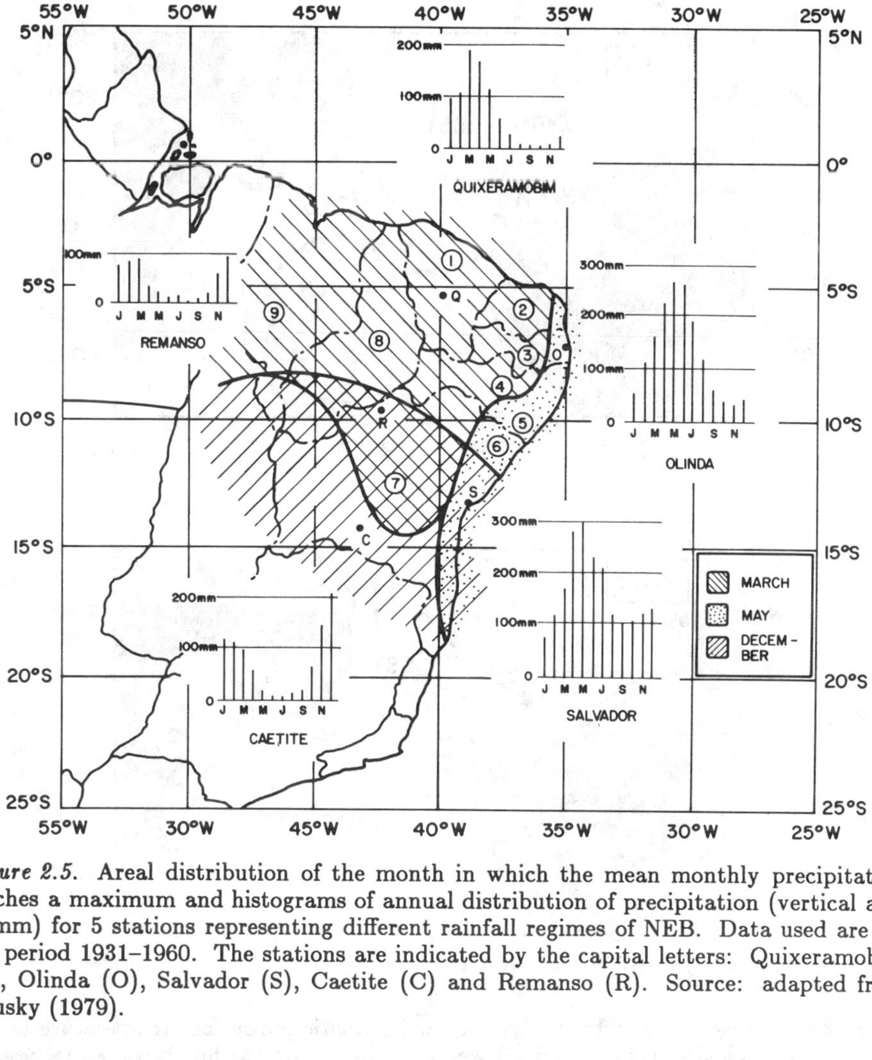

Figure 2.5. Areal distribution of the month in which the mean monthly precipitation reaches a maximum and histograms of annual distribution of precipitation (vertical axis in mm) for 5 stations representing different rainfall regimes of NEB. Data used are for the period 1931–1960. The stations are indicated by the capital letters: Quixeramobim (Q), Olinda (O), Salvador (S), Caetite (C) and Remanso (R). Source: adapted from Kousky (1979).

scale atmospheric circulation conducive to high precipitation rates also lies over NEB, primarily over the northern part of it, resulting in plentiful rainfall.

The ITCZ is part of the general circulation of the atmosphere. Therefore, the reasons for the variations in its strength and position from year to year, which cause the enormous interannual variability in NEB's rainfall, must be sought in the planetary-scale atmospheric circulation. Indeed, several observational studies have indicated the existence of a link between rainfall anomalies in NEB and anomalies in the atmospheric circulation on a much larger scale than that of NEB itself (Namias, 1972; Hastenrath and Heller, 1977; Nobre *et al.*, 1984). These studies have shown that in most of the years presenting large rainfall fluctuations about the mean, the atmospheric circulation over the entire Atlantic was changed. In wet years, the North Atlantic Subtropical Anticyclone (Azores High) is stronger than normal and displaced southward of its usual

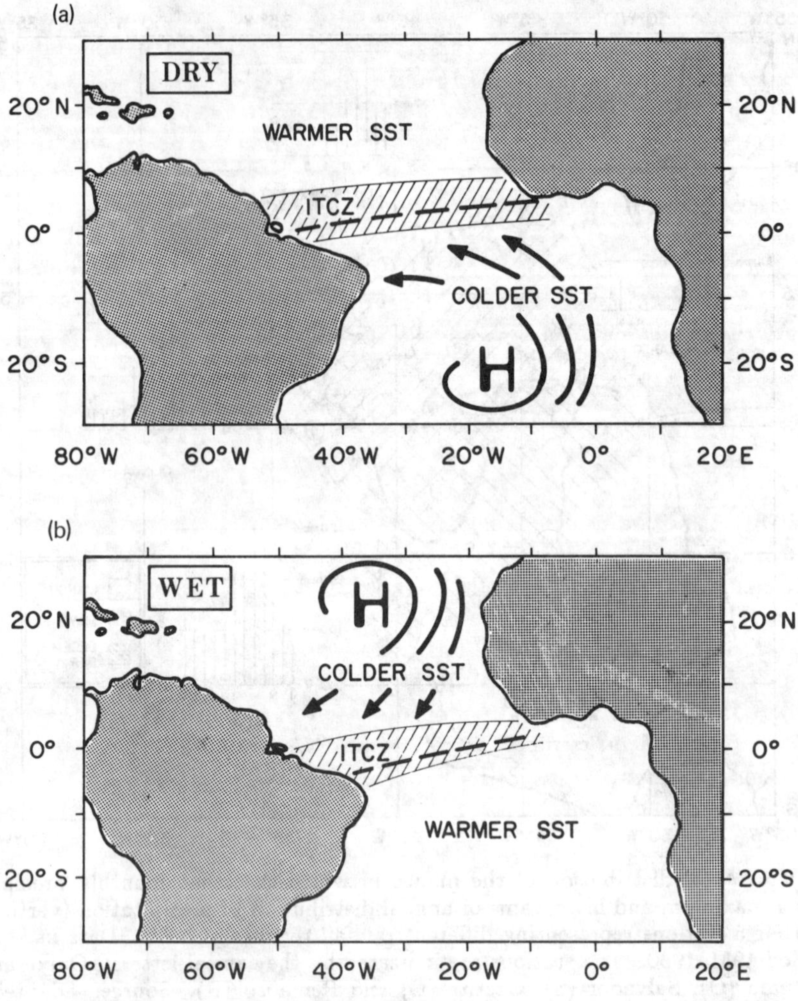

Figure 2.6. Large-scale circulation features and Atlantic sea-surface temperature (SST) distributions typical of (*a*) dry and (*b*) wet years in NEB. The hatched area represents the mean ITCZ position, with the dashed line indicating the axis of confluence of the trade winds from both hemispheres.

position. The North Atlantic trade winds are stronger, and the ITCZ is displaced farther south. A mirror reflection of this situation about the equator happens in dry years: the South Atlantic Anticyclone is more intense and displaced equatorward resulting in stronger South Atlantic trades, and the ITCZ is found north of its normal position.

Although the ITCZ is embedded in the general circulation of the atmosphere, it is also closely linked to the surface temperature of the tropical oceans. It generally lies over or near the area of highest sea-surface temperatures (SST). Therefore a relationship would be expected to exist between the general SST distribution in the tropical Atlantic (which, to a certain extent, determines the southernmost or northernmost positions of the ITCZ), and rainfall over NEB. Indeed, this relationship seems to hold for most years (Markham and McLain, 1977; Hastenrath and Heller, 1977; Moura and Shukla, 1981). Warmer (colder) SSTs in the tropical South Atlantic and colder (warmer) SSTs in the tropical North Atlantic are associated with wet (dry) years in NEB. In wet years (e.g., 1964, 1967, 1984 and 1985) the SST anomalies were positive in the tropical South Atlantic and negative or near zero in the tropical North Atlantic. In contrast, in dry years (e.g., 1951, 1953, 1958, 1970 and 1979) the SST anomalies were negative in the tropical South Atlantic and positive in the tropical North Atlantic. *Figure 2.6* shows in schematic form the large-scale circulation features, ITCZ positions and Atlantic SST distributions associated with wet and dry years in NEB.

2.2.2. Frontal influences

Another important rain-producing mechanism for NEB arises from the penetration of frontal systems well into the tropical latitudes of that region. Cold fronts or their remnants moving north from high latitudes of the southern hemisphere may reach NEB (as far north as 15° S along the east coast of Brazil) when the circulation patterns of subtropical latitudes are favorable. They may enter NEB at any time of the year, and they play an important role in the December–January maximum in monthly rainfall experienced in the southern part of the region (*see Figure 2.5*), and are also associated with increased precipitation along the coast from 5° to 18° S (Kousky, 1979). These frontal systems also generate and enhance convective precipitation inland over the southwestern and western portions of NEB (actually, this enhancement extends to Amazonia) as shown schematically in *Figure 2.7*. These events seem to be quite frequent from November through February, and they give rise to a large interannual variability (Santos, 1986).

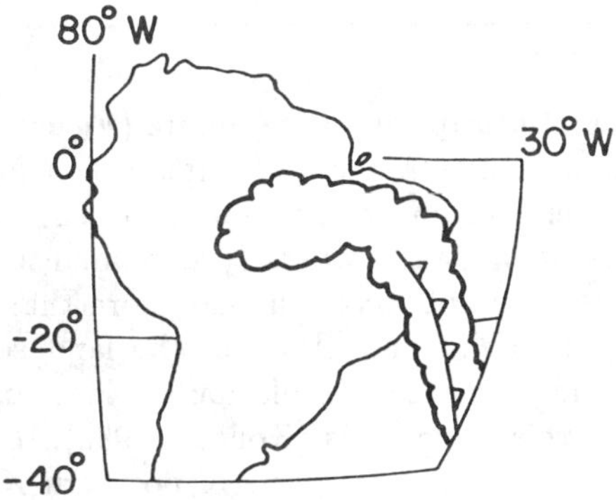

Figure 2.7. **Schematic diagram indicating the northernmost position of frontal systems that affect NEB and the convective cloudiness associated with them extending to the western part of NEB and to Amazonia. Source: Santos (1986).**

Probably the most important reason for the observed large interannual variability in the frequency of these events is the occurrence of blocking patterns in the tropospheric flow over South America and adjacent oceans. The atmospheric blocks prevent the penetration of rain-bearing frontal systems into NEB, keeping them stationary over the southern states of Brazil, Uruguay and northern Argentina where they cause high rainfall totals and floods.

2.2.3. Tropical meso-scale mechanisms

There are different types of meso-scale dynamic mechanisms that may produce precipitation in NEB, generally associated with large-scale circulation patterns. One of the most important involves the occurrence of cold upper tropospheric cyclonic vortices, which form over the Atlantic to the east of NEB. They are called "cold upper" vortices because the core is colder than the surrounding atmosphere, and their cyclonic circulation is generally present only above 500 mb. The vortices have been observed to form chiefly in advance of strengthening southern hemisphere mid-latitude frontal systems that have penetrated very deeply into the subtropics, mainly during southern hemisphere summer. The annual distribution of occurrences of such vortices (*Table 2.1*) clearly demonstrates their predominant appearance during the summer months from November through February. They move westward at a speed of about 4 to 6 degrees of longitude per day and promote intense rainfall in central and southern areas of NEB (Virji, 1981; Kousky and Gan, 1981).

Table 2.1. Number of upper tropospheric cyclones over the South Atlantic Ocean, and number of days with well defined cloud patterns for the period September 1974–August 1981. Source: adapted from Gan (1983).

	Jan	*Feb*	*Mar*	*Apr*	*May*	*Jun*	*Jul*	*Aug*	*Sep*	*Oct*	*Nov*	*Dec*	*Total*
Cyclones	18	15	8	5	0	0	0	0	2	1	11	10	70
Vortex days	93	82	53	10	1	0	0	0	5	3	42	68	357

The May–June coastal precipitation maximum (*Figure 2.5*) is probably due to other meso-scale mechanisms. A possible explanation for the occurrence of that maximum is that there would be a region of ascending motion along the coast due to convergence of the trade winds (predominantly onshore flow) with the nocturnal land breeze. The observation that more than 50% of the rain in that region falls during the night, and also that the land–sea temperature contrast – which drives the land–sea circulation – is maximum also during May–June, helps support this hypothesis (Kousky, 1980). Instability lines associated with the sea breeze along coastal parts of northern NEB seem also to be an important rain-producing factor for those areas (Cavalcanti, 1982).

Meso-scale circulations induced by topographic features are also important in determining the complex, small-scale, spatial structure of precipitation in NEB. For instance, some of the driest locations in NEB, which receive about

300 mm annually, are valleys located downwind (in the "rainshadow") of topographic barriers reaching up to 1000 m. On the other hand, locations upwind from such barriers are wet and suitable for agriculture. Also the diurnal valley–mountain circulation seems to be an important rain-causing mechanism for some interior valleys (Ramos, 1974).

In this section, the macro- and meso-scale rain-causing mechanisms have been described. It should not be forgotten, however, that local moist convection may occur due to surface heating during the day and the availability of moisture near the surface. Such micro-scale convection can always produce rain when the mechanisms of larger scales are not inhibiting it.

2.3. Drought Predictability, El Niño and the Drought of 1983

Severe droughts impose tremendous social and economic costs on the local population. It is therefore not surprising to find in the literature, as early as the 1920s, attempts to develop methods to forecast droughts far in advance. There have been two main classes of methods. One attempts to predict the character of the rainy season a few months in advance and is based on the knowledge of the physical processes, trying to identify large-scale parameters both in the troposphere and the oceans that are indicative of the rainfall regime over NEB. This may involve the use of statistical analysis (multiple regression). The other class uses statistical methods based on the existence of apparent periodicities in the historical rainfall series of a station and attempts to predict drought many years in advance by extrapolation of such periodicities into the future.

2.3.1. Physical methods

Recent studies have revealed a relationship between planetary-scale tropospheric circulation anomalies and rainfall fluctuations in NEB (Nobre *et al.*, 1985a and 1985b). The study of the autumn and winter northern hemisphere circulation anomalies at several levels in the troposphere revealed the appearance of characteristic trains of alternating cyclonic and anticyclonic circulation centers from the tropical North Atlantic across the North Pole to East Asia, forming a wave-like teleconnection pattern. The circulation anomalies were defined as the wind deviation from the mean for a given month, i.e., observed monthly mean wind minus the 18-year mean wind (1964–1981) for that calendar month.

Figure 2.8 shows a composite of circulation anomalies at 200 mb for the northern hemisphere for several months that preceded drought years in NEB. The pattern that precedes wet years is similar except that the circulation sense of these centers is reversed. That is, cyclonic (anticyclonic) circulation for a dry year becomes anticyclonic (cyclonic) circulation for a wet year. Besides this phase reversal from dry to wet years, of interest to drought forecasting is the fact that usually these patterns are established three to four months in advance of the rainy season in NEB, and are seen most frequently in December–January. Analysis of 22 years (1963 through 1986, except 1982 and 1983) has shown that

DRY

Wind anomalies 200mb

Dec/69, Nov/71, Dec/75, Jan/79, Jan/80.

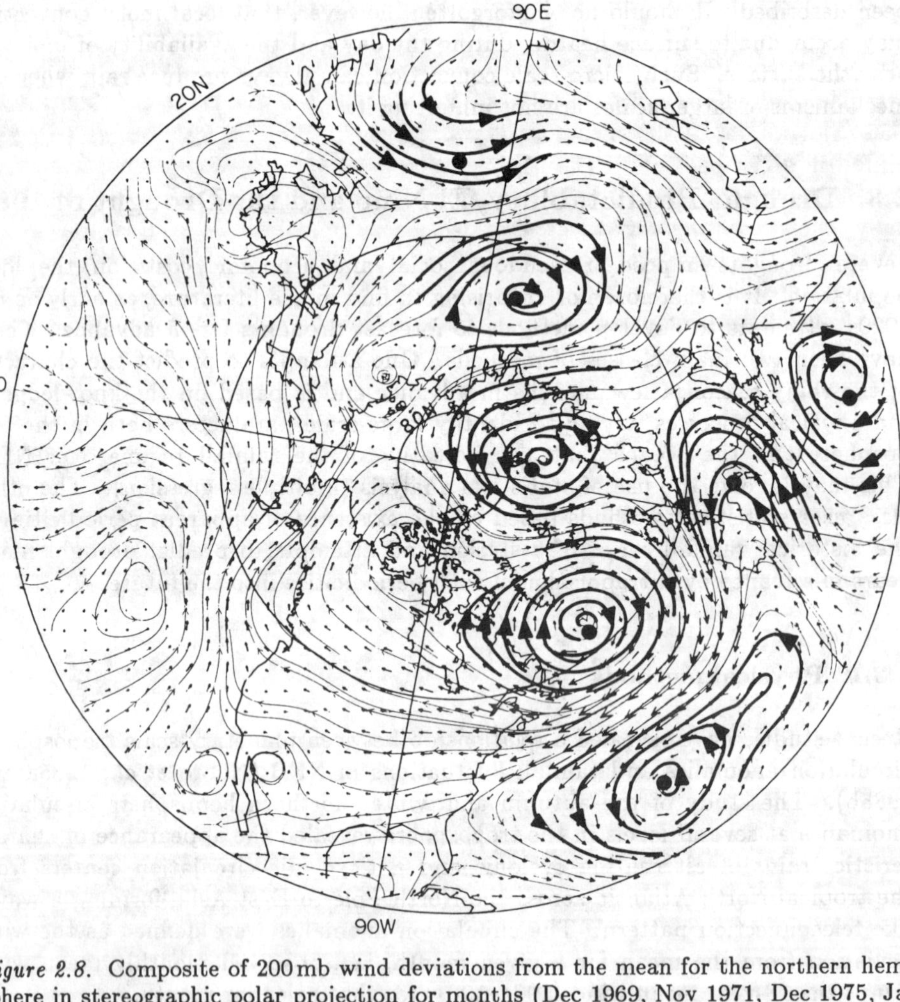

Figure 2.8. Composite of 200 mb wind deviations from the mean for the northern hemisphere in stereographic polar projection for months (Dec 1969, Nov 1971, Dec 1975, Jan 1979, Jan 1980) that preceded dry years in NEB. Source: Nobre *et. al.* (1985b).

this wave-like pattern appeared with the right phase (the one shown in *Figure 2.8* for dry years in NEB or with the sense of circulation centers reversed for wet years) in 15 out of 22 years. It appeared with the wrong phase in 2 years and did not appear at all in 5 years.

The distribution of SSTs in the tropical Atlantic also has some predictive value. In approximately 50% of the years, a pattern of warmer waters in the tropical North Atlantic and colder waters in the tropical South Atlantic, or vice versa, emerges two to three months in advance of the March–May season. As mentioned in Subsection 2.2.1, these SST patterns are associated with dry and wet years, respectively, in NEB.

2.3.2. Multiple regression

A number of studies have made use of multiple regression analysis to develop formulas to predict rainfall in NEB (Walker, 1928; Sampaio Ferraz, 1929; Serra, 1956; Shugahara, 1983; Hastenrath *et al.*, 1984). The general idea is to relate rainfall in all or part of NEB with meteorological and oceanographic (SST) data for points far away from NEB and for periods preceding the rainy season. Generally, the predictors are selected on the basis of diagnostic studies that relate features of the large-scale atmospheric circulation to rainfall in NEB. These features were reviewed in Subsection 2.2.

Table 2.2 lists four of these formulas from different authors. The percentage of the rainfall series' variance (given by the square of the total correlation coefficient, R) explained by these formulas is around 70%. This figure seems to be relatively high, but the multiple regression formulas are capable only of indicating qualitatively whether the rainy season will be above or below the normal and cannot give accurate predictions about the amount of rainfall.

Although these formulas are potentially valuable for drought forecasting, they have not been tested by applying them to new data not used in their derivation, and they have not been put to practical use due to the difficulty of obtaining meteorological and oceanographic data for distant points on a quasi-real-time basis. It is likely that such use will become feasible in the near future, because countries such as Brazil are developing the capabilities to acquire and process global data in real time.

2.3.3. Apparent periodicities

Some long rainfall series, such as that of Fortaleza going back to 1849, show an apparent quasi-periodic behavior at low frequencies, corresponding to periods of approximately 13 years and 24 to 26 years. Sampaio Ferraz (1950), who was the first to notice the existence of these apparent periodicities in the long-term series of rainfall annual totals at Fortaleza, suggested that they might be related to sunspot cycles. More recent studies have confirmed the existence of such periodicities and that they are statistically significant (Markham, 1974; Nobre *et al.*, 1984; Mesquita and Morettin, 1984). By extrapolating on the basis of those apparent periodicities, Girardi and Teixeira (1978) predicted the occurrences of droughts during the period 1979–1985. *Figure 2.2* shows that during that period four years could be considered dry (1979, 1980, 1982 and 1983) and three years were wet (1981, 1984 and 1985). However, a drought-forecasting model constructed on the basis of the two periodicities described above explained only 24% of the rainfall series' variance. It can therefore be concluded that the methods

Table 2.2. Multiple regression formulas relating rainfall in NEB to meteorological variables for antecedent periods at distant points.

Equation	(1)	(2)	(3)	(4)
Reference	*Walker (1928)*	*Sampaio Ferraz (1929)*	*Shugahara (1983)*	*Hastenrath et al. (1984)*
Variables[a]				
Y	Jan–Jun Ceará rainfall[b]	Jan–Jun Ceará rainfall[b]	Mar–May NEB rainfall[c]	Mar–Sep NEB rainfall index[d]
X_1	Jun–Aug Santiago pressure	Jun–Aug Punta Galera pressure	July S. India pressure	Jan Eq. Atlantic zonal wind
X_2	Jun–Nov Honolulu pressure	Jul–Nov S. Rhodesia rain	May–Jul Horta pressure	Jan NEB rain
X_3	Sep–Nov Cape Town pressure	Sep–Nov Cape Town pressure	Dec Accra pressure	Jan S. Oscillation index
X_4	Jul–Nov S. Rhodesia rain	Sep–Nov Samoa temperature	Dec NW USA temperature	
X_5	Sep–Nov St. Helena wind		Dec Honolulu pressure	
a_0	–	–	–	+0.013
a_1	+0.44	+0.67	−0.06	+1.09
a_2	+0.20	−0.39	+0.22	+0.54
a_3	−0.10	−0.16	−0.45	+0.37
a_4	−0.42	−0.13	−0.44	–
a_5	−0.22	–	−0.25	–
R	0.82	0.88	0.85	0.82

[a] Equations are of form $Y = a_0 + a_1X_1 + \ldots a_5X_5$. There is no a_0 for equations (1–3), and no a_5 or a_4 in some cases.
[b] "Ceará rainfall" was from one station (Quixeramobim).
[c] "NEB rainfall" is average for NEB in equation (3).
[d] "NEB rainfall index" in equation (4) is Mar–Sep rainfall departures from long-term average in northern NEB.

based on these apparent periodicities are not suitable for forecasting the years of occurrences of droughts in NEB (Nobre *et al.*, 1984).

2.3.4. El Niño/Southern Oscillation and the drought of 1983 in NEB

El Niño/Southern Oscillation (ENSO) episodes are part of a phenomenon affecting the weather and climate on a global scale, but particularly in the tropics (*see* Part I, Section 3, of this volume). Thus, it is reasonable to expect that

NEB's rainfall might also be affected in some way in ENSO years. In *Figure 2.2* solid bars denote ENSO years. For northern NEB most of the ENSO years coincided with years with rainfall below average during the March–May rainy season. On the other hand, for southern NEB, during the period 1911–1972, approximately half of the ENSO years were dryer, and the other half wetter, than average. Although *Figure 2.2* would seem to indicate a relationship between ENSO episodes and rain-deficient years in northern NEB, a definite link between the two events is not firmly established yet.

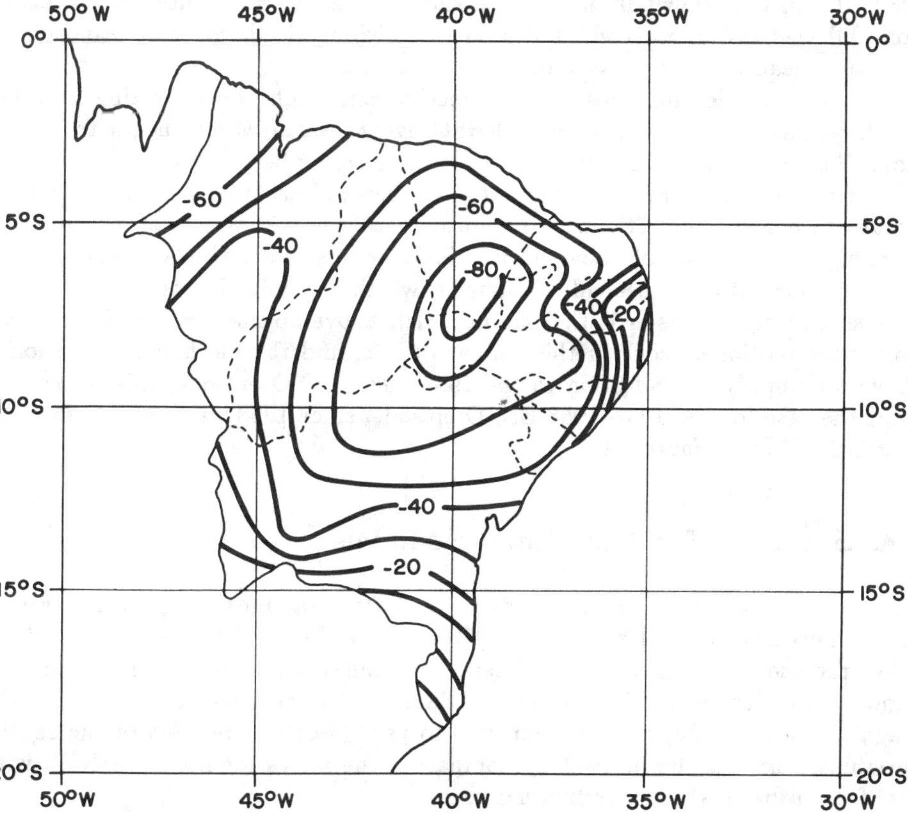

Figure 2.9. Rainfall departures from the mean (expressed in percentages) for the rainy season (March–May) of 1983 in NEB. Source: Rao *et al.* (1986).

The ENSO episode of 1982–1983 was among the most pronounced ever recorded and certainly the best documented to date. The year 1983 was the driest of the last 25 years in NEB. *Figure 2.9* shows rainfall deviations with respect to the 20-year mean rainfall totals for the March–May rainy season. Rainfall in the central part of NEB was as low as 20% of the mean (deviation of −80% in the figure) and less than half of the mean for most of the region. The physical cause of this severe drought seems to have been associated with the ENSO episode of 1982–1983 (Kousky *et al.*, 1984; Nobre and Rennó, 1985). The hypothesis is that strong convection, and thus precipitation, was generated over the anomalously warm waters of the central and eastern equatorial Pacific Ocean in a region where there is very little precipitation in the non-ENSO years. The ascending branch of the Amazonian Hadley–Walker cell (*Figure 2.4*), was displaced from its normal location over Amazonia to a new location farther west over the Pacific and enhanced by the strong convection. The descending branch then extended over the central and northeastern Amazon, NEB, and the tropical Atlantic Ocean, and reached the western coast of Africa. Even the ITCZ over the Atlantic was weakened due to this sinking motion during the early part of 1983. Thus, this unusually intense subsidence over NEB inhibited cloud formation and was the major contributor to establishing the large-scale atmospheric conditions leading to the severe drought of 1983.

It is possible that this type of mechanism might be operating in other ENSO years as well. For a given ENSO event, warm waters begin to appear along the west coast of South America during December, followed by an entire year of above-normal SST throughout the equatorial eastern and central Pacific. Return to near-normal SST occurs approximately 15–18 months after the initial warming. This period of above-normal SST includes two rainy seasons in NEB, but it is normally the second of these in which precipitation is most reduced. This second rainy season is the one in which above-normal SST, and enhanced convective motions, occur in the central Pacific, and the mechanism mentioned above may apply. In addition to the 1982–1983 ENSO episode, this seems also to be the case for the 1957–1958 ENSO episode, since 1958 was a severe drought year in NEB (*see Figure 2.2*).

2.4. Scenarios for Agroclimatic Models

The series of standardized rainfall departures from the mean, depicted in *Figure 2.2* for northern NEB (Hastenrath, 1986) and southern NEB (Hastenrath *et al.*, 1984) represent the best historical series of rainfall currently available for that region. An idea of the interannual variability can be obtained by noting the magnitudes of the departures from year to year. Some indication of the spatial variability can also be derived by comparing the series for northern NEB with that for southern NEB for individual years.

A definition of drought (or of very wet years) based solely on the rainfall departure indices of *Figure 2.2* is not a precise one. The distribution pattern of rainfall with time through the rainy season – which is often more important to agriculture than the total amount of rainfall in the season – is not taken into account by such definitions. Also, the energy and water balances at the surface, which indicate the effective amount of water left for agricultural purposes after runoff, evaporation and infiltration/percolation are not reflected in such simple indices. To the authors' knowledge, no comprehensive agricultural drought index has been developed for NEB so far (*see* Part I, Section 3, of this volume for a discussion on the definition of drought). Thus, although rainfall departure indices (*Figure 2.2*) will be used here to define scenarios for agroclimatic modeling of crop yield, their limitations must be kept in mind.

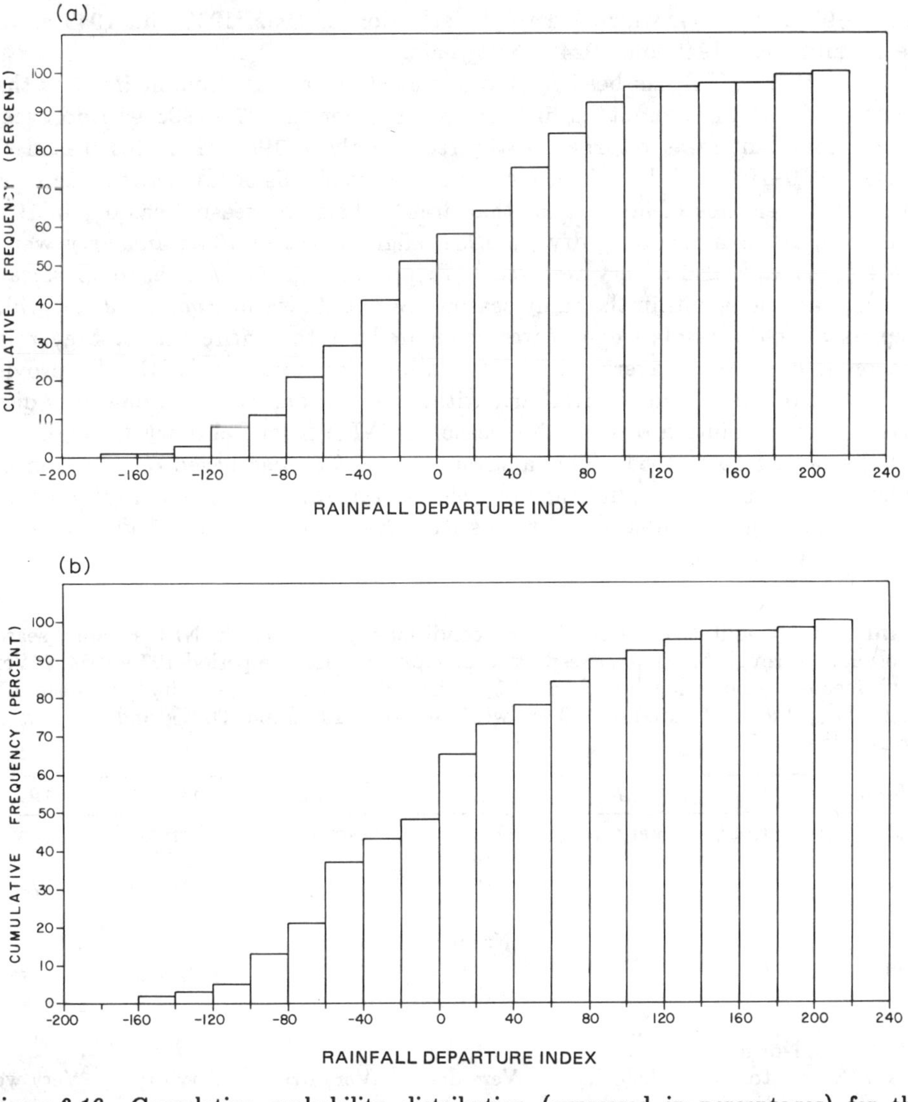

Figure 2.10. Cumulative probability distribution (expressed in percentages) for the rainfall departure indices of ***Figure 2.2*** for (*a*) northern NEB and (*b*) southern NEB.

The approach used involved determining cumulative probabilities of the rainfall departure indices (of ***Figure 2.2***), and then selecting as "dry" scenario years those with departures near the tenth percentile value, and as "wet" scenario years those with departures near the ninetieth percentile value. For northern NEB, for 10% of cases departure indices were below about −80, while for 90% of cases they were below about +105 (***Figure 2.10***). For northern NEB, 1980 and 1983 had departure indices close to −80 (***Figure 2.2***) and were selected for dry year scenarios; and 1984, 1967, 1965 and 1960 had indices near +105 and were selected as "wet" years. For southern NEB the 10 and 90 percentile limits were −90 and +100, which led to the selection of 1962, 1952 and 1942 as dry years, and 1969, 1945 and 1924 as wet years.

Northern NEB has been far more subject to serious drought impacts than southern NEB throughout its history, and the period 1979–1983 was no exception. Abundant rains returned to the region only in 1984. For this reason the period 1979–1984 will be taken, for the purposes of this study, as representative of several scenarios of interest for that area. The rainy season was dry in 1979, 1980 and 1983; moderately dry in 1982; slightly wet in 1981; and very wet in 1984 (1985 was also a very wet year) [*Figure 2.2(a)*]. Of course, temporal and spatial variations within the rainy season are not shown in *Figure 2.2 (a)*, which depicts only the rainfall departures calculated for the entire March–May rainy season and spatially averaged for 30 stations for northern NEB. To provide some idea of the temporal variability within the season, *Table 2.3* describes qualitatively the evolution of the rainy season in NEB from February through May for the period 1979–1984. This analysis (*Table 2.3*) was based on the monthly rainfall departure maps published in *Boletim Agroclimatologico Mensal* (Monthly Agroclimatological Bulletin) by the Instituto Nacional de Meteorologia (Brazilian Meteorological Office).

Table 2.3. Qualitative evaluation of conditions prevailing in NEB's rainy season (February through May) with respect to precipitation for the period 1979–1984. "Very dry" means precipitation less than 50% of the 1931–1960 average; "dry", between 50% and 100%; "normal", around 100%; "wet", between 100% and 200%; and "very wet", greater than 200%.

Month	*1979*	*1980*	*1981*	*1982*	*1983*	*1984*
Feb	Normal	Very wet	Very dry	Normal	Normal	Dry
Mar	Dry	Dry	Wet	Dry	Dry	Wet
Apr	Normal to dry	Dry	Normal to dry	Dry	Very dry	Wet
May	Normal to dry	Very dry	Very dry	Very dry	Very dry	Very wet

The years 1979 and 1980 were both drought years and thus are suitable to be used as a back-to-back drought event for agroclimatic modeling studies. In 1981 the total seasonal (March–May) rainfall was slightly above normal, but its temporal distribution was not particularly good for agriculture. Rainfall was very abundant in March, but the next two months were very dry. In 1982, on the other hand, total seasonal rainfall was below normal and all rainy season months were drier than normal; but the temporal distribution of the rainfall was more uniform than in 1981 so the effects on agriculture were less adverse. The following year, 1983, can be taken as an extreme case since it was the most severe drought in 25 years. Almost the opposite conditions prevailed in 1984. It had the wettest rainy season since 1974, and it rained abundantly in NEB as a whole from March through June. It can be said that the period 1979–1984 probably encompasses most of the likely scenarios for the meteorological variables in NEB.

2.5. Concluding Remarks

The present knowledge of NEB droughts suggests that they are caused primarily by features of the large-scale tropospheric circulation and probably enhanced by local secondary causes. Attempts to increase precipitation through modern techniques such as cloud seeding would thus probably be only of marginal value since these methods would have very little effect on such underlying causes of the drought as the descending branch of the Hadley–Walker circulation located over NEB during its rainy season. Recognition of this meteorological reality leads one to conclude that other approaches must be sought for mitigating the always adverse effects of severe droughts. It is becoming increasingly clear that these measures will have to rely heavily on water management.

With our present knowledge it is not possible to forecast droughts several years in advance. However, the analysis of anomalies in the large-scale circulation of the atmosphere and in the distribution of sea-surface temperatures in the tropical Atlantic and Pacific Oceans seems to promise the possibility of forecasting the character of the rainy season – whether dry or wet – two to three months in advance. This information can be very useful for short-term planning in preparation for a drought (or, alternatively, for floods). We believe that multiple regression formulas relating rainfall in NEB with meteorological and oceanographic conditions at distant points can be improved to the point of predicting rainfall amounts two to three months in advance of the rainy season with sufficient accuracy to be of practical operational value.

SECTION 3

The Effects on Crop Yields: A Simulation Approach

3.1. Introduction

In 1978 a simulation model of the water–plant–soil type was developed by researchers of the Brazilian Agricultural Research Enterprise (EMBRAPA), (Garagorry *et al.*, 1982). This model was the main element of a computerized system for analysis of agroclimatic data, used to meet the needs of researchers of the Center for Agricultural Research in the Semi-Arid Tropics (CPATSA, located in Petrolina, State of Pernambuco). These needs were related to:

(1) The determination of the optimal planting period for annual crops in NEB.
(2) The production of risk estimations associated with the several crops.
(3) The preparation of maps of agroclimatic zones.
(4) The appraisal of potential in different projects for the use of excess water in the soil.
(5) The development of proposals for improved agricultural systems at the farm level.

The outcome of previous experiments aimed at attaining the above results had been unsatisfactory, owing to the unavailability of a sufficiently large sample of years to provide the basis for estimations of the risks involved in determining the optimal time of the year for planting. On the other hand, computerized experiments with the above-mentioned model had been carried out for some time with beans (*Phaseolos*); cowpeas (*Vigna Unguiculata*) – a type of short-cycle bean (60–70 days), very resistant to moisture stress; corn and soybeans. The results of these experiments were considered satisfactory by CPATSA scientists (Garagorry *et al.*, 1982; Garagorry and Porto, 1983).

Two important aspects must be taken into consideration with respect to the use of a simulation model of the water–soil–plant type, rather than dealing simply with climatic data. Firstly, the monthly average or annual precipitation data used in the more traditional studies are of little use for ascertaining what really occurs at the farm level, due to the extreme variability of precipitation in NEB. The simulation model, for instance, may be able to explain why a certain crop had a lower yield in one year than another in spite of the fact that the total amount of rainfall in the former was larger. Several examples researched with the use of the model show that these types of situations can occur quite frequently.

Secondly, the five main objectives listed above are interrelated in one way or another. Generally speaking, it is extremely difficult to devise a model that is adequate for a range of different objectives unless these objectives are closely interrelated. This section focuses on the study of agricultural systems at the farm level, particularly with respect to the selection of optimal planting periods and optimal water management. The simulation model has clearly shown that the best planting period may vary significantly from one place to another, even on farms located in the same state (Garagorry *et al.*, 1982). In other words, NEB comprises several semi-arid tropic types, so an agroclimatic zonation is mandatory before any appropriate recommendations can be made to farmers.

To date, all studies of climatic zoning based solely on precipitation data have given unsatisfactory results. These results are very different from the ones obtained by the use of interrelated data from climate, soil and plants. An agroclimatic zonation will be one of the model's outputs for each crop. Moreover, the model takes into account the excess or deficit of water in the soil, which provides an estimation of the risk associated with these variables, and from this a probability distribution is obtained for a yield index for each crop.

In this section we use the modeling approach described above to estimate the impacts caused by climatic variations on productivity in NEB. With this end in mind, six municipalities representative of the semi-arid tropic zone of NEB were selected for study, based upon historical meteorological and yield data for these locations.

Although the scenarios suggested in Section 2 of this case study were used as far as possible, it should be noted that:

(1) While the year 1983 has been characterized as a year of meteorological drought, it was not necessarily a year of extensive agricultural drought. Although drought was meteorologically extreme and affected the entire NEB, in some localities the distribution of rainfall allowed for the maintenance of significant agricultural activity. That, for example, was the case for Petrolina (where CPATSA is located), which is included in the present study.

(2) While some drought scenarios have been selected from the post-1979 period, much of the climatic and yield data that were suitable for the model were available only up to the year 1978.

Because of these constraints, and some specific characteristics of the model (to be discussed later), other similar scenarios were substituted, using criteria based on the analysis of empirical data related to the agricultural years, which do not always coincide with the calendar year.

3.2. The Model and the Data

3.2.1. Outline of the agroclimatic model

This subsection provides an overview of the agroclimatic model for the semi-arid tropic zone of NEB. The model considers not only precipitation and evapotranspiration, but also plant development as a function of the soil's capacity to retain water. A detailed analysis using this system requires daily rainfall data for a series of years, and daily potential evapotranspiration data obtained from the monthly averages of potential evapotranspiration estimated by Hargreaves (1974) from the monthly averages of temperature and relative humidity. For the unit of time in the simulation analysis, the five-day period was selected as best reflecting the variations that occur in the availability of water for plants. The year is thus divided into 73 five-day periods. Precipitation and potential evapotranspiration are then expressed in terms of the five-day totals.

The territory within the boundary of a municipality is taken as the geographical unit for the purpose of applying the model, because of the unavailability of climate data at the farm level. The soil water-holding capacity was estimated for the prevailing type of soil in each municipality.

For each five-day planting period being considered, the water balance is computed through the growing season for the crop started in that five-day period. To estimate the initial moisture status at planting, water balance computations are performed for the 20 preceding five-day periods. For example, for a January 11–15 planting period in 1981, with a crop that would mature around June 1, the water balance would be computed for the 20 five-day periods October 3, 1980 to January 10, 1981, and then on as far as the five-day period including June 1. Note that the model would thus require precipitation data for parts of two years, 1980 and 1981, even for an analysis relating to a single planting period early in 1981.

After computing the water balance through the growing season for a crop started in a particular five-day period, the model estimates that crop's yield index, *YI*, which is defined as the ratio between the actual yield and the potential yield. *YI* is computed as a function of the relative evapotranspiration (here defined as the ratio between actual evapotranspiration and the maximum expected evapotranspiration of that particular plant), and the yield response factor (estimated empirically by FAO, 1979).

More specifically the yield index (*YI*) used in this section is computed as follows:

$$YI = \prod_{i=1}^{n} \left\{ \max \left[0; 1 - Ky_i \left(1 - \frac{ETA_i}{ETM_i} \right) \right] \right\} \tag{3.1}$$

where

n = total number of phenological periods (here $n = 4$: seeding, growing, pod formation and ripening of grains)

Ky_i = yield response factor for the ith phenological period

ETA_i = actual evapotranspiration of the crop during the *i*th phenological period

ETM_i = maximum evapotranspiration of the crop during the *i*th phenological period

The "max" function is explained under equation (3.2), below.

As the equation above indicates, *YI* is estimated using the maximum and the actual crop's evapotranspiration and the associated yield response factor (*Ky*) for each phenological period in turn (*Figure 3.1*). The contribution to *YI*, in a given phenological period, is at its maximum if the plant suffers no water shortage, that is, if $ETA = ETM$. Thus, the value of *YI* is a function of the value of *Ky* and of relative evapotranspiration ETA/ETM. *Figure 3.2* provides an idea of the variation of *YI* as a function of relative evapotranspiration.

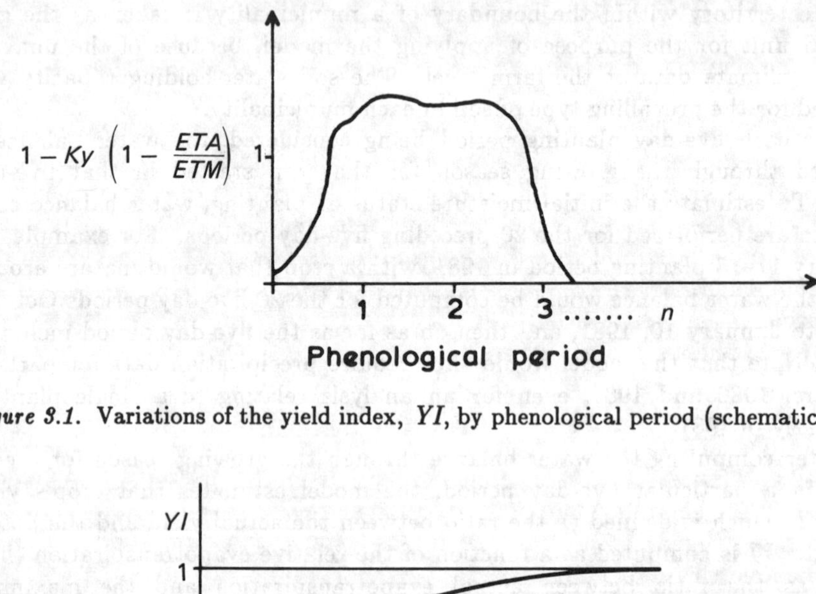

Figure 3.1. Variations of the yield index, *YI*, by phenological period (schematic).

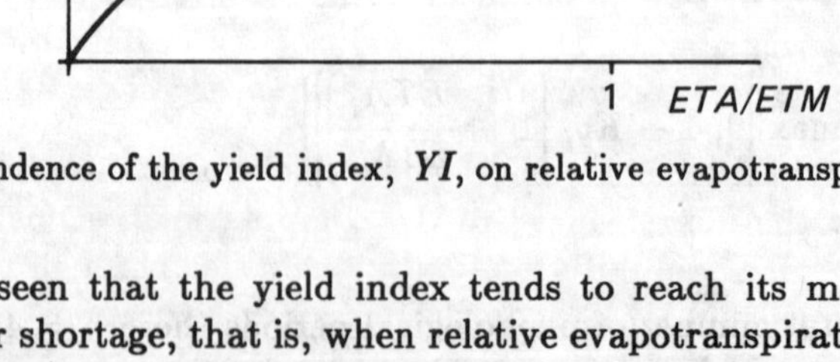

Figure 3.2. Dependence of the yield index, *YI*, on relative evapotranspiration.

It can be seen that the yield index tends to reach its maximum value if there is no water shortage, that is, when relative evapotranspiration tends to 1.

Equation (3.1) is based on the following equation derived by FAO to quantify the effect of water stress on yields:

$$(1 - YA/YM) = Ky\ (1 - ETA/ETM) \tag{3.2}$$

where

YA = actual crop yield
YM = maximum (or potential) crop yield
Ky = crop yield response factor
ETA = actual crop evapotranspiration
ETM = maximum crop evapotranspiration.

From equation (3.2), YI was redefined in equation (3.1) in order to specify the effects of phenological periods and to assure a positive yield index. The function "max" gives the greater value among several values provided as arguments. For instance, given two arguments, numbers 7 and 3; then max (3; 7) = 7. Its use in conjunction with a "0" argument in equation (3.1) avoids negative results. If $Ky = (1 - YA/YM) / (1 - ETA/ETM)$ were greater than 1, the use of equation (3.2) to obtain YI could give a negative result.

The program next classifies the yield index into three categories of results: good, fair, and poor. This classification is done according to values supplied by the user (who has the final decision over the planting of a certain crop). For instance, the farmer may have established that for the yield index to be considered good, it must be equal or greater than 80%; for fair, equal to or greater than 50% but less than 80%; and for poor, less than 50%. Thus, the values of 0.8 and 0.5 must be supplied to the program by the user so that the classification can be completed.

Besides classifying the yield index for each crop growing season as good, fair or poor, the program also accumulates the incidence of water excess or deficit.

After completing the calculation of yield indices and water excesses and deficits as outlined above, for each of the selected localities, over the whole series of years, on the basis of five-day time periods, the program provides the relative frequency of the good, fair and poor results, and the relative frequency of acceptable results (defined as the sum of the good and fair results). In addition, it prints the following indicators for each five-day period:

(1) Mean value, standard deviation, and coefficient of variation of the yield index, if the five-day period was one of the selected planting periods.
(2) A frequency distribution and mean value of water excess.
(3) A frequency distribution and mean value of water deficit.
(4) The difference between mean values of water excess and deficit.

The operation of the model is shown schematically in *Figure 3.3.* The items framed by the dotted lines are the computational elements of the model, where the water–plant–soil interrelation is quantified through the water balance and yield index calculations, and the necessary operations for obtaining the above-mentioned statistics are performed. These procedures are carried out for the crop corresponding to each selected five-day planting period for every year of that series, and for each municipality.

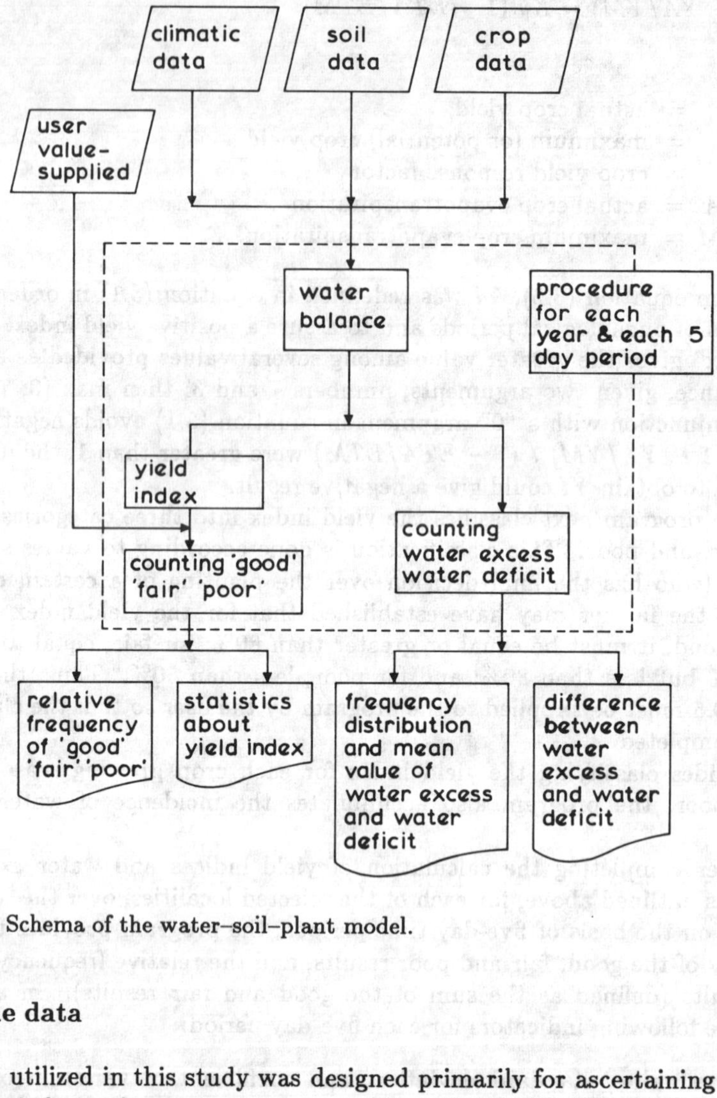

Figure 3.3. Schema of the water–soil–plant model.

3.2.2. The data

The model utilized in this study was designed primarily for ascertaining the best planting period as a function of the expected yield value, which, in turn, is computed from an expected rainfall distribution. A precipitation data series of at least 15 years is necessary in order to obtain useful results. It is essential that the data record be sufficiently long to allow for the estimation of the associated risk in determining the best planting period through the assessments of the standard deviation and the coefficient of variation of the expected yield index value.

For the climate impact analysis of the present study, a sample of six stations located in representative municipalities of the NEB semi-arid zone was chosen. Their locations, as well as the lengths of the available data series, are shown in *Figure 3.4* and *Table 3.1*. The daily precipitation data were provided by SUDENE and CPATSA. The remaining data required for the simulation were the following:

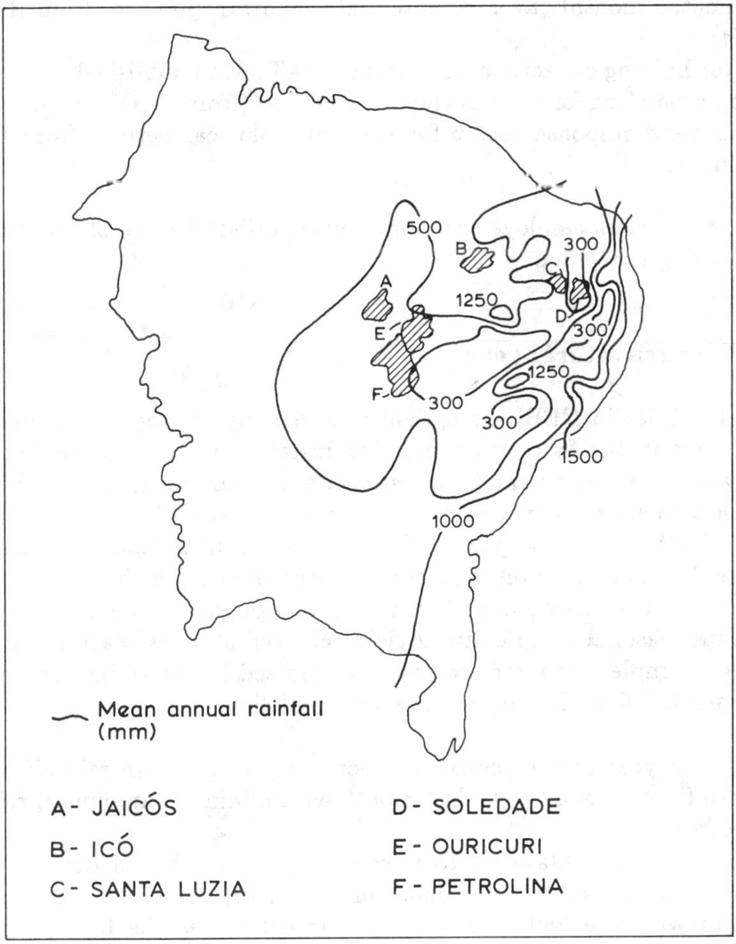

Figure 3.4. Locations of stations and municipalities referred to in the text.

Table 3.1. Locations and main climatic characteristics of the stations.

Location	*Geographic coordinates* *Lat.*	*Long.*	*Altitude* (m)	*Number of years (start–end)*	*Mean annual rainfall* (mm)	*Mean annual PET* (mm)
Jaicós	07° 22′S	41° 08′W	255	41 (1913–1977)	684	1882
Icó	06° 25′S	38° 51′W	160	51 (1913–1975)	720	2020
Santa Luzia	06° 52′S	36° 56′W	290	63 (1912–1978)	537	1811
Soledade	07° 07′S	36° 22′W	517	56 (1913–1980)	366	1614
Ouricuri	07° 53′S	40° 04′W	432	44 (1914–1978)	559	1965
Petrolina	09° 09′S	40° 22′W	355	20 (1965–1984)	577	2080

Source: Data obtained from SUDENE/CEPATSA.

(1) Estimated monthly average potential evapotranspiration (from Hargreaves, 1974).
(2) Water-holding capacity of soil (from CPATSA, unpublished).
(3) Crop coefficient for each phenological period (from FAO, 1979).
(4) Crop yield response factor for each phenological period (from CPATSA, unpublished).

The data series employed are for similar periods for five of the stations, the exception being Petrolina.

3.2.3. The agricultural year

In many localities in NEB, the agricultural year usually begins in mid-calendar year and ends in the following year. For instance, in some places the planting season starts in November and harvest occurs in March. Thus, the total annual rainfall for a calendar year may not be a good indicator of climate conditions for rainfed agriculture in some places. This limits the utilization of scenarios based on the total annual precipitation for a calendar year, such as the year 1983. That year was considered one of widespread drought, in meteorological terms, though some reasonable agricultural yield occurred at a few places.

A very simple computer program was devised to determine the agricultural year for each locality. The procedures were as follows:

(1) For each year of the series, the month with minimum rainfall was determined (in case of a tie the last month with minimum amount of rainfall was selected).
(2) For each month (January to December), the number of times it had been selected as the minimum rainfall month in (1), above, was noted, and the month with the highest frequency was defined as the final month of the agricultural year (if there was a tie, the last month with the largest frequency was chosen).

The agricultural year is then the 12-month period ending with the final month as obtained in step (2) above. The results obtained by this method were found to be more satisfactory than those obtained by other procedures (such as, for example, simply defining the final month of the agricultural year as the month with minimum precipitation). In our case, the beginning of the agricultural year coincides with the beginning of the agricultural season in each locality. The following months were thus defined as the beginning of the agricultural years:

Icó and Soledade:	January
Jaicós:	October
Petrolina and Ouricuri:	November
Santa Luzia:	December

3.3. Simulating the Effects of Climatic Variations on Cowpea Yields

3.3.1. Simulation design

This simulation exercise sought to analyze the possible effects on agricultural yield of a variation of ± 10% in precipitation and potential evapotranspiration (PET). To achieve this, the precipitation and potential evapotranspiration data were multiplied by 0.9, 1.0 and 1.1, producing a simulation of 3 × 3 factorial, and providing the following situations:

(1) Precipitation and PET multiplied by 1.0, i.e., with unaltered original data and representing a climate called "normal".
(2) "Normal" precipitation, with PET decreased by 10% (i.e., multiplied by 0.9).
(3) "Normal" precipitation, with PET increased by 10% (i.e., multiplied by 1.1).
(4) 10% increase in precipitation, and "normal" PET.
(5) 10% increase in precipitation, and 10% decrease in PET. (This is the most favorable situation for a semi-arid climate, because it improves the moisture situation through both an increase in the supply of water and a decrease in the moisture demand as reflected by the potential evapotranspiration.)
(6) 10% increase in both precipitation and PET.
(7) 10% decrease in precipitation with "normal" PET.
(8) 10% decrease in both precipitation and PET.
(9) 10% decrease in precipitation and an increase of 10% in PET. (This is the most unfavorable situation for plant growth in semi-arid regions, because the supply of water diminishes as the plant's need for water increases.)

These situations may be expressed by a pair (x_1, x_2), where x_i is the value by which the precipitation will be multiplied and x_2 is the value by which potential evapotranspiration will be multiplied. For example (1.0, 1.0) represents the first situation, described above as "normal" whereas (1.1, 0.9) represents the most favorable situation, as described in (5) above. The simulations were performed without altering either the soil and plant data, or the procedures for calculating the yield index, water excess or water deficit.

3.3.2. Types of results obtained

In order to assess the impacts caused by climate variations, estimations for expected yield are needed. Since they are expected values, there is a degree of variability associated with them, and it is important to have an understanding of this variability. To design a project for water and soil management that can respond adequately to eventual water needs at critical stages of plant growth, thus improving yield, it is necessary to have a thorough knowledge of the estimation of water excess and deficit, combined with knowledge of the expected values of the yield estimation.

Several indicators selected from the computer output are discussed below:

(1) Maximum probability (*PA*) of an acceptable relative yield, as indicated by the yield index defined in Subsection 3.2.1. *PA* is utilized to determine the best planting period in the following way: cutoff values at 10% intervals are assumed. When the maximum value of *PA* is found, the next lowest cutoff value is then selected for that case; then the five-day period corresponding to the first occurrence of a *PA* value greater or equal to that cutoff level is defined as the initial period for the best planting season; and the five-day period corresponding to the last occurrence of a *PA* value greater than or equal to that cutoff level will be the final period for the most suitable planting season. For example, if the maximum value for *PA* is 63.2%, the cutoff level will be 60. The initial period for the best planting season will be the one corresponding to the first occurrence of a value greater than or equal to 60; the final period will be the last one where *PA* is greater than or equal to 60.

(2) Maximum mean value of the crop yield index (*MYI*) within the best planting season, i.e., the growing season of a crop planted in the best planting period.

(3) Coefficient of variation of the crop yield index (*CVMYI*) corresponding to the value obtained in (2). If there is a tie in (2) for more than one five-day period, the maximum coefficient of variation will be chosen.

(4) Maximum value of the mean water excess (*EXC*) in millimeters within the best planting season as defined in (2) above.

(5) Maximum value for the mean water deficit (*DEF*) in millimeters within the best planting season as defined in (2) above.

3.3.3. Analysis of the obtained results

Tables 3.2–3.6 show the results obtained from the procedures described above. *Table 3.2* indicates the likelihood of obtaining acceptable relative yield (*PA*). These values may be seen as percentage probabilities of achieving a fair or good yield. The first three municipalities clearly have the best chance of obtaining acceptable yields, with probabilities greater than 60% for all the simulated situations. For the other three municipalities, the table indicates a need for supplementary irrigation under existing precipitation levels. At these locations the best chance of obtaining a good or a fair yield is at Ouricuri, with the precipitation level increased by 10% and crop demand for water through evapotranspiration decreased by 10%.

Table 3.2. Percentage probabilities (PA) of acceptable yields, for each municipality, according to the simulated climatic situation.

	Climatic situation[a]								
Location	*1*	*2*	*3*	*4*	*5*	*6*	*7*	*8*	*9*
Jaicós	85	90	80	90	93	85	80	88	78
Icó	88	92	86	90	92	88	88	88	80
Santa Luzia	70	75	65	73	76	68	67	71	63
Soledade	33	36	29	33	40	31	31	33	26
Ouricuri	40	49	35	45	54	38	35	35	35
Petrolina	42	47	37	42	47	42	42	42	32

[a]The simulated climatic situations are defined by pairs (x_1, x_2) in the text: situation 1 = (1.0, 1.0); 2 = (1.0, 0.9); 3 = (1.0, 1.1); 4 = (1.1, 1.0); 5 = (1.1, 0.9); 6 = (1.1, 1.1); 7 = (0.9, 1.0); 8 = (0.9, 0.9); 9 = (0.9, 1.1).

Table 3.3. Maximum yield index (MYI) in percentages.

	Climatic situation[a]								
Location	*1*	*2*	*3*	*4*	*5*	*6*	*7*	*8*	*9*
Jaicós	69	72	66	69	74	68	66	66	62
Icó	70	73	67	72	75	69	68	71	65
Santa Luzia	60	63	55	62	65	57	56	60	54
Soledade	30	33	26	33	35	30	27	30	26
Ouricuri	42	46	36	43	48	40	39	43	32
Petrolina	38	42	35	40	42	36	36	40	33

[a]The simulated climatic situations are defined by pairs (x_1, x_2) in the text: situation 1 = (1.0, 1.0); 2 = (1.0, 0.9); 3 = (1.0, 1.1); 4 = (1.1, 1.0); 5 = (1.1, 0.9); 6 = (1.1, 1.1); 7 = (0.9, 1.0); 8 = (0.9, 0.9); 9 = (0.9, 1.1).

Table 3.4. Coefficient of variation for the maximum yield index ($CVMYI$) in percentages.

	Climatic situation[a]								
Location	*1*	*2*	*3*	*4*	*5*	*6*	*7*	*8*	*9*
Jaicós	36.1	35.6	38.8	34.2	31.9	36.8	40.6	32.6	41.3
Icó	33.4	31.3	34.9	31.8	30.4	33.8	34.7	33.2	36.6
Santa Luzia	48.8	45.9	52.0	47.4	44.7	50.2	51.0	47.6	54.6
Soledade	91.0	84.6	100.7	87.3	81.1	92.8	95.7	88.8	102.1
Ouricuri	64.3	58.3	69.5	61.4	55.5	66.7	67.4	61.4	72.7
Petrolina	74.3	69.1	78.5	71.5	68.6	75.8	77.3	72.1	81.3

[a]The simulated climatic situations are defined by pairs (x_1, x_2) in the text: situation 1 = (1.0, 1.0); 2 = (1.0, 0.9); 3 = (1.0, 1.1); 4 = (1.1, 1.0); 5 = (1.1, 0.9); 6 = (1.1, 1.1); 7 = (0.9, 1.0); 8 = (0.9, 0.9); 9 = (0.9, 1.1).

Table 3.3 provides the values for the maximum expected yield index, for the different localities and climatic situations. One may observe that these results agree with those in *Table 3.2*, in that for all the situations yield performance is shown as better for the first three municipalities than for the last three. The most favorable result is the expected maximum yield index of 75% for situation (1.1, 0.9) at Icó.

Table 3.4 shows the values of the coefficient of variation for the maximum yield index, for the six localities and nine climate situations. Again, these results are consistent with those in the preceding paragraphs, with the first three municipalities showing the best results among the six. Their coefficients of variation are much smaller than those for Soledade, Ouricuri, and Petrolina. Soledade, which had the poorest results in *Tables 3.2* and *3.3*, also presents the greatest variability in its *MYI* values, as indicated by the *Table 3.4* results. This suggests that crop production there would be particularly unstable.

As was shown in *Table 3.1*, the total annual average precipitation for Santa Luzia, Ouricuri and Petrolina are similar, although potential evapotranspiration is somewhat lower at Santa Luzia. With respect to the variability of *MYI*, however, Santa Luzia shows a much smaller figure than the other two (*Table 3.4*). This implies a better agricultural performance for that municipality.

From the results of *Table 3.4* it can be seen that the variability is much smaller for the first three municipalities, which indicates more stability and reliability in the rainfall distribution from year to year for those localities. Such results are useful for agroclimatic zoning, in that they provide a means of comparing the suitability of the different locations for achieving acceptable yields. The procedures may also be seen as providing a simple type of risk analysis, because the coefficient of variation is related directly to the standard deviation of the yield index; the smaller the standard deviation, the smaller the coefficient of variation, and consequently, the smaller the risk.

Table 3.5 shows the maximum soil water excess, within the season with the best planting period, for the several localities and climatic situations. The importance of these water excess values stems from the fact that they may help to indicate the needs with respect to the management of water to improve yields. This excess water may be stored in small reservoirs, by the use of *in situ* collecting technologies, already developed for NEB semi-arid areas by CPATSA.

It will be noted that the first three municipalities, together with Petrolina, offer the best conditions for excess water utilization, because there would be relatively large volumes of excess water during the crop season that could be saved for later use.

Table 3.6 provides the maximum values for average water deficit during the best crop season. When one compares these results with those obtained in *Table 3.5*, it can be seen that the deficit is smaller than the excess for all the municipalities except Soledade. Where this is the case, it may be concluded that water management may appropriately depend solely on local rainfall. However, in the case of Soledade, where the deficit values are greater than the excess values, additional water from other sources would be needed or agricultural yields would suffer.

Table 3.5. Average water excess (EXC) in mm.

	Climatic situation[a]								
Location	*1*	*2*	*3*	*4*	*5*	*6*	*7*	*8*	*9*
Jaicós	185	196	176	214	229	208	153	163	144
Icó	202	206	191	237	241	226	166	177	163
Santa Luzia	164	170	156	191	200	185	135	141	126
Soledade	51	59	52	67	71	58	31	43	41
Ouricuri	93	100	87	110	117	106	74	80	56
Petrolina	102	108	96	123	130	116	88	94	83

[a]The simulated climatic situations are defined by pairs (x_1, x_2) in the text: situation 1 = (1.0, 1.0); 2 = (1.0, 0.9); 3 = (1.0, 1.1); 4 = (1.1, 1.0); 5 = (1.1, 0.9); 6 = (1.1, 1.1); 7 = (0.9, 1.0); 8 = (0.9, 0.9); 9 = (0.9, 1.1).

Table 3.6. Average water deficit (DEF) in mm.

	Climatic situation[a]								
Location	*1*	*2*	*3*	*4*	*5*	*6*	*7*	*8*	*9*
Jaicós	28	19	34	25	16	31	26	25	32
Icó	26	18	32	25	17	30	28	23	37
Santa Luzia	28	30	45	31	29	43	41	30	44
Soledade	56	52	73	60	50	63	55	48	76
Ouricuri	77	64	78	56	43	76	83	58	78
Petrolina	69	59	84	67	56	79	72	60	85

[a]The simulated climatic situations are defined by pairs (x_1, x_2) in the text: situation 1 = (1.0, 1.0); 2 = (1.0, 0.9); 3 = (1.0, 1.1); 4 = (1.1, 1.0); 5 = (1.1, 0.9); 6 = (1.1, 1.1); 7 = (0.9, 1.0); 8 = (0.9, 0.9); 9 = (0.9, 1.1).

3.4. Impacts under Specified Climatic Scenarios

3.4.1. Analysis of the scenarios proposed

The selection of climatic scenarios should take the impact model's major characteristics into account. For the model used here, there was a need to know the amount of water contained in the soil during the planting season, and to make yield estimations after the calculation of the water balance for all the five-day periods. These aspects, combined with the constraints mentioned in Subsection 3.2.2, made it inappropriate to use this model with all of the scenarios suggested in Section 2.

Of the six municipalities involved in the analyses here, only Petrolina and Soledade had daily precipitation data available after 1978. *Table 3.7* shows the results of the effects of a back-to-back drought in the years 1979 and 1980. The apparent contradictory nature of the results in this table may reflect the model's sensitivity to rainfall distribution through the year, and the fact that the amount of water contained in the soil in the first five-day period in the given year is computed from the rainfall data for the previous 100 days.

Table 3.7. Effects under the back-to-back drought scenario for Petrolina and Soledade (1979 and 1980).

	Long-term[a]			*1979*		*1980*	
Location	*Mean rainfall* (mm)	*Mean PET*[b] (mm)	*Sim*[c] *MYI* (%)	*Rainfall* (mm)	*MYI* (%)	*Rainfall* (mm)	*MYI* (%)
Soledade	366	1614	30	195	50	302	26
Petrolina	577	2080	38	508	52	536	80

[a] Long-term data are from *Table 3.1*.
[b] PET = potential evapotranspiration.
[c] Sim = simulated (data from *Table 3.3*).

Rainfall distribution through the year in those municipalities is quite variable, which may help to explain why, in 1979, the computed yield for Soledade was greater than that in 1980, despite the fact that the annual rainfall was less than in 1980. What happened was that rainfall was better distributed throughout 1979. In 1980 rainfall was concentrated in the months of February, March and April, with heavy rainfall occurring on isolated days widely separated from one another during those months.

With respect to 1983, which was considered in Section 2 as a year of extreme drought, the model was applied to Petrolina, the only municipality where daily data were available. The computed yield index was 71%, which is much greater than the expected index (38%) that had been computed from the long series of years, although the annual precipitation (540 mm) was much less than the long-term annual average of 577 mm. This suggests that identifying extreme drought in NEB on the basis of total annual rainfall alone can be seriously misleading.

3.4.2. Climate impact assessment for the specified scenarios

The selection of climatic scenarios was based on agricultural years rather than calendar years, and considered every year available in the data series for each municipality. The fact that agricultural years are more representative of agricultural activity in each locality was also taken into account, and it was assumed that the greater the number of years in the series, the more precise the average estimated yield would be. *Table 3.8* describes some of the statistical characteristics of annual rainfall, by agricultural years, in each of the municipalities.

The first three localities present a greater absolute variability from year to year in the occurrence of annual precipitation, as indicated by the standard deviation, although the relative variability may be smaller. At Jaicós and Icó the coefficients of variability, 33% and 32%, respectively, are smaller than at Soledade and Ouricuri (51% and 40%), indicating lower relative variability. The variability within the same year may be lower for the first three municipalities than for the others.

Table 3.8. Mean quantity, distribution and variability of the total annual rainfall (mm) in each locality studied.

Location	*Mean*[a]	*Min*[b]	*P10*[c]	*Q1*[d]	*Med*[d]	*Q3*[d]	*Max*[e]	*Range*[f]	*STD*[g]	*P90*	*CV*[h] (%)
Jaicós	684	303	420	526	652	832	1282	979	223	993	33
Icó	720	166	447	570	687	864	1259	1093	233	1076	32
Santa Luzia	537	90	198	308	531	670	1230	1143	246	834	46
Soledade	366	93	182	210	318	448	1035	942	185	641	51
Ouricuri	559	166	240	435	565	664	1234	1068	223	820	40
Petrolina	577	239	334	484	585	669	852	613	166	828	29

[a]Mean = total annual average for the corresponding municipality.
[b]Min = minimum annual total that occurred in the municipality in the series.
[c]P10 = annual total that occurs at 10% probability – that is, the probability of rainfall being no more than that amount is 10%. P90 = annual total that occurs at 90% probability – the probability of rainfall being that much or less is 90%.
[d]Q1 and Q3 = totals corresponding to the first and third quartiles, respectively; occurrence of annual totals equal to or less than Q1 and Q3 are at 25% and 75% probability, respectively; Med = the annual total at 50% probability (median).
[e]Max = maximum annual total that occurred in the series.
[f]Range = Max − Min.
[g]STD = standard deviation of annual totals, for each series and for each municipality.
[h]CV = coefficient of variability.

The 1-in-10 analysis (*Table 3.9*), was performed by taking the annual rainfall closest to the 10% and 90% levels, i.e., the P10 and P90 values from *Table 3.8*, and calculating the maximum yield index (*MYI*) for the corresponding agricultural year.

As expected, the *MYI* values in *Table 3.9* are quite high for wet years, indicating that the yield should tend to approximate the potential yield for each locality in such years. For the 1-in-10 dry years, *MYI* values for four of the other municipalities are distinctly low. The high *MYI*s obtained for these years at Jaicós and Icó reflect some anomalous features of these particular years in those localities, probably in the within-year rainfall distributions.

Extreme years were selected from among all the years in the analyzed series (*Tables 3.10* and *3.11*). *Table 3.10* shows values obtained for the yield index in a year of extreme drought and, for comparison, the long-term values based on the whole data series. One may note that even in the best case, that for Icó, the productivity level indicated was so low as to be considered catastrophic. The better performance indicated at Icó than elsewhere may be related to the fact that its calculation included consideration of the final 100 days of the year 1918 (a year for which total rainfall was 1078 mm), since Icó has an agricultural year which coincides with the calendar year.

In *Table 3.11 MYI* data indicate that, for very wet agricultural years, the yield would approach the potential yield, except at Petrolina, where it would apparently only reach two-thirds of potential. That may be related to rainfall variability within the year at Petrolina.

To summarize the foregoing, *Table 3.12* shows the percentage change in yield estimated in the several localities for extreme years, for 1-in-10 events, and for the back-to-back scenario proposed in Section 2. The figures for extreme

Table 3.9. Maximum yield index (*MYI*) for the 1-in-10 scenarios.

	Wet[a]			Dry[a]		
Location	*Year*	*Rainfall* (mm)	*MYI* (%)	*Year*	*Rainfall* (mm)	*MYI* (%)
Jaicós	1976–1977	931	83	1971–1972	413	82
Icó	1940	1031	92	1932	444	70
Santa Luzia	1970–1971	806	86	1952–1953	188	5
Soledade	1972	639	75	1946	164	47
Ouricuri	1946–1947	786	89	1952–1953	238	13
Petrolina	1974–1975	828	73	1969–1970	334	30

[a] In this table a "wet" year is that one with annual rainfall closest to the P90 values of *Table 3.8*, while a "dry" year is that with annual rainfall closest to the P10 level.

Table 3.10. Maximum yield index (*MYI*) for the extreme drought scenario compared with long-term values.

	Long-term			Extreme drought		
Location	*Mean rainfall*	*Mean PET*	*Simulated MYI* (%)	*Year*	*Rainfall*	*MYI* (%)
Jaicós	684	1882	69	1914–1915	303	22
Icó	720	2020	70	1919	166	43
Santa Luzia	537	1811	60	1918–1919	96	21
Soledade	366	1614	30	1958	93	7
Ouricuri	559	1965	42	1929–1930	166	8
Petrolina	577	2080	38	1975–1976	239	14

Table 3.11. Maximum yield index (*MYI*) for the extreme wet year scenario compared with long-term values.

	Long-term			Extreme wet year		
Location	*Mean rainfall*	*Mean PET*	*Simulated MYI* (%)	*Year*	*Rainfall*	*MYI* (%)
Jaicós	684	1882	69	1923–1924	1282	98
Icó	720	2020	70	1917	1259	97
Santa Luzia	537	1811	60	1963–1964	1239	94
Soledade	366	1614	30	1924	1035	92
Ouricuri	559	1965	42	1934–1935	1234	88
Petrolina	577	2080	38	1968–1969	852	67

drought indicate yields generally more than 50% below the long-term average, demonstrating the futility of trying to produce a cowpea crop in anomalously dry years. Results for Santa Luzia, Ouricuri and Petrolina also indicated very low yields for the driest 10% of years.

Table 3.12. Relative impacts (%) of different scenarios on *MYI* values in the simulation of cowpea yield. The scenario *MYI* values are shown expressed as percentages of the long-term *MYI*s.

		Scenarios					
		Extreme		*1-in-10*		*Back-to-back drought*	
Location	*Base years*	*dry year*	*wet year*	*dry*	*wet*	*1979*	*1980*
Jaicós	100	31	142	119	120	–	–
Icó	100	61	138	100	131	–	–
Santa Luzia	100	35	156	8	143	–	–
Soledade	100	23	306	156	250	166	87
Ouricuri	100	19	209	31	212	–	–
Petrolina	100	36	176	79	192	137	210

3.5. Model Validation

As mentioned by Garagorry and Porto (1983), when the model was applied according to its original objectives, it was well accepted among professionals and farmers. The results are being confirmed by feedback in response to inquiries made to farmers of the NEB region.

Some comparison was also made between the model outputs and IBGE data. The values obtained from the simulation of yield index (%) were compared to the present yield (t/ha) provided by IBGE. Since current yield data are available only for Petrolina, and only from 1975 on, it was not possible to carry out validation tests for the other municipalities.

A regression analysis was performed, where the dependent variable was observed yield, and the independent one was yield index. If the yield index was a good indicator, one would expect the determination coefficient and the correlation coefficient (r^2 and r) to be high. However, those values were very small. This could indicate that the yield index may not be a good indicator, but it could also be a reflection of inadequate sample size for analysis. Pending further research, these results should be considered as inconclusive.

3.6. Concluding Remarks

This study demonstrates that a simulation model can be utilized not only for its main original purpose of helping to select best planting times, but also for estimating the impacts of climatic variations on crop yields. In the latter case, the model's usefulness stems from the fact that the results indicate the probabilities of obtaining good crops. It can thus be useful as a component of measures to lessen the risks associated with climatic variations. Although there would still be a risk of crop loss, the model provides for the estimation of that risk. For example, in Jaicós (Piauí State), Icó (Ceará State) and Santa Luzia (Paraíba), the probability of obtaining good crops is 93%, 92% and 76%, respectively, when precipitation is 10% greater than average and potential evapotranspiration is 10% below average (*Table 3.2*). At Soledade (Paraíba), Ouricuri (Pernambuco) and Petrolina (Pernambuco), the probability of achieving acceptable crops under the same conditions is 54% at the most. In these cases, the chances of loss are great, even if the best planting season is chosen.

We may summarize the interaction between knowledge of climatic risk and the economics of agricultural production as follows. At present, planting decisions based on a limited knowledge of climatic risk place farmers at point A on the supply diagram (*Figure 3.5*). However, a more precise knowledge of risk may cause them to alter their decision making and to move them to B, implying an increase in product supply (A to B). Under these semi-arid conditions, even if the best planting seasons are known, there are still substantial risks of loss. An irrigation policy could move the supply curve to C, which represents a situation where climatic risks due to moisture constraint are largely eliminated. The benefits derived from this policy are translated into a larger supply product.

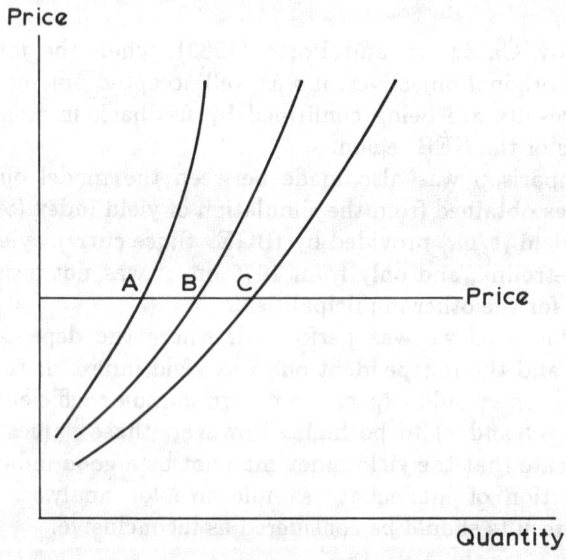

Figure 3.5. Supply curves of agricultural production. A = initial situation; B = situation with decisions based on knowledge of optimal planting times; C = situation after introduction of irrigation.

Another implication arising from the analyses made with this model is that, to maintain the water–soil–plant balance so as to minimize damage caused by droughts, there must be not only optimum utilization of water, but also a guaranteed water supply. Both water conservation, and practices aimed at improving the soil's capacity to retain water, are needed. These techniques will reduce the damage caused by consecutive years of droughts, which have been shown to have very adverse effects on the population and the economy of Northeast Brazil.

SECTION 4

The Effects on Agricultural Production and Yields

4.1. Introduction

The droughts that occurred in Northeast Brazil (NEB) during the 1979–1983 period brought serious consequences to the economy and population of that region. The extreme drought of 1983, which affected 88% of NEB, reduced the regional gross product by 15.8%. It was worse than in 1979 and 1980, when production growth rates were zero and –0.9%, respectively (MINTER, 1973). In 1983, some 12.5 million people received assistance in coping with the drought, but it has been estimated that the total number affected reached 12 million people (Section 1).

There is evidence that the industrial sector is less affected by drought, partly because some of its raw materials can be obtained from other regions (MINTER, 1973). The most severe impacts of droughts are on the agricultural sector, especially on subsistence crops. As noted in Section 1, subsistence crops are more affected than cattle, and small farmers are more affected than large landowners. Figueroa (1977) reports that, for medium and large farm owners, drought is mainly a problem of production, while for small farmers it involves family subsistence.

Examination of the distribution of workers engaged in emergency public works (*Table 4.1*) reveals that in these projects landless farmers predominated (75.1%), followed by landowners. Together, they represented 95.2% of the sample analyzed by the Joaquim Nabuco Foundation (*see* Subsections 1.4.4 to 1.4.6). This labor force came mainly from farms of 100 ha or less. *Table 4.2* shows that these farms supplied 69.7% of the people enlisted in the work projects. The remainder came from farms larger than 100 ha.

Among landholders, 92.2% owned a mean area of land up to 100 ha or less, and the majority of farmers (59.9%) owned a mean area of 20 ha or less. This

Table 4.1. Main occupation and other occupations of workers enlisted in emergency public works, by type of occupation, 1978.

Type of occupation	*Main occupations* (%)	*Other occupations* (%)
Landholder	20.1	0.8
Landless farmer	75.1	11.4
Herdsman	1.2	2.9
Tradesman	0.2	1.2
Urban worker	1.2	3.8
Other	1.8	2.6
Without occupation	0.4	77.3
Total	100.0	100.0

Source: Nabuco (1983).

Table 4.2. Percentage of those enlisted in public works, classified by size of their farms, 1978.

Area intervals (ha)	≤20	21–100	101–500	≥501	Total
Enlisted workers (%)	32.1	37.6	22.0	8.3	100.0

Source: Nabuco (1983).

size of landholding was considerably smaller than the mean farm area in NEB, which was 36.5 ha in 1980 (IBGE CAB, 1984). The landless farmers have been responsible for very small areas of land, and in the sample taken by the Joaquim Nabuco Foundation, 73.8% were responsible for areas up to 5 ha at most. Taking these area limits into account and comparing them with the distribution of food production in NEB, it can be seen that the farmers owning 100 ha or less were responsible for almost all the food production in the region. In 1980 they produced 76.3% of the rice, 82.0% of the regional production of beans and 94.0% of the manioc (IBGE CAB, 1984). Crop production is mainly concentrated in the semi-arid zone, where half of the NEB rural population is concentrated. Its mean annual precipitation varies from 400–800 mm, and the variability of precipitation reaches 60% (Ferraz, 1925).

The technical basis of food production in NEB is severely limited. From the distribution of the cultivated area according to technological groups (*Table 4.3*), it can be seen that 70.8% of the corn area, 74% of the rice area and 61.1% of the beans area are grown under a system where only common seeds are used, i.e., seeds stored by the farmer from his own previous crop. Also, under this system the seeds are planted by simply scattering them on the ground and counting upon the help of nature to produce a crop. *Table 4.3* shows that only a very small proportion of the food-cropped area is cultivated using modern technology. Irrigation, which should be a fundamental practice in NEB, is used in only 0.54% of the corn area, 0.43% of the rice area and 1.88% of the beans area. Although cotton and sugarcane represent the largest proportion of cultivated area involving modern techniques, even they are still mainly grown by traditional methods. Both for food and nonfood crops, agriculture in NEB is almost entirely based on

Table 4.3. Distribution of the harvested area by technological group, NEB, 1980.

	Area distribution by technological group (%)				
Applied technology	*Corn*	*Rice*	*Beans*	*Cotton*	*Sugarcane*
I. Use of selected seeds and:					
Irrigation, pesticides and fertilizers	0.094	0.129	0.439	0.648	2.825
Irrigation and pesticides	0.034	0.014	0.213	0.224	-
Irrigation and fertilizers	0.002	0.001	0.016	0.004	0.433
Irrigation	0.007	0.007	0.008	0.043	0.000
Pesticides	1.717	0.841	1.949	23.447	0.041
Fertilizers	0.147	0.944	0.174	0.293	3.610
Pesticides and fertilizers	0.319	4.350	0.393	1.561	13.446
Subtotal	2.320	6.285	3.192	26.220	20.355
II. Selected seeds only	23.408	0.898	5.362	5.940	0.184
III. Use of common seeds and:					
Irrigation, pesticides and fertilizers	0.105	0.100	0.422	0.105	7.127
Irrigation and pesticides	0.090	0.037	0.319	0.055	0.004
Irrigation and fertilizers	0.037	0.001	0.144	0.010	0.813
Irrigation	0.174	0.144	0.316	0.114	0.063
Pesticides	20.856	17.270	23.594	38.436	0.358
Fertilizers	0.857	0.388	2.169	0.579	33.490
Pesticides and fertilizers	1.318	0.808	3.373	2.351	33.750
Subtotal	23.437	18.748	30.345	41.650	75.605
IV. Common seeds only	70.834	74.069	61.100	26.190	3.856
Technology not specified	0.001	0.000	0.001	0.000	0.000
Total	100.000	100.000	100.000	100.000	100.000

Source: IBGE CAB (1984).

the use of common seeds. In export and raw materials crops this may occur due to the relative prices of selected versus common seeds; but with food crops, it may arise from attitudes about protection against risk.

4.2. Aims

The main objective of this section is to analyze the effects of drought on food production and yields in NEB, without neglecting either the importance of small farmers with respect to this food production, or the economic fragility of the region. Although food production depends on different factors, including prices and those of a socioeconomic nature, one may assume that drought is the most restrictive factor. This assumption has some empirical support. In Canindé, State of Ceará, about 130 small farmers cited climate as the greatest source of risk. No farmer considered changes in prices as an important factor in making decisions regarding his crop (Dillon and Mesquita, 1976).

Berry and Cline (1979), among others, concluded that the agrarian structure of NEB affects yields. They showed that the productivity per unit area is reduced when the size of the farm increases, even after taking the quality of the soil into account. It is important to stress that the results of regression analyses conducted by these authors only indicate part of the relationship between agrarian structure and yield. Perhaps the largest part, which has not been reflected in the analyses, is concerned with social relations established as a result of the high concentration of land ownership – these are fundamental in determining production and yield levels.

According to Correia de Andrade (1970), other factors besides agrarian structure should be considered as agrarian problems in NEB. These include the natural conditions, social structure, technocultural patterns, basic infrastructure, and political–administrative organizations.

The 1981–1983 Brazilian economic recession must also have affected the NEB economy and its production and yield levels. Although we are not aware of any study of the effects of the restriction on credit imposed by the recession, that restriction may well have reduced growth rate in the region. Between 1981 and 1983, the real value of rural credit financing fell to 43.7%, and the number of lending contracts was also considerably reduced (IBGE AEB, 1985). The reduction of resources may have been reflected in the extent of the cultivated area and the demand for industrial inputs and industrial labor. If this was the case, agricultural production and yield would have been affected.

Although the scope of this section is limited to the effects of drought on food, important industrial crops have also been considered. If we compare the distribution of food and industrial crops, it is apparent that food production is largely located among those areas in which annual droughts of 6 to 11 months' duration occur quite frequently (FENAME, 1983). Sugarcane, one of the main export crops, is cultivated in the more humid areas and the forest zone (Zona da Mata), where lack of rain is, in general, not a problem and the drought period lasts three months a year or less.

Data sources include the Brazilian Institute for Geography and Statistics (IBGE), the Superintendency for the Northeast Development (SUDENE) and the Technical Assistance and Rural Extension Enterprise of Rio Grande do Norte (EMATER-RN). The analysis covers the period from 1960 to 1983, with some data also for 1984. Two scenarios from this period have been considered: one for the years 1979 and 1980, and the other for 1983. These scenarios were chosen on the basis of the discussion in Section 2, which suggested that 1979 and 1980 were successive drought years that might be used to exemplify a back-to-back drought event, and that 1983 could be adopted as an extreme drought scenario since this embraced the severest drought of the last 25 years.

4.3. Production and Yield Behavior

On the basis of a paper by Evenson and Kislev (1973), Silva (1984) estimated the effects of climatic variations in the State of São Paulo, analyzing data relating to forests and rainfall inadequacies. Yield was taken as a function of investment in research, investment in extension, frost, and rainfall shortages. The results showed that the climatic variables had a significant impact on yield.

In analyzing the effects of drought on production and yield, the present study assumes that these variables, expressed in physical units, are functions of trend and annual rainfall. Trend was introduced to estimate the effects of variables that were not included, but which can affect production and yield over a period of time. Emphasis has been given to combining graphic analysis with the use of linear regressions. This should not only make comprehension easier, but it should also make the analysis more illustrative.

4.3.1. Trends and drought impacts for NEB crops

Figures 4.1 to *4.3* show the production and yield curves for food and raw material crops, for a period of 24 years, in several states of NEB. Except for beans in Ceará, the food crop curves appear to show a general production increase from 1960 to 1983. The trend of production increase was significant at the 5% and 10% levels for the t-test. The trend in yield was generally downward, or not significant if increasing (for example, rice in Maranhão). The diagrams indicate that no food crop increased its production without an area increase. During the period analyzed, food production increased only due to an areal increase, since the yields decreased or remained constant.

Cotton and sugarcane behaved quite differently from the food crops. Sugarcane increased production and yield throughout the period. Herbaceous cotton, on the other hand, suffered significant reductions not only in production, but also in yield. Unfortunately, problems with the cotton data hinder further analysis for that crop.

As already mentioned, 1983 was the worst year of the 1979–1983 period. In most cases it was the year in which production and yield had the lowest levels of the series. For food crops the lowered production in 1983 was due to reductions in both yield and area. Of the 1979–1983 period, 1983 was the year with the greatest reduction in area. During the other years of that period, there was an expansion in cultivated area. This suggests that an area reduction response only occurs in extreme cases. Apart from such exceptions, the reaction to the yield decrease trend has been an increase in the cultivated area for food crops, to try to maintain production levels even in drought years. It should be noted, however, that the area of expansion occurred mainly in Maranhão and Bahia, two states that suffered the least during the drought periods, and which still have an expanding agricultural frontier.

Besides the efforts of the affected families to respond to changing conditions through adjustment of cultivated area, another kind of change occurred regarding the distribution of production. It is instructive to consider 1975 as a "normal" year, when production and yield levels were generally equal to or above the mean (*see Figures 4.1* to *4.3*), and to examine the considerable changes in production between that year and 1980. *Figure 4.4* shows that the proportion of the production kept for subsistence increased by about 10 percentage points between 1975 and 1980. Uncertainty induced by drought led the families to search for more protection, as reflected by the increasing proportion of food kept for subsistence. The corollary to this was a reduction in the proportion of food products available for sale, which thus decreased the food supply in the domestic market.

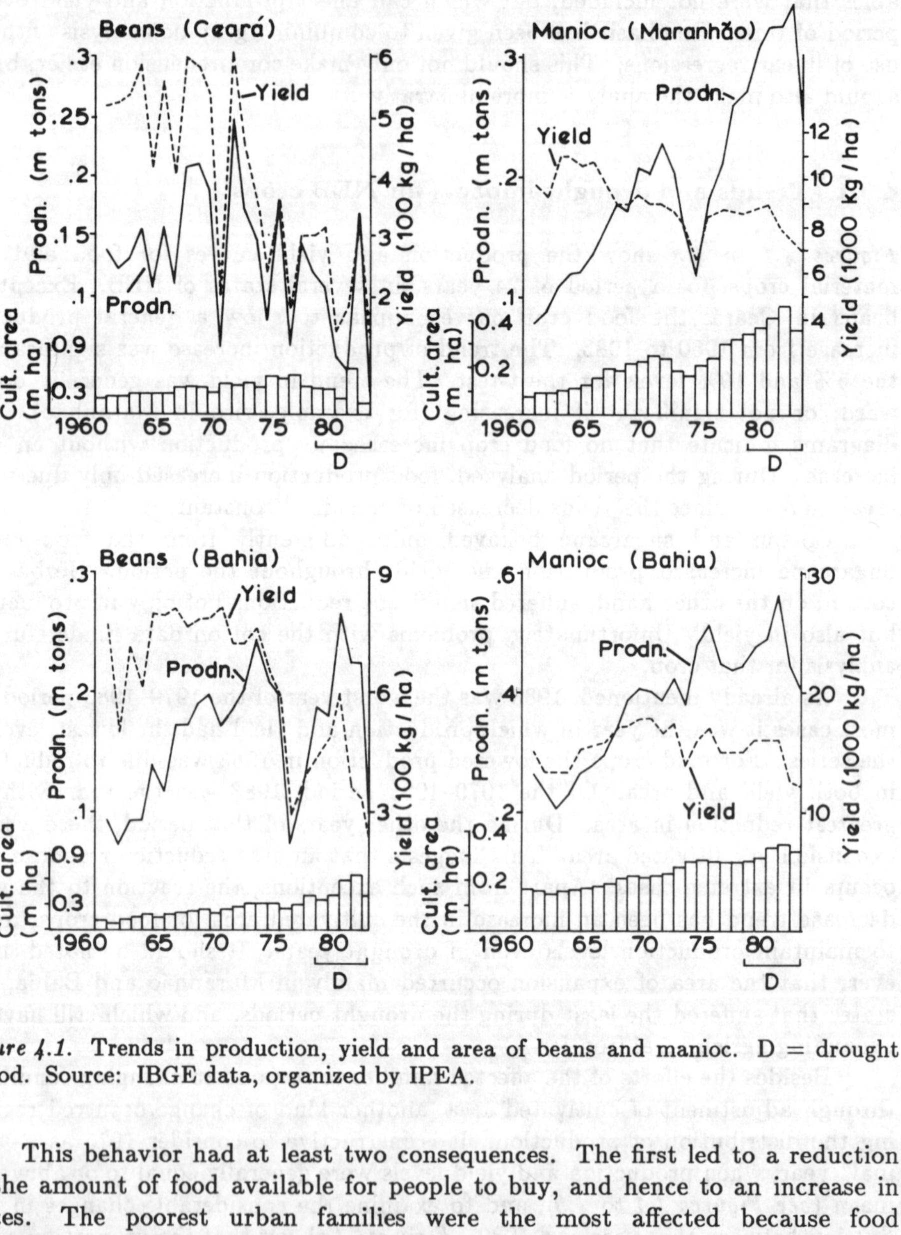

Figure 4.1. Trends in production, yield and area of beans and manioc. D = drought period. Source: IBGE data, organized by IPEA.

This behavior had at least two consequences. The first led to a reduction in the amount of food available for people to buy, and hence to an increase in prices. The poorest urban families were the most affected because food represents the major portion of their total expenses. The other consequence of the change in production distribution was related to domestic market demand, considering farmers as consumers. If total sales fall during drought periods because some part of NEB production is almost nonexistent, the demand for industrial goods is also reduced. This translates into a reduction in the amount of food, clothing, furniture, medicine, etc., which is purchased.

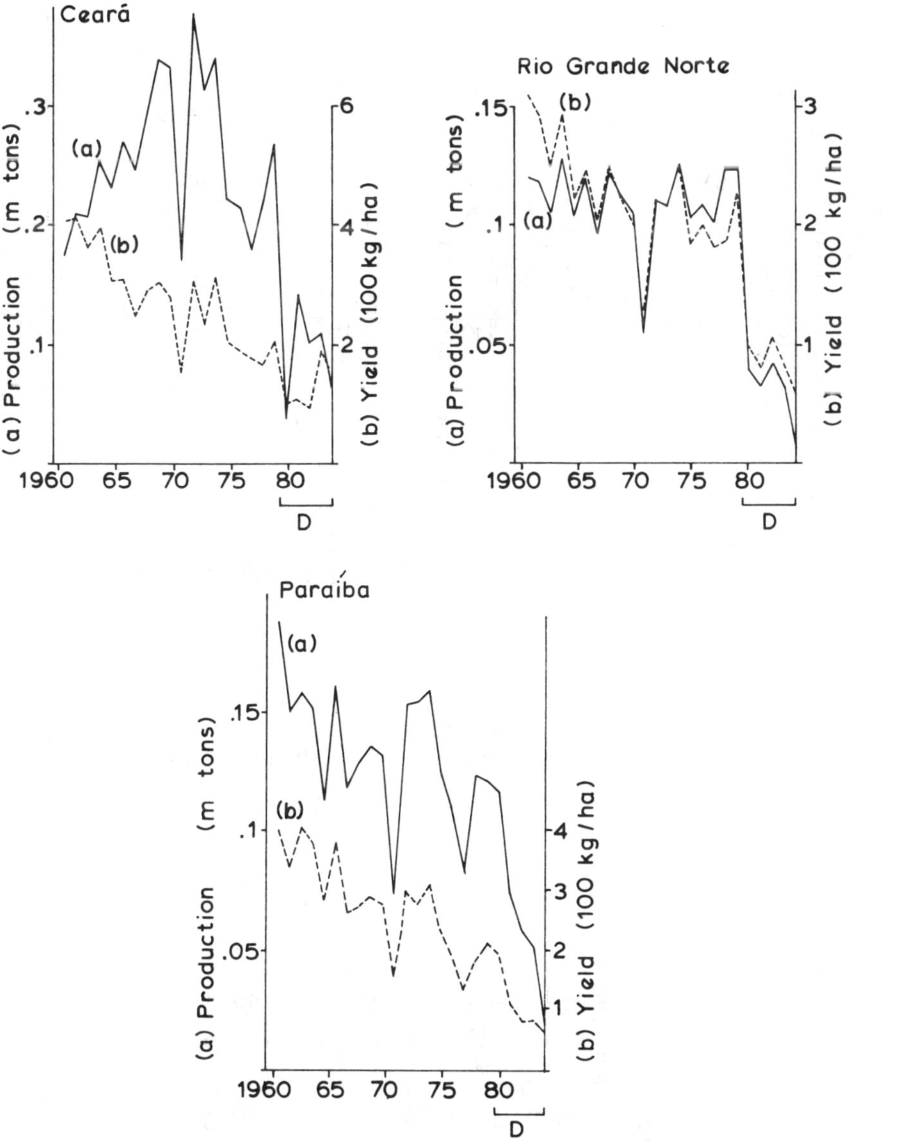

Figure 4.2. Trends in production and yield of herbaceous cotton. D = drought period. Source: IBGE data, organized by IPEA.

To illustrate these effects, the global expenses elasticity was calculated for low-income rural families in NEB (*Table 4.4*). This coefficient shows the variation in consumption of different kinds of goods, when global expenses vary. Assuming that global expenditure is a function of income, if income decreases by 10%, expenses for cereals and by-products will fall 12.7%, all other expenses remaining constant. Agricultural production in NEB fell by 24.4% in 1983, and

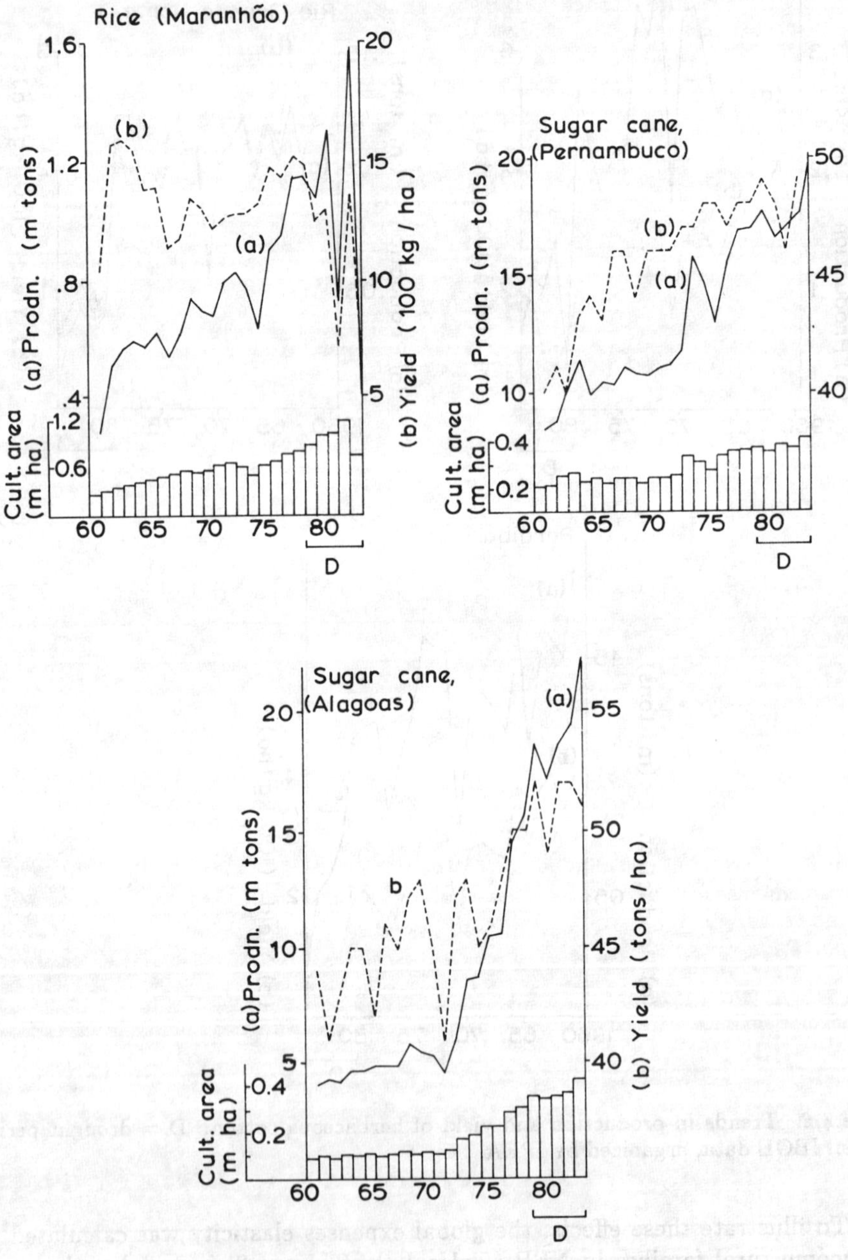

Figure 4.3. Trends in production, yield and area of rice and sugarcane. D = drought period. Source: IBGE data, organized by IPEA.

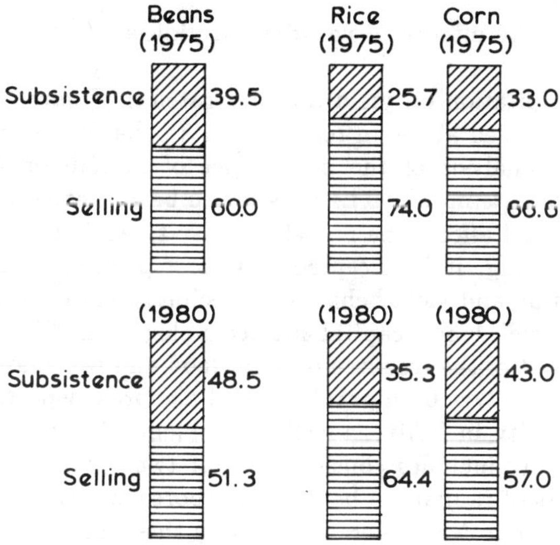

Figure 4.4. Distribution of production in normal year (1975) and drought year (1980). Source: IBGE.

Table 4.4. Global expenses elasticity for low-income rural families of NEB[a].

Category of goods	*Elasticity*[b]
Food	1.01
Cereals and by-products	1.27
Tubers and roots	0.82
Meat and fish	1.10
Eggs and dairy products	1.48
Legumes and oil seeds	0.73
Clothing	1.35
Furniture and cleaning materials	1.07
Labeled medicines	1.24
Rent and taxes	0.39
Transportation	1.25

Source: IBGE (1977; 1978).
[a] Low-income families were considered to be those earning up to two minimum wages a year, equivalent to US$75.
[b] Elasticity was calculated using the $y = ax^b$ function, where y = consumption by category of goods, x = global expenses and a,b are the parameters. Transforming these functions into the logarithmic form, the estimate of b defines the elasticity. The b parameter indicates the percentage variation that occurs in consumption when global expenses vary by 1%.

the effect on the domestic market has been estimated very approximately. The results indicate that the change in real consumption of cereals and by-products was −30.98%; for consumption of eggs and dairy products, it was −36% and for clothing, −32.9%. These consumption variations emphasize how the vulnerability of food production to the effects of drought can compromise stability and damage the growth of the domestic market for goods.

4.3.2. Analyzing the effects of rainfall on yields

The effects of precipitation on yield are illustrated in *Figures 4.5* and *4.6*. The yield trend was discussed above in relation to its behavior during drought years, but a more direct analysis of the importance of rainfall on agricultural yield changes will now be attempted. The question to be addressed is how introducing rainfall into the model affects the model's ability to explain yield variations. As noted at the beginning, it is accepted that precipitation is a key variable in explaining production and yield behavior. Precipitation can be considered a production factor, like soil, labor, capital and technology. In NEB it assumes a fundamental role as part of the resource base for food and raw material production.

The analysis involves both graphical and algebraic aspects. Because of the great climatic diversity in NEB, as well as problems that can occur with aggregated data, the study includes consideration of two specific localities or micro-regions. The effects of rainfall on beans and arboreous cotton at Caicó and Alexandria in Rio Grande do Norte were analyzed in these local studies.

Yield and rainfall over 25 years can be seen in *Figure 4.5*. The fluctuations in both variables are quite marked. With respect to 1983, which was considered a year of extreme drought in Section 2, the observations confirm the severity of that drought in both micro-regions. It can be seen that precipitation in 1983 was considerably lower than in 1979 and 1980 (back-to-back drought years), and was also much lower than that of 1970, another drought year. The diagrams show that precipitation and yield levels correspond only in a general way. The lowest yield for beans was 1980, not 1983. This finding is consistent with the simulation model proposed in Section 3, which shows that the yield index is a function of the amount of water available for the plant. As 1979–1980 were years of back-to-back droughts, the amount of water available in 1980 must not have been sufficient for the crops, and this had a substantial effect on the yield during that year. Thus, yield in 1980 was even lower than that of 1983, which was a single-year drought.

The broken lines in *Figure 4.5* represent the yield trend. Free-hand drawing would not necessarily have produced better lines, although lines drawn free-hand for graphical analysis can avoid routine errors of fitting linear relations that may or may not be appropriate (Waugh, 1973).

Analysis of bean production in Caicó and Alexandria shows that there are great deviations between the observed yields and any point on the trend lines. In general, the greater the deviations, the smaller the proportion of total yield variance explained by the estimated trend function, and the smaller the correlation between observed and trend yields as well. The correlation coefficients were thus quite low: 0.36 for Caicó and 0.31 for Alexandria (*Figure 4.6*). This suggests that a great proportion of yield variance cannot be explained by the trend variable alone. Strictly speaking, the proportion of yield variance that is explained by trend is given by the coefficient of determination. The correlation coefficient is the square root of this coefficient.

The correlation coefficient for arboreous cotton was higher than for beans – in the order of 0.77 (*Figure 4.6*). In this case the deviations were smaller, so that

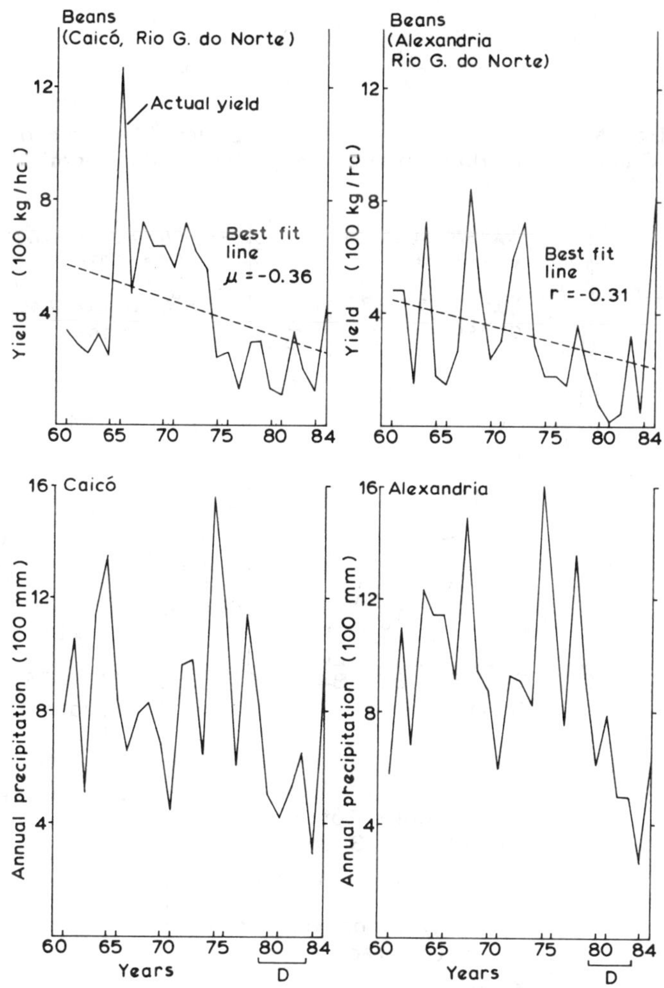

Figure 4.5. Annual precipitation and bean yields, 1960–1984. Source: EMATER-RN (n.d.).

the adjusted straight line accounts for a greater proportion of yield variance than in the previous case.

The proportion of unexplained yield variation can be reduced when new variables are introduced that account for an important phenomenon not previously considered, perhaps because its importance had not been recognized or because of lack of data. Thus, it was decided to introduce precipitation, with the object of trying to reduce the deviations or residuals, i.e., the differences between the observed values and the estimated yields.

The multiple regression equations, which contain precipitation as well as trend, have been represented algebraically and graphically (solid lines in *Figure 4.6*). For comparison, equations without precipitation are also shown in

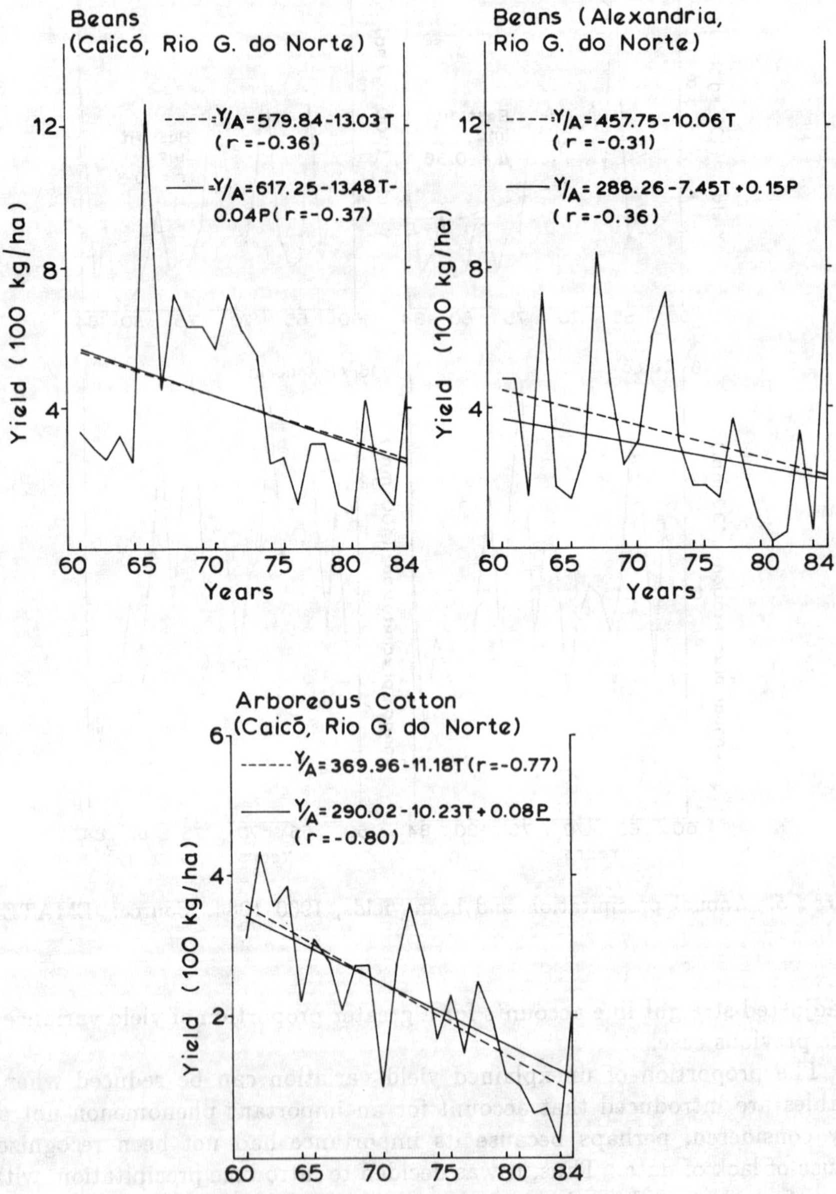

Figure 4.6. Observed yields of beans and of arboraceous cotton, and regression lines accounting for trend with (solid lines) or without (broken lines) precipitation. Source: EMATER-RN (n.d.).

algebraic and graphic forms (broken lines). Comparing the equations, there was not much reduction in unexplained yield variation when precipitation was introduced as an additional explanatory variable. In the graph it can be seen that the multiple regression curve follows the trend line rather closely, and reflects the year-to-year fluctuations in actual yields to only a small extent. For beans in Caicó, for example, the correlation coefficient changed only from 0.36 to 0.37. The deviations between estimated and actual yields remain high even after introducing precipitation. This may mean that other factors besides precipitation have affected yield when a period as long as 1960–1984 is being considered. This could have happened because this period included a variety of events in which there were years of drought, abnormally wet years, and normal years. It might be possible to improve the model by adding a quadratic term for precipitation, and perhaps including a simple potential evapotranspiration term. Also it may not be adequate to assume a single straight-line time trend through the period, as was done here.

The results obtained for arboreous cotton in Caicó (*Figure 4.6*) indicate that the model used here performed better for arboreous cotton than for beans, since the deviations were smaller. To test the validity of this model for arboreous cotton a regression equation was derived, using data for 1960–1981, and then used to make estimates for subsequent years. The estimates of the equation parameters were:

$$\begin{aligned} \text{Yield} &= 312.09 - 10.53T + 0.06P \\ r &= 0.77 \qquad (-4.69) \quad (1.26) \end{aligned} \tag{4.1}$$

where the values in brackets are estimates of the t-test, r is the correlation coefficient, T is the trend and P is the annual precipitation. Using this equation, estimated yield values were obtained for the rest of the period, 1982 to 1984, as shown in *Table 4.5*.

Table 4.5. Validation for arboreous cotton – Caicó.

Trend (i.e., year number)	*Annual rainfall* (mm)	*Estimated yield values* (kg/ha)	*Observed yield values* (kg/ha)	*Residuals (obs – est)* (kg/ha)
23	648	108.70	69	−39.70
24	300	77.37	20	−37.37
25	933	106.14	20	−93.86

Sources: EMATER-RN and SUDENE.

In the test, the deviations or residuals obtained by subtracting the model estimates from observed yields were rather large, suggesting that the model is biased and gives estimates that are too high, at least for recent years. A more comprehensive model should be developed. Examination of the arboreous cotton graph in *Figure 4.6* reveals that yield had a declining trend since 1960, perhaps induced by other variables, such as market conditions or governmental policies that deemphasized this crop.

The above results should not be taken as implying that rainfall is unimportant in the analysis of the problem. *Figure 4.5*, representing precipitation in Caicó and Alexandria, shows that rainfall did not have a uniform pattern throughout the 1960–1984 period. Starting in 1974, precipitation levels showed a general decreasing trend. This in turn seems to have been accompanied by yield reductions, which, between 1974 and 1984, showed similar deviations to those observed in the 1960–1973 period.

Table 4.6. Estimated regressions, correlation coefficients and marginal contributions of rainfall, 1974–1984.

Estimated equations	*Correlation coefficients*	*Marginal contribution of rainfall* (%)[a]
Caicó beans		
1. Yield = 208.44 + 4.88T	0.14	
r^2 = 0.02 (0.438)		
2. Yield = 89.77 + 22.66T − 0.24P	0.65	41%
R^2 = 0.43 (1.95) (2.42)		
Alexandria beans		
1. Yield = 107.74 + 16.48T	0.27	
r^2 = 0.07 (0.864)		
2. Yield = 600.71 + 67.43T + 0.48P	0.64	35%
R^2 = 0.42 (2.35) (2.15)		
Caicó arboreous cotton		
1. Yield = 218.24 − 12.64T	−0.56	
r^2 = 0.31 (−2.049)		
2. Yield = 59.64 − 3.19T + 0.13P	0.76	27%
R^2 = 0.58 (−0.48) (2.26)		

Sources: EMATER-RN and SUDENE.
[a]Difference between determination coefficients of the regression with P and the regression without P; the result was multiplied by 100.

To analyze this relation between rainfall and yield for 1974–1984, similar procedures were followed. Initially, some linear functions were derived for which yield depended only on trend, and others in which it also depended on rainfall. The resulting parameters of these functions and their respective correlation coefficients are shown in *Table 4.6.* For beans, the correlation coefficients with respect to the model, including both trend and precipitation, are higher than those obtained for the 1960–1984 period. For cotton, however, correlation for that model for 1974–1984 was almost identical to that obtained for the 1960–1984 period.

To try to assess further the effects of rainfall during 1974–1984, the marginal contribution of this variable was calculated (*see Table 4.6*). This was obtained using procedures described by Theil (1971), who showed that to obtain this individual contribution of each variable, the difference is calculated between the coefficient of determination (R^2) of the model in which this variable is included, and the coefficient of determination (r^2) for that same model but without that variable. The difference between the coefficients of determination is the marginal contribution of that variable. For *Table 4.6* the marginal contribution of rainfall on Caicó beans, for example, was calculated by subtracting 0.02 from 0.43 and multiplying by 100. The resulting value, 41%, indicates that when rainfall was introduced, the proportion of total yield variance explained by the regression increased by 41% in comparison to the regression that did not contain rainfall. The marginal contributions indicate that the model has been greatly improved by introducing rainfall, although there still is a great proportion of the total variance that has not been explained by the regression equations, since the highest coefficient of determination is only 58%. Using similar procedures with the 1960–1984 results, and squaring the correlation coefficients to obtain coefficients of determination, results in $[(0.37)^2 - (0.36)^2] \times 100 = 0.7\%$ marginal contribution for Caicó beans; for Alexandria beans $[(0.36)^2 - (0.31)^2] \times 100 = 3.4\%$ and for Caicó arboreous cotton $[(0.80)^2 - (0.77)^2] \times 100 = 50\%$.

The results obtained can lead one to conclude that the climate represents only one of the problems that affect the NEB economy and population. When one considers long periods, such as decades, each of which includes both normal years and drought years, rainfall fluctuations have little importance in determining the interperiod behavior of production and yield. For shorter periods of time, such as 1979–1983 or the one- or two-year time units considered in selecting scenarios for drought, the role of the climatic factor increases considerably in the determination of yield variations. However, the fact that yield levels are low and are continuing to decrease is due mainly to factors other than rainfall. This means that irrigation policies in NEB should be formulated in conjunction with other policies that would act in a complementary manner – for example, price policies, rural credit, land reform policies and macroeconomic policies. The empirical results seems to suggest that comprehensive, integrated policy development can be much more beneficial than attempting to promote irrigation in isolation from other factors.

4.3.3. Examining regional effects of drought through micro-regional level studies

As an extension of the analyses discussed above, attention is now focused on selected crops in micro-regions in which their production is concentrated (*Figure 4.7*) in order to ascertain how droughts have affected the main production centers of each state. Beginning with the production in all micro-regions, those that had the greatest concentration of food and raw material production in each state were identified and analyzed for the period from 1974–1975 to 1982 (*Figures 4.8* to *4.12*). In each micro-region shown, the production and yield changes for the back-to-back drought period (1979–1980) were estimated in relation to the period without drought.

Among the food crops, the least affected was rice, for which production is concentrated in Maranhão, a state that suffered the effects of drought only in

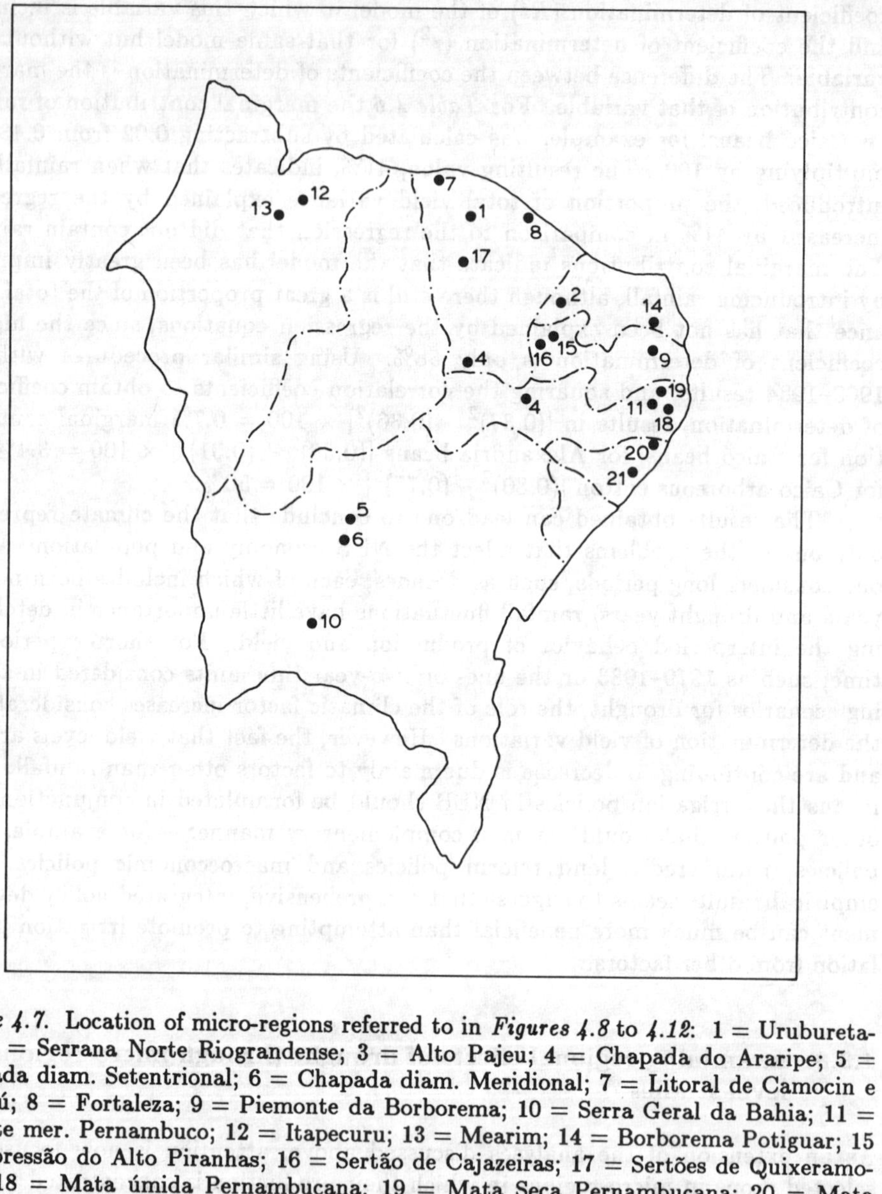

Figure 4.7. Location of micro-regions referred to in *Figures 4.8* to *4.12*: 1 = Urubureta-ma; 2 = Serrana Norte Riograndense; 3 = Alto Pajeu; 4 = Chapada do Araripe; 5 = Chapada diam. Setentrional; 6 = Chapada diam. Meridional; 7 = Litoral de Camocin e Acaraú; 8 = Fortaleza; 9 = Piemonte da Borborema; 10 = Serra Geral da Bahia; 11 = Agreste mer. Pernambuco; 12 = Itapecuru; 13 = Mearim; 14 = Borborema Potiguar; 15 = Depressão do Alto Piranhas; 16 = Sertão de Cajazeiras; 17 = Sertões de Quixeramo-bim; 18 = Mata úmida Pernambucana; 19 = Mata Seca Pernambucana; 20 = Mata Alagoana; 21 = Tab S. M. dos Campos.

1981. Beans and manioc were the most affected food crops. The state of Ceará, which is an important bean producer, had its production in drought years almost wiped out in some of its main micro-regions. Manioc was also very much affected in all micro-regions.

Among the raw materials, cotton suffered considerable damage from the droughts. All the micro-regions suffered a great decline of production as well as

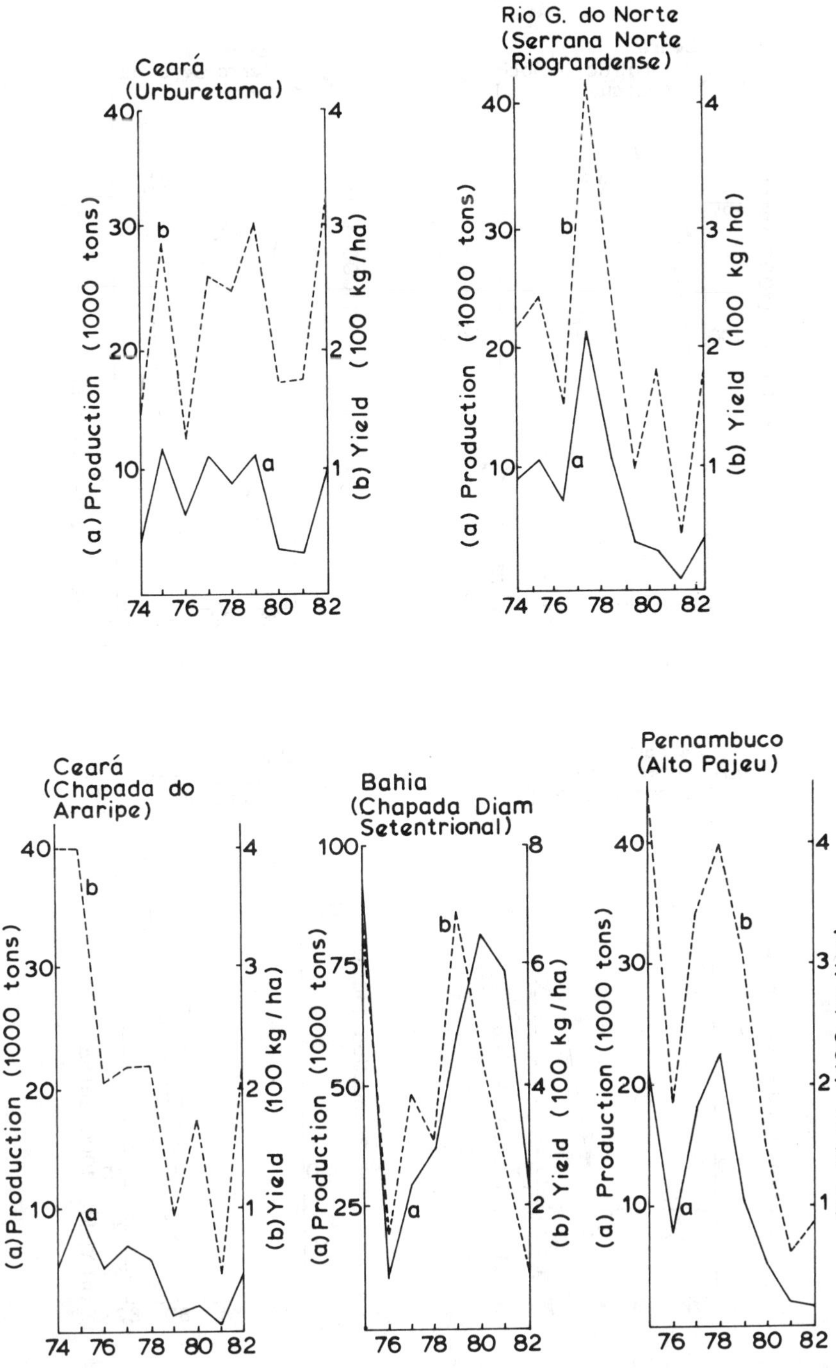

Figure 4.8. Beans: production and yield changes, 1974–1982. Source: IBGE.

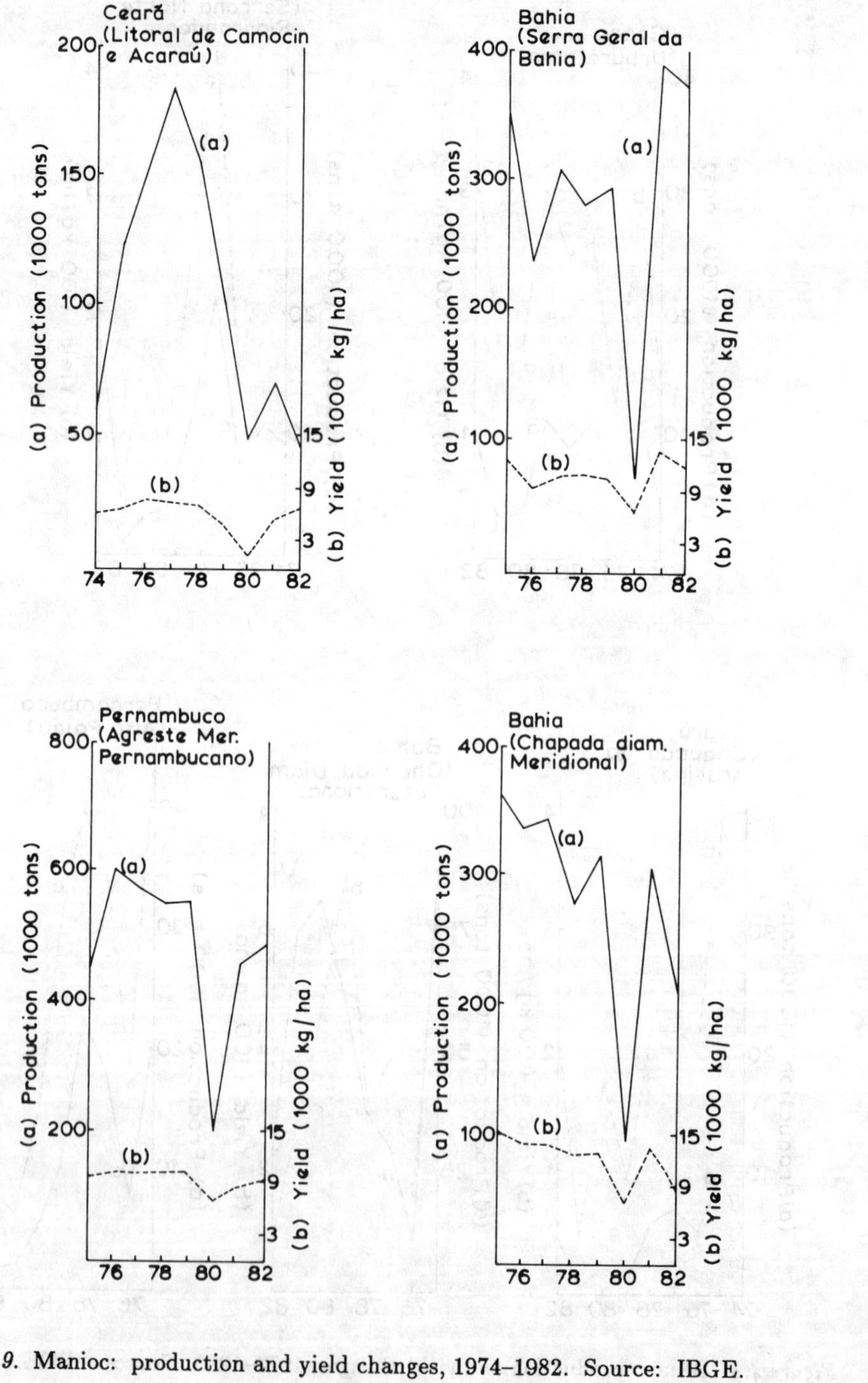

Figure 4.9. Manioc: production and yield changes, 1974–1982. Source: IBGE.

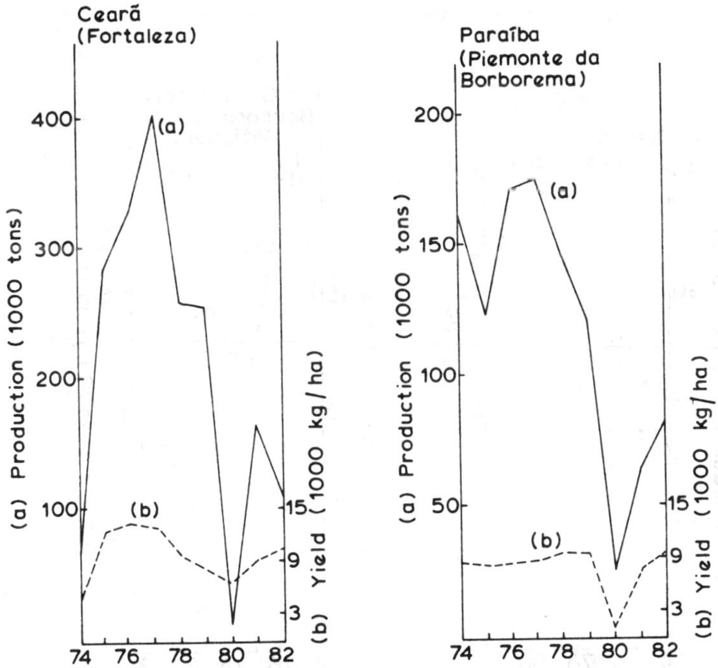

Figure 4.9 continued.

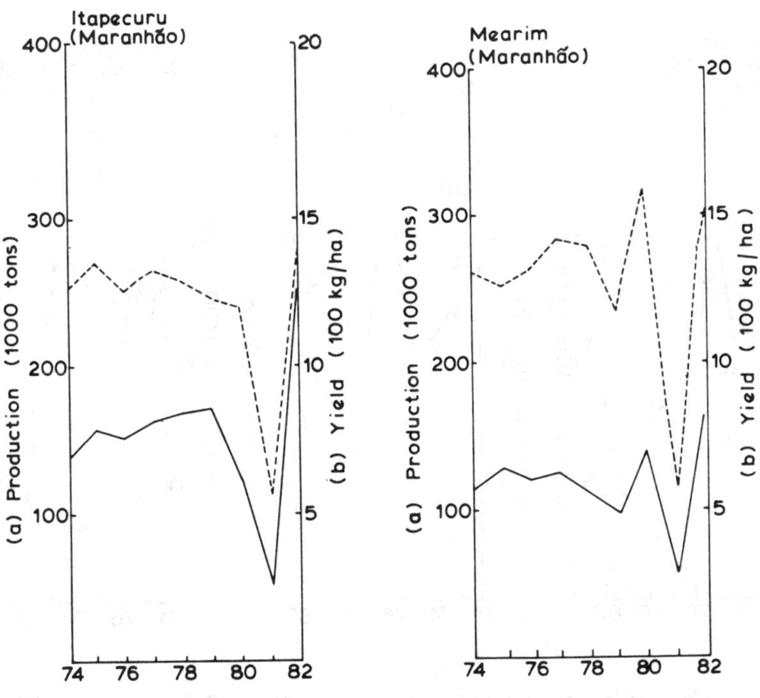

Figure 4.10. Rice: production and yield changes, 1974–1982. Source: IBGE.

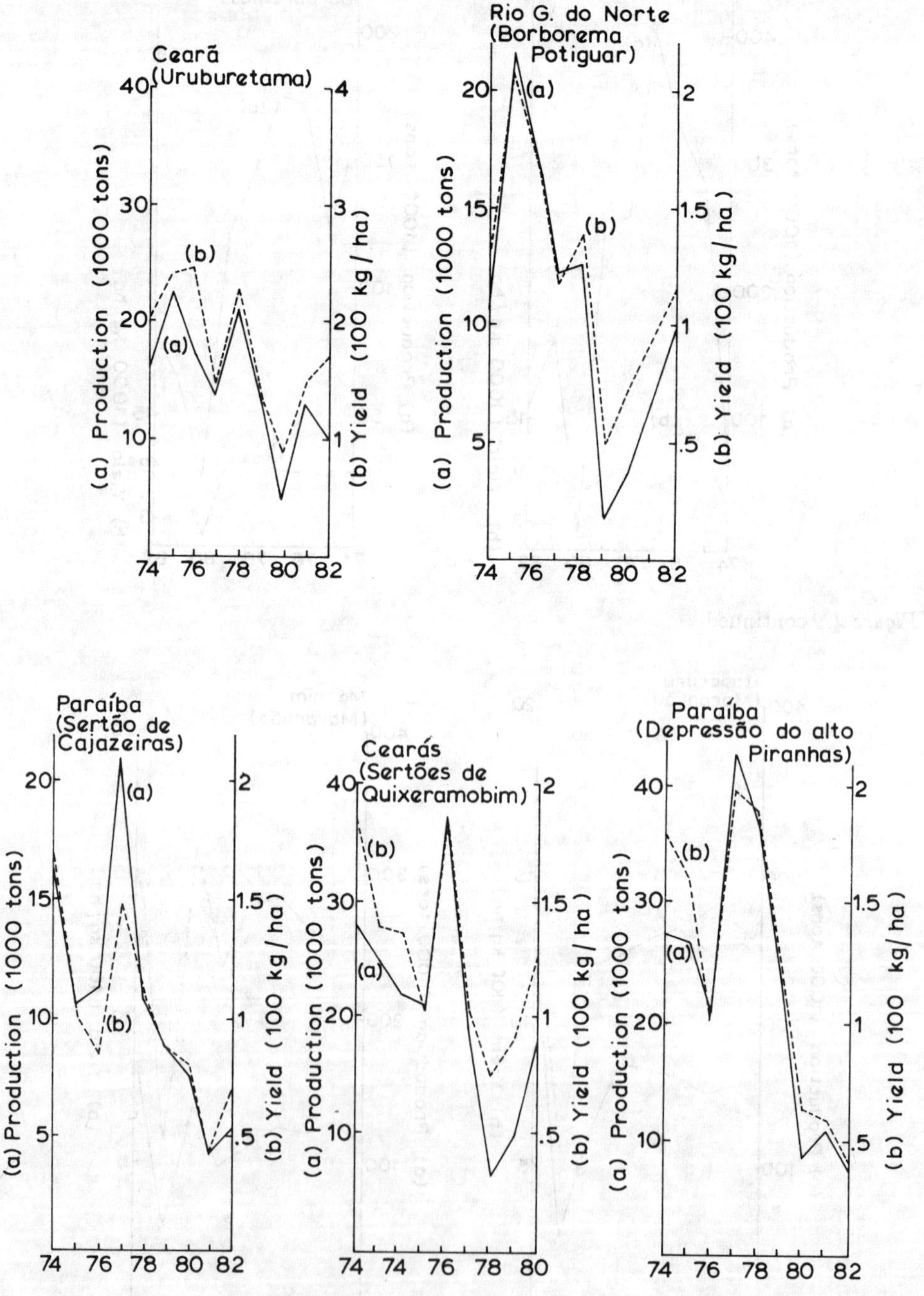

Figure 4.11. Cotton: production and yield changes, 1974–1982. Source: IBGE.

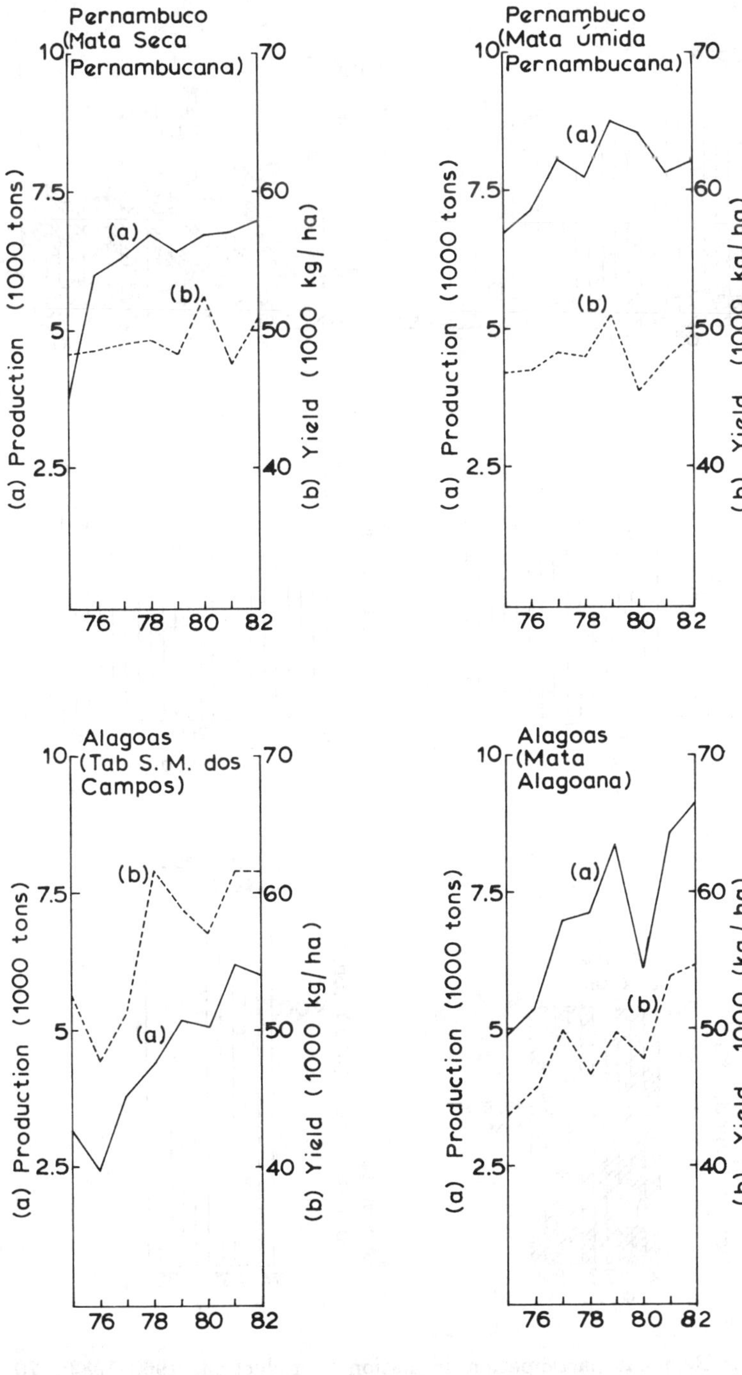

Figure 4.12. Sugarcane: production and yield changes, 1974–1982. Source: IBGE.

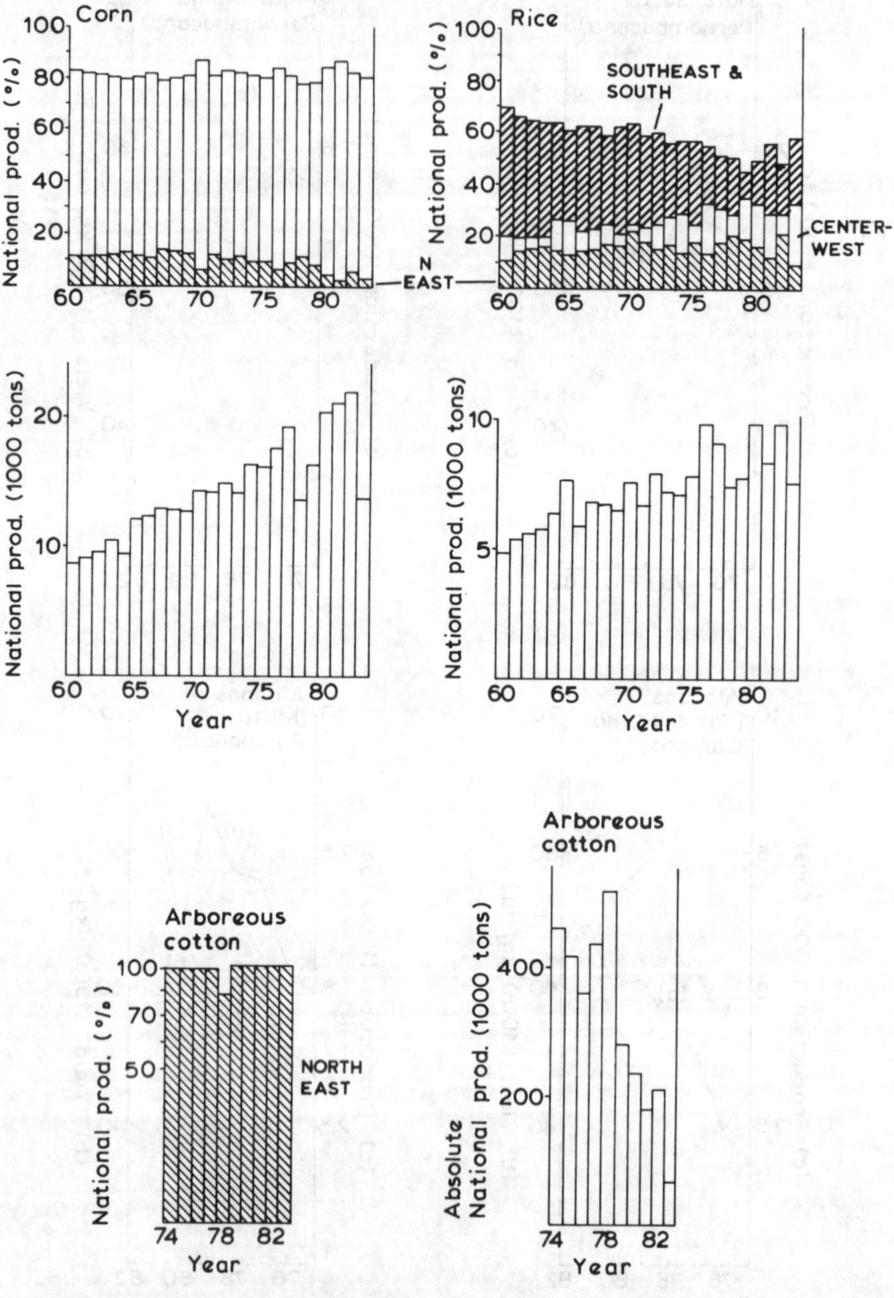

Figure 4.13. Regional participation in national production, 1960–1983: 70 = 1970 single-year drought; 79–80 = 1979, 1980 back-to-back drought; 83 = 1983 extreme drought. Source: IBGE data, organized by IPEA.

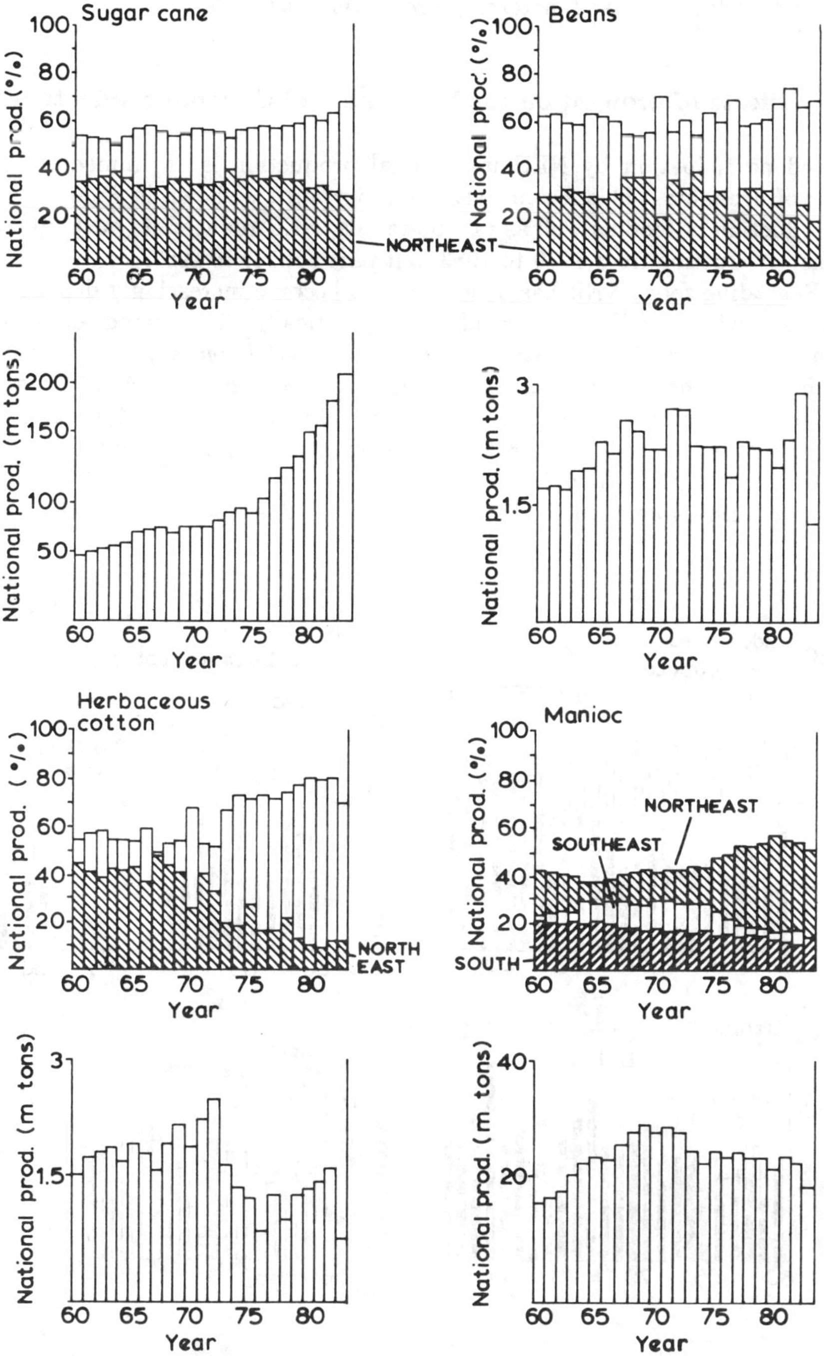

Figure 4.13 continued.

of yield. Although sugarcane was affected in 1980, this did not have much impact on its trend toward increasing production and yield.

4.3.4. Effects of drought on the NEB share of national production

Reduced participation by NEB in national production is an immediate consequence of droughts. This loss in relative importance is shown in *Figure 4.13*. In the production of food as well as raw materials, NEB reduced its participation in national production from 1979 to 1983, but particularly in 1983.

Regarding food, NEB has a tendency to become increasingly dependent on the other regions. NEB corn production practically disappeared as a part of national production in the extreme drought of 1983 (*Figure 4.13*). As for rice, NEB has been helped in maintaining its participation in national production because in Maranhão, which is a frontier state, the effects of drought were mainly confined to 1983. In the case of beans, a great part of production is concentrated in Ceará, which in normal years is an important producer, but is extremely vulnerable to drought (*Figure 4.14*). The participation of Ceará in

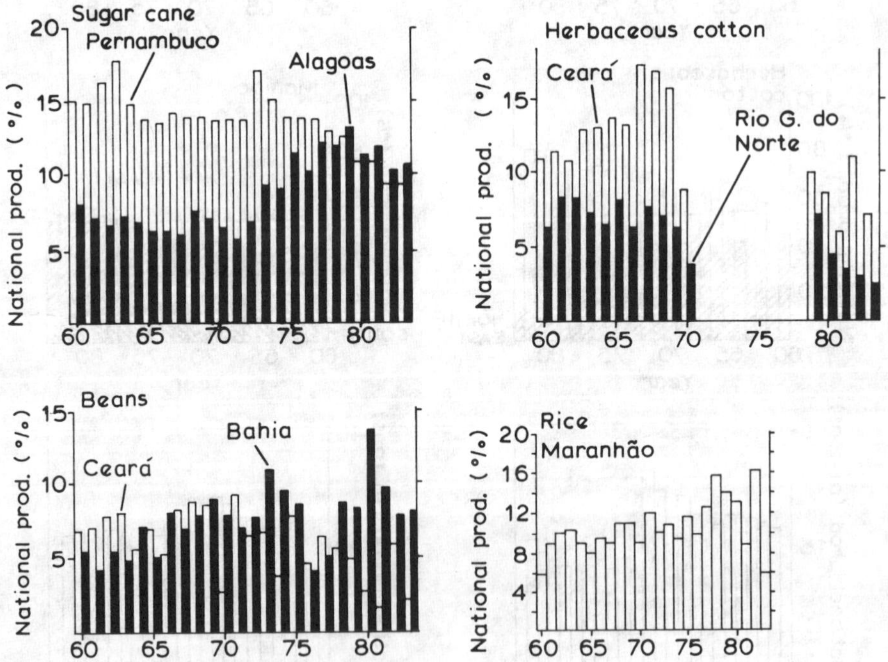

Figure 4.14. Participation of selected states in national production, 1960–1983: 70 = 1970 single year drought; 79/80 = 1979, 1980 back-to-back drought; 83 = 1983 extreme drought. Source: IBGE data, organized by IPEA.

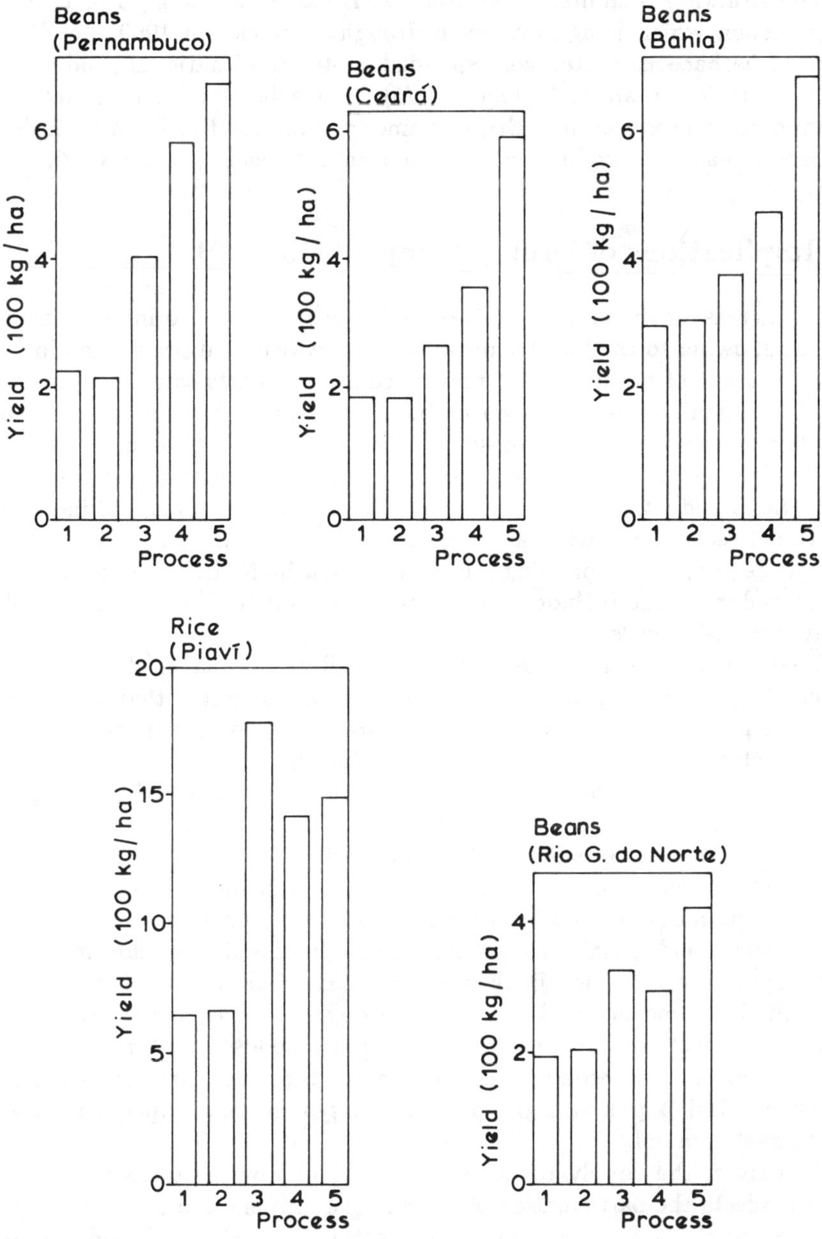

Figure 4.15. Processes of production and yield: 1 = common seeds only; 2 = common seeds and fertilizers; 3 = common seeds and irrigation; 4 = common seeds and irrigation, pesticides and fertilizers; 5 = selected seeds and irrigation, pesticides and fertilizers. Source: IBGE data, organized by IPEA.

national production in 1969, which may be taken as a normal year, was 8.54%. In 1970, a drought year, it fell to 2.67% and, in 1983, to 2%.

The situation is similar for cotton. NEB has been losing its relative share, a phenomenon which is aggravated in drought periods. In 1960 the NEB production of herbaceous cotton corresponded to 44.6% of national production. In 1983 it fell to less than 20% (*Figure 4.13*). For arboreous cotton, 100% of the national crop is produced in NEB, and since production there has been decreasing, absolute national production has also been decreasing (*Figure 4.13*).

4.4. Implications of Drought Impacts in NEB

Only by increasing the cultivated area has it been possible to increase food crop production, owing to the frail technical basis on which food production in NEB is based. Water is of fundamental importance in any efforts to change this technical basis. When supplemental water (i.e., by irrigation) is introduced into the production process, there is a considerable yield increase. Even if the other traditional technologies of the region, such as the use of common seeds, were maintained, the introduction of supplemental water could result in yield increases for beans of between 26 and 65%, and for rice of 171% (*Figure 4.15*). There is, however, a wide range of expanding food production in NEB. As shown in *Figure 4.15*, procedures that include water, pesticides and fertilizers may double the present food yield levels.

Nevertheless, other factors besides rainfall are crucial in determining production and yield behavior. The analysis here has suggested that variables such as price policies, credit and agrarian structure can be even more important than rainfall fluctuations. This leads to the conclusion that agricultural policies for NEB should include a combination of instruments in which irrigation, although an important factor, cannot be considered in isolation.

The impacts of drought on food production mainly affect small farmers because they produce most of the food, particularly beans and manioc, which are the food crops most affected by drought. As these people have a high potential as consumers of industrial wage goods, such as processed food, clothing, footwear and medicine, the reduction in their income during drought periods has a great impact on the domestic market. During drought periods, the reduction is proportionally greater for the consumption of these goods than for overall income itself. This affects the economy as a whole, for the small farmers in NEB make up a substantial population group, and changes in their expenditures greatly affect market demand.

Moreover, the supply reduction of food and raw materials resulting from droughts affects the participation of NEB in the national economy. This participation, particularly in rice, beans and cotton, is very much reduced during drought years. In the state of Ceará, for example, which is an important national beans and cotton producer, severe drought almost eliminates the production of these crops in its main producing areas. This has dramatically curtailed Ceará's participation in the national economy during drought periods. Consequently, the region is obliged to spend a great part of its income on food imports.

SECTION 5

Conclusions

5.1. Framework of the Scenarios

Though there is no established definition of drought in NEB (Northeast Brazil or the "Nordeste"), at least 10 extreme droughts and 33 partial droughts, including multi-year events, have apparently occurred in the past 400 years (*see* Section 1). Although information is not complete, there is ample evidence to show that droughts have caused great impacts upon the NEB economy and society.

There is also an apparent, but perhaps not real, tendency for these events to occur with increasing frequency. While five extreme events occurred over approximately 300 years (1603–1898), the other five occurred between 1900 and 1983 (*Table 1.5*). However, one cannot conclude that a change in climate is occurring, because some early events may not have been recorded. Also, population growth and the modification in the environment due to human occupation have increased the vulnerability of society to drought events. Thus, climatic conditions that would constitute an extreme drought event in 1983 might not have had such an extreme effect in the eighteenth century.

Based on the history of droughts, selected scenarios of drought were chosen for use in this case study. Fortunately (or unfortunately), the very recent period, 1979–1984, provided plenty of examples of extreme drought events (*see* Section 2). The year 1983 was chosen as a typical extreme event, while 1979–1980 provided the example of a back-to-back drought. However, due to lack of data or to special situations in some localities, and to the particular requirements of the impact models, other years have been alternatively used as drought scenarios (*see* Section 3).

5.2. Limitations of the Study

Some limitations of this case study must be considered. The first is the problem of the availability of data. There is a great volume of meteorological data, but they are not organized in a way to facilitate their use for such research. At present the National Institute for Meteorology (INEMET) is engaged in a project to improve the quality and availability of meteorological information in Brazil.

Also, there are no widely available agricultural productivity data. These data problems restricted the number of places that could be included in this study.

The second limitation relates to the validation of the models. Validation was attempted in Section 3, but results were not conclusive, primarily owing to the lack of data. Much work on this subject is still needed.

It was also beyond the scope of this case study to undertake any thorough investigation where the need was indicated. For example, in the multivariate regression analysis (Section 4), the distinction between the contributions of precipitation and of other factors could not be clarified, and more investigation is clearly needed on the apparent unimportance of weather in explaining yield variations.

5.3. Summary of the Results

The main results of this work are summarized in *Table 5.1.* It shows that the 1979–1983 droughts affected an average of 75% of NEB and involved the equivalent of US$1778 million of federal government expenditures in emergency programs. In 1983, a year of extreme drought, 88% of the area and 12 million people were affected, although only 2.5 million were directly assisted (Section 1).

Those droughts were not the severest in the history of NEB. However, in 1983, the deviation of rainfall in relation to the average was −57%. During 1979–1980, the deviation was −31% (Section 2). Such events greatly reduce the probability of farmers obtaining acceptable harvests. Although this probability is typically rather low in the semi-arid zone, it becomes still lower in a drought period, when the risk in agriculture considerably increases. Estimates for six municipalities (Section 3) have shown that, even with 10% less precipitation and more moisture demand, the probability of obtaining an acceptable bean harvest would still be at least 80% in Ceará (Icó municipality); for three of the municipalities the probability would range from 26% to 35%, and for another it would be 63%.

The results also showed that the droughts have a large impact on crop yields. Estimates made using a soil–water–plant type model indicated very low yields in extreme drought periods. It also appears that yield reductions of more than 60% in relation to the yield trend are likely in such periods (Section 3).

Among economic activities, food production is the most affected by drought. Beans and manioc are the most vulnerable crops. In the states of Ceará and Bahia, the main beans and manioc producers of NEB, the production of those crops was almost totally destroyed by drought during the 1979–1980 period. Among raw materials, cotton rather than sugarcane is the most affected because it is cultivated in the most drought-prone areas. For example, in the Borborema Potiguar area of the State of Rio Grande do Norte, bean production was reduced in 1979–1980 by almost 80% (Section 4). On the other hand, for sugarcane, which is cultivated in the Zona da Mata, a less drought-prone area, that period did not significantly disturb the trend toward increasing production in all the municipalities analyzed (Section 4).

These results have several implications. The first relates to the fact that, since food production is the most affected activity, the societal impact of drought falls most directly on the food producers, who are small farmers. As mentioned in Section 1, they are responsible for almost all food production and are the most vulnerable group. Another implication is related to the farmers' behavior when their survival is threatened. The data show that there are changes in the production distribution in drought years. In comparison to normal years, there is an increase of about 10 percentage points in the proportion of production that is kept for subsistence. This means less output is available for sale, which reduces the supply of food and increases food prices. The third implication of the effect of droughts is for the national market. Since the spending elasticity (the variation of spending for goods in response to income variation) of the rural population has generally been high, it was concluded that the fall of the gross agricultural production of 24.4% in 1983 led to the following changes in consumption: food, −24.64%; clothing, −32.94%; furniture and cleaning materials, −26.10%; labeled medicine, −30.25%; rent and taxes, −9.52% and transportation, −30.50%. Finally, drought had an adverse effect on the participation of NEB in the national economy. In 1983 the participation of NEB in the national production of corn, beans and cotton was reduced to half that of normal years (Section 4).

5.4. Strategy to Increase Resilience with Respect to Drought

The impact of drought in NEB is magnified because the region is economically underdeveloped and the majority of the inhabitants are poor and have no means of subsistence when the drought causes them to lose their employment. Economic development, with an appropriate distribution of income and technologically adequate management of the environment, must be promoted if the capacity of the population to resist adverse climatic events is to be improved. The results of this case study suggest the formulation of a strategy for NEB, with three main components:

(1) Elaboration of a plan, to be applied in drought years, aimed at reducing the direct effects of drought on the population. The application of knowledge about the atmosphere and oceans discussed in Subsection 2.5 might enable this plan to be put into action about three months in advance, in the case of extreme droughts, which would allow sufficient time to develop an operational plan, mobilize resources and adopt preventive measures. The principle aims of this plan should be the creation of employment, the provision of food and medical supplies, and the assurance of a good-quality water supply. It should include the design of a program of public works with the following characteristics: maximization of the number of people employed; adequate spatial distribution to avoid the need for large-scale population movements, and implementation of works and measures that could help to serve the established objectives of the existing rural development programs. Thus, the hydraulic and infrastructural works being planned under the

Table 5.1. Summary of results.

Northeast Brazil (NEB)	*Baseline*[b] *1979–1983*	*Scenarios*[a] *Normal year*[c]	*Back-to-back*[d]	*Extreme*[e] *event*	*1-in-10 event*
Indicators					
Population	38.1 million[f]				
Rural population	18.7 million[f]				
Total annual rate of population growth	2.18%[g]				
Literacy rate	50.7%[h]				
Life expectancy	52.6 years[h]				
Infant mortality rate	92.6/1000 live births[h]				
Underemployment rate	47.7%[h]				
Gross product	US$30 782 million[i]				
Percentage of area yearly affected by drought	75%		58%	88%	
Assisted people/year			584 840	2 502 300	
Financial resources spent for the emergency	US$1778 million				
Precipitation	1439 mm[j]		−31%[k]	−57%[k]	
Forecast of drought occurrence in NEB	2–3 months[l]				
Maximum probability of acceptable bean harvest					
Jaicós (Piauí)		0.85[m]		0.78[n]	
Icó (Ceará)		0.88		0.86	
Santa Luzia (Paraíba)		0.69		0.63	
Soledade (Paraíba)		0.33		0.26	
Ouricuri (Pernambuco)		0.40		0.35	
Petrolina (Pernambuco)		0.41		0.32	
Impact of drought on simulated yield index (*MYI*) for beans (% baseline)					
Jaicós	100[p]			31%[r]	118%[r]
Icó	100			61%	100%
Santa Luzia	100			35%	8%
Soledade	100		26%[q]	23%	156%
Ouricuri	100			19%	31%
Petrolina	100		80%	36%	79%
Distribution of production					
Subsistence–beans		39.5%[s]	48.5%[t]		
For sale–beans		60.0%	51.3%		
Subsistence–rice		25.7%	35.3%		
For sale–rice		74.0%	64.4%		
Subsistence–corn		33.0%	43.0%		
For sale–corn		66.6%	57.0%		
Effects of drought on domestic market					
Food				−24.60%[u]	
Clothing				−32.94%	
Furniture and cleansing material				−26.10%	
Labeled medicine				−30.26%	
Rent and taxes				−9.52%	
Transportation				−30.50%	
Impact of drought on bean production					
Uruburetama–Ceará			−12.0%[v]		
Chapada do Araripe–Ceará			−74.0%		
Serrana Norte R. Grandense – R.G.N.			−69.7%		
Alto Pajeú–Pernambuco			−54.2%		
Chapada Diam. Setentrional–Bahia			69.0%		

Table 5.1. continued.

Northeast Brazil (NEB)	*Baseline*[b] *1979–1983*	*Scenarios*[a] *Normal year*[c]	*Back-to-back*[d]	*Extreme*[e] *event*	*1-in-10 event*
Impact of drought on manioc production					
Litoral de Camocim e Acaraú–Ceará			−43.50%[v]		
Agrest Mer.Pernambucano Pernambuco			−31.12%		
Serra Geral da Bahia–Bahia			−34.12%		
Chapada Diam. Meridional–Bahia			−37.55%		
Piemonte da Borborema–Paraíba			−52.54%		
Fortaleza–Ceará			−50.60%		
Impact of drought on rice production					
Mearim–Maranhão			−0.84%[v]		
Itapecuru–Maranhão			4.52%		
Impact of drought on cotton production					
Uruburutama Ceará			−49.67%[v]		
Sertões de Quixeramobim–Ceará			−52.43%		
Sertão de Cajazeiras–Paraíba			−40.70%		
Borborema Potiguar–R.G.N.			−78.70%		
Depressão do Alto Piranhas–Paraíba			−45.00%		
Impact of drought on sugarcane production					
Mata Seca Pernambucano–Pernambuco			14.3%[v]		
Mata Omida Pernambucano–Pernambuco			15.8%		
Tab. S.M. dos Campos–Alagoas			49.4%		
Mata Alagoana–Alagoas			18.5%		
Share of NEB in national production					
Corn		16.60%[s]	15.17%	8.45%	
Beans		31.00%	25.46%	18.33%	
Cotton		36.75%	22.91%	19.90%	

[a] 1979–1980 = back-to-back drought years; 1983 = a year of extreme drought (Section 2).
[b] Unless otherwise indicated.
[c] Defined when necessary.
[d] 1979–1980 unless otherwise indicated (Section 2).
[e] 1983 unless other observations are made (Section 2).
[f] Population in 1985 (Section 1).
[g] Relative to the 1970 and 1980 period (Section 1).
[h] Relative to 1980 (Section 1).
[i] Refers to 1983 (Section 1).
[j] 1849–1984 period for Fortaleza (Section 2).
[k] Precipitation deviations from the average calculated for 30 northern stations (Section 2).
[l] Between two and three months in advance of the event (Section 2).
[m] Precipitation and potential evapotranspiration remain unaltered (Section 3).
[n] Indicates the situation in which precipitation is decreased by 10% and potential evapotranspiration is increased by 10% (Section 3). The authors considered this to be the worst year.
[p] Baseline: 1913–1976 for Jaicós; 1913–1975 for Icó; 1912–1978 for Santa Luzia; 1913–1980 for Soledade; 1914–1978 for Ouricuri and 1965–1984 for Petrolina. The yield index (*MYI*) in each period was taken as base 100 (Section 3).
[q] Yield index (*MYI*) in 1980 in relation to the baseline.
[r] 1914–1915, Jaicós; 1919, Icó; 1918–19, Santa Luzia; 1958, Soledade; 1929–30, Ouricuri; 1975–76, Petrolina. The percentages refer to the yield index values (*MYI*) in these years in relation to the baseline indicated in note p.
[s] Normal year, 1975 (Section 4).
[t] 1980 (Section 4).
[u] Expense elasticity. The fall of gross production in agriculture in NEB in 1983 by −24.4% was taken as the basis for calculations, as well as the elasticities in *Table 4.4*.
[v] Calculated taking the 1975 and 1978 crops as normal and the 1979–1980 production as back-to-back scenario years.

rural development programs could be taken into account. Recognition should also be given to the needs of other works of community interest, including urban infrastructure in towns that are far from the cities. In the future, emergency programs could incorporate greater planning efforts, and more attention could be given to making them compatible with permanent programs.

(2) Development of a set of permanent programs aimed at promoting rural development, to render the agricultural activity and the population less vulnerable to drought.

These programs are included within the Northeast Project (Projeto Nordeste), which is being put into place at the time of writing and will operate until the year 2000. Due to the underdevelopment and poverty that affects about 80% of the rural inhabitants of NEB, great emphasis should be given to programs aimed at directly benefiting this population, especially agrarian reform programs complemented by the provision of services to promote production, commercialization, cooperative efforts and productivity increases. A great effort should be developed to take advantage of the irrigable potential of NEB, estimated at 3 million hectares. Besides the impact on production, irrigation projects would create employment for the low-income population.

The programs aimed at creating employment, increasing production and yield, reducing dependency caused by drought and protecting the environment should be complemented by further action in the fields of education, health and basic sanitation. Such complementary programs help to provide the necessary conditions for improving the quality of life for the population and increasing their working productiveness.

(3) Considering the strong interrelation between the rural and urban sectors, and between agriculture and industry, it is necessary to continue giving support to the industrialization process, which would include continuation of programs that protect agroindustry and the small industries of upcountry areas. Well balanced development of the several rural and urban areas and of the many economic sectors would make the NEB economy more resistant to climatic crises.

Besides the development of specific programs, it is still necessary to improve the several kinds of sectoral governmental interventions, to make the different policies available to the NEB population. The emphasis is on those aimed at the promotion of agrarian reform, technological development, guaranteed prices and credit policies.

An important question that has been the object of attention for years and remains unsolved is the question of managing water resources. NEB lacks a comprehensive policy for the collection, storage and integrated use of water. As one reads the sections of this case study, it becomes clear that the definition of such a policy is a necessary measure to support the actions aimed at reducing the impact of droughts on society and on food production. The rational use of water, at the farm, municipality and state levels and for the whole region, is fundamental for improving the efficiency of this production and consumption factor.

This implies water storage by small dams, to improve the hydric balance for individual rural properties, and building large dams and even providing for long-distance movement and diversion of water among large drainage basins to supply the water needs to those areas in NEB that have water deficits.

Finally, there is evidently a lack of integrated climate–economy–society studies in Brazil. Given the magnitude of climatic impacts on the Brazilian economy and society, especially in Northeastern Brazil, the development of further research in this field is recommended, with particular emphasis on research that will lead to policies that increase resilience to drought.

Acknowledgments

We would like to express our gratitude to Norma Stenzel for help in manuscript preparation and translation, and to Nilton de O. Renno and Lourdes B. Silva for assistance in the preparation of several figures. Dr. V.E. Kousky provided valuable comments and suggestions with respect to Section 2.

Abbreviations and acronyms

BNB	Banco do Nordeste do Brasil (Bank of Northeast Brazil)
BNB/ETENE	Departamento de Estudos Econômicos do Nordeste (Northeast Economic Studies Department of the BNB)
CODEVASF	Companhia de Desenvolvimento do Vale do São Francisco (São Francisco Valley Development Company)
CEPATSA	Centro de Pesquisa Agropecuária do Trópico Semi-Árido (Semi-Arid Tropic Agricultural Research Center)
DNOCS	Departamento Nacional de Obras Contra as Secas (National Works Against Drought Department)
DNOS	Departamento Nacional de Obras de Saneamento (National Sanitation Works Department)
EMBRAPA	Empresa Brasileira de Pesquisa Agrocpecuária (Brazilian Agricultural Research Enterprise)
FJN	Fundação Joaquim Nabuco (Joaquim Nabuco Foundation)
IBGE	Fundação Instituto Brasileiro de Geografia e Estatística (Brazilian Institute for Geography and Statistics)
IPEA	Instituto de Planejamento Econômico e Social (Economic and Social Planning Institute)
IPLAN	Instituto de Planejamento do IPEA (Planning Institute of IPEA)
INEMET	Instituto Nacional de Meteorologia (National Institute for Meteorology)
INPE	Instituto de Pesquisas Espaciais (Space Research Institute)
NEB	Northeast Brazil (the "Nordeste")
MINTER	Ministério do Interior (Ministry of the Interior)
SUDENE	Superintendência do Desenvolvimento do Nordeste (Northeast Development Superintendency)
SUDENE/CPR	Coordenadora de Planejamento Regional da SUDENE (Regional Planning Coordination of SUDENE)

References

Alves, J. (Ed.) (1982). História das Secas (sec. XVII–XIX), *Coleção Mossorenze*, V. ccxxv, 2a.

Banco do Nordeste do Brasil (1973). *Solo e Água no Polígono das Secas*, 4th ed.

Berry, R.A. and Cline, W.R. (1979). *Agrarian Structure and Productivity in Developing Countries*. Johns Hopkins University Press, Baltimore, MD.

Carvalho, J.O. de *et. al.* (1973). *Plano Integrado para o Combate Preventivo aos Efeitos das Secas no Nordeste*. Ministério do Interior, Brasília.

Cavalcanti, I.F.A. (1982). *Um Estudo Sobre Interacões Entre Sistemas de Circulação de Escala Sinoótica e Circulacões Locais*. MSc Thesis INPE-2494-TDL/097, São José dos Campos, S.P., Brasil, 133 pp.

Charney, J.G. (1975). Dynamics of desert and drought in the Sahel. *Q. J. Roy. Meteor. Soc.*, **101**, 193–202.

Correia de Andrade, M. (1970). *Nordeste, Espaco e Tempo* [Northeast, space and time]. Ed. Vozes, São Paulo.

Dillon, J.L. and Mesquita, T.C. (1976). *Attitudes dos Pequenos Agricultores do Sertão do Ceará Diate do Risco* [Small farmers' attitudes toward risk in the inland of Ceará]. Fortaleza.

Duque, J.G. (1973). *O Nordeste e as Lavouras Xerófilas*. Banco do Nordeste do Brasil.

EMATER-RN – *Empresa de Assistência e Extensão Rural do Rio Grande do Norte* [Technical assistance and rural extension enterprise of Rio Grande do Norte]. Yield statistics for micro-regions (n.d.).

Evenson, R.E. and Kislev, Y.V. (1973). Research and productivity in wheat and maize. *J. Pol. Econ.*, **81**(6), 1309–1329, Nov/Dec.

FAO (1979). Yield response to water. In J. Doorenbos and A.H. Kassam (eds.), *Irrigation and Drainage*, Paper No. 33. Food and Agriculture Organization, Rome.

FENAME (1983). *Atlas Geográfico* [Geographic atlas]. Fundação Nacional de Material Escolar, Rio de Janeiro.

Ferraz, J.S. (1925). *Causas Prováveis das Secas do Nordeste* [Possible causes of droughts in the Northeast]. Conference held at the Clube de Engenharia, Dec. 20. 1924, Rio de Janeiro.

Figueroa, M. (1977). *O Problema Agrário no Nordeste do Brasil* [The agrarian problem in Northeast Brazil]. HUCITEC/SUDENE, São Paulo, 272 pp.

Fundaçao Joachim Nabuco (1983). Unpublished field survey data of the 1979–80 drought in Northeast Brazil. Fortaleza.

Furtado, C. (1964). *Formação Econômica do Brasil*. Editora Fundo de Cultura, Rio de Janeiro.

Furtado, C. (1980). *Formaçáo Econômica do Brazil*. Editora Fundo de Cultura, Rio de Janeiro.

Gan, M.A. (1983). *Um Estudo Observacional Sobre as Baixas Frias de Alta Troposfera, nas Latitudes Subtropicais do Atlântico Sul e Leste do Brasil*. São José dos Campos, INPE, Rel. INPE-2595-TDL/125.

Garagorry, F.L. and Porto, E.R. (1983). Agroclimatic modeling as research tool in Brazil. In D.F. Cusack (ed.), *Agroclimatic Information for Development (Reviving the Green Revolution)*. Westview Press, Boulder, CO.

Garagorry, F.L., Porto, E.R., Moita, A.W., Silva, A.S. and Chaib Filho, H. (1982). A study of sorghum production potential in the semi-arid tropic of Brazil. *Tech. Rep. DMQ/A/62*. EMBRAPA, Brasilia.

Gastwirth, J.L. (1974). The estimation of the Lorenz and Gini index. *Rev. of Econ. and Stat.*, **56**(3).

Girardi, C. and Teixeira, L. (1978). Prognóstico de período de seca para o Nordeste Brasileiro. *IAE/CTA Relatório Técnico ECA-06/78*, São José dos Campos, S.P., Brasil, 18 pp.

Gomes Filho, M.F. (1979). *Um Estudo Sobre a Influência do Albedo Diferencial e da Orografia na Circulação Atmosférica: Uma Aplicação para o Nordeste Brasileiro.* MSc Thesis INPE-1640-TDL/015, São José dos Campos, S.P., Brasil, 86 pp.

Hargreaves, G.H. (1974). *Precipitation Dependability and Potentials for Agricultural Production in Northeast Brazil.* Utah State University, Logan, UT.

Hastenrath, S. (1986). Rainfall departure indices for northern Nordeste for the period 1912–1985 (personal communication).

Hastenrath, S. and Heller, L. (1977). Dynamics of climate hazards in Northeast Brazil. *Q. J. Roy. Meteor. Soc.*, **103**, 77–92.

Hastenrath, S., Wu, M.C. and Chu, P.S. (1984). Towards the monitoring and prediction of Northeast Brazil droughts. *Q. J. Roy. Meteor. Soc.*, **110**, 411–425.

IBGE (various dates). *Produção Agrícola Municipal, Vários Anos* [Municipal agricultural production, various years]. Fundação Instituto Brasileiro de Geografia e Estatística, Rio de Janeiro.

IBGE AEB (1981). *Anuário Estatístico do Brasil de 1980* [Brazilian statistical yearbook, 1980]. Fundação Instituto Brasileiro de Geografia e Estatística, Rio de Janeiro.

IBGE AEB (1985). *Anuário Estatístico do Brasil de 1984* [Brazilian statistical yearbook, 1984]. Fundação Instituto Brasileiro de Geografia e Estatística, Rio de Janeiro.

IBGE CAB (1980). *Censo Agropecuário do Brasil, 1975* [Brazil's agricultural census, 1975]. Fundação Instituto Brasileiro de Geografia e Estatística, Rio de Janeiro.

IBGE CAB (1984). *Censo Agropecuário do Brasil, 1980* [Brazil's agricultural census, 1980]. Fundação Instituto Brasileiro de Geografia e Estatística, Rio de Janeiro.

IBGE ENDEF (1977). *Estudo Nacional da Despesa Familiar – Consumo Alimentar* [National study on family expenses – food consumption]. Fundação Instituto Brasileiro de Geografia e Estatística, Rio de Janeiro.

IBGE ENDEF (1978). *Estudo Nacional de Despesa Familiar – Despesas das Famílias* [National study on family expenses – family expenses]. Fundação Instituto Brasileiro de Geografia e Estatística, Rio de Janeiro.

IPEA (various dates). Setor de agricultura e abastecimento. – Basic statistics. Instituto de planejamento econômico e social.

IPEA IRN (1985). Indicadores regionais: Nordeste. Unpublished paper. Instituto de Planejamento Econômico e Social, Brasilia.

Jäger, L. (1976). Monatskarten des Niederschlags für die ganze Erde. *Ber. Deutsch. Wetter.*, **18**, No. 139. Im Selbstverlag des Deutschen Wetterdienstes. Offenbach, West Germany.

Jatibá, J. (Ed.) (1985). *Politica de emprego para e Nordeste.* Recife.

Kousky, V.E. (1979). Frontal influences on Northeast Brazil. *Mon. Wea. Rev.*, **107**(9), 1140–1153.

Kousky, V.E. (1980). Diurnal rainfall variation in Northeast Brazil. *Mon. Wea. Rev.*, **108**, 488–498.

Kousky, V.E. and Chu, P.S. (1978). Fluctuation in annual rainfall in Northeast Brazil. *J. Meteorol. Soc. Japan*, **56**(5), 457–465.

Kousky,V.E. and Gan, M.A. (1981). Upper troposheric cyclonic vortices in the tropical south Atlantic. *Tellus*, **33**, 538–551.

Kousky, V.E. Kagano, M.T. and Cavalcanti, I.F.A. (1984). A review of the Southern Oscillation: oceanic atmospheric circulation changes and related rainfall anomalies. *Tellus*, **36A**(5), 490–504.

Lima Viana, M.O. de (1982). *A Unidade de Produção Agropecuária, Serrões Semi-áridos do Nordeste.* Banco do Nordeste do Brasil S/A–ETENE, Fortaleza.

Magalhães, A.R. (1983). *Industrialização e Desenvolvimento Regional: a Nova Indústrial do Nordeste.* IPEA/IPLAN, Serie estudos para o planejamento, 24. Rio de Janeiro.

Markham, C.G. (1974). Apparent periodicities in rainfall at Fortaleza, Ceará, Brazil. *J. Appl. Meteor.*, **13**, 176–179.

Markham, C.G. and McLain, D.R. (1977). Sea surface temperature related to rain in Ceará, Northeastern Brazil. *Nature*, **265**, 320–323.

Marques, V.S., Rao, V.B. and Molion, L.C.B. (1983). Interannual and seasonal variation in the structure and energetics of the atmosphere over Northeast Brazil. *Tellus*, **35A**, 136–148.

Mesquita, A.R. and Morettin, P.A. (1984). Interannual variations of precipitation at Fortaleza, Ceará, Brazil. *Trop. Ocean Atmos. Newsletter*, **9–10**, September.

Miller, D.B. and Feddes, R.G. (1971). *Global Atlas of Relative Cloud Cover 1967–1970.* US Department of Commerce NOAA and US Air Force (AWS), Washington, DC.

MINTER (1973). *Plano Integrado para o Combate Preventivo aos Efeitos das Secas no Nordeste* [Integrated plan for the preventive combat against the droughts in Northeast]. Organized by José Otamar de Carvalho. Ministério do Interior. Brasília, 267 pp.

MINTER–SUDENE (1967). *Uma Política para o Desenvolvimento do Nordeste*, 2nd ed. Grupo de trabalho para o desenvolvimento do Nordeste, Recife.

MINTER–SUDENE (1984). *Aspectos gerais da agropecuária do Nordeste.* Ministério do Interior, Série Projeto Nordeste, Recife.

MINTER–SUDENE (1985). *Recursos naturais do Nordeste. Investigação e potencial (Sumário).* Recife.

Moura, A.D. and Shukla, J. (1981). On the dynamics of droughts in Northeast Brazil: observations, theory and numerical experiment with a general circulation model. *J. Atmos. Sci.*, **38**(12), 2653–2675.

Nabuco, J. (1983). *A Seca Nordestina de 79–80*, Vol 1. (Visão Geral). Recife.

Namias, J. (1972). Influence of Northern hemisphere general circulation on drought in northeast Brazil. *Tellus*, **24**, 336–343.

Nobre, C.A. and Rennó, N.O. (1985). *Droughts and Floods in South America Due to the 1982–83 El Niño/Southern Oscillation Episode.* Proceedings of the Sixteenth Conference on Hurricanes and Tropical Meteorology. American Meteorological Society, Boston, pp. 128–130.

Nobre, C.A., Yanasse, H.H. and Yansse, C.C.F. (1984). Previsão de secase no Nordeste pelo método das periodicidades: usos e abusos. Second WMO Symposium on Tropical Droughts, *TPM Report Series No. 15*, pp. 113–116.

Nobre, P., Moura, A.D. and Nobre, C.A. (1985a). Planetary-scale circulation anomalies associated with droughts over Northeast Brazil. *Trop. Ocean Atmos. Newsletter*, **13–15**, March.

Nobre, P., Nobre, C.A. and Moura, A.D. (1985b). *Large-Scale Circulation Anomalies and Prediction of Northeast Brazil Droughts.* Proceedings of the Sixteenth

Conference on Hurricanes and Tropical Meteorology. American Meteorological Society, Boston, pp. 125–127.

Oven, R. von (1983). *Analise Comparativa e Recommendacões para os Programas Especiais.* Projeto Nordeste, unpublished paper. Recife.

Ramos, R.P.L. (1974). *Precipitation Characteristics in the North Brazil Dry region.* Atmospheric Science Paper 224. Colorado State University, Fort Collins, CO.

Ramos Viana, P.J. (1981). Nordeste: a 79a. Naçao. *Revista de Economia do Nordeste*, **12**(1), Jan/Mar.

Rao, V.B., Satyamurti, P. and Brito, J.I.B. de (1986). On the 1983 drought in Northeast Brazil. *J. Climatol.*, **6**, 43–51.

Sampaio Ferraz, J. (1929). Sir Gilbert Walker's formula for Ceará droughts; suggestion for its physical explanation. *Meteor. Mag.*, **64**, 81–84.

Sampaio Ferraz, J. (1950). Iminência duma grande seca no Nordeste. *Rev. Bras. Geogr.*, **12**, 3–15.

Santos, A. de O. (1986). *Meridional Penetration of Frontal Systems in South America and its Relation to Organized Convection in the Amazon.* MSc. Thesis, INPE, São José dos Campos, SP, Brazil.

Serra, A.B. (1956). *As Secas do Nordeste.* Servico de Meteorologia, Ministério da Agricultura, Rio de Janeiro.

Shugahara, S. (1983). *Um Estudo Sobre a Viabilidade de Prever Chuvas no Nordeste do Brasil com Antecedência de um a Nove Meses.* MSc Thesis INPE-2628-TDL/111, 91 pp. São José dos Campos, SP, Brasil.

Silva, G.I.S.P. (1984). *Produtividade Agrícola, Pesquisa e Extensão Rural* [Agricultural productivity, research and rural extension]. FIPE/USP, São Paulo.

SUDENE (1984). *Projecto Nordeste: Concepção Básica.* Superintendência de Desenvolvimento do Nordeste, Recife.

SUDENE AQSN (1985). *Aspectos do Quadro Social do Nordeste.* Superintêndencia do Desenvolvimento do Nordeste, Recife.

SUDENE BNB (1986). *Relatório e avaliação do FINOR.* Superintendência de Desenvolvimento do Nordeste, Recife/Fortaleza.

SUDENE PFB (1979). *Produto e Formaçao Bruta de Capital Nordeste do Brasil, 1965–1977.* Superintendência de Desenvolvimento do Nordeste, Recife.

SUDENE PNUD Banco Mundial (1984). *Estimativas de Populaçao do Nordeste – 1985–2000.* Superintêndencia de Desenvolvimento do Nordeste. Recife.

SUDENE PSM (various dates). Precipitation statistics for micro-regions. Superintendência de Desenvolvimento do Nordeste, Recife.

Theil, H. (1971). *Principles of Econometrics.* John Wiley, New York.

Virji, H. (1981). A preliminary study of summertime tropospheric circulation patterns over South America estimated from cloud winds. *Mon. Wea. Rev.*, **109**, 599–610.

Walker, G.T. (1928). Ceará (Brazil) famines and the general air movement. *Beitr. Phys. Frein. Atmos.*, **14**, 88–93.

Waugh, F.V. (1973). *Análise de Demanda e Precos na Agriculture* [Demand analysis and prices in agriculture]. Escola Superior de Agricultura Luiz de Queiróz, Piracicaba.

World Bank (1983). *Uma Avaliação Preliminar dos Programas de Desenvolvimento Rural para o Nordeste.* Washington, DC.

PART IV

The Effects of Climatic Variations on Agriculture in the Central Sierra of Ecuador*

*The authors and editors acknowledge with gratitude the support of the United Nations University in preparation of the Ecuador case study;

Contents, Part IV

List of Contributors

BRAVO, Mr. Raul E.
Proyecto DRI-Guamote
MAG
Ministry of Agriculture
Quito
Ecuador

CAÑADAS CRUZ, Ing. Luis
PRONAREG
Ministry of Agriculture
Quito
Ecuador

ESTRADA, Ing. Washington
PRONAREG
Ministry of Agriculture
Quito
Ecuador

HODGES, Dr. Tom
ARS
US Department of Agriculture
Prosser, WA 99350
USA

KNAPP, Dr. Gregory
Department of Geography
University of Texas
Austin, TX 78712
USA

RAVELO, Dr. Andres C.
Atmospheric Science Department
University of Missouri
Columbia, MO 65211
USA

PLANCHUELO-RAVELO, Dr. Ana M.
Atmospheric Science Department
University of Missouri
Colombia, MO 65211
USA

ROVERE, Mr. Oscar
PRONAREG
Ministry of Agriculture
Quito
Eduador

SALCEDO SOLIS, Mr. Tarquino
Proyecto DRI-Guamote
MAG
Ministry of Agriculture
Quito
Ecuador

YUGCHA P., Ing. Trajano
PRONAREG
Ministry of Agriculture
Quito
Ecuador

Abstract of Part IV

The agriculturally oriented economy of the central Sierra of Ecuador is highly vulnerable to annual variations in climate. Severe drought and occasional frost or flooding can lead to crop failures and economic losses. Recognizing these problems, this case study seeks (a) to describe the types of climatic variations occurring in the central Sierra, (b) to evaluate their possible effects on agriculture in the region, and (c) to consider the responses of indigenous communities to such effects.

Precipitation in the region is affected both by local processes and by global circulation phenomena: the El Niño/Southern Oscillation and the North Atlantic Oscillation. Dry multi-year periods associated with changes in global circulation, such as occurred in the 1960s and late 1970s, tend to be widespread in their effects. Climatic scenarios of short-term and long-term drought and excess moisture were derived for this case study on the basis of historical records of meteorological observations. Later in the study estimates of effects under these scenarios are considered for crop yields, crop suitability and ecological zones.

A questionnaire survey was undertaken to determine local environmental perception, agricultural practices, crop yields and responses to hazards. Yields are low because of low levels of fertilization, although use of organic manure is sufficient to permit reasonably continuous cultivation. In general, potatoes do better in wet years and barley in dry years, the observed barley–potatoes crop mixture thus serving to smooth payoffs over time and minimize risk. Farming is practiced for subsistence, while temporary labor migration brings in cash and is an adaptive measure during drought years. Commercialization of farming is not pursued because of low prevailing prices of agricultural products. In the absence of commercialization, security of subsistence rather than maximization of yields is likely to remain the primary goal of farmers.

To make a preliminary estimate of the effects of variations of precipitation on barley yields, the CERES-Wheat growth simulation model was adapted for barley in the highlands of Ecuador. Data for adapting the model came from field surveys and weather data for the 1984–1985 crop growing season from the Riobamba area. The model was then run with weather data from the Riobamba station for seven "scenario years", with six planting dates from October 15 to December 30 and with long and short fallowing periods. In all years, the model estimated higher yields for middle and late planting dates than for the earliest planting dates. Under the "dry year" scenarios, yield estimates for the early planting dates were near zero (crop failures), owing to flowering and grain filling during periods of extreme drought stress.

An assessment is made of the effects of changes in frost risk on agriculture, employing a scenario of decreased seasonal minimum temperature to map the area potentially at risk. The results indicate that about 8% of cultivated land is at risk, particularly areas with peasant smallhold agriculture. Flats are a special instance of high-risk farming. Apart from the development of raised-field frost-averting technology for the flats, the scope of government proscriptive action for hazardous areas seems limited by the usefulness and fertility of marginal lands, uncertainty about processes setting the upper limit of cultivation, and the already high flexibility of local adaptations including temporary labor migration.

The sensitivity to climatic variations of zones of agroclimatic suitability is also considered. Zones of suitability for barley and potatoes have been delimited for Chimborazo Province based on annual precipitation receipt and frost risk. The changes in the distribution of these zones under different climatic scenarios are quite substantial. Preliminary indicators are that the potential area of good yields would contract, and the contraction may be greater during years with excess rainfall than during dry years.

The broad-scale distribution of terrestrial ecosystem complexes is determined to a large extent by climate and could be altered by long-term climatic change. Classifications that recognize the dependence of natural vegetation on climate provide one means of evaluating the possible impact of long-term climatic change on the geography of major vegetation zones. Regional mapping of the Holdridge Life Zone classification, for Chimborazo Province, is used to test the sensitivity of the distribution of ecosystems to simulated changes of precipitation. The shift of life zones under relatively small changes in precipitation is quite substantial.

The conclusion is that short-term fluctuations in precipitation are of major significance to areas of good yield, while longer-term climatic changes pose threats to more limited areas. We argue against evicting farmers from potentially threatened areas. Farmers are already coping in an appropriate manner with climatic variability, and government policy should not attempt to change peasant agricultural behavior (planting dates, crops, capital inputs, etc.). Rather it should work toward (1) in the short term, improving adaptability to climatic anomalies using traditional technology and (2) in the long term, improving the opportunities of farmers, through decentralized education, health care, road construction and attention to the welfare of migrant laborers.

SECTION 1

Introduction: Vulnerability to Climatic Variations

1.1. Introduction

Late in 1976, Ecuador entered a drought that was to last until 1980. Widely separated stations (Quito, Riobamba, Guayaquil – *see Figure 1.1*) recorded below-normal annual precipitation for all years between 1977 and 1980. According to some estimates (USDA FAS, 1977–1981), hard maize production declined by 16%, barley production by 62% and wheat production by 48% during this period. Effects of the drought were felt in almost every part of the country, with depressed output of feed and livestock as well as cereal crops (USDA FAS, 1981). The FAO World Food Program responded by increasing annual food aid from 947 metric tons in 1975 to 4859 mt in 1978 and 10430 mt in 1979 (PMA, n.d.). Food donations from the USA were also distributed through CARE and Catholic Relief Services – about 8400 mt in fiscal year 1979 alone (USDA FAS, 1979).

Little is known about the effects of the 1976–1980 drought on farmers in Ecuador. The previous major drought (in 1968) contributed to a permanent migration of farmers from affected areas, particularly from Loja in the southern Sierra of Ecuador (Duverneuil, 1983; Gondard, 1983). However, the absence of an effective agricultural extension service or program for gathering data from small farmers in the countryside contributes to our ignorance, as has the scarcity of studies of peasant technology and adaptation in the equatorial Andes. The present case study represents an interdisciplinary effort to improve our understanding of the effects of climatic variability on agriculture in the central Sierra of Ecuador.

1.2. The Design of the Study

1.2.1. The research problem

Ecuador is a diverse land with extensive areas of lowland tropical wet-and-dry and tropical wet climates, in addition to a lengthy if slender highland area in the

Andean chain extending from the Colombian border in the north to the Peruvian border in the south. About 46% of the nation's economically active population is engaged in agriculture. Perhaps half of the highland farmers are ethnically Indian (Quichua) smallholders, producing for family subsistence as well as for sale. These farmers have only recently been divorced from various forms of non-capitalist labor relationships with haciendas (*see* Subsection 1.6.2), due in part to a partial land reform and in part to the development of new labor-saving and capital-intensive hacienda forms oriented to the growing urban markets. Ecuadorian indigenous farmers are thus both ethnically distinctive and socioeconomically marginal.

The valley interiors of the central Sierra display a variety of regional climates. Some are moist valleys, some are drought-prone according to their disposition to prevailing rain-bearing winds (*see* Section 2 below). Farmers in the dry valleys can escape drought by location higher up the valley sides because, in general, rainfall increases with elevation. But at these higher locations temperatures are generally lower, and there is a greater risk of frosts that can damage or kill crops. Facing a higher risk of drought in the valley bottoms and a higher risk of frost on the higher valley sides, farmers need to consider what crops or combinations of crops represent the best means of combating the effects on agriculture of year-to-year variations in rainfall and temperature. Such is the setting for this case study. It offers the opportunity for us both to learn from the adaptive strategies of indigenous farmers and to consider how these might be improved.

1.2.2. The research approach

The case study focuses specifically on the impact of climatic variations on indigenous farming communities at high elevations (2800–3800 m). The problem is approached on four different scales of analysis: the Ecuadorian highlands as a whole, the central region of these highlands, the province of Chimborazo and the peasant communities of Chauzan and Galte (in Guamote *canton* of Chimborazo province) (*Figure 1.1*). The aim has been to focus closely on a "laboratory" region of manageable size, but it should be remembered throughout that statements made on one scale or level of generalization may not always be true on other scales. The study communities of Chauzan and Galte were chosen on the advice of Proyecto Desarollo Rural Integral de Guamote (DRI-Guamote). The DRI-Guamote project was initiated in October 1981, and subsequently published a three-volume report on the region (SEDRI, 1983). Two field workers from the DRI-Guamote project, who led a farm study for DRI in 1981–1982, also conducted a comparable survey for the present study in 1984, with additional questions related to agricultural terminology, agricultural practices and the like.

Each section of this study reports results from one of the four scales of analysis. As in the case studies reported elsewhere in this volume, attention is given first to the selection of climatic scenarios. Changes in precipitation and temperature under these scenarios are then used as inputs to a range of impact models or climatic indices – in this case, a crop growth model for barley, and

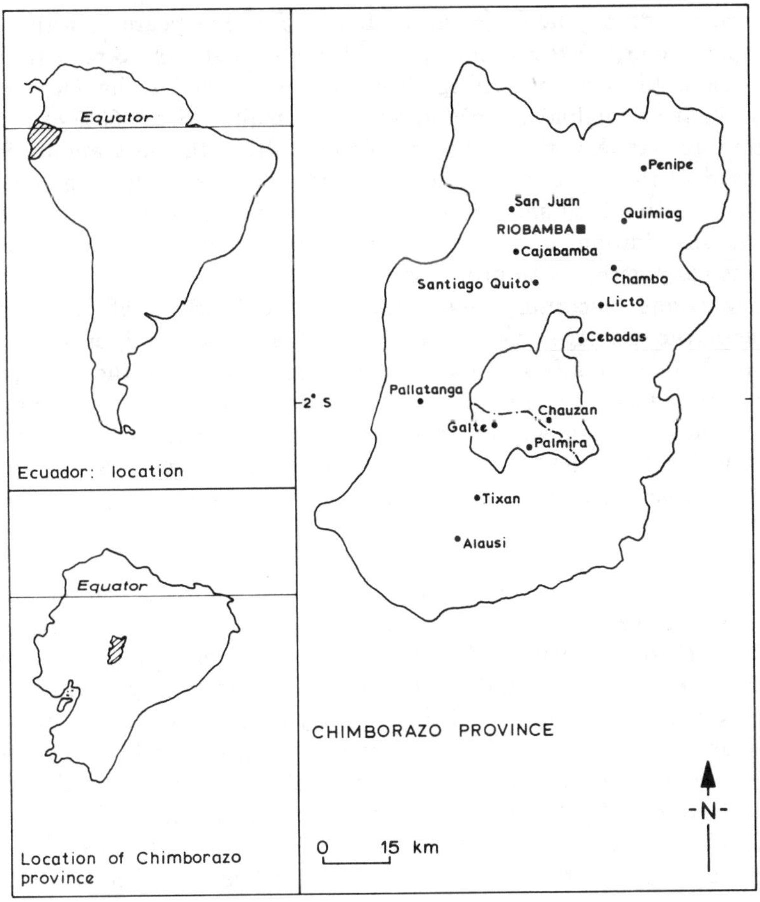

Figure 1.1. Location of Chimborazo Province in Ecuador.

indices of frost risk and of ecological life zones. The climatic scenarios adopted here are intended to be compatible with those in the other case studies, and thus include (a) a recent and extreme drought event, (b) the 90% and 10% rainfall events and (c) the "back-to-back" occurrence of two successive drought years. Selection of the climatic scenarios is discussed in Section 2.

1.3. The History of Impacts from Drought, Excess Rainfall and Frost

The historical record, both oral and written, indicates that drought has long been a common phenomenon in the central Sierra of Ecuador.

The *Libros del Cabildo de Quito*, which relate to the period 1644–1835 for the northern Sierra, reveal the kinds and frequencies of climatic anomalies experienced in the highlands (Bonifaz, 1980). For 187 years in which the data are extant, 57 refer to excess rainfall, 43 refer to drought and 3 refer to frosts.

The annotations concerning drought refer only to the October–March period, the normal highland growing season for maize. Some 63% of the annotations occur in January and February alone. Although most annotations refer merely to drought or fears of crop loss without specifying human deaths, in several cases actual food shortages or deaths are mentioned: in 1651, 1776, 1809 and 1815. In addition, in the years 1693, 1700, 1719 and 1791, epidemics are reported in conjunction with droughts.

Annotations concerning excess rainfall occur every month of the year. In many cases the damage from excess rainfall affected structures only (roads, houses, etc.); in others it is difficult to determine whether the damage was to structures or to farms or to both. If we exclude those annotations that explicitly refer to structural damage, we are left with 44 annotations, of which 55% occurred during the June–September (dry "summer") period. Although in many cases crop damage is mentioned, in only one case is famine explicitly mentioned – in 1724 when frost, rather than excess rainfall, was responsible for most of the damage.

In modern times, oral tradition as well as the Quito rainfall record indicate that 1926 was a major drought year with widespread hunger ("La Hambruna"). Droughts in 1968 and 1976–1980 stimulated international aid, but the extent of impact in terms of scarcity and mortality is uncertain largely because the Ecuadorian government in general and the Ministry of Agriculture in particular have traditionally placed a low priority on monitoring agricultural production. Retail food prices were published for Quito, Guayaquil, and Cuenca until March 1982 by INEC (*Indice de Precios al Consumidor*), but the quality of the data is uncertain. Production data are collected by the Ministry of Agriculture on the basis of telephone calls to regional offices. These data are published in Ecuador and reported to FAO, though (particularly for food staples) they are considered unreliable (USDA FAS, 1985). Similar information from the Foreign Agricultural Service of the United States Department of Agriculture is largely unverified. Efforts are now being made to determine crop production by a sampling procedure (INEC, 1984).

Thus, it is difficult to determine the effects even of the most recent (1976–1980) drought. There was a substantial reduction of highland food production, but the exact quantities and geographic distributions of shortages remains in doubt. The consumer price index in Quito rose 25% from April 1976 to April 1979 (INEC, 1976–1979). During the same period the price of peas rose 88%, sweet corn (choclo) 23% (April–September averages), beans 13%, broad beans 46%, and bread (which is price-controlled) only 5%. The (admittedly unreliable) USDA data indicate that the harvested area of wheat in 1979 was 52% less than in 1976, while yields remained roughly unchanged. The barley-harvested area decreased by 51% and yields declined by 21%. Maize yields decreased by 18% while the harvested area remained unchanged; potato yields and area harvested remained constant.

1.4. Existing Responses at the National and International Levels

Recent responses by national government and international agencies to the effects of climatic variability have included the following:

Government involvement in food sales. The ENPROVIT stores were established during the Roldos administration (1979–1981) to market staple food, especially grains and pulses, at a low price. However, the stores maintain low inventories and often run out of goods. On weekends, long lines are common and the types of food stocked (e.g., rolled oats) may not match peasant food preferences (Di Vincenzo, 1984).

Agricultural extension and small-farm aid. While there is no national agricultural extension service in Ecuador, a number of separate agencies have developed rural projects, e.g., (a) FODERUMA, a rural development agency of the Central Bank, which implemented a number of local projects in the early 1980s and subsequently created its own cadres of development experts, but which has reduced its activity since the Febres Cordero government came into office in 1985; (b) SEDRI (Secretaria de Desarollo Rural Integral) an agency for integrated rural development charged with the task of coordinating personnel from other agencies facilitating agricultural extension in regions designated for special attention.

Ecuadorian nongovernmental organizations. CAAP (Centro Andino de Acción Popular) has developed a program for studying and providing aid to small farmers in selected project sites (Field and Chiriboga, 1984). This project, with substantial participation by anthropologists, geographers and historians, is attempting to discover forms of development appropriate to Andean peasants and respectful of their traditions.

Department of Civil Defense. This agency spent 72.7 million sucres (69 sucres = US$1, 1985) on emergency projects between 1979 and 1983. Over half of these funds went to construction of wells and to the rehabilitation and improvement of canals. The funds came from external sources and involved food-for-work programs.

World Food Program (FAO). This began operations in Ecuador in 1965 and has disbursed substantial quantities of food. Projects have generally involved food-for-work, reforestation or construction projects, and are administered by various Ecuadorian agencies. There have also been four emergency food programs responding to the 1968 drought, the 1975 earthquake, the 1978–1979 drought (especially in the southern highlands) and the torrential rains in 1982–1983.

Food aid from donor countries. The United States has provided Ecuador with favorable credit terms for the importation of wheat. As a result of this and national policy to favor the noodle industry and keep urban food prices down, wheat imports to Ecuador have increased from 118 000 mt in 1973 to an estimated 380 000 mt in 1985 (USDA FAS, 1977 and 1985). It is clear that if the United States were to pull out of this market, other wheat exporters, such as Australia, would step in with almost equally favorable terms. The result, of course, has been to discourage Ecuadorian wheat production and encourage substitution of wheat for maize in the diet, discouraging rural maize production. Production of wheat, barley and soft corn in Ecuador all peaked in the late 1960s and have decreased since.

Until about 1975, United States aid agencies provided Ecuador with free food grains. This program was discontinued, in part because of unfortunate side effects on local grain production. In recent years US food aid has been channeled through Catholic Relief Services and CARE in food-for-work programs. Aid through these programs reached a peak in the drought years of 1978 and 1979.

From July 1983 to July 1984, 36% of food aid (by weight) came from the World Food Program and 29% from the United States (AID, CARE, Catholic Relief Services). About 19% came from the European Economic Community (excluding Italy), 13% from Italy, and 3% from other sources (World Food Program, 1985).

1.5. Existing Responses at the Farm Level

To succeed and to survive most farmers have, over the centuries, developed a set of strategies for responding to climatic anomalies. Andean farmers may have been particularly successful in this respect since there is no oral tradition of famine in the Inca realm (Murra, 1985). In the 187-year record analyzed by Bonifaz (1980), only in two of the 37 drought years is famine explicitly mentioned; in eight other years disease is mentioned, which may be connected to harvest failure. In only one of the 43 excessively wet years in this period is famine explicitly mentioned; in one other year epidemic is mentioned without directly linking disease to hunger. The last true famine in highland Ecuador occurred after the 1926 drought.

This is not to cast doubt on the impact of climate anomalies on production, but rather the impact of production losses on hunger – a matter that depends crucially on forms of peasant storage, sharing and cooperation, as well as on market and redistributive reactions to food shortage.

According to one recent assessment, "hunger does not appear to be an apparent problem in Ecuador – even during the 'Niño' floods food was still plentiful. It was just the transportation and distribution which were the main problem" (USDA FAS, 1985). This suggests that peasant farming systems in Ecuador are quite resilient to climatic anomalies as a result of a set of strategies that includes intercropping, cultivation of several fields in different locations, low capital investment in farming, temporary labor migration, reliance on both livestock and production, and the use of networks of family, neighbors and friends for aid in adversity.

A further adaptive measure, particularly in Chimborazo Province, has been the formation in recent years of village communes (*comunas*, which may or may not control communal land), and cooperatives that exploit some land communally while leaving other land under the control of members. In the cooperatives families accept and abide by the decisions of the directorate and receive in kind a small wage (equivalent to US$0.66 a day in 1985) plus the right of grazing their cattle in the communal paramolands (high grasslands). The profits generated by the production of the cooperative are contributed to the community fund and are reinvested in farm equipment and other inputs for agricultural production.

These societal adaptations are in addition to the technical adjustments that farmers may employ to reduce hazards (irrigation, drainage, mounding, use of staggered planting dates, etc.).

1.6. Background to the Study Region

1.6.1. The Ecuadorian Andes

Ecuador possesses a relatively short and narrow, but densely populated, portion of the Andean chain. The Andes here consist of an eastern cordillera, or chain (*Cordillera Real*), and a western cordillera that enclose a series of lower basins (the inter-Andean basins) separated from each other by lateral ranges (*nudos*). The inter-Andean basins drain either westward to the Pacific or eastward to the Atlantic via the Amazon system.

Terminology varies, but the Ecuadorian Andes (or *Sierra*) may be thought of as divisible into two parts: the southern Sierra (Loja, Azuay and Cañar provinces) with a population of about 1 million in 1982; and the central and northern Sierra with 2.7 million, including the capital city, Quito (population 900 000), once the northern capital of the Inca empire. Many although not all peasants (*campesinos*) of the central and northern Sierra are Quichua-speaking people, commonly called Indians (or in government parlance, "indigenous" people).

At one time the Indian population in Ecuador was sufficiently numerous and well-organized to challenge the Inca empire. Over the centuries, however, it has been increasingly "marginalized" by the intrusion of Inca, Spanish colonial, and post-colonial governments. Until very recently, most development programs have assumed that Indians are backward people, with an inadequate technology and poor knowledge of their environment, who need to be modernized and incorporated into a modern "mestizo" nation-state. It has become increasingly clear, however, that indigenous science is both sophisticated and appropriate (Knapp, 1984), and that nonmestizo ethnic identity and values can be consistent with prosperity (Salomon, 1985).

This is not to say that peasant farmers do not face problems of an environmental nature or that governmental policy cannot contribute to rural well-being. A search for such policy is a fundamental aim of this case study.

1.6.2. Chimborazo Province

Chimborazo Province covers an area of almost 8000 km^2 (*Figure 1.1*). Elevations vary from under 2000 m on the valley floors to 6675 m (the volcanic peak of Chimborazo). The upper limit of cultivation lies at around 3800 m.

Rainfall varies markedly with elevation and topographic position. The driest areas are valley bottoms (around Riobamba and Guamote), not only because of the rain shadow effect, but because local circulation cells occur with stagnant descending air in the valley interiors and rising air at valley margins. Precipitation thus increases away from the valley centers and reaches a maximum on the slopes of the volcano Chimborazo and on the higher slopes of the eastern cordillera.

Most soils are derived from volcanic ash. Between 2500 m and 3500 m, one typically finds a thick layer of loessal volcanic ash (called *cangahua*) with an overlying layer of more recent ash deposits. At lower, drier elevations, the slopes have often been subjected to extreme water and wind erosion and the underlying cangahua is often exposed and gullied. The soil under these conditions is thin and sandy and thus more drought-prone. Elsewhere, where moister conditions prevail, the cangahua is more intensively weathered, forming a deep, black prairie-type soil, sometimes over hardpan.

The population of Chimborazo Province was about 330 000 in 1982, with 22% (73 000) in the main town, Riobamba, 16% in other similar towns and 62% in the surrounding countryside.

In the northern and eastern parts of the province, most land is still in large farms (*haciendas*) where it is used for ranching and dairy operations, as well as for some grain and potato production. In the west and the south, there has been considerable redistribution of hacienda land to peasants, who practice a combination of subsistence agriculture and commercial production. Despite this redistribution, however, land ownership in the province was still highly unequal in the 1970s, with 85% of the farms being minifundias of under 5 ha and controlling only 16% of the farm land (PRONAREG-ORSTOM, 1979a).

About one quarter of the land area is cultivated, the principal crops being maize (especially at lower elevations), barley, potatoes, wheat and broad beans (*habas*). Onions are the main cash crop, as well as barley for beer (*Table 1.1*).

Maize, which requires higher temperatures and tolerates dry conditions better than potatoes, is the major crop at lower elevations (below 3300 m). It also requires less labor per calorie than potato production at equivalent levels of soil fertility (Knapp, 1984).

At higher elevations, potatoes, barley and broad beans are grown. Local wheat is supplemented by purchased wheat products. Other highland crops are quinoa (*Chenopodium quinoa*), oca (*Oxalis crenata*), ullucu (*Ullucus tuberosus*), and tarwi (*Lupinus mutabilis*). Maturation times for potatoes and barley are closely related to air temperature, which in turn is closely related to altitude. Barley requires at least 10 months to mature at 3200 m. Potato maturation time increases from 5–6 months at 2800 m to 9–10 months at 3800 m.

Planting times tend to be earliest at the highest elevations. Sowing at 3800 m may occur in August, while below 3000 m it may occur as late as

Table 1.1. Areas and yields of principal crops, Chimborazo Province, 1983.

Crop	*Area sown* (ha)	*Percent irrigated*	*Yield* (kg/ha)
Soft corn	17 900	28	400
Barley	15 300	12	284
Potatoes	9 500	25	7085
Wheat	9 100	32	750
Broad beans (*habas*)	5 100	6	382 (dry) 1200 (fresh)

Source: INEC (1984).

February (Cañadas, 1983). Multiple planting times (*suertes*) also help to limit effects of early season frosts.

Cattle and sheep are a significant component of small-farm agriculture in Chimborazo Province. Most pasture is in the high grassland (*paramo*) above the limit of cultivation, but some lower land is also grazed, including most of the land in large haciendas and much of the land on frost-prone flats. Animals also feed along roadsides, in fallowed fields, in the stubble of recently harvested fields, and along field borders. Here fodder includes field weeds, unused parts of crop plants (e.g., maize tassels and stalks), grass harvested from natural pastures, and sown fodder crops (alfalfa, ryegrass) (Kirkby, 1981). The animals provide valuable dung for fertilization, traction for plows, a source of cash income and a form of savings for emergencies. Wool and milk are used for subsistence as well as for sale. Costs may be low due to complementarity of feed production with agriculture, reliance on fallow and natural pasture, and the use of children to herd the stock. In addition to cattle and sheep, pigs and guinea pigs are an important source of protein and manure.

Farming in Chimborazo is based on an affordable and simple life style that uses both human and animal power. While there is some mechanization on large farms, elsewhere crops are normally cultivated and harvested manually. The lack of readily available credit and the uncertainty of cash income make farmers wary of spending money on purchased inputs such as chemical fertilizers, pesticides, herbicides and machinery.

1.7. The Galte–Chauzan Area

The county (*canton*) of Gaumote (population 25 000), which lies in the center of Chimborazo Province, faces a particularly extreme set of environmental conditions. It is both high (lying mainly above 3000 m) and dry, with the result that agriculture is prone to losses both from drought *and* frost.

Farmers in the county are especially disadvantaged. Of the population over 10 years of age, 74% are illiterate, and the infant mortality rate is 157/1000 (SEDRI, 1983). Only 17% of individual land holdings have irrigation.

The value of all agricultural and livestock production in the county in 1982 was estimated to be US$7m or US$280 per capita, of which 30% was from livestock (SEDRI, 1983). In 70% of households one or more members migrate temporarily to work in the larger cities in order to boost household income, which is limited mainly by farm size. Most farmholdings are of 1 to 3 ha, and the average has 3 cattle, 12 sheep, 11 guinea pigs, and 5 chickens (SEDRI, 1983).

Within Gaumote county two communities, composed of several hundred farm households, were chosen for detailed investigation: Galte and Chauzan (*Figure 1.1*). The region of Chauzan was once part of the Totorillas hacienda, but now includes the peasant organizations of Chauzan–Totorillas, Atapo–Santa Cruz, and Santa Teresita. Of these, Santa Teresita has the least effective social organization and is the most depressed farming organization in the study area. The area of Galte was once part of a state (social security) hacienda. It includes the peasant organizations of Galte–Laime, San Francisco–4 Esquinas, and San Miguel de Pumachaca.

1.8. Structure of the Study

This part of the central Sierra of Ecuador is clearly characterized by extreme marginality: it lies near the limit of frequent killing frost, yet is also prone to severe drought; the farms are small and undercapitalized; the Indian farmers are isolated from the centers of political influence in Ecuador and are not strongly organized in communes. *Prima facie*, these communities are thus especially vulnerable to variations of climate.

In the next section the most significant climatic variations will be analyzed. Changes in precipitation and temperature under particular climatic scenarios will then be input to simple impact models in order to assess their effects on Indian farming and to consider appropriate responses to mitigate such effects.

SECTION 2

Selection of the Climatic Scenarios

2.1. Introduction

In this section we describe the most important characteristics of Ecuadorian highland climate and construct plausible scenarios of short-term climatic variability and long-term climatic change for the central highlands. These scenarios will then be used in subsequent sections of this case study to estimate (a) the effects of climatic variability on barley yields (Section 4), and on crop zones (Section 6), and (b) the possible effects of long-term climatic changes on ecological zones (Section 7). We will be focusing here on variations in precipitation. Variability of temperature and frost risk will be discussed in more detail in Section 5.

2.2. Climatic Characteristics of Ecuador

Ecuador is located within the zone of influence of the intertropical convergence zone (ITCZ) and is therefore affected by changing amounts of uplift and rainfall and humidity characteristics as the ITCZ moves northward and southward during the year. In the highlands, the influence of the ITCZ and associated lowland air masses is supplemented by local convectional processes affected by solar position and topography (Frere *et al.*, 1975; Johnson, 1976; Pourrut, 1983).

In general, the Amazon basin portion of Ecuador is always under the influence of moist tropical air masses and has no pronounced dry season; usually all months receive at least 250 mm of precipitation. The coast is strikingly dissimilar. In April, May and June the ITCZ retreats to the north, to approximately 10° N. Then dry air masses from the south influence the coast until December. Dryness may be further enhanced by atmospheric flow from the east (below the 300 mb level), placing the region in the Andean rain shadow. Another factor is the temperature of ocean water off the Ecuadorian coast. From June to December much of the Ecuadorian coast is affected by the cool Peru or Humboldt Current flowing from the south, which contributes to aridity. Beginning at the end of December, warmer waters from the north penetrate southward, displacing the Peru Current to the west and promoting convection.

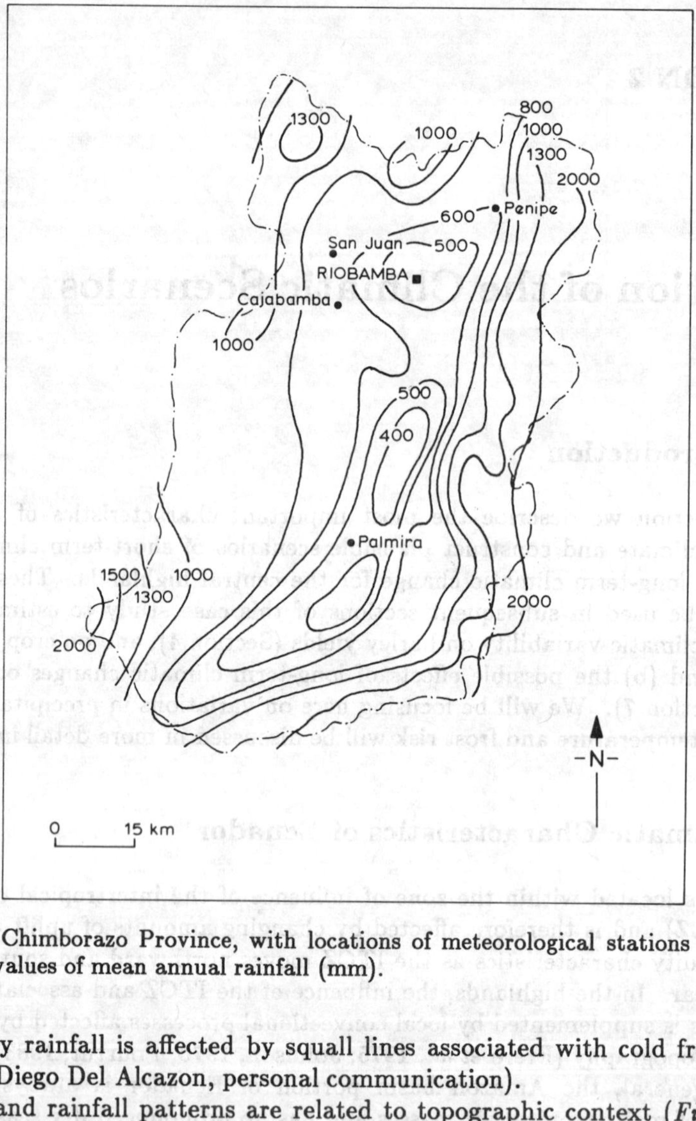

Figure 2.1. Chimborazo Province, with locations of meteorological stations mentioned in text and values of mean annual rainfall (mm).

Occasionally rainfall is affected by squall lines associated with cold fronts from the south (Diego Del Alcazon, personal communication).

Highland rainfall patterns are related to topographic context (*Figure 2.1*). Where wind patterns are favorable and topographic barriers do not intervene, lowland conditions may influence rainfall in the highlands. Where topographic barriers do intervene, orographic rainfall on the outer flanks gives way to rain shadows on the inner flanks and valley interiors. Localized circulation cells may also be set up with uplift on basin flanks and subsidence in the basin interiors. Convection is an important cause of rainfall at stations such as Quito, where the highest percentage of rainfall occurs from noon to 4 pm when temperatures are highest (Ferdon, 1950).

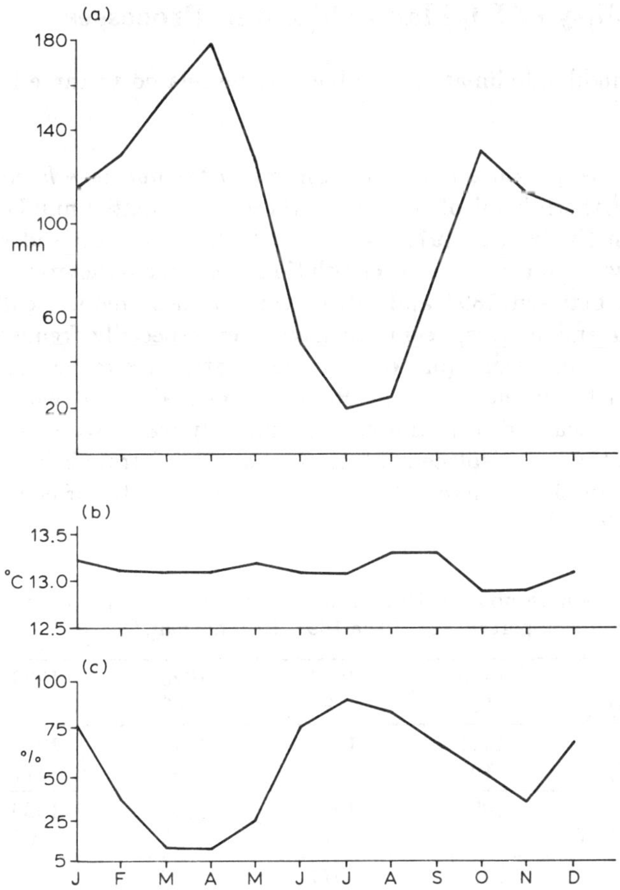

Figure 2.2. Seasonal climatic characteristics of Quito: (*a*) precipitation (average for 1891–1983), (*b*) temperature (average for 1949–1968), (*c*) probability of frost (data for 1949–1968).

Seasonal patterns of highland rainfall differ from those in adjacent lowlands and are probably correlated with seasonal variations in wind directions and strengths, and with variations in insolation due to varying altitude of the sun (Ferdon, 1950). Data for Quito and Riobamba show a double-peaked pattern of seasonal rainfall, the peaks coinciding with the equinoxes (high-sun months) (*Figure 2.2*). The dry season associated with the June solstice (summer, *verano*) is more intense and prolonged than that associated with the December solstice (little summer, *veranillo*). This difference may be due both to the greater influence of the high-pressure cell over Peru and Bolivia after the June solstice, and to the dryness of the adjacent western coastal air masses. (In highland Peru and Bolivia, there is a single dry season about the June solstice).

2.3. Variability of Highland Climate: Processes

Variability of highland climate may plausibly be related to three kinds of global events:

(1) *Variations in pressure and wind systems in the northern hemisphere.* During the northern hemisphere winter, there are a limited number of modes of circulation (Kutzbach, 1970; Namias, 1963). Van Loon and Rogers (1978) have shown that only four circulation modes characterized the bulk of Januaries between 1899 and 1975. One of these modes, called "GA" by Van Loon and Rogers, seems to have been especially frequent during the 1960s. GA conditions (or Northern Oscillation Extremes) involve a polar jet stream to the north of the Rockies, unusually cold winters in eastern North America and in Scandinavia, unusually warm winters in Greenland, and a southward displacement of the equatorial trough in Andean South America. In Quito these GA seasons are significantly drier than other seasons (*Table 2.1*).

Table 2.1. Northern Oscillation Extremes (GA years), 1891–1980, and January–April Quito precipitation as a percentage of the 1891–1979 normal.

GA year	1916	1924	1926	1929	1940
Precipitation (%)	114	90	38	124	73
GA year	1941	1942	1948	1953	1958
Precipitation (%)	105	84	91	117	100
GA year	1959	1960	1963	1966	1967
Precipitation (%)	62	74	113	65	76
GA year	1969	1977	1979	1980	Mean
Precipitation (%)	110	64	64	83	87

Sources: Knapp (1980, 1984); precipitation data from the Instituto Nacional de Meteorologia e Hydrologia, Quito.

(2) *Variations of insolation due to dust content of the atmosphere.* Insofar as local precipitation is convective and driven by insolation, changes in atmospheric dust content (e.g., due to volcanic eruptions) may be expected to affect local temperature and rainfall. Variations in the solar constant might be expected to have a broadly similar effect. However, until now no convincing evidence has been presented to demonstrate the influence of either effect on climate in Ecuador.

(3) *Variations associated with the Southern Oscillation and El Niño.* In general, the atmospheric circulation in the Pacific involves a high-pressure cell in the East Pacific and a low-pressure cell in the West Pacific. Pressure differences drive the easterly trade winds. The magnitude of pressure differences between East and West Pacific varies over time, this variation in pressure being known as the Southern Oscillation. One extreme state is a relatively weakly developed West Pacific Low coupled with a relatively

weak East Pacific High. Such a state, if pronounced and long-lasting, may be called the Southern Oscillation Extreme; extreme years are listed in *Table 2.2* together with dates of inception of cycle extremes.

Under normal conditions, strong trade winds in the Pacific are associated with an accumulation of warm surface water in the tropical West Pacific, while in the East Pacific cool subsurface waters are close to the surface. During a Southern Oscillation Extreme, weakening of the easterly trades may result in a return flow of warm surface water from the West Pacific to the East Pacific. Widespread warm water in the East Pacific eliminates the normal temperature inversion along the central South American coast and leads to torrential rains – the phenomenon of El Niño (Caviedes, 1982).

Table 2.2. Southern Oscillation Extremes, 1939–1983.

		Annual precipitation (% of normal)		
Initiation dates[a]	*Extreme years*[b]	*Guayaquil*[c]	*Lambayeque*[d]	*Quito*[e]
May 1939	1940	97	66	86
	1941*	148	57	88
September 1945	1946	68	–	80
January 1951	1952	51	36	100
	1953*	162	111	134
November 1956	1957*	144	254	96
	1958	133	88	83
February 1962	1963	60	95	102
September 1964	1965*	122	239	103
	1966	89	72	80
June 1968	1969	87	122	111
November 1971	1972*	119	376	114
October 1975	1976*	169	57	84
	1977*	56	170	74
February 1982	1983*	–	–	126

[a] The month preceding an extreme year when the Southern Oscillation Index (or a normalized proportion of the long-term monthly mean) began to fall.
[b] Years (1) with at least one month with a Southern Oscillation Index one standard deviation below normal; (2) with no months with a Southern Oscillation Index one standard deviation above normal; and (3) preceded by another extreme year or a year including one month with an index one standard deviation above normal. El Niño years are starred.
[c] 1915–1980 mean = 1094 mm.
[d] 1929–1977 mean = 19.4 mm.
[e] 1891–1983 mean = 1235 mm.
Source: Historical Southern Oscillation Index data supplied by Lockwood (personal communication); manuscript precipitation schedules in Quito and Lima.

However, Southern Oscillation Extremes do not necessarily always lead to El Niño events. Since 1940, there have been 15 Southern Oscillation Extreme years, but only 8 of these have been associated with increases in rainfall of 40% or more in coastal Ecuador (Guayaquil) or Peru (Lambayeque) (*Table 2.2*). These 8 years are commonly identified as "true" El Niño years in the literature, although variations occur depending on varying definitions of the phenomenon (Smith, 1983).

El Niño has a limited effect on adjacent highland areas. Quito precipitation deviations are listed in *Table 2.2.* There is no evidence that Southern Oscillation Extremes or El Niño years are different on average from normal years at that station. The most recent El Niño year, however, was clearly wetter than normal at Quito. At certain stations in the highlands with a strong exposure to western air masses, record extremes of precipitation were recorded during this event.

2.4. Variability of Highland Climate: Patterns

We will focus on the historical records of three stations to help select scenarios of climatic variations. Quito Observatory (which we will call Quito) is outside Chimborazo Province, but has the longest precipitation and temperature record in Ecuador. Riobamba Airport (which we will call Riobamba) is centrally located in Chimborazo Province. Palmira is a station close to the Chauzan–Galte study ara of Chimborazo Province (*Figure 2.1*).

Correlations of annual rainfall between the three stations are reasonably high, and the most recent drought (1976–1980) was characterized by similar low rainfall amounts at all three stations (depressions of 15–20% below normal). This is not to imply that identical processes or identical air mass influences operate in all three cases. In general, we will use the Quito record to study long-term variations, the Riobamba record to construct scenarios for Chimborazo Province as a whole and the Palmira record to construct scenarios for the Chauzan–Galte area.

2.4.1. Variations in temperature

Figure 2.3(a) indicates mean annual temperatures during the last 90 years in Quito. There appears to have been a warming throughout the period, probably due to the urban heat island effect. Quito's population increased from 400 000 in 1940 to about 1 000 000 in 1984, and this growth was associated with an increase in urban area, industrial development and transport. Unfortunately, we lack long records from nearby rural areas against which to compare this possible urban effect.

Frost frequencies vary considerably through the year (*Figure 2.2*). Since most frosts in highland Ecuador are radiation frosts, cloud cover associated with precipitation lowers frost frequency. Wet seasons have low frost risk, while the prolonged summer dry season is the period of greatest risk. Between 1891 and 1982, over half the 882 recorded frosts occurred in the summer (June–September).

Long-term trends in absolute minimum temperatures are indicated in *Figure 2.3(b)*. Frosts are seldom severe, and in a few years no frosts were recorded.

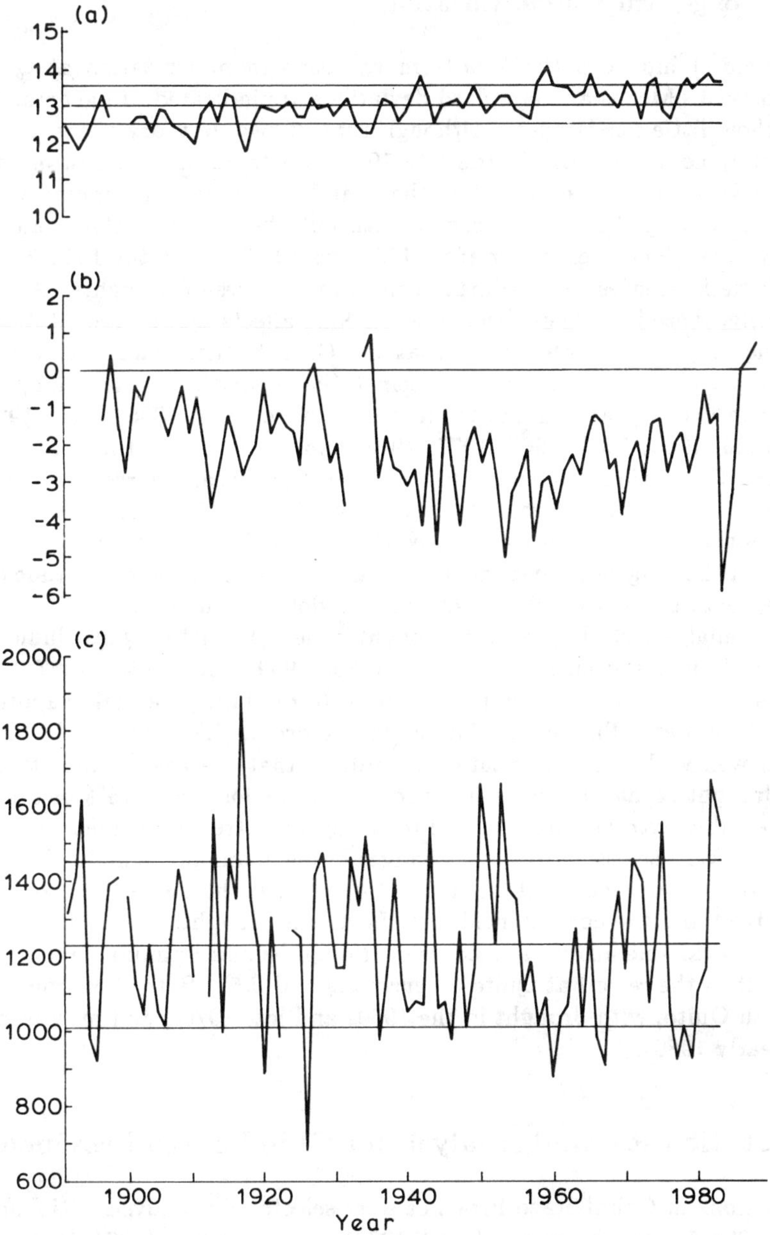

Figure 2.3. (*a*) Annual mean temperature, (*b*) annual absolute minimum grass temperature and (*c*) annual precipitation at Quito (1891–1983).

There is no clear trend of frost severity, nor can frost severity be clearly related to mean annual temperature.

2.4.2. Long-term rainfall variation

Figure 2.3(c) indicates the long-term variation in precipitation at Quito from 1891 to 1983. Means and standard deviations are indicated. In general, very dry years show little persistence. Although runs of two very dry years in a row are moderately common, only in the late 1970s did there occur a run of three consecutive very dry years. On the other hand, the decade beginning in the late 1950s was clearly drier than normal, with only two years wetter than the long-term average. The average for 1958–1967 was 14.4% below the 1891–1983 mean.

Some inferences may be made concerning causes of droughts. As indicated previously, there is little evidence that El Niño affects Quito precipitation, inducing either drought or excess precipitation. On the other hand, the occurrence of Northern Oscillation Extremes do seem to be related to Quito droughts (*Table 2.1*). The driest years on the Quito record were 1894, 1895, 1920, 1922, 1926, 1936, 1946, 1960, 1966, 1967, 1977, 1978, and 1979. Six of these 13 years were Northern Oscillation Extremes. By far the driest year on record, 1926, was a Northern Oscillation Extreme. Only 21% of all years (1891–1979) were Northern Oscillation Extremes. On average, Northern Oscillation Extreme years are 13% drier than the long-term average, giving an indication of the magnitude of potential change to be associated with changes in global circulation.

An analysis of the Palmer Drought Index (PDI) for Quito indicates that extreme droughts occurred in 1905, 1926, 1959, 1963–1965, 1967–1970, and 1976–1980. The PDI uses as inputs monthly rainfall, potential evapotranspiration and soil data (Palmer, 1965); yet occurrence of PDI droughts in these years accords well with the precipitation record, so that the use of the latter to generate drought scenarios seems justified. Over the long run, 16% of the years at Quito exhibit extreme droughts. Frequency of extreme droughts according to the PDI is very sensitive to slight changes in mean annual rainfall. For example, the 1958–1967 decade, with rainfall 14.4% below the 1891–1980 average, had over three times the normal incidence of extreme drought.

The available historical records of Riobamba and Palmira rainfall are much shorter than the record at Quito (*Figures 2.4* and *2.5*). Both show rough similarities with Quito, with drought in the 1960s and late 1970s, and increased rainfall in the early 1980s.

2.5. Station Rainfall Analysis for Chimborazo Province

Five stations in Chimborazo Province were selected for analysis: Riobamba Airport (1964–1982), Palmira (1964–1982), Cajabamba (1964–1982), Penipe (1965–1982) and San Juan (1964–1982) (*see Figure 2.1* for locations).

A seasonal rainfall analysis of data from these and other stations indicates that in all cases precipitation is low after June and rises to a peak in October, after which it declines to December (*Figure 2.6*). There are differences in the precise timing of the precipitation peak in the first half of the year. Penipe and

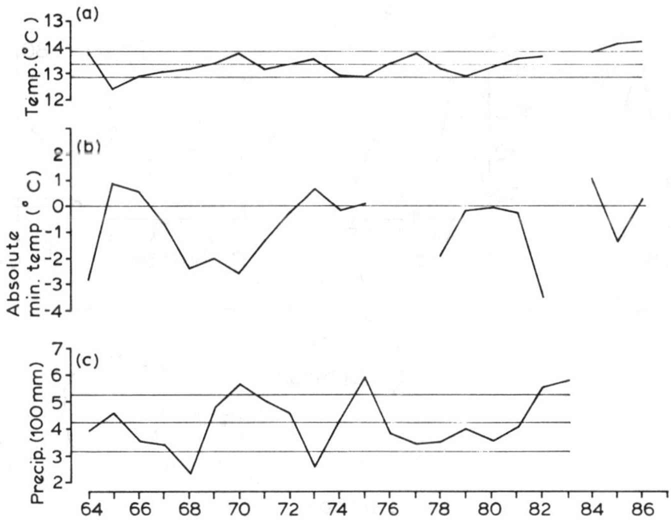

Figure 2.4. (*a*) Annual mean temperature, (*b*) annual absolute minimum temperature and (*c*) annual precipitation at Riobamba (1964–1983).

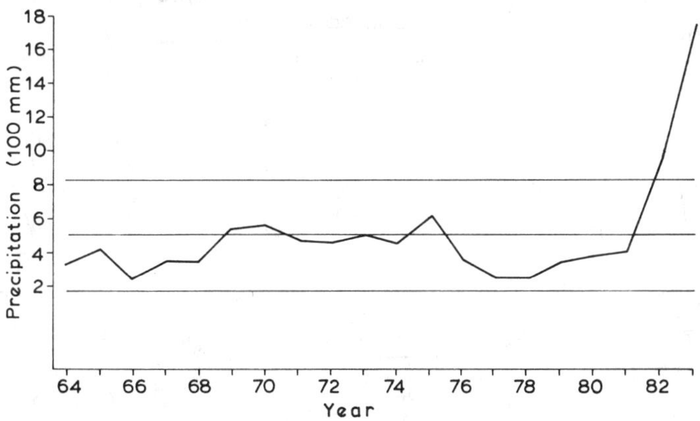

Figure 2.5. Annual precipitation at Palmira (1964–1983).

Alao in the eastern part of the province peak in June, while the stations in the western part of the province peak in April. The difference probably reflects respective influences from Amazon and coastal air masses.

During the 1976–1980 drought, average annual rainfall at Riobamba was depressed by 15% below the 1964–1981 average and at Palmira by 20%. The corresponding values for the other stations are: Penipe, −23%; San Juan, −11%; and Cajabamba, −8%. Although there is considerable local variation, the drought registered at Riobamba was also experienced elsewhere in the province, and the average rainfall depression for the four other stations closely matched the depression at Riobamba.

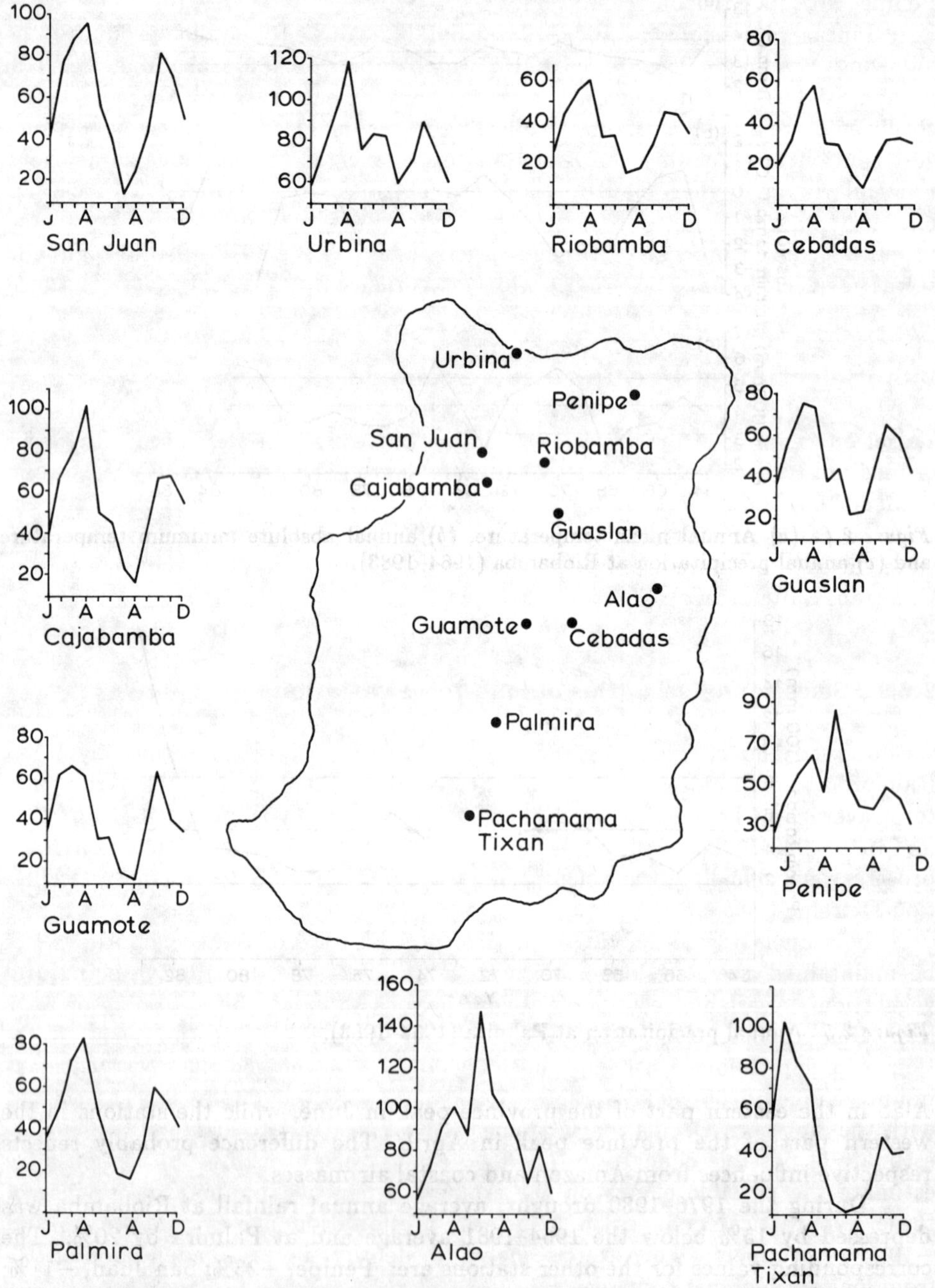

Figure 2.6. **Seasonal patterns of rainfall, Chimborazo Province.**

The 1982–1983 El Niño year was wetter than normal at all stations except Penipe, with its Amazon orientation.

Linear regressions of 1964–1981 rainfall against Riobamba produced the following r values: Palmira, +0.715; Cajabamba, +0.485; San Juan, +0.479; Penipe, +0.674. Although it is unclear to what extent low correlation coefficients are due to faulty data collection at regional stations, these values suggest caution in extending the Riobamba experience to the province as a whole. Particularly at higher elevation stations shielded from lowland tropical air masses (e.g., San Juan and Cajabamba), interannual variability may be related to quite localized processes. Broader events (such as the 1976–1980 drought) do seem, however, to have been more widely experienced.

2.6. Selection of Precipitation Scenarios

To select scenarios of precipitation variation, three stations were used. Scenarios of medium-term variation (i.e., *ca.* 30 years) were generated using Quito station data, since this station has the longest record. Riobamba data were used to generate scenarios of short-term variation for Chimborazo Province. Palmira data were used to generate scenarios for the crop–climate model in the Chauzan–Galte area.

2.6.1. Selection of scenarios of long-term variation

As observed above, the 1960s and late 1970s were drier than normal at Quito, Riobamba and Palmira. The magnitude of depression of rainfall below the long-term average in both cases was about 15% at Quito (14.4% for 1958–1967; 17% for 1976–1980). These dry periods seem to be associated with global-scale processes, specifically an augmentation of the frequency of the Northern Oscillation Extreme (*Table 2.1*).

It is unclear whether depressions of precipitation of these magnitudes can be maintained over the longer term so as to affect the distribution of natural vegetation. Available records of 30-year periods show a rainfall decline of 7.1% from 1891–1920 to 1951–1980. We chose our scenarios of medium-term alteration in precipitation in a compromise fashion. We averaged the observed maximum change in 30-year averages (7.1%) and observed depressions associated with the Northern Oscillation Extreme (13%, *Table 2.1*) to arrive at a plausible estimate of possible changes in the 30+ year average of ±10%. In Section 6 we estimate the possible effects on the ecological map of Chimborazo Province of these hypothetical changes in long-term precipitation averages of +10% and −10%, respectively. This exercise helps to identify the area potentially at risk from long-term climatic change.

2.6.2. Selection of scenarios of short-term variation

A well-adapted farming system may be expected to cope with extreme events that occur more frequently than 10% of the time. Therefore, one of the scenarios common to all case studies in this volume is the 1-in-10 rainfall event (*see* Part I, this volume).

Riobamba was chosen as an appropriate station to calculate the 1-in-10 scenario. The station is centrally located in Chimborazo Province; and although correlations between this station and others in the province are often poor, the relative rainfall depression during 1-in-10 droughts at Riobamba is similar to that at other stations. Thus, a map based on a 1-in-10 drought scenario at Riobamba, while not representing a 1-in-10 probability year for the entire province of Chimborazo, does indicate areas at risk from such events at the local level.

Precipitation values during the period 1964–1983 at Riobamba were ranked, cumulated and plotted on normal probability paper. The 1-in-10 dry year (10 percentile rank) and the 1-in-10 wet year (90 percentile rank) were interpolated from a straight line that was fitted to the plotted points. The interpolations were as follows:

(1) *Baseline*: This was determined from the 50 percentile rank value to be 445 mm, close to the calculated 1964–1983 mean of 443.8 mm.

(2) *1-in-10 dry year*: A dry year with 295 mm of precipitation is observed only once every 10 years in the Riobamba record and corresponds to 295/445 = 66% of the baseline. The 1-in-10 dry year is therefore defined as a year where precipitation is 66% of the baseline total, regardless of location in Chimborazo Province.

(3) *1-in-10 wet year*: A wet year with 595 mm precipitation has a 90 percentile rank and corresponds to 595/445 = 134% of the baseline. The 1-in-10 wet year is therefore defined as a year in which precipitation everywhere is 134% of the baseline total. Similar calculations at Palmira indicated that the 1-in-10 years are 61 and 139% of the baseline value, respectively, which is acceptably close to the Riobamba-derived values.

2.6.3. Selection of scenarios of extreme rainfall events

To determine areas subject to rare but catastrophic events, the Riobamba record was examined for extreme maximum and minimum precipitation during the 1964–1983 period. These extremes are based on the data set, rather than interpolated from the cumulative frequency distribution.

(1) *Dry extreme*: 1968 was a dry extreme at Riobamba with only 253.1 mm of precipitation, which was 57% of the baseline. The dry extreme scenario is therefore defined as a year where precipitation is 57% of the baseline value.

(2) *Wet extreme*: 1975 was the wet extreme with 621 mm, which is 140% of the baseline. The wet extreme scenario is therefore defined as a year where

precipitation is 140% of the baseline value. Use of similar methodology at Palmira results in extremes of 61 and 360% of the baseline, respectively. The former value is acceptably close to the Riobamba value. The latter value represents the 1982–1983 El Niño event. This event resulted in extreme rainfall only at Palmira and Cajabamba, that is, in western Chimborazo Province, so it is justifiable to discount it in mapping an extreme scenario for the entire province of Chimborazo.

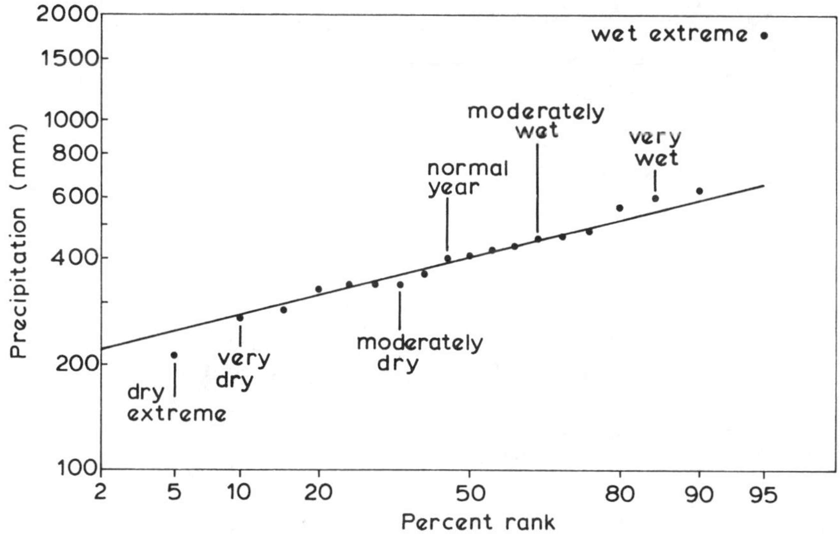

Figure 2.7. Scenarios of climatic variation at Palmira. Cumulative frequency of October–June rainfall, 1964–1983, showing the scenarios chosen for this study.

2.6.4. Selection of climatic scenarios for Palmira

In Section 6 of this case study, a barley yield model will be applied to the Chimborazo small-farm situation. This model requires daily rainfall and temperature data for the growing season and thus cannot be used with the Riobamba scenarios based on annual data. For this reason, planting season (October–June) rainfall data for Palmira were plotted on log normal paper and a line fitted to the resulting distribution (*Figure 2.7*). For each scenario, the year with precipitation closest to the target interpolated value was selected for use with the model, thus:

(1) *Baseline*: 50 percentile rank value = 416 mm.
(2) *1-in-10 dry year*: precipitation is 65% of the baseline.
(3) *1-in-10 wet year*: precipitation is 40% above the baseline.
(4) *Dry extreme*: 1979–1980, with October–June precipitation of 210 mm.
(5) *Wet extreme*: 1982–1983, with October–June precipitation of 1744 mm.

In addition, two further scenarios were selected: a moderately wet year (65 percentile) and a moderately dry year (35 percentile). Note that the 1-in-10 wet year is identified with the 85 percentile rank year, since the 90 percentile year lies above the fitted curve.

In subsequent sections of this case study, we shall consider the response of crop yields, farming systems and the spatial pattern of climatic resources in Ecuador under these scenarios of climatic variation.

SECTION 3

The Vulnerability of Indigenous Farming Systems to Climatic Variations

3.1. Introduction

In this section we consider the significant environmental constraints and opportunities for highland farmers and characterize the technical options that farmers may use to reduce risk. Our aims are to assess the sensitivity of agriculture to climate, and to identify points of vulnerability, so as to begin to identify appropriate policies. The focus of attention is on the results of a survey of indigenous farmers in the Galte–Chauzan area.

3.2. The Chauzan–Galte Area

3.2.1. Environment and land use

The areas of Chauzan and Galte contain several hundred farm families concentrated in six peasant organizations (*Figure 3.1*). Most of the area is above 3200 meters elevation (*Figure 3.2*). Soils, vegetation and crops are largely determined by elevation and topographic position (*Figure 3.3*). According to the Holdridge (1967) classification of life zones, most of the region is Montane Moist Forest, varying from 3000 to 4000 meters in elevation and with 500 to 1000 millimeters of rainfall per year. The valley bottom, however, is in the drier Montane Steppe life zone.

Soils are sandy (Vitrandepts and Psamments). Some areas, such as around Palmira, are wind-blown dunes and are thus unsuitable for cultivation. There are also less sandy soils, such as the Eutrandepts (fertile volcanic soils) around Galte and high-elevation Dystrandepts. Of the county of Guamote about 20% of the surface is eroded and another 20% in process of erosion (SEDRI, 1983). The vegetation of the area has been extensively modified by human activity. High elevations exhibit a paramo grassland (*Stipa ichu*, etc.). Shrubs and thorny

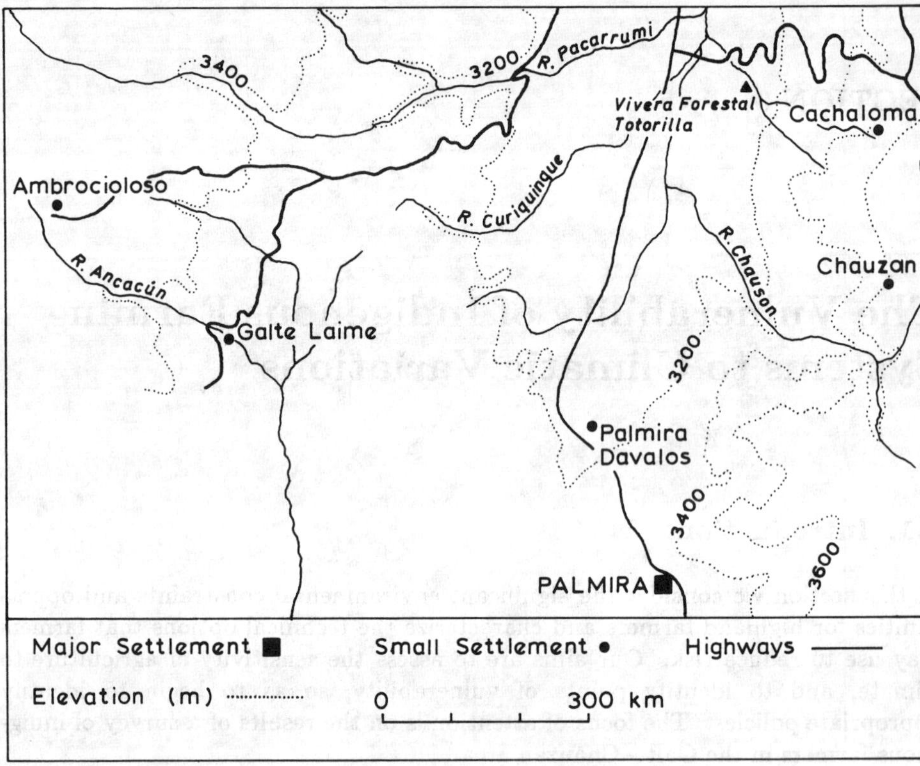

Figure 3.1. Galte and Chauzan: location and topography.

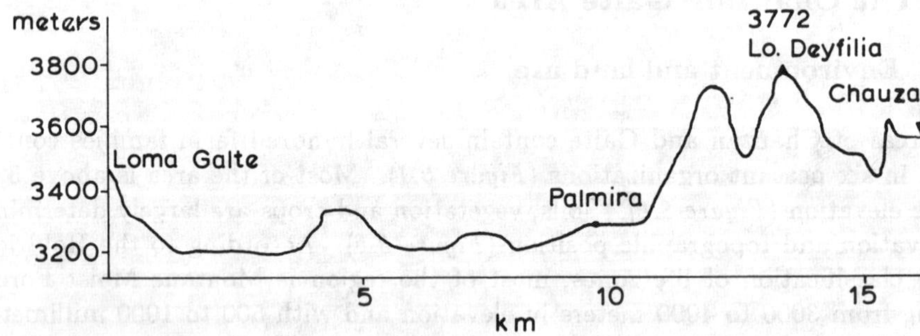

Figure 3.2. Topographical transect from Galte to Chauzan.

vegetation (*chilca*, *agave*, *tuna*, etc.) are common at lower elevations, especially in hedgerows.

In the county of Guamote, 65% of land is owned communally and 35% individually (SEDRI, 1983). The communally owned land is mostly in natural pasture. Some 81% of the individually owned land is cultivated, the remainder being pasture or unproductive land. Most of the cultivation is for broad beans,

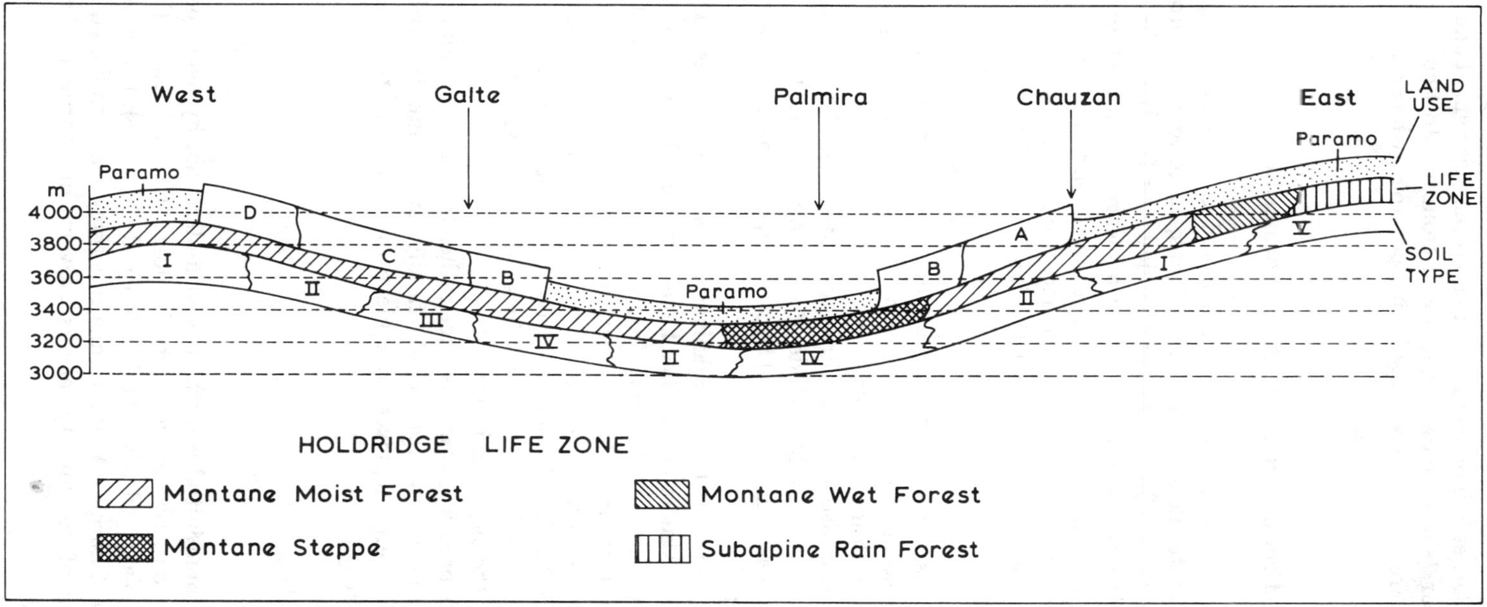

Figure 3.3. Generalized soils, Holdridge life zones and land use. A = barley, potatoes, broad beans, rye grass, lentils; B = potatoes, lentils, forest (cypress, eucalyptus, pine); C = barley, broad beans, rye grass, melloco, oca; D = barley, lentils, fallow, forest. Soil types: I = Dystrandept, II = Eutrandept, III = Vitrandept, IV = Ustipsamment, V = Hydrandept.

potatoes and barley. Other minor Andean crops are also grown. Land may be cropped year after year (short fallow) or fallowed from one to two years (long fallow). Broad beans are often intercropped with potatoes, while barley is typically grown alone.

3.2.2. The social and economic context of production

Until the late 1970s the communities of Chauzan and Galte were dependent on either the private or state hacienda systems. Thus, large areas of productive land and almost all rent benefited either one landowning family or the government. Peasants laboring on these lands gained a meager reward either in kind (i.e., harvested crops) or in the use of some hacienda resources: irrigation water, grazing lands, firewood, and the right of use of access paths to their own lands.

These haciendas were broken up and the land redistributed by the Ecuadorian land reform agency. The region of Chauzan now includes the peasant organizations of Chauzan–Totorillas, Atapo–Santa Cruz, and Santa Teresita. Of these, Santa Teresita has the least effective organization in the area. Chauzan–Totorillas, with a population of 437 in an area of 27.7 square km, is organized as a juridicial community with majority rule.

The region of Galte also includes three organizations: Galte–Laime, San Francisco–4 Esquinas, and San Miguel de Pumachaca. Galte–Laime, with a population of 620 persons in an area of 30 square km, is organized as a cooperative with 110 members.

All of the members of these organizations are indigenous (Quichua-speaking) smallholders. Almost all families have between 0.1 and 10 ha of land. The economy is mixed: almost all production is for on-farm consumption, yet most farmers resort to temporary labor migration to provide an additional cash income. The reason most often given for the lack of commercial production is low prices (SEDRI, 1983). In addition, the population has suffered from a lack of technical assistance, social security and education. In Guamote county, 74% of the population over the age of 10 is illiterate.

Since the farmers produce for subsistence rather than sale, the aim is not to maximize profit but rather to feed one's family and minimize risk. The knowledge and the strategies deployed to achieve this aim were the prime subjects of the present survey.

3.3. The Survey

Prior to this survey, some initial work had been undertaken by using personnel and resources of FAO and two private organizations: CESA (Centro Ecuatoriano de Servicios Agricolas) and AIG (Accion Integral Guamote), the latter sponsored by the Catholic Church. A questionnaire survey was conducted in 1982 with the participation of community leaders. Over time, through the use of lectures, meetings and film shows, it proved possible to improve local understanding of the aims of SEDRI and, in particular, for two of the present co-authors (T. Salcedo and R. Bravo) to establish a rapport with smallholders.

For the purposes of the present project, a new questionnaire was formulated and tested in different communities with individual peasants and with meetings and work groups of villagers. One difficulty was that, while women perform most of the agricultural tasks in the villages (the men being generally occupied in labor migration), they do not usually speak Spanish but rather Quichua, the indigenous language. Since the questionnaire was written in Spanish and the surveyors were monolingual, it proved necessary to direct all questions to the male heads of the family. The surveyors feel, however, that the survey results do generally reflect current practices and beliefs, with perhaps a 90% confidence level.

Ten family heads were to be interviewed in each of the six peasant organizations in the area, the 10 representing various altitudinal zones in each community. Attempts were made to find these individuals in their homes or fields in each community. In some cases the persons could be found near the village itself; in others, it was necessary to try to find acquaintances to locate the informants. Because of the problems of mistrust, it proved possible to interview only 50 of the hoped-for 60 informants. The community of Atapo Santa Cruz provided the lowest number of informants, with only 6 of the selected 10 responding.

The questionnaire employed included the following sections:

(1) *Environmental terminology.* An effort was made to determine local terms for selected environmental characteristics and hazards.

(2) *Adaptive strategies.* Farmers were asked about altitudinal zonation of crops and length of fallow.

(3) *Soil preparation practices.* Farmers were asked how they prepare soil for cultivation in brushland and paramo, after long fallows and after short fallows.

(4) *Crop calendar.* Farmers were asked about the times of sowing and of harvest, and crop combinations in use.

(5) *Yields.* Farmers were asked to estimate yields for crops and crop combinations during two recent agricultural years: 1981–1982, an unusually dry year and 1983–1984, an unusually wet year. This difference in weather provided a convenient indication of the effects of rainfall variations on output. Information was gathered on quantity of seed sown per hectare and yield per unit seed for long and short fallows, both with and without organic fertilizer (sheep dung).

(6) *Specific practices.* In addition to asking about fertilizer use, questions about variety of seeds, pests and pesticides, specific sequences of rotation, irrigation and movable corrals were included.

(7) *Hazards.* Farmers were asked to determine the months of greatest risk of frost, wind damage and rain damage for various crops. They were also asked to state adjustments they made to reduce frost and wind damage.

3.4. Interpretation of Results

3.4.1. Perception of physical constraints

In the following discussion, Spanish terms will be identified with the abbreviation (Sp.) and Quichua terms with the abbreviation (Q.).

Farmers recognize at least two altitudinal zones: a lower elevation zone used for cultivation, and a higher elevation zone used for pastures. When asked to provide a term for the latter high zone, *urcu* (*jagua urcu, jatun urcu*) (Q.), *cerro* (Sp.) or *pajonal* (Sp.) were the most common responses.

There is a much richer vocabulary of terms relating to different topographic and soil types. The most fundamental distinction for peasants here, as elsewhere in highland Ecuador, is between sloping land and concavities or flats in the landscape. Slopes are known by the name *loma* (Sp.) or *quingre* (Q.). Concavities in the landscape are known by the names *hondon* (Sp.), *bajio* (Sp.) or the Quicha terms *pugru* or *fundu huaycu*. Flats are known as *pamba* (Q.) or *plano* (Sp.). The importance of this distinction between slope, concavity and flat is that the different zones have different soils and different degrees of susceptibility to frost.

Farmers distinguish between good soil (*alli allpa*, Q.) and bad soil (*mana allpa*, Q.) or thin soil (*tullu allpa*, Q.). They recognize rich, black soil (*yana allpa*, Q.), sandy soil (*tiu allpa*, Q.) and various kinds of hard or eroded soil: duripan, *cangahua* (Sp.), wind-eroded soil, *guaira allpa* (Q.), stony soil, *rumi allpa* (Q.), hard clay soil, *grida allpa* (Q.) and hard soil, *sinchi allpa* (Q.).

There seems to be no strongly held view that increasing elevation brings reduced yields. Instead, farmers focus on topographic and soil factors related to fertility and frost protection to maximize yields. Drought (*usiag, chaquishca*, Q.) is a recognized problem, as is frost (*rasu*, Q.) and hard rain (*tamia sinchi*, Q.). Fortunately, peasants are not without strategies to deal with these problems.

3.4.2. Technical options

The majority of respondents (62%) denied that altitude was a consideration in choosing which crop would be grown. Those who did take altitude into consideration were in general agreement that potatoes and *habas* (broad beans) were suitable for higher elevations, while lupines (*chocho, tarwi*, Q.) were suitable for lower elevations. A few informants in two communities felt that barley was suitable for higher elevations, while all the other informants felt that barley did well at all elevations.

Farmers recognize that slopes are relatively more frost-free than flats, and they also recognize that deep soils with dark topsoil – found in flats and concavities – result in better yields than eroded soils. Farmers know that frequently cultivated soils as a rule yield less than soils that are allowed to rest; that rotation improves yields; and that green manure animal manure and fertilizers improve yields (Knapp, 1984).

In the Chauzan–Galte area most farmers plant in October–December, with some plantings in August and September. The peak harvest period is August–September, with some harvest also taking place in June–July and December–January. The long maturation times are typical of high elevations.

Table 3.1. Forms of management (values are numbers of respondents to questionnaire).

	Community						
	Chauzan–Totorillas	*San Francisco*	*Atapo*	*Galte–Laime*	*Santa Teresita*	*San Miguel*	*Total*
Fallow length							
1 year	10	10	5	9	5	6	45
2 years	1	2	0	0	1	1	5
Barbecho length							
1 month	1	1	0	0	0	0	2
2 months	4	5	1	1	2	6	19
3 months	6	7	5	7	3	3	31
Fertilizer applications							
Barley:							
22–44 kg/ha	2	0	0	2	0	0	4
Potatoes:							
44–88 kg/ha	1	0	0	2	0	0	3
88–176	0	0	0	1	3	0	4
176–264	2	1	2	1	0	1	7
264–352	5	3	2	1	0	0	11
352–440	0	6	1	4	0	0	11
Use animal fertilizer?							
Yes	10	10	6	9	6	6	47
Sheep	10	10	1	8	4	4	37
Cattle	7	5	0	1	0	1	14
Horses	5	8	0	3	2	2	20
Use pesticides?							
On broad beans	3	2	0	1	0	1	7
On potatoes	9	10	0	5	0	1	25

In these communities, irrigation is not available. Although cultivation results in the formation of lynchets or semi-terraces, true terracing with retaining walls is rare. There is no access to flat land; all the land is gently to steeply sloping. The major options for agriculture therefore involve manipulation of crop plants (selection of varieties, interplanting, rotation) and manipulation of soil (fallowing, incorporation of manure and organic matter).

Soil fertility management

Fallowing is widely practiced to improve soil fertility. When farmers do fallow, they generally do so for one rather than two years (*Table 3.1*). Fallowing is less likely to follow barley cultivation than potato cultivation (cf. *Tables 3.5* and *3.6*).

The same crop is seldom grown two years in a row. Barley is usually followed with habas (broad beans) (cf. *Table 3.5*); broad beans are usually followed by potatoes (cf. *Table 3.7*); and potatoes, in turn, usually by barley again or by fallow (cf. *Table 3.6*).

Communities differ in their fallowing and rotation practices. Fallowing is most often practiced in San Francisco; least often, in Santa Teresita.

Chemical fertilizer is often used with potatoes; application rates are about 264–352 kg per hectare (median) (*Table 3.1*). Rates are highest in San Francisco and lowest in Santa Teresita. Other crops are seldom fertilized with chemical fertilizer.

Animal fertilizer is almost universally used (*Table 3.1*). Sheep manure is the most important, followed by horse and cattle manures. Other studies (Knapp, 1984) have indicated the importance also of night soil, guinea pig manure and kitchen wastes.

Table 3.2. Movable corrals (*talanqueras*).

	Community						
	Chauzan–Totorillas	*San Francisco*	*Atapo*	*Galte–Laime*	*Santa Teresita*	*San Miguel*	*Total*
Use *talanqueras?*							
Yes	6	5	3	5	2	2	23
No	4	4	3	5	4	6	26
Size							
5–10 m^2	0	0	0	1	1	0	2
10–20 m^2	0	0	1	0	1	2	4
20–30 m^2	0	1	0	3	1	0	5
30–40 m^2	2	1	1	0	0	0	4
No. of animals							
20–30	1	2	2	1	2	3	11
30–40	0	0	1	2	0	1	4
40–60	1	1	0	2	1	0	5
Time in *talanquera*							
3 nights	5	4	0	3	1	0	13
4 nights	1	2	0	2	0	0	5

Even when there is no year-long fallow, there is often a two- or three-month rest (*barbecho*) between harvest and planting (*Table 3.1*). During this time crop residues, green manure, dung and kitchen wastes are incorporated into the subsoil to rot and contribute to soil fertility. A similar turning-under of organic material (*tzauma*) is practiced when fields are created in grassy or brushy areas.

During the time of rest, animals are folded at night in movable corrals (*talanqueras*) by about half the population (*Table 3.2*). Generally speaking, 30 animals spend 3 nights in a movable corral about 20 m^2 in size (*Table 3.2*). Other studies have shown that sheep produce about 0.2 kg/night/animal of manure (dry weight equivalent) (Knapp, 1984). Thus the movable corrals deposit $0.2 \times 30 \times 3 = 18$ kg of manure on $20\,m^2$ – an overall rate of 9000 kg of manure per hectare.

All these fertility management practices explain the ability of farmers to cultivate with minimal fallowing without major long-term yield deterioration.

Pest management

Farmers occasionally use pesticides for broad beans and frequently for potatoes (*Table 3.1*). The major problem afflicting potatoes is *lancha*, a fungus (*Table 3.3*). Grubs (*gusano blanco*) are also a problem. One reason for avoiding fallowing is the belief that certain pests multiply in fallowed fields.

Hazards

Most farmers assess the weather in recent years (i.e., 1981 to 1984) as normal or bad (*Table 3.4*). Although farmers claim to be unable to adapt to frost and drought risk, in fact they recognize that planting on slopes, especially east-facing slopes to catch the morning sun, minimizes frost risk. Cultivation during the rainy season minimizes not only frost, but also drought. Some farmers note that building potato ridges up and down slope also reduces frost risk, but at the cost of increasing erosion.

Wind is a recognized hazard, especially in summer. A common response to this is *shigua paja*, which involves minimal working of the soil and the spreading of grass as a mulch over the surface to limit wind and water erosion of seed, soil and fertilizer.

3.4.3. Yields

Individuals were requested to report yields during the recent dry year (1981–1982) and the recent wet year (1983–1984). Reported yields varied greatly from individual to individual, and average yields varied equally greatly from community to community. It is unclear to what extent these variations are real as opposed to variations in perception, memory and trust on the part of informants.

Barley

Yields in wet years were frequently three times as large as those in dry years. Yet the difference in yield between communities is greater than the difference between wet and dry years. Increased fallow length *per se* apparently had a minor effect on improving barley yields in both dry or wet years (*Table 3.5*). Much more important was the effect of organic fertilizer, which approximately doubled yields in both wet and dry years (*Table 3.5*).

Table 3.3. Hazards.

	Community						
	Chauzan–Totorillas	*San Francisco*	*Atapo*	*Galte–Laime*	*Santa Teresita*	*San Miguel*	*Total*
Principal pests							
Barley:							
Roya	8	9	6	8	6	8	45
Roya Amarilla	3	1	1	1	0	0	6
Lancha	2	3	0	2	0	1	8
Gusano	1	0	0	0	0	0	1
Amarillamiento	1	2	2	4	3	4	16
Potatoes:							
Lancha	10	9	6	9	6	6	46
Gusano Blanco	8	10	5	9	6	6	44
Cutzo	2	7	2	2	3	0	16
Mosco	2	1	0	0	0	0	3
Ayabala	4	1	2	6	5	1	19
Pulgrilla	3	0	0	0	0	0	3
Gusano Negro	0	4	1	0	0	0	5
Frost damage							
Barley:							
Jan–Feb	2	0	0	0	0	0	2
Mar–Apr	1	0	0	0	0	0	1
May–Jun	8	7	5	1	0	1	22
Jul–Aug	0	1	1	0	0	0	2
Sep–Oct	0	0	0	0	0	0	0
Nov–Dec	1	0	0	2	0	0	3
Potatoes:							
Jan–Feb	5	9	5	1	0	0	20
Mar–Apr	1	0	0	1	1	1	4
May–Jun	9	13	8	16	8	15	69
Jul–Aug	1	0	3	1	2	1	8
Sep–Oct	0	0	6	1	5	8	20
Nov–Dec	10	10	7	16	12	15	70
Habas:							
Jan–Feb	1	1	2	0	0	0	4
Mar–Apr	0	0	0	0	0	0	0
May–Jun	7	10	2	4	0	3	26
Jul–Aug	0	0	0	1	0	0	1
Sep–Oct	0	0	0	0	0	0	0
Nov–Dec	1	0	0	2	0	1	4

Potatoes

There was little relation between fallow length and potato yields, perhaps because short-fallow potatoes received more fertilizer (*Table 3.6*). Organic fertilizer application had a major effect on yields, particularly when yields are expressed net of seed. Gross yields generally increased fivefold with organic fertilization; net yields (in wet years) increased by 31 times.

Table 3.4. Hazard control.

	Community						
	Chauzan–Totorillas	*San Francisco*	*Atapo*	*Galte–Laime*	*Santa Teresita*	*San Miguel*	*Total*
Opinion of recent weather (1981–1984):							
Good	2	1	0	0	0	0	3
Bad	7	1	3	8	1	4	24
Normal	0	5	3	2	4	3	17
What is the effect of frost?							
Lose everything	9	8	6	8	5	6	42
Do you adapt to frost/drought risk?							
Yes	4	3	0	2	0	0	9
No	4	5	1	6	3	1	20
How?							
Use slopes	4	2	0	1	0	0	7
Cultivate in rainy season	1	3	0	0	0	0	4
Do you adapt to risk of wind damage?							
Yes	5	0	0	3	1	0	9
No	1	4	1	1	0	0	7
How?							
Shigua paja	5	0	0	3	1	0	9

The potential improvement under fertilization is even greater than *Table 3.6* suggests, for in the study area yields of 10 to 20 metric tons per ha are possible with optimal fertilization. Yields in wet years are about four times those in dry years. In communities with better fertilization and management practices this difference is somewhat greater.

As with barley, there was a considerable difference between community-average yields. San Francisco, the community most likely to fallow potatoes and to apply large quantities of fertilizer, had high yields, while Santa Teresita, which uses the least amount of fertilizer and pesticide, had the lowest yields.

In general, it appears that management and fertilization are major limiting factors. This is not to suggest, however, that current levels are suboptimal or backward, since optimality is a function of the goals of cultivators. Under subsistence conditions it is seldom necessary or desirable to maximize yields, particularly if such maximization would require costly inputs.

Broad beans

Broad beans (habas) produce relatively low yields in calorie terms, but they supply protein to the diet and nitrogen to the soil. The latter may account for their relatively high yields without fertilization (*Table 3.7*).

As with other crops, yields are improved in wet years, with long fallow, and in communities with superior cultivation practices. Habas are usually followed by potatoes in the rotation sequence.

Table 3.5. Management and levels of yields for barley.

	Community						
	Chauzan–Totorillas	*San Francisco*	*Atapo*	*Galte–Laime*	*Santa Teresita*	*San Miguel*	*Total*
Varieties (no. responses):							
Dorada	7	9	4	8	1	1	30
Duchicela	3	0	0	0	0	0	3
Comun	0	0	2	1	5	6	14
Rotations (no. responses):							
Barley–potatoes	0	1	0	2	2	1	6
Barley–broad beans	8	6	0	7	3	2	26
Barley–fallow	2	2	1	0	0	0	5
Barley–barley	0	0	0	0	0	0	0
Use of plow (no. responses):	0	0	1	1	0	0	2
Seed used (kg/ha)	79.2	86.0	79.1	78.3	79.2	83.6	79.2
Average yields in dry year (kg/ha):							
By fallow length							
2–3 months							187
1–2 years							233
By organic fertilizer application:							
With							311
Without							151
Overall average	246	394	301	116	70	157	212
Average yields in wet year (kg/ha):							
By fallow length							
2–3 months							630
1–2 years							665
By organic fertilizer application:							
With							826
Without							428
Overall average	641	1172	818	335	244	624	645

Calories

Since barley and potatoes are produced for direct consumption, it is valuable to calculate yields in calorie terms. Averaging the actual yields (net of seed) of the two most productive communities, we find that barley produces 3.2 million calories per ha (MCPH) in wet years and 0.9 MCPH in dry years. Potatoes produce 3.9 MCPH in wet years and 0.6 MCPH in dry years. In wet years, these crops would support 3–4 persons per ha (at one million calories/person/year). Broad beans produce only 0.7 MCPH in wet years and 0.3 MCPH in dry years.

3.4.4. Choice and combination of adaptive options

If we ignore, for the present, the differential inputs of fertilizer and labor, the calorie payoffs of potatoes and barley suggest that farmers should plant potatoes

Table 3.6. Management and levels of yields for potatoes.

	Community						
	Chauzan–Totorillas	*San Francisco*	*Atapo*	*Galte–Laime*	*Santa Teresita*	*San Miguel*	*Total*
Varieties (no. responses):							
Ubilla	8	9	2	9	6	4	38
Santa Catalina	3	0	3	0	0	0	6
Rotations (no. responses):							
Potatoes–barley	3	3	1	6	3	0	16
Potatoes–fallow	5	6	0	2	0	0	13
Potatoes–broad beans	2	1	0	0	3	1	7
Potatoes–potatoes	0	0	0	0	0	0	0
Usual preceding land use (% of responses):							
Potatoes							4%
Habas							50%
Barley							23%
Fallow							23%
Seed used (kg/ha)	526	693	625	746	567	629	629
Average yields in dry year (kg/ha):							
By fallow length:							
2–3 months							939
1–2 years							854
By organic fertilizer application:							
With							1170
Without							301
Overall average	953	1082	1163	592	414	783	897
Average yields in wet year (kg/ha):							
By fallow length:							
2–3 months							3250
1–2 years							3435
By organic fertilizer application:							
With							4216
Without							743
Overall average	2283	5104	6002	2384	867	2472	3334

in wet years and barley in dry years. Since farmers cannot predict rainfall, however, we would expect potato monoculture only in areas that are predictably wet. Planting in other areas should be determined by the nature of farming strategy. If farmers seek to minimize their maximum possible loss, they should plant barley wherever there is a risk of a dry year. Since pests are also a factor, and pests may strike either barley or potatoes, a mixture of the two crops may be most desirable in practice. Toward the dry limit of cultivation, a potato crop enables a potential windfall in a wet year. Moreover, average yields of potatoes and barley are so similar across wet and dry years that there is little yield reduction involved in a mixed strategy. This probably accounts for the potato–barley rotation pattern that is widely practiced in the Galte–Chausan area.

Potatoes may require more labor than barley to cultivate and to harvest (Knapp, 1984). They also seem to demand more fertilizer to enable yields well

Table 3.7. Management and levels of yields for broad beans (habas).

	Community						
	Chauzan–Totorillas	*San Francisco*	*Atapo*	*Galte–Laime*	*Santa Teresita*	*San Miguel*	*Total*
Varieties:							
Guagrahaba	6	5	2	5	2	2	22
Comun	3	4	0	1	0	2	10
Fina	2	0	0	0	0	0	2
Verde	0	0	3	3	4	2	12
Rotations:							
Habas–potatoes	9	9	0	6	3	0	27
Habas–barley	1	1	0	1	3	1	7
Habas–fallow	0	0	0	0	0	2	2
Habas–habas	0	0	0	0	0	0	0
Seed used (kg/ha)	56	72	74	78	67	74	66
Average yields in dry year (kg/ha):							
By fallow length:							
2–3 months							231
1–2 years							291
By organic fertilizer application:							
With							335
Without							194
Overall average	300	357	320	216	320	157	260
Average yields in wet year (kg/ha):							
By fallow length:							
2–3 months							567
1–2 years							627
By organic fertilizer application:							
With							681
Without							429
Overall average	663	725	691	515	328	533	594

above replacement of seed. On the other hand, barley tends to be grown immediately after potatoes and probably benefits from the cultivation and residual fertility of the potato crop. Potatoes have a higher retail value than barley and may well be more highly valued in the diet.

3.4.5. Relationship of payoffs to scenarios of wet and dry years

Farmers identified the recent dry year as 1981–1982 and the recent wet year as 1983–1984. At neighboring Palmira, 1983–1984 indeed was a wet year – too wet for the record to be credible. Seasonal 1983–1984 rainfall was recorded as 1693 mm – comparable to the 1982–1983 El Niño year. However, 1983–1984 was not an El Niño year, and no other station in Chimborazo Province recorded comparable values.

As a check of the Palmira record, a regression was performed between the average precipitation at the bracketing stations of Guasuntos and Cebadas (GC)

and P, Palmira rainfall, for the period 1973–1982. The relation

$$P = -105.3 + 1.425\ GC \qquad (3.1)$$

accounted for 87% of the variance (r^2). Applying this equation to the 1984 value of GC, Palmira should have received 608 mm of rainfall in the calendar year; in fact, 1914 mm was recorded. Local informants also verify that this amount could not have been received at Palmira. We conclude that although Palmira data are acceptable up to 1982, after that date they are suspect.

To estimate what the real 1983–1984 precipitation might have been at Palmira, available data at Guasuntos and Cebadas were used to estimate GC, and the above equation used to estimate predicted Palmira rainfall (P). (For three months in this year, only Guasuntos data were available; these were assumed to be equal to corresponding predicted Palmira rainfall for those months.) The result was a calculated Palmira rainfall value of 648 mm for the 1983–1984 growing season.

While the 1981–1982 crop season was quite wet at Palmira, interviews at Chauzan–Galte (and the low reported yields) indicate that this season was dry in those areas. Use of the above regression equation for Guasantos and Cebadas data produces a result corroborating the Palmira value. The same data, however, indicate a strong decline in precipitation from south to north: Guasuntos, 834 mm; Palmira, 584 mm and Cebadas, 268 mm. Furthermore, at Cebadas 59% of the precipitation arrived in the last two months of the rainy season. Chauzan is closer to Cebadas than Galte, but both Chauzan and Galte experienced comparably poor harvests in 1981–1982. This suggests that both areas experienced precipitation totals more akin to those at Cebadas than to those at Palmira.

If this were the case, 1981–1982 precipitation might have been similar to the 1966–1967 pattern at Palmira, when 337 mm of rain fell. This was a 35% probability event there (1982 rainfall experienced at Cebadas was a 27% probability event).

If our interpretations are valid, the scenarios of recent wet year and recent dry year at Chauzan–Galte are roughly equivalent to 650 mm and 300–400 mm precipitation totals, respectively. For these communities close to the 500 mm isohyet, agriculture is clearly marginal. The practice of farming, however, continues, indicating the resilience of traditional adaptive strategies.

It should be clear that the "dry limit" of cultivation is not that point at which yields drop to zero, but rather that point at which the expected yields over the whole range of variation of precipitation are economically and culturally unacceptable, given a strategy of crop rotation and agricultural management. The data provided by the present study are inadequate to define such a limit precisely. The communities involved must, however, be fairly close to the dry limit, since barley yields are quite frequently only two to four times the quantity of seed input, due to dry conditions.

Farmers are able to survive dry years by a variety of means. Since they invest little cash in farming and are generally not seriously in debt, a failed harvest does not require sale of assets to pay off loans. Temporary labor migration is probably the key to continued survival. Always practiced by the majority, earnings from migration can be directed into subsistence purchases and the scale of migration stepped up.

3.5. Conclusions

The major problems for farmers in Chauzan and Galte are the maintenance of soil fertility, pest control and coping with drought, frost and wind damage. Low soil fertility, rather than drought or frost, accounts for low overall yields in the Chauzan–Galte area.

Farmers have a variety of options for the improvement of yields. These include the use of chemical fertilizers, movable corrals, other forms of application of animal manures, use of green manure, fallowing and crop rotation. They may also mix crops to minimize long-term gain and/or minimize risk from drought, frost and pest incidence.

In practice, barley is rotated with broad beans and potatoes with fairly frequent fallowing. Evidence given above indicates that barley is less sensitive to drought than potatoes, while potatoes yield better (in calorie terms) in wetter years.

By combining (female) farming with (male) labor migration, peasants provide a cash flow plus a survival option for bad farming years. The farming component provides for family food and shelter with normally minimal cash outlay; this permits farmers to accept temporary urban jobs at lower levels of pay than urban residents. Employment is thus readily available for temporary migrants.

It is probable that a series of drought years would reduce the ability of migrants to survive on low-paying jobs and increase pressure on other family members to leave the countryside. Massive outmigration on a permanent basis – such as that observed during the Loja drought – could ensue.

During a recent dry year (1981–1982), when the rainfall total at Chausan–Galte was probably between 300 and 400 mm, the potato crop clearly failed (i.e., barely exceeded seed replacement), while the barley harvest was marginal (2:1 to 5:1 per unit seed). In the following section, crop–climate modeling techniques will be used to estimate barley yields for various scenario years.

SECTION 4

The Effects of Climatic Variations and Agricultural Practices on Barley Yields*

4.1. Introduction

In most agricultural regions of Ecuador, the year-to-year variations in agricultural output are largely determined by climatic variability, and farm productivity is at severe risk from climatic extremes such as drought, flooding and frost. For example, during the 1979–1980 crop growing season, severe drought conditions led to widespread crop failures and economic losses. A report from the Food and Agriculture Organization stated: "Dry conditions prevailing since September 1979 severely affected agricultural production in large areas of the Coastal and Sierra regions. Food crops sown in October 1979 are almost entirely lost and about 1 million people need assistance ..." (FAO, 1980). In the 1982–1983 season, coastal crops were lost due to El Niño-event flooding. Most recently, in January 1985, a dry spell adversely affected sowing of many crops in the coastal area of Ecuador (MAG-INAHMI, 1985).

Agriculture is the main economic activity in the Galte–Chauzan area (*see* Section 3); however, the ecosystem is only marginally suited for commercial agriculture, and local farmers operate with a low level of technology. Major crops for the area are barley, wheat, potatoes, broad beans and lupines. Barley is an important source of protein and carbohydrates for the indigenous communities of the area under study. In 1983, an estimated 17 000 hectares were sown with barley in Chimborazo Province. This represents about 50% of the total area cultivated with barley in Ecuador. The average yield (1961–1983) for commercial and subsistence barley at the provincial level is 635.8 kg/ha (MAG, 1984).

Crop yield simulation models have proved to be useful tools for assessing climatic impact on crop productivity. Ritchie (1984) and French and Hodges (1985) described applications of wheat models in agriculture. Hodges and Botner (1985) demonstrated the practical use of a maize model to predict crop yield in the Corn Belt of the USA.

---

*Two authors of this section, Dr. A. Ravelo and Dr. A. Planchuelo-Ravelo, are in the Atmospheric Science Department, University of Missouri, Columbia, Missouri, USA; Dr. Tom Hodges is with the US Department of Agriculture, Prosser, Washington, USA.

A barley yield model was adapted from the CERES-Wheat model developed by Ritchie and Otter (1985) to simulate responses to extreme climatic conditions. These responses can help establish new strategies in agricultural management and planning to prevent or reduce the probability of crop failures.

4.2. Model Description

The model simulates the following processes: seedling emergence, leaf growth and senescence, root growth, biomass accumulation, soil evaporation, transpiration, soil moisture distribution, phenological development, tillering and grain yield component determination. Each process is simulated in daily time steps using the previous values to complete a cycle. *Figure 4.1* shows the daily cycle for the barley model. The model estimates final grain yield as a function of crop, soil and weather data, but does not account for yield-reducing factors such as weeds, insects, nutrient deficiencies and diseases. Soil nutrient effects have not been modeled, owing to lack of information on initial extractable ammonium and nitrate, soil organic carbon content and pH.

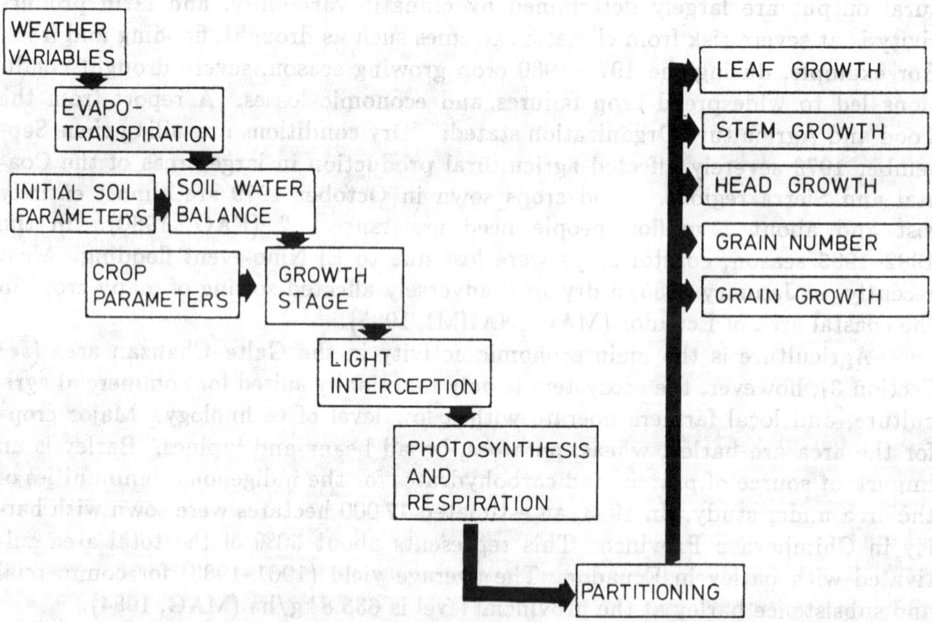

Figure 4.1. Daily cycle of barley yield model.

The model input data are:

- Latitude
- Daily solar radiation (cal/cm^2/day)

- Daily maximum and minimum temperatures (°C)
- Daily rainfall (mm)
- Initial soil parameters
- Plant genetic parameters
- Sowing dates (day of the year)
- Plant density (plants/m^2)
- Dates of irrigation (day of the year) and amounts (mm).

The major components of the barley model are:

- A daily soil moisture balance
- A crop growth and development model.

The crop growth stages considered in the model are presented in *Table 4.1.*

Table 4.1. Growth stages used in the barley yield model and plant components growing during each stage (in parentheses).

Growth stage	*Description*
7	Presowing
8	Sowing to germination
9	Germination to emergence (roots, leaves)
1	Emergence to terminal spikelet (roots, leaves)
2	Terminal spikelet to end of vegetative growth (roots, leaves, stems, heads)
3	End of vegetative growth to end of pre-anthesis ear growth (roots, leaves, stems, heads)
4	End of pre-anthesis ear growth to beginning of linear grain fill (roots, stems, heads)
5	Grain filling (roots, grain)
6	End of grain filling to harvest

4.3. Model Calibration

Model calibration consisted of developing coefficients for the crop and soil parameters. The model was run for the test year and varietal coefficients were adjusted, first to correct model flowering and maturity dates and then to correct model yield components.

Soil profile coefficients were based on soil type, depth and water-holding capacity obtained from FAO (1975) and Ravelo (1985).

Field surveys of barley plots for the 1984–1985 crop growing season in the vicinity of Riobamba provided information on growth stage dates, plant density, kernel weight, and kernel number per ear at harvest time. Barley yields were obtained by random sampling of selected crop fields. The surveyed fields were sown in early October 1984, with a plant density of 35 plants/m^2.

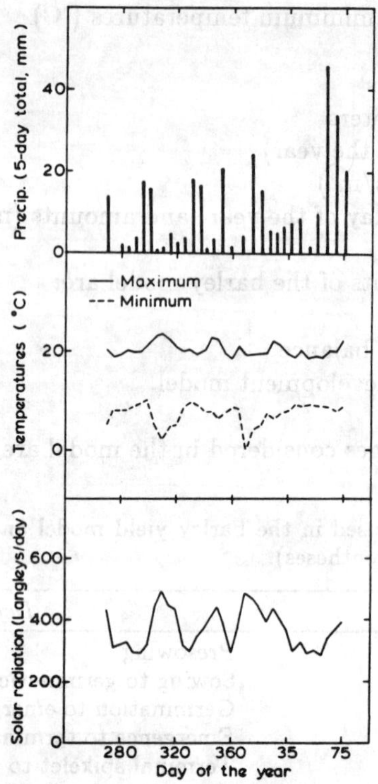

Figure 4.2. Precipitation (5-day totals), maximum and minimum temperatures, and solar radiation (5-day averages) during the 1984–1985 crop growing season at Riobamba.

The meteorological data needed to run the model were obtained from the Riobamba station. *Figure 4.2* shows precipitation, temperature and solar radiation data used in the calibration of the barley model. Precipitation was low during late January, February and early March. Two periods with minimum temperatures between 0° and 2 °C were observed in early November and early January, but no crop damage was reported.

Figure 4.3 shows the model's estimates of potential and actual evapotranspiration and moisture content in the four soil layers (0–5, 5–15, 15–33 and 33–51 cm depths) considered in the analysis. Potential evapotranspiration is estimated using an equilibrium evaporation concept as modified by Priestley and Taylor (1972), and actual evapotranspiration is calculated using procedures developed by Ritchie (1972). The soil moisture balance was run from 20 days before sowing to maturity stage. During the dry period in February and March, soil water reserves were depleted in all layers, and actual evapotranspiration was drastically reduced.

Daily model output for root weight, stem weight, leaf weight, grain weight and leaf area index (LAI) are shown in *Figure 4.4*. Observed grain weight is shown for comparison. Observed and estimated grain yields are 365 kg/ha and 388 kg/ha, respectively. This indicates that the model can be calibrated to

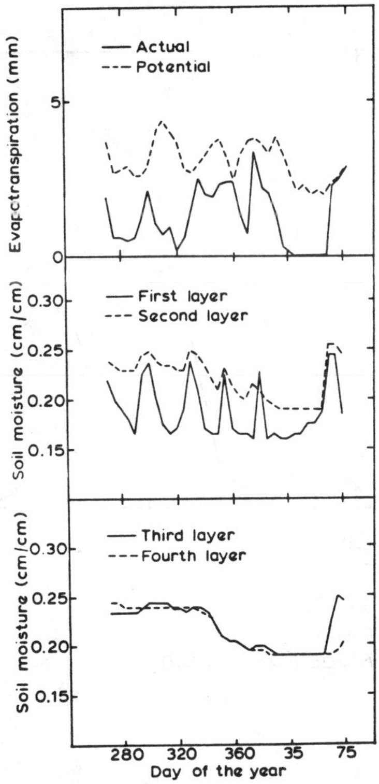

Figure 4.3. Evapotranspiration and soil moisture content during the 1984–1985 growing season at Riobamba.

estimate grain yield in these dry conditions fairly accurately. The low yield recorded is a clear reflection of the below-normal precipitation during critical crop reproductive stages. A crop sown early, for example, in September, may have avoided the adverse impact of water deficit during the critical moisture-sensitive stage.

In spite of the encouraging results for the 1984–1985 crop growing season, the model needs further testing with additional crop data before it can be used operationally.

4.4. Agroclimatic Data

Daily rainfall data were obtained for Palmira (1° 44′S, 78° 39′W, 3240 m altitude) for the period 1964 to 1983. No temperature data were available for this station. Therefore, daily maximum and minimum temperatures for the same period of record were obtained for Riobamba (1° 39′S, 78° 39′W, 2769 m) and adjusted by −2.3 °C (Cañadas, 1983) to account for the elevation difference between Palmira and Riobamba. Daily solar radiation was estimated by the method described by Richardson (1981), using daily temperatures and rainfall data.

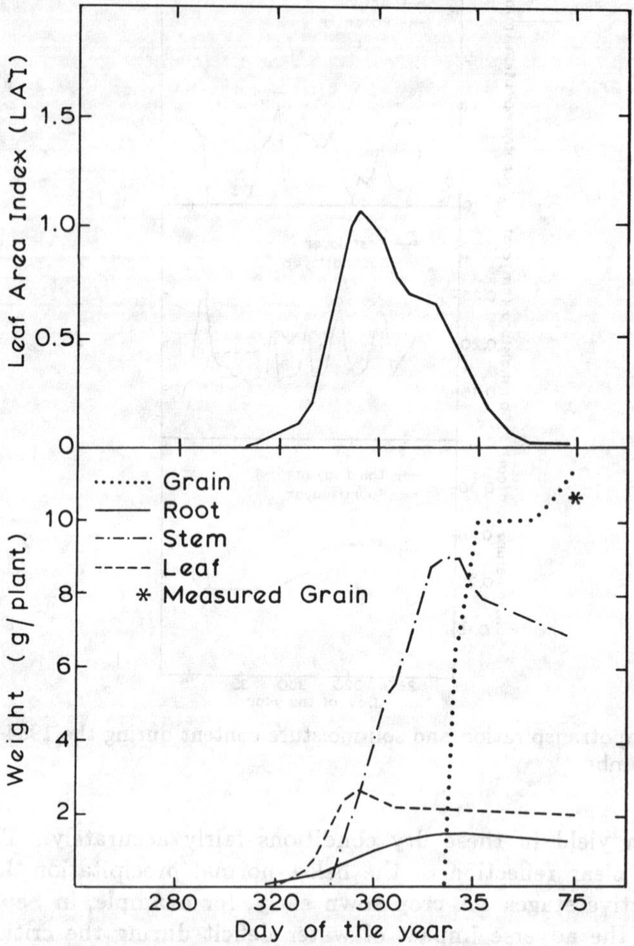

Figure 4.4. Daily model output for the 1984–1985 crop growing season.

Crop calendar information for the area is very limited. Barley sowings normally take place between October and December. Harvest occurs 5 to 8 months after sowing (Banco Nacional de Fomento, 1982). A longer crop growing season, 7 to 11 months, has been reported for the Chauzan–Galte area in Section 3 of this case study.

Three barley varieties are commonly cultivated in the area. Two improved varieties are "Dorada", cultivated on a large scale for human consumption, and "Dulchicela", cultivated on a lesser scale for both human consumption and beer production. An unimproved barley variety, "Riobambena comun", is also cultivated. Grain yield, plant density and phenological information were obtained from field surveys (Ravelo, 1985).

4.5. The Choice of Scenarios

Two groups of scenarios were selected to assess the respective impacts of climatic (specifically, precipitation) variability and of agricultural practices on barley yields.

4.5.1. Precipitation

Historical seasonal precipitation (October–June) was computed and ranked in percentiles. The percentile ranking was used to select seven crop growing seasons, ranging from very wet to very dry.

4.5.2. Agricultural practices

(1) *Sowing dates and plant density.* Six sowing dates at 15-day intervals, starting October 15, and 5 plant densities (25, 35, 45, 55 and 65 plants/m^2) were analyzed for the selected crop growing seasons. These plant densities represent the real range.

(2) *Fallowing.* Short (less than 12 months) and long (longer than 12 months) fallowing practices were considered for two "dry" seasons.

4.6. Analysis

4.6.1. Precipitation scenarios

Seasonal rainfall variability is the major factor determining the success or failure of subsistence agriculture in the highlands of Ecuador. Rainfall amount and distribution affect every crop grown in the farming communities of Galte and Chauzan. *Figure 4.5* shows the historical seasonal rainfall for Palmira and Riobamba.

The crop life cycle from sowing to harvest is complete when the plant growth stage requirements are fulfilled. Each growth stage has specific water needs; when the environment fails to provide optimum water supply, metabolic activities such as transpiration, respiration and photosynthesis are affected. Under water stress conditions, the crop undergoes incipient, transient or permanent wilting. Deficient rainfall during periods with high evaporative demand may lead to plant water stress and crop failure or reduction of yields. In the other extreme, excess rainfall may damage crops by flooding, soil erosion or reduced oxygen uptake by plant roots due to soil saturation. The model does not account for the effects of water excess. In addition, the prolonged cloud cover associated with above-normal rainfall reduces solar radiation and photosynthetic rates. Photosynthesis rates are affected more by reduced radiation during the early growth stages than during the senescent and ripening phases.

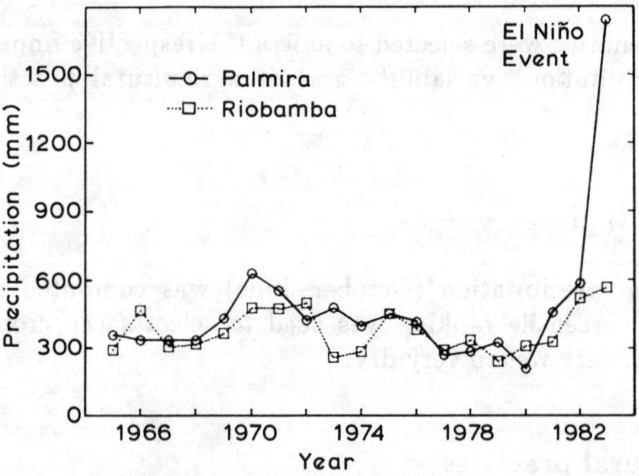

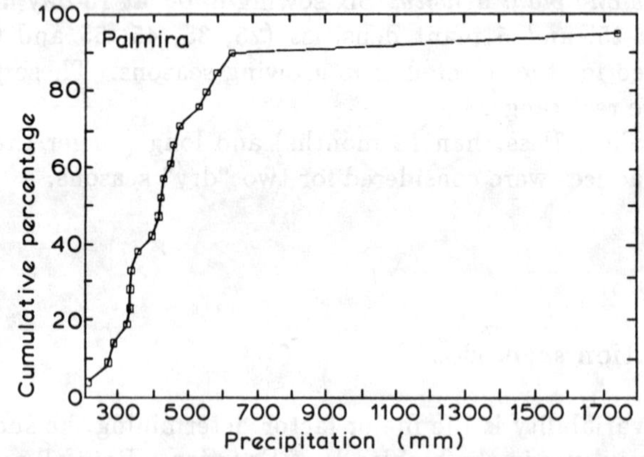

Figure 4.5. (*a*) Seasonal precipitation (October–June) for Riobamba and Palmira, 1965–1983. The extreme amount recorded in 1982–1983 at Palmira is coincident with the occurrence of a pronounced El Niño event at that time. This is discussed in Section 3. (*b*) Cumulative percentage curve for seasonal precipitation at Palmira.

The impact of rainfall on crop yield components was analyzed using the barley yield model. The parameters used as inputs to the model were six sowing dates at 15-day intervals starting October 15, and a plant density of 35 plants/m^2. Initial soil moisture was arbitrarily set at 35% of field capacity, and computation of the soil moisture balance was begun three weeks before sowing date.

Table 4.2 shows the selected crop growing seasons and model results for major barley yield components. Seven rainfall scenarios ranging from extremely wet (1982–1983) to extremely dry (1979–1980) were analyzed. Barley yield and components of the yield clearly reflect the impact of rainfall amount and distribution on crop productivity.

Table 4.2. Barley yield components under selected precipitation scenarios for Palmira. Values are averages of six sowing dates except where indicated.

Scenarios	*Seasonal (Oct–June) rainfall* (mm)	*Percentile rank of seasonal rain*	*Crop growing season (Oct–June)*	*Mean biomass at maturity* (g/m^2)	*Mean number of ears per* m^2
Extremely wet	1744	95	1982–83	515	91
Very wet	584	85	1981–82	576	112
Moderately wet	451	65	1974–75	461	123
Normal	416	50	1975–76	394	176
Moderately dry	337	35	1966–67	343	156
Very dry	269	10	1976–77	186	105
Extremely dry	210	5	1979–80	89	28

Scenarios	*Mean number of grains per* m^2	*Mean kernel weight* (g)	*Early sowing date (Oct. 15) yield* (kg/HG)	*Late sowing date (Dec. 30) yield* (kg/ha)	*Mean yield* (kg/ha)
Extremely wet	3667	0.042	1480	1827	1540
Very wet	4061	0.043	1352	1898	1746
Moderately wet	3347	0.038	867	671	1272
Normal	2597	0.038	231	1931	987
Moderately dry	1676	0.024	615	456	402
Very dry	1055	0.028	88	1120	295
Extremely dry	825	0.020	0	528	165

The model shows high yields for all wet scenarios. The 1981–1982 growing season had the highest modeled average yields (1711 kg/ha), due to more appropriate rainfall amount and distribution. Excessive rainfall was reported during the 1982–1983 season when El Niño occurred (*Figure 4.5*). The Palmira area is under the influence of maritime air masses from the Pacific Ocean, which caused the above-normal rainfall during 1982–1983. During this extremely wet season, the modeled photosynthetic rate was diminished by prolonged cloud cover, and the numbers of ears and grains per square meter were reduced.

The 1975–1976 crop season appears to have been a "normal" year (50 percentile ranking on rainfall amount) with an average yield of 1042 kg/ha. There is a large yield variability for the different sowing dates, ranging from 231 to 1931 kg/ha. These yield variations are a result of the distribution of rainfall in relation to the different crop growth stages.

The dry scenarios show low modeled yields. Low rainfall decreased the values of crop yield components, particularly the number and weight of grains. The 1979–1980 season shows a complete crop failure for the early sowings. The 1976–1977 season shows very low yield except for the latest sowing date. For the first five sowing dates, crops were subject to severe water stress during the moisture-sensitive crop growth stages of flowering and grain filling [*Figure 4.6(a)* and (*b*)]. The index ranges from 0 (no stress) to 1 (maximum stress). For the latest sowing date, crops exhibit no water stress during the sensitive growth stages of flowering and grain filling.

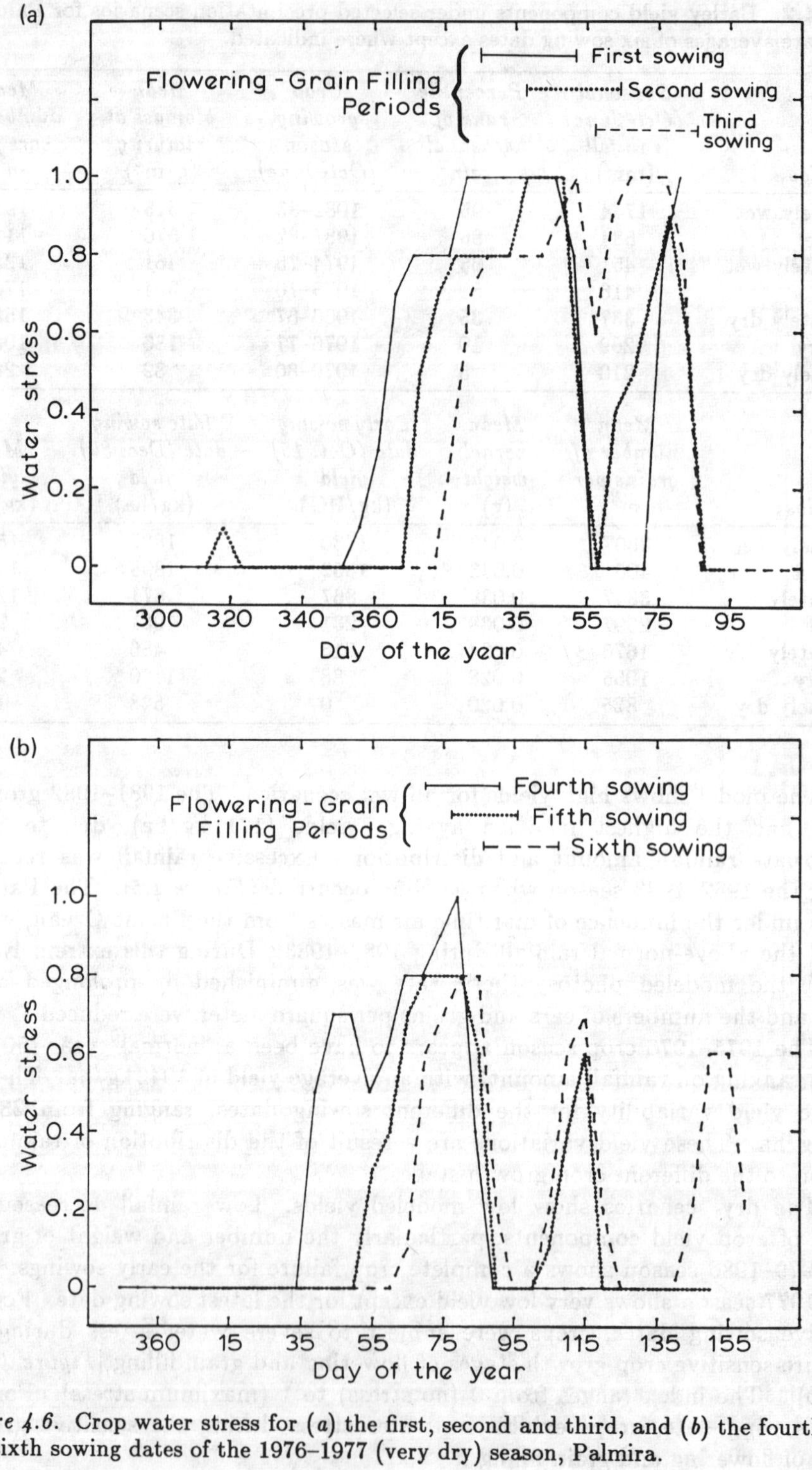

Figure 4.6. Crop water stress for (*a*) the first, second and third, and (*b*) the fourth, fifth and sixth sowing dates of the 1976–1977 (very dry) season, Palmira.

4.6.2. Agricultural practices

Sowing dates and plant density

Sowing date and plant density are variable in the farming communities of Galte and Chauzan. Sowing occurs sometime in October, November or December. Plant density varies from about 25 to 65 plants per square meter. Both sowing date and planting density depend on plot location, seed and work force availability, and traditional practices. Six sowing dates at 15-day intervals, starting October 15, and for 5 plant densities (25, 35, 45, 55 and 65 plants/m^2) were analyzed for wet and dry seasons to assess the impact of sowing dates and plant density on barley yield.

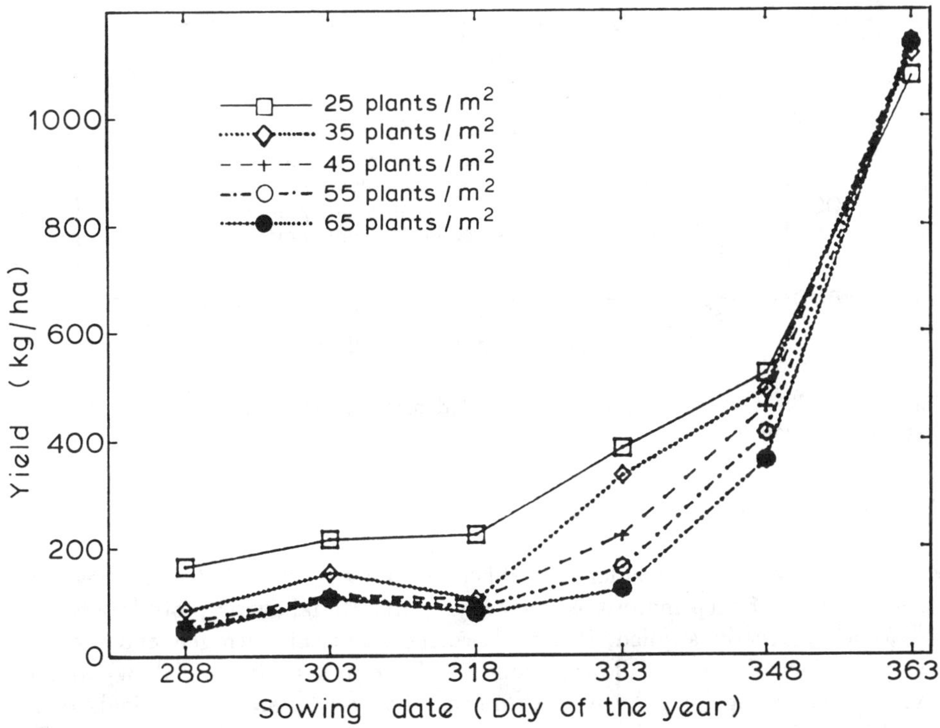

Figure 4.7. Effect of planting date and plant density on barley yield for the 1976–1977 (very dry) season, Palmira.

The responses of yield to sowing date and plant density were different for very dry (1976–1977) and very wet (1981–1982) seasons (*Figures 4.7* and *4.8*). Model results for the 1976–1977 season indicate that the plots with the lowest plant density, 25 plants/m^2, yielded more than plots with high plant density for the early sowing dates. The model also indicates that the latest sowing date (December 30) has a considerably higher grain yield for all planting densities.

The highest modeled yields occurred at higher plant density in 1981–1982. The crop sown on October 15 had the lowest yield for all planting densities. When water is not a limiting factor, high plant densities can increase crop yields by 30 to 40% (*Figure 4.8*).

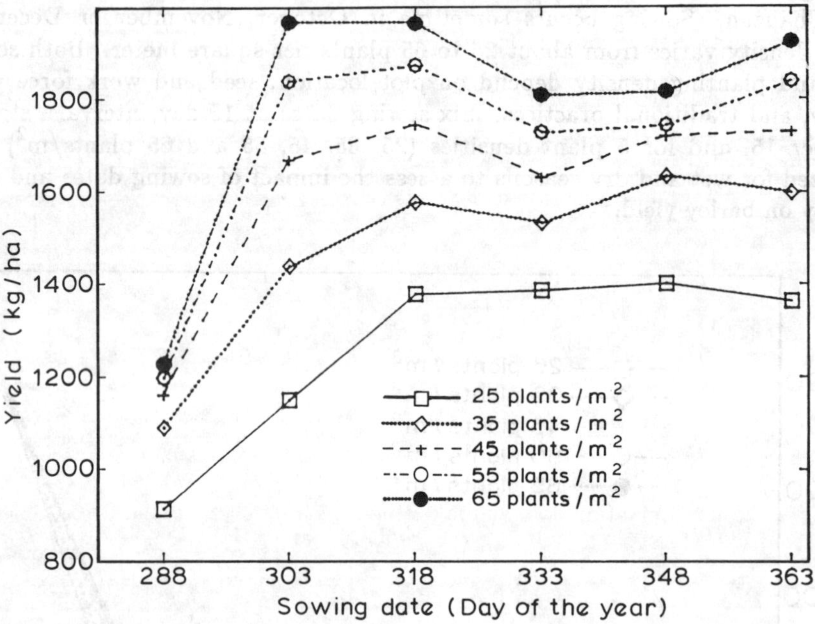

Figure 4.8. Effect of planting date and plant density on barley yield for the 1981–1982 (very wet) season, Palmira.

Fallowing practices

Farmers in the region considered in this study may leave their fields fallow for a year or longer. Sheep manure is normally applied to fallowing plots (*see* Section 3). When correctly applied, this practice increases soil nitrogen and moisture reserves. Two different fallowing schemes for one year (short fallow) and for several years (long fallow) were examined. The model simulated yields of fallowed crops by assuming increased soil moisture at sowing. The nutrient effect of manure was not included. The soil moisture balance was computed for four soil layers. The upper layer (5 cm depth) was assumed to be at 50% of field capacity for either fallowing practice. The three lower layers were assumed to be at field capacity for the long fallow and 70% of field capacity for the short fallow. The lack of information on amounts of organic nitrogen applied to plots prevented the use of the model to assess nitrogen fertilization effects on barley yields. Therefore, modeled yields are probably underestimating reality.

The response of yield to sowing date was different for a very dry (1976–1977) and an extremely dry (1979–1980) season (*Figures 4.9* and *4.10*). Yield obtained from crops with no fallowing are shown for comparison. In all

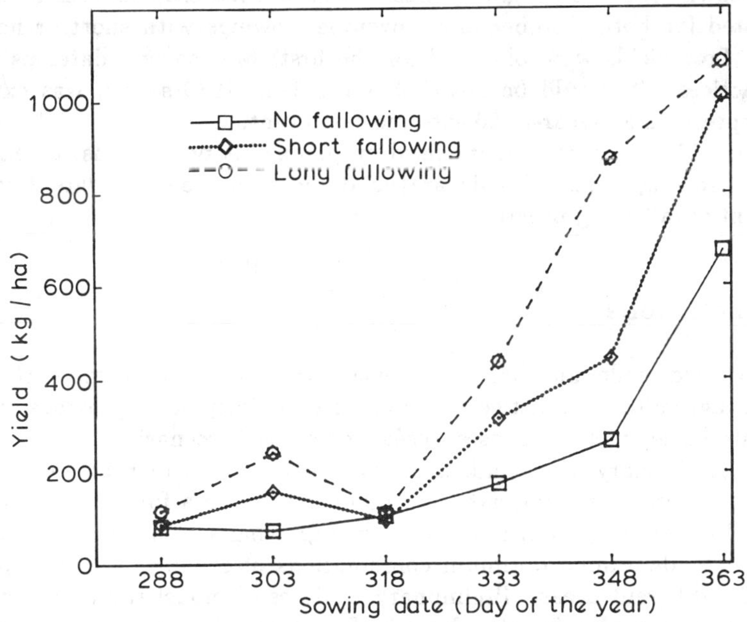

Figure 4.9. Effect of planting date and fallowing on barley yield for the 1976–1977 (very dry) season, Palmira.

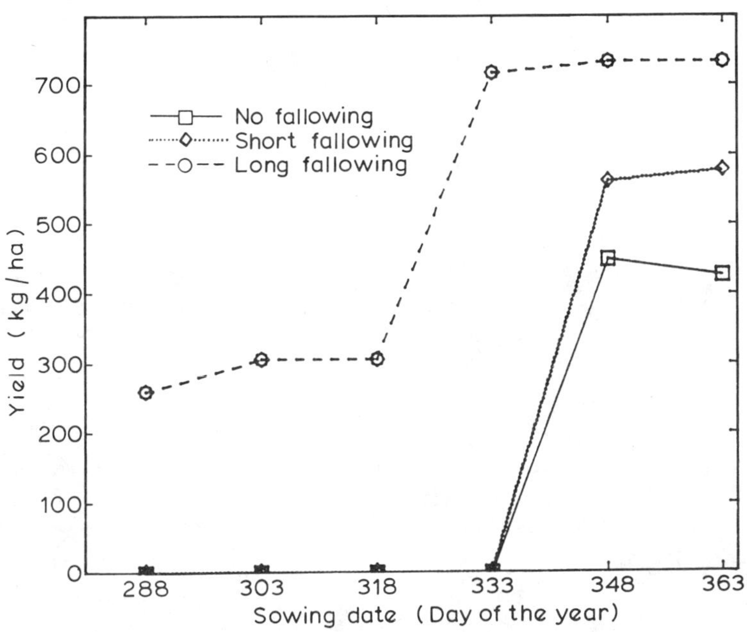

Figure 4.10. Effect of planting date and fallowing on barley yield for the 1979–1980 (extremely dry) season, Palmira.

cases late-sown crops had higher yields. In 1979–1980 crop failures or no yield are indicated for both October and November sowings with short or no fallow practices. Low yields were obtained for the first three sowing dates using long fallow practices. It should be noted that the 1979–1980 season was extremely dry and represents a 1-year-in-20 extreme case event.

In the 1976–1977 season, modeled crops exhibit poor yields for early and mid-season sowing dates. Yields are good for crops sown in late December, independent of fallowing practices.

4.7. Conclusions

The barley yield model appears to be a useful tool for assessing the effects of climate and agricultural practice on crop productivity and provides valuable insights into barley yield responses to selected climatic scenarios.

This preliminary work indicates that there is considerable potential for determining optimum sowing date and plant density as a function of initial soil moisture. Future analysis of additional crop growing seasons and locations will be required to determine optimum combinations of sowing dates and planting densities for different areas. Preliminary analyses of model results show that it is possible to reduce significantly the risk of crop failure by using the appropriate plant density, implementing fallowing practices and sowing at optimum dates.

Although the test results for the 1984–1985 crop season were encouraging, additional testing and calibration will be required before the model is used operationally.

SECTION 5

The Effects of Variations in Mean Temperature and Frost Risk

5.1. Aims

Villagers identify frost as one of the major hazards in the Ecuadorian highlands (*see* Section 3), and a considerable area is uncultivated owing to excessive frost risk or excessively low temperatures. This section explores the vulnerability of traditional agriculture to possible variations in mean temperature and frost risk. After an initial discussion of farmer adaptation to frost, meteorological records are analyzed to determine the patterns of fluctuation of frost risk and mean temperature. A plausible scenario of increased seasonal frost risk will be defined and used to determine areas potentially at risk in Chimborazo Province.

5.2. Previous Studies

Biogeographers have long been intrigued by the altitudinal zonation of crops and vegetation in tropical mountains. Each crop and plant species has an observed upper limit to its distribution, although there are disagreements on the precise climatic factors responsible for setting the upper limit of tree growth. Troll (1968) felt that the critical factor was risk of nocturnal frost. Walter (1973) emphasized soil temperature; and Ellenberg (1979), low effective radiation. The Holdridge (1967) system of ecological classification places emphasis on mean annual temperature. All of these factors are, of course, closely correlated with elevation. Frost risk and radiation are also affected by topographic position, while soil temperatures are related to soil properties.

Local correlations of altitude with crop distributions are so striking that studies of Andean agriculture are usually couched in an altitudinal framework (Brush, 1977; Gade, 1967; Guillet, 1983; Mayer, 1979). The notion of "verticality" which provided the framework for much ethnography and ethnohistory in the 1970s, explicitly assumed that the most significant agricultural problems and opportunities in the Andean environment were posed by factors associated with altitude. It was assumed that Andean farmers strive to maximize access to ecological, altitudinal zones (Brush, 1977; Masuda *et al.*, 1985; Murra, 1972; for a critique of this position *see* Sanchez, 1976).

Gade (1967) makes a useful distinction between the effective and absolute limits of cultivation in the Andes. The effective limit is the upper limit of the "agricultural region continuously enclosed by fields", e.g., of the region where people are critically dependent on their harvests. In the Vilcanota Valley of Peru, this limit was where hardy crops yielded "satisfactorily" in at least 60% of the years. Above this effective limit, one finds scattered fields of agro-pastoralists not critically dependent on good harvests. These fields are found up to the "absolute limit" of cultivation, which in this valley was 420 meters higher than the effective limit.

There has been only limited attention given to fluctuations in the upper limit of cultivation. Fossil cultivation ridges may exist up to 4300 m elevation in the Vilcanota Valley of Peru; this is above the current effective limit of cultivation but below the current absolute limit (Gade, 1967). At the Lauricocha Plain and environs (3950–4000 m) in Peru, farmers stated that cultivation was very risky 30–40 years ago due to frost, but satisfactory in the early 1970s. Nearby walled parcels of land suggest that the upper limit of cultivation was even higher – about 4100 m – during the Early Intermediate Period (200 BC – 700 AD) (Cardich, 1975). Gondard and Lopez (1983) noted that abandoned terraces or semi-terraces in northern highland Ecuador are found at higher elevations that the currently observed upper limit of cultivation. Abandoned field ridges were found in flats at 3150 m, higher than the current limit of cultivation in flats.

These studies are difficult to evaluate because, first, we do not know whether the apparent shifts in upper limit of cultivation were due to changes in social factors (effective limit) or climatic factors (absolute or effective limit) and, secondly, the changes may have been due to use/abandonment of frost management technology (e.g., walls, ridges, etc.).

Little work has been done on the geography of frost hazard in the Andes. Lauer (1982) has demonstrated patterns of frost hazard as a function of relief in Bolivia, but the strongly seasonal temperature regime there is very different from the equatorial temperature pattern in Ecuador.

5.3. Data Sources

In this study we will rely for the most part on data collected by national meteorological stations in order to estimate patterns of temperature and frost risk. Monthly temperature data (means, absolute minima, etc.) for a limited number of highland Ecuadorian meteorological stations are available in the INAMHI (Instituto Nacional de Meteorologia e Hidrologia) *Meteorological Annuals* (*Anuarios Meteorologicos*), available as of this writing for the years 1959 to 1981. More recent data are available on computer printouts or on the original station report cards on file at INAMHI. The data have been subjected to limited quality checks. Meteorological stations were visited in the mid-1970s, and the state of the instruments, the location and the reliability of the observers were noted (PRONAREG-ORSTOM, 1976, 1977a). Partially corrected data are found on photocopied tabular sheets in the archives of PRONAREG (Programa Nacional de Regionalizacion Agricola).

Especially useful for our purposes are the readings from thermometers in the station shelters (2 m above the ground), and (in selected cases) from thermometers at ground level. The latter thermometers, found at Quito observatory and some other stations, are lying horizontally on a grass surface. Temperature readings are published for 7:00 am, 1:00 pm and 7:00 pm. Mean daily temperatures are the mean of these three values, and absolute minimum temperatures reflect the minimum of these three values (normally the 7:00 am value). In a few cases (e.g., at Quito Observatory), readings are made 24 times a day, and averages and extremes are based on all 24 readings.

As in Section 2, we will use data from Quito Observatory (2828 m) for information on long-term trends in temperature and frost. This station has, by far, the longest run of such data in Ecuador.

For the purpose of studying altitudinal lapse rates in temperature and frost risk, seven stations from the central and northern highlands were chosen (cf. *Table 5.2.*). The stations are known to have good-quality data, have known topographic situations, and are not located in large cities. It would obviously be desirable to have a larger and better-controlled data base. The present study, therefore, should be considered a preliminary exploration.

As part of a larger project, the author (Knapp, 1984) collected a variety of data relevant to the determination of frost risks and adaptations in the Ecuadorian highlands. Information was derived from:

(1) Analysis of long-term hacienda records of yields in relation to environmental factors of precipitation, soil quality, slope and elevation (*Figure 5.1*).
(2) Interviews with small-farm operators in western Chimborazo and southern Imbabura Provinces.
(3) Archaeological study of fields and settlements.
(4) Experimental study of the possible frost-amelioration functions of traditional and prehistoric technologies.

5.4. Agricultural Adaptation to Low Temperature and Frost

Observation suggests that maize is grown to 3100–3400 m in the equatorial Andes, while other crops (potatoes, barley) are grown to 3600–3800 m. As altitude increases, mean annual temperature decreases, and maturation time for crops increases. Common maize and potato varieties require two more weeks of maturation time for every 100 m increase in elevation (Knapp, 1984). Long maturation times make it difficult to pursue a consistent cultivation cycle year after year, constrict the time available for harvest and field preparation for the succeeding crop, lengthen the time during which the growing crop must be protected against pests and theft, increase crop vulnerability to climatic hazards (wind, drought, excessive rain, hail, frost) and make it more difficult to match stages of crop growth with appropriate climatic periods. As elevation and maturation times increase, so does the incidence of the frost hazard.

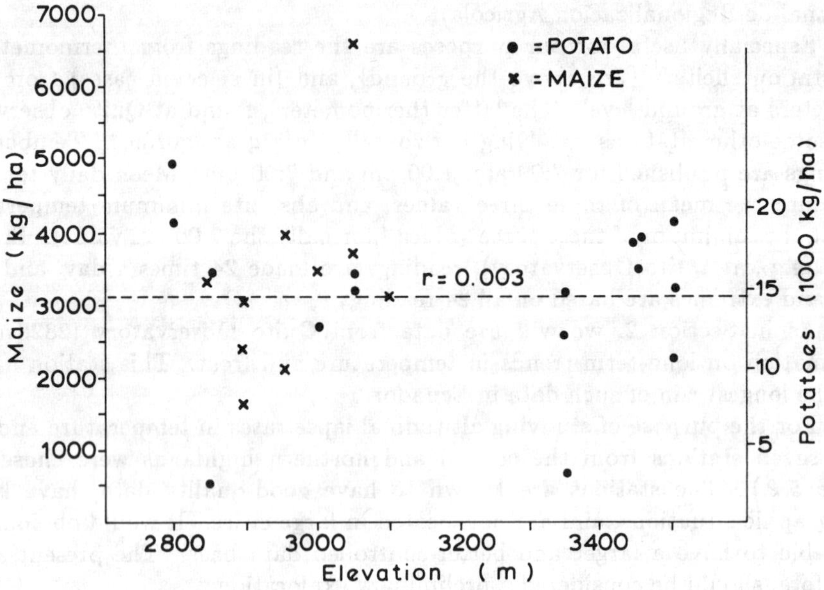

Figure 5.1. Relation between yields of potato and maize and elevation, northern highlands of Ecuador. Values of yields assume optimal fertilization. Source: Hacienda Record Books analyzed by Knapp (1984).

Furthermore, lowered temperatures may reach critical thresholds of plant functioning: maize, a C-4 plant type, is especially temperature-sensitive near 10 °C. Thus, farmers at higher elevations shift to crops with shorter maturation times or less sensitivity to lower temperatures.

In most cases, the high-elevation crops also do perfectly well at lower elevations. If there is a clear *lower* limit of cultivation of high-altitude crops (sometimes the case with potatoes, for example), this is often not due to maladaptation to warmth but rather to the existence of more attractive options. Maize, for example, requires less labor to harvest than potatoes, and subsistence farmers tend not to plant potatoes where maize can be grown.

Other things being equal, yields do not decrease with increasing elevation. Limited data from haciendas suggest that both maize and potato yields are constant throughout their altitudinal range (*Figure 5.1*). The same data source indicates that barley and wheat yields are also constant throughout their ranges (Knapp, 1984). Moreover, if maize and potato yields are expressed in calorie equivalents, calorie productivity per hectare is unchanged by the shift to potatoes (the *y*-axes for potatoes and maize in *Figure 5.1* are calibrated to be equivalent in calorie terms).

Since potatoes require more labor than maize at equal levels of calorie/hectare yields, in practice farmers often equalize their labor efficiency across zones by fertilizing and fallowing potatoes more than maize. Thus, the upper limit of maize is often the upper limit of short-fallow or annual cropping. The higher elevations are marked by long-fallow cultivation and/or rotations of potatoes with low-yielding legumes and labor-efficient small grains (*see* Section 3).

A decline in mean annual temperature, therefore, should not normally reduce crop yields, which may even increase if high evapotranspiration is a problem (*see* Section 4). Increased maturation times or the approach of critical temperatures may, however, lower the upper limits of cultivation of crops. As a result of variations of temperature, land may thus be lost to any cultivation at the upper limit of cultivation. Land may also be lost to intensive maize cultivation and be subject to the "disintensification" associated with potato cultivation. This latter problem will not be discussed further here.

Frost does least damage to crop plants when the seedlings are still in the ground or when the seeds or tubers have already matured. Nocturnal frosts in Ecuador are most dangerous when young plants extend less than a decimeter from the surface (Field and Chiriboga, 1984). For this reason (among others) sowing is staggered across several months of plantings (*suertes*), which, depending on elevation, may extend into February. Later plantings are not feasible because critical phases of crop growth would then occur in the driest and most frost-prone time of the year. The economically most damaging frosts are those that occur during the middle and late stages of crop growth, when considerable investment in weeding, ridging, and pest control has already been committed to the planting, when it is no longer possible to plant another *suerte* and when the crop may be susceptible to severe damage (e.g., at anthesis).

The practice of planting *suertes* may be seen as an example of the "minimax strategy" in game theory, where an attempt is made to minimize the maximum possible loss. If this is so, then farmers will time crucial phases of crop growth to coincide with the least risky months, which in the Ecuadorian highlands are January to May and especially March to May (*Table 5.1*).

Table 5.1. Risk of ground frost[a] by month, Quito Observatory (1891–1980).

	Month											
	J	*F*	*M*	*A*	*M*	*J*	*J*	*A*	*S*	*O*	*N*	*D*
Months with frost (%)	30	20	15	6	16	37	67	66	48	30	37	38

[a] Daily minimum grass temperature $\leq -0.1^\circ C$.
Source: Manuscript records of Quito Observatory.

Evidence for the special significance of these low-risk periods is found in farmers' folk terminology and perception. In Chauzan and Galte, the most dangerous frosts are said to occur in May–June (Section 3). At San Pablo (Imbabura Province), frosts are named after the nearest religious holiday – the earliest being Epiphany frosts (occurring in January); and the latest, Palm Sunday frosts (April).

Both frequency of frosts and absolute maximum frost may be significant for crop yields. Because of the nature of the data base, we will focus on absolute maximum frost and assume that the March–May season (late winter or late growing season) is the period of most interest for the study of frost risk. We now look in more detail at the spatial variation of significant temperature-related problems for agriculture.

5.5. The Geography of Mean Annual Temperature and Late Winter Minima

Using mean data for highland stations in northern Ecuador for the 1964–1973 period, PRONAREG-ORSTOM (1977b) has determined that the normal lapse rate for temperature (T, in °C) as a function of altitude (A, in km) is:

$$T = 28.4 - 5.7A \tag{5.1}$$

Stations with mean annual precipitation under 600 mm had mean annual temperatures about one degree higher than would otherwise be expected. These results are in rough accordance with the correlations obtained by Frere *et al.* (1975). They differ only slightly from those obtained by Cañadas (1983), who included subtropical and tropical in addition to highland data in his regression to arrive at a somewhat lower (5 °C/km) lapse rate.

Because of its equatorial situation, highland Ecuador's temperature pattern varies diurnally more than it does seasonally. The correlation with elevation is normally so close that soil temperature has been used in the field to infer elevation. Drier areas are, however, somewhat warmer; and since the western cordillera is generally drier than the eastern (particularly the eastern slope of the western cordillera), the western areas are warmer and have a higher limit of cultivation (Cañadas, 1983; Gondard, 1984a).

We know much less about geographic patterns of minima or of frosts than we know about the patterns of mean annual temperature. An analysis of the lapse rates of mean daily minimum temperatures of 19 northern highland stations in Ecuador indicated the altitudinal lapse rate was about 4.6 °C per km, somewhat lower than the mean temperature lapse rate (Knapp, 1984).

Frosts in the highlands of Ecuador are almost entirely nocturnal (radiation) frosts, as cold fronts do not normally penetrate this far north. Conditions that favor frosts are clear, windless, cloudless nights and low vapor pressure. These occur with more frequency during the dry seasons, and it is at these times that nocturnal frosts are most likely (*Table 5.1*). In view of the discussion in Section 4, we will focus in particular on frosts during the least risky period, March–May.

Nocturnal frosts are highly correlated with relief and topography, and we can expect frosts to be more severe on flats than on slopes due to the action of cold air drainage (Geiger, 1965). To obtain a first approximation of relations between altitude and frost risk, we give in *Table 5.2* data from selected meteorological stations. The mean annual temperatures of the seven stations are approximately those expected for their altitudes. Regression of mean annual temperature in °C (T) against altitude in km (A) yielded the equation:

$$T = 29.1 - 5.9A \tag{5.2}$$

with lapse rate and sea-level values similar to those in equation (5.1).

In the last three columns we have calculated the average absolute late growing season (March–May) minimum temperature for each of the stations, together with measures of standard deviation and number of observations. There is some indication in the data that March–May absolute minima are log-normally distributed. Medians are typically 0.3–0.6 °C higher than arithmetic means. Calculations of geometric instead of arithmetic means would not, however, appreciably change the absolute or relative values in *Table 5.2*.

Table 5.2. Mean annual temperature and late growing season minimum temperatures at seven stations in the Ecuadorian highlands.[a]

Station	*Elevation* (m)	*Mean annual temperature* (°C)	*Situation*	*Average absolute late growing season minimum air temperature*[b] *Mean* (°C)	*Std. dev.* (°C)	*Number obs.*
Ambato (INERHI)	2540	13.0	Open flat	4.7	1.4	13
San Pablo	2700	13.6	Open flat	3.1	1.3	18
Riobamba Airport	2796	13.4	Open flat	2.3	1.8	17
Izobamba	3058	10.9	Open; almost flat	2.4	1.1	16
Pachamama	3600	7.7	Open flat	−0.2	1.4	12
Tabacundo	2955	12.9	Slope	4.3	1.4	14
Cotopaxi	3560	8.1	Slope	1.1	0.9	18

[a]For Tabacundo and San Pablo: data for 1962–1980; for all other stations: 1959–1962, 1964–1966, and 1968–1979.
[b]March–May; Meteorological Station Shelters.
Source: Data in INAHMI *Meteorological Annuals*; collected by G. Knapp, L. Cañadas and O. Rovere.

The absolute seasonal temperature minima are measures of most severe frost rather than frost frequency. The temperatures reflect conditions in the meteorological station shelter at 2 m elevation above the ground surface. On the coldest nights, there is likely to be a strong surface inversion due to net radiation loss at the surface and cold air pooling on flats. Thus, as much as 4 °C should be subtracted from the mean figures in *Table 5.2* to approximate the most severe frost at the surface (Knapp, 1984). With the exception of Ambato, all flats may be expected to experience some slight frost during an average March–May season.

The values in *Table 5.2* allow us to make a preliminary calculation of lapse rates of March–May absolute minima. If the minima (M, in °C) are regressed against altitude (A, in km), we arrive at the following equations:

$$M = 14.4 - 4.1A \text{ (flats)} \tag{5.3a}$$

$$M = 19.9 - 5.3A \text{ (slopes)} \tag{5.3b}$$

These have no statistical significance since the slope equation is based on only two observations. However, there is a suggestion that lapse rates of minimum temperatures are less than lapse rates for average temperatures, at least on flats. The minimum temperature at the slope station Tabacundo is about 2.0 °C warmer than would be expected from its elevation if it were a flat; and Cotopaxi is about 1.3 °C warmer than would be expected if it were a flat. Flats 488 m below Tabacundo (a slope location – *see Table 5.2*) have the same theoretical absolute minimum temperature in the March–May period.

The author conducted frost risk experiments in 1981 at 2800 m, with minimum-reading upright thermometers placed on an open grassy slope (4% gradient) and an adjacent open grassy flat. Readings were taken during one week in August and another week in December. The average slope minimum was 0.6 °C less than the average flat minimum, but the difference between slope and flat temperatures increased on colder nights; the difference between the absolute slope minimum and the absolute flat minimum was 1.6 °C (Knapp, 1984).

A number of other factors may also affect the temperature near the ground at night. Soils with low thermal conductivities, such as recently cultivated soils, dry soil and soils of low bulk density (e.g., volcanic soils), can produce mean minima one to two degrees colder in the plant layer than soils with high thermal conductivities (Geiger, 1965). Risk of frost also depends on the degree of protection from sources of cold air (through construction of hedges or forest belts) and the provision of means for the drainage away of cold air (through manipulation of the arrangement of crop plants) (Geiger, 1965).

5.6. Relation of Frost Risk and Mean Temperatures to Cultivation Limits

Figure 5.3 shows the upper limit of cultivation in Chimborazo Province in relation to the 3600 m contour. This does not delimit "potential" areas, such as those discussed in Sections 6 and 7, but observed cultivation derived from air photographs and field work in the late 1960s and early 1970s. Cultivated area, mapped by Gondard and his associates, includes fallow (Gondard, 1984b; PRONAREG-ORSTOM, 1983, 1984).

This map indicates that in eastern Chimborazo Province the upper limit of cultivation is below 3600 m, while in western Chimborazo Province the upper limit exceeds 3600 m. In the latter case, the limit at times aproaches 3800 m. This regional variation in upper limits may in part be due to geographical variations in land tenure, the organization of agriculture and land use. The upper slopes in much of eastern Chimborazo Province are controlled by haciendas that use these lands as cattle pasture. Elsewhere, the upper slopes are in small farms or farm communities. It is also possible that the differences reflect differences in rainfall: the western cordillera is drier than the eastern at equivalent elevations in Chimborazo Province, and thus should experience slightly higher mean temperatures. It will also be noted that west-facing slopes have low cultivation limits, perhaps due to slower warm-up during critical early morning hours, which increases the duration of frosts and decreases time available for plant growth.

At 3800 m elevation, both lapse rate equations (5.1.) and (5.2) suggest a mean annual temperature of about 7 °C. This is consistent with readings at nearby Pachamama (*Table 5.2*). Late growing season absolute minimum temperature at 3800 m can be estimated as −0.2 °C [equation (5.3b) for slopes], suggesting a 50% risk of frost at shelter height.

The upper limit of cultivation in flats is lower than the upper limit of cultivation on slopes. The highest intensively cultivated flat in Chimborazo is that at Llano Grande (3080 m), which corresponds with Gade's (1967) effective limit of cultivation. Some cultivated fields may be found as high as Colta (3300 m), which perhaps is the absolute limit.

Use of the lapse rate equation (5.3a) for flats indicates that March–May absolute minimum temperature should be about 1.8 °C at Llano Grande and 0.9 °C at Colta. These values are somewhat higher than those at the upper limit of cultivation on the slopes. The lapse rate equation for flats is, however, less well calibrated than the equation for the slopes. Also, it is probable that nocturnal lapse rates between the ground and shelter height are steeper on flats. It is possible that flats lack the benefits of early morning warming experienced on east-facing slopes. Finally, farmers have more attractive options for agriculture on flats than they do on slopes. Flats may be used as high-quality meadows for cattle grazing.

It is probable that the upper limit of economically acceptable cultivation on flats is set by frost risk, since mean annual temperatures are higher than at the upper limit of cultivation on slopes. On slopes it is much less clear what determines the upper limit. As discussed earlier, potato and barley yields do not diminish as the upper limit of cultivation is approached. If mean temperatures play a role, the effect is sudden rather than gradual (perhaps increased maturation times push critical crop phases into unsuitable seasons). In moist northern Imbabura Province, none of eight hacienda potato plantings in the upper 3450–3600 m slope interval, during 1969 and 1973–1978, was lost to frost (Hacienda Zuleta, unpublished data). These were, however, relatively low-frost years in the Quito record [*see Figure 2.6(b)*].

Later in this section it will be assumed that frost risks, particularly March–May minimum temperatures, do in fact play an important role in setting the upper limit of cultivation in the drier highlands. As will be seen, if this is not the case, there is little likelihood that climatic change will have a significant effect on the upper limit of cultivation.

5.7. Fluctuations in Frost Frequency and Mean Temperature

We will now attempt to determine plausible scenarios of medium-term change in climatic factors related to the upper limit of cultivation.

5.7.1. Variations in mean annual temperature

Interannual variability of mean annual temperature at Quito (altitude 2818 m) is small (*see* Section 2). Mean decadal temperature rose from 12.6 °C (close to the value expected for Quito's altitude) in the 1890s to 13.6 °C in the 1970s (*Table 5.3*). Some of this variation may be due to the growth of the Quito urban heat island. However, the peaks of temperature in the early 1940s, the 1960s, and the late 1970s correspond with global southward shifts of circulation features and concurrent declines in Ecuadorian rainfall (*see* Section 2). There is about a 0.5 °C difference between these peaks and other periods.

Table 5.3. Decadal variations in mean annual temperature at Quito (°C).

Decade	*Mean decadal temperature*	*Mean: coolest 3 years*	*Mean: warmest 3 years*	*Range of mean annual temperatures*
1891–1900	12.6	12.2	13.1	12.0–13.5
1901–1910	12.7	12.4	12.9	12.2–13.1
1911–1920	12.9	12.3	13.2	11.9–13.5
1921–1930	13.1	13.0	13.2	12.9–13.4
1931–1940	13.0	12.6	13.2	12.5–13.6
1941–1950	13.3	13.1	13.4	12.8–14.0
1951–1960	13.5	13.0	13.9	12.9–14.2
1961–1970	13.6	13.4	13.7	13.2–14.1
1971–1980	13.6	13.2	13.9	12.9–14.1

Source: Manuscript records of Quito Observatory.

During the last 40 years, mean decadal temperatures have been virtually constant at Quito (*Table 5.3*). During this time, the 1-in-10 cold years (coldest years per decade) have varied within the range of only 0.4 °C.

Whether we look at short-term (decadal) mean temperatures, or means associated with global circulation shifts, there is little evidence for substantial change at Quito. Changes of the order of 0.4–0.5 °C correspond with shifts of less than 100 m in elevation, relatively trivial in the Andean context. Thus, if mean annual temperature sets the upper limit of cultivation (through governing maturation times or other plant processes), there is little reason to expect substantial changes in location of the limit.

In generating scenarios of significant change we will, therefore, focus on variations in frequency of frost.

5.7.2. Variations in frequency of late growing season frost

Variability of absolute surface temperature minima at Quito has been greater than for mean temperature (*Table 5.4*). The long-term mean of late winter minima at Quito is 0.81 °C. This is 2.04 °C below the calculated minima for Quito's elevation based on equation (5.3a) because the former is derived from readings at ground surface rather than at shelter height (2 m). It will be recalled that the highest intensively cultivated flat in Chimborazo Province (Llano Grande) is at 3080 m, with a calculated late winter shelter minimum of 1.8 °C. If the surface minimum is about 2 °C less than the shelter minimum, Llano Grande would experience a surface minimum of about 0 °C.

Decadal means of late winter absolute surface temperature minima at Quito varied from −0.16 °C in the 1950s to 2.59 °C in the 1970s (*Table 5.4*). If 30-year means are calculated, these vary from 1.0 °C (1891–1920) to 1.2 °C (1921–1950) and 0.4 °C (1951–1980). At this time scale variation of about 0.8 degrees in late winter surface minima seems to be experienced. The recent 30-year period appears to have been relatively frost-prone.

Table 5.4. Decadal patterns of mean late winter (March–May), absolute minimum surface temperature, Quito Observatory.

	Late winter absolute seasonal surface minimum temperature				
Decade	*Mean* (°C)	*Standard deviation* (°C)	*Range* (°C)	*No. of frosts*	*10% probability minimum*[a]
1891–1900	0.63	1.47	−1.5 to 2.3	6	−1.25
1901–1910	1.28	1.60	−1.4 to 3.2	10	−0.77
1911–1920	0.97	1.26	−0.7 to 2.9	9	−0.64
1921–1930	2.59	1.98	−2.0 to 4.1	7	−0.06
1931–1940	0.61	1.37	−1.5 to 3.0	9	−1.14
1941–1950	0.36	0.92	−1.0 to 1.9	10	−0.82
1951–1960	−0.16	0.95	−1.4 to 1.5	10	−1.38
1961–1970	0.51	1.64	−3.2[b] to 2.5	10	−1.59
1971–1980	0.96	1.20	−1.5 to 2.4	8	−0.58
Whole period 1891–1980	0.81	1.55	−3.2 to 4.1	79	−1.17

[a]Mean minus 1.28 × standard deviation.
[b]−3.2 °C observed May 5, 1964. Excluding this value, the extreme frost would be −2.2 °C.
Source: Manuscript data, Quito Observatory.

In *Table 5.4* we have calculated two forms of the 1-in-10 frost risk for each decade. The observed minimum temperature for each decade is recorded. Except for the extreme −3.2 °C observed on May 5, 1964, these extreme decadal minima vary from −0.7 to −2.2 °C, or by about 1.5 °C. When average minima are calculated for 30-year periods the values are −1.2 °C (1891–1920), −1.5 °C (1921–1950) and −2.0 °C (1951–1980). The variation among these values is 0.8, about the same as for the variation among 30-year means.

For comparison, *Table 5.4* presents calculated 10% probability late winter surface minima for each decade, based on an assumption that minima are normally distributed. Calculated 10% minima vary from −0.06 to −1.59 °C, or by about 1.6 °C. This is similar to the interdecadal variation of observed 1-in-10 frosts.

5.7.3. Variations in number of frosts

In addition to degree of frost, the number of frosts may also affect crop success. *Figure 5.2* presents data on number of frosts per year for Quito (note that these data are not limited to the March–May late summer period). An increase in frost frequency is apparent for the 1940–1960 period, consistent with data presented in *Table 5.4*.

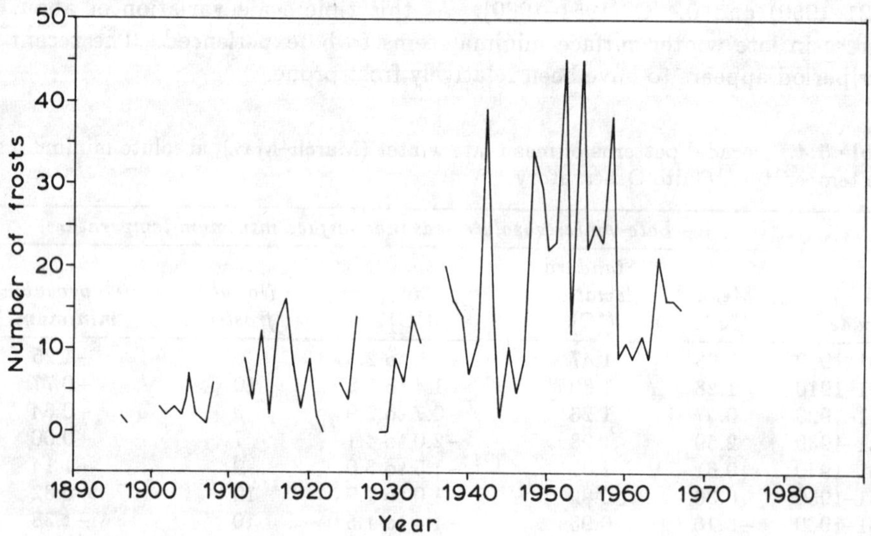

Figure 5.2. Number of ground frosts per year, Quito. Source: Manuscript data, Quito Observatory.

5.7.4. Conclusions

From the foregoing discussion it is evident that there is general agreement between the various measures of late winter frost risk. Calculated 10% probability temperature minima are about 2 °C below the long-term mean and are less variable from decade to decade than means, the former varying by about 1.5 °C between decades. On a 30-year basis, mean and 10% probability events both vary by about 0.8 °C. We will use the latter value to develop our scenarios of change in frost risk.

5.8. Scenarios of Changes in Frost Risk

Farmers at the upper limit of cultivation are no strangers to frost risk. As is the case with drought, risk-averting strategies and acceptance of occasional losses probably push the limit into areas with considerable chance of frost.

If farmers are prepared for 1-in-10 late winter extreme frosts, the 10% probability frost is the relevant parameter for constructing a scenario of change. If farmers determine this parameter over the course of a generation (say, 30 years), then 10% probability rainfall over a 30-year period determines the upper limit of cultivation. Farmers might also adapt to mean or 40% probability frosts, as Gade (1967) suggests. As noted above, both values (means and 1-in-10 extremes) have changed by a maximum of 0.8 °C over three consecutive 30-year periods. We therefore adopt a 0.8 °C shift to indicate plausible medium-term changes in frost risk at the upper limit of cultivation. This corresponds to about a 200 m change in elevation of the upper limit of cultivation [equations (5.3) and (5.4)].

5.9. Evaluation of Areas Affected by Change in Frost Risk

5.9.1. Impact on slope agriculture of increased frost risk

Figure 5.3 indicates the distribution of cultivated land above 3600 m in Chimborazo Province based on aerial photographs and field survey during the 1960s and early 1970s. The total cultivated area in the province, including fallow land, is 193 000 ha, considerably more than is cultivated in a given year. This area, which includes fallow land, totals 16 000 ha, and amounts to 8% of the total arable land in the province. Since the greatest elevation of cultivation achieved anywhere in the region is 3800 m, the shaded area in *Figure 5.3* indicates the area that would be potentially lost to cultivation if there were a 0.8–1.0 °C drop in late growing season minimum temperatures (or in 10% probability temperatures calculated on a 30-year basis).

To estimate the number of people who might be affected by such a change, topographic maps that show individual rural buildings (based on 1959–1963 air-photos) were examined. The number of buildings above 3600 m is 890, almost all of which are probably dwellings since outbuildings, schools and churches are rare in these extremely peripheral locations. Assuming that 5 persons occupy each dwelling, the population of the affected area above 3600 m is 4450, most of whom are indigenous (Indian) farmers. Since cultivated land is generally located close to farm dwellings, it is fair to assume that the land upon which this population depends lies above, or at least very close to, the 3600 m contour. Furthermore, it is likely that the occupancy of high areas such as these has increased since 1956–1963, owing to population growth and agrarian reform. Uphill migration has increased in recent years.

Of course, other important factors would serve to modify the actual response to this possible change in frost risk. High lands brought into cultivation recently are often highly fertile and therefore are particularly useful to villagers. In addition, farmers possess a flexible tool of adaptation to crisis in the practice of temporary labor migration. For these reasons, the identification of a considerable group of farmers at risk should not suggest at this point the imposition of coercive programs to remove farmers from these areas.

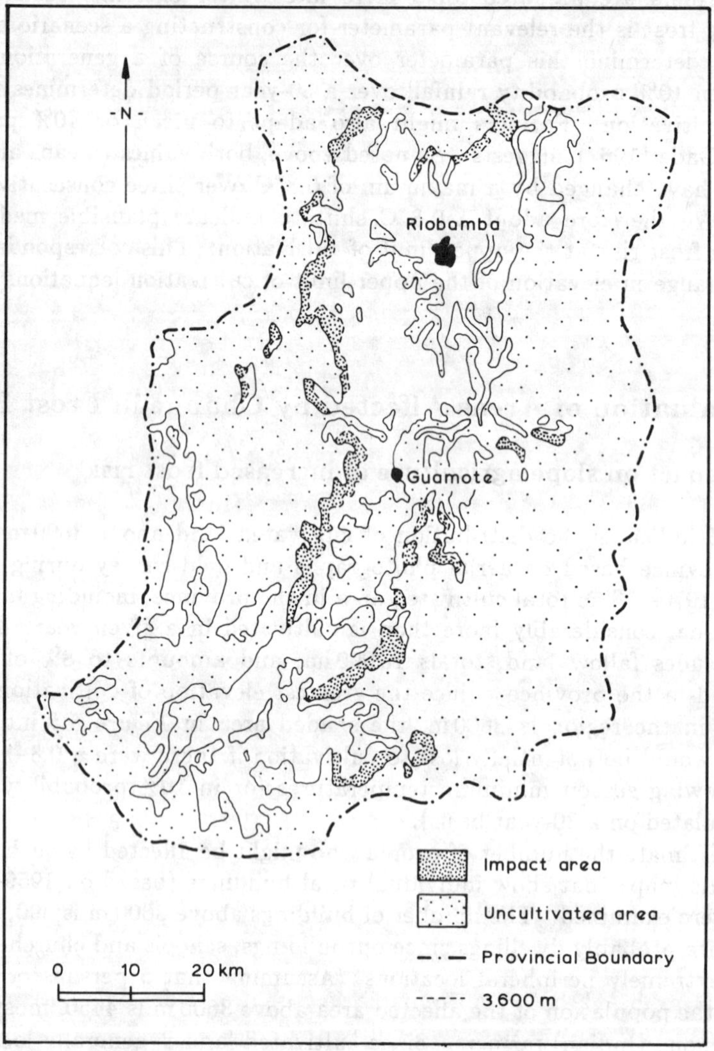

Figure 5.3. Impact of change in frost risk on probable limit of cultivation, Chimborazo Province.

5.9.2. Impact on flat agriculture of increased frost risk

Table 5.5 gives the area of flats and their present land use in Chimborazo Province. The highest intensively cultivated flat is Llano Grande at 3080 m. If the upper limit of cultivation of flats were to decline by 200 m, this flat containing 1800 ha (7% of the 26 600 ha of cultivable flats) could go out of production. Topographic maps indicate about 80 dwellings in the area, perhaps corresponding to 400 persons.

Table 5.5. Flats – areas, location, elevations and land use in Chimborazo Province.

Name	*Area* (km^2)	*Location* *Lat. (S)*	*Long. (W)*	*Elevation* (m)	*Land use*
–	2.9	1°37′	78°47′	3200	pasture
–	0.7	1°35′	78°15′	3200	pasture
Riobamba (N.)	5.2	1°38′	78°40′	2800	cultivated
Guano	1.3	1°37′	78°36′	2600	cultivated
Llano Grande	1.8	1°40′	78°45′	3080	cultivated
Cajabamba	1.4	1°42′	78°45′	3200	pasture
Colta	2.0	1°44′	78°45′	3300	pasture/lakeside cultivation
Riobamba (S.)	4.5	1°41′	78°37′	2740	cultivated
San Luis	8.6	1°44′	78°37′	2700	cultivated
Chambo	1.9	1°44′	78°36′	2740	cultivated
Colta Outlet	0.8	1°48′	78°45′	3260	pasture/totora beds
Punin	1.9	1°45′	78°49′	2760	cultivated
Pungala	0.4	1°49′	78°35′	2880	cultivated
Columbe	1.1	1°53′	78°44′	3140	pasture
La Clementina	2.5	1°52′	78°44′	3140	pasture
Galtepamba	2.2	2°01′	78°40′	3120	pasture
Palmira Davalos	2.2	2°03′	78°44′	3200	pasture
Pachamama	0.7	2°11′	78°46′	3600	paramo
TOTAL	48.1				

Source: PRONAREG-ORSTOM (1979b) soil maps and (1983, 1984) land use maps.

5.9.3. Impact on slope agriculture of decreased frost risk

If late winter absolute temperature minima were to increase by about 0.8 °C, thus *decreasing* frost risk, the upper limit of cultivation might be raised to 4000 m. This increase would have potentially major impacts: 153 400 ha of Chimborazo Province lie between 3500 and 4000 m. Even if 60% of this total is below 3800 m, 61 360 ha would remain for potential use. This potential increase correponds to 32% of the currently cultivated area.

The reason that the increase under this scenario is so much greater than the change (8%) in area under the increased frost risk scenario is that in this case we are considering changes in *potential* area while the preceding scenario considered *actual* cultivated area. Furthermore, there is abundant land above 3600 m that is not currently cultivated.

It is clear that potentially great increases in cultivable area could be achieved if frost risk were to fall or if frost-averting technology could be placed in use.

5.9.4. Impact on flat agriculture of decreased frost risk

If the upper limit of cultivation on flats were to increase 200 m, flats up to 3280 m elevation might be harvested intensively. About 18 400 ha of flats could go into production (*Table 5.5*), equivalent to 69% of the current cultivable area. Again this is a potential rather than actual increase, so is not directly comparable with the area of change under the increased frost risk scenario. The scenario does, however, indicate significant potential gains.

5.10. Conclusion

The upper limit of cultivation in the central Sierra of Ecuador appears to be governed, in part, by length of maturation time (related, in turn, to mean annual temperature). The upper limit of cultivation on slopes varies from 3600 to 3800 m, depending on rainfall receipt. Mean annual temperature at these elevations is approximately 7 °C. Interannual changes in mean annual temperature are slight and of themsevles should result in negligible shifts in cultivation limits.

Frost risk (as measured by late growing season absolute minimum temperature) also increases with altitude. Although farmers can cope with frost risk through staggering planting dates, planting on east-facing slopes, using crop ridges and employing cultivars with a short growing season, frost remains a serious perceived problem.

If the 3800 m limit of cultivation in Chimborazo Province can be attributed in significant measure to frost risk, this should be a fluctuating boundary. Late growing season minima and frost frequencies have varied during the present century, with the late 1940s and 1950s having been especially risky. The upper limit of cultivation in Chimborazo has apparently increased in elevation in the last two decades, but it is difficult to separate climatic factors from human factors: the last two decades have also been a time of agrarian reform involving the distribution of highlands to small farmers.

For a 0.8 °C decrease in late growing season temperature minima (with a corresponding downward shift of the upper limit of cultivation to 3600 m), it is suggested that 8% of Chimborazo's arable land could go out of production. These areas are primarily cultivated by indigenous smallholders of the western part of the province. These farmers are land-poor and hardly capable of shifting the potentially affected lands into less productive uses.

Flats are a special case. Analysis of late growing season temperature minima suggests that the upper limit of cultivation on flats should be at most 500 m lower than the upper limit of cultivation on slopes. In fact, in Chimborazo Province the difference in limits is closer to 720 m. This may be due to use of east-facing slopes, other frost-averting slope adaptations and the existence of attractive alternatives to cultivation on flats. If the cultivation limit on Chimborazo flats could be raised by only 120 m, the area of rich flat land in the province could be increased by two-thirds.

There is, in fact, evidence that flats have been cultivated in the past at higher elevations. Prehistoric raised fields have been found in flats of humid western Imbabura Province at elevations up to 3150 m where the upper limit of cultivation is at most 3600 m (Gondard and Lopez, 1983).

It is probably *not* desirable, however, to attempt to prohibit farming on high marginal slope areas. We know too little of the factors governing the upper limit of cultivation, and such policies would deny smallholders some of their most fertile and productive land. Policies that benefit small farmers in general should simultaneously seek to increase the ability of farmers to cope with the specific hazard posed by potential shifts in the upper limit of cultivation.

SECTION 6

The Effects on Zones of Agroclimatic Suitability for Crops

6.1. Introduction

In this section we examine the effects of short-term variability of precipitation on the potentially cultivable area of Chimborazo Province. After an initial discussion of the relation between climate and potential production, climatically induced shifts in areas suitable for certain crops are mapped for given scenarios of climatic variation. The areas delimited by these shifts represent areas of potential climate impact.

6.2. Climatic Limits to Cultivation

The amount of water required by a crop can be estimated by calculating potential evapotranspiration (PET). The investigation and evaluation of various methods to estimate PET in the central Sierra has indicated that the ratio between PET calculated using the Thornthwaite and Penman methods, respectively, is quite uniform at about 0.45 (Cañadas and Salvador, 1982). Further, analysis of Penman PET estimated for five locations at different altitudes suggested that mean annual Penman PET decreases only about 2 mm for every 100 m increase in elevation (Cañadas and Salvador, 1982). This is a relatively small altitudinal lapse rate. Work elsewhere in the temperate Andes tends to corroborate this result, which appears to be attributable to high net radiation in mountains even when temperatures are low (Frere *et al.*, 1975; Knapp, 1984).

In general, at higher elevations around the basin margins, precipitation is higher than in the basin centers, and crop water requirements are normally satisfied during the wet season (Cañadas and Salvador, 1982). In the drier basin interiors, precipitation totals often fall short of PET, and irrigation is often desirable or necessary, particularly for long-cycle crops such as maize.

In the central Sierra the variance of growing season rainfall is between 10 to 100 times larger than the variance of growing season PET (Cañadas and Salvador, 1982). This suggests that year-to-year variations in crop yield are more likely to be due to variations in precipitation than to variations in potential evapotranspiration.

So far there has been hardly any experimental study to determine the lower limit of rainfall for successful rainfed agriculture in the Ecuadorian Sierra. Not only the amounts but also the temporal distribution of rainfall are, of course, important. Yields are determined by the soil water available for the development of crops. Since highland soils are derived from volcanic ash, soil moisture-holding capacity can be related to soil depth, texture and structure. PRONAREG-ORSTOM has in fact produced maps of the limits of potential wheat production, based on the assumption that the dry limit of rainfed cultivation of wheat is the ustic-aridic soil moisture regime boundary – that is, the point where the "soil moisture control section" (USDA, 1975) is dry on average for six or more months a year. Wheat, however, is less tolerant of extremes than barley.

In order to determine empirically the wet and dry limits of cultivation in Chimborazo Province, maps of potential and of actual land use were consulted (PRONAREG-ORSTOM, 1977–1982; 1982a; 1982b). A comparison of these maps with rainfall receipt indicated that the dry limit of cultivation of barley corresponded approximately to the 400 mm/year isohyet. Similarly, the wet limit of actual and potential potato and cereal agriculture corresponded to the 1400 mm/year isohyet.

Previous studies of crop records on haciendas in Chimborazo Province indicate that potato yields in years with a rainfall receipt of 1400 mm are about 39% of optimal yield, and barley yields with annual rainfall under 400 mm are about 25% of optimal (Knapp, 1984; *see also* Section 4). An annual precipitation of less than 250 mm leads to complete loss in the sense that yields are then less than the seed input.

6.3. Areas of Suitability for Crops: Average Years

The area of Chimborazo Province under 4000 m elevation with mean annual precipitation between 400 and 1400 mm is 397 630 ha. This category of suitable land may be further subdivided in terms of drought/downpour risk and frost risk as follows.

The lower limit of rainfed cultivation of potatoes is defined by PRONAREG as the ustic-udic moisture regime boundary – that is, the point where the "soil moisture control section" is dry on average for at least three months a year. As part of the present project, the ustic-udic boundary in Chimborazo was taken from PRONAREG ORSTOM (1979c) soil maps and compared with the distribution of precipitation, indicating that this boundary approximately coincides with the 800 mm isohyet.

The actual land-use map was compared with the precipitation map to determine that no areas that receive more than 1200 mm of precipitation per year are currently cultivated for barley.

The precipitation interval of 400 to 1400 mm/year can be further subdivided into three subregions (*Table 6.1*): a barley subregion with 400–800 mm/year precipitation, a potato subregion with 1200–1400 mm/year precipitation, and a barley–potato subregion with 800–1200 mm/year precipitation.

Table 6.1. Typology of agroclimatic suitability zones.

	Mean annual climate		
Crop	*Precipitation* (mm)	*Temperature* (°C)	*Limitations*
Barley	400 – 800	6.0 – 8.4 8.4 – 18.0	Drought/frost Drought
Barley or potatoes	800 – 1200	6.0 – 8.4 8.4 – 18.0	Frost –
Potatoes	1200 – 1400	6.0 – 8.4 8.4 – 18.0	Frost/downpour Downpour

Table 6.2. Summary of the effects of climatic scenarios on agroclimatic suitability zones (ha).

	Scenarios				
	Dry			*Wet*	
	Extreme	*1-in-10*	*Baseline*	*1-in-10*	*Extreme*
Total area suitable for:					
Barley	309 830	340 460	346 120	163 570	155 685
Potatoes	5 300	70 140	274 015	128 000	131 371
Area with limitations:					
Frost	109 000	109 000	118 040	29 780	27 125
Downpour	304 530	270 320	123 665	93 200	75 960
Flood	–	–	43 020	44 620	39 711
Total cultivable area	309 830	340 460	397 630	223 170	207 331

In addition, following PRONAREG-ORSTOM (1982a, 1982b), we hypothesized that frost-prone areas correspond to soils of mean annual temperatures between 6 and 8.4 °C (elevations of 3600–4000 m). This range brackets the current upper limit of cultivation and will remain constant throughout the following scenarios.

The final map of crop potential (*Figure 6.1*) shows the land area of Chimborazo divided into these classes (*see also Table 6.2*, baseline scenario). About 118 000 ha are characterized by frost risk; 124 000 ha are suitable only for barley (due to drought risk); 43 000 ha are suitable only for potatoes (due to risk of excess precipitation).

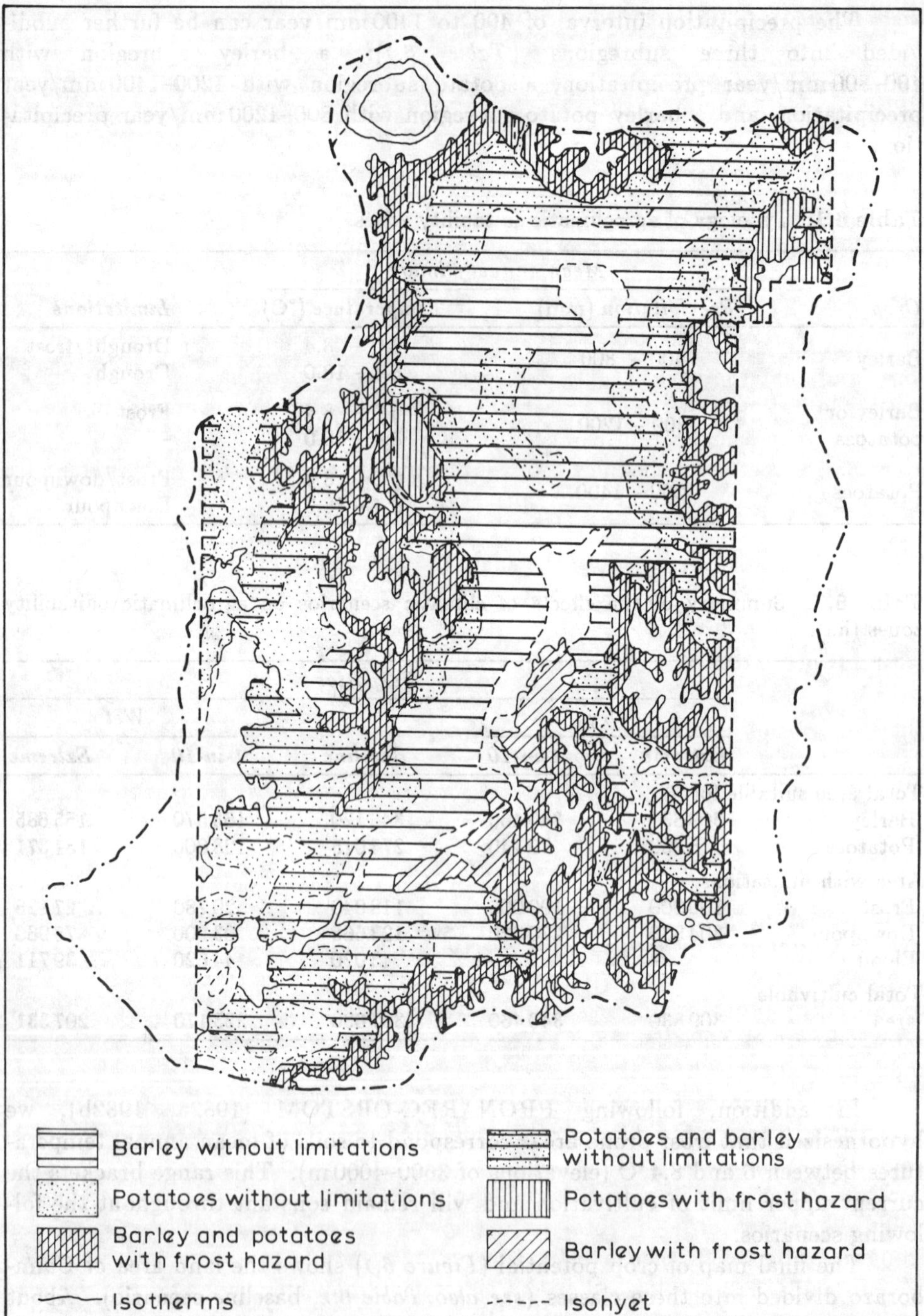

Figure 6.1. Agroclimatic suitability for crops: baseline scenario (1964–1973).

6.4. Effects of Climatic Variations on Crop Potential

We analyzed the effect of given variations of climate on the distribution of crop potential, using the scenarios of climatic fluctuation as defined in Section 2, viz:

- 1-in-10 dry year (65% of mean annual precipitation)
- 1-in-10 wet year (134% of mean annual precipitation)
- Extreme dry year (57% of mean annual precipitation)
- Extreme wet year (140% of mean annual precipitation)

Assuming that the same variations in precipitation occurred at all locations, and that these variations were not associated with changes in frost risk, new precipitation maps were created for each scenario. For example, in the 1-in-10 dry year, the 400 mm isohyet becomes the 260 (0.65×400) mm isohyet; interpolation between the new isohyets produces the new 400 mm isohyet to determine the new limit to barley cultivation.

Figures 6.2 to *6.5* show the shifts in zones under the different scenarios and *Table 6.2* indicates the extent of the zones in each case.

6.5. Discussion

The scenarios of precipitation change were derived from analysis of precipition at a single station. It is unlikely that the real extreme or 1-in-10 fluctuations in the isohyets are the same as those produced in our model by the expedient of multiplication of isohyets by percentages.

With simulated rainfall reduction (either 1-in-10 or extreme), the area with good potato yields drops to almost zero. This appears logical in view of the assumed 800 mm/year dry limit of potato cultivation and the relatively small area of Chimborazo Province with precipitation substantially above this value. The area with good barley yields suffers almost no change in comparison, since areas that drop out because of drought are replaced by areas which normally are too wet for good yields (some of these wet areas, of course, would not be planted under normal expectations).

With a simulated rainfall increase, the area of good potato yields increases. Nevertheless, total area with good yields (including the barley area) decreases from the baseline case, and is even less than during extreme dry years. This result – that dry Chimborazo Province has more to fear (in terms of area of good yields, if not in terms of total production) from rainfall increases than from drought – deserves further investigation.

In dry years, the lower limit to cultivation moves upslope and the proportion of potentially cultivable land with frost risk increases dramatically. The significance of frost would thus be greater than indicated in Section 5.

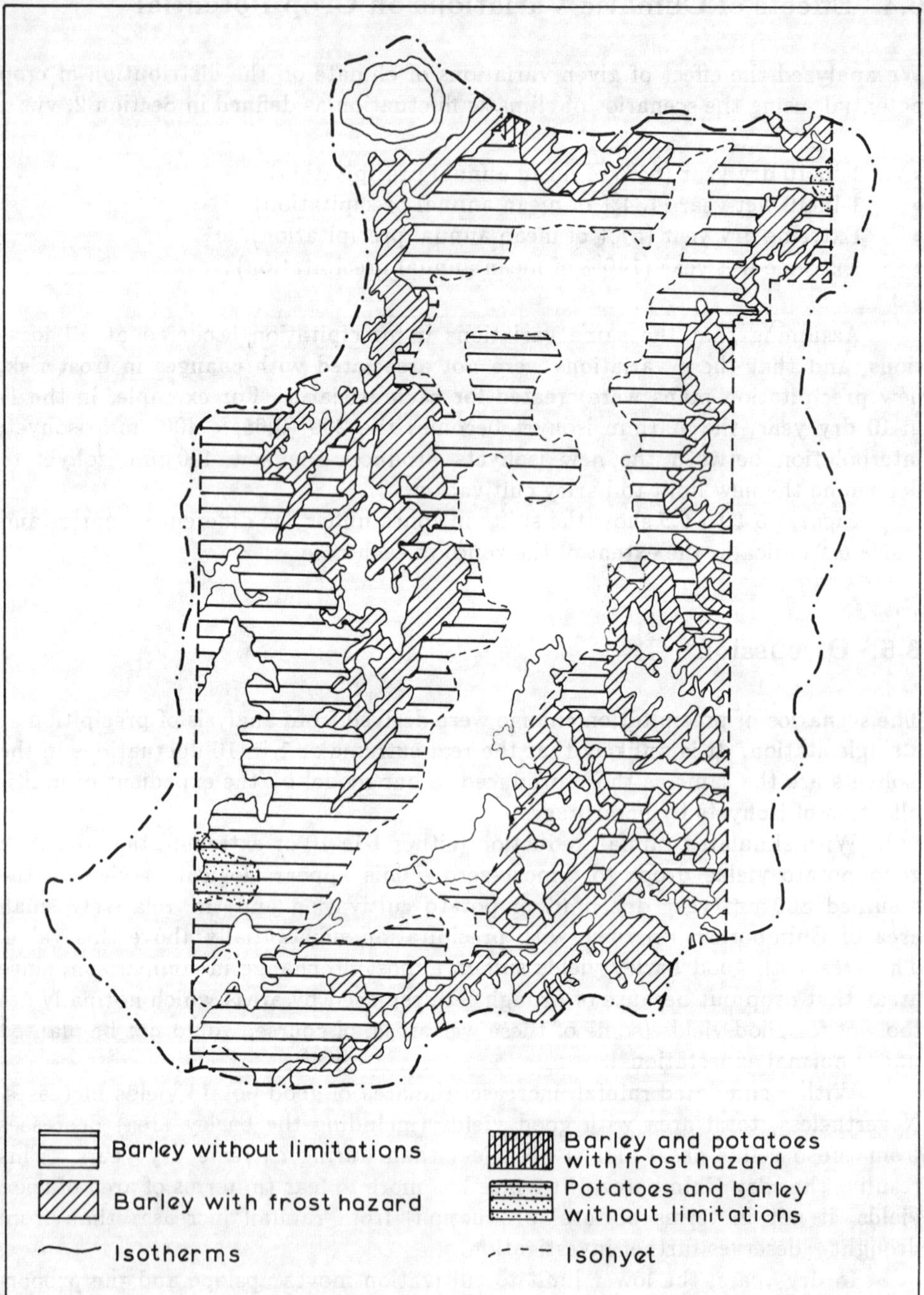

Figure 6.2. Agroclimatic suitability for crops: extreme dry year scenario.

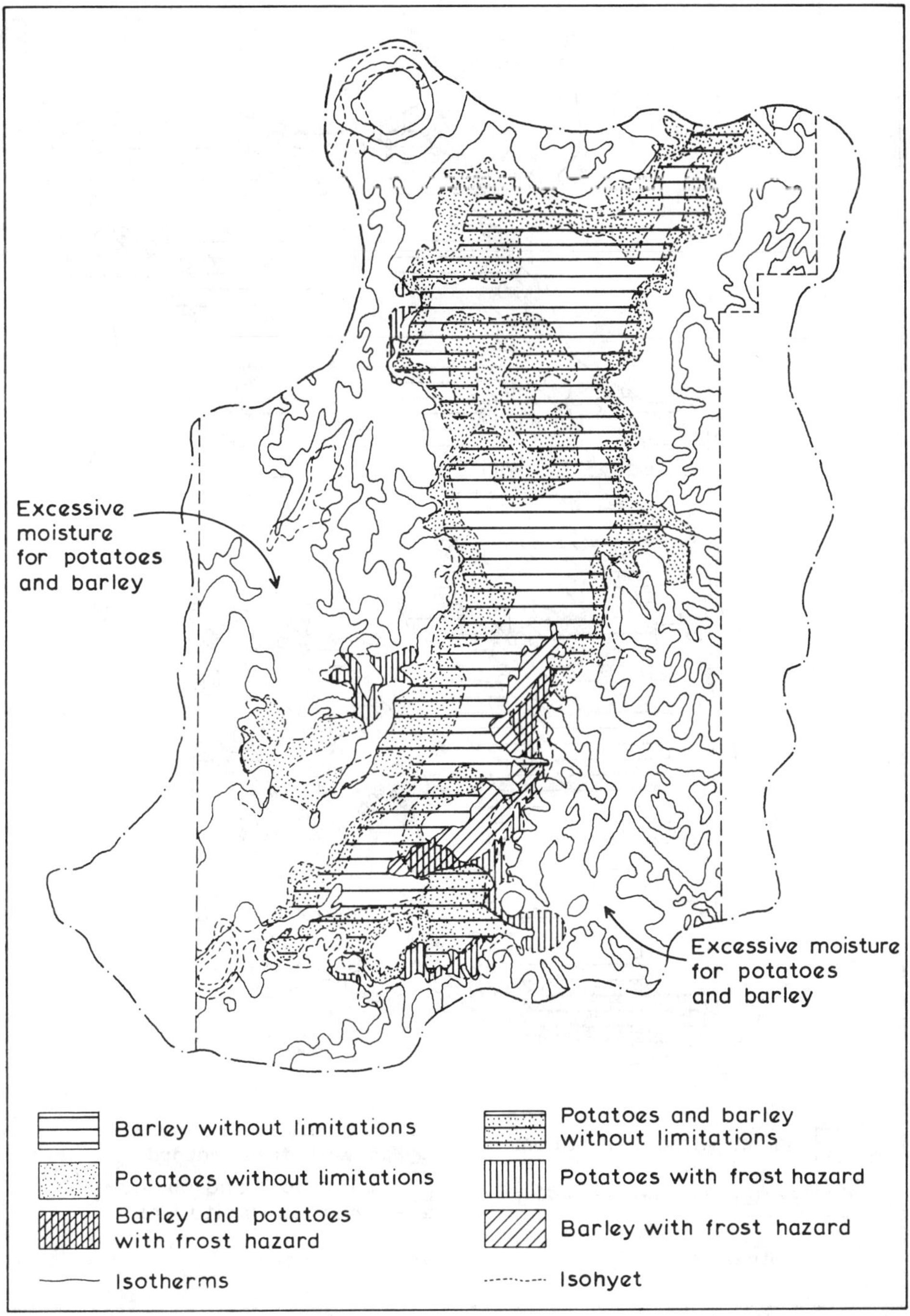

Figure 6.3. Agroclimatic suitability for crops: extreme wet year scenario.

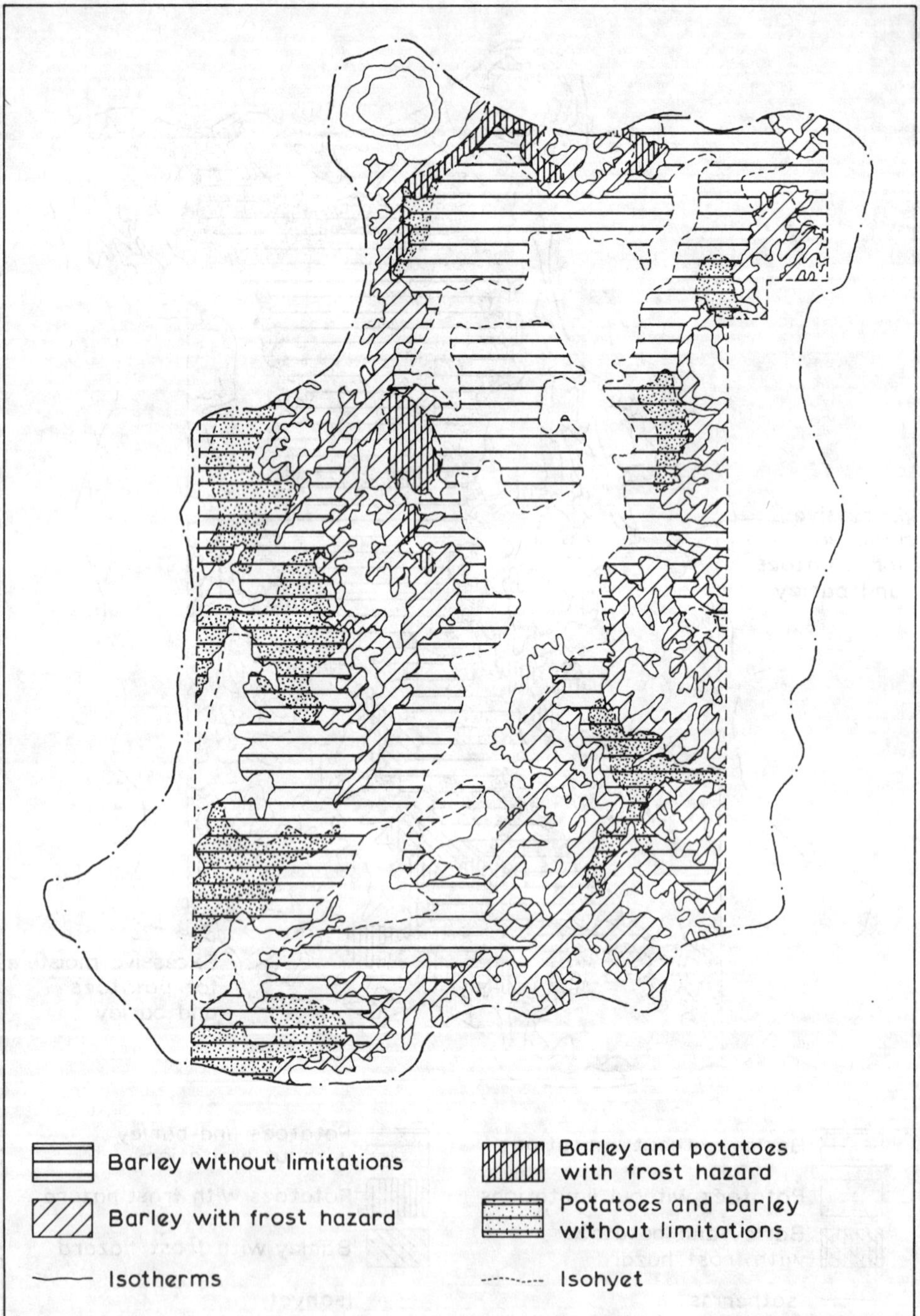

Figure 6.4. Agroclimatic suitability for crops: 1-in-10 dry year scenario.

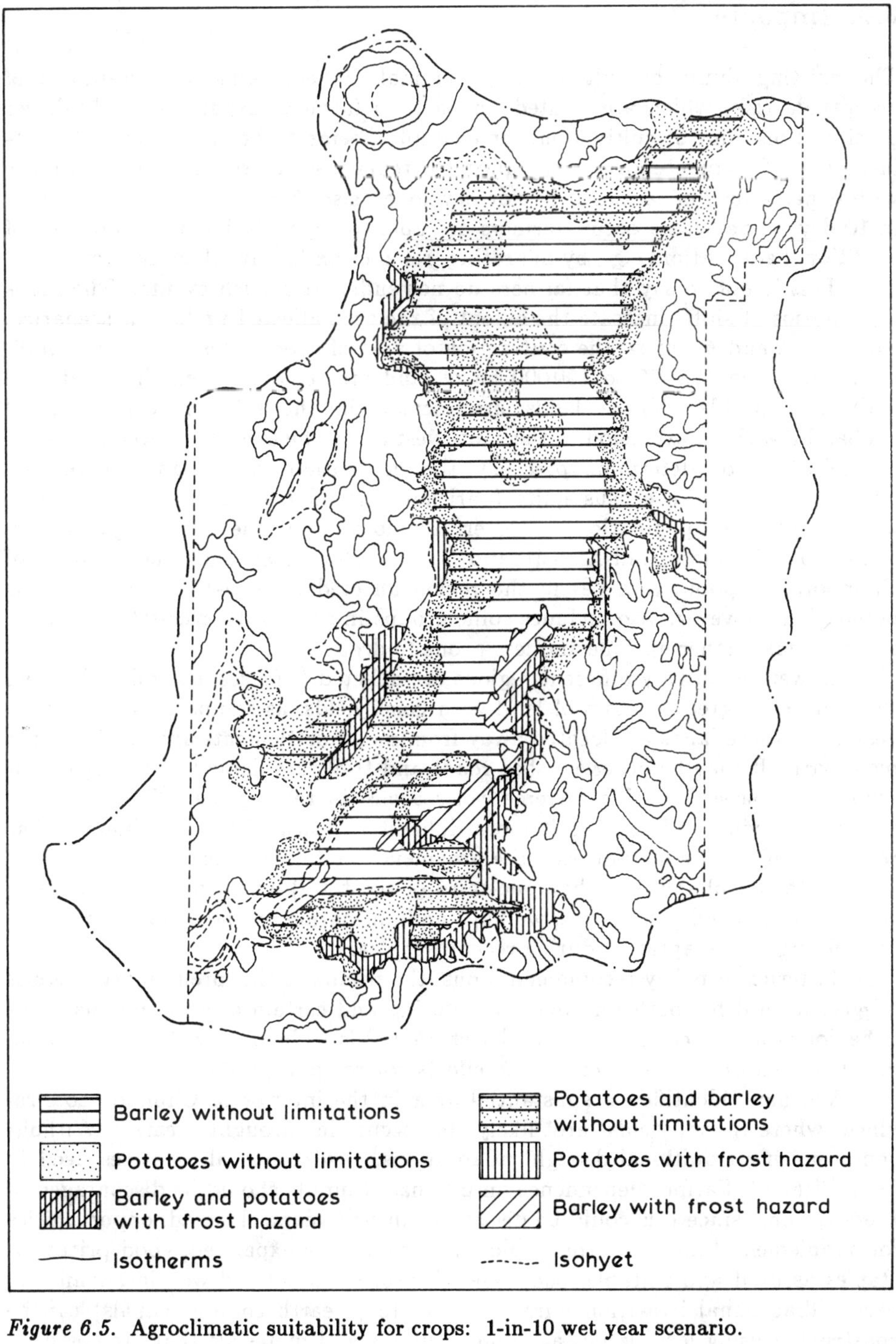

Figure 6.5. **Agroclimatic suitability for crops: 1-in-10 wet year scenario.**

6.6. Impacts

The existing limits of cultivation presumably reflect farmers' perceptions of acceptable risk, which are related, in part, to farmers' experience. The lower precipitation limit of cultivation, for example, seems to be associated with a 1-in-10 risk of complete crop loss (failure to replace seed), roughly in accordance with expectations of farmer behavior developed elsewhere. Presumably when 1-in-10 dry or wet years occur, farmers are able to cope within the parameters of traditional adaptation, e.g., by migrating, reliance on family alliances, etc.

This is not to say that farmers do not suffer from such events. The maps of scenarios of shifts indicate the extent of the area affected under our scenarios. *Figures 6.2* and *6.4* show the contraction of area of good yields under dry conditions. Between 175 000 to 190 000 ha of land are removed from the total area with good yields. These lands are located in the inter-Andean basin of Riobamba with an extension to the southwest along the axis of the Alausi basin.

We cannot determine precisely the population that might be affected because most of the census units (parishes) extend beyond the affected area. Many centers of population, however, are located in the area (e.g., Guano, Riobamba, Guamote, and Alausi), and it is therefore likely that the majority of Chimborazo's population lies in the region affected by 1-in-10 droughts. This population, however, also includes some groups in areas with access to irrigation and to lands further upslope and less prone to drought.

In wet years, very different regions are affected (*Figures 6.3* and *6.5*). Here there are even greater losses of land with good yields, as compared with normal years, but these lands are located away from major settlements and roads for the most part. It may be observed that roads and trails are liable to be damaged or rendered impassible during extreme wet seasons, increasing the difficulty of harvest or resowing and making it more difficult to learn of local problems. The regions involved are distant from the provincial capital. A substantial number of farmers (but probably less than one-third of the total) would be affected. As in the case with the dry scenarios, we have no information on the actual dimensions of suffering and adaptation during recent wet years.

In terms of policy recommendations, the results of this preliminary exercise suggest a need for both improved monitoring and certain specific actions. The behavior of actual communities under stress needs to be observed in connection, of course, with humane relief and aid efforts where appropriate.

We have identified a substantial area in the interior of Chimborazo Province where poor yields are likely to occur in drought years. Without compromising traditional drought adaptations, new coping alternatives need to be facilitated. Farmer dependence on purchased inputs should be discouraged in these regions, since this could be disastrous in bad years. Instead, opportunities for supplemental and off-farm employment should be expanded, good prices for staples assured and suitable food-for-work projects identified well in advance of need. Traditional irrigation infrastructures (e.g., earth contour canals) can be constructed using local labor, avoiding methods and materials that are costly to install or maintain. General improvement of health, education and community facilities would further expand options and increase resilience to inevitable hazards.

The same general prescriptions apply to areas identified as being prone to risk from excessive rainfall. In addition, such areas need good all-weather roads and trails to maintain access during wet seasons and wet years.

In general, Andean communities are located in regions with moderate precipitation receipt, about 900 mm being the average value for a sample of communities in Ecuador (Knapp, 1984). The same is true for Chimborazo, with many communities having access to both dry and wet land. Farmer access to multiple niches needs to be supported and improved, and a wide spread of planting dates encouraged to help reduce the risk of crop loss. These elements of diversity should not be compromised by emphasis on maximum yields.

In general, our recommendations are in accordance with traditional wisdom of the farmers themselves. Insofar as capital inputs are supplied to modernize agriculture, the preferred locations are those with least risk from drought or downpour – ideally, irrigated areas or areas with mean annual precipitation of about 900 mm.

SECTION 7

The Effects on Holdridge Life Zones

7.1. Introduction

This section considers the transitional areas between life zones of the Holdridge ecological classification system and identifies the location and extent of areas that might be affected under given scenarios of long-term climatic change.

Holdridge (1947, 1967) proposed a life zone classification to predict the potential vegetation of a region for values of climatic indices. The Holdridge classification is based on the principle that different types of vegetation have different, yet definite, climatic and edaphic requirements in which the climatic usually overshadow edaphic factors in conditioning large-scale patterns and sensitivity to change. In particular, the system is based on two factors: mean annual precipitation and mean annual biotemperature, the latter defined as the mean of unit-period temperatures with substitution of zero for all temperature values below 0 °C and above 30 °C, respectively. A "life zone" is objectively defined as an area that falls within specified ranges of mean annual precipitation and mean annual biotemperature, and which consists not only of vegetation, but also soils, geology, topography, atmospheric influences and animal activities (Holdridge, 1967).

A classification that recognizes the dependence of natural vegetation on climatic factors provides a basis for constructing maps to display the impact of climatic changes on the geography of the major vegetation zones (cf. Emanuel *et al.*, 1985). In this section we use the Holdridge system to consider the possible sensitivity of the distributions of terrestrial ecosystem complexes to changes in precipitation, especially in those areas where the shifts in their boundaries are likely to be significant agriculturally.

Three maps using the Holdridge classification will be compared for Chimborazo Province, Ecuador: one derived from soils data and representing recent conditions, one for a change of climate toward more moist long-term conditions, and one for a change of climate toward drier long-term conditions (these scenarios were discussed and defined in Section 2, above).

7.2. The Holdridge Life Zone Model

The underlying postulate of the Holdridge system is that the primary influences on vegetation are those elements that make up climate. In this respect, it is not unlike the systems of Köppen (1931), Clements (1920), Thornthwaite (1948), Gaussen (1959), Troll (1968) and others.

Holdridge's model divides the world into seven latitudinal regions (based on mean annual temperature at sea level). All of Ecuador falls into one of these regions – the tropical. The tropical region is divided into 37 life zones, of which 22 are found in the Ecuadorian Andes and the immediately adjacent lowlands. In addition, Ecuador has a wet high-elevation life zone, which falls outside the Holdridge classification system. Fortunately, this aberrant life zone is limited in extent.

The life zones are normally shown on a tri-axial diagram as hexagons (*Figure 7.1*), the boundaries of each being defined in terms of mean annual biotemperature (equivalent to elevation) and mean annual precipitation. The third axis of the diagram is determined by the other two; it represents the moisture regime, defined as a proportion between mean annual biotemperature and mean annual precipitation. Holdridge assumes that biotemperature is linearly related to potential evapotranspiration (which is not true for the Ecuadorian Andes; *see* discussion in the preceding section). The latitudinal regions and the altitudinal belts are defined in terms of mean annual biotemperature and not in terms of latitudinal degrees or meters of elevation.

To determine which life zone corresponds to a give site, the first step is to locate the point on the life zone chart that represents its mean annual precipitation and mean annual biotemperature. Generally, the point will fall within one of the hexagons. The name of the altitudinal belt associated with this hexagon, and the vegetation type name within the hexagon, are recorded.

The next step is to determine the latitudinal region in which the site occurs. This is done by converting the mean annual temperature at the site to a corresponding mean annual temperature and biotemperature at sea level, using the site elevation and a suitable temperature lapse rate for the area. This procedure, in essence, corrects for the different seasonalities encountered in different latitudinal regions.

The combined names of latitudinal region, altitudinal belt and vegetation type, in that order, provide the name of the life zone. For example, a site at 3000 m with a mean annual biotemperature of 8 °C and total annual rainfall of 2250 mm is the Tropical Montane Rain Forest Life Zone. Note that the "Tropical" modifier refers to the latitudinal region while the "Montane" modifier refers to elevation.

It should be noted that the latitudinal region and altitudinal belt bracketed by 12 °C and 24 °C are further divided by the "frost line" or critical temperature line into two parts: Warm Temperate–Subtropical and Premontane–Lower Montane, respectively. In this region or belt there is a very significant change in species composition due to the presence or absence of killing frost or near-freezing temperatures.

7.3. Maps of the Holdridge Classification

The first important feature of the Holdridge system, not found in other classifications, is that it truly recognizes that the tropics are complex. Holdridge devised a classification model with a greater number of life zones, specifically suitable for the analysis of complex tropical mountain vegetation, but applicable to other areas as well. (However, the use of a logarithmic scale for biotemperature does result in a relatively low diversity representation of tropical wet *lowland* environments.)

The system consists of three levels of classification. The climatically determined life zones, already defined, form the primary level. These are subdivided into associations that are differentiated on the basis of local environmental conditions. The third level of classification is the subdivision of the association on the basis of actual vegetation cover or land use (Holdridge *et al.*, 1971). For example, totora reed wetlands would be an edaphic/hydric association within the Low Montane Dry Forest Life Zone, defined in terms of potential vegetation, while areas actually drained and in kikuyu grass for dairy cattle pasture would be a subdivision of the association. So far, relatively little work has been done in the Andes to identify associations in the Holdridge system. Four basic types of association are defined in the Holdridge system: climatic (the name placed inside each hexagon in the life zone chart), edaphic, atmospheric and hydric. Edaphic associations are those in which azonal soils cause deviations from the climatic associations, e.g., because of sandy texture. Atmospheric associations are those associated with, e.g., unusual seasonal precipitation patterns, such as a Mediterranean climate or cloud forests. Hydric associations are associated with shallow water.

Most publications of the Holdridge classification refer only to the larger units, the life zones. Moreover, in Holdridge's scheme, each life zone can be further subdivided into a number of associations depending on "unusual" or "special" prevailing edaphic, hydric or atmospheric conditions and their combinations; but for each life zone only one association qualifies as climatic, and its name is lifted to become the name of the life zone. Further subdivisions of the associations may be the seral stages including modified natural vegetation as well as different patterns of human intervention (cropping, pastures, etc.).

Thus defined, life zones have been mapped in Central and South America, the eastern and central USA, Nigeria, and Thailand (Budowski, 1964; Holdridge *et al.*, 1971); and a broad-scale life zone map has been created based on data from about 8000 meteorological stations worldwide (Emanuel *et al.*, 1985).

Numerous applications of the life zone system have been derived especially in relation to the determination of potential evapotranspiration (Holdridge, 1959), studies of leaf physiognomy (Tasaico, 1959), relating the number of strata of natural vegetation associations to the height of dominant trees (Wadsworth, 1960), and the development of improved land-use systems for the tropics in general (Hunter, 1959) and more specifically for Peru (Lindsay, 1959; Tosi, 1960).

The Holdridge map of Ecuador was published by Cañadas (1983) at a scale of 1:1 000 000, together with a report that included discussion of climate, soils, vegetation and land use for the life zones of the country. The map was based on precipitation (1964–1973 data) and temperature maps and was refined using aerial photography and field surveys.

7.4. Life Zone Boundaries in Space and Time

Within each hexagonal boundary of the life zone chart, there are six small triangular sections (*Figure 7.1*). Within the limits of climate demarcated by the triangles, the vegetation possesses some of the characteristics of the adjacent zone or zones and is considered to be transitional. It must be emphasized that transitional in the life zone context indicates a change in the appearance of the vegetation on an areal basis rather than on a temporal basis (Holdridge *et al.*, 1971).

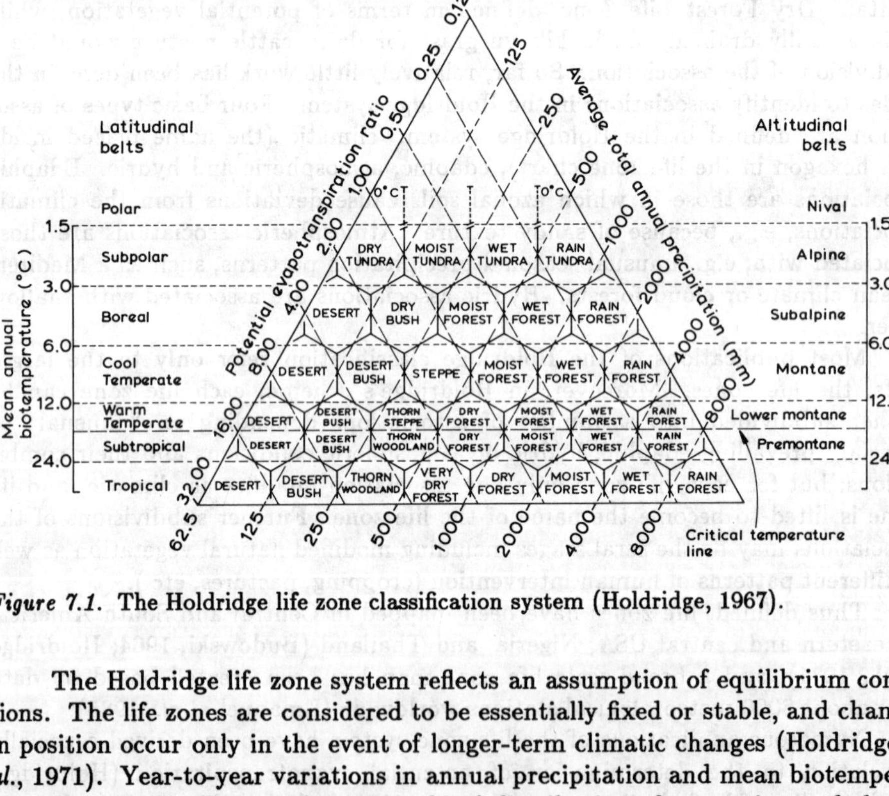

Figure 7.1. The Holdridge life zone classification system (Holdridge, 1967).

The Holdridge life zone system reflects an assumption of equilibrium conditions. The life zones are considered to be essentially fixed or stable, and changes in position occur only in the event of longer-term climatic changes (Holdridge *et al.*, 1971). Year-to-year variations in annual precipitation and mean biotemperatures have little effect on the geographical distribution of vegetation and should not affect the boundaries of life zones. Therefore, life zone determinations should be based on long-term records, or on filed observations of vegetation forms that reflect long-term influences.

The long-term association of a given climate and vegetation may produce a characteristic soil for the life zone, a "zonal" soil. Zonal soils are defined –

following the soil geograher Dokuchaiev and his followers (Glinka, 1927) – as soils that reflect the influence of the active factors of soil genesis: climate and organisms. Thus, the possibility exists of defining life zone boundaries using soil formation – an approach that will be used below.

Since the life zone represents an equilibrium condition, scenarios of climate change should be described as new equilibria. Effects might be perceived first in the transitional zones, where different vegetation types may already coexist and readily be able to respond to changing conditions.

In this section we will not be modeling long-term changes in temperature. The analysis performed in Section 5 suggested that such changes have been slight in recent Ecuadorian history and that changes in frost risk may have been more significant than changes in mean annual temperature in the medium-term. Global rises in CO_2 concentration may result in temperature increases of the order of 1 °C in the Andes (Emanuel *et al.*, 1985). Such changes could increase the upper limit of cultivation by 200 m if frost risks also improve; the results of such an increase were discussed in Section 5 in connection with a scenario of change of medium-term frost risk.

In this section we will consider effects under scenarios of long-term precipitation change developed using the precipitation record at Quito (Section 2). The scenarios were produced using historically observed changes in 30-year rainfall averages and observed correlations of Ecuadorian rainfall with the phases of the Northern Oscillation.

7.5. Significance of Life Zone Boundaries

The life zones are essentially defined in terms of climatic parameters. According to Holdridge (1967), verification that the boundaries of life zones correspond to actual vegetation patterns can be demonstrated both from field mapping experience and the correlation of objective measurements of forest stands with life zone units. This assertion has been questioned by other ecologists, but will not be discussed further here. The problem has been partly masked by using observed vegetation changes as evidence for the precise location of life zone boundaries, even where climatic evidence is contradictory.

Life zones have been used to indicate agricultural or land-use regions, both by the authors of ecological monographs in the Holdridge tradition and by social scientists (e.g., Brush, 1977). Others have pointed out that crop moisture and temperature requirements may have little relation to life zone boundaries, that actual land use is as dependent on prevailing economic rationality and the nature of alternatives as it is on ecological crop requirements, and that the life zone system may distract attention from variation in azonal soils, slope and water availability (Knapp, 1984).

In the previous section, the crucial precipitation values for highland barley and potato cultivation in Ecuador were determined to be 400, 800, 1200 and 1400 mm/year. In contrast, life zone boundaries are located at 250, 500, 1000 and 2000 mm/year. Probably only the shift of the 500 mm boundary is significant for highland farmers, since this is broadly coincident with the limit of low yields for barley cultivation. The wet limit of cultivation falls roughly midway between 1000 and 2000 mm/year, making the Holdridge system almost irrelevant for the study of that boundary.

Given these caveats, modeling variations in the life zones is an illuminating exercise contributing to an understanding of potential impacts of long-term climate change. At the present stage of knowledge, it is perhaps best treated as a general indication of magnitudes of potential change rather than as a means of identifying specific regions and land-use types that might be affected.

7.6. The Ecological Map of Chimborazo Province

Given the scarcity of reliable long-term climatic data for the Andes, it has frequently been necessary to identify life zones on the basis of observable vegetation patterns (Tosi, 1960). Unfortunately, the vegetation of much of the Andes, and certainly in Chimborazo Province, is in a disturbed condition; large areas have been converted to agricultural use. Other sources of data on life zone distribution are therefore required.

When the soils of Ecuador were mapped in the 1970s by Colmet-Daage (PRONAREG-ORSTOM, 1979b; 1979c), it was shown that climate and topography were the essential factors accounting for variations of these soils. This led Colmet-Daage to define "topo-climo-sequences" for the highlands and to suggest the elaboration of ecological maps at the scale of 1:50 000 on the basis of soils charts at the same scale.

Throughout the highlands, the distribution of volcanic ash can vary between tens and hundreds of meters thickness. It is necessary to distinguish between soft ash, which is relatively recent, and the hard cemented ash or cangahua, which is older. Erosion has, of course, been very active on the steep slopes and sometimes has produced a mixture of both old and new deposits. Topography thus plays a role in the depth of soft ash and, consequently, affects the water-holding capacity of the soil. Climate also plays a much greater role in soil differentiation than variations in the composition of the parent material, e.g., by controlling rates of erosion, leaching/concentration of soluble salts (duripan formation), etc. Consequently, climatic–ecological maps can be drawn from soils maps because of soil property variations due to climatic factors.

This approach was taken up by L. Cañadas, A. Jacome, and W. Estrada, who between 1977 and 1980 identified life zones in Ecuador according to the following procedure (Cañadas, 1983):

(1) The pattern of isotherms was determined from topographic maps based on the known lapse rate of temperature with elevation (*see* Section 5).
(2) Following the Holdridge scheme, the altitudinal belts were defined in terms of mean annual biotemperature.
(3) The altitudinal belts were plotted onto existing soil maps (PRONAREG-ORSTOM, 1979b; 1979c), together with whatever precipitation data were available.

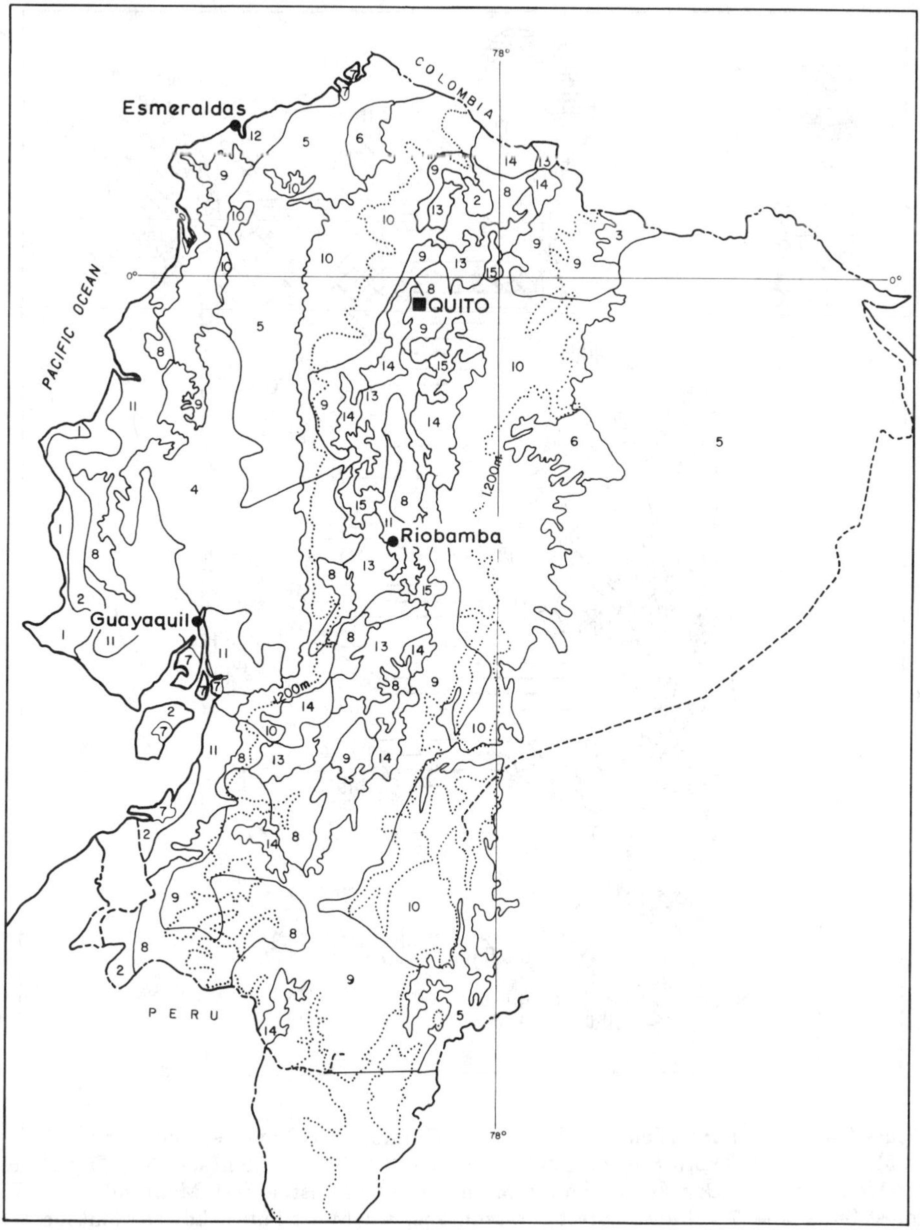

Figure 7.2. The distribution of life zones in Ecuador (Cañadas, 1983). Key: 1 = Desert Bush Tropical; 2 = Thorn Woodland Tropical – Pre-Montane; 3 = Very Dry Forest Tropical; 4 = Dry Forest Tropical; 5 = Moist Forest Tropical; 6 = Wet Forest Tropical; 7 = Mangroves; 8 = Dry Forest Pre-Montane–Lower Montane; 9 = Moist Forest Pre-Montane–Lower Montane; 10 = Wet Forest Pre-Montane–Lower Montane; 11 = Steppe Lower Montane; 12 = Steppe Montane; 13 = Moist Forest Montane; 14 = Wet Forest Montane; 15 = Wet Rain Forest Subalpine (Paramo).

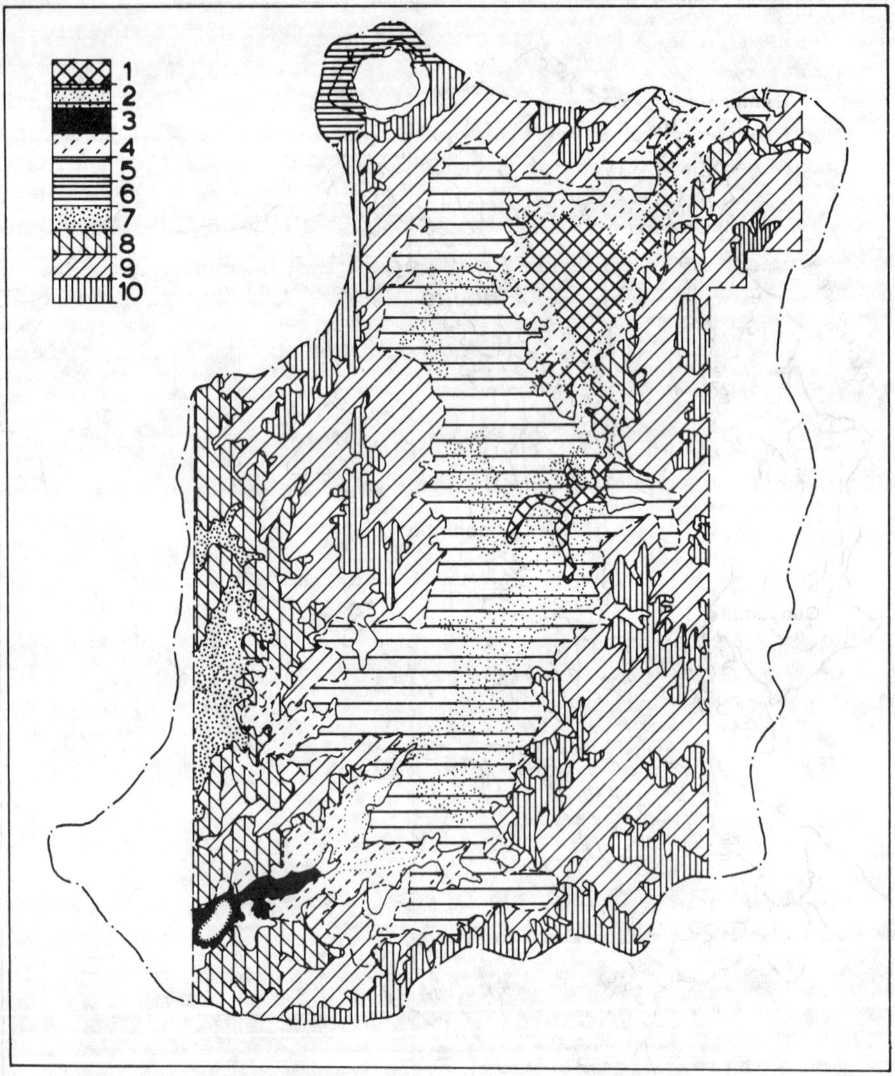

Figure 7.3. The distribution of Life Zones in Chimborazo Province, Ecuador (Cañadas, 1983). Key: 1 = Thorn Steppe Lower Montane; 2 = Steppe Montane; 3 = Dry Forest Pre-Montane; 4 = Dry Forest Lower Montane; 5 = Moist Forest Montane; 6 = Wet Forest Subalpine; 7 = Moist Forest Pre-Montane; 8 = Moist Forest Lower Montane; 9 = Wet Forest Montane; 10 = Rain Forest Subalpine.

(4) Zonal soils were identified for each life zone on the basis of their relationship to precipitation and biotemperature values. Azonal soils (i.e., those with special texture or topographic position) were also identified and were assigned if necessary to a drier or moister life zone than precipitation

suggested. The resulting map therefore reflects association patterns as well as life zone patterns.

(5) The life zone boundaries were drawn along the boundaries of the zonal and azonal soils.

The resulting maps of life zones in Ecuador as a whole (*Figure 7.2*) and of Chimborazo Province (*Figure 7.3*) reflect the present climate, which includes both the normal years and extreme events of recent decades. By using soil information, of course, we run the risk of drawing boundaries along extinct climatic ecotones, e.g., boundaries that under former climatic conditions were crucial for soil formation, but which are no longer in the same location.

7.7. Selection of Climatic Scenarios

As in Section 2, the following scenarios were selected to test the sensitivity of terrestrial ecosystem complexes to possible changes of precipitation:

(1) The present *baseline* climate, which is reflected in the present-day current vegetation distribution, which we assume is related to present-day soil boundaries.

(2) A *dry scenario*, based on historical experience of change, and with precipitation assumed to be 10% below the baseline.

(3) A *wet scenario*, with precipitation assumed to be 10% above the baseline.

The life zone map of Chimborazo Province (*Figure 7.3*) was assumed to correspond with the normal values of total annual precipitation.

Table 7.1. Areal extent of life zones in Chimborazo Province under various climatic scenarios (for definitions, *see* text).

	Extent of life zone under scenario:					
	Baseline		*Dry*		*Wet*	
Life zone	*ha*	*%*	*ha*	*%*	*ha*	*%*
Thorn Woodland Pre-Montane	3 080	0.62	1 650	0.33	960	0.19
Thorn Steppe Lower Montane	21 800	4.38	29 520	5.93	9 840	1.98
Steppe Montane	31 840	6.40	44 520	8.95	14 820	2.98
Dry Forest Pre-Montane	4 020	0.81	11 400	2.29	2 720	0.55
Dry Forest Lower Montane	30 680	6.17	46 470	9.34	45 830	9.21
Moist Forest Montane	111 370	22.39	193 940	38.90	106 630	21.44
Moist Forest Pre-Montane	14 030	2.82	5 420	1.09	16 390	3.29
Moist Forest Lower Montane	38 910	7.82	31 950	6.42	55 860	11.23
Wet Forest Montane	178 620	35.91	72 780	14.63	184 780	37.15
Wet Forest Subalpine	5 920	1.19	59 790	12.02	3 600	0.72
Rain Forest Subalpine	57 180	11.49	–	–	56 010	11.26
TOTAL		100.0		100.0		100.0

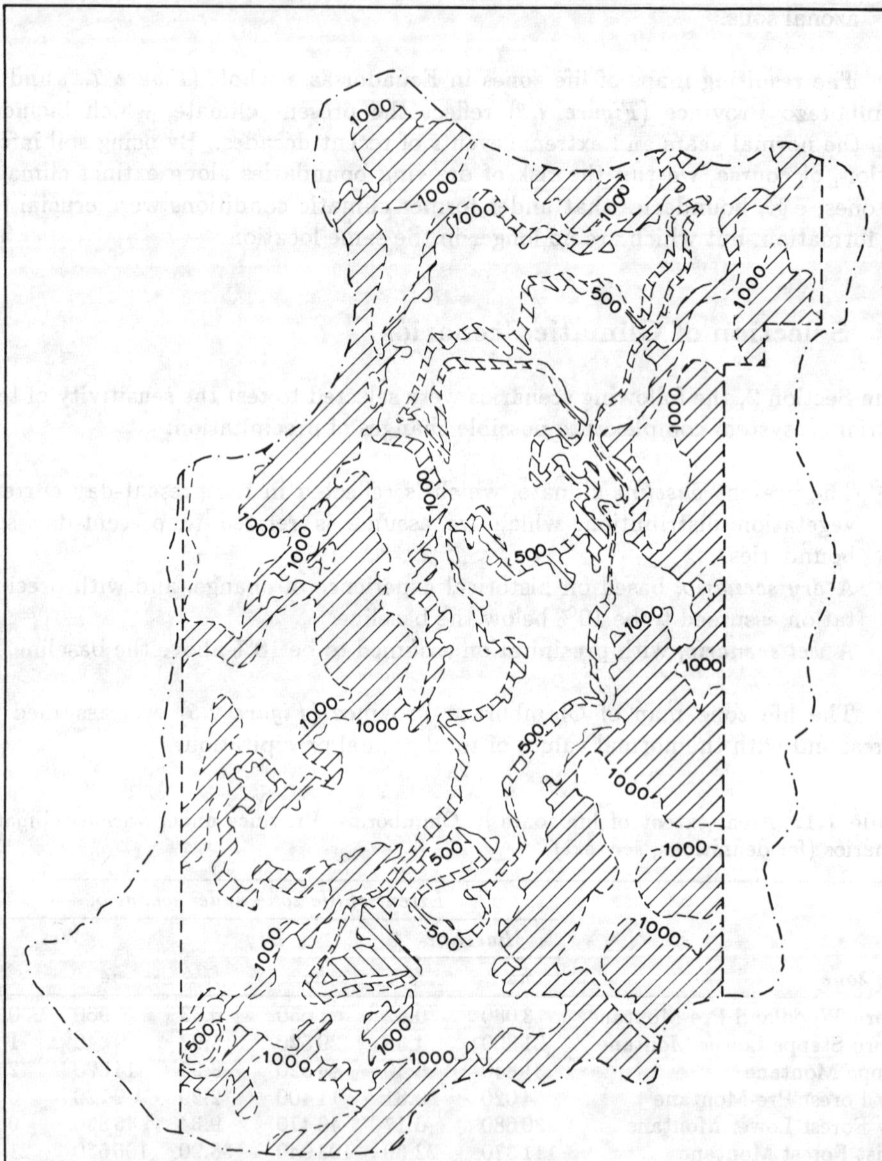

Figure 7.4. Hypothesized shift of 500 mm and 1000 mm isohyets under dry climate scenario.

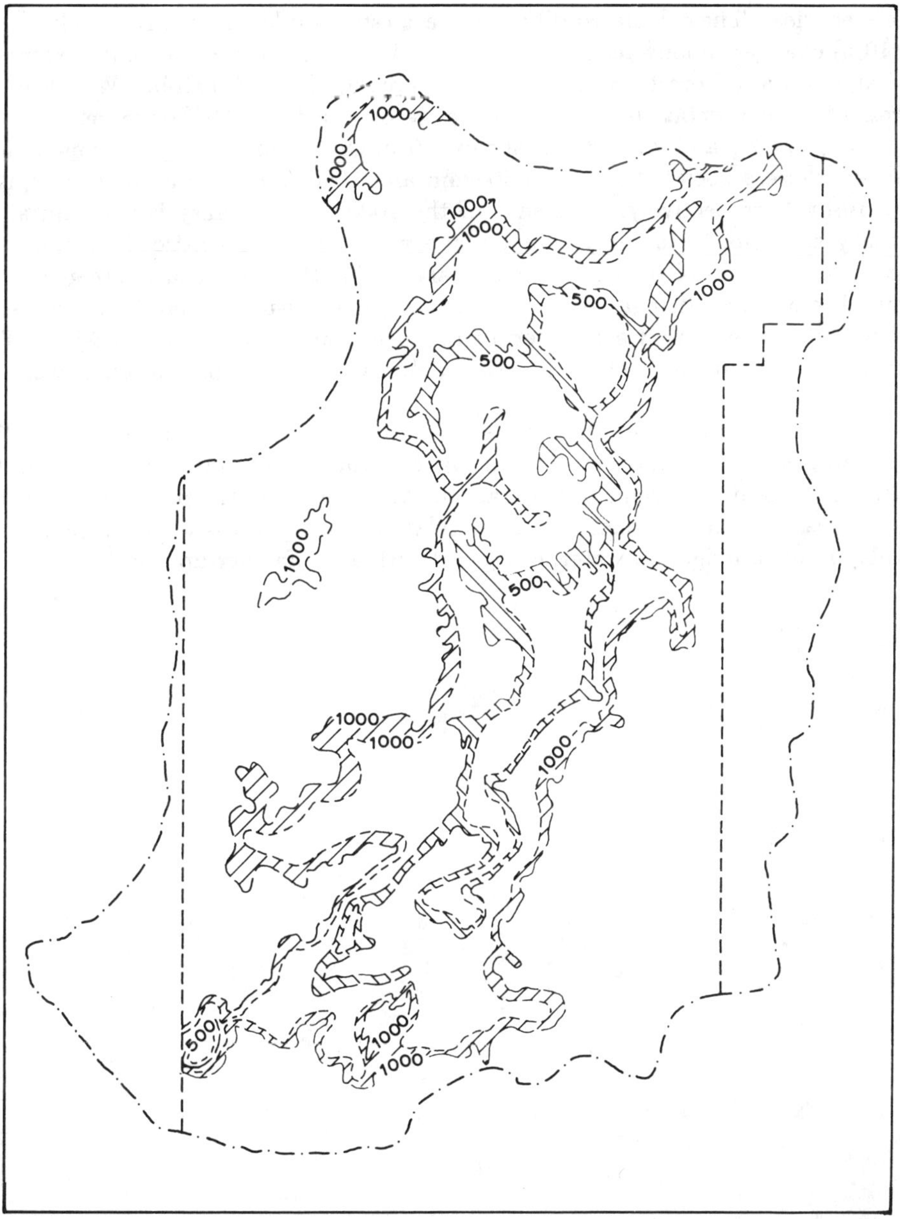

Figure 7.5. Hypothesized shift of 500 mm and 1000 mm isohyets under wet climate scenario.

7.8. Results

Table 7.1 indicates the differences in areal extent of the life zones associated with the scenarios. The calculated differences are remarkable for such relatively slight (±10%) changes in long-term precipitation. Under the dry scenario, for example, Subalpine Rain Forest disappears and is replaced by Subalpine Wet Forest. Areas of other vegetation types often change by a factor of 50% or more.

Figures 7.4 and *7.5* indicate the extent of change of two key life zone boundaries: those associated with the 500 mm and 1000 mm mean annual precipitation isohyet, respectively. The shift of the 1000 mm boundary is substantial in the dry scenario; fortunately, this boundary is of relatively little significance to agriculture. Shifts of the 500 mm boundary during the dry scenario are generally limited to a narrow fringe. This may suggest that dry-margin farmers have relatively little to fear from a long-term shift in precipitation. *Figure 7.5* identifies certain areas that may benefit if a slight increase in mean precipitation were to occur.

Thus, despite substantial shifts in relative areas and locations of the life zones in general, the current analysis suggests that fairly limited areas would be affected at the dry margins of cultivation. As observed in the preceding chapter, these areas are already ones that have relatively good access to settlements and roads, if not to irrigation water, compared with areas farther upslope.

SECTION 8

Conclusions and Implications for Policies of Rural Development

8.1. Past Interventions

The indigenous farmers of the Sierra of Ecuador have not always been a "marginal" group. Prior to the Inca and Spanish conquests, they had access to the best land and controlled the key cultural institutions. They used sophisticated systems of raised-field and canal irrigation agriculture to support populations locally more dense than those in existence today.

As a result of centuries of colonization and manipulation, current indigenous farmers find themselves on the worst land, with the least access to resources or political influence, of the various Ecuadorian ethnic groups in the highlands. The problem, therefore, is not one of instituting policies affecting indigenous people, but of undoing the effects of past conscious or *de facto* policies; of unshackling the potential of indigenous farmers to use their traditions and skills to shape their own futures.

As discussed in the first section of this case study, few institutions are concerned with monitoring or helping indigenous agriculture. There is no general extension service, no adequate monitoring system for yields, no information on hunger or famine, and no information on the impacts of droughts or downpours. There has been no effective national agency concerned with indigenous people, athough this may have been a blessing in view of the lack of trained personnel to staff such an agency. There have been a few trial projects to help traditional farmers, sponsored by the US Agency for International Development, the Ecuadorian government, the Catholic Church, and others. For a number of years now, the Ecuadorian government has channeled foreign researchers into Chimborazo Province. It is safe to say, however, that despite a bewildering profusion of reports and suggestions, the province remains desperately poor and without much to show in the way of improvements.

In this context the present study constitutes perhaps just one further compilation of information and listing of suggestions. It has the merit, however, of being the first to consider the potential impacts of climatic variations on traditional crop yields and distributions and to suggest policy responses to such impacts.

8.2. The Components of the Case Study

Good long-term meteorological data exist for the Quito Observatory meteorological station. Analyses of these data in Sections 2 and 5 indicated that some global climatic processes (particularly the Northern Oscillation) do have significant connections to Ecuadorian highland climate. Long-term scenarios of precipitation change were generated to reflect these processes.

Apart from long-term change, short-term variations of precipitation are also important in Ecuador. Scenarios of short-term precipitation change were generated using rainfall data from Riobamba and Palmira in Chimborazo Province.

Chimborazo Province was chosen as an appropriate subregion of the highlands for more detailed study. This province has a large population of poor, ethnically indigenous peasants who are ostensibly vulnerable to both long-term and short-term variations of climate.

The study of climate impacts involves a "chicken-and-egg" problem. The delimitation of significant environmental problems requires a knowledge of local adaptive strategies, while the study of local adaptation requires at least an initial guess as to the significant environmental problems. One solution to this quandary is to ask the peasants themselves what the significant problems are. Such an approach was followed in Section 3, the ethnographic portion of the project. The interviews conducted for this portion of the project also enabled collection of data on cropping practices and yields. These indicated the significance of nonagricultural strategies, such as temporary labor migration, to maintain resilience in the face of an uncertain environment. The farmers made it clear that low prices were a fundamental disincentive to commercial agriculture, and that farming was useful primarily for maintaining a low family cost of living.

Another approach is to attempt to correlate yield data with various environmental parameters (elevation, slope, temperature, precipitation, etc.). Unfortunately, the only time series of yield data that are currently available are for large rather than for small farms, and there are few even of those series. In this quandary agroclimatic modeling based on models developed elsewhere is an intriguing alternative, and the trial presented in Section 4 may serve to whet our appetite. Of course, such models at present are vulnerable to the frequently poor quality of meteorological data and the enormous spatial variation in climatic parameters in mountain environments. In addition, while soil fertility is seldom given much attention in developed country agricultural systems analysis (it being assumed that chemical fertilizers are readily available and relatively cheap), in Ecuadorian small farms fertility is the single most important limiting factor to increased production. In many cases, small-farm potato and maize yields could be increased many-fold with supplementary fertilization.

As a consequence of the aforementioned problems, we really do not know what climatic parameter(s) set the upper limit of cultivation in Ecuador. In Section 5 it was assumed that the worst frost of the late growing season set the upper limit of cultivation. The analysis of the consequences of a plausible 0.8 °C intensification of late growing season frost severity indicated that 8% of the arable land in Chimborazo, with over 4000 farmers, could go out of production. The most affected would be poor indigenous farmers on the west side of Chimborazo Province.

Sections 6 and 7 explored the consequences of drought and excessive precipitation. Section 6 investigated the likely impacts of short-term climatic fluctuations on the areas with good yields of potatoes and barley. The agroclimatic requirements of crops were inferred from the crop–climate study in Section 4, ethnographic material in Section 3 and prior empirical crop–climate studies. Actual land use was taken as an indicator of the limits of yield tolerated by traditional farmers.

The testing of the scenarios indicated that the potential area of arable land with good yields would decline both for 1-in-10 dry years and 1-in-10 wet years. Wet years were actually more damaging than dry years.

Especially interesting was the decline in area suitable for potatoes in dry years and the area suitable for barley in wet years. This result demonstrates the appropriateness of the current practices, which employ a mixed planting strategy to minimize maximum possible loss. Experiments with the crop–climate model developed in Section 4 also indicated the desirability of the (already well-established) practice of staggering planting dates.

Section 7 represents perhaps the first effort to study shifts in Holdridge life zones under conditions of precipitation variation in the Andes. Since Holdridge life zones are assumed to reflect long-term conditions rather than annual values, the exercise in Section 7 focused on long-term changes in average climate. Even under scenarios of modest changes in precipitation, areas of various life zones changed substantially. On the other hand, the life zone boundary most significant to agriculture (the 500 mm isohyet, close to the limit of cultivation) exhibited little change.

Section 7 is also noteworthy for its pioneering use of soil map data to define life zones in regions where the natural vegetation has largely been removed.

8.3. Synthesis and Policy Recommendations

We are, of course, not alone in studying peasant technology in Ecuador. The Centro Andino de Acción Popular (CAAP) has performed a valuable service in the cooperative study of peasant technology, as have a number of Ecuadorian and foreign anthropologists, agronomists and geographers. Nevertheless, most would admit that this type of study began recently and is still at a rudimentary state of development. Thus, our recommendations in some respects must be tentative.

A number of authors would argue that the fundamental problem of the poor farmers of Ecuador is their structurally determined or reinforced poverty, e.g., lack of power, and that any concern with purely environmental problems merely diverts attention from the real issues. The environment, however, clearly presents serious problems to agriculturalists, regardless of prevailing social structure, and a concern with these problems need not necessarily imply an ignorance of or diversion of attention from social issues.

This project has suggested that there is the possibility of significant shifts of arable land for specific crops under plausible scenarios of climatic variation – in precipitation or frost risk. The same analysis has suggested possible technical responses: staggered planting dates, cropping of cultivars with different moisture requirements and possible application of frost amelioration technology in flats. To a remarkable extent, farmers are already practicing these technical responses and others like them. The problem seems to be less one of introducing new practices than of preventing the introduction of risky new practices.

The use of marginal areas could be discouraged through government proscriptions or made more secure through insurance programs. Both alternatives, however, would actually limit peasant options by reducing the already small land base or requiring contributions of scarce cash.

In general, capital-intensive solutions are by definition inappropriate for small family farmers whose major resource is labor and who have virtually no access to credit. The best option for the short run is to work with, rather than against, peasant perception and technology, refining and facilitating the employment of traditional contour canal irrigation, crop rotation, soil fertilization, drainage and frost mitigation practices. This shift in emphasis is, of course, distasteful to those who have much money, but little field experience, and perhaps can never be fully realized in the current political context.

Agencies that cannot dedicate many years to respectful cooperation with small farmers may wish, however, to pursue the extension and refinement of peasant technology with particular emphasis on mixed traditional–modern technologies. Little attention has been paid, for example, to determining low-cost amendments to traditional manuring and mulching practices, to improving traditional small-scale irrigation features, to reviving the prehistoric ridged-field frost control strategy, or to diversifying and extending traditional multi-functional hedgerows.

In the long run, the ability of peasant farmers to cope with crisis depends not only on a diversified and functioning traditional technology, but also on a major commitment by a society to the well-being of rural people. This commitment is reflected in pricing policy, land reform policy, rural credit and even the treatment of temporary migrant laborers. Regardless of social structure, one can ask: to what extent are poor farmers enabled and encouraged to stand up, get the tools they need, and increase their adaptability?

References

Banco Nacional de Fomento (1982). *Guia Agricola, "Elementos Basicos"*. Quito, 203 pp.

Bonifaz, E. (1980). Estudio de las Lluvias en la sierra. *Boletin: Oregano de Informacion de la Camara de Agricultura de la Primera Zona (Quito)*, **27**, 3–7.

Brush, S.B. (1977). *Mountain, Field and Family. The Economy and Human Ecology of an Andean Valley.* University of Pennsylvania Press, Philadelphia, PA.

Budowski, G. (1964). *Holdridge World Classification of Life Zones. A Reappraisal.* Technical Paper, Inter-American Institute of Agricultural Sciences.

Cañadas, L. (1983). *El Mapa Bioclimatico y Ecologico del Ecuador.* Banco Central del Ecuador, Quito, 260 pp.

Cañadas, L. and Salvador, H. (1982). *Agrometeorological Assessment Models for Rural Development in the Central Sierra of Ecuador.* Final Report. Center for Environmental Assessment Services, Columbia, MO.

Cardich, A. (1975). Agricultores y pastores en Lauricocha y Limites superiores del cultivos. *Revista del Museo Nacional (Lima)*, **41**, 11–36.

Caviedes, C. (1982). Natural hazards: in search of a method and a theory. *Geojournal*, **6**(2), 101–109.

Clements, F.E. (1920). *Plant Succession: An Analysis of the Development of Vegetation.* Carnegie Institute, Washington, DC, 299 pp.

Di Vincenzo, J. (1984). *Middlemen and Peasants in Pimampiro: Changing Market Relations in a Highland Ecuadorian Canton.* Unpublished M.A. thesis, University of Texas, Austin, TX.

Duverneuil, F. (1983). Distribucion de la poblacion en al provincia de Loja y su evolucion entre 1950 y 1974. *Cultura*, **15**, 255–284.

Ellenberg, H. (1979). Man's influence on tropical mountain ecosystems in South America. *J. Ecology*, **67**, 401–416.

Emanuel, W., Shugart, H.H. and Stevenson, M.P. (1985). Climatic change and the broad-scale distribution of terrestrial ecosystem complexes. *Climatic Change*, **7**(1), 29–43.

FAO (1975). *Soil Map of the World*, Volume 4(1). Food and Agriculture Organization, Paris, 93 pp.

FAO (1980). *Global Information and Early-Warning System in Food and Agriculture.* 14 March 1980, Food and Agriculture Organization, Rome, 21 pp.

Ferdon, E.N. (1950). *Studies in Ecuadorian Geography.* School of American Research (Monograph No. 15), University of Southern California, Santa Fe, NM.

Field, L. and Chiriboga, M. (eds.) (1984). *Agricultura Andina: Propuesta de Investigacion.* Centro Andino de Accion Popular, Quito.

French, V. and Hodges, T. (1985). Comparison of crop phenology models. *Agron. J.*, **77**, 170–171.

Frere, M., Rea, J. and Rijks, J.Q. (1975). *Estudio Agroclimatologo de la Zona Andina.* Food and Agriculture Organization, Rome.

Gade, D. (1967). *Plant Use and Folk Agriculture in the Vilcanota Valley of Peru: A Cultural-Historical Geography of Plant Resources.* Ph.D. thesis, University of Wisconsin–Madison. University Microfilms, Ann Arbor, MI.

Gaussen, H. (1959). The vegetation maps. In *Institut Francais de Pondichery Travaus de la Section Scientifique et Technique*, **1**(4), 155–178.

Geiger, R. (1965). *The Climate Near the Ground.* Harvard University Press, Cambridge, MA.

Glinka, K.D. (1927). Dokuchaiev's ideas in the development of pedology and cognate sciences. In *Russian Pedological Invest. I. Acad. Sci.*, Leningrad.

Gondard, P. (1983). La utilizacion del suelo y los paisajes vegetatles en la provincia de Loja. Aproximacion a los sistemas de produccion agricola. *Cultura*, **15**, 285–330.

Gondard, P. (1984a). Agricultura de Altura. *Ecuador Debate*, **6**, 25–47.

Gondard, P. (1984b). *Inventario y Cartografia del Uso Actual del Suelo en los Andes Ecuatorianos.* PRONAREG-ORSTOM and CEPEIGE, Quito.

Gondard, P. and Lopez, F. (1983). *Inventario Arqueologico Preliminar de los Andes Septentrionales del Ecuador.* MAG, PRONAREG, ORSTOM, and Museum of the Central Bank of Ecuador, Quito.

Guillet, D. (1983). Toward a cultural ecology of mountains: The central Andes and Himalayas compared. *Current Anthropology*, **24**, 561–574.

Hodges, T. and Botner, D. (1985). Using the CERES-maize model to estimate maize production for the US corn belt. *Agron. Abstract*, Annual Meeting of American Society of Agronomy, Chicago, IL, 12 pp.

Holdridge, L.R. (1947). Determination of world plant formations from simple climatic data. *Science*, **105**, 367–368.

Holdridge, L.R. (1959). Simple methods for determining potential evapotranspiration from temperature data. *Science*, **130**, 572.

Holdridge, L.R. (1967). *Life Zone Ecology.* Tropical Science Center, San José, Costa Rica.

Holdridge, L.R., Grenke, W.C., Hatheway, W.H., Liang, T. and Tosi, J.A. (eds.) (1971). *Forest Environments in Tropical Life Zones: A Pilot Study.* Pergamon Press, Oxford, UK.

Hunter, L.R. (1959). A new guide to land use planning in tropical areas. *Ceiba*, **8**(2), 44–70.

INEC (various years, ending 1982). *Indice de precios al consumidor.* Published monthly, Instituto Nacional de Estadistica y Censos, Quito.

INEC (1984). *Sistema Estadistico Agropecuario Nacional per Muestreo por Areas.* Instituto Nacional de Estadistica y Censos, Quito.

Johnson, A.M. (1976). The climate of Peru, Bolivia, and Ecuador. In W. Schwerdtfeger (ed.) *Climates of Central and South America*, World Survey of Climatology, Volume 12, Elsevier, Amsterdam, pp. 147–218.

Kirkby, R.A. (1981). *The Study of Agronomic Practices and Maize Varieties Appropriate to the Circumstances of Small Farmers in Highland Ecuador.* Ph.D. thesis, University of Toronto. University Microfilms, Ann Arbor, MI.

Knapp, G. (1980). Cambios en el clima de los Andes Ecuatorianos: una tentativa explicacion genetica. *Revista del Centro Panamericano de Estudios e Investigaciones Geograficas*, **4**, 39–48.

Knapp, G. (1984). *Soil, Slope and Water in the Equatorial Andes: A Study of Prehistoric Agricultural Adaptation.* Ph.D. thesis, University of Wisconsin–Madison. University Microfilms, Ann Arbor, MI.

Köppen, W. (1931). *Gründe der Klimakunde.* Walter de Gruyter, Berlin.

Kutzbach, J. (1970). Large-scale features of monthly mean northern hemisphere anomaly maps of sea level pressure. *Mon. Weather Rev.*, **98**, 708–716.

Lauer, W. (1982). Zur Okoklimatologie der Kallawaya-Region (Bolivien). *Erdkunde*, **36**, 223–247.

Linsay, A. (1959). Ecology and land use in Peru. *Ecology*, **43**(2), 349–350.

MAG (1984). *Estadisticas Agricolas.* Ministerio de Agricultura, Division de Estadistica (unpublished data), Quito.

MAG-INAHMI (1985). *Boletin Agroclimatico Mensual del Ecuador*, Ministerio de Agricultura and Instituto Nacional de Meteorologia e Hidrologia, Quito, 12 pp.

Masuda, S., Shimada, I. and Morris, C. (eds.) (1985). *Andean Ecology and Civilization*. University of Tokyo Press, Tokyo.

Mayer, E. (1979). *Land Use in the Andes: Ecology and Agriculture in the Mantaro Valley of Peru with Special Reference to Potatoes*. Centro Internacional de la Papa, Lima.

Murra, J. (1972). El control vertical de un maximo de pisos ecologicos en la economia de las sociedades Andinas. In J. Murra (ed.) *Visita de la Provincia de Leon de Huanuco en 1562, Volume 2*, Universidad Nacional Hermilio Valdizan, Huanuco, Peru, pp. 429–476.

Murra, J. (1985). "El archipelago vertical" revisited. In S. Masuda, I. Shimada and C. Morris (eds.) *Andean Ecology and Civilization*. University of Tokyo Press, Tokyo, pp. 3–14.

Namias, J. (1963). The interaction of circulation and weather between hemispheres. *Mon. Weather Rev.*, **91**, 482–486.

Palmer, (1965). *Meteorological Drought*. Research Paper No. 45, US Department of Commerce, Washington, D.C., 59 pp.

Parry, M.L., Carter, T.R. and Konijn, N.T. (eds.) (1988). *The Impact of Climatic Variations on Agriculture. Volume 1. Assessments in Cool Temperate and Cold Regions*. Reidel, Dordrecht, Netherlands.

Pourrut, P. (1983). Los climas del Ecuador: fundamentos explicativos. In *Los Climas del Ecuador*, CEDIG (Centro Ecuatoriano de Investigacion Geografica), Documentos de Investigacion No. 4, pp. 7–41.

Priestley, C.H.B. and Taylor, R.J. (1972). On the assessment of surface heat and evaporation using large scale parameters. *Mon. Weather Rev.*, **100**(2), 81–92.

PMA (Programa Mundial de Alimentos) (n.d.). *Ayuda Alimentaria Proporcionada al Ecuador (toneladas y dolares)*. Photocopied sheet, Programa Mundial de Alimentos, Quito.

PRONAREG-ORSTOM (1976). *Estudio Hidro-Meteorologico e Hidrogeologico Prelimina de las Cuencas de los Rios Pastaza, Chimbo y Chanchan. Volumen Anexo. Anexo 1: Inspeccion de las Estaciones en el Campo*, Programa Nacional de Regionalización Agraria–Office de Recherche Scientifique et Technique Outre-Mer, Quito, pp. 2–15.

PRONAREG-ORSTOM (1977–1982). *Cartas de Uso Actual de la Tierra en la Sierra Ecuatoriana*. Programa Nacional de Regionalización Agraria–Office de Recherche Scientifique et Technique Outre-Mer, Ministerio de Agricultura y Ganadería, Quito.

PRONAREG-ORSTOM (1977a). *Informe de Visita a las Estaciones Meteorologicas y Pluviometricas en las Provincias de Carchi-Pichincha-Imbabura-Bolivar-Esmeraldas-Los Rios y Guayas. 13 July 1977*. Unpublished manuscript, Programa Nacional de Regionalización Agraria–Office de Recherche Scientifique et Technique Outre-Mer, Quito.

PRONAREG-ORSTOM (1977b). *Estudio Hidro-Meteorologico e Hidrogeologico Preliminar de las Cuencas de los Rios Esmeraldas y del Norte Ecuatoriano*, Programa Nacional de Regionalización Agraria–Office de Recherche Scientifique et Technique Outre-Mer, Quito.

PRONAREG-ORSTOM (1979a). *Diagnostico Socio-Economico del Medio Rural Ecuatoriano*. Programa Nacional de Regionalización Agraria–Office de Recherche Scientifique et Technique Outre-Mer, Documentos A-D (4 vols.), Quito.

PRONAREG-ORSTOM (1979b). *Mapas de Suelos por Regionalization*, Programa Nacional de Regionalización Agraria–Office de Recherche Scientifique et Technique Outre-Mer, Centre des Antilles-Bureau des Sols, Martinique and Quito.

PRONAREG-ORSTOM (1979c). *Mapas de Suelos Sierra*, Programa Nacional de Regionalización Agraria–Office de Recherche Scientifique et Technique Outre-Mer, Ministerio de Agricultura y Ganaderia, Quito and Martinique.

PRONAREG-ORSTOM (1982a). *Zonificación Potencial de la Papa*. Programa Nacional de Regionalización Agraria–Office de Recherche Scientifique et Technique Outre-Mer, Ministerio de Agricultura y Ganaderia, Quito.

PRONAREG-ORSTOM (1982b). *Zonificación Potencial de Trigo-Cebada*. Programa Nacional de Regionalización Agraria–Office de Recherche Scientifique et Technique Outre-Mer, Ministerio de Agricultura y Ganaderia, Quito.

PRONAREG-ORSTOM (1983). *Riobamba: Mapa de Uso Actual del Suelo y Formaciones Vegetales*. Programa Nacional de Regionalización Agraria–Office de Recherche Scientifique et Technique Outre-Mer, Quito.

PRONAREG-ORSTOM (1984). *Canar: Mapa del Uso Actual del Suelo y Formaciones Vegetales*. Programa Nacional de Regionalización Agraria–Office de Recherche Scientifique et Technique Outre-Mer, Quito.

Ravelo, A.C. (1985). *Field Surveys to Galte and Chauzan Area* (unpublished data).

Richardson, C.W. (1981). Stochastic simulation of daily precipitation, temperature and solar radiation. *Water Res. Res.*, **17**(1), 182–190.

Ritchie, J.T. (1972). Model for predicting evaporation from a row crop with incomplete cover. *Water Res. Res.*, **8**(5), 1204–1213.

Ritchie, J.T. (1984). A user-oriented model for the soil water balance in wheat. In W. Day and R.K. Atkin (eds.) *Wheat Growth and Modelling*. Plenum Publishing Corp., NATO ASI Series.

Ritchie, J.T. and Otter, S. (1985). Description and performance of CERES-wheat. In W.D. Willis (ed.) *A User-Oriented Wheat Yield Model*. USDA Wheat Yield Model. USDA-ARR Series.

Salomon, F. (1985). The dynamic potential of the complementarity concept. In S. Masuda, I. Shimada and C. Morris (eds.) *Andean Ecology and Civilization*, University of Tokyo Press, Tokyo, pp. 511–531.

Sanchez, R. (1976). *The Model of Verticality in the Andean Economy: A Critical Reconsideration*. Actes du XLII – Congres International des Americanistes, Vol. IV, Paris, pp. 213–231.

SEDRI (1983). Proyecto de desarollo rural integral Guamote. *Proyecto PNUD-FAO ECU -79-007*. 3 volumes, Secretaria de Deserollo Rural Integral, Quito.

Smith, R.L. (1983). Peru coastal currents during El Niño: 1976 and 1982. *Science*, **221**, 1397–1399.

Tasaico, H. (1959). *La Fisionomia de las Hojas de los Árboles en Algunas Formaciones Tropicales*. Unpublished thesis, Turrialba, Costa Rica. Inter-American Institute of Agricultural Sciences.

Thornthwaite, C.W. (1948). Approach towards a rational classification of climate. *Geographical Rev.*, **38**, 55–94.

Tosi, L.A. (1960). Zonas de vida natural en el Perú. *Boletin Técnico No. 5*, Inter-American Institute of Agricultural Sciences, Lima.

Troll, C. (1968). The cordilleras of the tropical Americas. Aspects of climatic, phytogeographical and agrarian ecology. In *Geo-Ecology of the Mountainous Regions of the Tropical Americas*, Colloquim Geographicum Band 9, Ferd, Dummlers Verlag, Bonn, pp. 13–56.

USDA (1975). Soil taxonomy. *Agriculture Handbook No. 436*, US Department of Agriculture, Soil Survey Staff. US Government Printing Office, Washington, D.C.

USDA FAS (1977–1985). *Ecuador Annual Situation Report.* US Department of Agriculture, Foreign Agricultural Service, Quito.

Van Loon, H. and Rogers, J.R. (1978). The seesaw in winter temperatures between Greenland and northern Europe. Part I: General description. *Mon. Wea. Rev.*, **106**, 296–310.

Wadsworth, F.H. (1960). *Datos de Crecimiento de Plantaciones Forestales en México, Indias Occidentales, Centro y Sur América.* Forest Service, US Department of Agriculture, Rio Piedras, Puerto Rico.

Walter, H. (1973). *Vegetation of the Earth.* Springer-Verlag, New York.

World Food Program (1985). Unpublished manuscripts reporting food donations to Ecuador. Quito.

PART V

The Effects of Climatic Variations on Agriculture in Dry Tropical Regions of India

Contents, Part V

List of Contributors

GADGIL, Dr. S.
Centre for Ecological Science
Indian Institute of Science
Bangalore 560 012
India

HUDA, Dr. A.K.S.
International Crops Research Institute
for the Semi-Arid Tropics (ICRISAT)
Patancheru P.O.,
Andhra Pradesh 502 324
India

JODHA, Dr. N.S.
International Centre for Integrated
Mountain Development
G.P.O. Box 3226
Kathmandu
Nepal

SINGH, Dr. R.P.
International Crops Research Institute
for the Semi-Arid Tropics (ICRISAT)
Patancheru P.O.,
Andhra Pradesh 502 324
India

VIRMANI, Dr. S.M.
International Crops Research Institute
for the Semi-Arid Tropics (ICRISAT)
Patancheru P.O.,
Andhra Pradesh 502 324
India

Abstract of Part V

This study relates to the dry tropical regions of India. The major problem facing the area is inter- and intrayear variability of rainfall. Farmers have adapted to this by building diversity and flexibility into their farming systems. Policy makers and planners have also attempted to reduce the instability of dryland agriculture and its impact on the rest of the economy. However, the measures adopted by them have often in the past been designed to protect against drought rather than to respond to both negative (drought) and positive (high rainfall) aspects of rainfall variability. More recently, advances in science and technology have provided new methods and data enabling improved understanding of the agroclimatic, agrobiological and other environmental variables affecting agriculture; and these have demonstrated the extent to which it is possible to obtain higher and more stable crop production. In addition, drought relief interventions, which originally were mainly short-term and purely welfare-oriented *ad hoc* activities, have recently acquired productivity orientation and long-term focus. Section 1 considers these developments in some detail.

In Section 2 a method of assessing the impact of rainfall variability on crop yields was demonstrated with respect to sorghum in a semi-arid tropical subregion in Karnataka State in South India. The large variation of the distribution of seasonal rainfall has led farmers to devote a large proportion of their annual crop production to post-rainy cropping, so that the crops are raised on residual moisture. This practice has led to increased soil erosion as bare lands are exposed during the rainy season.

Section 3 considers the variations in sorghum yields that occur under moist and dry scenarios. In areas where the annual rainfall exceeds 750 mm and the available moisture storage capacity of soils exceeds 150 mm, the cultivation of two crops, one in the rainy season and a sequential crop in the post-rainy season, is feasible. The study identifies the areas in which it would be worthwhile to collect surface runoff water for subsequent use for protective irrigation, to ensure crop survival during the dry season. Simulation of the response of sorghum to supplemental water provided at different growth stages during post-rainy seasons indicates that irrigations of 50 mm each at sowing and anthesis would increase sorghum yield in poor years by 136%, in average years by 38% and in good years by 17%.

Short-term responses take the form of adjustments of plant population, changes in levels of input such as the application of fertilizers and variations in other agronomic practices. Adjustments to severe drought situations often include curtailment of essential consumption and the disposal of productive assets.

The study concludes with a plea for dovetailing the application of technology with public interventions, and with farmers' traditional responses to climatic variability. The public interventions should be more in the nature of permanent drought-relieving assets rather than short-term measures to alleviate distress. Traditional adjustment mechanisms are losing their effectiveness in the face of rising population pressure and other forces of change. These adjustments can be modernized and strengthened with the help of new technological developments that have the potential for higher productivity under the same soil–rainfall conditions.

SECTION 1

Introduction

1.1. Background of the Case Study

India, owing to its size and geographical location, is characterized by extremely varied agroclimatic conditions, ranging from humid to arid and tropical to temperate. The most important climatic element for agriculture in the tropics is the rainfall. The dependence of Indian agriculture on monsoon rainfall is well known; in fact, the Indian economy has been described as a gamble on the monsoon rains. Any climatic change involving a large-scale change in the rainfall is therefore expected to make a significant impact on agricultural productivity in India.

However, analysis of the variation of rainfall over the country as a whole from 1871–1978 does not reveal any long-term changes in rainfall (Mooley and Parthasarathy, 1984). Even over the smaller regions, i.e., the meteorological subdivisions of the country, no long-term trends are found (Parthasarathy, 1984). Hence, in the Indian context, it will be more relevant to consider the impact of short-term variability of rainfall rather than of long-term climatic change.

The regions most susceptible to inter- and intrayear variability of rainfall are the arid and semi-arid tropics in India. The annual rainfall in these dry tropics ranges from less than 350 to 1500 mm and is accompanied by a high coefficient of variation. More importantly, evapotranspiration exceeds precipitation for 8 to 9 months in a year, which, in the absence of irrigation or moisture conservation measures, limits the period available for crop growth to roughly 50 to 200 days in most of the areas.

The variability of rainfall has not only shaped the farming systems in the regions, but has greatly influenced the public policies and programs for these areas. The only difference between the two is that, through diversification and flexibility of farming systems, the farmer tries to adapt both to good and to bad rainfall situations, whereas the policy maker and administrator often respond only to the negative side of the rainfall variability. Hence, historically, public interventions have largely been directed toward protection of the people from the

Table 1.1. Some agroclimatic features of locations included in the case study.

Locations	*Rainfall during growing season (Jun–Oct)* (mm)	*Potential evapotranspiration* (PET)	*Major soils*[a] *A*	*B*	*C*	*D*	*E*
	Areas with crop growing season of 200–260 days[b]						
Akola	801 (877)[c]	730				√	
Amravati	880 (975)	701				√	
Bangalore	671 (924)	570	√				
Indore	944 (1001)	733			√	√	√
	Areas with crop growing season of 145–199 days						
Aurangabad	718 (792)	683				√	
Dharwad	687 (889)	628				√	
Gulbarga	653 (753)	761			√		√
Hyderabad	593 (792)	715	√				√
Mahbubnagar	693 (792)	638	√				
Sholapur	652 (742)	707			√	√	√
	Areas with crop growing season of 50–100 days						
Anantapur	409 (527)	764	√				
Barmer	283 (310)	881		√			
Jodhpur	356 (382)	880		√			

Locations	*Major rainfed crops*[d]	*Return period of drought* (years)[e]	*Section of study covering the area*[f] *1*	*2*	*3*	*4*
	Areas with crop growing season of 200–260 days[e]					
Akola	C,S,L		√			√
Amravati	C,S,L		√	√		
Bangalore	P,F,L	4–5	√	√		
Indore	W,K,B		√		√	
	Areas with crop growing season of 145–199 days					
Aurangabad	S,P,L		√	√		
Dharwad	S,P,L		√		√	
Gulbarga	S,L,P	2–4	√	√		
Hyderabad	S,P,L		√			√
Mahbubnagar	S,P,L		√			√
Sholapur	S,L,P		√			√
	Areas with crop growing season of 50–130 days					
Anantapur	G,S,P		√		√	
Barmer	P,L	2–3	√			
Jodhpur	P,L,S		√	√		√

[a] Soils are as follows: A = Alfisol, B = Aridisols, C = Entisols, D = Inceptisols, E = Vertisols. Properties of these soil groups are described in Part I, Section 2 of this volume.

[b] Length of growing season (or water availability period) is calculated using rainfall variables and storage capacity of soils (Raman and Srinivasa Murthy, 1971).

[c] Figures in parentheses indicate total annual mean rainfall (mm).

[d] S = sorghum, P = pearl millet, G = groundnut, C = cotton, L = pigeonpea or other rainy season pulses, K = chickpea, W = wheat, F = finger millet, B = soyabean.

[e] Length of period (years) in which one drought year is expected (*see Table 1.3*).

[f] Locations covered by different sections of the Indian case study are indicated by the respective section numbers (1, 2, etc.)

impact of droughts. Harnessing the potential of good rain years did not constitute an integral part of public policy to the problem of rainfall variability in the past.

Similarly, while the adaptive strategies of individual farmers have generally increased the stability of agriculture in the dry tropics, these have also been unable to harness fully the potential of the environment.

The recent advances in the field of science and technology have helped in better quantification of agroclimatic, agrobiological and other environmental variables and their interactions. The resulting knowledge in the fields of agroclimatology, soil and water management, plant science and agronomy has shown that, even without any change in the basic pattern of rainfall, agricultural productivity can be substantially increased and sustained. This offers great scope for strengthening the conventional approaches both at the farm level and at the policymaking and planning levels to handle the problems associated with rainfall variability. The objective of this study is to demonstrate this, though our focus will be mainly on public interventions. This, in turn, offers an opportunity to identify and integrate new approaches into policies and programs to combat the impact of rainfall variability in the dry tropics.

The focus of the present study is on the quantification of environmental, particularly rainfall-related, variables and their interactions with crops, soils and other resource management factors. Selection of approaches taken in the study was guided by: (i) the extent to which they provided appropriate frameworks and methods for understanding the effect of rainfall on crop farming in the dry tropics; and (ii) their suitability for use in strategies for agriculture in the areas affected by low and variable rainfall. Inevitably, our choice was seriously limited by the range of data and models available and, consequently, it has been necessary to focus our studies sometimes on a variety of locations where such resources exist. Moreover, most of these locations are similar in agroclimatic terms; their agroclimatic characteristics are summarized in *Table 1.1.*

1.2. Aims and Approach

The purpose of the study is to:

(1) Highlight the features of current public strategies for combating the impact of rainfall variability on agriculture.
(2) Identify appropriate research strategies for improving our understanding of how agroclimatic and other variables affect crop farming in the dry tropics.
(3) Juxtapose (1) and (2) to indicate the scope for making fuller use of the agroclimatic environment and the farmer's resource base for improving and sustaining agricultural productivity in the dry tropics.

The study has four components. The present section, after introducing the theme of the case study, provides a brief description of extent and coverage of the dry tropics in India, their agricultural characteristics and the problem of drought in these areas. This is followed by a discussion of the government's present drought-combating policies and the potential contributions of new approaches.

Section 2 considers the value of identifying homogeneous agroclimatic clusters to facilitate area-specific interventions in different zones of the dry tropics. Kartanaka State is used as an illustration.

Section 3 has two parts. The first uses a Markov chain probability model and long-term weekly and daily rainfall data (in association with soils information) to demonstrate a methodology that can help assess the dependability of rainfall, and indicate the scope for double cropping strategies to cover both rainy (*kharif*) and post-rainy (*rabi*) seasons, in different parts of the dry tropics. The second part simulates varying crop performance under different moisture conditions in order to consider the opportunities for improving crop planning and cropping decisions in dry farming areas. The crop chosen for the purpose is sorghum, a major staple of the Indian dryland farmer.

Section 4, using village- and farm-level data, evaluates farmers' adaptive strategies for varying rainfall situations, emphasizing that public policies should be sensitive to such strategies, and can influence them by promoting the introduction of new knowledge.

Finally, Section 5 integrates the major issues emerging from the discussions in the earlier sections, in order to develop suggestions for possible improvements in public interventions in the drought-prone areas.

1.3. The Study Area

We may define the dry tropical regions of India as those districts (151 in all) where mean annual rainfall lies between 350 mm and 1500 mm (*Table 1.2*). These cover about 174 million hectares, accounting for about 53% of India's total geographical area (*Table 1.2*; *Figure 1.1*). The great majority of the districts do not have dependable irrigation for even 10% of their cropped areas. About 205 million, or 39% of India's total rural population, live in these regions. The fact that the dry tropics account for 55% of the country's total cropped area of about 94 million hectares is of special significance in such a predominantly agricultural country as India. Because of their agroclimatic features, particularly the rainfall pattern and soils, these regions constitute a natural habitat for crops, such as coarse cereals (e.g., pearl millet, maize) and pulses, that have low water requirements. About 79% of the cropped area devoted to coarse cereals in India is located in the dry tropics, as is about 92% of the sorghum acreage, and more than half the area of the edible oilseeds and fiber crops such as cotton.

The vulnerability of such a major agricultural tract to weather variability acts as a source of instability of Indian agriculture in general. Despite remarkable achievements through the green revolution in well watered areas, food grain production in India still fluctuates dramatically, due largely to the influence of weather in dry farming areas (*Table 1.3*).

The vulnerability of the dry farming areas to low and unstable rainfall ultimately manifests itself through low levels of productivity, with no significant upward trend. An analysis of agricultural performance of the 151 dry tropical

Table 1.2. Characteristics of dry tropical regions (151 districts) in India[a].

Variables for 151 dry districts	*Percentage*	
Contribution to all-India total[b]		
Geographical area	52.9	(173.9)[c]
Rural population	38.9	(204.6)
Total cropped area	54.6	(93.7)
Area of coarse cereals	78.9	(28.1)
Area of pulses	59.3	(6.2)
Area of edible oil crops	58.0	(8.7)
Area of sorghum	92.0	(14.8)
Distribution of agricultural productivity level (Rs/ha)[d]		
Less than 500	8.6	
500–1000	53.0	
1000–1500	21.8	
1500–2500	9.3	
2500 and above	7.3	
Distribution of compound rate of agricultural growth (1956–57 to 1978–79)[e]		
4% and above	6.6	
2 to 4%	28.5	
0 to 2%	62.3	
Negative	2.6	

[a]The table includes data on 151 districts with an annual rainfall of less than 350 mm to 1500 mm falling in arid and semi-arid tropical parts of India (*see Figure 1.1*).

[b]Rural population data relate to 1981. The area data relate to 1976–1979 averages, the latest period for which district level data for all areas are available in published form.

[c]Figures in parentheses indicate absolute values of concerned variable relating to 151 districts. (Area in million hectares, population in million people).

[d]Based on average of production and prices of crops for 1976–1979 (Rupees per hectare); 12.5 Rupees = US$ 1 (1980).

[e]Based on constant prices (1976–1979 average) of crop produce. It is quite possible that in a number of districts where new dry farming technologies have been adopted in recent years, the productivity might have increased since 1979. However, relevant data were not readily available to test this.

districts indicated the following (*Table 1.2*): over 8% of the districts had a productivity level of less than Rs. 500/ha, (less than US$ 40/ha). For more than half of the districts, the per hectare productivity level was between Rs. 500 and Rs. 1000. Only about 17% of the districts had agricultural productivity per hectare during the reference period (1976–1979) of Rs. 1500 or more.

Agricultural growth from 1956–1957 to 1978–1979 for the dry districts was not encouraging: 62% of the districts had a compound growth rate of less than 2% per year, and about 3% showed a negative growth rate. Only 7% of the districts had a growth rate of 4% or higher (*Table 1.2*).

This disappointing performance of agriculture in a major agroclimatic zone of the country is bound to concern national policy makers and planners. It also indicates that drought mitigation continues to be a major preoccupation of policy.

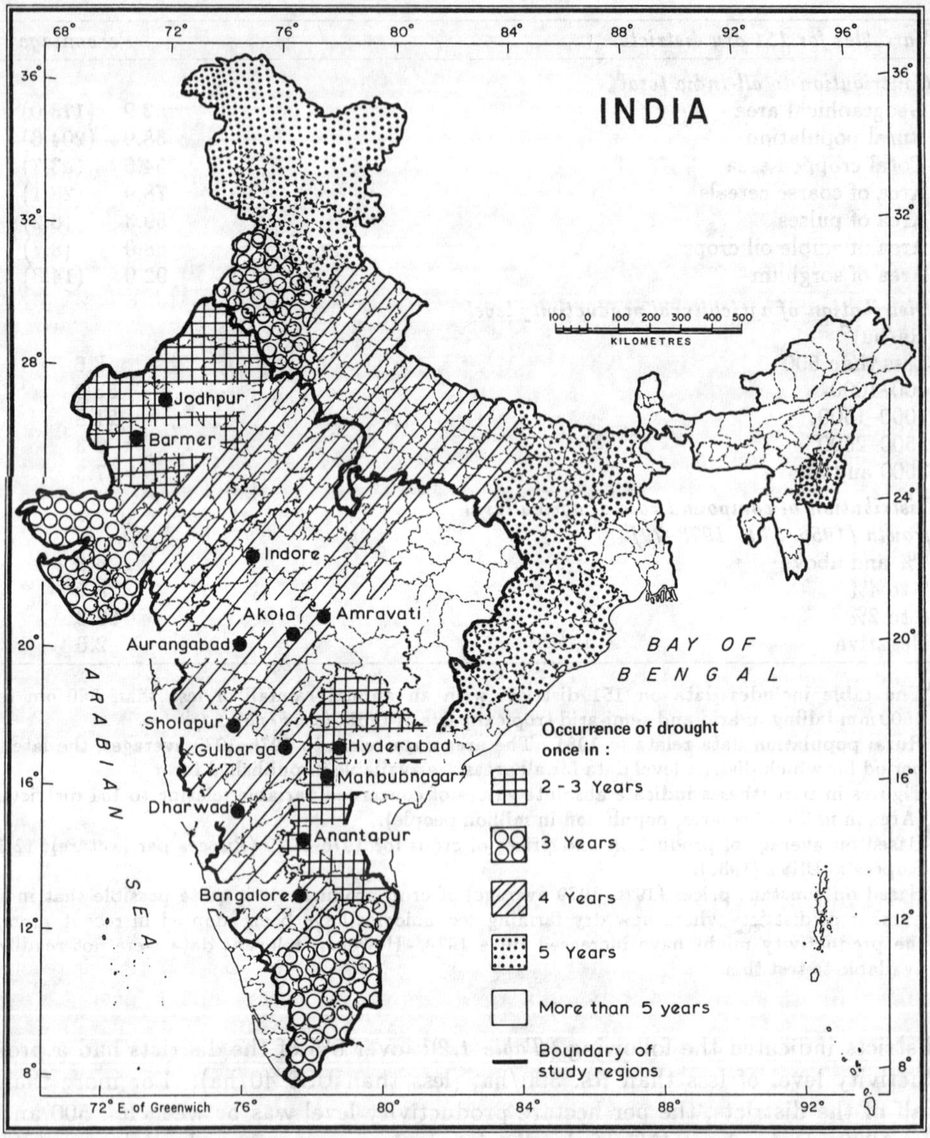

Figure 1.1. **The regions and locations included in the present study, and distribution of areas according to frequency of drought.**

Table 1.3. **Total Indian food grain production (million tons). Source: Swaminathan (1983).**

Period	*1961–65*	*1965–66*	*1970–71*	*1972–73*	*1974–75*	*1975–76*	*1978–79*	*1979–80*
Production	89.4	72.3	108.4	97.0	99.8	121.0	131.4	109.0

1.3.1. The incidence of drought

Depending on the purpose, drought can be defined in several ways. Broadly speaking, to the meteorologist a drought is a rainless or rain-deficient period during which some precipitation should have been received. The agriculturalist considers drought to be a condition of shortage of moisture for crop or animal production. The hydrologist views it as a phenomenon responsible for depression of surface and underground water levels and decline in stream flows. To an economist or administrator drought means a water shortage adversely affecting the established economy of the region, including an excessive rise in food prices.

We have adopted the definition of drought used by the Indian Planning Commission (1973). Accordingly, a drought is taken to occur if the annual rainfall of a region is at least 25% less than the long-term mean. The major limitation of this definition is that it completely disregards the distribution pattern of precipitation, which at times is as important as the total precipitation in arid and semi-arid regions. Its rationale, however, is that a rainfall deficit of 25% or more, irrespective of its distribution, often causes serious problems for the farmer. Furthermore, it is administratively simpler for the government to initiate relief measures on learning that such a deficit has occurred, than to undertake the time-consuming procedure of appointing fact-finding committees etc. For a drought (or crop failure) that results when the deficit is less than 25% but when the rainfall is poorly distributed, the response is more complicated, and involves a long procedure of verification, etc. In the present study we adhere to the definition based on annual rainfall, because our purpose is to examine the drought-combating policies and programs of India, which are designed in the context of what the Indian government considers to be a drought situation. Furthermore, a number of more comprehensive definitions, which take into account quantity and distribution of precipitation as well as their interactions with soils and crops at different stages of growth, are too sophisticated and complex to be of practical value where the necessary specialized data and expertise for applying them may not be available.

Long-term annual rainfall data for 60 or more years for the 31 meteorological subdivisions covering all India were analyzed. The results (*Table 1.4* and *Figure 1.1*) indicate that the area subject to drought once in 2 to 3 years constitutes more than 13% of India's area and about 19% of the dry tropical part. Another 16% of the dry tropics area has drought once in 3 years. More than half of the area of the dry tropics is affected by drought once in 4 years. The frequency of droughts in other parts of India is much lower. The last three columns of *Table 1.4* indicate that the bulk of the drought-prone areas of India fall in dry (arid and semi-arid) tropical regions of the country.

Historical evidence on famines in India during the period 1291 AD to 1979, summarized by Jaiswal and Kolte (1981), indicates that 46% of the total 119 famines during that period affected the areas lying within the dry tropical zone; another 31% affected this zone as well as some other parts of the country, and 21% of the famines also affected the parts of India outside the arid and semi-arid tropics.

Table 1.4. Distribution (in percentages) of geographical area, rural population and cropped area in India according to degree of recurrence of drought[a].

Frequency of droughts (return period)	Whole India: Geographical area	Whole India: Rural population	Whole India: Cropped area	Dry tropical regions[b]: Geographical area	Dry tropical regions[b]: Rural population	Dry tropical regions[b]: Cropped area
2–3 years	13.2	5.8	11.2	18.6	9.4	14.7
3 years	11.6	12.6	14.5	15.6	18.9	18.4
4 years	36.5	39.0	42.4	51.4	62.9	55.7
5 years	30.9	38.2	30.4	14.4	8.7	11.2
> 5 years	7.8	4.4	1.4	–	–	–
Total	100.0 (305.4)[c]	100.0 (525.5)	100.0 (174.7)	100.0 (216.6)	100.0 (325.4)	100.0 (133.1)

Frequency of droughts (return period)	Dry tropical regions relative to all India: Geographical area	Rural population	Cropped area
2–3 years	100.0	100.0	100.0
3 years	95.8	92.9	96.5
4 years	100.0	100.0	100.0
5 years	32.8	14.1	28.1
> 5 years	–	–	–
Total	70.9 (216.6)	61.9 (325.4)	76.2 (133.1)

[a]Calculated on the basis of details relating to districts falling under 31 different meteorological subdivisions in India, which are subject to varying frequency of occurrence of drought as reported by Planning Commission (1973). Drought is defined as a mean deficit of 25% or more in annual rainfall as compared to the long-term average.

[b]Based on the meteorological subdivisions broadly falling under arid and semi-arid tropical regions. The exclusion of specific districts with high rainfall or extensive irrigation facility (falling within these subdivisions) would reduce the extent of dry tropics reported here. This more precise demarcation of the dry tropics (using rainfall, irrigation and extent of dryland crops) forms the basis of *Table 1.2*, and explains the difference between the extent of the dry tropics reported in *Tables 1.2* and *1.4*.

[c]Figures in parentheses indicate actual values of area (in million hectares, 1981 census). The area figures relate to year 1978–1979, the latest year for which districtwise published data are available.

1.4. Drought as a Policy and Planning Issue

Public interventions regarding the drought problem are often undertaken by establishing inquiry commissions and by formulating both short-term relief schemes and long-term strategies to counter the impact of droughts. In this subsection we consider the development of different policy approaches.

1.4.1. Famine policies and programs of the colonial government

The first well documented and comprehensive evidence on policy makers' concern for drought-prone areas is revealed by the reports of the three Famine

Commissions set up by the British Government in India toward the end of nineteenth century. These commissions were set up in India in 1880, 1898, and 1901, following three devastating famines. The key features of their recommendations are summarized below (*see* Bhatia, 1967).

(1) *Adoption of a Famine Code.* For short-term relief, the measures suggested by the commissions to fight the consequences of famine included initiation of relief works and supplies, suspension or remission of land revenue and taxes, and provision of loans to facilitate reinitiation of cropping activity in the post-famine period. Furthermore, to facilitate and standardize the execution of short-term relief measures, as well as to monitor the emerging scarcities due to rain failures, the first Famine Commission report suggested a model Famine Code. This was adopted by the provincial governments, which executed the relief measures (Curry, 1983). Through occasional updating the Famine Code still continues to offer a general framework for famine relief administration in Indian states.

(2) *Improvement of the railway network.* One of the long-term strategies to expedite short-term relief measures was the extension of a transport network, particularly of railways. To the extent famine was treated as a temporary shortfall in supplies of food grains in specific areas, the commissions advocated provision of a better railway network to facilitate distributing food supplies within the shortest time, wherever a problem occurred.

(3) *Use of protective irrigation.* The most important recommendation of the first famine commission concerning the causes rather than the consequences of famine related to the irrigation policy of the British government in India. The commission recommended that irrigation should be given highest priority. Public irrigation works were classified into two categories: (a) productive works, which had to satisfy minimum net return criteria, and (b) protective works, which were undertaken mainly to afford protection for crops against drought and where commercial profitability was a secondary consideration.

In accordance with the above policy, a number of irrigation projects were undertaken in traditional drought-prone areas. By 1945–1946, protective irrigation works irrigated more than 26% of the total area served by public works in British India (Anon., 1972). After independence in 1947, the distinction between productive and protective irrigation schemes disappeared, but protection of drought-prone areas by irrigation continued to be an important concern of government.

Without belittling the importance of irrigation, it is worth mentioning that protective irrigation schemes that were intended to cover vast areas of crops with low moisture requirements, such as sorghum, millet and pulses, ultimately became intensive irrigation schemes serving small pockets of high water consumption crops, such as sugar cane, vegetable crops, paddy, etc. The final consequence was the creation of pockets of prosperity within vast drought-prone areas (Anon., 1973). This result was due to several factors related to hydrology, location specificities and design conditions, and the relative levels of private profitability of water use on different crops. In addition to these problems, the lack of sufficient water (usually from high rainfall regions) restricts the scope for extending irrigation to substantial parts of the dry regions (Jodha, 1979).

(4) *The Bombay dry farming system (BDF).* When the limitations of trying to cover vast tracts of dry areas by irrigation were recognized, attention was directed during the 1930s to improving agriculture through efficient use of local soil moisture resources. Research at five locations was initiated in 1933, to develop relevant technology for the purpose. It abruptly ended in 1943, owing to the shift in priorities to activities directly related to the war effort of the British government (Kanitkar *et al.*, 1960). Nevertheless, based on this brief dryland research effort, a package of practices was recommended to dryland farmers. It was called the Bombay Dry Farming (BDF) System, because one research location was in Bombay State. It emphasized contour bunding (for soil and moisture conservation), periodic deep plowing and some appropriate agronomic practices. However, the BDF did not have much impact on farming in drought-prone areas, for a number of reasons:

(a) The package did not contain any biological innovations comparable to the high-yielding varieties used during the subsequent green revolution period in India.
(b) Contour bunding violated farmers' property boundaries (or field boundaries) and, hence, did not get their support in implementation.
(c) The whole research effort had been too short-lived to provide a full understanding of the problem and to offer workable solutions based on local potential and designed to meet local needs.

The only component of BDF that was widely implemented was field border bunding, a poor substitute for contour bunding, and the impact of which on agricultural productivity and stability was quite mixed (PEO, 1964).

1.4.2. General programs during the post-independence period

The agricultural or rural development programs initiated in dry areas during the post-independence period can be grouped into general and specific programs. The former imply extension of general rural development programs to drought-prone areas. The latter include programs especially designed to help dry regions.

The general programs included activities such as the introduction of land reforms and the promotion of rural cooperatives, community development projects, etc. Most of these programs were initiated during the early 1950s, and they reflected the newly independent nation's enthusiasm to solve its rural problems through restructuring the rural economy. Since most of these programs had evolved with reference to well endowed areas, they proved to be insufficiently sensitive to the specific problems of drought-prone areas. The criteria and norms underlying the conception and execution of these programs were too standardized to be effective in the highly unstable environments of dry farming regions (Jodha, 1984). General programs had, therefore, to be supplemented or replaced by more specific programs.

1.4.3. Specific programs

Specific programs adopted for drought-prone areas included the following: *ad hoc* relief measures during droughts, soil conservation works and dry farming schemes, the Drought-Prone Area Program (DPAP) and technological transformation of dryland agriculture through development and diffusion of improved technologies.

Drought relief programs

Drought relief is one of the oldest and most consistently adopted programs in the drought-prone areas. The measures initially adopted and institutionalized following the report of the first famine commission, in 1883, have undergone considerable qualitative change since then (Curry, 1983; Torry, 1986). The sources of information are now more verifiable than was the case earlier. They provide data on rainfall, crop conditions, market prices, and the pressure of demand on public food distribution systems, as well as newspaper reports and reports of fact-finding missions, all of which are used as a basis for decisions. At the same time, the democratic process has caused pressure mechanisms to become an important factor in the whole exercise relating to the declaration of drought conditions and the allocation of relief resources. Consequently, drought has ceased to be a purely natural phenomenon. For these reasons as well as "leakages" associated with relief works, the amount of relief expenditure has substantially increased in recent years (Morris, 1975; Rangaswami, 1974).

To meet the demands for relief coming every year from one part of the country or another, the government now maintains a substantial buffer stock of food grains, currently about 30 million tons. In addition, an elaborate public distribution system through more than 25 000 fair price shops is maintained not only in urban, but also in rural areas throughout the country.

In addition, and especially during the 1960s, production components were increasingly added to drought relief programs, including various soil moisture conservation measures, percolation tanks, and works relating to minor irrigation, land shaping and afforestation (Torry, 1984).

Rehabilitation support, including input supplies for producing the next crop, relief employment through various programs, such as food for work, employment guarantee schemes, etc. during the post-drought period (i.e., until the arrival of next harvest), have also become an important part of drought relief programs, facilitated by the availability of buffer stocks of grain.

Drought-Prone Area Program (DPAP)

Faced with the very poor results for the general rural development programs in certain low rainfall regions, and alarmed by the continuing impact of recurrent droughts of 1965–1967 in these areas, the government considered special policies for drought-prone areas. In view of the shorter crop season and therefore longer duration of seasonal unemployment, a rural works program with direct central assistance was initiated as a part of the Fourth Five Year Plan in the chronically drought-affected areas (Jaiswal and Kolte, 1981).

Following the mid-term appraisal of the plan in 1971, an area dimension was added to the rural works program, and it was redesignated as the Drought Prone Area Program (DPAP). Activities such as animal husbandry, soil moisture conservation measures, minor irrigation, pasture development, etc., which were most suited to these regions, but which had been adopted on an *ad hoc* basis in the past, were integrated under this program. Planning of a number of activities on a watershed basis, heavy emphasis on development of infrastructure and use of improved dry farming technology were other features of the new program. Restoration of the ecological balance in these rapidly degrading areas was also the long-term objective of the program. The program had special exclusive funding from the central government in addition to the contribution of state governments.

The evaluation reports indicate that DPAP has helped greatly in developing the infrastructure and the asset situation of the rural people, with dairy development being the most successful component of the program. Furthermore, as well as increasing output in normal years, it has served to minimize the sharp drop in crop yields in the years of drought (Anon., 1984).

Important deficiencies of the program, particularly at the implementation stage, are inadequate integration of the activities and an inability to execute the program on a watershed basis (Jaiswal and Kolte, 1981). However, as it is one of the relatively successful anti-poverty programs started in 1970–1971, this program has been given support for the Seventh Five Year Plan period, initiated in 1985–1986. During the plan period, this program is placing greater emphasis on adoption of dry farming technology, use of the watershed concept and generation of employment.

Focus on farm technology

The circumstances that led to the initiation of the DPAP developed further, during the late 1960s, with the widening of inequalities between dry and wet areas. The latter experienced a green revolution based on new technologies. It was realized then that no program based on institutional or infrastructural measures could ensure growth and stability of drought-prone agriculture unless it was supported by appropriate technological developments. This thinking was further strengthened by the success of the improved technologies that had led to the green revolution in well watered areas. The experience of the BDF during the late 1930s, the limited impact of soil moisture conservation technologies developed during the late 1950s, and the development of hybrids of coarse cereals and their adoption in moist areas in the late 1960s, were other elements in the background for action to initiate a coordinated approach to the development and diffusion of relevant technologies for dryland agriculture.

The action took the form of the establishment, in 1970, of the All India Coordinated Research Project on Dryland Agriculture, AICRPDA (in 1983 renamed the Central Research Institute for Dryland Agriculture, CRIDA). The dryland project coordinates the adaptive research through a network of 23 research stations representing the main dry agroclimatic zones of tropical India. This effort at the national level was complemented by a massive international research effort, as part of which the Consultative Group on International Agricultural Research established the International Crops Research Institute for the Semi-Arid Tropics (ICRISAT) in 1972, with its headquarters near Hyderabad. The Institute has a mandate to serve all semi-arid tropical regions in the world by developing crop-based and resource-based technologies relevant to the agroclimatic and socioeconomic environments of the semi-arid tropics.

These efforts represent both a quantum jump in the research resource allocation to dryland agriculture and a basic change in the approach toward the problems of these dry regions. Their objective is to help to stabilize and develop dryland agriculture through the development and diffusion of relevant technologies. As part of this new approach, climate is considered as a resource, not merely a hazard. The present emphasis is on both exploiting the potential of good rain years and minimizing losses from poor rain years. This is attempted through various conservation measures and the development of suitable crop varieties as well as resource management practices. Input of scientific knowledge on agroclimatology is an important feature of the new research approach. In addition, we may note the following emphases:

(1) Integration of resource- and crop-centered technologies is focused on exploiting the complementarities between the new high-yielding varieties and moisture conservation measures.
(2) Multi-location testing is emphasized in order to evaluate the location specificity of technologies.
(3) On-farm testing of technologies through verification trials, pilot projects, etc., is provided for, in order to take account of farm-level realities.
(4) The present approach has a much stronger biological and scientific foundation than earlier efforts, in terms of a vast stock of germplasm of major dryland crops and better knowledge of soil, water and plant relationships.
(5) There is special emphasis on developing resistance to yield reducers such as drought, disease and insect pests.
(6) The watershed as a unit of resource assessment and management is central to the present research effort.

There are, however, several important limitations of the new technology-centered initiative (Jodha, 1983). First, there is a heavy emphasis on crop production, which ignores the fact that small farmers in the dry tropics have a very diversified agriculture, which may include activities based on livestock, bush/forest, etc. Secondly, new technologies are often too standardized and costly to fit the farmers' flexible and low-risk production systems. Thirdly, despite provisions such as multi-location trials, on-farm testing, demonstrations, operational research projects, etc., the mechanism for transfer of technology is the weakest component of the new approach. Finally, because of the factors mentioned above, the technology components are not fully integrated with the public measures designed to combat drought and accelerate development of dryland agriculture.

1.5. A Brief Evaluation of Public Interventions

In the preceding discussion we have considered the chronological development of public policies and programs toward drought. We can evaluate their successes and deficiencies by classifying some of their basic characteristics (*Table 1.5*).

1.5.1. Types of interventions

Three broad types can be recognized:

(1) *Food supply measures.* These involve either the intertemporal transfer of food by storage of buffer stocks, or the interregional transfer of food by trade, transport networks, etc. Mechanisms to distribute relief supplies to ultimate users directly or through employment and income generation activities, or what Sen (1981) calls "entitlement", are part of this category of measures.

(2) *Resource-oriented interventions.* These involve intertemporal water supply transfers, i.e., through conservation measures under the BDF, etc., and interregional transfers of water as under the protective irrigation approach. There are instances of this type of intervention being supplemented by food supply measures, as reflected by the productive use of relief supplies or works in support of conservation measures, afforestation, etc.

(3) *Crop-oriented measures.* These involve activities aimed at raising the productivity of crops and making them less sensitive to drought. Such measures are facilitated by infrastructural and institutional support, as under the DPAP, and a greater focus on development and diffusion of technology, as under AICRPDA.

Public interventions in drought-prone areas in India have thus tried to reduce the vulnerability of the local food system to droughts and to reduce the impacts of drought at the production or farm level. These are the key steps often suggested to cope with climatic variability in vulnerable areas.

1.5.2. Some important deficiencies

Table 1.5 summarizes a number of important limitations in past policy. Many of these have now been overcome, but some, in one form or another, still persist today.

Since the drought-prone areas catch the eye of policy makers largely during crisis periods (i.e., severe droughts), their deficiencies tend to dominate the policy makers' perception of these areas. Consequently, dry regions are often perceived as national liabilities. Because of such inaccurate perceptions, the proposed solutions often take the form of strategies and programs based on perpetual dependence on external resources. The most damaging consequence of this deficiency-induced perception is complete disregard of the potential of the areas and of approaches that could harness that potential (Jodha, 1984).

Many public measures to help dry regions were introduced in response to crisis situations. For example, the protective irrigation strategy emerged as the British government's response to the crisis generated by three severe famines toward the end of the nineteenth century. *Ad hoc* drought relief undertaken every year in one part or another of the dry tropics is a further illustration of crisis management (Morris, 1975). The introduction of the DPAP, and the initiation of research through AICRPDA since the early 1970s, as well as the very high priority given to dry farming in the Prime Minister's 20 point program in the early 1980s, again have their origin in the crisis generated by stagnation of dry areas and dry crops, and the widening gap between dry and wet areas.

Having originated in crisis, the public programs (particularly the relief measures) in dry tracts acquire some attributes of crisis management practices. Important among these attributes are the predominance of *ad hoc* measures, great and persistent dependence on resource transfers from other areas, inadequate attention to the basic causes of crises, and a tendency to introduce inappropriate and untested solutions, particularly those evolved in the context of stable environments. Periodic relief works and the introduction of general programs are characteristic of these interventions (Jodha, 1984; Morris, 1975).

Finally, there may be recognition of the need to prevent widening regional inequities, and this generates the occasional intense concern for the long-term development needs of the dry farming regions. Consequently, programs with a strong technological orientation – such as the BDF in the early 1930s, AICRPDA since the early 1970s, or programs with the institutional and infrastructural orientation such as the DPAP – are initiated. However, such programs often fail to get full support for sufficiently long periods to make their impact. This may occur for two reasons. First, the programs are initiated without sufficient understanding of their components and their overall implications, including the required resource support. Secondly, most of the agricultural policies and programs in India, including the norms and yardsticks that govern program formulation, execution and evaluation, are developed in the context of well endowed regions and are not always sensitive to the problems of the dry areas. Failures, or rather inappropriately judged levels of performance of the programs in dry farming areas, are often projected as deficiencies of these areas. These supposed deficiencies are then used as grounds for reducing support to the programs. Thus, the lack of a clear understanding of the characteristics of dry farming tracts, and the consequent failure to design development strategies suited to them, obstruct the sustained and long-term support to help these regions (Jodha, 1984).

Table 1.5. The character of public interventions in drought-prone areas in India.

	Major interventions/programs			
			Drought relief	
Features	*Protective irrigation (1)*	*BDF/Soil conservation (2)*	*Supplies (3)*	*Productive components*[a] *(4)*
A. General:				
Period of initiation	1890s	1940s/1950s	1890s	1960s
Gradual evolution				√
Coverage: All dry areas		√	√	√
Coverage: Specific areas	√	√		
Special funding provisions	√			√p
B. Major orientation:				
Welfare/relief/food supply			√	√p
Resource development/ management	√			√
Crop productivity	√			
Long-term focus				
Short-term *ad hoc* device			√	√
Technology oriented		√		
Administration oriented			√	√
C. Positive attributes:				
Local potential centered		√		√
Flexibility/responsive to changing situation				√
Scope for long-term sustainability				
Integrated approach–reorientation through new scientific knowledge				
D. Deficiencies/constraints:				
Perpetual dependence on external resources	√		√	
Indifferent to local potential	√			
Lacked productive technological component		√		√
Insensitive to farmer circumstances		√		
Institutional constraints		√		
Leakages/operational problems		√	√	√
Location specificity (physical constraints)	√			√
Inadequate information base and analysis for planning operations[b]		√		

Table 1.5. continued.

	Major interventions/programs			
	Technological transformation of agriculture			*Integrated strategy involving (4,5,6 & 7)*[a]
Features	*DPAP (5)*	*Crop-based (6)*	*Resource-based (7)*	*(8)*
A. General:				
Period of initiation	1970s	1970s	1970s	1970s
Gradual evolution	√			√
Coverage: All dry areas		√	√	√
Coverage: Specific areas	√			
Special funding provisions	√			√p
B. Major orientation:				
Welfare/relief/food supply			√	
Resource development/ management	√		√	√
Crop productivity	√	√	√	√
Long-term focus	√			√
Short-term *ad hoc* device				
Technology oriented		√	√	√
Administration oriented	√			
C. Positive attributes:				
Local potential centered	√	√	√	√
Flexibility/responsive to changing situation	√			√
Scope for long-term sustainability	√			√
Integrated approach–reoriented through new scientific knowledge	√			√
D. Deficiencies/constraints:				
Perpetual dependence on external resources				
Indifferent to local potential				
Lacked productive technological component				
Insensitive to farmer circumstances		√	√	√
Institutional constraints			√	√
Leakages/operational problems			√	
Location specificity (physical constraints)			√	
Inadequate information base and analysis for planning operations[b]	√	√	√	√

[a]p = partial applicability.
[b]Poor resource characterization and understanding of interactions of involved variables, due to which interventions fail to acquire relevant forms.

1.5.3. Evolution over time

Despite the deficiencies discussed above, some important positive developments in policies of intervention have occurred, the implications of which are considered below.

First, there has been a gradual change in policy makers' perception of climate as a factor influencing the economy of drought-prone areas. The climate, rather than being considered only as a source of distress, has been increasingly recognized as a potential factor for growth. Consequently, the total resource base and production potential of drought-prone areas, rather than the problem of drought *per se*, have received increasing attention. However, this change can become a matter of practical importance only if various climatic variables, particularly rainfall, are quantified and their stability as well as their interactions with soil, plants, etc., are understood. This understanding can help in matching the development interventions to the interregional heterogeneity of dry farming areas.

Purely welfare-oriented activities, such as providing relief supplies during droughts, are increasingly being complemented by productive activities leading to creation of social assets through afforestation, soil conservation works, etc. However, the appropriate choice of these works (e.g., percolation tanks, contour bunding, pasture or forest development, etc.) also requires quantification and understanding of the environmental and socioeconomic variables and their spatial patterns.

Closely related to the above is the shift from purely *ad hoc* measures to systematic and planned efforts to help drought-prone areas. The food supply-oriented, resource-oriented and crop-oriented interventions are becoming better integrated into long-term strategies. The composition of these strategies again needs to be varied to fit the different zones within the dry tropics, which again underscores the need for quantification of environmental and other variables.

An important feature, both of long-term development projects, such as the DPAP, and of production-oriented relief strategies, is a stronger focus on the technological aspects of agriculture in dry regions. The largely institutional and infrastructure-centered programs of the past are increasingly being complemented by technology-oriented measures.

A major outcome of these changes is a shift in the development programs for these regions from a largely external resource-based approach to an approach based on local resources. This approach recognizes the continued need for external resource support, but aims to reduce reliance on a perpetual flow of resources from outside.

Finally, what seems to be gradually evolving, with the transformation of the various interventions, is an integrated and comprehensive development strategy for drought-prone areas. The three key components of this emerging strategy are:

(1) The local resource-oriented and area-oriented development programs represented by the DPAP, etc.

(2) A sharper focus on technological transformation of dryland agriculture.
(3) An increasing production orientation of relief works.

This integrated approach places increased demands on available resources, on long-term planning and on commitment to implementation at the grassroots level. In addition, it requires much more comprehensive information on agro-climatic and technological variables than has hitherto been available. Detailed characterization of the agroclimatic environment is essential. The following sections describe how this can be achieved.

SECTION 2

Effects of Monsoon Variability on Agriculture

2.1. The Setting

In the tropics the most important climatic element for agriculture is rainfall. The dependence of Indian agriculture on monsoon rainfall is well known (Swaminathan, 1983); in fact, the Indian economy has been described as a gamble on the monsoon rains. We thus expect any climatic change involving a large-scale change in rainfall to have a significant impact on agricultural productivity. However, as pointed out in Section 1, there is no evidence for long-term increasing or decreasing rainfall, either for the Indian region as a whole or for meteorological subdivisions of the country. Thus, in the Indian context, it is more pertinent to analyze impacts of climatic variability rather than of climatic change.

The variability of rainfall in India is predominantly interannual, between years with adequate rainfall and years of drought (defined in this case study as years with rainfall less than 75% of the mean (*see* Section 1). The amplitude of these interannual variations over the country as a whole is not very large: the standard deviation is about 10% of the mean and the worst drought of the century involved a deficit of only 25% of the mean. However, the total rainfall over the country is not the most useful index, because even in a year in which Indian rainfall is normal, many regions suffer from droughts and some from floods. Over smaller subregions the coefficient of variation is much larger than 10%, as much as 60% in arid regions. In addition to the variation in the total amount of rainfall received, the rainfall pattern (i.e., the distribution within the year) also varies considerably. The aim here is to develop a method for assessment of the impact of this variability in the amount and pattern of the monsoon rainfall on agricultural productivity.

Since the agricultural productivity of India as a whole is the sum total of production in the different agricultural zones, the impact of the variability of rainfall over each of these zones has to be assessed. Clearly, the first task is the identification of the agricultural zones. These can be characterized by homogeneity of the crop complex grown and can be determined by classification or cluster analysis of cropping patterns over a suitable network of observations. Alternatively, if the nature of the soil required and the optimal rainfall as well as the tolerance range for each crop are known, the agricultural zones can be derived directly from an analysis of the spatial variation of the soil and rainfall.

If we assume that over historical time the traditional cropping pattern over any region has evolved so as to be the optimal pattern for the soil and climate (particularly the rainfall) of that region, then even in the absence of such detailed information about the climatic preferences and tolerance of the crops, it is possible to deduce the agricultural zones. Over a monsoonal region where there is considerable variation in the annual rainfall received as well in its distribution within the year, we can expect the cropping pattern to be the optimal one for the most probable amount and pattern of rainfall, i.e., the modes of the distributions of these two features. If the mode and mean coincide (as is found to be the case for both annual and monthly rainfall), the cropping pattern can be explained largely in terms of the climatological mean rainfall.

Regions over which the climatological mean rainfall patterns are similar will thus be characterized by similar cropping patterns, provided the soil is homogeneous. If we take climatic clusters to be regions that are homogeneous with respect to the interannual variation in the amount and pattern of rainfall, agroclimatic zones can be obtained by superposition of the climatic clusters on a soil map.

The impact of the variability of rainfall on the yield of an agricultural zone cannot be assessed by analysis of the mean rainfall over that zone because, although the mean rainfall pattern is similar over an agroclimatic zone, the rainfall variations are not necessarily coherent within each zone. In other words, in any given year only a part of the region may experience drought or an abnormal rainfall pattern. Thus, the impact of the variability of rainfall has to be assessed for coherent subregions within each agroclimatic zone.

Following a brief discussion of the nature of the rain-giving system and the spatial variation of rainfall, the climatic clusters of the Indian region, within each of which the mean annual rainfall and the rainfall patterns are similar, will be described. This provides some insight into the number of agroclimatic zones in the country. For the state of Karnataka, for which data on the cropping pattern are available for 175 subdivisions, the agricultural zones are derived using a clustering algorithm specially developed for the purpose. Focusing on an agroclimatic zone in the northern part of the state in the semi-arid zone that extends across the peninsula, a subregion over which rainfall variations are coherent has been identified. For this subregion, the impact of the variation of amount and pattern of rainfall is assessed qualitatively using available data on the yield of traditional varieties of the major crop grown (sorghum). The biological basis for the impact of rainfall variability, which is not considered in this approach, is dealt with in Section 3.

2.2. Monsoons, the ITCZ and the Spatial Variation of Rainfall

The monsoonal regions of the world are characterized by a seasonal migration of the equatorial trough or the intertropical convergence zone (ITCZ), which occurs in response to the seasonal variation in the latitude of maximum insolation. One manifestation of the seasonal migration of the ITCZ is the seasonal reversal of winds by which the monsoon is traditionally defined. The more important manifestation, however, is the seasonality of rainfall, and in fact in India the word monsoon is synonymous with rainfall.

The annual rainfall over the Indian region can be thought of as the integrated effect of the annual visit of the ITCZ to the region. The spatial variation of the annual rainfall is shown in Section 1. The west coast of the peninsula and the northeast parts of the country receive very high rainfall, while in the monsoon zone north of the peninsula the rainfall is maximum near the Bay of Bengal and decreases westward.

Given the seasonal migration of the ITCZ, we expect the rainfall in the region around the maximum latitude attained by the ITCZ in the summer, i.e., around 21° N, to be unimodal with a peak corresponding to the time at which the ITCZ is over the region. That is indeed what is observed in ***Figure 2.1***, which depicts the spatial variation of the mean rainfall patterns over the Indian region. In the region to the south of the monsoon zone, i.e., over the Indian peninsula, we expect the pattern to be bimodal with one mode corresponding to the advancing phase and the other to the retreating phase. However this is not exactly what is observed, because two factors – viz., orography and the differential response of the Arabian Sea and the Bay of Bengal to the summer monsoon – generate considerable east–west asymmetry across the peninsula. Generally in the monsoon season, in addition to the prominent ITCZ stretching across the monsoon zone, there is a cloud band along the west coast of peninsula where heavy rain is received upstream of and on the crest of the mountains known as the Western Ghats. This orographic feature accentuates the rainfall on this western part of the peninsula during the summer when the region lies upstream in the monsoon wind. In the autumn, during the retreat of the monsoon, the ITCZ is far more active over the warmer Bay of Bengal than over the Arabian Sea, and the heaviest rainfall occurs over the eastern part of the peninsula. Thus, there is considerable variation in the nature of the rainfall pattern across the peninsula.

2.3. Climatic Clusters of the Indian Region

The climatic clusters of the Indian region, derived by analysis of the mean monthly rainfall at 119 stations, are shown in ***Figure 2.2***. There is a close correspondence between these clusters and the distribution of the vegetation types in the country, which lends support to our assumption about the role played by the mean rainfall in determining the nature of the vegetation and, hence, the cropping patterns.

The clusters F, G, H, J, K, L have similar patterns and the amount of rainfall received increases eastward. These clusters lie along the monsoon trough, i.e., the mean summer location of the ITCZ, which is most intense near the head of the Bay of Bengal where most of the synoptic scale disturbances are generated. There is a considerable variation in the pattern of rainfall over the

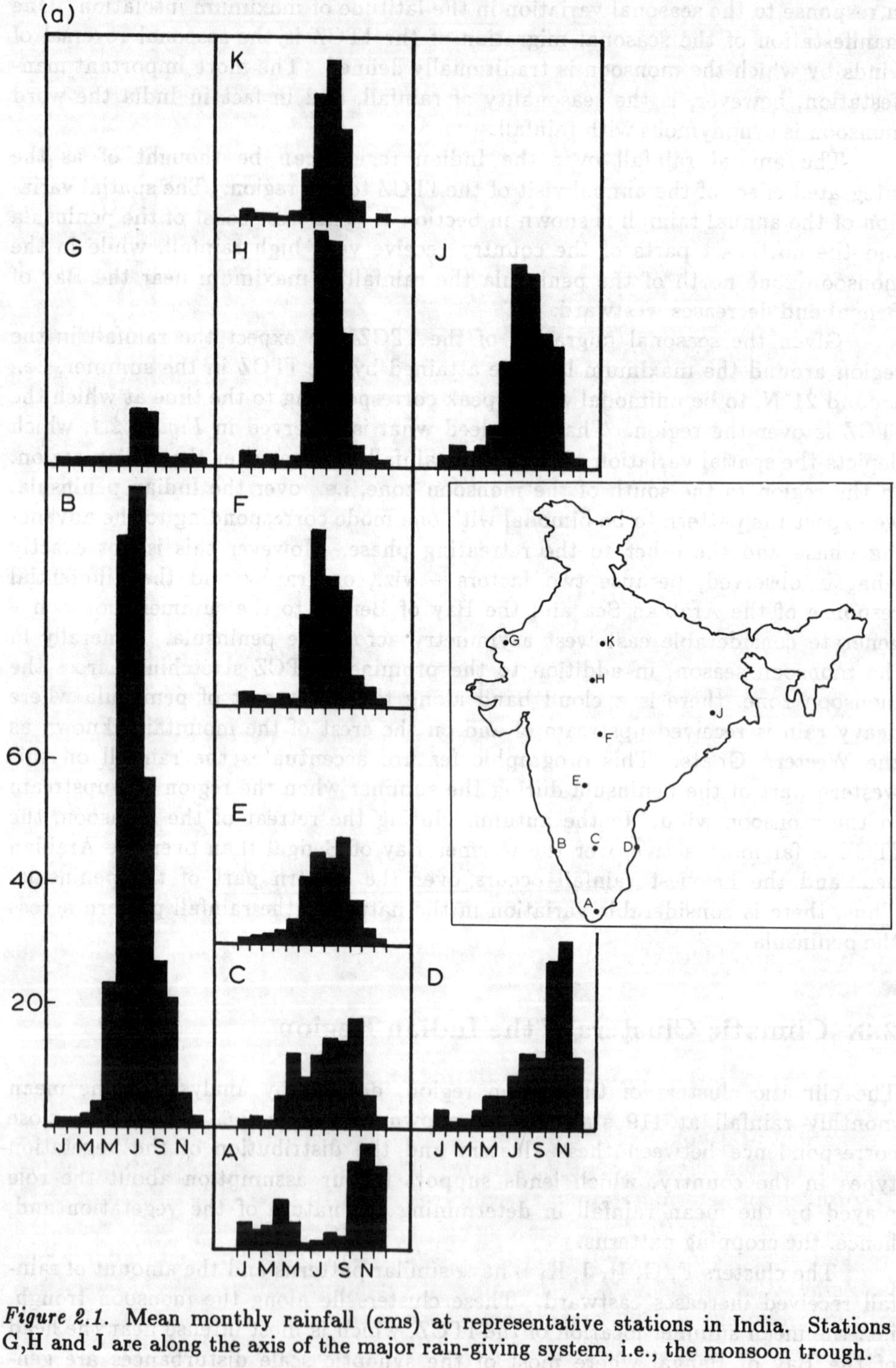

Figure 2.1. **Mean monthly rainfall (cms) at representative stations in India. Stations G,H and J are along the axis of the major rain-giving system, i.e., the monsoon trough.**

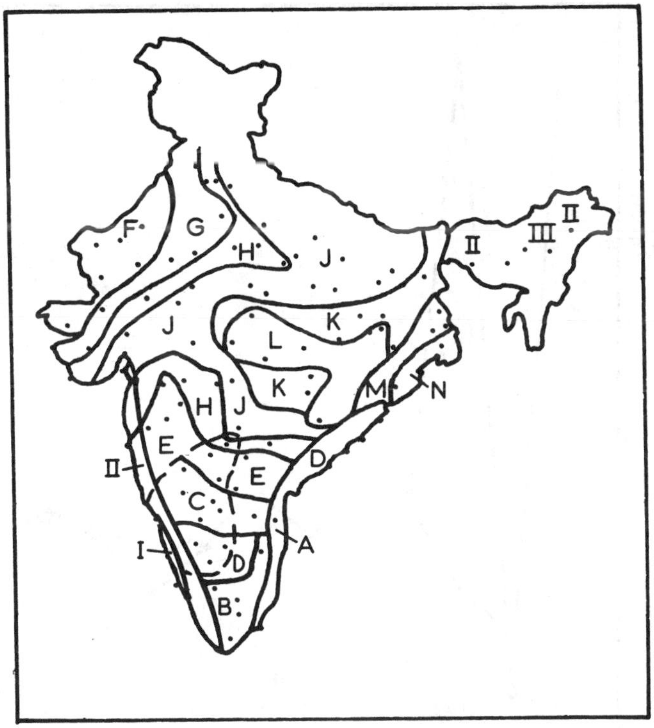

Figure 2.2. Clusters obtained from analysis of mean monthly rainfall patterns in the Indian region (Gadgil and Joshi, 1983).

peninsula. The west coast clusters have prominent peaks after the summer, while away from the coast the mode shifts from November in the southern tip to earlier in the year at higher latitudes. The northernmost part of the peninsula has two modes – in July and September. In this paper we focus on one particular state, Karnataka, the boundary of which is indicated in *Figure 2.2* by a broken line.

2.4. Agricultural Zones of Karnataka

The agricultural zones of Karnataka can be determined by an analysis of the spatial variation of the cropping patterns. There are two possible approaches to the problem of determination of agricultural zones. If the characteristics of each zone are specified in terms of the range of cropping patterns for each zone, the problem is reduced to one of classification in which the different subdivisions have simply to be assigned to the different categories/zones. Alternatively, the natural groupings or clusters of the cropping patterns can be determined, in which case the limits of each zone do not have to be specified *a priori*, but are

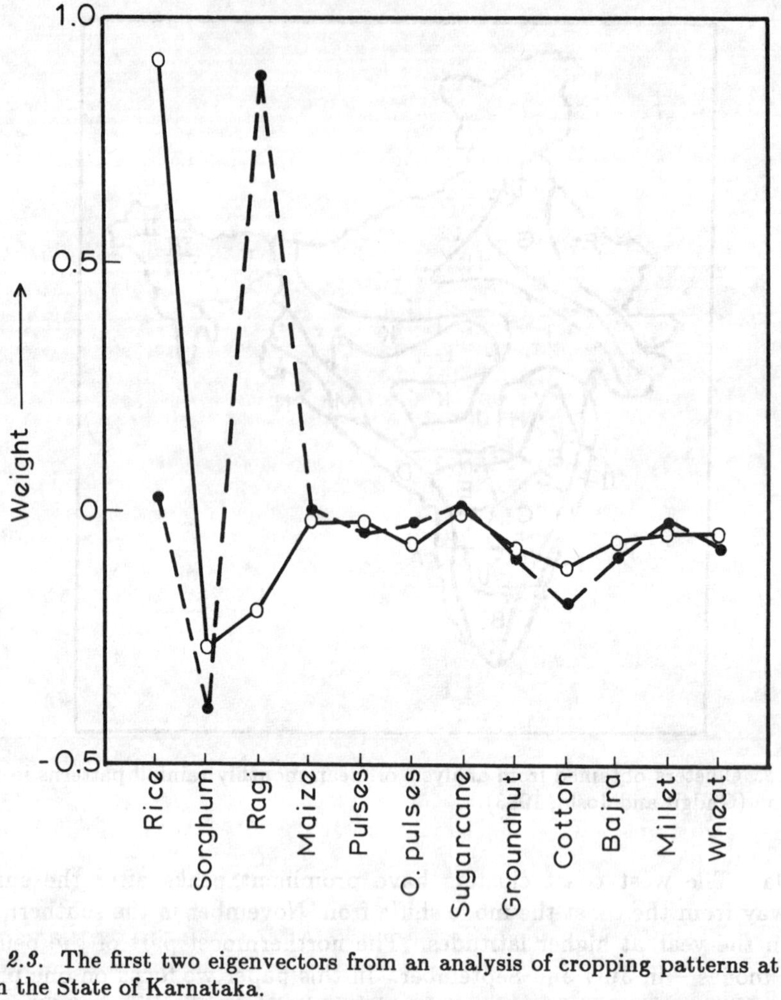

Figure 2.3. The first two eigenvectors from an analysis of cropping patterns at 175 stations in the State of Karnataka.

determined by the clustering algorithm itself. We have adopted the latter approach.

The cropping pattern is described in terms of the percentage area under 12 major crops for 175 Taluks (subdivisions) of the state. Clustering thus involves analysis in a 12-dimensional space. It is extremely useful if the variation in the 12-dimensional space can be effectively represented in a two-dimensional space, because then the human capacity for pattern recognition can be exploited in developing a suitable clustering algorithm. For such a representation, we have used principal component analysis or the method of empirical orthogonal functions (Lorenz, 1956). The first two eigenvectors explaining 84% of the variance are shown in *Figure 2.3.* The weight factors are the largest for the three major crops of the state: rice, sorghum and *ragi* (finger millet). Since the two leading eigenvectors explain a large percentage of the variance, the cropping

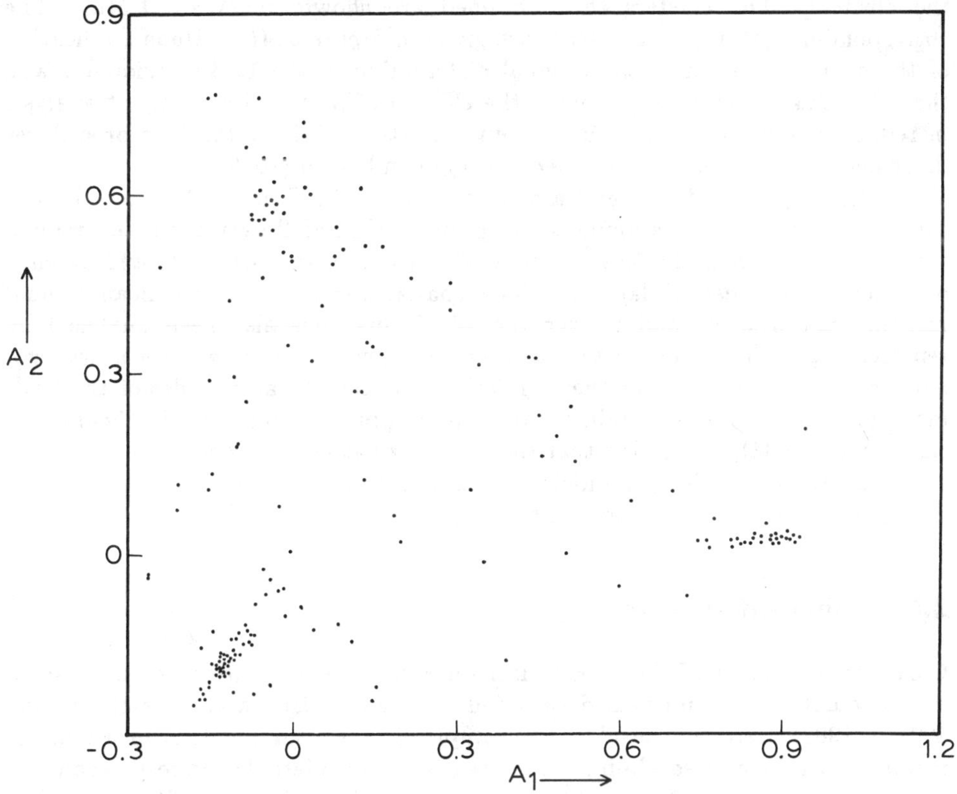

Figure 2.4. Distributions of the cropping patterns at the different stations in the amplitude plane of the eigenvectors in *Figure 2.3*.

pattern of any Taluk can be adequately represented as a point in the plane of the amplitudes A_1 and A_2 of these two eigenvectors. The distribution of the different Taluks in this amplitude plane is shown in *Figure 2.4*.

An algorithm was developed for cluster analysis of this type of distribution. Clusters, which can be thought of as dense clumps on the amplitude plane, are generally separated from one another by intervening regions with relatively low density. To facilitate the identification of these clusters, all points with very large nearest neighbor distances (NNDs) on the amplitude plane were dropped, the threshold NND being determined by the distribution of the NNDs. The chosen threshold implied dropping less than 10% of the points. From the remaining points, local maxima in density were identified as seed points, and each point was assigned to the nearest seed point to form the initial formation of clusters. The centroids of these clusters were calculated, and each point was again assigned to the cluster with the nearest centroid. This process was iterated until convergence was attained, a few iterations being sufficient for the purpose. Neighboring clusters were then merged if the separation between them was not sufficiently large in comparison with the grain-scale (i.e., mean NND) of the clusters. The clusters thus obtained are shown in *Figure 2.5(a)*. The corresponding geographical extents are given in *Figure 2.5(b)*. Upon application of the same algorithm to the original distribution in the 12-dimensional space, the same clusters emerged, proving the efficacy of principal component analysis in reducing computational time without distortion of the data. The procedures described here are discussed further in Gadgil and Joshi (1983).

The annual rainfall over Karnataka is shown in *Figure 2.6*. There is considerable variation in the amount and pattern of rainfall over the study region, which stretches from the west coast to the center of the peninsula and extends over several degrees of latitude. The spatial variation of the mean annual monthly patterns of rainfall over the 175 Taluks has also been analyzed by representing each rainfall pattern as a linear superposition of empirical orthogonal functions. We find that the correlation between the amplitudes of the leading eigenvectors for mean rainfall and the cropping patterns is significant and rather high (0.84), suggesting that the cropping pattern is indeed determined by the mean rainfall. Thus, it should be possible to estimate the effect of a given change in mean rainfall pattern on the cropping pattern.

2.5. Coherent Regions

Coherent subregions of the agricultural zones have been identified by analysis of monthly data at 50 stations distributed throughout Karnataka for the period 1901 to 1982. Here we consider the example of agricultural zone A in which the major crop grown is sorghum. This area is of considerable interest because it lies on the southern fringe of the monsoon zone in which the ITCZ is located during the peak summer months of July and August. Thus, we expect the variability in the pattern and amount of rainfall to be particularly large. Gulbarga (17.51° N, 76.51° E) was chosen as the central station.

To analyze the coherence in the amounts of rainfall received, we computed cross-correlations with respect to the central station for the annual rainfall as well as for the summer monsoon (June–September) rainfall and delineated the locations with cross-correlations greater than 0.5. Anomalies in the amount of rainfall, i.e., droughts or floods, would be expected to occur simultaneously over this region.

Next, the coherence in the monthly rainfall patterns was studied. For each year a disparity measure of the difference between the normalized rainfall patterns at Gulbarga and each of the other stations was computed. The normalized rainfall at any station in a given year was taken as the ratio of the monthly rainfall in that year to the mean annual rainfall of that station. The disparity index between the rainfall patterns of two stations, A and B, in a given year (or two years, A and B, at a given location) was defined as the Euclidean distance between the points representing the normalized rainfall patterns at A and B in the 12-dimensional space of monthly rainfall. In a given year, the disparity measure will be small between stations with similar patterns and large for stations with disparate patterns. From the distance statistics for 1901–1982,

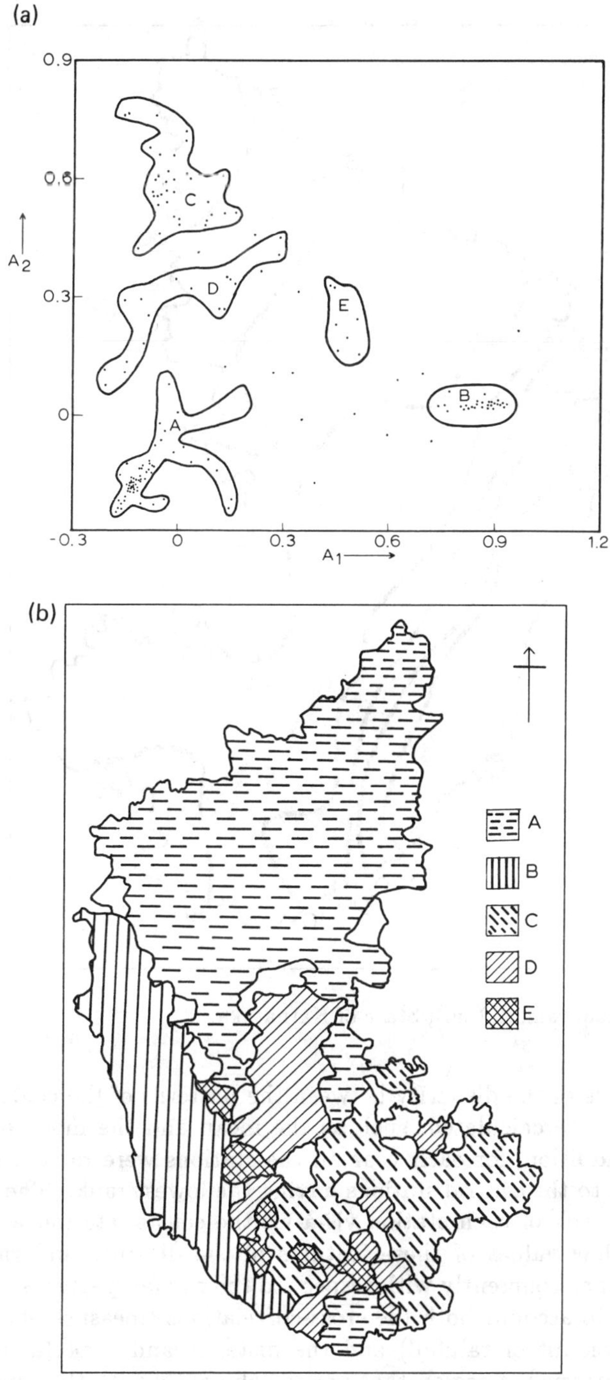

Figure 2.5(a) Clusters of the points in *Figure 2.4*. (*b*) Geographical location of the clusters in (*a*).

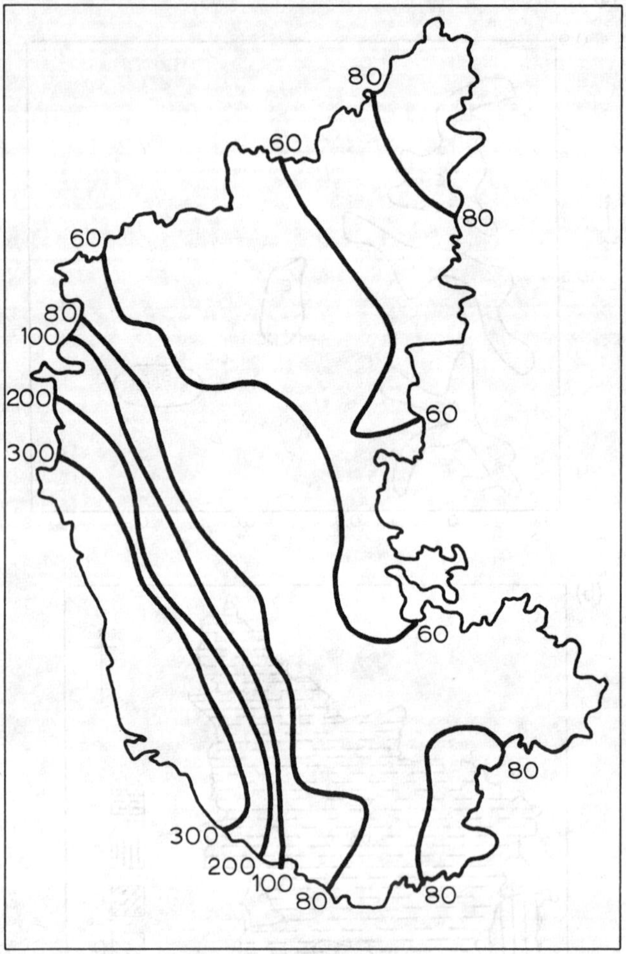

Figure 2.6. Annual rainfall (cm), State of Karnataka.

various measures of the disparity between the patterns of the central station and other stations were calculated, such as the mean and the mode of the distance statistics. In addition, for each year all the stations were ranked with the most similar pattern to the central stations having the lowest rank. The mean and the most probable rank of each station vis-à-vis the central station were calculated. Stations with low values of these features of the distance and rank were then considered to vary coherently with respect to the rainfall patterns.

Taking into account both the cross-correlations (measure of similarity variation in the amount of rainfall) and the distance and rank (a measure of the similarity of patterns), a region that varies coherently with the central station at Gulbarga was identified (*Figure 2.7*).

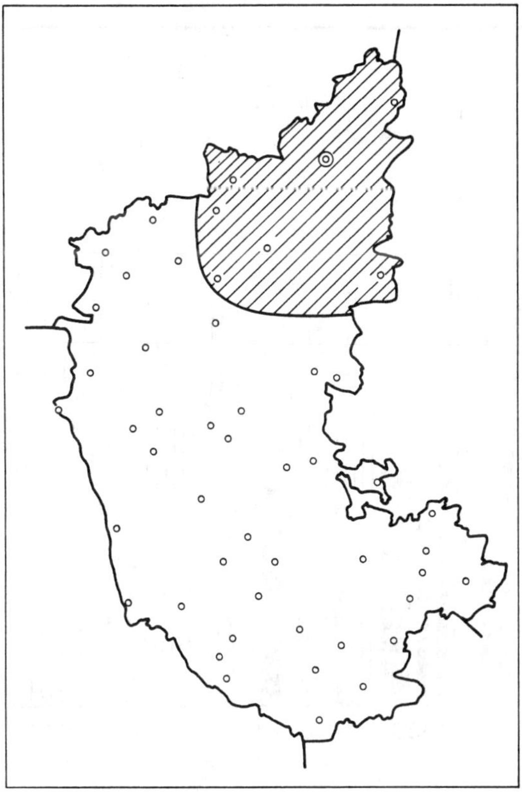

Figure 2.7. Coherent subregion of cluster A with central station at Gulbarga (circled).

2.6. Variability of Rainfall and Yield

The variation of the annual rainfall over the coherent subregion of cluster A during this century is shown in *Figure 2.8.* The mean annual rainfall for the region is 70.7 cm and the standard deviation is 17.7 cm, implying a large coefficient of variation. The years with rainfall deficit of more than 25% of the mean are marked in *Figure 2.8.* It is interesting that whereas a large number of droughts occurred during 1918–1941, only two occurred during the baseline period of 1941–1970. The frequency distribution (*Figure 2.9*) shows a mode around 70 cm, i.e., near the annual mean. For data relating to the period 1918–1970, there is a 1-in-10 chance of rainfall being less than 49 cm. However if only the baseline period (1941–1970) is considered (as in Section 3), a chance of 1 in 10 implies rainfall less than 59 cm. On the latter scale 1971 and 1972 were consecutive drought years.

The variation of the distribution of rainfall within years when the total rainfall is close to the most probable value is illustrated in *Figure 2.10*, in which

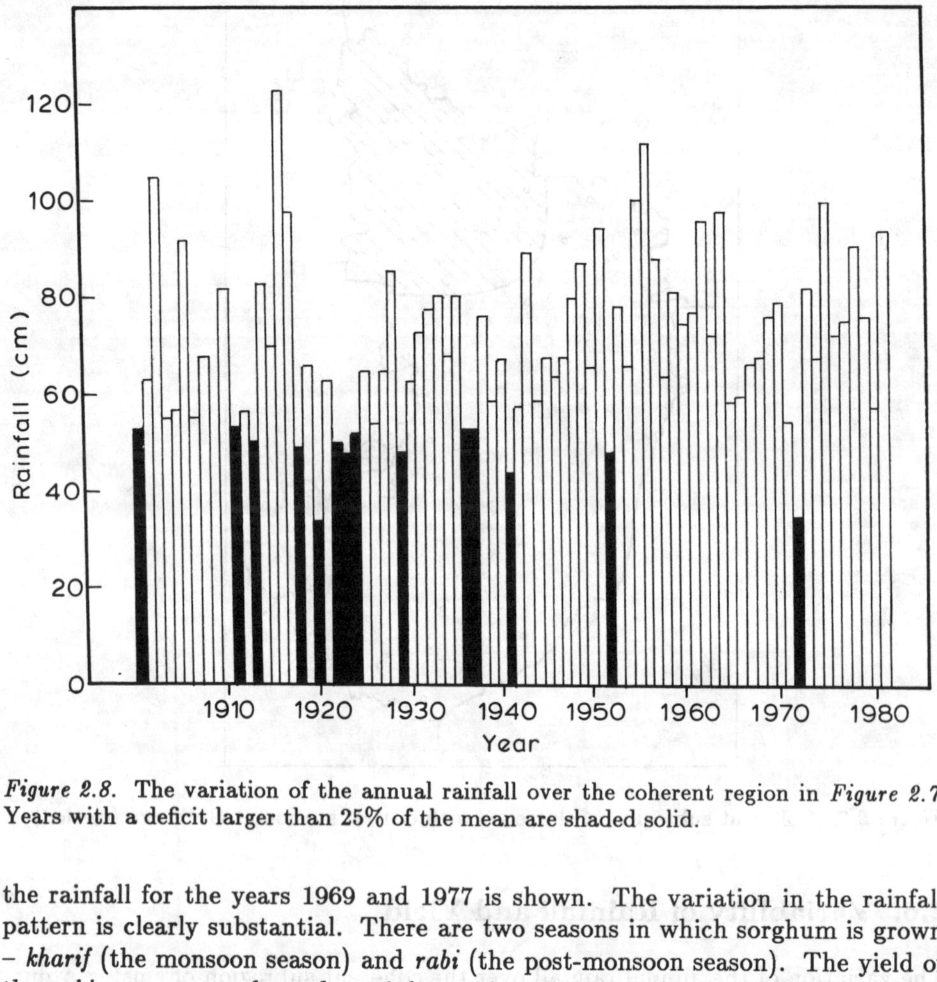

Figure 2.8. The variation of the annual rainfall over the coherent region in *Figure 2.7.* Years with a deficit larger than 25% of the mean are shaded solid.

the rainfall for the years 1969 and 1977 is shown. The variation in the rainfall pattern is clearly substantial. There are two seasons in which sorghum is grown – *kharif* (the monsoon season) and *rabi* (the post-monsoon season). The yield of the *rabi* crop grown depends mainly on the residual soil moisture at the end of the monsoon, which in turn is determined by the total rainfall received. The yield of *kharif*, on the other hand, depends not only on the amount, but on the pattern as well. Given the great variability of the rainfall patterns, it is not surpising that, on the average, twice as much land is devoted to cultivation of *rabi* as that of *kharif*.

We next illustrate the impact of variation of rainfall on the yields of the local varieties of sorghum, using the data on yields collected by the Government of Karnataka for 1955–1981. There is considerable variation in the rainfall pattern as well as in the annual totals in these 27 years for which the yield data are available. However, the yield data series is too small for a systematic analysis of the trends and of the variance that can be attributed to the variability of the rainfall. We therefore make some estimates of the magnitude of the impact of variability of the annual rainfall and the rainfall pattern on the yield by comparison of years with similar annual rainfall but different patterns (e.g., 1969,

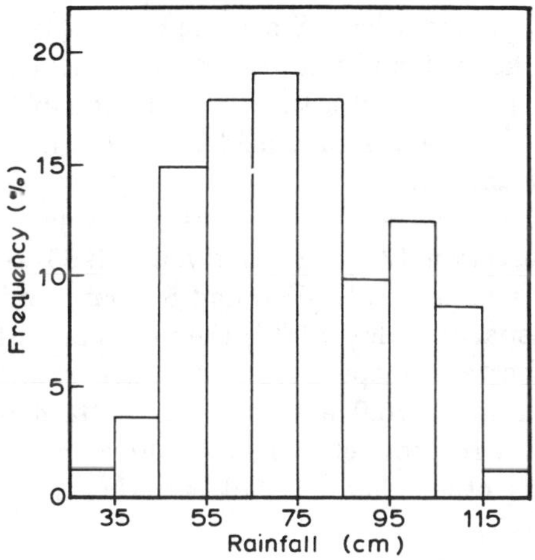

Figure 2.9. Frequency distribution of the annual rainfall over the study region, 1918–1970.

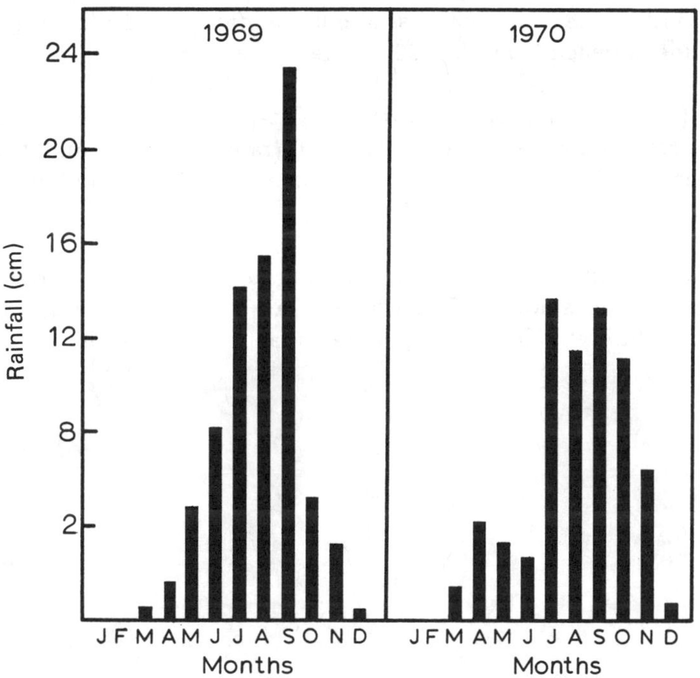

Figure 2.10. Rainfall patterns for 1969 and 1977, years with similar annual rainfall.

1977), on the one hand, and those with similar patterns but different annual totals (e.g., 1966, 1969), on the other. Since the general level of technology and availability of irrigation have steadily improved over the years, we have selected the contrasting years within as short a time span as possible to avoid bias introduced by these factors. We have also considered the impact of a back-to-back drought event, viz., 1971 and 1972.

The disparity between the rainfall patterns of any two years is determined as in the preceding subsection (2.5). In the years 1965 and 1966, which are characterized by similar annual totals (59.4 and 59.5 cm) and with patterns that are also quite similar (disparity index 0.169), the yields are rather similar as well: 0.55 and 0.51 tons per hectare, respectively. Another example is the pair 1969 and 1970 with annual rainfall 76.9 and 79.4 cm, pattern disparity 0.157 and yields 0.97 and 0.98 t/ha, respectively. This confirms our hypothesis that the variability due to factors other than rainfall is rather small if years within a short time span are compared.

We consider next the impact of variation of the rainfall pattern. The yield in the year 1977 (0.53 t/ha) was about 40% less than that of 1969, the annual rainfall totals being similar but the patterns markedly different (disparity index 0.30).

The impact of variation in the annual total is illustrated by comparison of 1969 with 1966. The yield in the year 1966, with a pattern very close to that of 1969 (disparity index 0.08) but much less annual rainfall, is 0.51 t/ha, implying a similar magnitude of deficit from the 1969 value.

The effect of a back-to-back drought event is reflected in the depression of yield in 1971 (i.e., the first of the two droughts; rainfall 54.7 cm and yield 0.82 t/ha) with respect to the previous year 1970 (rainfall 79.4 cm and yield 0.98 t/ha). This was followed by a major drought in 1972 (annual rainfall 35.2 cm) leading to a further depression of yield to 0.42, equivalent to about 50% of the yield of the first drought year.

These cases, though few in number, strongly suggest that the variability of the pattern and/or the amount of rainfall has a significant impact, and that back-to-back drought events have a very large impact on crop yields.

SECTION 3

Effects on Agricultural Productivity

3.1. The Setting

The dry tropics are characterized by a short and intense rainy season followed by a relatively long dry season. Generally the rainy season lasts from two to four months; the amount of rainfall received ranges between 400 and 1200 mm; the rainy season may have one peak (unimodal) or two peak periods of rainfall (bimodal). In the dry tropics, the rainy season is distinctly bimodal in areas lying between 5 and 12° N latitude, while it is generally unimodal in those areas located above 20° N latitude. The areas lying between 12 and 20° N show some tendency towards bimodality. The dry tropical region is also characterized by warm thermal regimes throughout the year. The annual potential evapotranspiration for most of the dry tropics ranges between 1500 and 2500 mm. The potential evapotranspiration is quite intense in the hot dry season preceding the rainy season. The US Class A open pan evaporation rates during this period range between 10 and 15 mm/day.

As was described in the previous section, the dry tropical areas show considerable variability of rainfall from year to year. The distribution of rainfall during the rainy season is quite erratic. Therefore, for developing alternate crop production strategies, it is important to consider the rainfall distribution pattern as well as its reliability. At ICRISAT (International Crops Research Institute for the Semi-Arid Tropics), it has been found that the rainfall can be characterized, in agronomic terms, most usefully by estimating the probabilities of rainfall and soil water in a stochastic process.

Markov chain procedures for estimating initial and conditional probabilities of rainfall can be used effectively for defining the dependability of rainfall during the rainy season (Robertson, 1976; Virmani *et al.*, 1982). The risks that sowing rains received by crops will not be enough to ensure survival, and the dependability of rainfall later at critical stages of crop development, can then be assessed through such analyses.

Water balance studies are important for characterizing the variations in availability of soil water for crop production, but few empirical data are available. The water-holding capacity of the soil has a considerable influence on the availability of soil water during the growing season. Information on daily soil water availability can be generated using soil water balance models. Such information is useful to agricultural scientists in devising cropping strategies for optimizing production in different types of soils, for evaluating risks and for optimizing inputs.

The objective of this section is to examine the variability of climate with particular emphasis on rainfall, and soil water availability, and their impact on agricultural productivity at 5 locations in India: Jodhpur (26°18′N, 73°01′E), Anantapur (14°41′N, 77°37′E), Hyderabad (17°27′N, 78°28′E), Dharwad (15°27′N, 75°00′E) and Indore (22°43′N, 75°48′E) (*see Figure 3.1*). These locations were chosen to represent a broad gradation of conditions from arid to wet semi-arid. This section first describes the agroclimatic characteristics, such as annual rainfall, weekly rainfall, soil water and runoff, for these locations. It then deals with the assessment and ranking of the locations in terms of productivity, using actual and simulated crop yield data.

3.2. Agroclimatic Description

3.2.1. Annual rainfall

The annual rainfall variability for the five locations is shown in *Figure 3.1*. The mean annual rainfall based on data from 1941–1970 ranged between 382 mm (Jodhpur) and 1001 mm (Indore), and the coefficients of variation ranged between 20 (Hyderabad) and 42% (Jodhpur). To calculate the frequency of occurrence of drought in these locations, we followed the definition of drought that the Indian Planning Commission (1973) used while evolving their relief and development approach toward drought-prone areas in the country. Accordingly, if the annual rainfall is 75% or less of the long-term average of the location, it is considered to be a drought situation (*see* Section 1). This definition has also been used in Sections 2 and 4 of this case study. The number of years in which rainfall fell below 75% of the corresponding average for the location was 8 in Jodhpur, 6 in Anantapur, 2 in Hyderabad, 4 in Dharwad and 6 in Indore. The lowest rainfall during 1941–1970 was 135 mm in Jodhpur in 1960, and the highest rainfall was 1743 mm in Indore in 1959.

The probability of occurrence of annual rainfall gives important preliminary information to help in developing alternative cropping systems for any region. The probability of receiving at least 800 mm of rainfall was 75% in Indore, 70% in Dharwad, 45% in Hyderabad, 7% in Anantapur and 3% in Jodhpur (*Figure 3.2*). In 70% of the years the rainfall received was at least 300 mm in Jodhpur, 470 mm in Anantapur, 700 mm in Hyderabad, 800 mm in Dharwad, and 860 mm in Indore (*Figure 3.2*). Farming systems research at ICRISAT has amply demonstrated that in the Hyderabad region a double crop system (e.g., intercrop, sequential or relay systems) is feasible if the water-holding capacity of soil exceeds 150 mm so that it can store sufficient available

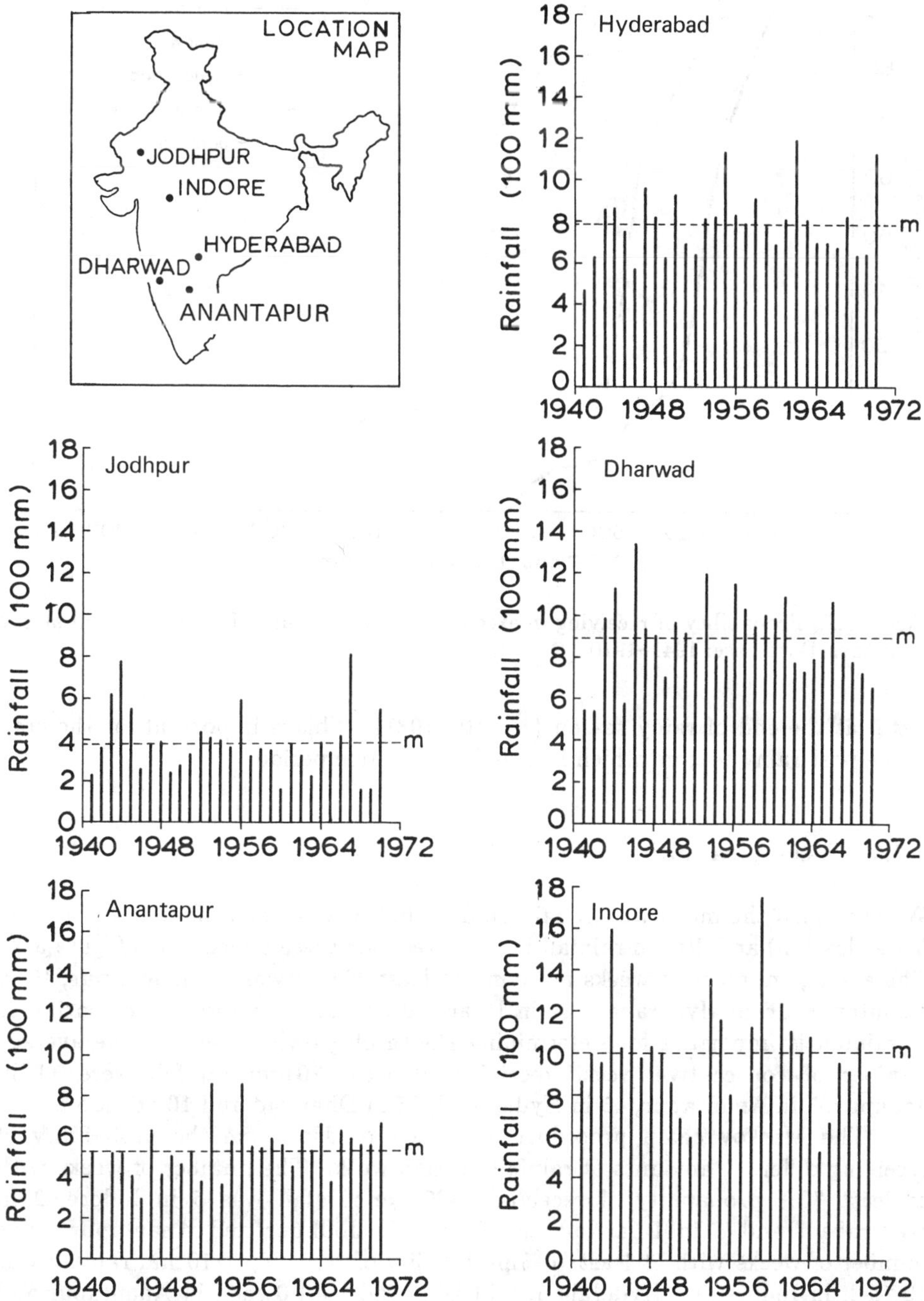

Figure 3.1. Location map and annual rainfall for five locations during 1941–1970 (m = mean annual rainfall).

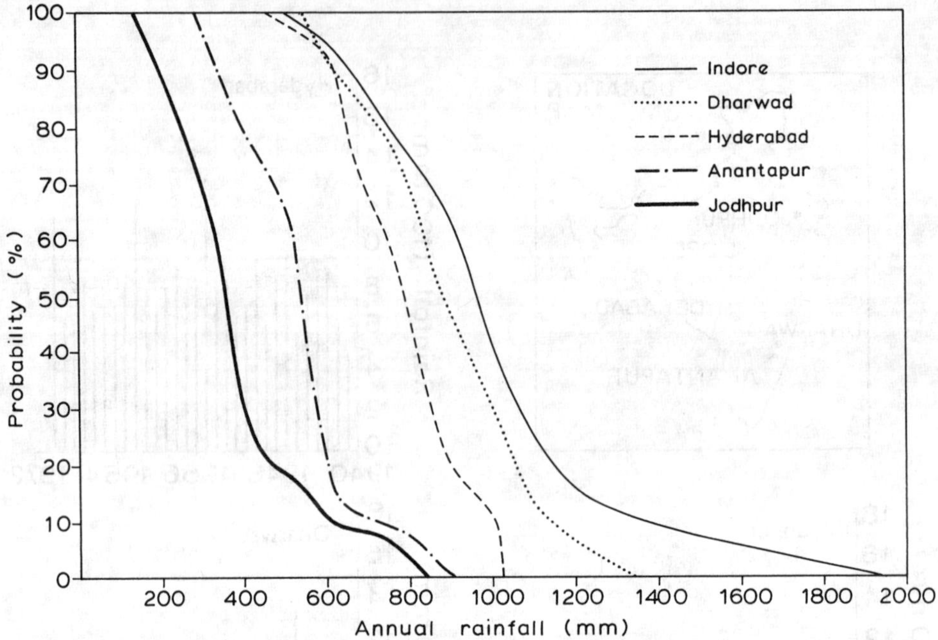

Figure 3.2. Probability of receiving in excess of a given amount of annual rainfall at five locations. Data base: 1941–1970.

water at the end of rainy season (Krantz, 1981). This is important for the crop in the post-rainy season, which is usually a rain-free period.

3.2.2. Weekly rainfall

We examined the mean weekly rainfall distribution and the probability of receiving at least 10 and 20 mm rainfall in each week for these 5 locations (*Figure 3.3*). The average number of weeks receiving at least 20 mm was 11 in Jodhpur, 10 in Anantapur, 16 in Hyderabad, 22 in Dharwad and 16 in Indore. The continuity of rainfall is important for determining length of growing season. The average number of consecutive weeks receiving at least 20 mm rainfall were 11 in Jodhpur, 5 in Anantapur, 15 in Hyderabad, 15 in Dharwad and 16 in Indore.

The Markov chain procedure was used to determine the probability of receiving different amounts of rainfall in each week. The number of weeks with at least 70% probability of receiving >20 mm rainfall was 5 in Indore, 2 in Hyderabad, and 1 in Dharwad and 0 each in Jodhpur and Anantapur. The number of weeks with at least 70% probability of receiving >10 mm rainfall was 10 each in Indore and Hyderabad, 9 in Dharwad, and 0 each in Anantapur and Jodhpur. These results indicate the relative risks of growing rainy season crops across the range of agroclimates represented by these locations.

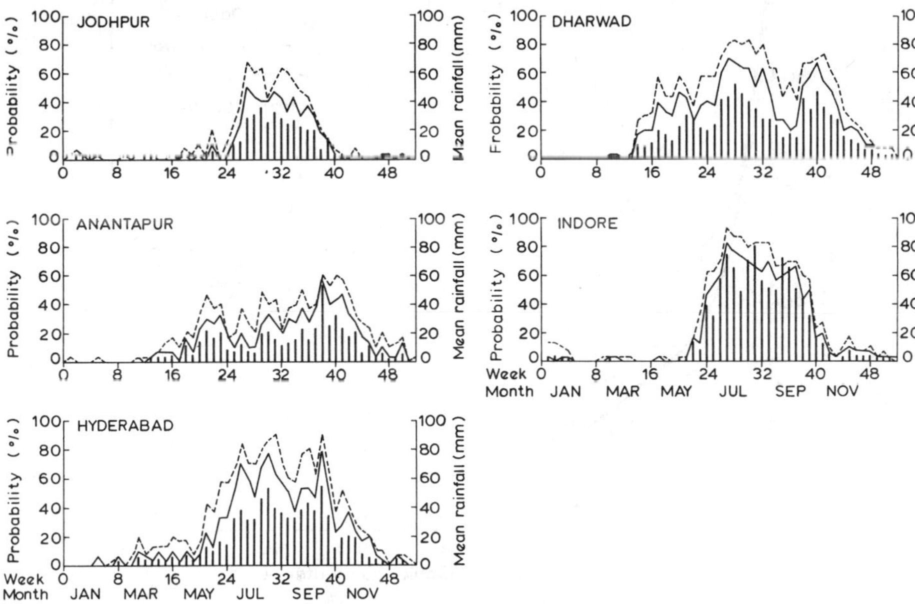

Figure 3.3. Mean weekly rainfall distribution (vertical lines) and the probabilities of receiving at least 10 mm (dashed line) and 20 mm (solid line) rainfall in a week for five locations. Data base: 1941–1970.

3.2.3. Soil water

Probabilities of available soil water (ASW) stored in the profile at the beginning of post-rainy seasons were simulated using the Ritchie (1972) soil water model to indicate the possibility of growing a second crop after rainy season cropping. This model simulated daily soil water using the information on initial ASW, available water-holding capacity of the soil, rainfall/irrigation and evaporative demand. In the model daily evapotranspiration (ET) was calculated by adding soil evaporation and transpiration. Potential evaporation below a plant canopy was calculated after simulating the potential evaporation from bare soil and using leaf area index (LAI) values. Potential evaporation from bare soil was calculated using net radiation as input data. Net radiation was calculated from albedo, maximum solar radiation reaching the soil surface and sky emissivity. Soil water at the beginning of the post-rainy season was simulated assuming that a sorghum crop had been grown in the rainy season and using climatic data from 1941 to 1970 (*Figure 3.4*). Available water-holding capacities (AWHC) of soils for Anantapur and Jodhpur were 50 and 100 mm, respectively. For the other locations, AWHC was 150 mm each. In 70% of the years, the simulated ASW was at least 5 mm each in Jodhpur and Anantapur, 75 mm in Dharwad, 105 mm in Indore and 110 mm in Hyderabad.

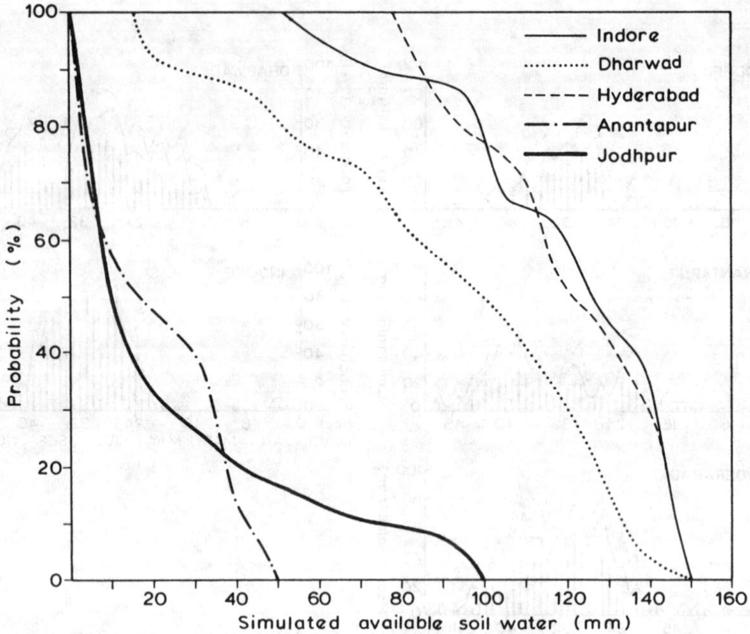

Figure 3.4. Probability of exceeding a given level of available soil water at the beginning of the post-rainy season after a rainy season cropping sorghum at five locations. Data base: 1941–1970.

These simulation results show that, after cropping rainy season sorghum, another crop can be grown in the post-rainy season in 70% of the years under residual soil water conditions in Indore, Dharwad and Hyderabad, where 50–75% of the soil profile was charged with available soil water at the beginning of post-rainy season. For Jodhpur and Anantapur, the soil was left with a negligible amount of water at the end of rainy season crop, and thus growing two crops in a year without irrigation involves high risk. This is not only because of the amount and distribution of the rainfall, but also because of the poor water-holding capacity of the soils.

3.2.4. Runoff

In the rainy season (June–October), much of the rainfall water runs off, but conservation of the runoff water can enable supplemental irrigation, either during intermittent droughts in the rainy season or for a post-rainy season crop. Thus, a quantification of the probability of runoff water is useful for developing watershed-based cropping strategies. Probabilities of simulated runoff for these locations, assuming a rainy season sorghum crop, during 1941–1970, are shown in *Figure 3.5*. Daily available soil water was first simulated using the Ritchie (1972) method. Daily runoff was simulated when the rainfall became excess to the capacity of the profile. Accumulated runoff was simulated at the end of each rainy season. *Figure 3.5* shows that in 70% of the years the simulated runoff was

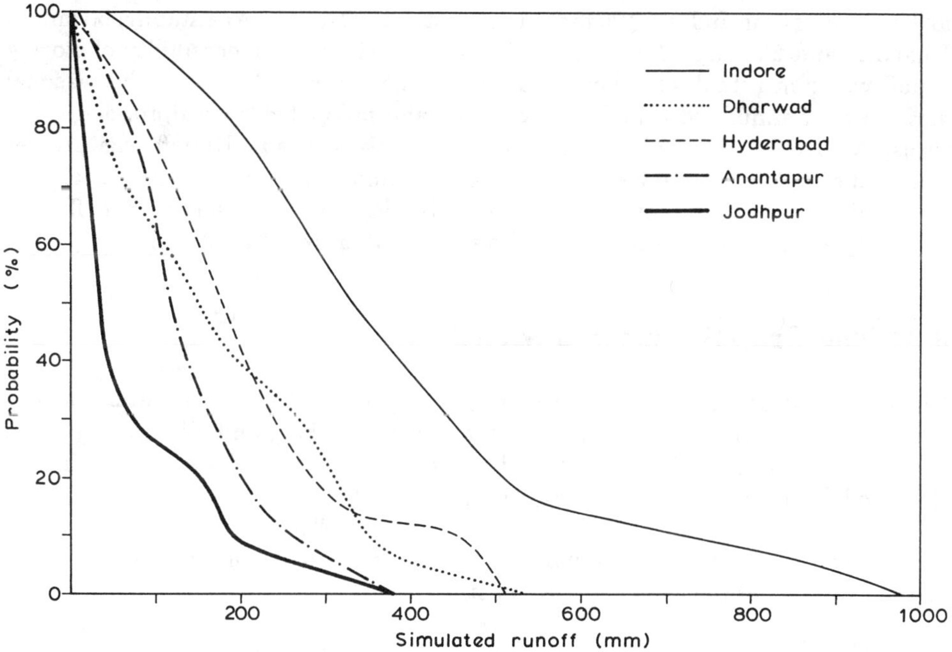

Figure 3.5. Probability of exceeding a given level of simulated runoff during June–October for five locations. Data base: 1941–1970.

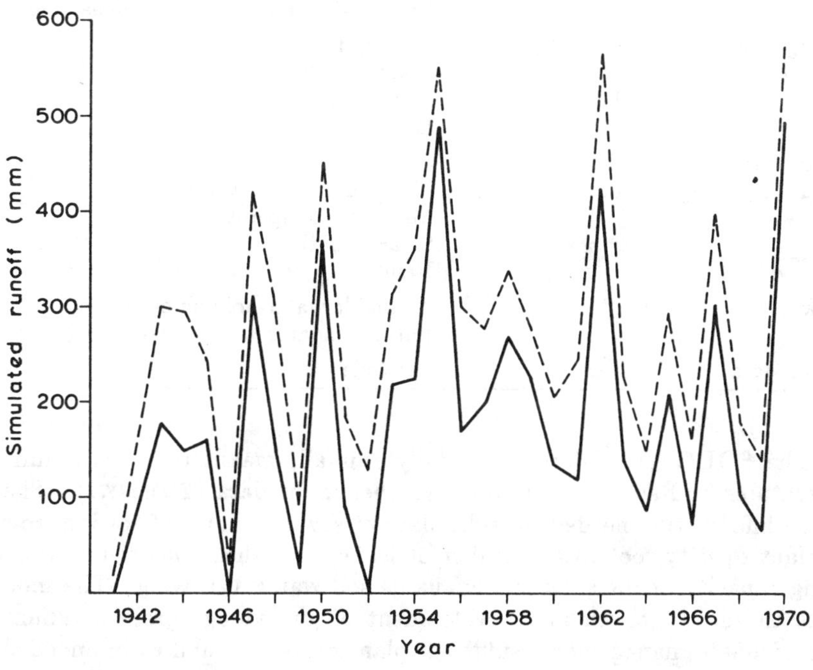

Figure 3.6. Seasonal variability of simulated runoff from sorghum (solid line) and fallow (dashed line) fields in rainy seasons at Hyderabad. Data base: 1941–1970.

at least 240 mm in Indore, 120 mm in Hyderabad, 90 mm in Anantapur, 60 mm in Dharwad and 20 mm in Jodhpur. This indicates the greater possibility of storing runoff water in a tank at Indore and Hyderabad. ***Figure 3.6*** shows the seasonal variability of simulated runoff from cropped and fallow fields (medium deep Vertisols, AWHC = 150 mm) in rainy seasons at Hyderabad. Runoff was greater from fallow fields. There were two years (1941 and 1946) when there was almost no runoff from either fallow or cropped fields. In 1952 there was no runoff from cropped fields, while 125 mm of runoff was estimated from fallow fields.

3.3. The Crop Simulation Model

Choosing crop production strategies using a crop model has the potential for providing useful quantitative information for decision making and eliminating much of the repetitive trial and error of field experiments. We used a sorghum simulation model, SORGF, developed by Arkin *et al.* (1976) and modified by Huda *et al.* (1984) for applications in the semi-arid tropics (SAT). Sorghum is one of the principal crops grown in the Indian dry tropics (Section 1), and we used sorghum as the example for our yield simulation study.

Table 3.1. Input data required for SORGF – a sorghum simulation model.

Category	*Specific input*
Plant data	Leaf number – total number of leaves produced Leaf area – maximum area of each leaf
Planting data	Sowing date Plant population Row width Depth of sowing
Climatic data (daily from sowing to maturity)	Maximum temperature Minimum temperature Solar radiation Rainfall
Soil data	Available water-holding capacity Initial available water content
Location data	Latitude

The SORGF model requires daily rainfall, maximum and minimum air temperatures and solar radiation as weather input data (*Table 3.1*). The plant and soil information needed includes date of sowing, depth of sowing, row spacing, plant density, potential number of leaves and their maximum size, water-holding capacity of the soils and available soil water at sowing. The model calculates the daily growth and development of an average grain sorghum plant under adequate management (sufficient plant protection and recommended doses of nutrient) in a field stand. It accounts for the physical and physiological processes, such as phenology, leaf area development, light interception and water use, which are independently computed and used as submodels. The potential dry matter is calculated from radiation intercepted, and the net dry matter is calculated accounting for water and temperature stress. Partitioning of dry matter into different plant parts is based upon the stage of development of the plant. The final grain yield per unit area is calculated by multiplying plant density by the grain weight per plant at physiological maturity. A flow diagram of the revised SORGF model is given in *Figure 3.7.* A brief description of the subroutines used in the model follows.

3.3.1. Emergence

Seedling emergence is computed in two steps. First, the date of germination is computed. This is assumed to occur when 18 growing degree-days (GDD) above a base temperature of 6.3 °C have accumulated, with at least 40% available soil water in the top layer (0.0–0.15 m) of the soil profile. The second stage of emergence is computed as the date when 50 GDD above a base temperature of 11.4 °C have accumulated, and it is also a function of depth of sowing: for each 0.01 m of additional depth, 2.8 extra GDD are required.

3.3.2. Leaf area

The model assumes a new leaf is initiated for every 50 additional GDD above a base temperature of 7 °C. Increase in individual leaf area is dependent on temperature as well. Leaf senescence is computed as a function of aging and soil water deficits. The leaf area of the senesced leaf is subtracted from the total plant leaf area to obtain daily leaf area increase per plant.

3.3.3. Phenology

The effects of temperature and day-length were included to determine the duration of three growth stages as defined by Eastin (1972). These stages are duration from emergence to panicle initiation (GS1), panicle initiation to anthesis (GS2), and anthesis to physiological maturity (GS3). Differences in the duration of GS1 and GS2 of a cultivar are accounted for by temperature and day-length. A threshold value of day-length was found to be 13.6 h. With the increase in day-length at emergence beyond this threshold, there was a linear increase in the duration of GS1 and GS2. However, the intercepts and slope of the equations were different for GS1 and GS2. The intercepts for GS1 and GS2 were 370 and 650 GDD above 7 °C, respectively. The slopes for GS1 and GS2 were 400 and 120 GDD for every hour increase in day-length. Differences in GS3 were accounted for as a temperature effect only. The duration of GS3 decreased with increase in mean air temperature up to 27 °C and increased above 27 °C.

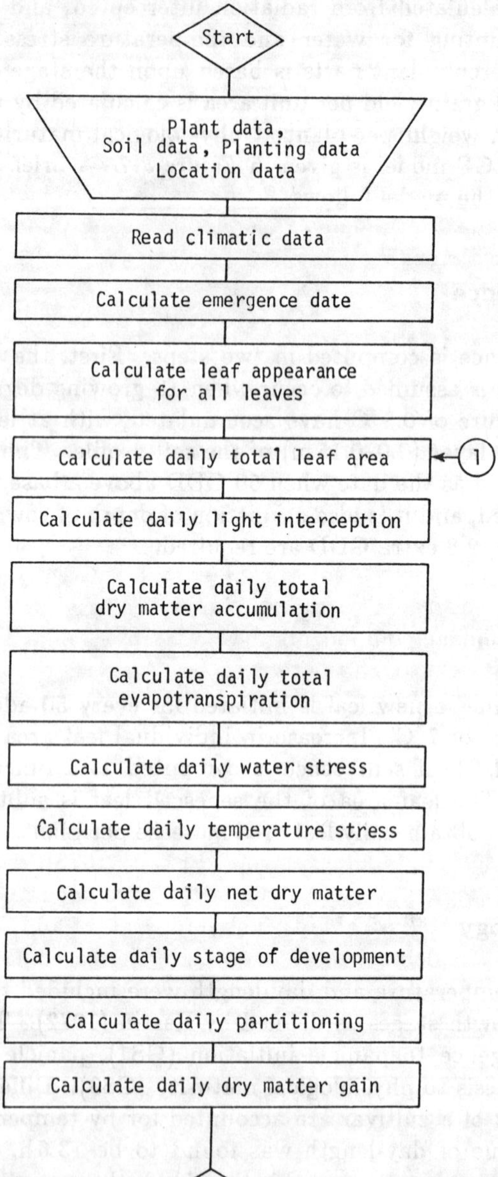

Figure 3.7. A flow diagram of the revised sorghum simulation model, SORGF.

3.3.4. Light interception

The light interception portion of the model simulates the relative quantum flux intercepted by a single plant. Intercepted photosynthetically active radiation (PAR) was calculated on an hourly basis following a Beer's Law relationship using solar radiation and light transmission values. Hourly solar radiation was computed from the input daily solar radiation and by accounting for hours of sunlight, which was calculated as a sine function of the local solar time and day-length. Light transmission was calculated from the relationship of extinction coefficients and maximum light transmission, using information on row spacings and LAI (leaf area index).

3.3.5. Soil water

The input data required for the computation of daily soil water for the entire soil profile were discussed earlier in this section. The revised SORGF model computes daily soil water for various layers in the profile using similar input data to those mentioned earlier. Daily evapotranspiration was simulated after Ritchie (1972) and was apportioned to the appropriate layers using a root growth model of Stinson (1979). Daily potential root growth was calculated using the ratio of cumulative daily leaf area to the maximum leaf area for the whole plant. The extraction of drainage components developed by Williams and Hann (1978) was used to predict flow through the root zone.

3.3.6. Dry matter accumulation and partitioning

The relationship between total dry matter and intercepted PAR was used following the approaches of Gallagher and Biscoe (1978) and Stapper and Arkin (1980). It was found that 3 g of dry matter were produced for each MJ of PAR intercepted for irrigated sorghum. The temperature and water-stress coefficients were then used to calculate daily dry weight increase. Partitioning of total dry matter to plant parts (e.g., leaf, culm, head and grain) varied according to the stage of plant development.

3.4. Crop Simulation Results

3.4.1. Agroecological capability analysis – average conditions

We used the 30-year period (1941–1970) averaged meteorological variables to simulate sorghum grain yield under high and low management across the selected locations (*Table 3.2*). Simulation under high management was based on a high plant density (180 000 plants/ha), and a high-yielding hybrid. Simulation under low management assumed low plant density (50 000 plants/ha) and a local variety. Using the 30-year averages of meteorological data resulted in simulated

Table 3.2. Rainfall, simulated and actual grain yield of sorghum in selected locations.

Management	*Measure*	*Locations* *Jodhpur*	*Anantapur*	*Hyderabad*	*Dharwad*	*Indore*
		Annual rainfall (1941–70)				
	Mean (mm)	382	527	792	889	1001
	CV (%)	42	25	20	22	27
		Simulated yield (mean climate, 1941–70)				
High[a]	Mean (t/ha)	2.7	3.4	5.0	5.4	6.4
Low[b]	Mean (t/ha)	0.5	0.8	1.5	1.6	2.0
		Simulated yield (individual weather-years, 1941–70)				
High[a]	Mean (t/ha)	3.1	2.6	4.6	5.7	6.4
	CV (%)	39	34	6	14	3
Low[b]	Mean (t/ha)	1.1	0.9	1.4	1.9	2.0
	CV (%)	34	32	5	9	3
		Actual yield (recorded, 1954–70)				
	Mean (t/ha)	0.1	0.4	0.5	0.6	0.7
	CV (%)	128	26	28	28	29

[a]High management is simulated with 180 000 plants/ha and using a hybrid.
[b]Low management is simulated with 50 000 plants/ha and using a local variety.

yields ranging from 2.7 to 6.4 t/ha under high management and 0.5 to 2.0 t/ha under low management across these locations.

The purpose of this exercise was to illustrate that, in the absence of individual-year weather data, it is possible to use averaged weather data to simulate the agroecological capability of various locations. However, year-to-year variability of simulated yield is important for many purposes, including the yield stabilization aspect, and this is discussed below.

3.4.2. Thirty individual years – productivity estimates

We collected the sorghum yields for 1954–1970 (data were available only from 1954) reported by the Indian Government Statistical Services, based on crop-cutting experiments for the selected locations in India from the season and crop reports of respective states. Each of these yields is the average for the whole district.

Figure 3.8(a) shows that the average yields obtained by the farmers were always less than 1.0 t/ha. In Jodhpur, greater than 0.2 t/ha sorghum yield was reported only in 3 out of 10 years. In 7 out of 10 years, sorghum yield was in the range of 0.3 to 0.6 t/ha at 4 other locations. This indicates that the farmers generally have adopted a very low level of management practices (e.g., little or no fertilizers, no plant protection measures, etc.). Singh and Das (1984) and Huda *et al.* (1985) reported that one can generally obtain 1.0 to 1.5 t/ha of sorghum grain yield from the plots without added fertilizers. Limited surveys conducted

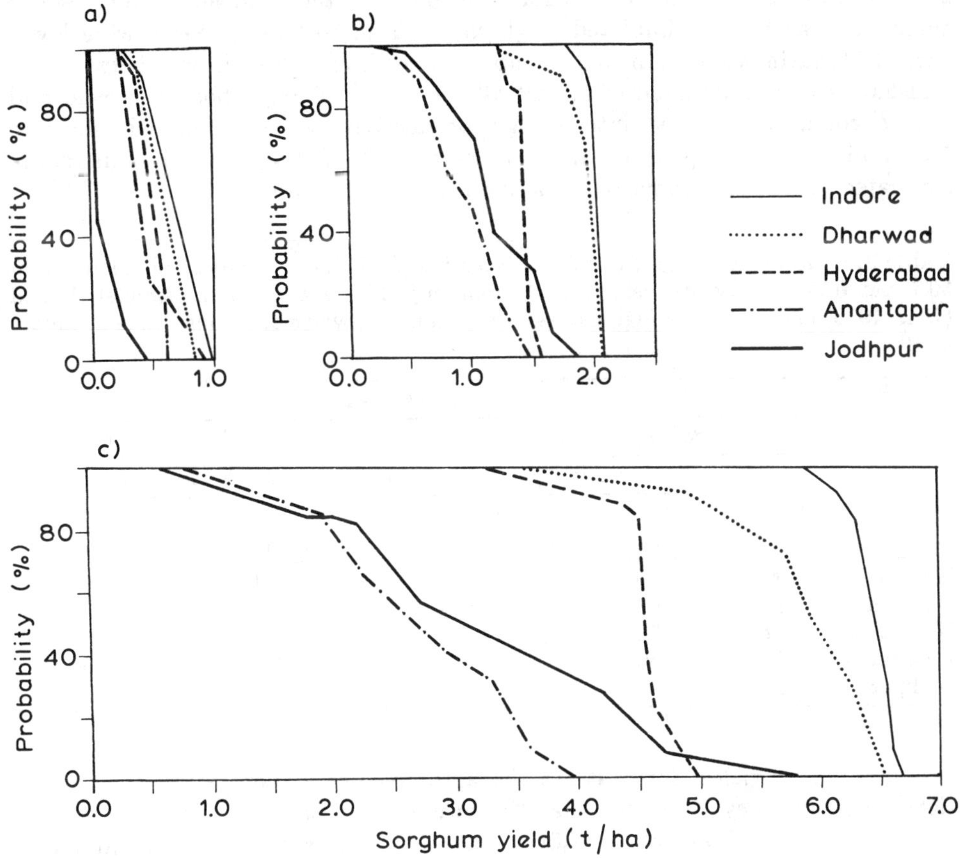

Figure 3.8. Sorghum yields: (*a*) actual, (*b*) simulated under low management and (*c*) simulated under high management.

on the use of fertilizers in the drylands of India also revealed that fertilizer consumption is only about 10 kg/ha (Tandon, 1981).

Sorghum grain yields simulated for low management (50 000 plants/ha and a local variety) using climatic data from 1941–1970 are shown in *Figure 3.8*(*b*). Results show that the five locations could be classified into three groups: low, medium and high productivity zones. For Anantapur and Jodhpur, the simulated sorghum grain yields of less than 1.2 t/ha in 50% of the years indicated a low productivity area. Hyderabad was in a medium productive area according to the simulated yields, which in almost all the years were in the range of 1.2 to 1.5 t/ha. Dharwad and Indore were in a highly productive area where simulated yields in 50% of the years ranged between 1.5 and 2.1 t/ha.

To examine the possible yields that might be attained in these regions, sorghum grain yields were simulated for high management (180 000 plants/ha, and a high-yielding hybrid) using climatic data from 1941–1970 [*Figure 3.8*(*c*)]. The order of the productivity zones for these selected locations remained the same as was observed under low management, but the yield levels were two or three times as large. Simulated sorghum yields in 50% of the years were lower than 3.0 t/ha in Anantapur and Jodhpur. The simulated yield was between 4.0 and 5.0 t/ha in Hyderabad in almost all the years. Though both Dharwad and Indore could be grouped into a high productivity zone, simulated yields in Dharwad ranged between 3.3 and 5.5 t/ha in 50% of the years, while in Indore simulated yields were above 6.0 t/ha in almost all the years.

Table 3.3. Ranking of the five selected locations in India in terms of productivity in 30% and 70% of the years, using actual sorghum yield (1954–1970) and simulated yield (1941–1970) under low and high management. Ranking was made using yields at Indore as 100.

	Ranking					
	30% of the years			*70% of the years*		
		Simulated yield			*Simulated yield*	
	Actual yield	*Low management*	*High management*	*Actual yield*	*Low management*	*High management*
Indore	100	100	100	100	100	100
Dharwad	89	98	97	93	95	89
Hyderabad	72	71	71	82	70	70
Anantapur	58	56	50	64	36	34
Jodhpur	19	70	62	5	49	37

Simulated yields demonstrate the agroecological potentials of these locations, and thus they are higher than the actual district average yields. *Table 3.3* shows the relative levels of these five locations in terms of actual and simulated yields in 30% and 70% of the years. Among these locations, both simulated and actual yields were maximum at Indore. Yields at Indore were taken as 100%, and the ranking of the other locations can then be made with reference to Indore. According to the actual yields, Indore and Dharwad could be grouped in a high productive zone, Hyderabad in a medium productive zone, and Anantapur and Jodhpur in a low productive zone. Simulated yields (both under low and high management) result in a similar ranking of these three groupings. However, simulated yields at Jodhpur ranked better than those for Anantapur, while the reverse was true for actual yields. Though the average annual rainfall (*Table 3.2*) at Jodhpur (382 mm) was lower than that of Anantapur (527 mm), the distribution of rainfall was better in Jodhpur for the rainy season crop (as indicated in *Figure 3.3*). For example, in Jodhpur 11 weeks receiving >20 mm rainfall were consecutive (02 July to 16 September) while in Anantapur, only five of the weeks that received >20 mm were consecutive weeks (10 September to 14 October). In addition, the available water-holding capacity of soils at Jodhpur (100 mm) is higher than that of Anantapur (50 mm).

In only 10% of the years does annual rainfall at Jodhpur and Anantapur exceed 700 mm, and in such years simulated yields of sorghum in Jodhpur – where AWHC is 100 mm – become comparable to those of Hyderabad. Though sorghum is not usually grown by the farmers in the Jodhpur region, experimental yields under high management in a good and well distributed rainfall year were as high as 3.0 t/ha (R.P. Singh, Director, Central Research Institute for Dryland Agriculture, Hyderabad, personal communication, 1985).

In spite of the discrepancy in the ranking of Jodhpur and Anantapur, the simulations clearly confirm that those locations represent less productive agroclimates for sorghum than do the other three locations. These results suggest that more drought-tolerant crops, such as millet, will be better suited for Anantapur and Jodhpur. The agroclimatic environments represented by Indore can be more profitably utilized by growing high-value crops, such as soybeans, and by adopting cropping systems (e.g., intercrop and sequential systems) capable of harnessing the better soil water availability of this location. The agroclimatic environments of Hyderabad and Dharwad are suitable for sorghum-based cropping systems. High-yielding cultivars of sorghum (e.g., CSH 5, CSH 6) give yields of 4–6 t/ha under experimental conditions in Dharwad and in the ICRISAT Center, Patancheru, near Hyderabad (ICRISAT, 1985).

3.4.3. Extreme case analysis

Table 3.4 shows the years in which simulated extreme (minimum and maximum) sorghum yields were obtained. Rainfall during the crop season in those years is also shown in *Table 3.4*. The simulated minimum yields corresponded to the extreme drought years, but the simulated maximum yields did not always occur in the highest rainfall years. Distribution as well as amount of rainfall were important in simulating maximum yields. These results agree with findings based on examination of the actual sorghum yield data reported by Statistical Services. For example, at Jodhpur there was no sorghum yield reported in 1960, when only 87 mm rainfall was received; while the maximum sorghum yield of 0.9 t/ha for Dharwad was in 1969, when 499 mm rainfall was received, although there were many years in which more than 800 mm of rainfall were received.

3.4.4. Back-to-back drought events

Climatic data for the years when drought occurred were analyzed as consecutive years and then as separate years for yield simulation. There was no difference in the simulated yields between back-to-back and separated drought events. This is because, in the semi-arid tropics, the growing season usually starts with a dry soil profile almost every year. However, two or three consecutive drought years will have a very great economic impact, and this aspect is discussed later in the case study (Section 4).

Table 3.4. Minimum and maximum simulated yields of sorghum under high management at the five selected locations in India. The years of occurrence of such events and the rainfall in the growing season for the corresponding years are also given.

Locations	*Minimum yield* (t/ha)	*Year*	*Seasonal rainfall* (mm)	*Maximum yield* (t/ha)	*Year*	*Seasonal rainfall* (mm)
Jodhpur	0.6	1960	87	5.8	1967	528
Anantapur	0.7	1944	112	4.3	1947	434
Hyderabad	3.4	1960	426	5.0	1961	489
Dharwad	3.3	1942	296	6.4	1953	725
Indore	5.9	1965	459	6.7	1967	814

3.4.5. Long-term changes in climate and some possible responses

We used the weather data for ICRISAT Center, Patancheru, from 1974–1985 to simulate the effect of long-term changes in climate on sorghum yield. Sorghum yields were simulated for the 12-year period using actual weather data for both rainy and post-rainy seasons for medium deep Vertisols (150 mm AWHC). The minimum, mean and maximum simulated grain yields using actual weather data were 4.2, 4.8 and 5.2 t/ha for rainy season sorghum, and 1.4, 2.9 and 4.2 t/ha for post-rainy season sorghum under residual soil water after a rainy season sorghum cropping.

Supplemental irrigation

The SORGF model was used to simulate the response of sorghum to supplemental irrigation assumed at different growth stages during post-rainy seasons after a rainy season sorghum crop (*Table 3.5*). The average rainfall between October and January during 1974–1985 was 123 mm. Irrigations of 50 mm each, assumed at sowing and anthesis, increased the simulation results for minimum yield by 136%, mean yield by 38% and for maximum yield by 17%.

Temperature and radiation

Long-term changes in daily maximum and minimum air temperatures, and intercepted solar radiation were assumed. The average weekly solar radiation, and maximum and minimum temperatures, are given in *Figure 3.9.* A combination of various hypothetical conditions was generated. *Figure 3.10* shows the minimum, mean and maximum simulated yields of sorghum corresponding to changes in daily temperature of 1° to 5 °C and daily intercepted solar radiation of 10% and 20% for both rainy and post-rainy seasons. The yields are expressed as percentages of the simulated yields for no change, i.e., for no change in temperature and radiation, the yield value is 100.

The simulated average duration to maturity for sorghum in the rainy and post-rainy seasons were 86 and 108 days after emergence, respectively. The

Table 3.5. Simulated grain yield for supplemental water assumed at different growth stages of sorghum during post-rainy seasons (sowing date assumed 1 October) after a rainy season sorghum cropping at ICRISAT Center, Patancheru. Simulation base: 1974–1985.

Supplemental water[a] *assumed at:*			*Simulated grain yield* (t/ha)		
Sowing	*Panicle initiation*	*Anthesis*	*Minimum*	*Mean*	*Maximum*
–	–	–	1.4	2.9	4.2
X	–	–	2.2	2.9	4.2
–	X	–	2.2	3.2	4.4
X	X	–	2.7	3.3	4.4
–	–	X	2.3	4.0	4.9
X	–	X	3.3	4.0	4.9
–	X	X	3.3	4.3	4.9
X	X	X	3.9	4.4	4.9

[a]X = 50 mm irrigation.

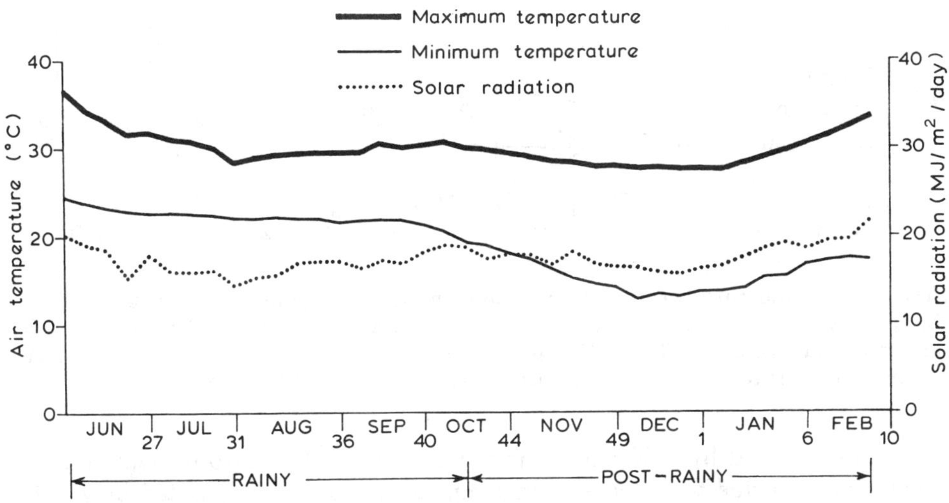

Figure 3.9. Weekly averaged daily maximum and minimum air temperatures and solar radiation at ICRISAT Center, Patancheru. Temperature data from 1974–1985 and solar radiation data from 1978–1985 were used to calculate average values.

longer duration in the post-rainy season is due to the relatively cooler temperature (*Figure 3.9*). Assuming a 5 °C increase in temperature, the simulated total growing period is shortened by 13 days in the rainy season and 28 days in the post-rainy season. Conversely, a decrease in temperature of 5 °C delayed the simulated duration by 31 days in the rainy season, and 45 days in the post-rainy season.

The manipulation of the temperature values in combination with the changes in intercepted solar radiation had significant impacts on the simulated yield. For example, in the rainy season, decreasing the temperature 5 °C while increasing intercepted radiation by 20%, increased the maximum yield by 29%,

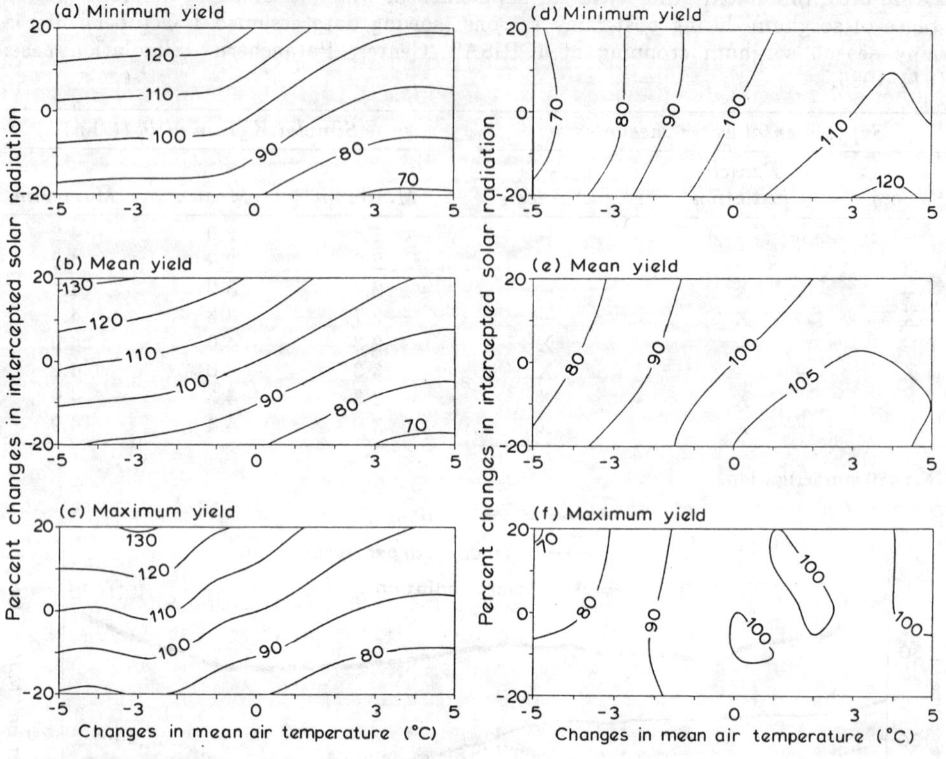

Figure 8.10. Minimum, mean and maximum simulated sorghum grain yields (shown as contours) as a function of assumed changes in mean air temperature (1 to 5 °C) and intercepted solar radiation (0 to 20%) during the rainy season (a, b, c) and the post-rainy season (d, e, f) at ICRISAT Center, Patancheru. Data base: 1974–1985.

the minimum yield by 29% and mean yield by 33%. It is also evident that the longer duration due to a decrease in temperature (5 °C) contributed only a 7% increase in minimum yield, a 13% increase in mean yield, and a 10% increase in maximum yield. The increase (20%) in intercepted radiation alone contributed an additional 20% improvement in minimum, mean, and maximum yield. This is because in the rainy season cloud cover limits higher yields.

The decrease in temperature by 5 °C in the post-rainy season reduced by 26% the simulated mean yields. This decrease in temperature coupled with a 20% increase in intercepted solar radiation in the post-rainy season reduced the minimum, mean and maximum yields, respectively, by 31, 29 and 31%. High radiation would cause more evapotranspiration losses, and low temperature would increase the duration of the growing period and thus expose the crops to drought stress for a longer period. By increasing the temperatures in the post-rainy season, and with a 20% reduction in intercepted radiation, though the maximum simulated yield was reduced by 8%, the minimum yields could be increased by 19% and the mean yields by 4%. The relatively shorter duration of the post-rainy growing season due to increased temperature and low radiation causing reduced evapotranspiration losses (reducing the period of drought stress) resulted in stabilization of the yields.

Though the global solar radiation could not be changed, the impact of higher solar radiation on evapotranspiration losses in the post-rainy season could be manipulated through agronomic practices, such as low plant density and organic mulches. In the rainy season the intercepted radiation can be increased by manipulating the canopy architecture and by choice of cultivars.

3.5. Conclusions

The dry tropics are noted for their variable rainfall and high potential evapotranspiration. Generalized agroclimatic analyses based on mean monthly rainfall provide insufficient agroclimatic detail within ecological regions for planning strategies to increase and stabilize crop production. Probability estimates of weekly rainfall, soil water at critical growth stages and seasonal runoff are much more sensitive parameters. This information is important for evaluating risks for crop production in dryland areas and for developing strategies to overcome them.

We have demonstrated the utility of a sorghum simulation model to assess yields associated with climatic variation and change, management levels and soil type. This model was used to delineate the areas in terms of growing a second crop in the post-rainy season in addition to rainy season sorghum. The use of such models for assessing crop yields due to simulated changes in climate has been illustrated, indicating that, for stabilizing yield, interception of solar radiation becomes an important factor in the rainy season while availability of soil water, particularly at crop establishment and grain filling stages, becomes important in the post-rainy season. In agroclimatic environments similar to that of ICRISAT Center, Patancheru, longer-duration cultivars with higher solar radiation utilization efficiency could stabilize sorghum yields in the rainy season. Shorter-duration sorghum in association with management practices for reduced soil evaporation losses and the use of runoff water for irrigation could stabilize sorghum yields in the post-rainy season. However, since the actual yields obtained by the farmers are much lower than the agroecological potentials of each region, it is suggested that adoption of appropriate management factors (e.g., timely field operations, establishing adequate plant stand, high-yielding cultivars, nutrient and pest management, etc.) would go a long way toward improving agricultural productivity in these regions.

causing reduced evapotranspiration losses (reducing the period of drought stress) resulted in stabilization of the yields.

Though the global solar radiation could not be altered, the impact of higher solar radiation on evapotranspiration losses in the postrainy season could be manipulated through agronomic practices such as higher plant density and organic mulches. In the rainy season the influence of radiation can be altered by manipulating the canopy [illegible].

3.5. Conclusions

The dry tropics are noted for their variable rainfall and high potential evapotranspiration. Computerized agroclimatic analyses based on long-term weekly rainfall provide important agronomic detail within ecological regions for planning strategies to increase and stabilize crop production. Probability estimates of weekly rainfall, soil water at critical growth stages and seasonal runoff are much more sensitive parameters. This information is important for evaluating risks for crop production in drylands and for developing strategies to overcome them.

We have demonstrated the utility of a sorghum simulation model to assess yields associated with climatic variation and change management practices [illegible] crops. This model was used to delineate the areas in terms of growing a second crop in the postrainy season in addition to the rainy season crop. The use of such models for assessing crop yields under simulated changes in climate has been illustrated, indicating that [illegible] the interaction of solar radiation becomes an important factor in rainy season, while availability of soil water particularly at crop establishment and grain filling stages becomes important in the postrainy season. In agroclimatic environments similar to that of ICRISAT Center, longer duration cultivars with higher radiation conversion efficiency could stabilize sorghum yields in the rainy season. Short duration sorghum in association with management practices for reduction of soil evaporation losses and the use of soil water for irrigation could increase sorghum yields in the postrainy season. However, since the actual yields obtained by the farmers are much lower than the projected potentials of such regions, it is suggested that adoption of appropriate management factors (e.g., timely field operations, establishing adequate plant stand, high yielding cultivars, nutrient and pest management, etc.) could go a long way towards improving agricultural productivity in these regions.

SECTION 4

Village-Level Farm Adjustment to Rainfall Variability

4.1. The Farmer Context

The preceding three sections described a broad framework, consisting of public policies and scientific approaches to manage and harness the agroclimatic environment of the dry tropics in India. However, the final agency to be affected by the issues and measures discussed is the farmer. Since the farmer's decisions and responses would largely determine the effectiveness of various measures discussed in the preceding sections, it is necessary to examine briefly the farmer's adaptation strategies against rainfall variability. The aims of this section are:

(1) To describe farmers' long-term and short-term adjustment mechanisms to deal with rainfall variability or the resultant instability and risk characterizing their occupation.
(2) To indicate the manner in which farmers' adaptation strategies could be potentially strengthened by improved understanding of the environment and by modifications of public policies and programs designed specifically for areas with high rainfall variability.

We will deal with these in turn. At the outset, however, it is important to recall the types of variability with which individual farmers must cope. Three such types are recognizable:

(1) Pronounced year-to-year variations in rainfall receipt, which may include substantial deficits from the seasonal normal.
(2) Less pronounced variations in the *distribution* of rainfall within the growing season, leading to important variations of soil moisture at certain critical stages of crop growth. Existing climatological data sets are rarely sufficiently detailed to identify this type of variability.
(3) The third dimension of rainfall variability relates to long-term pattern of rainfall, particularly the frequency of occurrence of droughts.

In the following section we consider, first, farmers' adaptation strategies to the perceived long-term variability of rainfall.

4.2. Long-Term Adaptations

The measures adopted by the farmer to cope with rainfall variability are reflected through different structural and operational features of his farming systems (Jodha, 1978; Walker and Jodha, 1986).

The structural features of a farming system represent the farmer's adaptation to long-term patterns of rainfall in a region. These include:

(1) The degree of diversification through mixed-farming/mixed-cropping flexibility in farm operations and practices.
(2) The systems of self-provisioning, on-farm storage and recycling procedures for both input and output.
(3) The choice and combination of farm enterprises, assets, etc.

On the basis of the relative degree and nature of these responses, the long-term adaptation of the farming system in one region can be differentiated from that in another (Spitz, 1980). For instance, the greater the weather-induced risk in a region, the greater will be the extent of diversification and flexibility in the farming systems, and *vice versa.* Not only do the diversification and flexibility equip the farmer to exploit the benefit of good rain years and to protect against losses during poor rain years, but they also determine the farmer's preparedness to adjust to the short-term variability of rainfall. An indication of such long-term adaptation strategies can be obtained by comparison of relevant features of farming systems and associated factors in three agroclimatic regions discussed below.

It should be noted that since a variety of factors collectively or individually play their role in shaping farming systems, it is not easy to pinpoint the exact role of agroclimatic conditions, especially rainfall, in determining the farmer's adjustment strategies. However, since rainfall plays a crucial role in shaping farmers' activities in the seasonally dry tropics, the existing farming systems in diverse agroclimatic environments can be viewed as the consequences of adjustment strategies in response to the long-term pattern of rainfall. Juxtaposition of relevant features of the farming systems in different agroclimatic environments can illustrate this.

Table 4.1 presents such information for three areas. The three areas, Akola, Sholapur and Jodhpur, belong to three zones that can be differentiated on the basis of normal annual rainfall, length of growing season and frequency of drought-related crop failures. (For the location of zones and study areas, *see Figure 1.1.*) Accordingly, Akola belongs to the relatively stable and high rainfall zone. This zone accounts for more than 14% of the total area of arid and semi-arid tropical regions in India. Sholapur represents the zone where the drought frequency is once in four years. This is the largest dry zone, accounting for over 51% of the arid and semi-arid tropical region of the country. Jodhpur represents

Table 4.1. Indicators of farmers' long-term strategies against weather risk in three districts with different degrees of weather risk in the dry tropical regions of India.[a]

	Situation in the villages of:		
Details of degree of risk and adaptations:	*Akola*	*Sholapur*	*Jodhpur*
A. Characteristics of weather risk:[b]			
Annual average rainfall (mm)	820	690	382
Probability of favorable soil moisture conditions for rainy season cropping	0.66	0.33	0.21
Length of growing season (days)	200	155	60
Incidence of crop failure (average of 3 years)[c]			
– plots with complete crop failure (%)	4	17	33
– plots with partial crop failure (%)	7	24	–
B. Indicators of spatial diversification:			
Scattered land fragments per farm (no.)	2.8	5.8	7.3
Split plots per farm (no.)	5.0	11.2	–
Fragments per farm by distance from village (no.)			
– Up to 0.8 km	1.5	1.4	1.2
– 0.8 to 1.6 km	1.1	3.4	4.3
– Above 1.6 km	0.2	1.0	1.8
C. Indicators of crop-based diversification:			
Extent of intercropping (%)	83	35	100
Total sole crops planted (no.)	20	34	12
Total combinations of mixed crops planted (no.)	43	56	30
D. Crop/livestock-based mixed farming:[d]			
Ratio of land and livestock values	93:7	91:9	63:37
Ratio of crop and livestock incomes	80:24	71:29	69:31
E. Occupational and institutional adjustments:			
Occupations (sources of income) per household (no.)	1.5	2.3	2.9
Households with more than 2 occupations (%)	14	18	39
Households with incidence of seasonal out-migration (%)	2	24	33
Cases of land tenancy induced by risk-sharing/management consideratons (no.)	9	66	–

[a]Table reformulated from Jodha (1967), Binswanger *et al.* (1980), Walker and Jodha (1986), Singh and Walker (1982), Walker *et al.* (1983), and other unpublished data from ICRISAT's village-level studies (Binswanger and Jodha, 1978).

[b]Information for the first three items is based on district-level details of soils and long-term average rainfall of the concerned districts. The rest of the information in the table relates to two villages in each of the districts. Number of sample households was 120 each in Akola and Sholapur. In Jodhpur it was 140.

[c]1975–1976 to 1977–1978 for Akola and Sholapur, 1963–1964 to 1965–1966 for Jodhpur.

[d]Assets (livestock/land) and income valued in terms of prices for 1976–1977 for all regions.

mainly arid tracts where the crop growing season is too short and the frequency of droughts is once in two or three years, and which account for over 18% of the dry tropical region of India.

The three areas represent three different risk environments, and the different farming systems can be considered as regionally differentiated indicators of farmers' adjustment strategies against climate-induced risk. The farmers in the three areas have incorporated varying degrees of diversification in their long-term strategies that depend on the degree of weather risk. Spatial diversification (*Table 4.1*, B) is thus an attempt, through scattered land fragmentation, to cope with the spatial variability of rainfall, even over short distances. Crop-based diversification, through seasonal subplotting of cropped land, intercropping and planting of different crop combinations, is an attempt to make fullest use of crop characteristics to gain protection against intraseasonal variations in the environment (*Table 4.1*, C). In the village studies by ICRISAT, as many as 84 crop combinations through intercropping were recorded in a single village (in this case in the Sholapur region) during one crop season (Jodha, 1980).

Crop- and livestock-based mixed farming, and occupational diversification through adoption of multiple sources of earning (including some having very low payoffs), are other measures used to adjust to high-risk environments. Similarly, the periodic resort to seasonal migration to escape local scarcities, dependence on common property resources (village pastures, forests, etc.) and the use of land tenancy as a risk-sharing device, etc., are other components of farmers' long-term strategies against weather risk. The extent of these practices increases with the increasing degree of weather risk (*Table 4.1*, D and E). Jodhpur, with the highest degree of weather risk, has the highest number of these risk-reducing features incorporated in the farming system. Akola, which has the lowest weather risk, has the smallest number of such features. Sholapur falls between these two.

The only exceptions to the general pattern of diversification observed in the three environments are: (i) the lower frequency of occurrence of intercropping in Sholapur villages, and (ii) the smaller number of crops and crop combinations in Jodhpur villages (*Table 4.1*, C). The former is due to the extensive practice of *rabi* (post-monsoon) cropping in the Sholapur villages. For *rabi* crops, the soil moisture is mainly that which is already in the soil and therefore known at the time of planting. The farmer therefore does not need intercropping as a device to cope with the situation caused by the uncertainty about moisture that arises in the case of rainy season cropping (Jodha, 1980). Regarding the second exception, it may be noted that Jodhpur has a smaller number of crops because only a limited number of crops are well adapted to that arid and sandy area.

4.3. Short-Term Adjustments

The operational features of farming systems reflect the farmer's short-term adjustment strategies against interyear or intrayear variability of rainfall and its interactions with other factors. These include the farmer's varying responses from year to year in terms of choice of crops, type and level of input use and resource management practices.

Such responses vary with the different types of rainfall variability mentioned earlier. The farmer must cope with different quantities of soil moisture at the time of crop planting, and the possibility of either normal rainfall or substantial deficit of rainfall causing drought. The farmer's responses in the event of drought can be further divided into: (i) those directed toward minimizing losses through various salvage operations, resource conservation and cost reductions, and (ii) those directed toward management of losses through curtailment of current commitments, and augmentation of income through various means, including disposal of assets and short-term outmigration for employment. *Tables 4.2* to *4.4* indicate various responses to these different situations.

4.3.1. Adaptation to different moisture situations

Cropping decisions by farmers are responses to the soil moisture availability rather than to the quantity of rainfall. However, since data on soil moisture are often not available, rainfall data are used to provide the context for studying farmers' responses. In the present case, data were available on both soils and daily rainfall recorded at the village level, which helped in quantifying soil moisture situations at different stages of crop growth. These data were collected under ICRISAT's village-level studies in three agroclimatic zones over a period of nine years (1975–1983). The availability of soil moisture was estimated by computer modeling considering precipitation, evapotranspiration and moisture retention capacity of soil (Keig and McAlpine, 1977).

To examine contrasts between the farmer's responses or adjustment measures under different soil moisture situations, two years with significantly different availabilities of soil moisture were selected out of the nine years for each of the villages. *Table 4.2* presents the soil moisture characteristics of the selected years as well as the details of farmers' decisions about cropping, input use, etc., during each of these years. Though moisture availability may not be the sole determinant of farmers' decisions and operations, the latter seem to differ considerably between years with different soil moisture situations. For instance, during the low soil moisture years, compared to high soil moisture years, the per household cropped area declined in all the three regions. The extent of intercropping as an insurance against bad weather also increased during the low soil moisture years in all the areas except the *rabi* (post-rainy) cropping tract of Sholapur. Similarly, the emphasis on short-duration crops and low water-requiring crops also increased during the low soil moisture years. In contrast, the long-duration crops and high moisture-requiring crops were allocated more area during high soil moisture years. The input use and related management practices also change according to moisture availability to crops.

Table 4.2. Changes in cropping decisions in response to soil moisture situation during planting period in the study villages in three districts in the dry tropical regions of India.[a]

	Changes in the proportion of area under crops during the years with: (I) high soil moisture and, (II) low soil moisture, during the planting period in the villages of:					
	Mahbubnagar		*Sholapur*		*Akola*	
	(I)	*(II)*	*(I)*	*(II)*	*(I)*	*(II)*
A. Soil moisture (mm) during 6-week planting period:[b]						
Mean per week	56.3	21.6	129.2 (175.2)[c]	36.0 (58.5)	68.6	22.7
Range during 6 weeks	48–99	21–28	67–150 (150–220)	28–41 (44–77)	63–74	12–59
B. Cropping decisions:[d]						
Cropped area per household (ha)	4.2	3.9	7.5	6.0	5.4	5.9
Proportion of the following in gross cropped area (%)						
– Intercropping	83	85	38	25	94	90
– Crops with low moisture needs[e]	29	36	54	62	21	23
– Crops with high moisture needs[f]	5	3	11	6	48	43
– Long-duration crops[g]	8	7	13	8	16	11
– Short-duration crops[h]	1	4	2	5	3	8

[a] Based on data from ICRISAT's village-level studies conducted during 1975 to 1985. Data relate to 30 farm households in one village in each of the three study districts.

[b] Based on daily rainfall recorded in each of the villages. Weekly soil moisture (cumulative moisture availability in soil) for successive weeks is worked out on the basis of a computer system, which takes into account precipitation, evapotranspiration, soil type, etc. For details, *see* Keig and McAlpine (1977). The rainy season crop planting period coincides with weeks 23 to 28 of the calendar year.

[c] For Sholapur, where the bulk of the area is planted in the post-rainy season, the main crop planting period covers weeks 41 to 46. The relevant soil moisture information for this period is indicated in parentheses.

[d] Refers only to rainfed crops.

[e] Includes castor seed in Mahbubnagar and sorghum in other areas.

[f] Includes groundnut, some other oilseeds and minor pulses; also rainfed wheat in Sholapur and cotton in Akola.

[g] Mainly pigeonpea.

[h] Includes mungbean and other minor pulses planted only when sufficient moisture is available at the beginning of the season.

4.3.2. Adjustment measures to minimize drought losses

Adjustment measures specifically directed against the short-term variability of rainfall can be identified by comparing the choice of particular farm operations during a drought year with that during a normal rain year. We illustrate the situation with relevant data from two villages in the Jodhpur district, which is faced with drought once in two to three years (drought being defined as a 25% or more deficit from normal, *see* Section 1).

Table 4.3 shows the extent of adoption of specific farm practices to minimize the extent of losses following the failure of rain soon after the completion of sowing operations during the year 1963–1964. When it was seen that there was no possibility of harvesting grain from the crops, the farmers concentrated on collecting fodder, fuel, etc. They also minimized the input use on crops, particularly the use of inputs involving out-of-pocket cost (e.g., hired labor). The extent of such measures during the normal rain year (1964–1965) was negligible, as indicated by the number of plots involved. The farmers did not need many of such salvage operations during a normal rain year.

Table 4.3. Loss-minimizing practices adopted by farmers during a drought year and a nondrought year in the study villages in Jodhpur district.[a]

	Situation during:	
Weather and farm operations	*Drought year* (1963–1964)	*Normal year* (1964–1965)
A. Characteristics of weather risk:		
Rainfall during the year (mm)	159[b]	377
Total rainy days (no.)	8	21
B. Coverage of plots by risk/loss-minimizing farm practices (no.):		
Collection of weeded material as fodder	53	5
Harvest of field borders for fodder	68	6
Harvest of premature crops	27	–
Harvest of crop by-product only	49	2
Harvest of mature crop	16	144
Interculturing	7	65
Weeding more than once	18	–
Thinning	37	–
Abandoning of post-sowing operations	36	–
Use of hired resources for post-sowing operations	2	24
Harvest of premature *Z. nummularia* (bush) for fodder	92	–
Lopping of trees for fodder/fuel	53	4

[a]Table based on data from Jodha (1967). The data relate to 160 plots belonging to sample households in two villages, acquired by observation of the plots during the reference years.
[b]Rains received only at the beginning of the crop season.

4.3.3. Management of drought losses

Despite adoption of various salvage operations and cost minimization measures, farmers have to face losses during a drought year. In attempting to live with the situation, they adopt a number of measures, some of which are drastic (e.g., the reduction of food consumption). Others include postponing current commitments, supplementing income through various means, using up past savings, and selling or mortgaging assets. Details of such loss management measures adopted by rural households during drought years are given in *Table 4.4* for two villages, one each from Jodhpur and Sholapur districts. Both of these districts fall into high-risk zones due to the higher frequency of drought years.

Results varied among groups and between districts, but regardless of these variations (*Table 4.4*) the asset position of the sample household deteriorated considerably due to sale or mortgage, or even loss of both productive (e.g., bullocks) and nonproductive assets (i.e., building) during the drought year. Indebtedness also increased substantially. *Table 4.4* also shows that a large proportion of households had to curtail their current commitments by postponing the payment of dues and withdrawing children from school, etc. Data reported elsewhere indicate significant declines in current food consumption by drought-affected households (Jodha, 1978).

Drought-year sustenance income ranged from a quarter to one-half of the normal-year income in most cases. However, the most significant aspect of the sustenance income relates to its sources. Nearly one-fourth to one-half of this income during drought years was realized through employment on public relief works. Sale of assets and earnings during outmigration (included with "others" in *Table 4.4*) were other important contributors to the sustenance income. The adverse impact on the economy and family of the farmer continued for several years.

4.4. Reduced Efficacy of the Strategies

The farmers' adjustment responses to rainfall variability have developed over several generations. More importantly, they have historically evolved in the context of a low-population, subsistence orientation of farming and agricultural practices that were, by nature, resource-extensive. Under the changed circumstances of today, with increased population pressure on land, the increased role of market forces, and a number of side effects of institutional and technological changes, the farmer's traditional adaptation strategies are losing their efficacy (Jodha and Mascarenhas, 1985; Walker and Jodha, 1986). Some examples are given below.

Farm-level diversification through crop- and livestock-based mixed farming is under severe strain because grazing lands, including rural common property resources, are fast shrinking due to the increased cultivation needed by a growing population (Jodha, 1986a). This has also adversely affected seasonal migration as a means to cope with local scarcities during droughts. The practices of on-farm storage and recycling are fast dwindling because of the increased operation

Table 4.4. Measures adopted by rural households to manage drought-induced losses in the study villages in Jodhpur and Sholapur districts.[a]

Details of situation during drought years[b] *(compared to pre-) drought year)*	*Districts and household categories:* Jodhpur			Sholapur		
	Small farm-ers	*Large farm-ers*	*All house-holds*[c]	*Small farm-ers*	*Large farm-ers*	*All house-holds*
A. No. of households	57	23	144	20	20	80
B. Decline in value of (%):						
Productive assets	37	25	37	31	5	16
Nonproductive assets	28	11	31	50	12	31
C. Increase in indebtedness (%)	45	187	192	75	33	63
D. Sustenance income[d]:						
Average per household (in 100 Rs)	26	42	33	25	48	29
Proportion (%) of sustenance income from:						
Crops	8	9	2	9	23	14
Livestock	4	14	10	4	–	1
Relief employment	47	23	25	49	35	47
Sale of assets	17	23	26	15	19	17
Borrowings	7	11	10	8	17	8
Others[e]	23	21	27	15	7	13
E. Proportion (%) of households that:						
Postponed payment of private dues[f]	69	19	49	33	10	25
Withdrew children from school	51	30	42	65	10	54
Affected by outmigration	48	18	37	20	5	19

[a]Source: Jodha (1978). The number of villages is two in each district.
[b]Drought years were 1963–1964 and 1972–1973 in the case of Jodhpur and Sholapur districts, respectively.
[c]All households include agricultural laborers and medium farmers besides small and large farmers. In the case of Jodhpur villages they also include artisans and other occupational groups.
[d]Sustenance income is defined as total inflow of cash and kind including borrowing, except long-term loans unrelated to sustenance during drought year. Value of sustenance income is expressed in terms of 1972–1973 prices.
[e]Includes income from other casual/agricultural employment, handicrafts, transport, remittance, free help from well-off relations, etc.
[f]The nonpayment of official dues (such as the land revenue, cooperative credit repayment installments, etc.) is excluded here because the government postponed recovery of such dues.

of market forces, which help to syphon off rural supplies, but often do not respond to rural scarcities during drought years in dry regions (Jodha, 1975). Collective risk-sharing arrangements have weakened due to rural factionalism, which has increased as a side effect of introduction of democracy at the village level (Jodha, 1975).

A major consequence of the reduced efficacy of the farmers' adjustment strategies has been increasing dependence on public relief during drought years. However the relief measures, until recently when they became more production-oriented, largely replaced rather than strengthened farmers' self-help strategies.

4.5. Complementary Roles of Technology and Public Interventions

Public relief, although an important welfare measure, cannot be a permanent solution to the problems of the drought-prone areas and their people. To avoid perpetual dependence on relief, and to prevent the latter from becoming an ever-increasing drain on public funds, it is essential that farmers' own adjustments be strengthened. Public measures can play a complementary role in this. As a result of the reorientation of public interventions in dry areas, and the generation of more productive options by new technologies, there is now the potential for developing an effective complementary role for public measures in strengthening farmers' coping strategies. *Table 4.5* summarizes the relevant details.

In the first place, improved understanding of rainfall probabilities, and the possibility of extending cropping seasons in specific soil and rainfall situations, can help generate more information for the farmer to plan his cropping strategies in both long- and short-term contexts. This can be achieved through the quantification of agroclimatic resources, illustrated in Section 2. Thus, what farmers have attempted by trial and error in the past can now be achieved using a more scientific approach.

Improved knowledge about climate and soils can be used more effectively with the the new technologies. For instance, the crop-centered technologies have offered new cultivars (e.g., hybrid sorghum), which are characterized by short-duration, high-yield potential and responsiveness to high input and management. They offer scope for high production during good rain years. Likewise, the enhanced growing season due to moisture conservation and water-harvesting techniques, coupled with short-duration crop varieties, offer opportunities for higher cropping intensity during good rain years (Jodha, 1986b). Use of the high production potential of good rain years can help cushion the impact of drought period scarcities, and thereby strengthen farmers' long-term preparedness against drought. This can be complemented by strengthening the farmers' resource base through emphasis on production-oriented components in drought relief interventions. Bunding, the reshaping of land on a watershed basis, and the creation of farm ponds, minor irrigation facilities, etc., are examples of this. In addition, infrastructural development, including social forestry, pasture development, percolation tanks, road networks, markets and support services, as envisaged under the development orientation of public interventions in drought-prone areas, may also indirectly strengthen farmers' capacities to withstand droughts. These measures can help them integrate with the mainstream of the rural development process in the country.

Table 4.5. Indicators of the potential role of technological and institutional intervention measures in strengthening farmers' adaptation strategies against rainfall variability.[a]

Key areas of adaptations	*Better understanding of environment: rainfall probability, crop–water–soil interactions, etc.*	*More productive technology options*				*Public interventions*			
		Crop varieties with:		*Moisture conservation for:*				*Development orientation*	
		Short duration	*High input response*	*Enhanced growing season*	*Increased usable moisture*	*Productive components*	*Post-drought rehabilitation*	*Infrastructure*	*Support services (input, etc.)*
A. Long-term preparedness through:									
- better use of good rain years to cushion impact of drought years	√	√	√	√	√			√	
- strengthening the resource base for higher, more stable productivity of farm						√			√
- reliance on market network, community assets								√	√
B. Loss minimization during low rainfall drought years through:									
- choice of short-duration crops, mid-season correction	√	√			√				√
- effective use of moisture	√	√			√				√
C. Drought loss management: through:									
- supplementary earning						√			
- sustenance relief						√	√	√	
- post-drought recovery							√		

[a]The actual extent of potential benefits indicated vary in different regions depending on the agroclimatic conditions, which partly determine the effectiveness of technological and institutional interventions.

The loss management or salvage operations undertaken by the farmer during low rainfall years or droughts, caused by late arrival or early termination of rains, can also be strengthened by new technologies. For example, new short-duration, photoperiod-insensitive cultivars and moisture conservation techniques can help ensure effective availability of moisture for crops. The different components of this new technology can enhance the flexibility of farming systems to withstand the intrayear fluctuations in precipitation. In addition, improved support services, particularly the provision of seed and other input supplies can complement the mid-season corrections that farmers make by switching to short-duration crops.

Current public relief measures, which, unlike those in the past, give more attention to post-drought rehabilitation, offer qualitatively different support to farmers' own loss management efforts during drought and the recovery in the post-drought period. For example, food-for-work and employment guarantee schemes can help particularly the rural poor to face drought and post-drought situations, enabling them to avoid more drastic measures, such as parting with their productive assets and thereby damaging their own productive capacity during good rain years.

SECTION 5

Conclusions and Implications

5.1. The Scope of this Study

This study relates to the dry tropical regions of India. These regions, which are composed of 151 arid and semi-arid tropical districts with limited possibility of irrigation, constitute 53% of India's total geographical area. Around 30% of India's rural population live in these areas, which have over 55% of the cropped area and 74% of the area of coarse grains in India. The bulk of the nationally (as distinct from regionally) important crops, such as pulses, oilseeds, and fiber (e.g., cotton) crops, are produced here.

The major problem of the seasonally dry tropical regions is interyear and intrayear variability of rainfall. Farmers have adapted to this by building diversity and flexibility into their farming systems. Policy makers and planners have also attempted to reduce the instability of dryland agriculture and its impact on the rest of the economy. However, the measures adopted by them have often in the past been designed to protect against drought rather than to respond to both negative (drought) and positive (high rainfall) aspects of rainfall variability.

Both farm-level adjustments and public interventions have imparted some degree of stability to dryland agriculture. However, they have both evolved through a process of trial and error, and without much input from modern science and technology, and have thus not fully exploited the potential within the agricultural environment for stability and growth in dryland agriculture. More recently, advances in science and technology have provided new methods and data enabling improved understanding of the agroclimatic, agrobiological and other environmental variables affecting agriculture, and these have demonstrated the extent to which it is possible to obtain higher and more stable crop production.

The opportunity offered by combining traditional and conventional approaches has been enhanced by recent changes in the perspectives of public intervention measures. Public measures for combating droughts have placed greater emphasis on the incorporation of productive components in relief measures. In area-based development strategies for dryland agriculture, complementary use of institutional, infrastructural and technological components is also increasingly emphasized. The present study demonstrated some approaches to help the new public policies and programs for dryland agriculture that reflect these changes in emphasis.

In the determination of what would be included in this case study, the availability of relevant material was a prime consideration. Hence, the experimental results presented in different sections of the study do not all relate to the same locations and periods. However, the locations are in the same broad agroclimatic zones and are closely related.

In Section 1 we concluded that the drought relief interventions, which originally were mainly short-term and purely welfare-oriented *ad hoc* activities, have recently acquired productivity orientation and long-term focus. Some form of integrated development strategy for drought-prone areas is taking shape, combining technological, institutional and infrastructural components. We also identified specific components that could be enhanced by the inclusion of scientific approaches and information discussed elsewhere in the case study. One of these approaches is a method of identifying agroclimatically homogeneous tracts to facilitate area-specific intervention measures and assessing the effects of rainfall variability on crop yields in these areas (*see* Section 2). In Section 3, we considered methods for assessing the dependability of rainfall for quantifying the interacting effects of the different variables that influence crop performance in various soil moisture situations. Section 4 evaluated conventional long-term and short-term adjustment strategies in coping with different levels of rainfall variability, and considered the opportunities by combining them with new agroclimatic knowledge and techniques, and with modern public intervention measures.

5.2. Summary of Results

The variability of agricultural production in India is directly related to the interannual variability of rainfall received in different parts of the country. A method of assessing the impact of rainfall variability on crop yields was demonstrated with respect to sorghum in a semi-arid tropical subregion in Karnataka State in South India (Section 2). The mean annual rainfall of the area is 710 mm and the standard deviation is 180 mm. Rainfall variability for two data periods, 1918–1941 and 1941–1970, showed that the frequency of droughts was greater in the earlier than in the later period. In the years 1941–1980, only two consecutive drought years (1971 and 1972) were identified. Homogeneous climatic clusters or cropping zones were identified, and the possibility of predicting crop performance in advance on the basis of the location and intensity of the inter-tropical convergence zone (ITCZ) was also considered. The rather large variation in the distribution of the seasonal rainfall seems to have led the farmers to devote a huge proportion of their annual crop production to post-rainy season cropping when the crops are raised on residual moisture. This practice, however, has led to increased soil erosion as bare lands are exposed during the rainy season.

Section 3 considered the variations in crop yield (sorghum) that occurred under various climatic scenarios. In areas where the annual rainfall exceeds 750 mm and the available moisture storage capacity of soils exceeded 150 mm, the cultivation of two crops, one in the rainy season and a sequential crop in the post-rainy season, is feasible. The study identified the areas in which it would be worthwhile to collect surface runoff water for subsequent use for protective irrigation to ensure crop survival during the dry season. Simulation of the response of sorghum to supplemental water provided at different growth stages during post-rainy seasons of 1974 to 1985 at ICRISAT Center, indicated that irrigations of 50 mm each at sowing and anthesis would increase sorghum yield in poor years by 136%, in average years by 38%, and in good years by 17%. With regard to possible impacts of climatic change on sorghum yield, the simulation results using 1974–1985 data from ICRISAT Center suggested that lower temperatures (resulting in longer duration), in association with higher intercepted radiation, would increase yields in the rainy season. The reverse was true in the post-rainy season, when the crop is raised on stored soil moisture.

The long-term adaptation of the farmers is manifested in agricultural diversification through adoption of mixed-farming and mixed-cropping systems, systems of self-provisioning and recycling, and combinations involving farm and off-farm enterprises (Section 4). The relative investment in different "drought proofing" mechanisms varies with the extent and nature of climatic variability. For example, in areas with highly variable rainfall, the dependence of the farmer on livestock and off-farm work is much greater than it is in areas with relatively stable rainfall. The extent of diversification, including having several plots some distance apart, also increases with increased variability and reduced dependability of rainfall.

In the short term, the responses of farmers to climatic variability involve adjustments of plant population, changes in levels of input such as the application of fertilizers and variations in other agronomic practices. The farmer's adjustments to severe drought situations often take the form of drastic actions, including curtailment of essential consumption, and the disposal of productive assets.

Section 4 makes a strong plea for dovetailing the application of technology with public interventions and with farmers' traditional responses to climatic variability. The public interventions should be more in the nature of permanent drought-relieving assets rather than short-term measures to alleviate distress. Traditional adjustment mechanisms are losing their effectiveness in the face of rising population pressure and other forces of change. These adjustments can be modernized and strengthened with the help of new technological developments that have the potential for higher productivity under the same soil–rainfall conditions.

5.3. Policy and Planning Implications

Table 5.1 summarizes conclusions of the case study and their operational utility. We conclude here by emphasizing some of the issues affecting their implementation, focusing, first, on the prerequisites for fuller use of the approaches and, secondly, on how the findings relate to specific components of the emerging integrated public interventions in drought-prone areas.

Table 5.1. Major findings of this study and their operational utility.

Major findings/ inferences	*Areas of operational utility*
Section 2. Effects of monsoon variability on agriculture	
Methods for analyzing spatial and temporal variability of rainfall	Screening of environments based on rainfall amount and distribution
Concept of climatic clusters of different Indian regions	Identifying homogeneous agroclimatic zones
Correspondence between climatic clusters and cropping zones.	Agricultural zoning based on integrated use of climatic and crop-related variables
Relating climatic fluctuations to crop yield variations	Improved understanding of climatic basis of agricultural instability
Preliminary attempts to predict performance of crops in different zones on the basis of location and intensity of ITCZ influencing the patterns of annual rainfall in different regions	Advance planning of seasonal cropping strategies as well as relief operations for different regions
Section 3. Effects of variations in climate and soil water on agricultural productivity	
Methods for analyzing spatial as well as temporal variability or dependability of rainfall, amount of runoff, and soil water	Screening of environments for stable arable farming; for potential runoff, its collection, and recycling and intensive cropping
Probability of rainfall on short-term basis	Clustering of analogues for area-based crop planning for development and diffusion of improved farming systems techniques
Simulation of crop yield probabilities based on soil, crop, weather and management data	Demarcation of high potential/low potential areas for different crops/cropping systems within the dry tropical regions
Areas with >750 mm rainfall and soils with storage capacity >150 mm found suitable for double cropping	Identification of high potential areas for intensive cropping
Low temperature in combination with higher radiation intercepted by cultivars increases and stabilizes crop yield in rainy season	Understanding of yield-influencing interactions between variables relating to microenvironment, crop and management
Section 4. Village-level farm adjustment to rainfall variability	
Farm-level diversity of agriculture is positively associated with long-term variability of rainfall	Understanding of regionally differentiated farming systems to help area-focused interventions in drought-prone agriculture

Table 5.1 continued.

Major findings/ inferences	*Areas of operational utility*
Section 4. Village-level farm adjustment to rainfall variability	
Interyear variations in rainfall (amount and distribution) induce farm-level adjustments through changes in cropping patterns, input use, etc.	Understanding need for and direction of introducing greater flexibility in farmers' cropping strategies through new technologies for adjustment to rainfall variability
Mid-season failure of rains gives rise to a number of salvage activities to minimize losses at farm level	Suggests the need and scope for more and better options through new technologies (i.e., short-duration, photoperiod-insensitive, drought-resistant cultivars) to strengthen farmers' adjustment to intraseasonal variability of rainfall
To combat impact of drought, rural households undertake quite drastic steps (including sale of productive assets and curtailment in consumption) with long-term production and welfare implications	Highlights the need for complementing farmers' adjustment measures by public intervention during droughts

5.3.1. Fuller use of the approaches

The studies reported here improve our understanding of the agroclimatic and socioeconomic environment of dry tropical areas and indicate the broad directions of interventions for growth and stability of agriculture in these areas. However, they relate to limited areas and crops, and it is important to extend such studies, to cover larger areas and a greater variety of different tracts/crops/farming systems in the dry tropics.

The logical and practical means to facilitate the above approaches require the development both of hardware and of software, as well as an expanded data base including as many locations as possible. The prospects for achieving this now seem quite encouraging. The government is giving major emphasis to computerization of agricultural information systems; fairly long-term rainfall data for several locations are now available; there is fairly disaggregated information on soils in different regions; and there are data on crop- and resource-centered technologies available with central and regional (as well as international) research organizations; and a sizeable volume of information from farm surveys. The crucial missing factor is the appropriately oriented manpower to use these facilities and approaches.

There is an absence of any mechanism for integrating the findings of scientific studies, such as this one, with the actual planning processes in drought-prone areas. Policy makers and planners at different levels need to be made aware of the utility and effectiveness of new approaches, particularly by demonstration through well-conducted pilot projects.

Table 5.2. Findings of this study and their potential relationship with specific features of public interventions in drought-prone areas.

Findings (section numbers in parentheses)	Intervention programs with their specific features[a]													
	Drought Prone Area Program (DPAP)					Technological transformation (AICRPDA/CRIDA etc.)						Drought relief programs		
	Area approach to integrated development	*Focus on new technology*	*Use of watershed concept*	*Restoration of ecological balance*	*Institutional infra-structural components*	*Climate as a source of growth*	*Integration of crop-centered and resource-centered technologies*	*Concern for location specificity of technology*	*Concern for farmer circumstances*	*Focus on watershed concept*	*Resistance to yield reducers*	*Integration of productive components*	*Emphasis on new technology*	*Relief as a long-term activity*
Basis for:														
Demarcation of homogeneous cropping zones (2)	√				√							√		√
Guiding spatial planning (2,3)	√		√	√	√	√						√		
Identification of areas for:														
Dependable cropping (2,3)	√							√						
Double cropping (2,3)	√		√					√		√		√		
Water harvesting, recycling (3)	√			√		√	√	√		√		√	√	√
Quantification of:														
Role of environmental factors in crop performance (3)	√	√		√			√	√			√		√	
Soil moisture–crop interactions (3)						√	√	√			√		√	
Soil moisture availability pattern (3)						√		√		√	√		√	
Improved technology and:														
Levels of agricultural risk (3,4)		√							√		√			
Role of farmer's resource position (4)			√	√			√		√		√			
Role of management factor (3)							√		√	√				
Farmer's risk management and:														
Additional options through relief programs (4)											√			√
New technology (4)		√							√		√		√	
Institutional changes (4)					√								√	

[a]For details of public interventions and their features *see* Section 1.

5.3.2. Relating the findings of the studies to public intervention measures

One way of making policy-makers aware of the value of the approaches discussed here is systematically to highlight the potential links between the latter and the specific components of the current development strategies for drought-prone areas. A number of these linkages are indicated in *Table 5.2.* This clearly demonstrates the considerable scope that exists for strengthening drought-combating strategies through the use of improved knowledge about the physical, agrobiological and socioeconomic aspects of drought-prone environments.

Acknowledgments

Thanks are due to J.G. Sekaran for assistance in analyzing the data, M. Shankaraiah and M. Ashok Reddy for cartographic work and K.P.C.S. Raju for typing the manuscript.

References

Anonymous (1972). *Report of the Irrigation Commission, Volume 1.* Ministry of Irrigation and Power, Government of India, New Delhi.

Anonymous (1973). *Report of the Fact Finding Committee for Survey of Scarcity Areas.* Maharashtra State, Government of Maharashtra, Bombay.

Anonymous (1984). *Report on Area Development Programs.* Report of the Working Group, Ministry of Rural Development, Government of India, New Delhi.

Arkin, G.F., Vanderlip, R.L. and Ritchie, J.T. (1976). A dynamic grain sorghum growth model. *Transactions of American Society of Agricultural Engineers*, **19**, 622–626.

Bhatia, B.M. (1967). *Famines in India: A Study in Some Aspects of the Economic History of India (1860–1965).* Asia Publishing House, Bombay, India.

Binswanger, H.P. and Jodha, N.S. (1978). *Manual of Instruction for Economics Investigators in ICRISAT's Village Level Studies.* Economics Program, ICRISAT, Patancheru, A.P., India.

Binswanger, H.P., Virmani, S.M. and Kampen, J. (1980). Farming system components for selected areas in India: evidence from ICRISAT. *Res. Bull. 2*, ICRISAT, Patancheru, A.P., India.

Curry, B. (1983). *Coping with Complexity in Food Crisis Management.* Food Crisis Management Group, School of Social Sciences, Flinders University of South Australia, Australia.

Eastin, J.D. (1972). Photosynthesis and translocation in relation to plant development. In N.G.P. Rao and L.R. House (eds.) *Sorghum in the Seventies.* Oxford and IBH, Bombay, India, pp. 214–246..

Gadgil, S. and Joshi, N.V. (1983). Climatic clusters of the Indian region. *J. Climatol.*, **3**, 47–63.

Gallagher, J.N. and Biscoe, P.V. (1978). Radiation absorption, growth and yield of cereals. *J. Agric. Sci., Camb.*, **91**, 47–60.

Huda, A.K.S., Sivakumar, M.V.K. and Rego, T.J. (1985). An analysis of agroclimatic data in relation to fertilizer use in dryland farming areas of India. *Fertilizer News*, **30**(2), 35–41.

Huda, A.K.S., Sivakumar, M.V.K., Virmani, S.M., Seetharama, N., Singh, S. and Sekaran, J.G. (1984). *Modeling the Effect of Environmental Factors on Sorghum Growth and Development.* Proceedings of the International Symposium on Agrometeorology of Sorghum and Millet in the Semi-Arid Tropics, 15–20 November 1982, ICRISAT, Patancheru, A.P., India, pp. 277–287.

ICRISAT (1985). *Annual Report 1984*, International Crops Research Institute for the Semi-Arid Tropics, Patancheru, A.P., India.

Jaiswal, N.K. and Kolte, N.V. (1981). *Development of Drought Prone Areas.* National Institute of Rural Development, Hyderabad, A.P., India.

Jodha, N.S. (1967). *Capital Formation in Arid Agriculture: A Study of Resource Conservation and Reclamation Measures Applied to Arid Agriculture.* Ph.D. Thesis, University of Jodhpur, Jodhpur, India.

Jodha, N.S. (1975). Famine and famine policies: some empirical evidence. *Econ. Pol. Weekly*, **10**(41), 1609–1623.

Jodha, N.S. (1978). Effectiveness of farmers' adjustment to risk. *Econ. Pol. Weekly (Quarterly Review of Agriculture)*, **13**(25), A38–A48.

Jodha, N.S. (1979). Dry farming technology: achievements and obstacles. In C.H. Shah and C.N. Vakil (eds.) *Agricultural Development of India: Policy and Problems*. Orient Longman, Bombay, India.

Jodha, N.S. (1980). Intercropping in traditional farming systems. *J. Development Studies*, **16**(4), 427–442.

Jodha, N.S. (1983). *Dry Farming Research: Issues and Approaches*. Paper presented at ISAE/ICRISAT/AICRPDA workshop on Technology Options and Economic Policy for Dryland Agriculture: Potential and Challenges, 22–24 August. ICRISAT, Patancheru, A.P., India.

Jodha, N.S. (1984). *Development Strategy for Rainfed Agriculture: Possibilities and Constraints*. Paper presented at Silver Jubilee National Seminar at the Institute of Economic Growth, New Delhi, 27–30 April, India.

Jodha, N.S. (1986a). Common property resources and rural poor in dry regions of India. *Econ. Pol. Weekly*, **21**(27), 1169–1181.

Jodha, N.S. (1986b). Research and technology of dryland farming in India: some issues for future strategy. *Indian J. Agricul. Economics*, **41**(3).

Jodha, N.S. and Mascarenhas, A.C. (1985). Adjustment in self-provisioning societies. In R.W. Kates, J.H. Ausubel and M. Berberian (eds.) *Climate Impact Assessment: Studies of the Interaction of Climate and Society*. John Wiley and Sons, New York, pp. 437–464.

Jodha, N.S., Asokan, M. and Ryan, J.G. (1977). Village study methodology and resource endowments of the selected villages. *Economic Program Occasional Paper-16*, ICRISAT, Patancheru, A.P., India.

Kanitkar, N.V., Sirur, S.S. and Gokhale, D.H. (1960). *Dry Farming in India*. Indian Council of Agricultural Research, New Delhi, India.

Keig, G. and McAlpine, J.R. (1977). *A Computer System for the Estimation and Analysis of Soil Moisture Regimes from Simple Climatic Data*. (2nd ed.) Techmemo, Division of Land Research, CSIRO, Australia.

Krantz, B.A. (1981). *Intercropping in an Operational Scale in an Improved Farming System*. Proceedings of the International Workshop on Intercropping, 10–13 January 1979, ICRISAT, Patancheru, A.P., India, pp. 318–327.

Lorenz, E.N. (1956). Empirical orthogonal functions and statistical weather prediction. *Sci. Rep. No. 1*, Dept. of Meteorology, Massachusetts Institute of Technology, Cambridge, MA, 49 pp.

Mooley, D.A. and Parthasarathy, B. (1984). Fluctuations in all India summer monsoon during 1871–1978. *Climatic Change*, **6**, 287–301.

Morris, D.M. (1975). What is a famine? *Econ. Pol. Weekly*, **9**(44), 1855–1864.

Parthasarathy, B. (1984). Interannual and long-term variability of summer monsoon rainfall. Proceedings of Indian Academic Sciences. *Earth Planet Sci.*, **93**, 371–384.

PEO (1964). *Study of Soil Conservation Program for Agricultural Land*. Program Evaluation Organization, Planning Commission, Government of India, New Delhi.

Planning Commission (1973). *Integrated Agricultural Development in Drought Prone Areas*. Report of Task Force on Integrated Rural Development, Government of India, New Delhi, India.

Raman, C.R.V. and Srinivasa Murthy, B. (1971). Water availability periods for crop planning. *Scientific Report No. 173*, India Meteorological Department, Pune, India.

Rangaswami, A. (1974). Financing famine relief: calling a bluff. *Econ. Pol. Weekly*, **9**(45 and 46), 1885–1888.

Ritchie, J.T. (1972). Model for predicting evaporation from a row crop with incomplete cover. *Water Resource Research*, **8**, 1204–1213.

Robertson, G.W. (1976). Dry and wet spells. UNDP/FAO, Tun Razak Agriculture Research Center, Sungh: Tekam, Malaysia, Project Field Report, *Agrometeorol. A-6*, 15 pp.

Sen, A. (1981). *Poverty and Famines: An Essay on Entitlement and Deprivation.* Clarendon Press, Oxford, UK.

Singh, R.P. and Das, S.K. (1984). Nutrient management in drylands with special reference to cropping systems and semi-arid red soils. *Project Bull. No. 8*, All India Coordinated Research Project for Dryland Agriculture, Hyderabad, India, 56 pp.

Singh, R.P. and Walker, T.S. (1982). Determinants and implications of crop failure in the semi-arid tropics of India. *Economics Program Progress Report-40*, ICRISAT, Patancheru, A.P., India.

Slater, L.E. (1981). Dimensions of the world food and climate problem. In W. Bach, J. Pankrath and S.H. Schneider (eds.) *Food–Climate Interactions.* D. Reidel Publishing Company, Dordrecht, Netherlands.

Spitz, P. (1980). Drought and self-provisioning. In J. Ausubel, and A.K. Biswas (eds.) *Climatic Constraints and Human Activities.* Pergamon Press, Oxford, UK.

Stapper, M. and Arkin, G.F. (1980). CORNF: A dynamic growth and development model for maize [*Zea Mays (L.)*]. *Program and Model Documentation No. 80-2*, Texas Agricultural Experiment Station, TX.

Stinson, D.L. (1979). *A Simulation-Based Soil and Water Resource Evaluation for Ratoon Cropping Grain Sorghum in the Central Blacklands of Texas.* M.Sc. Thesis, Texas AM University, College Station, TX.

Swaminathan, M.S. (1983). *Science and the Conquest of Hunger.* Concept Publishing Company, New Delhi, India.

Swindale, L.D., Virmani, S.M. and Sivakumar, M.V.K. (1981). Climatic variability and crop yields in the semi-arid tropics. In W. Bach, J. Pankrath and S.H. Schneider (eds.) *Food–Climate Interactions.* D. Reidel Publishing Company, Dordrecht, Netherlands.

Tandon, H.L.S. (1981). Research and development of fertilizer use in dryland agriculture. *Fertilizer News*, **26**(6), 25–34.

Torry, W.T. (1986). Drought and the government–village emergency food distribution system in India. *Human Organization*, **45**(1), 11–23.

Virmani, S.M., Sivakumar, M.V.K. and Reddy, S.J. (1982). Rainfall probability estimates for selected locations in semi-arid India. *Res. Bull. No. 1* (2nd enlarged edition), ICRISAT, Patancheru, A.P., India.

Walker, T.S. and Jodha, N.S. (1986). How small farm households adapt to risk. In P. Hazell, C. Pomareda and A. Valdes (eds.) *Crop Insurance for Agricultural Development: Issues and Experiences.* Johns Hopkins University Press, Baltimore, MD.

Walker, T.S., Singh, R.P., Asokan, M. and Binswanger, H.P. (1983). Fluctuations in income in three villages of peninsular India. *Economics Program Progress Report-57.* ICRISAT, Patancheru, A.P., India.

Williams, J.R. and Hann, R.W. Jr. (1978). Optimal operation of large agricultural watersheds with water quality constraints. *TR-96*, Texas Water Resources Institute, Texas A & M University, College Station, TX.

PART VI

The Effects of Climatic Variations on Agriculture in the Semi-Arid Zone of the European USSR

Contents, Part VI

B. THE SARATOV REGION

List of Contributors

NIKONOV, Acad. A.A.
President
Lenin All-Union Academy of
Agricultural Science
21 Bolshoi Kharitonievsky
Pereulok
Moscow 107814
USSR

PETROVA, Dr. L.N.
Stavropol Research Institute
of Agriculture (CHUUCX)
Shpakovskij Region
356200 Stavropolskij Kraj
USSR

STOLYAROVA, Dr. H.M.
Computing Center of the
USSR Academy of Sciences
Vavilova Str. 40
117333 Moscow
USSR

LEBEDEV, Dr. V.Yu.
Computing Center of the
USSR Academy of Sciences
Vavilova Str. 40
117333 Moscow
USSR

SIPTITS, Dr. S.O.
All-Union Research Planning and
Technological Institute of Cybernetics
Moscow
USSR

MILYUTIN, Dr. N.N.
All-Union Research Planning and
Technological Institute of Cybernetics
Moscow
USSR

KONIJN, Dr. N.T.
Department of Soil Science and Geology
University of Wageningen
PO Box 37
NL 6700 AA Wageningen
The Netherlands

PITOVRANOV, Dr. S.E.
International Institute for
Applied Systems Analysis
Schlossplatz 1
A-2361 Laxenburg
Austria

MAXIMOV, Dr. A.D.
Institute of Economics, Labor
and Management in Agriculture
Moscow
USSR

SIROTENKO, Dr. O.D.
All-Union Hydro-Meteorological
Research Institute
12 Pavlik Morozov Str.
123376 Moscow
USSR

ABASHINA, Dr. E.V.
All-Union Hydro-Meteorological
Research Institute
12 Pavlik Morozov Str.
123376 Moscow
USSR

PAVLOVA, Dr. V.N.
All-Union Hydro-Meteorological
Research Institute
12 Pavlik Morozov Str.
123376 Moscow
USSR

CARTER, Mr. T.R.
Atmospheric Impacts Research Group
Department of Geography
University of Birmingham
PO Box 363
Birmingham B15 2TT
United Kingdom
and
International Institute for
Applied Systems Analysis
Schlossplatz 1
A-2361 Laxenburg
Austria

Abstract of Part VI

A. The Stavropol Territory

The Stavropol territory is a large agricultural region in the southern European USSR, situated between the Black Sea and Caspian Sea and to the north of the Caucasus mountains. The climate over much of the territory is characterized by unpredictable precipitation and quite frequent droughts presenting a considerable hindrance to agricultural development in the region.

This study has three main aims:

(1) To evaluate the present-day sensitivity of crop production to climatic variations in the Stavropol territory.
(2) To assess the likely effects on crop production of a shift to a drier or to a wetter climatic regime.
(3) To examine a range of technological and management adjustments that could be implemented to achieve the multiple objectives of maximized total production, maximized stability of production and minimized expenditures under different climatic scenarios.

The climate of the region is classified into four characteristic types of weather-year (WT) on the basis of the hydrothermal coefficient (HTC): WT1 is cool and wet, WT2 is normal, WT3 is dry and WT4 is very dry. The frequency of occurrence of these types in the territory as a whole over the 11-year period 1970–1980 is used to describe the present-day (baseline) climate. By substituting dry year-types for wet year-types in the baseline scenario, or *vice versa*, four climatic scenarios (S) are identified: S1 (baseline climate), S2 (drier climate), S3 (driest climate) and S4 (wetter climate).

The study involves the linking of two types of mathematical model: a crop production model for estimating winter wheat yields as a function of climate, soil and nutrient supply; and a crop allocation model for calculating the optimum area given over to a certain crop rotation to fulfill specific production objectives for a given technology, while minimizing the total expenditures on elements of that technology.

Results indicate that the use of fallow preceding a winter wheat crop is the most effective way of achieving production targets in the drier agricultural zones of the territory. Following a change in climate toward a yet drier regime, increased fallow allocation is one, obvious, farming adjustment. Alternatively, in those regimes where the area of fallow is already close to a maximum, another strategy for maintaining production levels is to use improved technologies, such as more intensive fertilizer applications.

This option involves increased expenditures, however, and its effectiveness may be limited in the absence of sufficient moisture in the soil.

The results, while preliminary, suggest that a modeling approach such as this, if developed further, could serve as a useful tool for agricultural planning in the Stavropol territory and in other semi-arid regions of the Soviet Union.

B. The Saratov Region

The Saratov region lies about 800 km to the north of the Stavropol territory, and marks the transition zone between a cold forest climate to the north and a steppe climate to the south. As in the Stavropol territory, agricultural production in the Saratov region is highly variable, due in large measure to the effects of frequent and unpredictable droughts.

The two main aims of this study are:

(1) To investigate the sensitivity of grain production to present-day climatic variations in the Saratov region, and to estimate the likely effects of longer-term climatic change on crop productivity.
(2) To explore the potential benefits of accurate seasonal weather forecasts both for scheduling farm level activities and for adjusting regional crop allocation patterns.

Two approaches are adopted to tackle these problems. In Section 6, a dynamic crop–weather model is used to provide estimates of spring wheat yields at the Ershov location in the dry steppe zone of the region. Spring wheat is the major grain crop grown in this zone, although it is near its southern limit in the Saratov region (replaced by winter-sown cereals to the south). In Section 7, an optimization approach is used to compute the allocation of crops that is most appropriate for meeting production targets under a range of climatic conditions.

In both sections, climatic variability is characterized by a set of representative weather-years selected from the historical climatological record as “climatic scenarios”. The effects of scenario weather conditions on spring wheat yields and on crop production for a given regional allocation have been calculated with, as a reference, the effects of computed average or “baseline” conditions.

Results indicate that spring wheat yields are reduced significantly by below-average seasonal precipitation, and that this effect is more pronounced if low precipitation is combined with above-average seasonal temperatures. However, the regional optimization model results suggest that, while regional production is reduced markedly during drought conditions, there is considerable scope for improving the level of production (by perhaps more than double) in dry years, by reallocating certain areas to more drought-resistant crops. Moreover, the availability of accurate seasonal weather forecasts would provide opportunities both for scheduling the best use of scarce water supplies for irrigation, and for planning the crop allocation that optimizes production under the forecasted climate.

A. The Stavropol Territory

SECTION 1

Introduction

1.1. The Effects of Drought on Agriculture in the European Part of the Soviet Union

In the European part of the Soviet Union, large areas are characterized by a climate that is unfavorable for agricultural development. In the north of the area, as reported in the companion to this volume (Pitovranov *et al.*, 1988), temperature is a limiting factor for many crops. In the south, one of the most detrimental factors is drought.

It is difficult to define the boundaries of areas where drought is a recurrent problem rather than an infrequent, seasonal anomaly. However, in the regions of interest in this study, to the north of the Caucasus mountains and adjacent to the Caspian Sea, droughts are a characteristic feature of the climate. Each decade has three to five years that can be considered dry, exhibiting various degrees of drought severity. Moreover, two to three dry years often come in succession.

Historically, these unfavorable conditions were so detrimental as to cause major disruption and frequent fatalities. Today, the severity of the droughts has not decreased, and the frequency of their occurrence appears to be increasing. However, the social and economic structure of Soviet society, combined with the current scientific and technical potential within agriculture have made it possible to alleviate and sometimes to overcome the most severe consequences of droughts.

A national long-term agricultural development project has been elaborated and is being put into practice. The project provides for extensive measures to be taken to reduce the vulnerability of agriculture to variability in the natural environment and to establish zones of guaranteed production of major agricultural products. Measures include land reclamation, increased use of chemicals, and the mechanization and electrification of agriculture. Along with the national program, considerable emphasis is placed on the development and introduction

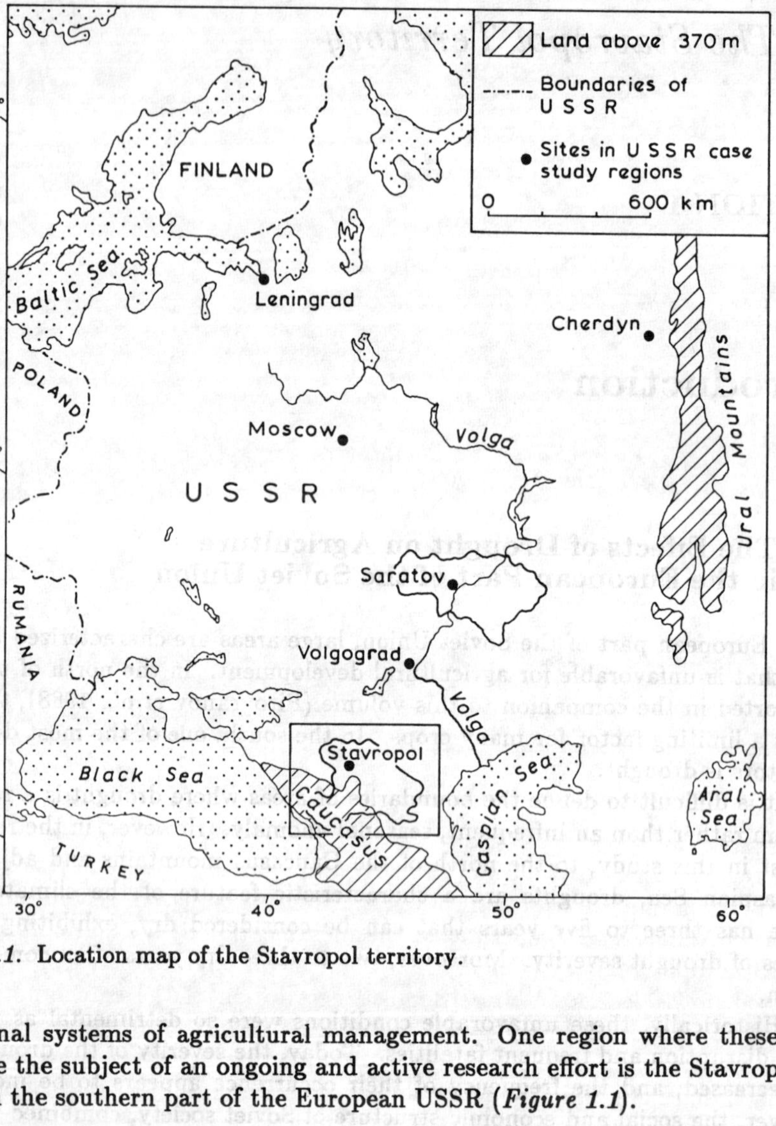

Figure 1.1. Location map of the Stavropol territory.

of regional systems of agricultural management. One region where these questions are the subject of an ongoing and active research effort is the Stavropol territory in the southern part of the European USSR (*Figure 1.1*).

1.2. Characteristics of the Stavropol Territory

The physiography of the Stavropol territory is extraordinarily diverse. Within a distance of only 300–400 km, there occur sharp contrasts in geomorphological, soil and climatic conditions – from the semi-desertic Caspian lowlands to the Caucasus mountains, which include the highest peaks in Europe. The region is thus an appropriate natural laboratory for studying the effects of diverse soil and climatic conditions on agricultural production.

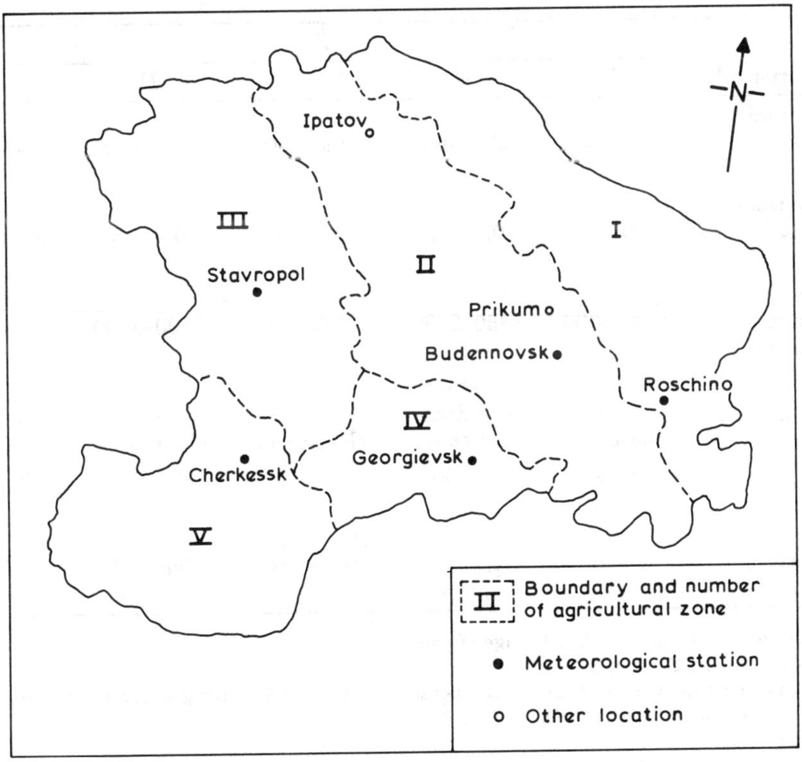

Figure 1.2. Agricultural zones in the Stavropol territory.

Previous detailed studies of soil cover, of climate, of agricultural enterprise types and of the distribution of agricultural production sectors and their efficiency (Nikonov, 1980) permitted the identification of five agricultural zones in the Stavropol territory (*Figure 1.2*). Some of the characteristics of these zones are listed in *Table 1.1*. A more detailed description of the agroclimatic conditions in the territory is provided in Section 2.

More than half of the territory is occupied by dry steppes accounting for the major share of grain production and sheep breeding. In these regions the frequent occurrence of droughts is a characteristic feature of the local climate. Conditions during the winter period are also rather severe. An unreliable and shallow snow cover, and frequent thaws followed by severe frosts (with recorded temperature minima as low as −36° C) create unfavorable conditions for winter crops.

The effect of climate on the principal crop in the region, winter wheat, can be readily detected from observed interannual variations in yield (*Figure 1.3*). As a general rule, a severe drought year in the semi-arid zones of the Soviet Union can be so characterized when crop yields fall below 75% of the long-term average. Thus, of the 29 years depicted in the period 1956–1984, there were four

Table 1.1. Agricultural zones of the Stavropol territory.

Characteristics[a]	*Zone* I	II	III	IV	V
Mean annual precipitation (mm)	260–400	370–450	450–530	480–600	400–1700
Hydrothermal coefficient (HTC)[b]	0.5	0.5–0.9	0.7–1.1	0.9–1.5	above 1.5
Effective accumulated temperatures[c]	3500–3600	3350–3500	3000–3400	2700–3400	1200–2700
Prevailing soil types	light chestnut	chestnut and dark chestnut	chernozem	chernozem	mountain soils
Altitude (m)	0–80	80–500	100–800	200–900	500–3000
Prevailing types of collective and state farms	sheep grain/ sheep	grain/ livestock grain/ sheep	grain/fruit/ livestock	grain/ livestock	livestock

[a]Values represent the geographical range of conditions.
[b]For explanation *see* Section 2.
[c]The annual accumulated sum of air temperatures above 0°C during those days when daily mean air temperature is greater than 10°C.

years of severe drought, including two years when yield reductions of 40% and 42% occurred, respectively. However, both minimum (in dry years) and maximum values of winter wheat grain output (per five-year period) exhibited a consistent increase during this period (*Figure 1.4*).

Two agricultural zones in the region (I and II) account for almost half of the total area and have the greatest share of land under winter wheat. Due to frequent occurrences of drought, interannual yield variability is considerable in these zones compared with the rest of the territory. In years with severe droughts, winter wheat yields in zone I can be as low as 400 to 800 kg/ha and in zone II, as low as 1000 to 1200 kg/ha. In favorable years, the yields may be 3.5 to 7 times larger in zone I and 2 to 3 times larger in zone II, respectively. The coefficient of variation of yields (C_v) averaged over the whole Stavropol territory is about 32%. In the moderately humid zones (IV and V), C_v is about 15–17%, in zone III about 25%, and in zones I and II as high as 52–63%.

Generally, the differences in maximum yields between zones in the Stavropol territory are within a narrow range of 13–26%. In contrast, minimum yield levels in the dry zones are 9 to 15 times less than those in the most favorable zones.

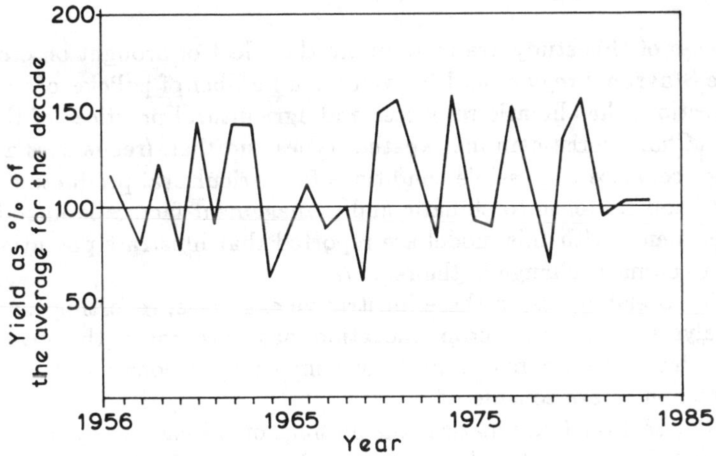

Figure 1.3. Annual winter wheat yields in the region, 1956–1984 (values = percentage of the average yields for the preceding decade).

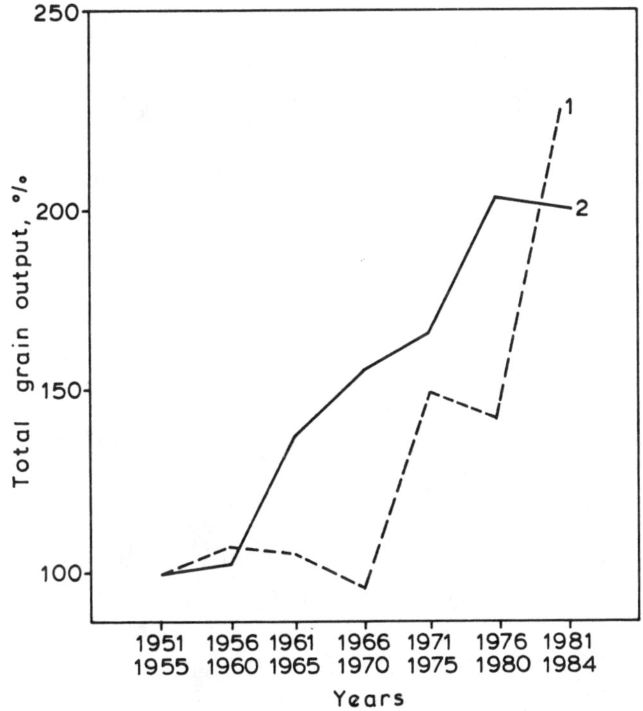

Figure 1.4. Minimum (1) and maximum (2) total winter wheat grain output in the region by five-year period, 1951–1984 (percentages of the 1951–1955 values).

1.3. Aims and approach

The purposes of this study are to estimate the effect of drought on crop production in the Stavropol region, and to evaluate a number of policies of response. In the next section, the climatic resources and agricultural practices of the area are described. Characteristic annual weather types and their frequencies are selected as climatic scenarios of possible conditions for agricultural production. A model of crop and soil response to climatic and management factors is introduced, and some experiments with this model are reported that illustrate potential adaptive strategies to climatic change in the region.

Section 3 elaborates on these illustrative examples, reporting more detailed model analyses of optimal crop allocation patterns under the given climatic scenarios. Possible economic impacts of climatic variations and the likely technological responses are considered.

Finally, Section 4 summarizes the findings of the case study, discusses some implications for policy and makes recommendations for further research.

SECTION 2

Background to the Methods, Models and Scenarios

2.1. Classification of the Agroclimate and Development of the Climatic Scenarios

2.1.1. Droughts and drought classification

In order to characterize the range of severity of moisture conditions in the Stavropol region, a classification of drought has been devised that relates specifically to winter wheat, the most important crop in the area.

The vegetation period of winter wheat extends throughout much of the year. On the average it lasts 277 days, within a range from 267 to 287 days. Droughts may occur in the autumn, spring and summer. During these dry periods the effect of different meteorological parameters upon crop growth varies greatly. In autumn droughts are usually related to insufficient stored soil water rather than to high temperatures. In that period soil water storage can be controlled through the selection of an appropriate preceding crop and the use of an adequate soil management technology.

In the drier regions, the most favorable conditions for winter wheat development are created when it is preceded by fallow. In this case the amount of water stored is approximately double the amount stored after a winter wheat crop has been cultivated. Usually the amount of water stored increases during the winter, although the amount of water available to plants when the growing season starts can vary considerably from year to year. Thus, initial water storage in combination with climate conditions are considered to be key factors determining crop yields.

A classification of the weather conditions during the spring–summer period in the Stavropol territory has been presented by Petrova (1986), based on the results of an analysis of meteorological data (including temperature and precipitation), spring soil water storage, and winter wheat yields for the period 1951–1980. The study showed a clear relationship between winter wheat yields and water stored in autumn. The analysis also failed to reveal any consistency

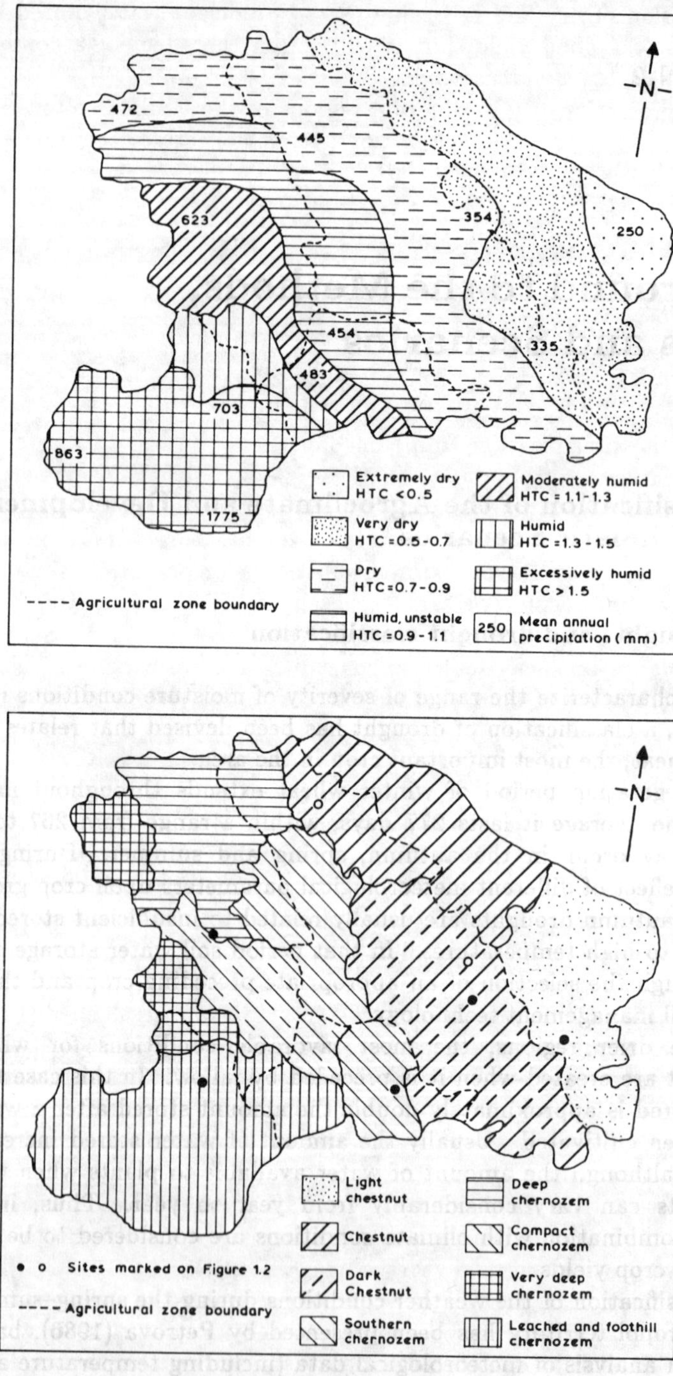

Figure 2.1. **The Stavropol territory: (*a*) agroclimatic regions, (*b*) agricultural soil classes.**

in the succession of dry and favorable years. Sufficient water stored in spring is no guarantee of a good yield, but it will soften the effect of a coming drought period.

The degree of aridity in different regions of the USSR is often expressed as a ratio of the total seasonal precipitation over the effective accumulated temperature – the hydrothermal coefficient (*HTC*):

$$HTC = \frac{\sum_{i=1}^{n} P_i \cdot 10}{\sum_{i=1}^{n} T_i > 10^{\circ}\mathrm{C}} \tag{2.1}$$

where P_i is the precipitation (mm) on day i, and $\sum T_i > 10^{\circ}$C is the sum of mean temperatures T above 0°C on those days (i) when T exceeds 10°C (the effective accumulated temperature). The summation is conducted for the period (n days) of interest. Effective accumulated temperature is reduced by a scaling factor of 10 to serve as an approximate substitute for potential evaporation.

The distribution of *HTC* values has been used to distinguish seven agroclimatic regions in the Stavropol territory [*Figure 2.1*(*a*)]. These regions bear some resemblance to the agricultural zones described in Section 1. They are also related to the pattern of dominant soils in the region, which are shown in *Figure 2.1*(*b*).

2.1.2. Climatic typology

As part of the analysis of weather conditions affecting yields of winter wheat, four types of weather-years (WT) were identified for the Stavropol territory. A qualitative description of the typical weather conditions in zones III and IV (which exhibit a varying degree of aridity) is presented in *Table 2.1*, with more quantitative information shown in *Table 2.2*. In the driest zones (I and II), the meteorological conditions are much more harsh: growing season precipitation is lower and temperatures are higher, there are more days with a relative humidity below 30%, and a limited amount of water is stored in the soil.

However, average seasonal weather is less important for crop growth than the pattern of weather during the growing season. *Figure 2.2* presents the monthly course of the *HTC* values in zone III averaged over those years in the period 1951–1981 that fell in each of the four categories of weather-year (WT), along with values for selected years during the period 1971–1980 (listed in the caption to *Figure 2.2*). In all cases the four months April through July are presented. Four *HTC* curve-types are presented for conditions in zone III. The first type (WT1) can be considered as representing the most favorable conditions for crop growth. In the second type of year (WT2), the aridity pattern for the spring–summer growth period is also favorable, with *HTC* values seldom below

Table 2.1. Meteorological conditions (N = normal) for four types of weather-year (WT).

Type of weather-year	*Typical years during 1971–80*	*Drought conditions*	*Temperature*	*Rainfall*	*HTC*	*Soil water storage: Fallow crop*	*Soil water storage: Preceding crop*
WT1: Favorable	1971 1973 1978	Slight: at end of growth period	N or < N	Close to N or slightly below	Close to N, < N at end of growth period	> N	> N
WT2: Favorable (humid, cool)	1974 1977 1980	Of short duration, favorable moisture conditions in spring and at end of growth period	< N both in spring and at end of growth period	> N	> N	Close to or < N	< N
WT3: Dry (recurrent droughts)	1972 1975	Recurrent, prolonged	> N	< N, >N in May or June	< N in May or June	Close to or >N	< N
WT4: Very dry (severe droughts)	1976 1979	Severe, of increasing intensity	N or > N	Considerably < N	< N for prolonged period	< N	< N

Table 2.2. Growing period (April–July) weather conditions in zones III and IV of the Stavropol territory during four types of weather-year (based on data from Stavropol meteorological station, 1951–1981).

Type of weather-year[a]	*Precipitation* (mm)	*Effective accumulated temperature*	*HTC*[b] *for April–July*	*Soil water storage* (mm)[c] *Fallow*	*Soil water storage* (mm)[c] *Non-fallow*	*Transpiration* (mm)	*Number of days with relative humidity < 30%*
WT1	190	1893	1.0	324	228	194	18
WT2	259	1887	1.4	253	187	161	12
WT3	216	2317	0.9	293	165	185	33
WT4	135	2002	0.7	254	182	205	29
Normal	243	1943	1.3	276	195	181	-

[a] Weather-year types are defined in *Table 2.1*.
[b] Mean value.
[c] Soil water stored (available moisture) is given for the period 1971–1981.

1.0. Years with recurrent periods of drought (WT3) are characterized by greater *HTC* fluctuations and rather low *HTC* values during the spring or the summer period. Years of severe drought (WT4) are characterized by a long period

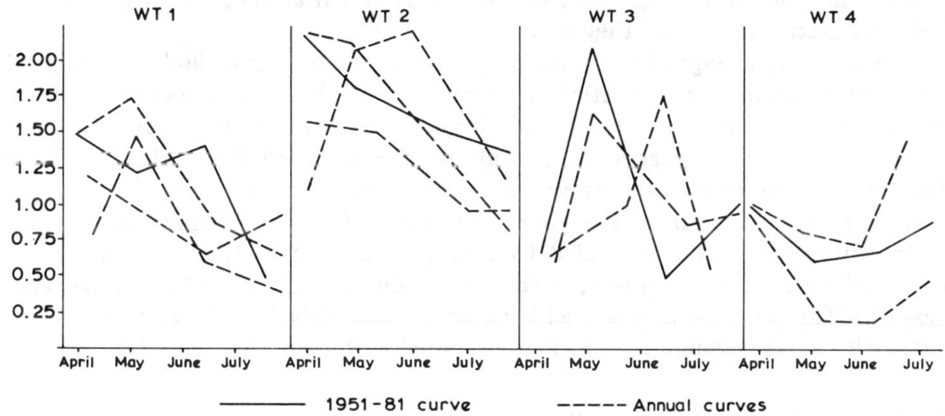

Figure 2.2. Monthly values of the hydrothermal coefficient (*HTC*) in agricultural zone III during the spring–summer period in four types of weather-year (WT). Solid lines link the average values for all years in the period 1951–1981 that fall in a particular category. Dashed lines link the values for individual years in the period 1971–1980: WT1 (1971, 1973, 1978), WT2 (1974, 1977, 1980), WT3 (1972, 1975), WT4 (1976, 1979).

Table 2.3. Winter wheat yield as related to weather conditions in the dry zone (II) of the Stavropol territory[a] (Prikum Experimental Station, 1961–1981)

Type of weather-year	*Number of years of observations*	*Frequency of occurrence* (%)	*Yield* (t/ha)
WT1 (favorable)	6	30	2.4
WT2 (humid and cool)	8	40	1.6
WT3 (dry)	3	15	0.9
WT4 (very dry)	3	15	0.6

[a]No clean fallow used in the rotation and no fertilizers applied.

without rain, resulting in high water deficits. Under such conditions, the *HTC* may drop to below 0.2.

The frequency of occurrence of certain types of weather-year varies by zone. In zone II, for example, the frequency of WT1 years is about 30% of the total number, the frequency of WT2 years 40%, and of WT3 and WT4 years 15% each (*Table 2.3*). In zone I, the number of years in the WT2 category decreases with a corresponding increase in the number of dry years (i.e., WT3 and WT4).

In addition, it should be mentioned that, even in the more favorable years, droughts of short duration may occur. Thus, in zone III in 60% of the years during 1951–1981 with a WT1 climatic pattern, a 5- to 15-day period without rain occurred, most often at the end of the growing season. In years with the WT2 weather pattern, a short period of drought occurred in only 18% of the years, though in the eastern part of the territory 25% of the years experienced such a period. In general, drought periods have a detrimental effect on winter wheat yield, no matter how short they are.

Moreover, drought is not the only climatic factor responsible for reduced crop yields. Significant reductions in winter wheat yields have been observed in the years with the WT2 (cool and humid) weather pattern, resulting from diseases, lodging and harvesting difficulties. Studies carried out at the Prikum Experimental Station of the Stavropol Agricultural Institute (zone II – *see* Section 1, *Figure 1.1*) during the period 1961–1981 (*Table 2.3*) have shown that when winter wheat was preceded by a crop in a rotation (i.e., with no fallow used and no fertilizer applied), yields were reduced by 31% in humid and cool years (WT2), by 60% in years with recurrent droughts (WT3), and by 75% in years with severe droughts (WT4) as compared with favorable years (WT1).

Table 2.4. The effect of clean fallow and weather conditions on winter wheat yield in zone II of the Stavropol territory[a] (Ipatov region, 1971–1980).

	Yields (t/ha)	
Type of weather-year	*Wheat crop preceded by clean fallow*	*Wheat crop preceded by a rotating crop*
WT1 (favorable)	3.35	2.29
WT2 (humid and cool)	2.70	1.77
WT3 (dry)	1.73	1.19
WT4 (very dry)	1.92	1.01

[a]At low fertilizer application rates ($N_{35}P_{20}$).

Table 2.5. The effect of various technologies and weather conditions on winter wheat yield in zone III of the Stavropol territory (1965–1985).

	Experimental farm: wheat crop preceded by a rotating crop		*Experimental plot: cropped fallow, low fertilizer application rate*		*Experimental plot: clean fallow, fertilizer application*	
Type of weather-year	t/ha	%	t/ha	%	t/ha	%
WT1 (favorable)	3.6	100	5.5	100	6.1	100
WT2 (humid and cool)	3.2	88	5.0	92	6.1	100
WT3 (dry)	2.7	74	4.0	73	5.5	91
WT4 (very dry)	2.5	68	4.4	80	5.3	87

Various measures to mitigate such negative effects include the use of fallow periods (*Table 2.4*), application of fertilizers (*Table 2.5*), as well as the use of improved soil management and crop cultivation practices. In zones with a more humid climatic regime, but affected by recurrent droughts, winter wheat should be preceded by fallow or peas with an optimal mineral and organic fertilizer application. Thus, the establishment of a favorable nutritive and water regime in soils of the different regions makes it possible not only to stabilize, but also to increase winter wheat yields in spite of the dry conditions. This is the problem that is addressed in the following discussion and in subsequent sections of this case study.

2.1.3. Development of the climatic scenarios

The classification of the weather conditions described above is used for the construction of the weather scenarios that underlie the climate impact studies. To analyze the possible effects of climatic variations on winter wheat production, four scenarios were developed (*Table 2.6*). The following series of selected types of weather-year (WT) was adopted as the baseline scenario (S1): 1, 1, 4, 2, 2, 4, 3, 2, 1, 3, 1. This series closely represents the annual events during the 11-year period 1971–1981. Although a longer-term series would have been preferable, data constraints restricted the analysis to an 11-year period. This was still considered sufficient, however, to represent long-term conditions in the region. Through substitution of favorable years in the historical sequence by the dry year types (WT3 and WT4), two scenarios of increased aridity were obtained (S2 and S3). Finally, dry year types in the baseline sequence were replaced by weather types (WT1 and WT2) in a fourth scenario (S4).

Table 2.6. Frequency of occurrence of various weather-year types under the four climatic scenarios (in percentages).

Type of weather-year	*Climatic scenarios*			
	S1 (baseline)	*S2*	*S3*	*S4*
WT1 (favorable)	36.4	18.2	9.1	45.5
WT2 (humid and cool)	27.3	18.2	9.1	36.4
WT3 (dry)	18.2	36.4	45.5	9.1
WT4 (very dry)	18.2	27.3	36.4	9.1

2.2. A Crop Production Model

A crop production model (CPM) is used in this study to estimate crop yield. It is composed of two submodels that run at different time steps. The first submodel considers the effects of the radiative, temperature and water regimes on a crop at 10-day intervals during the growing season. The crop yield thus estimated is termed the "water-limited yield" as water is, in most years, the major limiting climatic factor for growth, and as nutrients are assumed to be abundantly available. This submodel is further documented in Konijn (1984) and is also used in the Kenya case study (*see* Part II, Section 4, this volume).

The second submodel determines to what extent the water-limited yield is further constrained by insufficient nutrient supply. This submodel operates on an annual basis, keeping track of changes in the nutrient status of the soil from one year to the next.

The results generated by these two submodels include the information on crop yields in various production environments that is required as an input to optimization models used later in this section and in Section 3.

2.2.1. Yields determined by radiative, temperature and water regimes

The photosynthetic biomass production of a crop is closely related to the prevailing radiative, temperature and water regimes. Crop responses to varying light intensity can be estimated using an approach developed by de Wit (1965) and revised by Goudriaan and van Laar (1978). The approach distinguishes between two major groups of plants, the C-3 and C-4 plants, that are characterized by different biochemical pathways for assimilating carbon dioxide. Winter wheat, the crop selected for this study, belongs to the so-called C-3 group, which performs better in cool, temperate climates.

The production of biomass depends on the development of plant leaf area. Only a part of the carbon dioxide assimilated during photosynthesis will be converted into additional leaf area. The remaining fraction is required to build up the root, stem and grain. This partitioning changes during the development of the crop. For instance, just after germination most of the assimilates go to the roots and the leaves, whereas during grain filling an increasing proportion goes to the grain. The parameters that describe this partitioning during the growing season are crop-specific.

Partitioning is accompanied by the conversion of the assimilates into dry matter, a process that causes weight loss to the plant. Different crops each have a different dry matter composition, and those that ultimately contain products with a high energy content generally have a low dry matter weight. The process of conversion is often called growth respiration. Loss in weight will occur also because the plant needs to expend energy in maintaining various metabolistic functions, such as the preservation of cell structure. The effect on production of the maintenance and growth respiration can be expressed in the following equation:

$$\Delta Y_n = CV(\Delta Y_g - mr \cdot Y) \tag{2.2}$$

where

ΔY_n = gain in net dry matter production over a 10-day interval
CV = gain in assimilates over a 10-day interval
Y_g = growth respiration
Y = standing dry matter
mr = maintenance respiration

Water stress on plants often limits dry matter production. In semi-arid regions, water stress is the main cause of production losses. During periods of limited available water, a plant, in order not to dry out, reduces water loss through transpiration by closure of its stomata. However, closure stops not only the transpiration, but also the intake of carbon dioxide, and under these conditions no photosynthesis is possible.

The reduction in biomass production is proportional to the crops actual evapotranspiration compared to the potential for a well-watered crop (de Wit, 1958):

$$\Delta B_{wl} = \frac{ET_{wl}}{ET_{pot}} \cdot \Delta B_{pot} \tag{2.3}$$

where

ΔB_{pot} = change in potential biomass over a 10-day interval
ΔB_{wl} = change in water-limited biomass over a 10-day interval
ET_{wl} = actual evapotranspiration
ET_{pot} = potential evapotranspiration

How far transpiration is limiting depends on the water available in the soil. A simple water balance helps to determine the water availability to plants:

$$S_{t+\Delta t} = S_t + P_{\Delta t} - R_{\Delta t} + I_{\Delta t} - ET_{\Delta t} - D_{\Delta t} \tag{2.4}$$

The soil water content at the beginning of each 10-day period (S_t) may increase during that period (Δt) because of precipitation (P_t) or irrigation (I_t). The amount of water that enters the soil through precipitation depends on the surface characteristics of the soil, and a proportion may run off (R_t). Finally, the water that infiltrates will partly be lost through evapotranspiration (ET_t) or at high soil moisture content levels may percolate (D_t) to below the rooting zone. Groundwater levels in the Stavropol region are assumed to be too deep to contribute water to plants through capillary rise.

The crop yield that is finally determined at this stage is termed the water-limited yield (Y_{wl}), but this may well be further constrained by limited nutrient availability (see below).

2.2.2. Yields determined by limited nutrient availability

The effect of nutrient availability on crop yields is determined on an annual basis. This estimation is accompanied by an updating of the soil nutrient status. Yield responses to phosphorus and to nitrogen are simulated in the model. The effects of potassium are not modeled, however, for soils are rich in potassium in the Stavropol territory, and in the majority of cases winter wheat (the crop examined in this case study) does not respond to potassium application.

Phosphorus

Phosphorus is the most limiting nutrient in the region, and the phosphorus content of the soil is subject to continuous changes, depending on the original phosphorus content of the soil, its supply from organic and inorganic sources, and its recovery by the crop being grown. The following equation describes the dynamics of soil phosphorus:

$$P_s(i+1) = P_s(i) + \frac{\xi_{PH} \cdot H(i) + P(i) - V_P(i)}{100 \cdot \gamma \cdot H_\pi} \tag{2.5}$$

where

$P_s(i)$ = phosphorus content of the soil in year i, according to the so-called Machigin method
ξ_{PH} = available phosphorous content of organic fertilizer
$H(i)$ = amount of organic fertilizer applied
$P(i)$ = amount of mineral fertilizer applied
$V_P(i)$ = phosphorus recovery by the crop
γ = bulk density of the plowed layer
H_π = depth of the plowed layer

The recovered phosphorus can come from three different sources: the soil itself, mineral fertilizer and organic fertilizer. It is assumed that the phosphorus uptake is proportional to the accumulated dry matter. Therefore the factor most limiting to dry matter accumulation determines the final amount of phosphorus recovered. This can be expressed as an equation:

$$V_P(i) = \frac{Y_{\min}(i)}{Y_P(i)} \Big\{ 100 \cdot P_s(i) \cdot \gamma \cdot H_\pi \cdot k_P^S + k_P^M \cdot P(i) + \xi_{PR} \cdot H(i) \cdot k_P^H \cdot 10 + R(i) \cdot \xi_{PR} \cdot k_P^H \cdot 10^{-2} \Big\} \tag{2.6}$$

where

$Y_{\min}(i)$ = wheat yield determined by the most limiting factor
$Y_P(i)$ = wheat yield determined by phosphorus as the most limiting factor
k_P^S, k_P^M, k_P^H = coefficients for phosphorus recovery for soil, mineral fertilizer and organic fertilizer, respectively
$R(i)$ = residue of the preceding crop (stubble and roots)
ξ_{PR} = phosphorus content of the residue

The recovery coefficients are climate-dependent and will thus be different for each of the five agricultural zones in the Stavropol territory. The application of mineral fertilizer is of special importance in explaining changes in phosphorus content from year to year. The recovery coefficient for mineral fertilizer is independent of soil moisture content, being determined by the amount of fertilizer applied:

$$k_P^M = 0{\cdot}195 - 0.077 \cdot \exp\{-19.543/P(i)\} \tag{2.7}$$

The equation describing phosphorus recovery from organic fertilizers and crop residues is similar to equation (2.7), depending on the agricultural crop in question. If no fertilizers are applied, inorganic phosphorus in the soil is the main source of phosphorus.

Nitrogen

In contrast to phosphorus, it is the organic sources that determine the availability of nitrogen to a high degree. Fluctuations in organic matter content in the soil occur thus:

$$G(i+1) = \varepsilon G(i) + \kappa\{R(i)+10 \cdot H(i)\} \tag{2.8}$$

where

$G(i)$ = humus content of the plowed layer (%)
ε = humus mineralization coefficient
κ = organic matter humification constant
$R(i)$ = residue of the preceding crop
$H(i)$ = amount of organic fertilizer applied

The mineral nitrogen fertilizers applied are assumed to have no residual effect, part of the application being recovered by the crop and the rest considered to be leached or volatized. Nitrogen recovery by the crop for the different nitrogen sources in the soil can be described by:

$$V_N(i) = \frac{Y_{\min}(i)}{Y_N(i)} \Big\{0{\cdot}005\ (1-\varepsilon)\ G(i) \cdot \gamma \cdot H_\pi \cdot k_N^S + R(i) \cdot \xi_{NH} \cdot k_N^S + 0{\cdot}05\ H(i) \cdot k_N^H + 0{\cdot}01 \cdot N(i) \cdot k_N^M\Big\} \tag{2.9}$$

where

$Y_N(i)$ = wheat yield determined by nitrogen as the most limiting factor
k_N^S, k_N^M, k_N^H = coefficients for nitrogen recovery for soil, mineral fertilizer and organic fertilizer, respectively
ξ_{NH} = nitrogen content of organic matter
$N(i)$ = amount of mineral nitrogen applied

Other symbols are as for equation (2.6).

The recovery coefficients, k_N^M and k_N^H, have a specific (constant) value for each agricultural crop. Recovery from soil organic matter, in contrast, is a function of the soil organic matter and the soil moisture content:

$$k_N^S = 1 - 0{\cdot}849 \cdot \exp\{-0{\cdot}6143/G(i)\} + \frac{0.32}{1 + \exp\,(-44.23\,\tilde{W} + 27{\cdot}81)} \quad (2.10)$$

where $\tilde{W}$ = the average soil moisture content for growing period.

Nutrient-limited yield

Equations (2.6) and (2.9) describe the change in nutrient status from year to year. The yield that will be generated for the optimization model is determined by the most limiting factor, which can be expressed as:

$$Y_{\min}(i) = \min\left\{ Y_N^*(i)\ ,\ Y_P^*(i)\ ,\ Y_{wl}(i) \right\}\ ,$$
$$Y_N^*(i) = Y_N(i)/\xi_{YN}\ ,\ Y_P^*(i) = Y_P(i)/\xi_{YP} \quad (2.11)$$

where

$Y_{wl}(i)$ = limited wheat yield (*see* Subsection 2.2.1, above)
ξ_{YN} = average nitrogen content of winter wheat
ξYP = average phosphorus content of winter wheat

Other symbols are explained above.

Following the typology of weather-years in Subsection 2.1.2, the variable $Y_{wl}(i)$ can be represented as:

$$Y_{wl}(i) = \left\{ Y_{wl}^j \right\},\ j = \overline{1,4} \quad (2.12)$$

where

j = type of year [based on the *HTC* value for a particular location – equation (2.1)]
Y_{wl}^j = average yield of winter wheat in the jth type of year

2.2.3. Model input requirements

Data

Model simulations are conducted for agrometeorological conditions and crop rotations that are characteristic of each particular agricultural zone. The agrometeorological input information is ascribed to one representative meteorological station in each zone (*see Figure 1.2*). These are located at Roschino (Zone I), Budennovsk (Zone II), Stavropol (Zone III), Georgievsk (Zone IV) and Cherkassk (Zone V).

The crop rotations included the following crops (in addition to winter wheat): winter barley, spring barley, corn (maize) for grain, corn for silage, sunflowers, potatoes, peas, and fallow (no crop). Areas sown to winter wheat after various preceding crops were derived for each zone from averages (by administrative division) for the most recent five-year period.

Options for the model user

Several possibilities are available to the model user:

(1) Multi-year simulations can be conducted according to the sequences of year-types described in Subsection 2.1.3, either for the baseline climatic scenario, for the three other climatic scenarios or for any other combination of years.
(2) Calculations can be made for individual zones or for all five zones together.
(3) Any specified crop rotation can be assumed, and any number of these simulated.
(4) One of three rates of nitrogen, phosphorus and organic fertilizer application can be selected, and in any combination.

2.3. Model Simulations of Long-Term Changes in Soil Fertility

Model simulation runs have been conducted to examine the long-term dynamics of humus and phosphorus content in the plowed layer for various levels of mineral fertilizer application. The experiments extended over a simulation period of between 20 and 80 years, considering in addition the effects of nutrient status on crop yields.

Table 2.7 shows the changes in soil fertility status that could be expected under the characteristic crop rotation system in each of the agricultural zones. In zone I all winter wheat is preceded by fallow, while in zone II, fallow → winter wheat → winter wheat is the common rotation. In the other zones 6-, 8-, 10- and 12-year crop rotation systems prevail, including various crop combinations.

The table depicts the actual (present-day) levels of soil fertility and the levels projected by the model following long-term adherence to the present crop rotations. Results indicate that the application of $N_{40}P_{40}$ or $N_{60}P_{60}$ with 3.5 to 4.0 t/ha of organic fertilizer would not significantly change the humus content in the two driest zones over a long period. In zone III the same fertilizer applications lead to a significant decrease in humus content in the plowed layer. The other zones show some accumulation of organic matter content, probably because legumes are present in the rotations and residue production is greater. The rate of change of organic matter content in the soil is closely related to the hydrothermal conditions that control the rate of decomposition and

Table 2.7. Present-day levels and long-term model projections of soil fertility by agricultural zone, given existing winter wheat rotation systems and two rates of mineral fertilizer application.

Fertilizer application rates for wheat and preceding crop	*Soil fertility ingredients*	*Agricultural Zones*									
		I		*II*		*III*		*IV*		*V*	
		Actual (A)	*Projected (P)*	*A*	*P*	*A*	*P*	*A*	*P*	*A*	*P*
$N_{40}P_{40}$	Humus in plowed layer (%)	1.9	1.8	3.3	3.2	4.8	4.5	3.5	4.0	7.7	7.7
	P_2O_5 (mg/kg)[a]	2.1	3.1	1.7	2.4	1.8	2.6	2.0	3.1	2.1	1.7
$N_{60}P_{60}$	Humus in in plowed layer (%)	1.9	1.8	3.3	3.2	4.8	4.5	3.5	4.2	7.7	8.3
	P_2O_5 (mg/kg)[a]	2.1	4.5	1.7	3.7	1.8	3.6	2.0	4.1	2.1	2.6

[a]Machigin method.

mineralization of organic matter. Under the warmer and drier conditions of zones I and II, the rate of humus mineralization is higher than the decomposition.

The behavior of the soil phosphorus content at the various fertilizer application rates is less complicated. Phosphorus accumulates in all soils in the Stavropol territory when it is applied at rates exceeding 60 kg/ha of P_2O_5 (*Table 2.7*). At a rate of 40 kg/ha the accumulation is much lower, with even a decrease in available phosphorus observed in zone V, due to greater phosphorus recovery at the higher yield levels [(*see* equation (2.6)]. The time period of simulation required for the model estimates of the final phosphorus content in the plowed layer to reach a stable equilibrium varied from zone to zone. However, similar patterns of available phosphorus have been observed at the Stavropol Research Institute of Agriculture, which reinforces confidence in the model's ability to imitate the dynamics of the nutrient status in soil.

2.4. The Effects of Climatic Change on Winter Wheat Yield

2.4.1. Climatic change and maximum yield

In a second series of model experiments, the response of winter wheat yields to various levels of nutrient supply and to different degrees of aridity were investigated for each agricultural zone. Differences in aridity are represented by three of the climatic scenarios described earlier in Subsection 2.1.3: the baseline scenario S1, and two scenarios (S2 and S3) with an increased frequency of dry types of weather-years.

Table 2.8. Model estimates of winter wheat yield (t/ha) in each agricultural zone under three climatic scenarios of increasing drought frequency (S1, S2 and S3, respectively), for various levels of mineral nutrient supply and for different crop rotations.

		Agricultural zones									
		I		*II*		*III*		*IV*		*V*	
		Winter wheat preceded by:									
Mineral nutrient application rate	*Climatic scenario*	*Fallow (F)*	*Rotating crop (RC)*	*F*	*RC*	*F*	*RC*	*F*	*RC*	*F*	*RC*
$P_{40}P_{40}$	S1	2.78	–	3.17	–	4.04	3.56	3.69	3.40	4.14	3.41
	S2	2.47	–	2.83	–	3.72	2.51	3.68	2.64	3.91	3.10
	S3	1.86	–	2.15	–	2.82	1.74	2.88	1.90	3.44	2.57
$P_{60}P_{60}$	S1	3.13	–	3.69	–	4.76	3.76	4.79	4.01	5.76	4.28
	S2	2.71	–	3.27	–	4.31	2.62	4.36	2.90	5.47	4.04
	S3	2.01	–	2.47	–	3.32	1.84	3.34	2.09	4.18	2.91

Table 2.8 shows estimates of average winter wheat yields in rotations with fallow and other crops. In zones III, IV and V, where fallow and peas occurred in the rotation, they were considered simply as clean fallow. The other preceding crops were grouped together as nonfallow preceding crops, in spite of their different properties and contrasting effects on wheat yield.

Model runs for the 11-year scenario periods involved a number of repeated iterations, depending on the number of crops included in a rotation, in order to obtain comparable results. Thus, the rotation fallow → wheat was run for 22 years, while an 8-course rotation required an 88-year simulation run.

The results in *Table 2.8* illustrate how the maximum estimate of winter wheat yield varies by zone and depends on the preceding crop, the amount of fertilizer applied and the climatic conditions. The findings correspond closely to observed yields at experimental stations.

2.4.2. Climatic change and yield stability

In order to assess the stability of winter wheat production in the Stavropol territory, the coefficient of variation C_v (standard deviation/mean) has been used, where stability is represented by $(100-C_v)$. *Figure 2.3* shows the results of model runs that aimed to evaluate the stability of the zone-wise wheat production under present-day fertilizer applications (a rate of $N_{40}P_{30}$ annually), present-day rotations, and under the four climatic scenarios (S1–S4). There is a consistent increase in the stability of the total grain output as one moves from zone I to zone V.

Since phosphorus fertilizers are known to induce the greatest yield response, a series of simulation experiments was conducted to study the stability of wheat production as a function of phosphorus applications (*Figure 2.4*). The stability increases with increased phosphorus application, but the maximum increase is only about 6%, indicating that the present level of application is

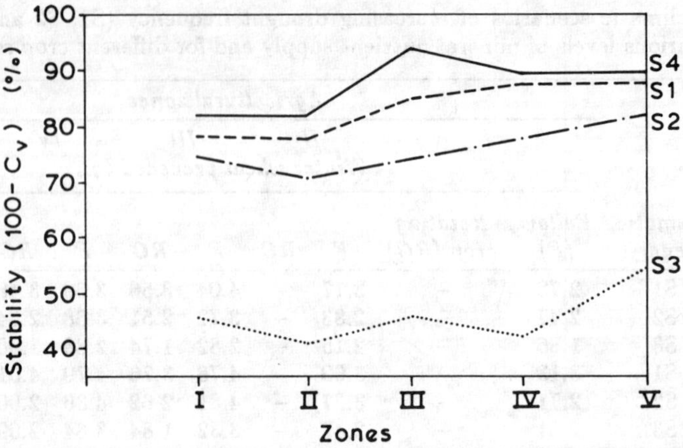

Figure 2.3. **Stability of total winter wheat production, by agricultural zone, under the four climatic scenarios.**

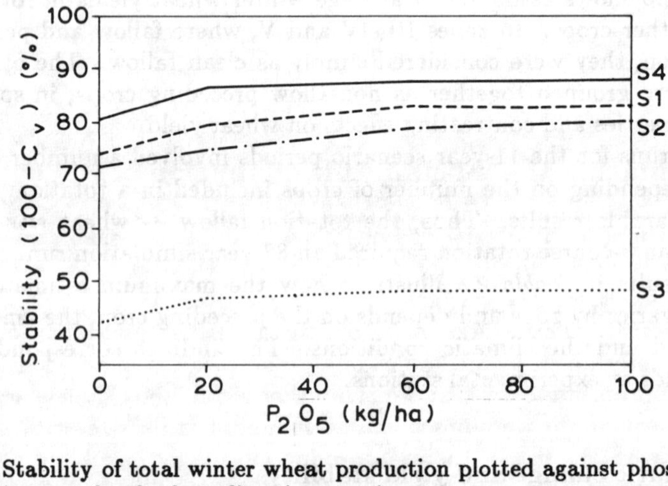

Figure 2.4. **Stability of total winter wheat production plotted against phosphorus fertilizer application rate for the four climatic scenarios.**

already approaching the maximum potential. This means that with increasing aridity (as illustrated by scenario S2 and S3) hardly any stability can be gained from additional phosphorus applications.

Figure 2.5 **emphasizes that the stability of total production does not depend merely on a single determining production factor, but a whole complex of agrometeorological conditions. The figure shows the stability–yield relationship for two climatic scenarios: the baseline scenario (S1) and the scenario with the highest frequency of very dry weather-years (S3). An increase in aridity (i.e., moving from the S1 line towards the S3 line on *Figure 2.5* results not only in a decrease in long-term average yield, but also in a substantial decrease in stability.**

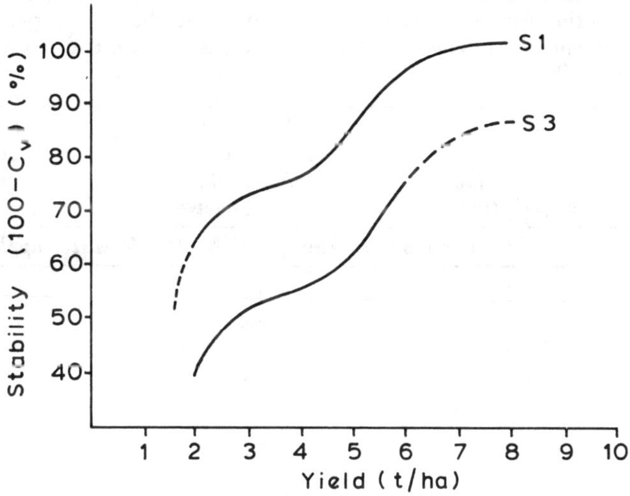

Figure 2.5. The stability of total wheat production as a function of long-term average yield for scenarios S1 (baseline) and S3 (high drought frequency).

2.5. Estimating the Optimum Distribution of Fallow

In the Stavropol territory the use of fallow is an indispensable requirement for optimal winter wheat production, particularly in the driest zones (I and II). In this connection the zonal pattern of fallow distribution is of great interest, for this characteristic affects total wheat production. The total fallow area required, however, is determined by a complex of agrometeorological and economic factors.

The appropriate extent and distribution of fallow is that which maximizes the total wheat production under limited mineral fertilizer reserves. This can be determined with the use of the crop production model, involving multiple iterations according to a so-called "configuration method" of non-linear searching. *Table 2.9* gives the model results for the optimal fallow distribution under the baseline climatic scenario (S1) and under the scenario of greatest drought frequency (S3) that maximize total wheat production, as well as the stability of wheat yields, in each zone.

Computations are carried out for two levels of fertilizer application, and the table also includes a relative index of the total production obtained under each fallow distribution.

Under a low level of fertilizer application ($N_{40}P_{40}$), the maximum production obtained with an adjusted fallow distribution is only 5% higher than the present-day production level, while maximum yield stability could be achieved with a radical change in fallow distribution in most zones, with average production unchanged from the present.

Table 2.9. Optimum fallow distribution (expressed as a percentage of the area sown to winter wheat) for maximized total wheat production and for maximized yield stability. Total wheat production for each fallow distribution is also shown as a proportion of production for the present-day fallow distribution, baseline climate and low rate of mineral fertilizer application ($N_{40}P_{40}$).

Agricultural zones	*Actual fallow distribution*	*Fallow distribution maximizing total wheat production*	*Fallow distribution maximizing yield stability*
	Scenario S1 (baseline) with $N_{40}P_{40}$ *fertilizer application*		
I	49.1	33.2	50.0
II	34.3	33.3	50.0
III	24.4	0.0	0.0
IV	17.1	0.0	0.0
V	0.0	0.0	0.0
Total wheat production	1.00	1.05	1.00
	Scenario S1 with $N_{60}P_{60}$ *fertilizer application*		
I	49.1	33.2	50.0
II	34.3	24.9	50.0
III	24.4	0.0	0.0
IV	17.1	0.0	0.0
V	0.0	0.0	0.0
Total wheat production	1.09	1.18	1.11
	Scenario S3 with $N_{40}P_{40}$ *fertilizer application*		
I	49.1	50.0	50.0
II	34.3	50.0	50.0
III	24.4	50.0	50.0
IV	17.1	0.0	0.0
V	0.0	0.0	0.0
Total wheat production	0.60	0.65	0.64

For a higher level of applied fertilizer ($N_{60}P_{60}$), the optimal fallow distributions are very similar to those in the previous case, but maximum production would be 8% above the present, while total production under maximum stability might increase 2% above the present-day value.

Of special importance is the determination of the area under fallow for the unfavorable (drier) weather conditions (scenario S3). In this case the optimum fallow distribution is achieved at an area share of 50% in the first three zones. Zones IV and V do not need fallow at all. The maximum total wheat production would be 8% higher than levels expected at present under these conditions, while yield stability could be maximized at a level of total production 7% higher than at present. These results were obtained at a fertilizer supply rate of ($N_{40}P_{40}$). Further increases of fertilizer application under these weather conditions do not lead to any significant change in total output and yield stability or of the optimum fallow distribution for the different zones.

Further analysis of *Table 2.9* does not reveal any similarity between the scenarios for fallow distribution at maximized production and yield stability under the given weather conditions. Increasing inorganic fertilizer applications does not change the trend. In case of an unfavorable climatic variation – for example, toward more arid conditions – the area under fallow in the first three zones would already be close to the maximum. Maximization of the total production and yield stability would not, therefore, lead to a marked change in the fallow distribution pattern. Thus, the possibilities of controlling production and yield stability through changes in the fallow distribution pattern are limited. Further increases in the efficiency of agriculture would instead call for active monitoring of the hydrothermal conditions during crop growth and development.

Other possible areas of improvements include soil management, methods of fertilizer application, pest and disease control and the introduction of improved (drought-resistant) winter wheat varieties. However, the use of these technologies involves costs that must be weighed carefully against the expected returns. This is taken up in the next section, which adopts an optimization approach to estimate the costs of applying different levels of technology to wheat production in the Stavropol region under different climatic conditions.

SECTION 3

The Effects on Crop Production and Crop Allocation

3.1. Introduction

In Section 2 it was shown that moisture conditions in much of the Stavropol territory are limited and unreliable, the high variability of the interannual and seasonal precipitation leading to large variations in agricultural production. This is an important factor to reckon with in the context of a centrally planned economy, such as that of the Soviet Union. The effects of droughts have been severe in the past, and there is evidence to suggest that the frequency of drought occurrence may even have increased in recent decades. So there are good reasons to look at the implications of changing patterns in drought frequency and their effect on agricultural production.

In recent years, a long-term program for national agricultural development has been devised and is currently being implemented, which includes measures designed to reduce the dependence of agricultural production on the weather. The program makes a distinction between the areas that guarantee a certain level of production and the more risky, dry regions. In these dry areas one aim is to stabilize economic conditions through the introduction of more advanced and adapted technologies. Moreover, attention has focused not only on the national "all-union" agricultural programs, but also on the introduction of improved agricultural management at a regional level.

Mathematical modeling is a very useful tool in regional agricultural management, and two such models are used in this section to evaluate the effects of climatic variations on agricultural production in the Stavropol territory: a crop production model (introduced in Section 2), and a crop allocation model.

Four climatic scenarios have been developed to represent both the present-day (baseline) climatic conditions, and plausible conditions under changed climatic regimes (e.g., drier or moister than present). These are described in detail in Section 2.

The crop investigated in this study is winter wheat, the principal crop in the region. In this section various technologies will be considered and two kinds of rotation for winter wheat:

(1) Continuous winter wheat.
(2) An alternating fallow → winter wheat rotation.

In a planned economy, such as the one in Stavropol, the detrimental effects of climate on planned production targets are of major concern. These can sometimes be offset by the deployment or reallocation of resources and technologies, but this can involve costs that far exceed the benefits reaped from increased or stabilized production. In the following subsections we investigate some of the options available to agricultural planners in the face of climatic variation.

3.2. Methods

3.2.1. Models

An essential aspect of the study is the linkage between a crop production model and a crop area allocation model. The former estimates the effects on crop yields of different climatic scenarios. The latter estimates the costs of achieving certain production targets through adaptations in technology and management, under different climatic scenarios.

For each of the climatic scenarios, the crop production model generated winter wheat yield estimates for a given agricultural zone, technology and rotation. Because of the large number of yield values generated, a special procedure to average yields was developed. The averaged winter wheat yields were required as inputs to the allocation model. That model computed simultaneously the optimal allocation of arable land in the territory for winter wheat and the optimal winter wheat technology required to achieve a given production target while minimizing the total expenditures.

The crop area allocation model is described in detail below. A schematic description of the crop production model is given in Section 2.

3.2.2. Simulating baseline yields

The Stavropol territory, as we have seen, is characterized by high rainfall variability. Seven different agroclimatic zones were identified in Section 2 on the basis of the hydrothermal coefficient (*HTC*). Values of the *HTC* showed a close relation to winter wheat yields, and following that finding four major types of weather-year were defined (*see Table 2.2*).

Yield estimations with the crop production model were conducted initially for all possible combinations of the four typical weather-years. These combinations were assumed to have an equal probability of occurrence.

The total number of model runs does not depend merely on the type of weather-years, however. In each of the five agricultural zones (*see Table 1.1*), two kinds of rotations were considered: winter wheat after fallow, and winter wheat after winter wheat.

The crop production model was run for both rotations over a sequence of 16 years composed of the four recognized standard types of weather-year (WT): 1, 4, 2, 2, 4, 3, 2, 1, 3, 1, 2, 3, 3, 4, 4, 1. In this sequence each of the four types is preceded by a year of the same type and years of the three other types (the first entry, WT1, is assumed to be preceded by the sixteenth entry, also WT1). Thus, we obtain four yield estimations for each type of weather-year. Yields are accompanied by the soil moisture content as a proportion of the field capacity.

Table 3.1. Yields (t/ha) and the soil moisture content (proportion of field capacity) for the rotation winter wheat after fallow. (Location: climatological station Roschino, zone I.)

Preceding year-type	*Weather-year type and representative year*			
	WT1 (1978)	*WT2 (1977)*	*WT3 (1972)*	*WT4 (1979)*
WT1	4.480 (0.80)	3.903 (0.62)	1.849 (0.62)	0.963 (0.61)
WT2	5.816 (0.79)	5.906 (0.80)	3.206 (0.78)	2.889 (0.74)
WT3	4.378 (0.70)	3.357 (0.69)	2.021 (0.62)	1.516 (0.60)
WT4	6.466 (0.80)	6.465 (0.80)	3.639 (0.74)	3.355 (0.77)

Some results are presented in *Table 3.1* for the climatological station Roschino (for location, *see Figure 1.2*). The table shows yields in metric tons per hectare for the weather-year types WT1, represented by 1978, WT2 (1977), WT3 (1972) and WT4 (1979). The preceding years are indicated in the first column.

3.3. The Crop Area Allocation Model for Different Technologies

A "technology" is defined here as a given combination of resource elements, each allocated to a region at a specified level of quantity and cost. The technologies used in this study have been identified at the Stavropol Institute for Agriculture. They were selected on the basis of experiments with different equipment, different tillage practices and varying applications of fertilizers.

Each technology is characterized by 10 resource elements (l). These elements can exhibit quite different values in the Stavropol territory, depending on the regional environmental conditions. However, within each of the five agricultural zones, the technologies are considered homogeneous.

The 10 elements of each technology are:

(1) Fuel and lubricating materials.
(2) Labor.

(3) Biocides (e.g., insecticides and herbicides).
(4) Organic fertilizer.
(5) Nitrogen fertilizer.
(6) Phosphorus fertilizer.
(7) Potassium fertilizer.
(8) Tractors.
(9) Motor lorries.
(10) Harvesters.

For both rotations – winter wheat after fallow and winter wheat after winter wheat – five technologies have been distinguished, characterized by the relative expenditure allocated to the ten elements in each. In *Table 3.2* the five technologies for the fallow → winter wheat rotation are presented, in terms of the expenditure for each of the 10 technology elements in agricultural zone III. The second technology turns out to be the most expensive one, mainly due to the high volume of organic fertilizer included in this technology.

Table 3.2. Specific expenditures of resources for fallow technologies in zone III.

Resource element	*Technologies*				
	T1	*T2*	*T3*	*T4*	*T5*
1	78.070	104.610	49.950	47.260	50.830
2	24.400	24.200	22.000	19.800	26.000
3	1.500	1.500	1.500	1.500	1.500
4	10.000	20.000	0.000	0.000	0.000
5	0.000	0.150	0.200	0.300	0.600
6	0.200	0.300	0.200	0.300	0.600
7	0.000	0.000	0.050	0.100	0.200
8	0.480	0.500	0.460	0.480	0.550
9	114.500	217.500	14.500	10.800	19.700
10	0.154	0.154	0.154	0.154	0.154

Each of the technologies can be described by its total cost (C_T) calculated as the sum of the costs per resource element. If a_l is the cost per unit of the lth resource element, the total cost can be described by:

$$C_T = \sum_l a_l \cdot C_{ijkl} \tag{3.1}$$

where

i = the agricultural zone
j = the technology
k = the rotation

A typical feature of socialist countries is the planning of the economy. In this study the planning concentrates around production targets for the Stavropol territory. Let P be the target for total wheat production (for example, in thousands of tons). Y_{ijk} is the yield and X_{ijk} the area of winter wheat in agricultural zone i corresponding to the jth technology and rotation k. Given that the production target is for the whole territory, the following inequality needs to be fulfilled to meet the target:

$$\sum_{ijk} Y_{ijk} \times X_{ijk} \geq P \tag{3.2}$$

Combining equations (3.1) and (3.2) makes it possible to calculate the total expenditure (J) based on the required target within the constraints fixed:

$$J = \sum_{ijkl} a_l \cdot C_{ijkl} \cdot X_{ijk} \tag{3.3}$$

The objective is to minimize the total expenditure while meeting the production target. This problem is solved through linear programming using the MINOS package. The input and output are created using the software package GEMINI (Lebedev, 1984). Model runs that minimize total expenditure have been carried out for 15 different production targets, four types of weather-year, two rotations and five different agricultural zones. An example of model input and output for a sample minimization run is presented in *Table 3.3*.

3.4. Model Linkages, Scenario Analysis and Results

In order to estimate the likely effects of climatic variations on optimum crop area allocation, four climatic scenarios were selected (*see* Section 2), each reflecting a different frequency of certain types of weather-year (cf. *Table 2.6*).

The average yield for each of the scenarios (s = S1, S2, S3 and S4) is determined as follows. The mean winter wheat yield (Y^r) for each type of weather-year (r) has a frequency of occurrence of p_r^s. If we denote the average scenario yield as Y^s, the estimation can be symbolized as:

$$Y^s = \sum_{i=1}^{4} p_r^s \cdot Y^r \tag{3.4a}$$

The crop production model can provide two kinds of yield estimation: the water-limited yield and nutrient-limited yield. This is described in more depth in Section 2. The nutrient-limited yield is required as an input to the crop allocation model to allow studies of the effect of fertilizer use under different climatic scenarios to be conducted, for, of the various technology elements, only fertilizer

Table 3.3. An example of model input and output for a minimization run.

	Technologies									
	Nonfallow					*Fallow*				
	T1	*T2*	*T3*	*T4*	*T5*	*T1*	*T2*	*T3*	*T4*	*T5*
					Zone I					
Costs	52.71	57.58	63.49	73.31	82.03	91.94	128.54	66.81	72.78	91.00
Yields (t/ha)	0.95	1.07	1.16	1.23	1.28	2.04	2.58	1.75	1.89	2.23
Optimum area for wheat (%)	0	0	0	0	0	0	0.15	0	0.071	0.28
					Zone II					
Costs	52.99	57.85	63.81	73.61	82.34	92.26	129.19	66.97	72.91	91.07
Yields (t/ha)	1.24	1.34	1.48	1.54	1.59	2.37	3.24	1.91	2.03	2.30
Optimum area for wheat (%)	0	0	0	0	0	0	0.50	0	0	0
					Zone III					
Costs	54.27	59.25	65.12	75.03	83.80	92.78	129.80	67.95	73.87	91.97
Yields (t/ha)	2.54	2.76	2.81	2.98	3.07	2.89	3.86	2.90	3.00	3.22
Optimum area for wheat (%)	0	0.52	0	0	0	0	0.15	0.09	0	0
					Zone IV					
Costs	54.33	59.20	65.31	75.43	84.26	92.15	129.84	68.04	74.07	92.19
Yields (t/ha)	2.60	2.71	3.00	3.38	3.54	2.25	3.90	3.00	3.20	3.44
Optimum area for wheat (%)	0	0	0	0.62	0	0	0.15	0	0.04	0
					Zone V					
Costs	55.20	60.09	66.04	75.95	84.74	93.62	130.66	68.83	74.79	93.05
Yields (t/ha)	3.48	3.62	3.74	3.91	4.03	3.75	4.73	3.80	3.93	4.31
Optimum area for wheat (%)	0	0	0.70	0	0	0	0.15	0	0	0

use is simulated in both the crop production model and in the crop allocation model.

To evaluate the role of fertilizers, winter wheat yields (Y) were estimated for each agricultural zone (i), technology (j) and rotation (k). Since there are four types of weather-year (r = WT1, WT2, WT3 and WT4) and there are four possible preceding years, yields for each type of weather-year were averaged over the four possibilities. To arrive at the yields for each scenario (s = S1, S2, S3 and S4), the frequency of occurrence (p) of each weather-year per scenario needed to be estimated. The yield estimation can thus be summarized as:

$$Y_{ijk}^{s} = \sum_{r=1}^{4} p_r^s \cdot Y_{ijk}^{r} \tag{3.4b}$$

Yields determined by equation (3.4b) were used as inputs for the allocation model.

In *Table 3.4* the present allocation to winter wheat (*see also Table 2.9*) is compared with results for the optimum allocation of winter wheat required to achieve a production target 50% greater than the average annual 1951–1955 production (used as a reference in this study) under the baseline climate scenario, S1. Results are expressed as percentage fallow of the area sown to winter wheat. The results can be considered a satisfactory representation of reality for most of the zones and, further, suggest that the present allocation to winter wheat in the Stavropol territory is close to optimal for a range of different production targets.

Table 3.4. A comparison between the present allocation to winter wheat and model estimates for the optimal allocation to achieve a production target 150% of the average annual 1951–1955 production under the baseline climatic scenario, S1.

Agricultural zones	*Existing allocation (fallow area, %)*	*Optimal allocation with 150% plan and weather-type W1 (fallow area, %)*
I	49.1	49
II	34.3	36.4
III	24.4	24
IV	17.1	19
V	0	15

Yield estimates by agricultural zone, for five technologies and the four climatic scenarios, are presented in *Table 3.5.*

3.5. Interpretation of Results

The best preceding land use for winter wheat under semi-arid conditions in the Stavropol area is fallow. This has been confirmed repeatedly in field trials, in particular in the two driest agricultural zones I and II, and is reinforced by the results given in *Table 3.5.* In fact, there is a direct relationship between the area under fallow in these zones and total wheat production.

Table 3.6 shows that in zones I and II the area under fallow increases to its maximum either when the planned production increases (from 125% to 175% of 1951–1955 production) or when the climate gets drier (scenarios S2 and S3). The response in the wetter zones III, IV and V to these changes is small by comparison. The costs of production also increase with higher planned production and/or drier climate. *Figure 3.1* is plotted on the basis of multiple model runs for a range of planned production target, including those shown in *Table 3.6.* It can be used to evaluate the cost of achieving a specified production target under a given climate scenario in the Stavropol territory. The short vertical lines on the curves delimit the areas required under fallow. Each curve refers to one of the four scenarios.

Up to this point, the level of technology has been assumed to be constant. *Table 3.7* demonstrates how technological transformations might alter crop

Table 3.5. Estimated winter wheat yields by agricultural zone for two rotations, five technologies and four scenarios (t/ha).

Zone	Preceding land use	Technology T1	T2	T3	T4	T5
				Scenario S1		
I	Nonfallow	1.08	1.21	1.30	1.41	1.50
	Fallow	2.19	2.92	2.01	2.16	2.50
II	Nonfallow	1.37	1.48	1.58	1.71	1.81
	Fallow	2.62	3.51	2.26	2.38	2.59
III	Nonfallow	2.73	2.90	3.02	3.19	3.31
	Fallow	3.19	4.16	3.00	3.08	3.71
IV	Nonfallow	3.21	3.36	3.41	3.47	3.63
	Fallow	3.44	4.59	3.50	3.66	3.78
V	Nonfallow	3.70	3.83	3.95	4.12	4.24
	Fallow	4.04	5.02	4.00	4.24	4.68
				Scenario S2		
I	Nonfallow	0.95	1.07	1.16	1.23	1.28
	Fallow	2.04	2.58	1.75	1.89	2.23
II	Nonfallow	1.24	1.34	1.48	1.54	1.59
	Fallow	2.37	3.24	1.91	2.03	2.30
III	Nonfallow	2.54	2.76	2.81	2.98	3.07
	Fallow	2.89	3.86	2.90	3.00	3.22
IV	Nonfallow	2.60	2.71	3.00	3.38	3.54
	Fallow	2.25	3.90	3.00	3.20	3.44
V	Nonfallow	3.48	3.62	3.74	3.91	4.03
	Fallow	3.75	4.73	3.80	3.93	4.31
				Scenario S3		
I	Nonfallow	0.86	0.99	1.06	1.11	1.14
	Fallow	1.94	2.34	1.58	1.71	2.05
II	Nonfallow	1.16	1.25	1.33	1.41	1.44
	Fallow	2.20	3.03	1.67	1.79	2.09
III	Nonfallow	2.42	2.55	2.67	2.84	2.91
	Fallow	2.67	3.65	2.70	2.90	2.99
IV	Nonfallow	2.45	2.66	2.94	3.31	3.45
	Fallow	2.75	3.70	2.95	3.35	
V	Nonfallow	3.33	3.47	3.59	3.76	3.89
	Fallow	3.54	4.52	3.50	3.70	4.06
				Scenario S4		
I	Nonfallow	1.16	1.22	1.39	1.53	1.65
	Fallow	2.29	3.15	2.18	2.34	2.68
II	Nonfallow	1.45	1.57	1.67	1.83	1.96
	Fallow	2.79	3.74	2.49	2.62	2.80
III	Nonfallow	2.86	3.04	3.16	3.33	3.48
	Fallow	3.39	4.37	3.20	3.40	4.03
IV	Nonfallow	3.31	3.40	3.45	3.54	3.71
	Fallow	3.60	4.70	3.60	3.90	4.10
V	Nonfallow	3.84	3.98	4.10	4.27	4.39
	Fallow	4.25	5.23	4.20	4.47	4.93

Table 3.6. Optimum allocation of winter wheat (percentage fallow area) for a constant technology to achieve three planned production targets ($P = 125\%$, 150% and 175% relative to present mean production) while minimizing resource costs (expressed as percentages relative to the costs of plan $P = 150\%$ under the baseline scenario, S1). Computations are for the four climatic scenarios S1, S2, S3 and S4 (*see Table 2.6*).

	Scenario S1	*Scenario S2*	*Scenario S3*	*Scenario S4*
		Production target, P = 125%		
Relative costs (%)	86	92	98	83
		Allocation (% of fallow)		
Zone I	49	49	49	49
Zone II	29	29	32	29
Zone III	24	24	24	24
Zone IV	19	19	19	19
Zone V	15	15	15	15
Whole territory	32.5	32.5	34	32.5
		Production target, P = 150%		
Relative costs (%)	100	111	123	94
		Allocation (% of fallow)		
Zone I	49	50	50	49
Zone II	36.4	50	50	29.4
Zone III	24	24	24	24
Zone IV	19	19	19	19
Zone V	15	15	15	15
Whole territory	35.8	42	42	32.5
		Production target, P = 175%		
Relative costs (%)	121	139	159	110
		Allocation (% of fallow)		
Zone I	50	50	50	49
Zone II	50	50	50	50
Zone III	24	24	24	24
Zone IV	19	19	24	19
Zone V	15	15	15	15
Whole territory	42	42	42	41.8

allocation requirements under the four climatic scenarios and for the given planned production targets.

It was shown in Subsection 3.3 and *Table 3.2* that the costs of the nonfallow technologies increased gradually with the increase of mineral fertilizer applied. The first two fallow technologies (T1 and T2) include organic fertilizer (resource element 4) and are thus rather expensive; the other fallow technologies (T3, T4 and T5) include only mineral fertilizer and are comparatively cheap.

The results imply that the cheaper technologies are favored as long as the required total grain production is low. When the plan requires a higher grain

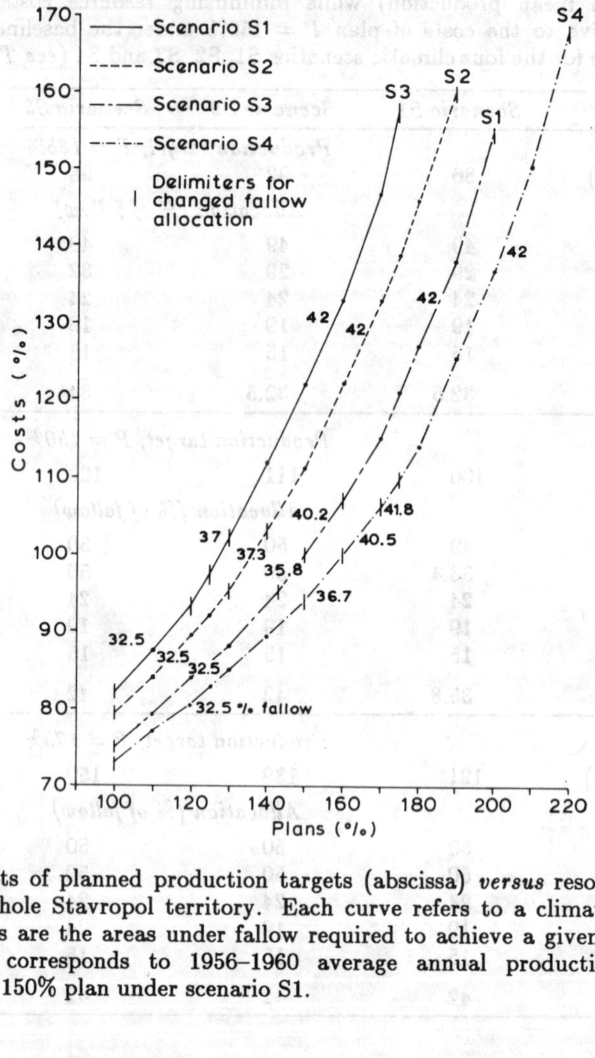

Figure 3.1. Plots of planned production targets (abscissa) *versus* resource costs (ordinate) for the whole Stavropol territory. Each curve refers to a climatic scenario, and values on curves are the areas under fallow required to achieve a given planned target. A 100% plan corresponds to 1956–1960 average annual production; 100% costs correspond to a 150% plan under scenario S1.

production, more fallow is included in the rotation and expenditures increase together with the application of technology (T2) that includes the greatest application of organic fertilizer. It is interesting to note that, under a drier climate (e.g., scenarios S2 and S3), the maximum achievable production target decreases.

Figure 3.2 shows that when planned production targets are relatively modest, the optimum area under fallow remains at a relatively low level (32.5%) for all four climatic scenarios. With a further increase in planned production, the area under fallow increases linearly until it reaches saturation (42%). This point of saturation is approached at the 175% plan for the most favorable weather scenario (S4), but at the 140% plan under the driest weather scenario (S3).

Table 3.7. Optimum allocation of winter wheat (percentage fallow area) to achieve a range of planned production targets (percentage relative to the 1951–1955 average annual response) for different fallow technologies and for the four climatic scenarios (entire Stavropol region).

	Scenario S1					*Scenario S2*				
P	*T1*	*T2*	*T3*	*T4*	*T5*	*T1*	*T2*	*T3*	*T4*	*T5*
100%	20.5	4.7	7.3	0	0	20.5	4.7	7.3	0	0
110%	20.5	4.7	7.3	0	0	20.5	4.7	7.3	0	0
120%	20.5	4.7	7.3	0	0	20.5	4.7	7.3	0	0
125%	20.5	4.7	7.3	0	0	20.5	4.7	7.3	0	0
130%	20.5	4.7	7.3	0	0	11.4	9.3	11.8	0	0
140%	3	13.5	16	0	0	8.3	10.8	17.8	0.3	0
150%	0	15	20.8	0	0	8	11	17	6	0
160%	0	15	25.2	0	0	0	21.3	10.8	9.9	0
170%	0	15	8.3	17	1.7	0	28.7	3.4	9.9	0
175%	0	18.5	0	21.8	1.7	0	30.3	1.7	2.2	7.8
180%	0	22.8	0	17.4	1.8	0	32	0	0.3	9.7
190%	0	30.3	0	5.7	6	0	42	0	0	0
200%	0	42	0	0	0					
210%										
220%										

	Scenario S3					*Scenario S4*				
P	*T1*	*T2*	*T3*	*T4*	*T5*	*T1*	*T2*	*T3*	*T4*	*T5*
100%	20.5	4.7	7	0.3	0	15.1	7.4	10	0	0
110%	20.5	4.7	7	0.3	0	12.2	8.8	11.5	0	0
120%	20.5	4.7	5.2	2	0	0.5	14.7	17.3	0	0
125%	8.3	10.8	12.8	2	0	0	15	17.5	0	0
130%	8.3	10.8	15.7	2	0	0	15	17.3	0	0
140%	8.3	11.7	20	2	0	0	15	17.5	0	0
150%	8.3	19	12.7	2	0	0	15	17.4	0.1	0
160%	8.3	26.2	4.2	3.3	0	0	15	21.4	0.1	0
170%	0	32	0	0.3	9.7	0	15	25.5	0.1	0
175%	0	42	0	0	0	0	15	26.1	0.1	0
180%					0	15	3.2	2.2	1.8	
190%					0	22.2	0	18	1.8	
200%					0	30.3	0	7.6	4.1	
210%					0	34.4	0	0	7.6	
220%					0	42	0	0	0	

3.6. Conclusions

Two completely different models were linked to study the required changes in the area under fallow to achieve higher grain production levels under four climatic scenarios. The models were apparently able to describe the complex interdependence of changes in the climate, production costs, levels of technology and total grain production.

One conclusion that can be drawn is that a further increase in production is achievable only through technological development, for example, through

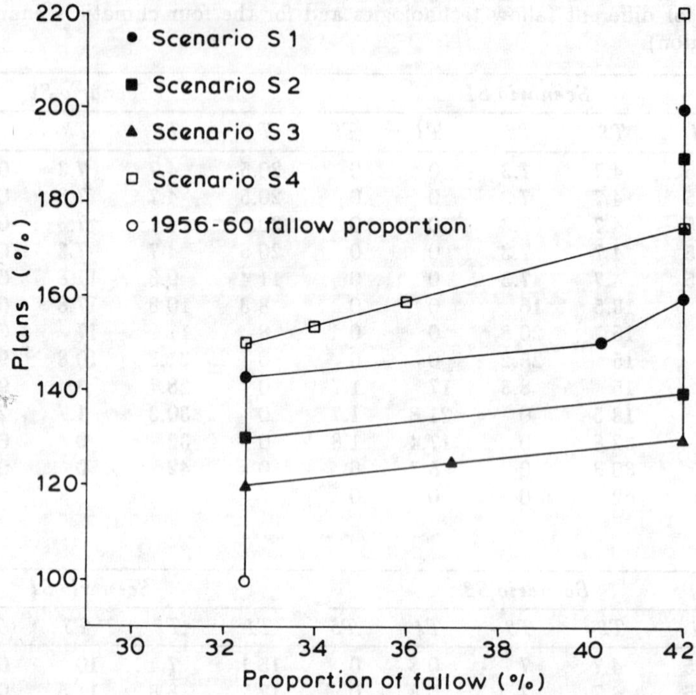

Figure 3.2. Plots of optimal fallow area (abscissa) *versus* planned wheat production (ordinate) under four climatic scenarios. A 100% plan corresponds to 1951–1955 average annual production.

improved tillage techniques and more appropriate levels of fertilizer application. This conclusion is derived from the observation that in the driest zones the maximum possible area under fallow is already obtained under a climatic scenario of slightly drier conditions than present (S2). Clearly, the allocation of the area is closely related to the regional climatic conditions.

The foregoing represents only a preliminary attempt to link two models in order to analyze the effect of variations of climate in the Stavropol territory. However, it shows how useful the approach can be for illuminating the interdependences of physical and economic factors that determine agricultural production.

SECTION 4

Conclusions and Implications for Agricultural Planning and Policy

4.1. The Experiments

In order to investigate the effects of climatic variations on production and allocation of winter wheat in the Stavropol territory, two large series of optimization experiments have been conducted. The first set of experiments, described in detail in Section 2, involved simulations with a crop production model. A nonlinear optimization procedure was employed to compute the optimum share of fallow in each of the five agricultural zones in the territory along with the optimum applications of nitrogen and phosphorus fertilizers required to maximize the stability of yields or to maximize the average total production, under different climatic scenarios. The second series of optimization experiments was devoted to determining winter wheat allocation in a wider sense, by selecting optimum technologies and rotations that would minimize total expenditures on grain production under the same climatic scenarios.

4.2. The Importance of Fallow in Dry Regions

The results of the optimization experiments in Section 2 served both to evaluate the influence of climate on yields and soil fertility, and also to assess model adequacy. The dependence of soil fertility on climate was investigated on the basis of imitation experiments.

The trends of humus and phosphorus content in the plowed layer were determined in long-term simulation runs of 20, 40 or 80 years duration (Subsection 2.3, *Table 2.7*). Another set of imitation experiments investigated the climate dependence of yields under different levels of nutrition (*see Table 2.8*). The results in *Table 2.8* demonstrate that the maximum achievable level of winter wheat yield differs between agricultural zones, preceding land uses, fertilizer application rates and climate.

Of particular importance is the difference between winter wheat yields after fallow and after nonfallow growing seasons – a difference that increases significantly with increasing aridity. This implies that in planning grain cultivation in these regions, special consideration should be given to evaluating the degree of aridity, for this offers an important indicator of the fallow requirement. It is clear that, should there be a trend toward greater aridity, then the areas under fallow will become more essential, even in those zones that are at present favorable for agriculture.

4.3. Measures to Stabilize Yields and Maximize Production

Bearing in mind the importance of fallows in dry climatic conditions, further experiments were carried out to study the influence of fallow areas and their allocation on yield stability and total production (*see* Subsection 2.5). The results of the experiments, conducted with the crop production model, are summarized in *Table 4.1.* Note that some, but not all, of these results are also shown in *Table 2.9.* The values illustrate that it is possible to increase average production a little under the existing climate and rate of fertilizer application, but only at the expense of decreasing yield stability. By increasing the application of fertilizers, however, it is possible to reduce the little area under fallow slightly, and to redistribute the fallow areas between zones in order to maximize total production under the baseline climate. The grain in production is then about 18% (*Table 4.1*). If the goal is to maximize yield stability under the same conditions, then total grain production is still 11% higher than the reference level (1951–1955 annual average).

Table 4.1. Summary of experiments to determine fallow allocations required to maximize total production and to stabilize yields for two levels of fertilizer application and two climatic scenarios.

Agricultural zone	*Present-day fallow allocation* (%) *and fertilizer rate* ($N_{60}P_{60}$)	*Allocation maximizing total production; high fertilizer rate* ($N_{60}P_{60}$) *Baseline climate* (S1)	*Increased aridity* (S3)	*Allocation maximing yield stability; high fertilizer rate* ($N_{60}P_{60}$) *Baseline* (S1)	*Increased aridity* (S3)
I	49.1	33.2	50.0	50.0	50.0
II	34.3	24.9	50.0	50.0	50.0
III	24.4	0	50.0	0	50.0
IV	17.1	0	0	0	0
V	0	0	0	0	0
Relative total production (%)	100	118	63	111	62

It follows from these calculations that higher levels of nutrition cannot effectively substitute for the efficiency of fallows because the main limiting factor under the conditions in the Stavropol region is water. However, through the judicious use of fertilizers that makes the best use of water accumulated by fallows, it is possible to increase yields by a certain amount.

The selection of an appropriate fallow allocation assumes even greater importance if the climate becomes more arid in the future. In this case, as the calculations show for scenario S3, the optimum fallow allocation would require a similar share of fallow in zones I, II and III. Under these circumstances it would be necessary to increase the area under fallow throughout the whole Stavropol territory, and restrict the cultivation practices in most areas to a simple fallow → wheat rotation.

4.4. Technological Adjustments to Minimize Expenditures

The second series of optimization experiments dealt with the linkage of the crop production model to an allocation model to compute total expenditures required to achieve planned production targets (*see* Section 3). Although the approach, model and experimental criteria are quite different from those discussed in Section 2, the results obtained are in good agreement – namely, that the allocation of fallows is directly related to total grain production. The existing fallow allocation is close to optimal for minimizing total expenditures on grain production (*see Tables 3.4* and *3.6*). Under climatic scenarios of increased aridity, an increase of fallow area in every zone is required to maintain the current level of total production. Expenditures also increase, however, with a drier climatic regime.

4.5. Recommendations

The results of this investigation demonstrate the great potential of a model-based approach to agricultural planning. Although the models described above are in an early stage of development and far from perfect, it would seem prudent to expand on these preliminary experiments, so that the approach can quickly be adopted for use as a direct planning tool. Even at this stage, however, the model framework offers an opportunity to provide sound judgments about the likely effects of climatic variations on the level and stability of crop production and on soil fertility, and about the effectiveness of certain technological and mangement adjustments as a buffer against adverse climatic conditions.

B. The Saratov Region

SECTION 5

Introduction

The Saratov region is situated in the southwestern part of the East European plain (***Figure** 5.1*), covering a total area of 100,200 km^2. It lies approximately between longitudes 42°30′ and 51°E, and between latitudes 50° and 52°30′N, and about 80% of the territory falls in the steppe zone. Most of the area (about 85%) is agricultural land, predominantly arable, while approximately 5% of the region is covered by forests. The Volga River (the largest in Europe) dissects the region for a distance of 420 km from northeast to southwest as it flows toward the Caspian Sea. The river thus presents a convenient geographical reference that can be used in delimiting territorial subregions (*see* Subsection 5.2, below).

5.1. Main Regional Climatic Characteristics

The region marks the transition between a cold forest climate to the north and a steppe climate to the south, following Köppen's (1931) classification. The climate of the region is continental, characterized by cold winters with moderate snow cover and warm, dry summers. The coldest month is January (mean monthly temperature about −11°C in the southwest of the region; −14°C in the northeast); and the snow cover usually reaches a maximum depth in March, averaging 30–60 cm in the northwest and 20–40 cm in other areas of the region. Spring tends to be short and dry, and the warmest month in summer is July (mean monthly temperature about 20°C in the northwest and 24°C in the southwest of the region). Mean annual precipitation varies regionally from about 250 mm to 450 mm. Annual evaporation exceeds precipitation, and evaporation during the warm period of the year (April to October) varies from 450 mm to 750 mm. The high frequency of dry winds and droughts is an important feature of the climate for agriculture.

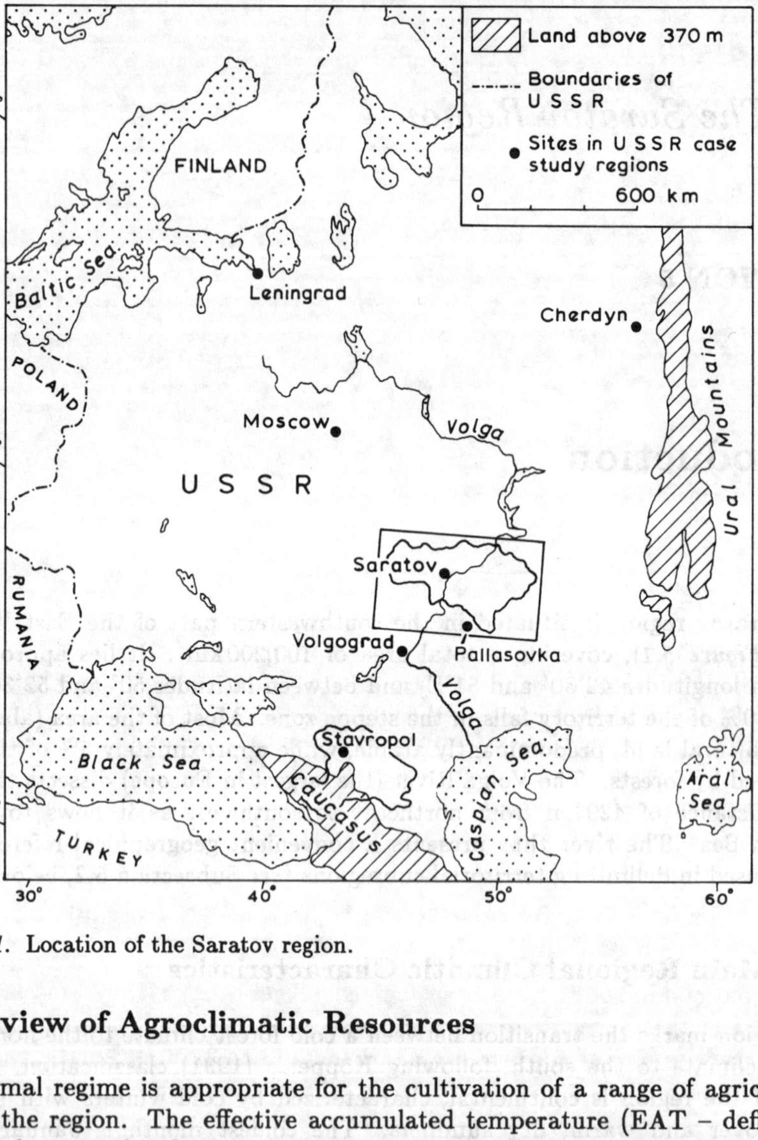

Figure 5.1. Location of the Saratov region.

5.2. Review of Agroclimatic Resources

The thermal regime is appropriate for the cultivation of a range of agricultural crops in the region. The effective accumulated temperature (EAT – defined as the annual accumulated sum of air temperatures above 0° C during the period when daily mean temperature is greater than 10° C) varies between about 2400 degree-days in the northeast and 3000 degree-days in the southeast of the region.

The most limiting factor for agriculture is water. Years with a sufficient water supply during the whole growing season are relatively infrequent (only 28 such years during the 90-year period 1891–1980). The most important period for the growth of grain crops is from May to June, and dry weather conditions have been registered in this period in 40 years of the 90, among them 17 years with severe drought.

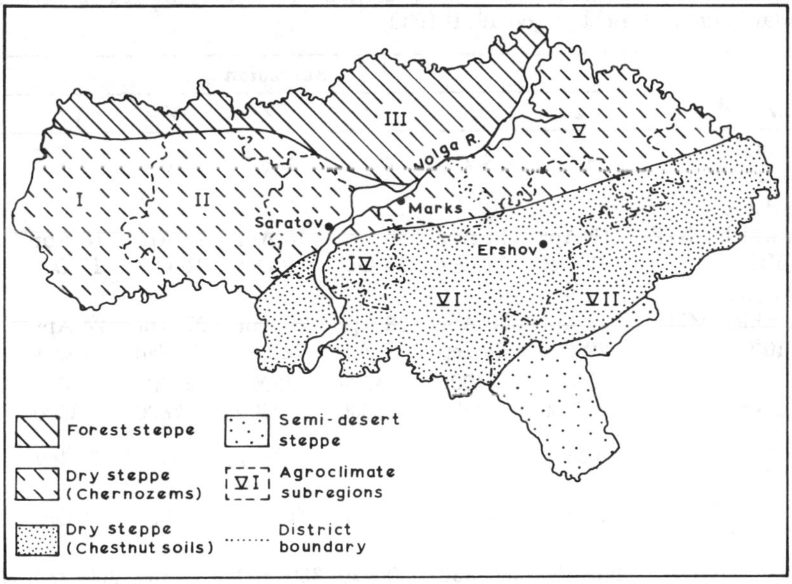

Figure 5.2. Soil types and agroclimatic subregions in the Saratov region.

Four different soil–climate subzones can be identified in the region (*Figure 5.2*):

(1) Dry steppe with chernozem soils.
(2) Dry steppe with chestnut and dark chestnut soils.
(3) Semi-desert steppe.
(4) Forest steppe.

In order to characterize the pattern of agroclimatic resources in the region, seven agroclimatic subregions have been defined (*Figure 5.2*). They are classified on the basis of the thermal and moisture regimes, as well as types of soils, agricultural activities and administrative districts. The main agroclimatic characteristics of the subregions are shown in *Table 5.1* and described below.

I – Western subregion

This is typical dry chernozemic steppe with a wetter and less continental climatic regime compared with other microzones of the region. Spring water storage is considerable: the probability of a level in any one year exceeding 150 mm is 75–77%. Levels decrease to about 60–70 mm by the grain harvesting period. The combination of abundant thermal resources and a high probability of a favorable moisture regime in the latter half of the summer provides an opportunity for high yields, both for early and late crops.

Table 5.1. The main agroclimatic characteristics of the Saratov region (averages for the standard climatological period 1951–1980).

Characteristic[a]	*Subregion* *I*	*II*	*III*	*IV*	*V*	*VI*	*VII*
Mean annual temperature (°C)	4.5	5.2	4.2	4.1–5.7	5.0	4.8	4.9
Average dates between which MDT exceeds 5°C	15 Apr 15 Oct	14 Apr 17 Oct	16 Apr 15 Oct	15 Apr 19 Oct	15 Apr 18 Oct	15 Apr 17 Oct	15 Apr 18 Oct
Average dates between which MDT exceeds 10°C	27 Apr 29 Sep	30 Apr 28 Sep	28 Apr 30 Sep	26 Apr 28 Sep	27 Apr 29 Sep	24 Apr 2 Oct	24 Apr 2 Oct
EAT (degree-days)	2400–2800	2400–2800	2400–2600	2600–2800	2600–2800	2800–3000	2800–3000
Average date of last spring frost	8 May	5 May	7 May	3 May	1 May	2 May	2 May
Average date of first autumn frost	23 Sep	30 Sep	28 Sep	30 Sep	30 Sep	1 Oct	27 Sep
Mean annual precipitation (mm)	400–500	350–450	400–420	350–400	320–360	300–330	255–326
Mean April–October precipitation (mm)	250–310	250–310	250–295	225–265	180–203	190–217	160–200
Precipitation during period when MDT exceed 10°C (mm)	225–251	183–230	183–236	180–201	160–180	141–155	112–128
Spring soil moisture storage in the top 1 m (mm)	150–175	125–165	140–160	120–150	125–150	100–125	70–90
Average water storage in snow under the maximum snow cover (m^3/ha)	560–1100	560–1100	800–1150	520–1000	520–1000	500–800	500–800
Mean maximum snow depth (cm)	20–30	20–40	30–60	20–40	20–40	20–30	20–30
Number of days with a dry wind	13	14–25	15	21	22–27	25–28	27–36
Hydrothermal coefficient (*HTC*)[b]	0.8–0.9	0.7–0.8	0.7–0.8	0.6–0.7	0.6	0.5	0.4–0.5

[a] Abbreviations: MDT = mean daily temperature, EAT = effective accumulated temperature.
[b] $HTC = Pn \cdot 10/\text{EAT}$ where *Pn* is total precipitation (mm) and EAT is effective accumulated temperature (*see* Subsection 5.2).

II – Central right-bank subregion

Subregion II is bordered to the east by the Volga River, hence the "right-bank" designation. Also located in dry chernozemic steppe, this subregion records on average less precipitation in summer than subregion I. A soil moisture storage of greater than 150 mm in early spring is recorded in 5–7 years of every 10, which implies favorable conditions for winter cereals, sunflowers, sugar beet and other crops that are highly responsive to soils that are deep and thoroughly moist in springtime. The abundant thermal resources offer the possibility, as in subregion I, for supplementary agricultural production of heat-loving and late-ripening crops when the precipitation regime is favorable in the second half of the summer.

III – Northern right-bank subregion

Subregion III lies in the forest–steppe transition zone. As in subregions I and II, spring moisture storage is plentiful, levels above 150 mm occurring in 75% of years. However, the summer months receive slightly less precipitation than in those subregions. The climatic conditions are favorable for winter-sown crops and sunflowers. The soils in the subregion vary from light grey forest soils to different types of leached chernozems. The depth of chernozemic topsoils varies from 20 to 60 cm, and the humus content of the soils in the subregion varies from 5% to 7.6%.

IV – Suburban subregion

This subregion is so labeled because it surrounds the urban center of Saratov. Areas on the right bank of the Volga are warm in summer, with occasional droughts; areas on the left bank are warmer still and more drought-prone. In the subregion as a whole, spring soil moisture storage exceeds 150 mm in 40–50% of years. Thermal resources are considerable, especially in the areas close to the river, which are favorable for growing fruit and vegetables. Soils are southern chernozemic and dark chestnut, with a humus content of 3–5% extending over a profile some 50–60 cm deep.

V – Northern left-bank subregion

Subregion V has a drier and more continental climate than the subregions to the west. As a result, water from the spring thaw normally penetrates to a depth of only about 70 cm in the soil. Soil moisture storage is therefore only occasionally above 150 mm (in 3 out of every 10 years). The relatively deep snow cover (20–40 cm), nevertheless, is often insufficient to protect crops from severe winter frosts. The area is referred to as the most frost-susceptible in the region, with an average absolute minimum winter temperature of about −34°C. Most of the subregion has southern chernozemic soils, while the remainder has meadow chernozems, dark chestnut or saline soils. Humus content in the southern chernozemic and dark chestnut soils varies from 3–4.5%.

VI – Central left-bank subregion

High temperatures and low precipitation during the growing season combine to produce very dry conditions in Subregion VI. These factors help to explain the dominance of chestnut soils, which replace the chernozems found further west

Table 5.2. Agricultural land in the Saratov region: soils, topography, drainage, runoff and erosion.

	Subregion						
Descriptor	*I*	*II*	*III*	*IV*	*V*	*VI*	*VII*
				Agricultural land (1000 ha)			
Total area	1091.7	1190.5	939.3	603.7	1200.5	1415.6	2046.7
Arable area	875.3	933.2	701.4	456.8	927.0	1140.2	1341.1
				Major soils (% of agricultural land)			
Leached chernozem	5.0	–	16.5	–	–	–	–
Typical chernozem	38.4	11.6	–	–	–	–	–
Common chernozem	47.9	48.3	29.1	7.6	–	–	–
Southern chernozem	–	9.5	8.5	21.2	51.1	11.7	–
Chernozem on bedrock alluvium	–	20.4	44.1	19.7	–	–	–
Dark chestnut	–	–	–	29.4	41.0	42.0	16.7
Chestnut	–	–	–	–	3.4	35.3	43.4
Light chestnut	–	–	–	–	–	–	9.4
Salt marsh	–	–	–	–	2.9	6.1	21.1
Other	8.7	10.2	1.8	22.0	1.6	4.9	9.4
				Slopes of arable fields (classed frequencies, %)			
< 1°C	49.7	40.4	22.3	55.7	51.0	60.6	57.6
1–3°	39.0	46.4	43.4	34.6	36.0	33.9	34.6
3–5°	9.5	11.5	20.8	7.6	11.0	4.8	6.3
> 5°	1.8	1.7	13.5	2.1	2.0	0.7	1.5
				Drainage			
Dissection by natural water courses (km/km^2)	0.5–0.6	0.6–0.7	0.6–0.9	0.3–0.6	0.3	0.2–0.4	0.2–0.4
Depth of local erosion (m)	50–100	50–200	50–270	50–250	0–150	0–100	0–100
Number of rivers per km^2	3–7	4–7	5–6	1–6	2	1	1
				Runoff (mm)			
From rain	30–45	30–40	20–45	15–30	15–25	10–15	10–15
From snow	40–60	45–60	45–55	25–45	30–40	20–30	15–25
Over plowed fields	10–16	10–18	25	9–15	7–10	7–10	6
Over winter-cultivated fields	50–70	50–85	70	40–60	50–80	35–80	30–60
				Erosion			
Arable land subject to erosion (%):							
by water	44.9	59.1	59.5	44.6	45.3	51.8	54.5
by wind	2.3	3.5	5.1	6.5	1.2	2.3	0.7
Average no. of days with dust-storms in warm period of year	0.6–1.5	0.6–1.5	0.8–2.5	0.9–8.1	0.4–3.3	3.3–6.1	6.0–12.9
				Soil conservation			
Woodland (%):							
natural	6.8	9.8	15.8	10.9	3.9	2.1	0.6
shelter belts	2.3	1.9	2.3	2.6	1.9	1.3	0.9
Protective vegetation (% arable):							
fields	0.8	0.6	0.5	0.7	0.7	0.6	0.4
river valleys	0.9	0.9	0.9	0.7	0.3	0.1	0.1
sandy areas	0.04	0.2	0.4	0.5	0.3	0.02	0.06

under a wetter regime. Spring soil moisture storage rarely exceeds 125 mm in the subregion and, along with the common occurrence of dry winds during the early summer in this predominantly arable zone, contributes to a high frequency of dust storms and a significant level of wind erosion of soil.

VII – Southeastern subregion

The greater part of Subregion VII is characterized by a very dry climate. The average depth of wetting in early spring is 55–70 cm, and the probability of spring soil moisture storage exceeding 150 mm (usually associated with a depth of wetting of more than 90 cm) is 6–18%. The water deficit is especially noticeable during the growing season when evaporation is twice as great as precipitation. Droughts during the growing season are observed in 82% of years. The soil type in the subregion is a relatively homogeneous dark chestnut type, with a humus content of 3–4%.

The characteristics of the soils and some other environmental parameters of the region are shown in *Table 5.2*.

5.3. The Specialization of Regional Agriculture

The agroclimate resources of a subregion define, to a large extent, its specialization and allocation of crops – in particular, the proportion between winter and spring crops. In subregions I, II and III, cultivation of grain crops and cattle breeding are the most important activities. The specialization of the suburban subregion V is cattle breeding and poultry. In Subregions V, VI and VII, cattle breeding, sheep breeding and grain cultivation are the dominant agricultural activities. The share of agricultural production (by value) between different activities is given in *Table 5.3*, based on data from 1981–1984.

Table 5.3. Share of agricultural production by value (%) averaged over 1981–1984.

	Subregion						
Share of production by value (%)	*I*	*II*	*III*	*IV*	*V*	*VI*	*VII*
Grain	19.2	25.0	23.0	13.3	34.6	38.1	43.1
Sunflowers	5.0	4.4	8.2	–	–	–	–
Sugar beet	4.3	–	–	–	–	–	–
Vegetables	–	–	–	7.6	–	–	–
Beef cattle	40.3	43.0	59.6	30.2	34.8	30.5	26.8
Pigs	13.0	8.6	7.5	–	20.4	7.0	–
Sheep	–	–	–	–	6.5	7.0	16.9
Poultry	–	–	–	19.6	–	–	–
Total	(81.8)	(81.0)	98.3)	(70.7)	(96.3)	(82.6)	(86.8)

5.4. Outline of this Study

The role of water as a limiting factor for agriculture in the Saratov region is a central theme of the next two sections. The study focuses on the sensitivity of agriculture (specifically, of grain production) to climatic variations, and considers the range of options for coping with drought both at the farm level and in a regional planning context.

In the following sections, representative weather-years (scenarios) have been selected from the historical climatological record as indicative of the characteristic range of climatic conditions in the study area. The effects of these scenario conditions on spring wheat yields are evaluated, in Section 6, using a dynamic crop–weather model for individual sites in the region. In Section 7 a crop allocation model is used to assess the maximum economic returns of crop production under different climatic scenarios both for the actual present-day crop allocation pattern and for an optimal allocation appropriate to the scenario climate. In addition, the effect of a possible longer-term change in global climate on spring wheat yields in the region is investigated in Section 6. The value of seasonal weather forecasts for agriculture is also demonstrated, both as an aid for irrigation scheduling (explored using the crop–weather model in Section 6), and as an indicator of the most appropriate crop allocation strategies for optimizing regional production under the forecasted climate (described in Section 7).

In order to provide a reference with which to compare the results of scenario experiments, a baseline climate has been selected, represented either by the standard 30-year climatological averaging period 1951–1980 (in Section 6) or by a representative "normal" weather-year (in Section 7).

SECTION 6

The Effects on Spring Wheat Yields

6.1. Introduction: Alternative Approaches to Crop Yield Estimation

In this section a dynamic crop–weather model is used to examine the sensitivity of spring wheat yields to short-term climatic variability and long-term climatic changes in the Saratov region. A dynamic modeling approch based on the theory of production in agricultural ecosystems (Sirotenko *et al.*, 1984) is adopted in preference to other less process-oriented approaches to crop yield estimation: expert judgment (e.g., NDU, 1978 and 1980), regression models (e.g., Scammell, 1981) or physical statistical models (e.g., Menzhulin and Savvateev, 1980; Monteith, 1981; Terjung *et al.*, 1976). A more complete description of these alternative approaches can be found in the companion volume to this book (Part V, Section 3).

6.2. The Dynamic Crop–Weather Model

A detailed description of the dynamic crop–weather is given in Sirotenko (1981) and Sirotenko *et al.* (1982). A brief description is also available in Pitovranov *et al.* (1988). In terms of mathematics, the model is a combination of two interdependent sets of ordinary differential equations incorporating the nonlinearities that govern the dynamics of plant growth and soil water content during the growing season. These include the partitioning of phytomass to specific plant organs (e.g., stem, roots, leaves and grain) and the water status in 10 cm layers of the soil down to a depth of 1.5 m. The model operates with a one-day time step for a calculation period from emergence to maturity of a particular crop.

The model has been used to simulate the growth and yield of a number of crops, mainly cereals. Independent testing of a version for spring wheat has shown that the model explains about 80% of the variance of actual yields. It has been validated for a range of environmental conditions in the USSR, extending from the humid forest zone (toward the northern limit of spring wheat cultivation) to the dry steppe zone examined in this section. Plots of model-derived and actual (experimental station) annual yields for the period 1972–1981 were

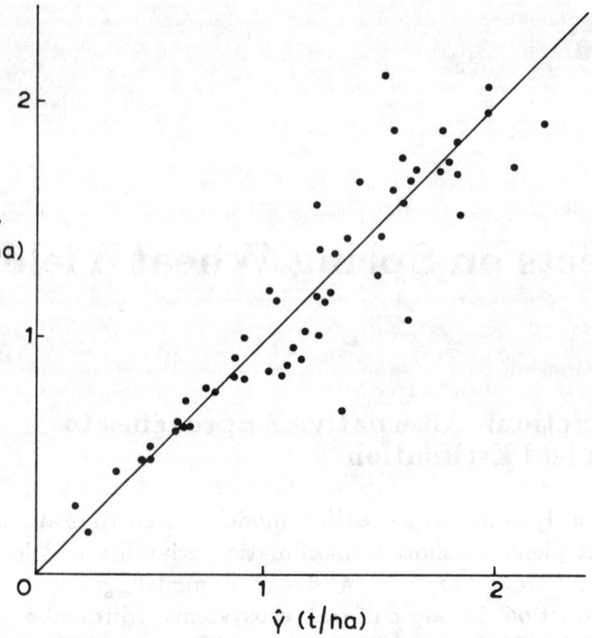

Figure 6.1. Relationship between actual (Y) and estimated ($\hat{Y}$) mean regional spring wheat yields in the Volga basin, 1975–1982 (correlation coefficient, $r = 0.92$; significant at the 1% level).

presented in the companion volume (Pitovranov *et al.*, 1988) for two sites (for locations *see Figure 5.1*): Cherdyn in the Perm region, northern European USSR (latitude 60.4° N) and Ershov in the present study region of Saratov (latitude 51.3° N). A further plot of estimated versus actual yields at several sites in the Volga basin is reproduced here to show the close agreement of model estimates to observed yields in the present study area (*Figure 6.1*). The capabilities of the model for estimating crop yield in semi-arid conditions are further illustrated below, for the Ershov location in the Saratov region.

6.3. Background to the Study Area

Ershov is situated approximately 100 km east of the left bank of the Volga River in agroclimatic subregion VI of the Saratov region (*see Figure 5.2*). As noted in Section 5, the region is characterized by a dry continental climate, owing to both the remoteness of the Atlantic and to the proximity of the Central Asian deserts. Ershov itself is located in the dry steppe zone, on soils that are predominantly of the dark chestnut type. Grain production is the major branch of agriculture, the leading crop being spring wheat, which occupies 60% of the sown area. Only about 12% is under winter cereals.

Annual precipitation at Ershov is about 330 mm, with an average of 88 mm recorded during the period from sowing to maturity of spring wheat. At the time of sowing, the available water in the top 1 m of soil averages 106 mm. During the growing season for spring wheat, actual evapotranspiration averages 182 mm, which is only about 35% of the optimal water demand. The hydrothermal coefficient *HTC* (*see Table 5.1*) for the area is 0.5 to 0.6. The duration of the period with mean daily temperatures above +5°C varies from 184 to 189 days, and the effective accumulated temperature during this period ranges from 3050 to 3250 degree-days.

Among the adverse weather conditions that occur in the region, special reference should be made to the dry winds that occur in June and July nearly every year. In July, the probability of medium-intensity dry winds is 100% at Ershov, while that of very intense dry winds (causing severe leaf wilting and grain capture) is 45%.

The agroclimatic conditions of the area are thus characterized by an adequate heat supply for early spring crops, but by sharply deficient amounts of precipitation in 6–7 years out of 10. Further information on the agricultural and environmental characteristics of the Ershov area (subregion VI) is given in *Tables 5.1* and *5.2*.

6.4. Estimating Yield Responses to Observed Climatic Variability

Two approaches may be distinguished in the application of dynamic crop production models to climatological studies:

(1) Estimating the effects of the observed *variability* of climate on crop yields, whether short-term variations (monthly or seasonal), medium-term (up to 2–3 years) or longer-term (i.e., as observed in the instrumental climatic record). The effects of these are examined in this subsection.
(2) Estimating yield responses to possible longer-term *climatic change*. These are considered in Subsection 6.5, below.

6.4.1. Responses to interannual climatic variability

Table 6.1 presents the results of calculations for four years that represent the interannual variability of weather conditions typical for the steppe zone, ranging from the dramatically dry year, 1975, to the abundantly wet year, 1978. The first two columns in *Table 6.1* give a further indication of the validity of the crop–weather model. Three types of reference yield are depicted in *Table 6.1*: under an ideal water regime, under an ideal level of mineral nutrition, and under both an ideal water regime *and* an ideal level of mineral nutrition.

These calculations confirm that in many dry regions (including the steppe zone) the moisture conditions generally determine the "first minimum" level of yields. The optimization of mineral nutrition proves to be effective only in years

Table 6.1. Actual spring wheat yields (t/ha) and those simulated by the crop–weather model[a] for Ershov, Saratov region.

			Potential yield		
Year	*Actual yield*	*Simulated yield*	*Ideal water regime*	*Ideal nitrogen nutrition*	*Ideal water regime and nitrogen nutrition*
1975	0.4	0.1	7.0	0.2	9.4
1976	1.2	1.2	7.3	2.0	9.4
1977	0.7	0.9	7.0	1.2	8.9
1978	3.0	3.1	7.0	4.1	9.0

[a]The values in *Tables 6.1* and *6.2* have been calculated using a new version (1985) of the crop–weather model that incorporates the dynamics of available forms of nitrogen in soil.

of fair (1976) to good (1978) moisture conditions. The calculation of potential yields demonstrates that under irrigation the edaphic and climatic conditions of Ershov would make it possible to obtain stable yields of about 7.0 metric tons/ha.

6.4.2. Responses to seasonal weather variability

Figure 6.2 depicts plots of within-season yield estimates for the same four years as in *Table 6.1.* Each point on a curve in *Figure 6.2* (e.g., point X) corresponds to the calculated *final* yield, based on the actual (observed) weather conditions prior to that date and long-term average (1951–1981) weather conditions following the date. Yields are given relative to the long-term mean yield. The initial point of any curve fixes the yield level according to the observed water content at the date of emergence and average climatic conditions, while the final point represents the yield level under the observed weather conditions of the given year. In fact, *Figure 6.2* represents detailed agroclimatic portraits of individual years. For example, in 1975 the meteorological conditions during the growing season progressively deteriorated for spring wheat production, the poor initial state of estimated yields (71% of the long-term mean) quickly reduced by the end of the growing season to a pitifully low estimate (8%). In contrast, the 1978 situation was quite different, the starting and final estimates being 140% and 202%, respectively.

If the appropriate data are available, a detailed analysis of the agroclimatic conditions of a specific year can be conducted. An example is illustrated in *Figure 6.3*, which shows estimates of the individual effects on yield of temperature, air humidity, growing season precipitation and soil water content at the time of sowing. These have been obtained by replacing the corresponding long-term averages (i.e., the baseline scenario – *see* Section 5) by the observed values for each year during the period 1951–1981.

As expected, the water regime greatly affects crop yields under the dry conditions of the Volga basin. Thus, the yield estimates based on observed values of

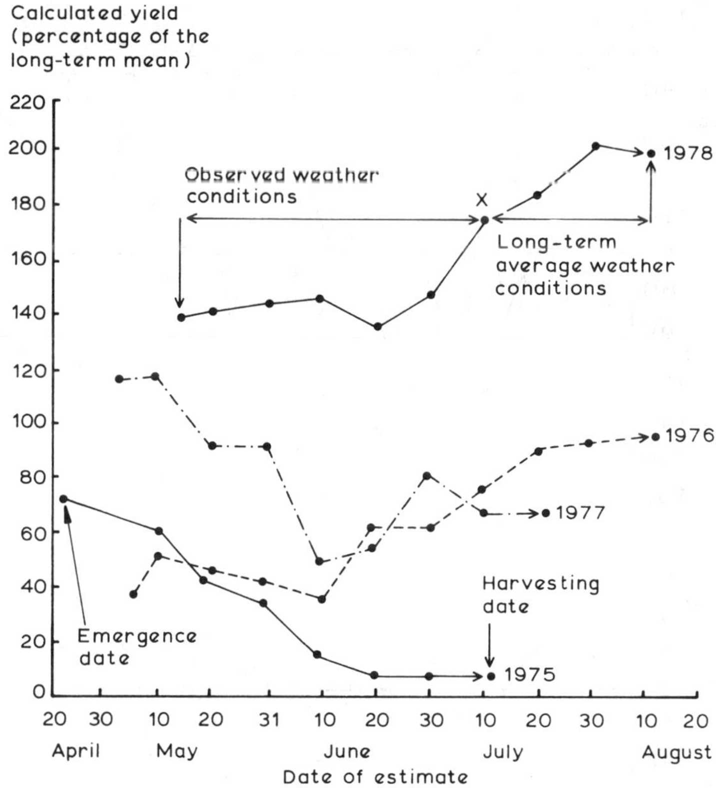

Figure 6.2. Harvest yields of spring wheat in four years at Ershov (percentage of the long-term mean yield) from model calculations at different dates in the growing season. For explanation, *see* text.

air humidity [*Figure 6.3*(*a*)] vary from 65% (1957) to 175% (1978); those based on observed precipitation [*Figure 6.3*(*b*)], from 57% (1957) to 166% (1958); and those based on the observed initial soil water content [*Figure 6.3*(*c*)], from 53% (1972) to 142% (1953).

Over the same period, temperature conditions did not affect estimated yields as appreciably as precipitation or spring soil water storage. In particular years, however, such as 1954 or 1972, temperature anomalies resulted in a 20–25% yield decrease [*Figure 6.3*(*d*)].

It is also possible, with this approach, to interpret the combined climatic influence on the crop in terms of its individual components. For example, in 1962 exceptionally favorable climatic conditions emerged, thanks to the fortunate coincidence of good moisture conditions during the growing and pre-growing seasons, their respective estimates being 160% (precipitation) and 125% (soil water content) relative to the baseline. In contrast, conditions in 1976 were close to normal as a result of an abnormally wet and cool spring and summer, compensating for an exceptionally low spring soil water storage (55% of the normal). Estimates for the temperature and precipitation conditions in 1976 are 115% and 130%, respectively.

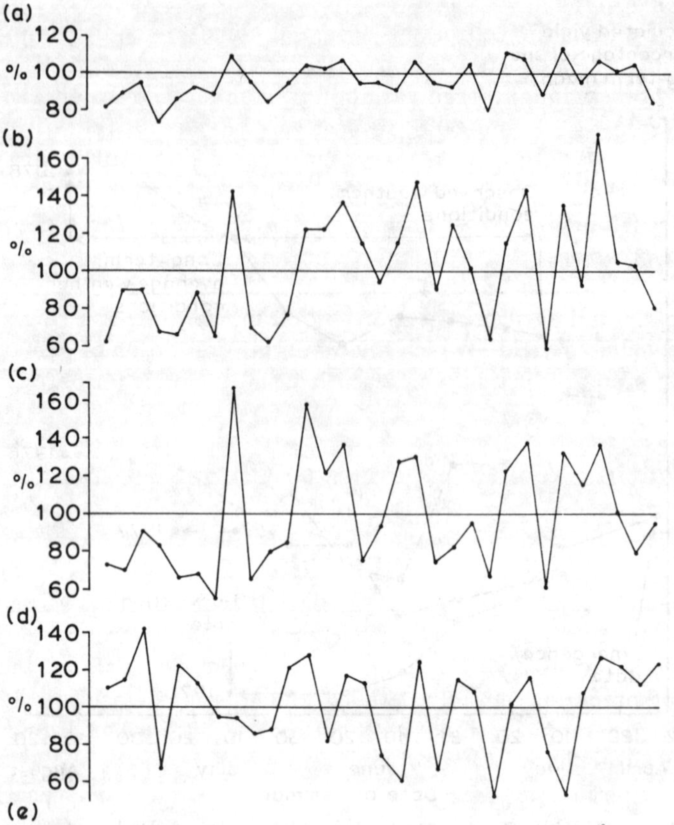

Figure 6.3. Estimated effects of climatic conditions on spring wheat yields in the dry steppe zone (Ershov agrometeorological station). Influence during the growing season of (*a*) mean air temperature, (*b*) mean air humidity and (*c*) precipitation; influence at emergence date of (*d*) soil water storage; and (*e*) combined influence of all modeled climatic variables.

Finally, by including all of the observed climatic conditions in the model simulations, ***integrated*** effects on crop yields can be represented. The 1951–1981 series for Ershov exhibits a large variability: standard deviation and coefficient of variation were 64.3% and 69%, respectively [***Figure 6.3(e)***]. The variations were especially large during the most recent decade, with estimates ranging from 13% to 246% of the long-term mean yields. During the whole period 1951–1981, yield estimates fell below 50% of normal on nine occasions, and exceeded 150% of normal on six occasions.

Thus, it is possible on the basis of the dynamic crop–weather model to combine climatic and agroclimatic information in the form of a single integrated index – the calculated yield – which allows an interpretation of climatic variability in terms of crop production, i.e., essentially in economic terms.

6.5. Estimating Yield Responses to Climatic Change

Until quite recently, dynamic models have not been used for estimating yield responses to climatic change. This is due, first, to the fact that there are relatively few models that can be applied to this problem, and secondly, to the difficulties of interpreting the rather general information contained in many scenarios of long-term climatic change. Some of these difficulties, together with procedures for preparing suitably detailed model input data from the often limited data sets generated in climatic scenarios, are outlined in Pitovranov *et al.* (1988) and in Sirotenko *et al.* (1984).

Model "experiments" were conducted for three climatic scenarios for the Saratov region: synthetic, empirically derived and general circulation model (GCM)-derived scenarios. These are of the same type as those conducted for the Cherdyn region in the companion volume to this collection (Pitovranov *et al.*, 1988).

6.5.1. Experiments with synthetic scenarios

In order to estimate the response of spring wheat yields to a range of climatic conditions within the proven range of the model's validity, growing season temperatures and precipitation were adjusted by synthetic increments on either side of their long-term mean values. ***Table 6.2*** shows the yield estimates obtained for 25 different combinations of these arbitrary adjustments in temperature and precipitation (synthetic scenarios). The calculations were all made using long-term average values of initial soil water storage, and are for Palassovka in the Volgograd region (just to the south of the Saratov region – ***see Figure 5.1***).

Under dry steppe conditions such as these, a decrease in temperature prolongs the growing season, contributing to an increase in crop yields that is particularly noticeable with increased precipitation. Conversely, higher temperatures, unless they occur in conjunction with increased precipitation, are detrimental to yields (***Table 6.2***).

Table 6.2. Response of spring wheat yield (as percentages of the long-term mean) to variations in air temperature (ΔT) and precipitation (ΔP) during the growing season (Palassovka, Volgograd region).

	ΔT (°C)				
ΔP (mm)	−1.0	−0.5	0	+0.5	+1.0
−40	79	79	76	76	76
−20	92	92	89	89	89
0	104	103	100	100	99
+20	115	114	110	109	108
+40	125	124	120	118	117

6.5.2. Experiments with an empirically derived scenario

The model has been used for estimating possible changes in spring wheat yields under the assumption of a northern hemisphere warming of 0.5° C. Scenarios of regional climatic change were obtained from published empirical relationships between mean annual northern hemisphere temperatures, and mean monthly regional temperatures and precipitation (Vinnikov and Groisman, 1979; Groisman, 1981). As part of a much wider study using agroclimatic data from 300 stations in the USSR grain belt, model estimates of spring wheat yields were obtained for the Ershov location.

According to the scenario, a warming of 0.45°C would be expected at Ershov, with concurrent decreases in precipitation during the autumn–winter period and during the growing season of 10% and 5%, respectively. The change in the hydrothermal conditions of the autumn–winter period would result, according to our calculations, in a 6% decrease of water content in the top 1 m of soil by the date of emergence. Calculations using the crop–weather model demonstrate that spring wheat yields would decrease by about 10% under this scenario.

6.5.3. Experiments with a GCM-derived scenario

A climatic scenario for a doubling of atmospheric carbon dioxide (CO_2) concentrations, obtained from the Goddard Institute for Space Studies (GISS) general circulation model (GCM), suggests a more radical reconstruction of climatic conditions than the above changes. The GISS GCM generated estimates of climatic conditions under both present-day ($1 \times CO_2$) and doubled ($2 \times CO_2$) concentrations of CO_2 (Hansen *et al.*, 1983, 1984). A detailed description of the GISS $2 \times CO_2$ scenario, along with many examples of experiments performed using the scenario, are reported in Volume 1 (Parry *et al.*, 1988).

The scenario was constructed for Ershov by adding the difference between GISS-generated $1 \times CO_2$ and $2 \times CO_2$ temperature and precipitation values to the mean growing season values at Ershov for the period 1951–1981. The scenario implies an increase in the growing season temperature and precipitation of 3.3° C and 22%, respectively. Calculations show that the spring wheat yield response consistent with this climatic change (assuming present cultivars and management) is an increase of 14% relative to long-term average yields.

It is also possible to obtain a *combined* estimate of the effects of changes in hydrometereological conditions along with the *direct* effects of doubled atmospheric CO_2 concentrations on plant photosynthesis [modeled using the semi-empirical Rabinowitch–Chartier approach – *see* Rabinowitch (1951); Chartier (1970)]. Accounting for this, an increase in yield of 80% is obtained though the estimate is not reliable, since the validity of the carbon dioxide response model has not been tested using direct experimental data.

6.6. Drought Control Measures

We consider here the use of climatic information and simulation models for scheduling irrigation to mitigate drought effects. *Table 6.3* gives calculated spring wheat yield increments (i.e., yield additional to that obtained for a rainfed crop) at Ershov under four irrigation schedules:

(1) Spring water charge (the first column).
(2) Spring water charge and single irrigation on one of four different dates (the second to fifth columns).
(3) Spring water charge and repeated irrigation to maintain the soil water content at 70% of the field capacity.
(4) Optimal irrigation (the final column), where the soil water content is maintained at a level of 70% of the field capacity throughout the soil profile.

Table 6.3. Spring wheat yield increments (t/ha) at Ershov under different irrigation schedules, calculated by the dynamic crop–weather model.

	Irrigation (IRR) schedule						
		SWC[a] + *single IRR*[b]				*SWC + IRR at 70% field*	*Constant 70% field capacity*
Year	*SWC*[a]	*1/3* W_R	*2/3* W_R	W_R	*1/3* W_B	*capacity*	*(optimal)*
1975	0.50	0.61	0.88	0.96	1.06	3.41	6.90
1976	1.86	2.08	2.73	2.83	2.70	3.67	6.10
1977	0.44	0.66	0.99	1.14	0.59	3.89	6.00
1978	0.63	0.84	0.95	1.23	2.15	2.90	3.90

[a]SWC = spring water charge.
[b]W_R and W_B are the effective accumulated temperatures on days with a mean temperature exceeding 5°C during the emergence-heading and heading-wax maturity periods, respectively.

Noteworthy is the large interannual variability in efficiency of the water-charging method of irrigation, the yield increments ranging from 0.44 to 1.86 t/ha, i.e., more than fourfold. The greatest yield increments were obtained when the spring soil water content was low. This regularity makes is possible to predict the efficiency of water-charging irrigation and other water conservation practices from spring water storage data. The data in *Table 6.3* show that the efficiency of irrigation is largely time-dependent. In particular, if the water charge is abundant, irrigation should not proceed during the growing season prior to the crop heading stage.

Finally, the model also facilitates the simulation of the traditional method of water application in order to maintain a critical soil water content, for example, at 70% of the field capacity. Calculations show that high yield increments can thereby be attained (2.9 to 3.9 tons/ha). For the purpose of illustration, *Table 6.4* gives the irrigation schedule calculated by this method for 1976, in terms of the timing of the growing season water applications.

Table 6.4. Calculated irrigation schedule to maintain soil water content at 70% of field capacity at Ershov in 1976.

Days after emergence	29	38	50	55	66
Irrigation amount (mm)	34.5	36.5	38.5	34.1	35.4

The yield increments obtained by this method of water application are large, but still do not exhaust the complete potential of irrigation (*see* the final column of *Table 6.3*).

6.7. Conclusions

In this section we have demonstrated the use of climatic information and dynamic models not only for estimating the influence of possible climatic changes and variability on agriculture in semi-arid regions, but also for assessing the efficiency of various protective measures against adverse climatic conditions. Adaptive measures are also considered in the next section, but at the district scale, where strategies for increasing agricultural production under variable climatic conditions are explored using a crop allocation optimization model.

SECTION 7

Optimal Strategies for Allocating Land to Grain Crops

7.1. Introduction

Losses to agriculture in the USSR caused by unfavorable weather conditions have been assessed at 1.2–2.0 billion roubles per year. A proportion of these losses can be attributed to poor adaptation of regional agricultural activities to the prevailing climatic conditions.

Over the next 10 years it is likely that the major advances in agriculture will occur not through biological or technological development, but rather through the optimal use of more reliable climatic information (Borisenkov, 1981). The aim of this section is to review some different economic strategies for adaptation in agriculture, based on climatic and weather data. In particular, the potential benefits of such strategies for grain production in the Marks administrative district of the Saratov region are examined.

One general economic strategy is considered in this section: the optimization of regional grain crop allocation. The effectiveness of three specific strategies is compared based on the principles of:

(1) Maximizing the minimum benefit.
(2) Minimizing the dispersion of benefit.
(3) Maximizing the average benefit.

The potential benefit of growing season weather forecasts for timely adjustment of the strategies is also considered.

7.2. Climate and Agriculture in the Marks District

The Marks administrative district is located in the central part of the Saratov region, about 65 km to the east of Saratov itself, on the left bank of the Volga (*Figure 5.1*). Only a brief climatic and environmental description of the region will be given in this section. More detailed background information can be found in Section 5, particularly *Table 5.1*.

7.2.1. Climate

The climatic characteristics of the district are typical of those of subregion V. Mean annual temperature is 5.1°C, with the effective growing season (days with mean temperature above 5°C) extending, on average, from 15 April to 19 October. The effective accumulated temperature during the period when mean daily temperatures are above 10°C averages 2600 degree-days.

The main limiting factor for agriculture in the region is deficiency of water: mean annual precipitation is only 360 mm, and growing season (April–October) totals average only 200 mm. Growing season moisture conditions for crops can be classified into three types: dry, limited water, and sufficient water. The frequency and some characteristics of these weather-types are shown in *Table 7.1*.

Table 7.1. The frequency of different types of weather-year (% of years during the period 1894–1984) and their moisture characteristics in the Marks district.

	Weather-year types		
	Dry	*Limited water*	*Sufficient water*
Frequency (% of years)	63	18	18
Spring water storage in top 1 m of soil (mm)	<125	125–150	>150
Growing season precipitation (mm)	<150	150–250	>250

7.2.2. Agriculture

There are 201,700 ha of arable land in the district, 31% of which is irrigated. The average cropped area, yields and production for the period 1981–1984 are given in *Table 7.2*, both for rainfed and for irrigated crops.

The values shown in *Table 7.2* are averages, however, which mask significant differences in productivity in different years. These differences are mainly attributable to variable weather conditions. *Table 7.3* illustrates how yields of different grain crops and legumes varied during three recent years. Each year falls into one of the weather-type categories described above. Changes in management and technology between the years were minimal.

Throughout the remainder of this section, the years 1981, 1982 and 1978 will be regarded as representative of the three weather-types, providing a basis for the construction of climatic scenarios of altered climatic conditions. The grain crops are chosen for optimization experiments because of their dominant share of arable area and their importance in the region.

Table 7.2. Different uses of arable land in the Marks district and areas, yields and production of rainfed (R) and irrigated (IRR) crops. Values in parentheses are percentages of the total arable area. Data are averages for 1981–1984. Source: A. Maximov, personal communication.

	Area (1000 ha)			*Yield* (t/ha)		*Total production*
Land use	*Total*	*R*	*IRR*	*R*	*IRR*	(1000 tons)
Arable land	201.7	138.3	63.4			
Fallow	13.5 (6.7)	13.5	–			
Grain and leguminous crops	132.9 (65.9)	104.0	28.9			134.07
Winter wheat	14.6	10.2	4.4	0.74	2.17	17.10
Winter rye	4.9	4.4	0.5	1.30	1.62	6.53
Spring wheat	34.0	30.7	3.3	0.61	1.14	22.49
Barley	51.9	45.0	6.9	1.06	1.54	58.33
Millet	22.7	13.4	9.3	0.59	1.22	19.25
Maize	4.6	0.1	4.5	0.90	2.25	10.22
Peas	0.2	0.2	–	0.75	–	0.15
Industrial crops	10.7 (5.3)	7.4	3.3			5.51
Sunflowers	8.5	7.4	1.1	0.36	1.36	4.16
Flax	0.1	–	0.1	–	0.70	0.07
Soyabeans	2.1	–	2.1	–	0.61	1.28
Potato, vegetable and melon fields	2.1 (1.0)	0.2	2.0			25.00
Potatoes	1.0	0.02	0.9	1.5	4.38	4.15
Vegetables	0.9	–	0.9	–	15.32	14.25
Melons	0.2	0.1	0.1	–	–	6.60
Fodder crops	42.5 (21.1)	13.3	29.2			519.55
Maize for silage and green feed	15.7	4.9	10.8	8.62	24.43	306.08
Perennial herbs	14.4	3.6	10.8	–	–	162.28
Annual grass crops	6.7	3.6	3.1	–	–	51.19

Table 7.3. Yields of grain crops and legumes in the Marks district for three years: 1981 (dry), 1982 (limited water) and 1978 (sufficient water). Source: A. Maximov, personal communication.

	Yields (t/ha)					
	1981 Dry		*1982 Limited water*		*1978 Sufficient water*	
Crop	*Rainfed*	*Irrigated*	*Rainfed*	*Irrigated*	*Rainfed*	*Irrigated*
Winter wheat	0.50	2.54	0.77	1.93	0.72	1.52
Winter rye	0.88	2.60	1.31	2.17	0.94	1.73
Spring wheat	0.85	1.07	0.95	2.19	1.96	2.34
Barley	0.44	2.60	1.80	2.50	2.04	2.56
Millet	0.27	0.83	0.99	1.48	1.42	1.82
Maize	–	2.03	–	1.81	–	2.68
Peas	0.80	–	1.30	–	1.67	–

7.3. Optimal Allocation of Grain Crops

If we consider the optimal allocation of crops to be that which maximizes the return from grain crop production in a region, then the optimal strategy can be described by:

$$F = \max F(S) \tag{7.1}$$

where S is the crop allocation and $F(S)$ is the objective function. This function takes the following form:

$$F(S) = \sum_{i=1}^{n} (K_i y_i - \alpha_i) s_i + (K_i Y_i - \beta_i) S_i \tag{7.2}$$

where $i = 1,n$ is the set of all agricultural crops, K_i is the purchase price of the ith crop, s_i and S_i are the respective areas and y_i and Y_i are the yields of the ith crop on rainfed and irrigated land, and α_i and β_i are the costs per hectare of producing the ith crop.

The optimization problem is solved with the constraints:

$$\sum_{i=1}^{n} s_i = A_1 , \sum_{i=1}^{n} S_i = A_2 \tag{7.3}$$

where A_1 and A_2 are the total areas of rainfed and irrigated land.

Another set of constraints is produced by decision makers operating at supra-regional levels. These constraints may be formulated as follows:

$$c_i \geq s_i \geq \alpha_i , \quad C_i \geq S_i \geq \beta_i , \quad i = 1,n \tag{7.4}$$

Solving (7.1)–(7.4) gives the optimal crop allocation that maximizes the return from crop production for given amounts of yield. It should be noted, however, that in practice there are certain obstacles to the application of such solutions. For example, a switch from grain crops to industrial or fodder crops would entail investments in different types of machinery, different crop storage facilities, etc. Consideration in this analysis, however, is restricted to grain crops, which are most readily interchangeable. The information required for solving problems (7.1)–(7.4) is given in *Table 7.4*.

It can be seen that the highest purchase prices are for spring wheat. The constraints on allocation to less valuable crops, such as millet, are considerable.

As indicated above, the variability of annual weather conditions has different effects on the performance of different crops. However, this procedure enables us to determine the optimal allocation, of land for a variety of types of

Table 7.4. The constraints on crop allocation (area), purchase prices and production costs in the Marks district (1981–1984 average). Source: A. Maximov, personal communication.

	Constraints on cropped area (1000 ha)				*Production costs* (roubles/ha)		
	Rainfed		*Irrigated*				*Purchase prices*
Crop	*Upper*	*Lower*	*Upper*	*Lower*	*Rainfed*	*Irrigated*	(roubles/t)
Winter wheat	9.0	2.0	35.0	10.0	38.0	146.7	120
Winter rye	2.0	0.5	12.0	4.0	38.8	–	90
Spring wheat	10.0	3.0	50.0	15.0	38.9	145.4	150
Barley	9.0	3.0	45.0	15.0	37.5	145.4	100
Millet	9.5	2.0	18.0	13.0	35.4	143.3	80
Maize	6.0	4.5	0.2	0	–	247.6	80
Peas	0.3	0	2.0	0.2	32.8	–	70

Table 7.5. The optimal allocation of land (1000 ha) in the dry year (1981), limited water year (1982) and sufficient water year (1978). Source: A. Maximov, personal communication.

	1981 (dry)		*1982 (limited water)*		*1978 (sufficient water)*	
Crop	*Rainfed*	*Irrigated*	*Rainfed*	*Irrigated*	*Rainfed*	*Irrigated*
Winter wheat	35.0	9.0	10.0	2.9	10.0	2.0
Winter rye	12.0	1.4	4.0	0.5	4.0	0.5
Spring wheat	27.0	3.0	31.8	10.0	50.0	7.8
Barley	15.0	9.0	4.5	9.0	26.8	3.0
Millet	13.0	2.0	13.0	2.0	13.0	2.0
Maize for grain	–	4.5	–	4.5	–	4.5
Peas	2.0	–	0.2	–	0.2	–
Total	104.0	28.9	104.0	28.9	104.0	28.9

Table 7.6. The actual and optimal return from grain crop production in the three different years (million roubles).

1981 (dry)		*1982 (limited water)*		*1978 (sufficient water)*	
actual	*optimal*	*actual*	*optimal*	*actual*	*optimal*
1.712	4.303	12.562	14.020	20.156	23.790

weather. The optimal allocation of land for three types of weather-year (1981, 1982 and 1978) are given in *Table 7.5*.

A comparison of the optimal structure for different years (*Table 7.5*) with the actual allocation (*Table 7.2*) shows that actual allocation (averaged over 1981–1984) closely resembles the optimal allocation for the year with limited water supply (1982). A further comparison can be made between the actual return on grain crop production in the district and the return for an optimal allocation of crops in the three years (*Table 7.6*). An optimal allocation produces an additional return of from 1.5 to 3.5 million roubles in the different years.

7.4. Climatically Optimal Economic Strategies

In order to develop optimal economic strategies that account for variation in weather and climate, a benefit function B can be defined:

$$B = B(W,D) \tag{7.5}$$

where D is one of a set of possible decisions (Ω_D) made in response to W, one of the set of possible weather conditions (Ω_W). The function (7.5) is defined for all possible W and D, viz, $\Omega_W \times \Omega_D$. If the set $\Omega_W \times \Omega_D$ is discrete, the function can be written as a matrix.

In this study, the number of possible weather conditions is equal to three (the three types of weather-year described above). Let us assume, in addition, that the number of possible economic decisions is also three – namely, the choice between the optimal crop allocations for the dry, limited water and sufficient water weather-years. The benefit matrix can then be written as:

$$B_{jk} = \begin{bmatrix} 4.303 & 1.971 & 0.334 \\ 10.587 & 14.020 & 11.170 \\ 14.909 & 20.624 & 23.790 \end{bmatrix} \tag{7.6}$$

where the elements of the matrix are the return (in millions of roubles) obtained under weather condition j for a crop allocation oriented to weather condition k. Then, where j matches k [the leading diagonal in (7.6)], the crop allocation is optimal for maximized return (cf. *Table 7.6*).

7.4.1. Cautious strategies

In the absence of any information on the probability of certain weather conditions, several alternative economic strategies can be followed for allocating crops in the region. One is the strategy of *maximizing the minimum return*, so that:

$$B_{\max} = \max_k \min_j B_{jk} \tag{7.7}$$

It is evident that for the benefit matrix (7.6) the minimum return is maximized in the first column of the matrix, i.e., the optimal strategy is that one constantly oriented to dry weather conditions. It is thus a strategy of maximum caution and does not use any information about the probability of different climatic events.

Another cautious strategy is one of *minimizing the dispersion of return* which, following (7.5), can be denoted by:

$$G = M_D\,[(B(W,D) - \overline{B(W,D)})^2] \qquad (7.8)$$

where M_D is the probabilistic averaging of the set of possible weather conditions W under the chosen strategy D. Given the probability of different types of weather in the district (*Table 7.1*), it is easy to show that the optimal strategy for benefit function (7.8) is again one of a constant optimal dry weather allocation.

7.4.2. Higher-risk strategies

A third, less cautious strategy is one that involves *maximizing the (statistically) average return*, such that:

$$B_D = M\,[B(W,D)] \qquad (7.9a)$$

or, for the case of a discrete number of decisions and weather conditions,

$$B_{Dk} = \sum_{j=1}^{n} B(W_j,D_k) \cdot D(W_j) \qquad (7.9b)$$

where the notation is as above. The average benefit calculated for the three different strategies in matrix (7.6), knowing the probability of each weather type, is:

(1) Decision based on dry year allocation $B_{D1} = 7.44$ m roubles.
(2) Decision based on limited water year allocation $B_{D2} = 7.69$ m roubles.
(3) Decision based on sufficient water year allocation $B_{D3} = 6.72$ m roubles.

So, in order to maximize the average benefit over a given time period, it is necessary to choose a strategy oriented to limited water conditions. It should be noted that the average return for the actual 1981–1984 allocation B_{act} was even lower than the dry year strategy, at 7.17 m roubles.

7.4.3. Altering the constraints

It is clear that the choice of strategy for maximizing average benefit depends on the benefit matrix – in particular, on the constraints included in the matrix. Let us suppose that we can relax the constraints (7.4) connected with a given structure of land use for crops. Then the benefit matrix is as follows:

$$B_{jk} = \begin{bmatrix} 6.748 & 0.914 & 1.798 \\ 8.002 & 19.913 & 16.013 \\ 5.956 & 23.059 & 40.720 \end{bmatrix} \tag{7.10}$$

In this case the average return for the three strategies is:

(1) $B_{D1} = 6.82$ m roubles.
(2) $B_{D2} = 8.54$ m roubles.
(3) $B_{D3} = 11.75$ m roubles.

Here, the most profitable strategy is oriented to sufficient water conditions.

For a situation where changes in allocation are permissible only for spring crops, while winter crops are allocated according to their present-day areas, the benefit matrix is then:

$$B_{jk} = \begin{bmatrix} 2.35 & 1.85 & 1.65 \\ 12.45 & 13.80 & 11.20 \\ 19.10 & 20.66 & 23.10 \end{bmatrix} \tag{7.11}$$

The returns for the three strategies are thus:

(1) $B_{D1} = 7.15$ m roubles.
(2) $B_{D2} = 7.57$ m roubles.
(3) $B_{D3} = 7.44$ m roubles.

Most favorable in this case, is the limited water strategy.

7.5. The Use of Weather Forecast Information in Strategy Formulation

In this subsection, we explore how weather forecast information can be incorporated into the choice of appropriate economic strategies.

Let us consider n possible weather forecasts $F_l (l = 1, n)$. For each forecast there is a probability of a correct or of an incorrect prediction of weather-type W_j. Thus, the complete forecast is given as a set of combined probabilities $P(W_j, F_l)$. These, can be displayed in a matrix P_{jl}, which is known as the "matrix of conjugation". For the case where the number of possible forecasts are equal to the number of weather types (i.e., equal to three in this study), the matrix of conjugation is as follows:

	F_1	F_2	F_3	$\sum_l P_{jl}$
W_1	P_{11}	P_{12}	P_{13}	P_1
W_2	P_{21}	P_{22}	P_{23}	P_2
W_3	P_{31}	P_{32}	P_{33}	P_3
$\sum_j P_{ij}$	$\bar{P}_1$	$\bar{P}_2$	$\bar{P}_3$	

(7.12)

where P_j $(j = 1, 2, 3)$ is the probability of the different types of weather occurring (*see Table 3.1*). For an assessment of the utility of the forecast, we have used the Baijes approach to deteriving average statistical benefit:

$$B = M_D\,[B(W,D)] \tag{7.13}$$

The algorithm for defining the optimal economic strategy for a given forecast, according to the Baijes approach, is formulated as follows:

(1) The average utility, corresponding to different economic strategies after obtaining the forecast F_l is defined by:

$$B\,(D_k,F_l) = \sum_{j=1}^{n} B_{jk}\,P_{jl} \tag{7.14}$$

where $P_{j|l} = P(W_j \mid F_l)$ is the conditional probability of weather-type W_j occurring after obtaining forecast F_l. These probabilities are connected with the combined probability P_{jl} as follows:

$$P_{j|l} = P_{jl} \mid \bar{P}_l \tag{7.15}$$

(2) The strategy D_l, which maximizes the benefit function, is given as:

$$\max_k B(D_k \mid F_l) \tag{7.16}$$

(3) This procedure is reiterated for all possible forecasts. The result of this procedure is a set of optimal economic strategies corresponding to different weather forecasts.

It is easy to estimate under which circumstances following a strategy of "believing" a weather forecast is more profitable than adopting a climatological strategy based on long-term probabilities of weather types. For the case of the matrix (7.12), it would be profitable to believe a dry year forecast if the inequality:

$$B_{11}P_{1|1} + B_{21}P_{2|1} + B_{31}P_{3|1} > B_{12}P_1 + B_{22}P_2 + B_{32}P_3 \tag{7.17}$$

is satisfied, wheres the right-hand side of (7.17) is the prefered climatological strategy of allocation according to the limited water conditions.

Obviously, it would be possible to select optimal strategies of crop allocation that differed from the climatological ones if forecasts were available by, at latest, the early spring about the type of weather to be expected in the growing season. To illustrate this, one early indicator of growing season moisture conditions in a semi-arid region such as the Marks district, is the spring water storage in the top 1 m of soil in autumn-plowed land. Using a method developed by Fedoseev (1971) for this region, and forecasting seasonal weather-types according to spring water storage, the following matrix of conjugation for different predicted yields (which can be attributed to different weather-types W_j) is obtained:

	F_1	F_2	F_3	Σ
W_1	0.45	0.10	0.08	0.63
W_2	0.05	0.10	0.03	0.18
W_3	0.03	0.05	0.11	0.19

(7.18)

If we use the benefit matrix (7.11) where changes in allocation are only possible for spring grain crops, then according to formulae (7.14)–(7.16), a strategy of confidence in the forecast is markedly more beneficial than a climatological strategy, the average benefit B equal to 17.6 million roubles.

7.6. Conclusion

The foregoing results for the Marks district illustrate the complexities of choosing appropriate crop allocation strategies under a highly variable climate. Factors that determine an optimal strategy include:

(1) The farming objectives: for example, goals of production maximization, production stability or risk avoidance.
(2) The constraints imposed on agriculture by, for example, production quotas, input costs, labor, machinery, purchase prices, and land suitability.
(3) The expected weather conditions.

Allocation strategies that are optimal under specific types of weather conditions have been defined and their performance (in terms of value of crop production) compared with that obtained under existing allocation patterns. The results indicate that:

(1) An optimal allocation determined on *climatological* grounds (based on a knowledge of the probability of certain weather-types) would yield higher returns than the existing allocation.

(2) The benefits of adjusting allocation patterns according to *seasonal weather forecasts* are potentially great; experiments in the Marks district suggest an increased return on production of more than double the existing levels.

An important next step, therefore, is to evaluate the *levels of confidence* in seasonal weather forecasts that would be required before forecasts could be used operationally to determine optimal allocation patterns, in place of the climatological strategy.

SECTION 8

Conclusions and Recommendations

8.1. Synopsis of the Major Findings

The Saratov region, like much of the steppe zone in the European USSR, is characterized by relatively fertile soils, but a highly variable climate, particularly with respect to seasonal precipitation. Therefore, the planning and conduct of a weather-sensitive activity, such as agriculture, calls for considerable ingenuity and enterprise at both farm and regional levels. This study has examined the sensitivity of crop production to climatic variations on both scales of analysis.

8.1.1. The farm level: spring wheat yields

In Section 6, a dynamic crop–weather model was employed to evaluate yield responses of spring wheat to a number of climatic scenarios. The results indicated that:

(1) Yields of rainfed crops display a positive relationship to growing season precipitation, but a negative relationship to seasonal air temperatures (because of increased water loss through enhanced evapotranspiration).

(2) Due to the conflicting projections of future climatic change in the region, the implications for future spring wheat productivity are contradictory:

   (a) An empirical technique that relates regional temperatures and precipitation to mean annual northern hemisphere temperatures (Vinnikov and Groisman, 1979) suggests that a future hemispheric warming of 0.5°C would be mirrored by a growing season warming in the Saratov region of 0.45°C accompanied by decreased precipitation (−5%). Spring wheat yields under this scenario would be *reduced* by about 10%.

   (b) Conversely, the climatic change estimated to occur in response to a doubling of atmospheric carbon dioxide concentrations by the Goddard Institute for Space Studies general circulation model (the GISS 2 × CO_2 scenario) consists of a considerable warming during the growing season in the Saratov region (+3.3°C) but accompanied, in this case, by an increase in precipitation (+22%). A 14% *increase* in mean yield was estimated for this scenario, assuming present cultivars and management.

The assumption that crop management activities would remain unchanged over the long time scale represented by a CO_2-induced climatic change is clearly unrealistic, as farmers are continually seeking new approaches to optimize crop productivity under variable climatic conditions. Further experiments with the crop–weather model indicated the large gulf that exists between the actual yields observed in the field, and the optimal yields that could be obtained under ideal moisture and nutrient conditions. The results, summarized in *Table 8.1*, demonstrate that significant progress could be made towards narrowing this gulf through improved irrigation scheduling and fertilizer application.

Table 8.1. Estimated spring wheat yields for rainfed conditions during three types of weather-year and potential improvements in yields obtainable with fertilizer applications and irrigation. Source: *Tables 6.1* and *6.3*.

	Weather-year types		
Model estimates	*Dry (1975)*	*Average (1976)*	*Wet (1978)*
Yield – rainfed (t/ha)	0.1	1.2	3.1
Yield increments (t/ha):			
Optimal fertilizer application	+0.1	+0.8	+1.0
Irrigation:			
Spring charge[a]	+0.5	+1.9	+0.6
Early season[b]	+0.6	+2.1	+0.8
Late season[c]	+1.1	+2.7	+2.2
70% field capacity[d]	+3.4	+3.7	+2.9
Optimal water	+6.9	+6.1	+3.9
Optimal water + fertilizer	+9.3	+8.2	+5.9

[a] Based on stored water available before planting.
[b] Spring charge plus single irrigation following emergence stage.
[c] Spring charge plus single irrigation following heading stage.
[d] Spring charge plus repeated irrigation to maintain soil water content.

8.1.2. The regional level: crop allocation

A wide range of crops is grown in the Saratov region, each exhibiting its own particular sensitivity to climatic variations. With a centrally planned system of agriculture, in order to maintain satisfactory levels of regional production while keeping costs low, the allocation of land to different crops is strictly controlled. However, the results in Section 7 for the Marks district suggest that the average returns on crop production (7.17 m roubles per year) are not as high as they might be if the crop mix were more closely attuned to the variability of regional climate (*see Table 8.2*). Moreover, if some of the present-day constraints on cropped area were relaxed, mean returns could be further improved by altering the regional crop allocation.

Table 8.2. Summary of crop allocation experiments in the Marks district: Expected long-term minimum and mean returns on crop production under present-day climatic conditions for crop allocation patterns optimized according to specific types of weather-year and with different constraints on allocation. Allocation strategies yielding the maximum return are indicated in bold type. Also shown is the average return on crop production from an allocation strategy based on forecasted moisture conditions. Source: Benefit matrices and results in Section 7.

		Returns (m roubles) on allocation optimized for:			
		Single weather-type			
Constraints on crop allocation[a]	*Nature of long-term return*	*Dry*	*Limited water*	*Sufficient water*	*Forecasted moisture conditions*[b]
Present-day constraints	Minimum	**4.30**	1.97	0.33	–
	Mean	7.44	**7.69**	6.72	–
No area constraints (all crops)	Minimum	**5.96**	0.91	1.80	–
	Mean	6.82	8.54	**11.75**	–
No area constraints (spring cereals only)	Minimum	**2.35**	1.85	1.65	–
	Mean	7.15	**7.57**	7.44	17.6

[a]Based on areas, production costs and purchase prices of crops.
[b]Based on spring soil water storage.

8.2. The Role of Seasonal Weather Forecasts

A topic that was addressed in both model-based assessments was the potential role that accurate weather forecasts could play in the planning of crop management activities and crop allocation. The benefits of such information include higher crop yields, more efficient conservation and use of irrigation water, and improved returns on regional crop production through appropriate crop mixes designed to exploit the forecasted weather conditions.

In both Sections 6 and 7, examples were presented of forecasts of growing season moisture conditions based on spring soil water storage. Encouraging as these preliminary experiments are, the prospects of obtaining accurate *meteorological* forecasts of seasonal weather conditions in the foreseeable future remain poor. Some numerical experiments by Marchuk *et al.* (1979) have related summer temperature anomalies in the European part of the USSR to ocean temperature anomalies in the Iceland region, and investigations of ocean–atmosphere processes appear to hold promise of some progress toward adequate seasonal forecasts. However, at present the period of effective forecasts based on numerical modeling of atmospheric processes is limited to a few days or, at most, one week.

8.3. Recommendations

Several themes have emerged from this study that merit further attention:

(1) *Dynamic crop–weather models* analogous to the spring wheat yield model can be very useful for examining crop management options as well as for evaluating yield sensitivity to climate. Models should be developed for other crops in order to compare the differential effects of climate and management factors on different crops in the same region. These models could also provide useful input information for allocation models.

(2) *Improved probabilistic information* on regional climate is required to provide a firmer basis for long-term planning of crop allocation.

(3) An evaluation of the *level of confidence* in a seasonal weather forecast, which would be required before it could be used as an operational tool, would be useful to provide a marker against which to gauge improvements in the reliability of weather forecasts.

Clearly, while the problems of sustaining agricultural production in such a variable environment are many, they are not all insurmountable, and the potential benefits to be gained from further research are considerable.

References

Borisenkov, E.P. (1981). The study of climate and its applied aspects. *Meteorologia i Gidrologia*, **6** [in Russian].

Chartier, P. (1970). A model of CO_2 assimilation in the leaf, in I. Setlik (ed.) *Prediction and Measurement of Photosynthetic Productivity*, Pudoc, Wageningen, Netherlands, p. 307.

Fedoseev, A.P. (1971). The inertial forecast of the water storage in a soil and its economic efficiency. *Meteorologia i Gidrologia*, **6**, 111–120 [in Russian].

Goudriaan, J. and van Laar, H.H. (1978). Calculation of daily totals of the gross CO_2 assimilation of leaf canopies. *Neth. J. Agric. Sci.*, **26**, 373–382.

Groisman, P.Ya. (1981). Empirical estimates of the reaction of precipitation in the USSR to global warming and cooling processes. *Izvestia Academii Nauk SSSR, Ser. Geograph.*, **5**, 86–95 [in Russian].

Hansen, J., Lacis, A., Rind, D., Russell, G., Stone, P., Fung, I., Ruedy, J. and Lerner, J. (1984). Climate sensitivity: Analysis of feedback mechanisms. In J. Hansen and T. Takahashi (eds.) *Climate Processes and Climate Sensitivity*. Maurice Ewing Series, **5**. pp. 130–163. Amer. Geophys. Union, Washington, DC.

Hansen, J., Russell, G., Rind, D., Stone, P., Lacis, A., Lebedeff, S., Ruedy, R. and Travis, L. (1983). Efficient three-dimensional global models for climate studies: Models I and II. *Mon. Wea. Rev.*, **3**, 609–622.

Köppen, W. (1931). *Grundriss der Klimakunde*, Walter de Gruyter, Berlin.

Konijn, N.T. (1984). *A Crop Production and Environment Model for Long-term Consequences of Agricultural Production.* Working Paper WP-84-51. International Institute for Applied Systems Analysis, Laxenburg, Austria, 37 pp.

Lebedev, Yu.V. (1984). GEMINI system of MPS-file generation basing on formalized tasks. *Proceedings: All Union Research Institute for Systems Studies*, **2**. Moscow [in Russian].

Lebedev, Yu.V., Stolyarova, H.M. and Petrova, L.N. (1986). Economic consequences of climate impacts on production and allocation of winter wheat in the dry region. *Vestnik s.-kh Nauki*, **6** [in Russian].

Marchuk, G.J. (1979). The modeling of climatic change and the problem of long-term weather forecasting. *Meteorologia i Gidrologia*, **7**, 25–35 [in Russian].

Menzhulin, G.V. and Savvateev, S.P. (1980). Influence of present-day climatic changes on crop yields. In *Problems of Atmospheric Carbon Dioxide*. Gidrometeoizdat, Leningrad [in Russian].

Monteith, J.L. (1981). Climatic variation and the growth of crops, *Q. J. Roy. Met. Soc.*, **107**, 749–774.

NDU (1978). *Climate Change to the Year 2000.* National Defense University, Fort McNair, Washington, DC.

NDU (1980). *Crop Yields and Climate Change to the Year 2000.* National Defense University, Fort McNair, Washington, DC.

Nikonov, A.A. (Ed.) (1980). *System of Agricultural Management in the Stavropol Territory.* Moscow [in Russian].

Parry, M.L., Carter, T.R. and Konijn, N.T. (Eds.) (1988). *The Impact of Climatic Variations on Agriculture. Volume 1: Assessments in Cool Temperate and Cold Regions.* Reidel, Dordrecht, Netherlands.

Petrova, L.N. (1986). Weather conditions typology to be used in grain production simulation models. *Vestnik s.-kh Nauki*, **6** [in Russian].

Petrova, L.N., Siptits, S.O. and Milyutin, N.N. (1986). An imitation model of winter wheat grain production in the Stavropol territory, *Vestik s.-kh Nauki*, **3**, 15–24 [in Russian].

Pitovranov, S.E., Iakimets, V.I., Kiselev, V.I. and Sirotenko, O.D. (1988). The effect of climatic variations on agriculture in the subarctic zone of the USSR, Part V in M.L. Parry, T.R. Carter and N.T. Konijn (eds.) *The Impact of Climatic Variations on Agriculture. Volume 1: Assessments in Cool Temperate and Cold Regions*, Reidel, Dordrecht, Netherlands.

Rabinowitch, E. (1951). *Photosynthesis and Related Processes.* Vol. II, Part 1. Interscience, New York.

Scammell, R.P. (1981). Climate and crop yields. *Climate Monitor*, **10**(1), 12–17.

Sirotenko, O.D. (1981). *Mathematical Modeling of Hydrothermal Regimes and the Productivity of Agroecosystems.* Gidrometeoizdat, Leningrad [in Russian].

Sirotenko, O.D., Abashina, E.V. and Pavlova, V.N. (1984). An estimate of the influence of possible climate variations and changes on the productivity of agriculture. *Izvestia Akademii Nauk SSSR, Ser. Physik Atmosph. i Ocean.*, **20**(11), 1104–1110 [in Russian].

Sirotenko, O.D., Abashina, E.V. and Pavlova, V.N. and Dolgiy-Trach, V.A. (1982). *A Technique for Quantitative Estimates of the Influence of Dry Conditions on Spring Wheat Yields.* Gidrometeoizdat, Moscow [in Russian].

Terjung, W.H., Louie, S.S.-F. and O'Rourke, P.A. (1976). Toward an energy budget model of photosynthesis predicting world productivity. *Vegetatio*, **32**(1).

Vinnikov, K.Ya and Groisman, P.Ya. (1979). An empirical model of present-day climatic changes. *Meteorologia i Gidrologia*, **3**, 25–28 [in Russian; English translation available in *Soviet Meteorol. Hydrol.* **3**].

Wit, C.T. de (1958). Transpiration and crop yields. *Agric. Res. Rep.*, **646**, Pudoc, Wageningen, Netherlands, 88 pp.

Wit, C.T. de (1965). Photosynthesis of leaf canopies. *Agric. Res. Rep.*, **663**, Pudoc, Wageningen, Netherlands, 57 pp.

PART VII

The Effects of Climatic Variations on Agriculture in the Australian Wheatbelt

Contents, Part VII

List of Contributors

HOBBS, Dr. Jack
Department of Geography and Planning
University of New England
Armidale, New South Wales 2351
Australia

ANDERSON, Professor Jock R.
Department of Agricultural Economics
and Business Management
University of New England
Armidale, New South Wales 2351
Australia

DILLON, Professor John L.
Department of Agricultural Economics
and Business Management
University of New England
Armidale, New South Wales 2351
Australia

HARRIS, Dr. Hazel
International Center for Agricultural
Research in the Dry Areas (ICARDA)
PO Box 5466
Aleppo
Syria

Abstract of Part VII

Climatic variation, especially as manifested in droughts, is a feature of the Australian agricultural scene. Section 1 of this case study summarizes the place of agriculture in the Australian economy, with particular reference to wheat. The study area stretches from central Queensland, through New South Wales, Victoria and South Australia, and into Western Australia. Its general characteristics and the nature of the impacts of rainfall variability are summarized. The questions and problems presented to planners and decision makers are reviewed, together with coping strategies and possible future directions.

These themes are developed in the following sections, examining climatic impacts in a variety of scenarios and looking at the implications of such impacts for planners and decision makers at all levels.

In Section 2 the major controls on Australian climatic patterns are outlined, including the subtropical high-pressure zone, the intertropical convergence zone, the mid-latitude westerlies and the southeast trades. The role of atmospheric teleconnections is also considered, and apparent relationships are examined between the Walker circulation–Southern Oscillation/El Niño complex and the occurrence of drought in Australia. Evidence for climatic fluctuation during the period of instrumental records is summarized. Much of this evidence points to changes in the seasonal distribution of rainfall in the wheatbelt, although the detailed picture is complex. Attempts to identify drought cycles and to predict drought in Australia are reviewed, but some skepticism still prevails. Rainfall probabilities, particularly the one-year-in-ten event, have been calculated for the baseline (1951–1980) and other periods. Considerable spatial and temporal variability of risk is apparent throughout the wheatbelt.

In Section 3 attention is turned to the socio-economic impacts of climatic variability. A framework is sketched for exploring the pervasive impacts of variations in climate on the uncertainties and instabilities of the Australian rural sector. Low rainfall is identified as the dominating climatic variable. Effects of climatic variation are investigated and measured at farm, regional, industrial, sectoral and national levels.

The final section of this case study looks at the implications of climatic variability for policy makers and planners. Most farmers develop successful strategies to cope with most droughts. Some interventions made during extreme episodes include partial subsidies on inputs, such as purchased fodder, and interest rates. These have serious flaws in meeting goals of efficiency, equity and welfare support. Alternative policies to provide incentive for rational private direct welfare support are discussed and advocated as a preferable approach for future intervention.

SECTION 1

Introduction: Policy and Planning Issues for the Australian Wheat Industry

1.1. Agriculture in the Australian Economy

The emphasis in Australian agriculture is on large, extensive farming operations, reflecting adaptation to both the physical and economic environments. In general terms, Australia is a flat, dry country with very variable weather conditions, where soils are not highly fertile and the availability of water is a major limiting factor. Such conditions, combined with a small domestic market and the need to export a large proportion of the country's agricultural products, favor large farms with an emphasis on extensive grazing and cereal growing.

The total number of agricultural establishments has shown a long-term downward trend, although in 1980–1981 rural holdings covered 499 million hectares, about 7% more than in 1959–1960. The average size of Australian rural holdings has risen significantly over the last 40 years, to the present average farm size of about 2800 hectares. A relatively large proportion of the total area of all holdings is, however, occupied by a few very large properties, mostly in pastoral and more arid regions with relatively low carrying capacities. In 1977–1978, 18.7% of farms surveyed by the Bureau of Agricultural Economics were smaller than 200 ha, while 18.1% were larger than 2000 ha (BAE, 1983a).

Agriculture is playing a progressively smaller role in the Australian economy (*Table 1.1*). The contribution made to export earnings, however, continues to be very significant. Up to the early 1950s, it was metaphorically true that Australia "rode on the sheep's back". More recently, however, beef and (very recently and most significantly) wheat have become very important in terms of both gross value (*Table 1.2*) and export earnings.

The importance of wheat in the Australian agricultural economy, and the fact that much of it is grown in semi-arid, marginal areas, make it the clear choice for the primary focus of this case study. Wheat is Australia's most important crop in terms of area, value of output and value of exports. In the three seasons to 1980–81, wheat averaged 60% of the area sown to annual crops, its

Table 1.1. Agriculture in the Australian economy. Source: BAE (1985).

	Gross domestic product (GDP)[a] (A$ billion)	*Contribution by agriculture, fishing and forestry to GDP* (%)	*Total exports* (A$ billion)	*Contribution by agriculture, fishing and forestry to exports* (%)	*Index of prices paid*[b]	*Index of prices received*[b]
1953/4[c]	8	19	2	84	37	19
1963/4[c]	15	13	2	79	36	24
1973/4[c]	40	8	6	54	42	35
1976/7[c]	64	6	10	47	60	59
1979/80[c]	91	6	15	46	78	79
1980/1	116	6	19	46	100	100
1981/2	131	6	20	43	99	111
1982/3	142	4	22	37	105	123
1983/4	161	6	25	37	111	133

[a] At factor cost.
[b] BAE indices based on 1980/1 averages = 100.
[b] Averages for the 3 years ended in the designated fiscal year ending 30 June.

Table 1.2. Gross value of Australian rural production (GVRP) and its major relative contributions. Sources: Stoeckel and Miller (1982) and BAE (1985).

	GVRP (A$ billion)	*Wheat* (%)	*Wool* (%)	*Beef* (%)	*Other cereals* (%)
1951/2[a]	2.0	13.4	43.1	7.7	3.4
1959/60	2.6	10.4	29.5	14.2	3.6
1964/5	3.4	15.1	24.6	14.1	3.9
1969/70	3.7	14.2	19.7	16.8	4.0
1974/5	5.9	21.4	16.2	8.9	7.5
1977/8	7.0	13.3	17.2	16.8	5.8
1978/9	10.3	22.2	13.4	21.0	6.3
1979/80	11.7	20.7	13.9	10.2	6.9
1980/1	11.5	14.6	14.5	17.9	7.3
1981/2	12.6	10.6	14.2	15.0	7.2
1982/3	11.6	13.5	15.1	17.8	5.6
1983/4	15.5	23.3	13.1	13.3	9.3
1984/5	15.6	22.2	14.6	13.3	8.7

[a] Average of 3 years ending in 1951/2.

gross value averaged A$1903 million (41% of the gross value of all crop production), and its export value averaged A$1345 million (just over 50% of the total value of crop exports and about 20% of the total value of all rural exports) (BAE, 1983a). Major markets for Australian wheat and flour in recent years have been China, Egypt and other Middle Eastern countries, and Japan, with the USSR being an occasional large purchaser.

1.2. The Study Area

The Australian wheatbelt stretches from central Queensland, through inland New South Wales, into northern Victoria and South Australia, and then into parts of southwestern Western Australia. The area sown to wheat reached 11.3 million ha in 1980 81. Western Australia now has the largest area sown to wheat, but production in New South Wales has generally been greater as wheat cultivation has expanded into more marginal areas in the west. New South Wales and Western Australia together account for about 60% of the total Australian wheat production from about 70% of the area sown. The number of farms growing wheat has remained around 48 000 during the last 20 years. Relatively few farms are devoted entirely to wheat, which is usually combined with other crops and/or livestock. It was estimated in 1980 that most farms growing wheat devoted less than 30% of their areas to this activity.

Mainly because of its latitudinal extent, the wheatbelt shows considerable regional variability in climatic characteristics. Summer rainfall dominates in central and southern Queensland and into northern New South Wales, but the remainder of the wheatbelt experiences winter-dominated rainfall patterns.

Rainfall is the most important single climatic variable affecting the Australian community. Australian wheat yields have always fluctuated greatly, mainly because of erratic rainfall. Annual rainfall patterns in Australia show a basically concentric zonation of rainfall, decreasing rapidly inland toward a large arid interior zone with rainfall less than 250 mm. The present-day wheatbelt mostly lies roughly between isohyets of 250 and 635 mm, with a shift to wetter areas (500–760 mm) in the warmer, summer rainfall northeastern section.

Representative annual rainfall patterns are shown in *Figure 1.1(a)* for locations from central Queensland (Clermont), through northern New South Wales (Inverell), central New South Wales (Wellington), to South Australia (Warooka). Within each region and in the Western Australian sector, variations are also apparent from east to west as well as north to south (for example, *see Figure 1.1(b)*).

Some idea of rainfall variability is given in *Figures 1.2* and *1.3*. *Figure 1.2* shows isohyets of the 10th and 90th percentiles of annual rainfall. *Figure 1.3* shows the variability of Australian annual rainfall based on an index relating the difference between 90th and 10th percentiles to the median. The region with greatest variability lies mostly in the summer rainfall area of the arid zone. Closely related to this index is one that assesses the incidence of rainfall deficiency, rather than general variability. Analysis of the distribution of annual rainfall less than the median provides a better index for consideration of droughts. Hence, *Figure 1.4* shows the distribution of an index of drought risk.

Rainfall is the best single index of water availability; and monthly rainfall totals are sufficient for most studies of the occurrence, spread and breaking of drought (Reynolds *et al.*, 1983). Rainfall data generally provide a degree of uniformity, reliability and completeness than cannot be obtained with other criteria used as bases for the study of drought, and these attributes also make rainfall data the most useful climatic data base when considering climatic limitations to yield.

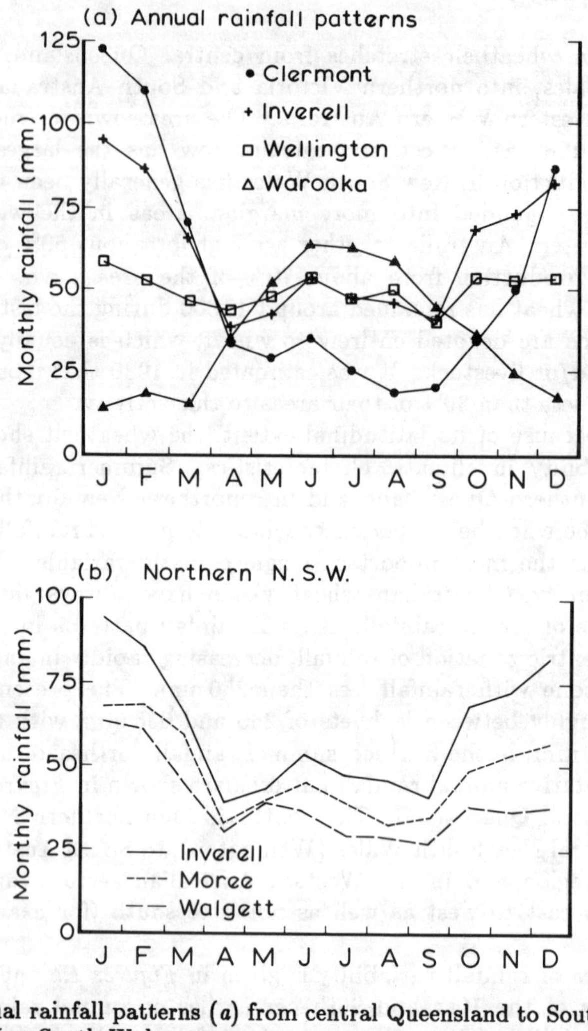

Figure 1.1. Annual rainfall patterns (*a*) from central Queensland to South Australia; (*b*) across northern New South Wales.

Throughout the Australian wheatbelt, the seasonal water regime is the main control on the timing and duration of the crop cycle. Other major constraints are the timing of sowing rains, the duration of the mid-winter depression of temperature and radiation values, the length of the frost period, and the rapid increase in temperature and evaporation rates during the late spring and early summer months. These constraints are generally closely related to the large-scale controls on Australian climate (*see* Section 2).

An indication of the timing of sowing rains may be given by the occurrence of median monthly rainfall equal to or exceeding 25 mm in April, May and June. Only the extreme southern and eastern fringes of the wheatbelt satisfy this condition in April, but by June all of the wheatbelt meets the criterion. Thus, there

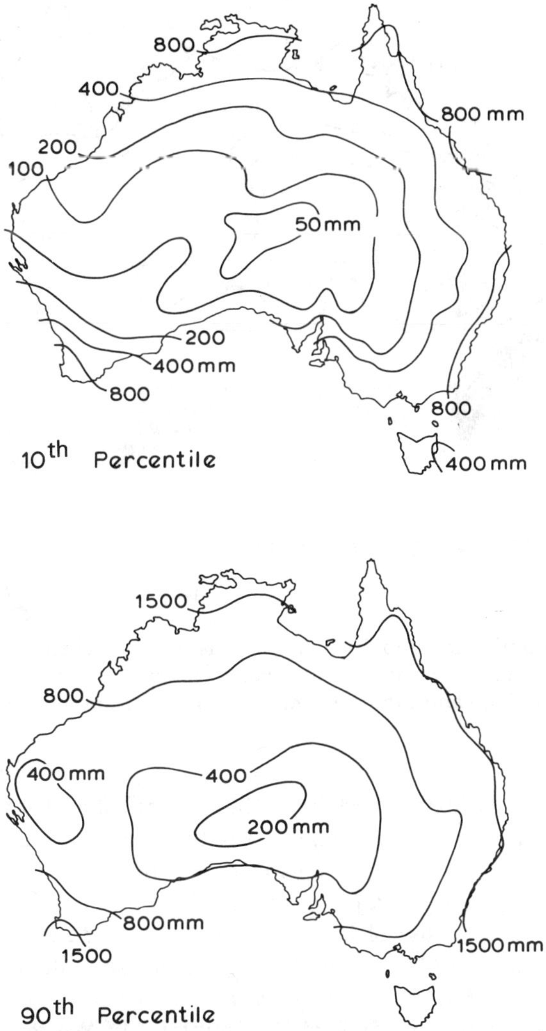

Figure 1.2. Isohyets (mm/y) of tenth and ninetieth percentiles of annual rainfall. Source: Ceplecha (1971).

is some early sowing in April, but mostly it takes place in May or June, although excessively dry or excessively wet conditions can delay sowing until July or August. Frost during elongation and heading is a major constraint throughout most parts of the wheatbelt, especially in the more southern areas. In the westernmost parts of the wheatbelt in Western Australia and the northernmost parts of the belt in central Queensland, approximate mean earliest safe-ear-emergence dates are in the first half of August, but for most of the remainder of the wheatbelt dates range through to early October, generally being later further south or at higher altitudes.

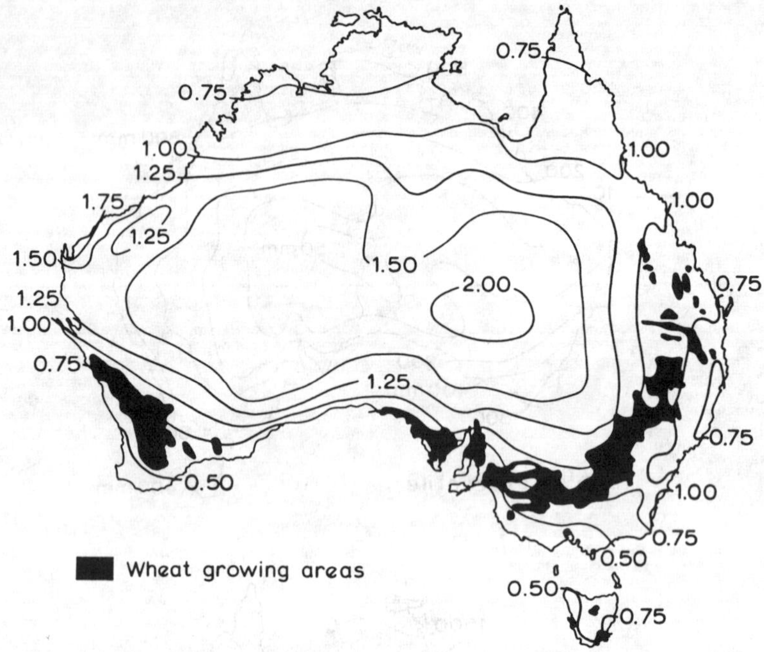

Figure 1.3. Variability of annual rainfall in Australia. Variability is represented by index values calculated according to the formula: index = 90 percentile − 10 percentile/50 percentile. Low index values indicate relatively low variability. Source: Reynolds *et al.* (1983).

The work of Gibbs and Maher (1967) established that annual rainfalls in the first decile give the nearest equivalent to drought conditions as specified in published reports in terms of various criteria. In the period from 1864 to 1984, there have been at least 10 major droughts affecting more than 20% of Australia, and at least eight droughts of lesser severity affecting extensive areas. In every 100 years, 21 years are likely to be free of major droughts, 62 will be marked by drought over 20–50% of the continent, and two years will have a widespread drought covering more than 50% of the country. In Western Australia, drought conditions might be expected in 54 years out of every 100, with incidence in other states less than this, down to Tasmania with droughts in 26 out of every 100 years. Large and irregular rainfall variations, spatially and temporally, are an integral part of Australian climate and must be expected to continue, although climatic fluctuations could lead to shifts in the locations of high-risk zones.

In terms of latitude, Australia starts growing wheat where the main wheat producers in other parts of the world stop, so that Australian production is generally from relatively low-latitude regions. This places Australian wheat outside the main areas of adaptation, and this is clearly reflected in yields. This extension of the crop to relatively low latitudes means not only that it is affected by

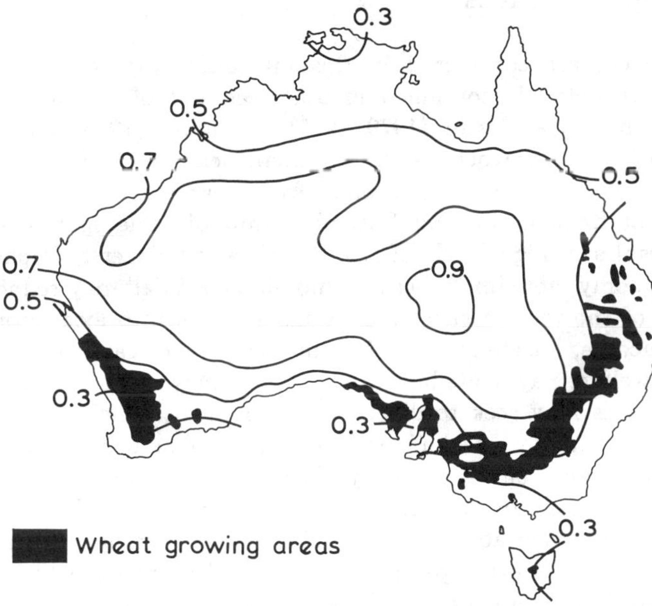

Figure 1.4. Drought risk in Australia. Susceptibility to drought based on annual rainfall totals is shown by index values calculated according to the formula: index = 50 percentile − 10 percentile/30 percentile. Low index values indicate relatively low risk. Source: Reynolds *et al.* (1983).

the inherent rainfall variability, but also by photo-period effects and generally higher temperatures than in most other major producing regions.

Relatively low yields per hectare in Australia are also reflections of a dominance of old, nitrogen- and phosphorous-deficient soils and of a concentration of breeding efforts into disease- and drought-resistant strains, which has limited the ability to select for high-yielding varieties. On the positive side, however, terrain in the wheatbelt is generally level or undulating, facilitating a high degree of mechanization and minimizing the erosion effects of cultivation, although wind and water erosion are problems in some areas.

It needs to be emphasized that the situation in Australia is perhaps different from other regions covered by case studies in this volume. Australian wheat production should not be viewed in subsistence terms, to the extent that people starve if the crop fails. Rather, production is for export and has to be viewed in the broader context of international supply and demand. The viability of Australian wheat producers depends upon being able to satisfy traditional markets and upon having sufficient reserves from good years to meet demands in bad years. Further, this viability depends very much on the nature of international demand, on the success or failure of crops elsewhere, and perhaps also on political forces and decisions elsewhere, as well as upon yields in Australia in a particular season.

1.3. Drought Impacts

Drought impacts are considered in this introductory section only in general terms. Greater detail is contained in other sections of the case study and in work such as that by Anderson (1979, 1985), Callaghan (1973) and Gibbs (1984).

Much of the wheatbelt is in marginal, semi-arid climatic regions that experience marked variability in rainfall from year to year (*see* Section 2). A combination of frequent low rainfalls plus unpredictable spatial and temporal variability results in high levels of risk for wheat producers. Impacts permeate the whole economy, not simply because unreliable rainfall may reduce yields, but also because of the related effects. Less money goes into local economies, local people lose income, businesses may fail, infrastructures can crumble, unemployment grows, welfare payments increase and there may be outmigration. Farmers and their families may seek to diversify employment, which may mean moving and may lead to family breakup with a variety of associated social problems. Clearly, impacts can occur at a variety of levels – at farm level, at industry level, at the regional level, or at the national, perhaps even international level. Impacts of climatic variability on local farms and agricultural industries can reverberate strongly in other parts of the regional economy, which in turn may have significant impact upon the national economy.

Gibbs (1968) reported that the 1965–1967 drought in eastern Australia, for example, reduced farm income by over $1000 million. More recently, the Bureau of Agricultural Economics (BAE, 1983b) estimated that the net value of rural production in 1982–83 was $2300 million compared with an average of over $4000 million for the previous four years.

In terms of wheat production, ***Figure 1.5*** shows the variations for New South Wales, Victoria and Queensland from 1885 to 1983 in relation to drought severity based on the percentage of the area of each state that experienced rainfall in the lowest 10% of occurrences. This is only a crude measure of drought severity, since the calendar year does not coincide with the growing period for wheat, and because whole states rather than just the wheatbelt regions are considered.

Droughts in New South Wales in 1888, 1902, 1914, 1919, 1940, 1945, 1957, 1965, 1972 and 1982 are reflected by a reduction of wheat production to less than 50% of that of the preceding years. The graph for Victoria points to even more severe reductions in wheat production in association with major droughts. In Queensland, however, the correspondence is generally less marked, probably because a drought index calculated on the basis of the whole of such a relatively large state is less reliable than for a small state of which the wheatbelt occupies a relatively large area.

Data for wheat production on a state-by-state basis are included in *Table 1.3.* These data provide a clear indication of the impact of the 1982–83 drought, and also point to the considerable regional variations within the wheatbelt. National wheat production in 1982–83 was only 63% of the previous five-year average, with the greatest reduction in Victoria (16%) and New South Wales (29%). In 1982–83 New South Wales and Victoria together produced only 20% of the Australian total, compared with an average 54% over the previous five

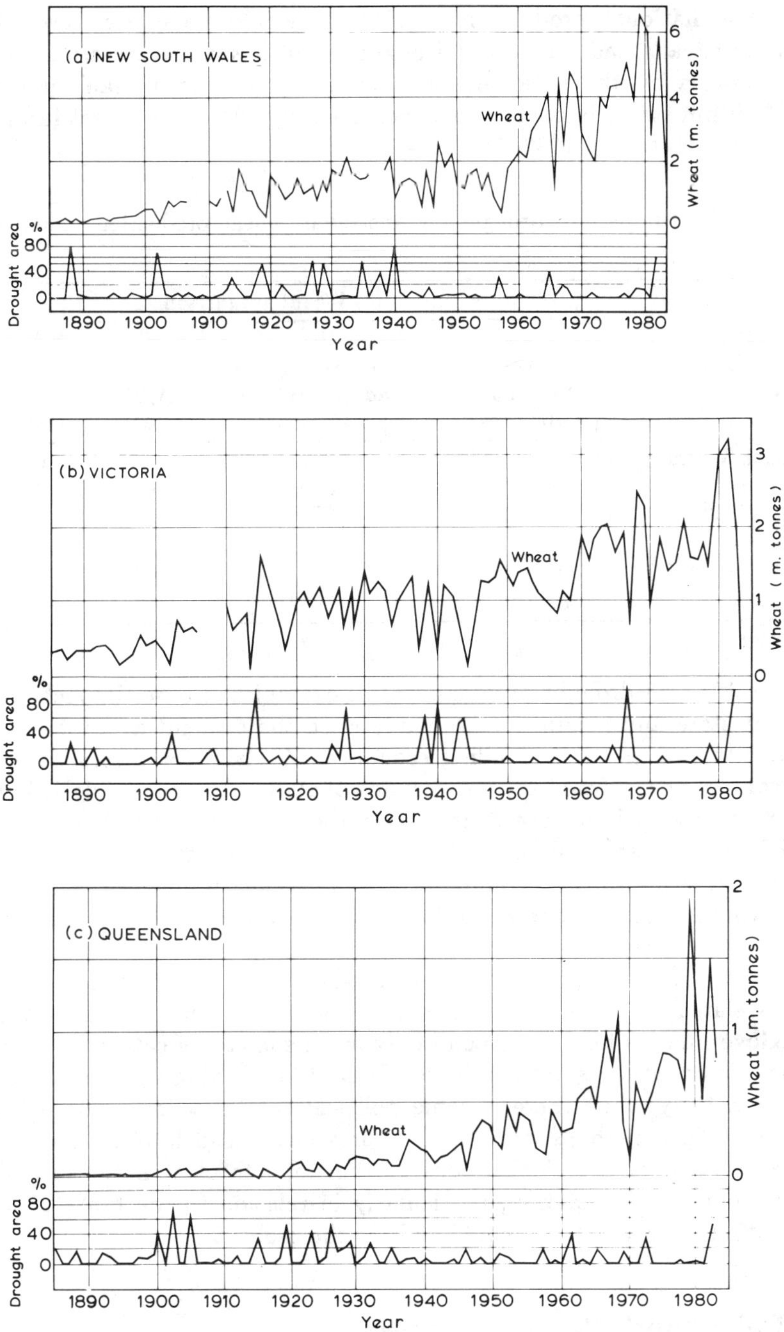

Figure 1.5. Drought intensity and wheat production in New South Wales, Victoria and Queensland, 1885–1983. Drought intensity is indicated by area of the state experiencing annual rainfall in the lowest 10% of occurrences. Source: Gibbs (1984).

years. The national production in 1982–83 would have been even lower if Western Australia had shared the drought conditions of the eastern states. Generally, impacts on the national scale are less pronounced than for individual states. Within states impacts also varied greatly, with some wheat farms having no wheat production in 1982–83.

Table 1.3. The impact of drought on Australian wheat production. Source: Gibbs (1984).

	Wheat production				
	A *1982–83* ($\times 10^3$ tons)	*B* *Average, 5 years ending 1981–82* ($\times 10^3$ tons)	*A/B* (%)	*A/C* (%)	*B/D* (%)
New South Wales	1450	5052	29	16	36
Victoria	400	2550	16	4	18
South Australia	680	1658	41	8	12
Queensland	760	1069	71	8	8
Western Australia	5700	3834	149	64	27
Australia	8990*C*	14 166*D*	63	–	–

The 1982–83 drought provides several examples of other impacts. Political leaders at state and federal levels readily cited the drought as a major factor in the downturn of an already ailing economy. A claim for increased wages for rural workers was refused on the grounds that the drought made it impossible for farmers to afford increases. At the same time, large amounts of money were included in state and federal budgets to provide drought relief. A variety of businesses catering to rural communities suffered sharp downturns in trade, and state railways suffered large reductions in revenue because there was less wheat to transport (Gibbs, 1984).

In the context of the Australian wheatbelt, problems are not always drought-related. They may be related to too much rain or rain at the wrong time. Governments tend to respond only when something catastrophic happens. At other times action is not considered helpful for gaining votes and therefore is not taken. Only when there is some political capital to be made will governments act. This, perhaps cynical, view provides a lead into a consideration in the next section of the implications of the impacts of climatic variability for planners and decision makers, particularly in relation to the types of problems and questions to which they should address themselves.

1.4. Policy and Planning Issues

1.4.1. Climatic variability and agricultural risk

It can be argued, and indeed is a popularly held view by foreigners as well as Australians, that risk-taking to some extent is part of the way of life in

Australia. The majority of Australian farmers seems to subscribe to the view that rainfall variability, in particular rainfall deficiency, is merely an unfortunate occasional abnormality of the environment; and that when drought does occur government assistance will aid survival. Such attitudes are widely held by the general community and by administrators and politicians, and it may be that farmers actually consider the best strategy for coping with drought is not to prepare for it, but to apply for government assistance when drought occurs (Gibbs, 1984).

Climatic variability has been and continues to be a major influence on the structure of Australian agriculture. Farmers in Australia arrived from Europe 200 years ago, and later from other countries, to search out the environment for niches favorable to their economic welfare. They did this in a relatively unconstrained manner, free to choose, to live, and in some cases to die as a consequence of nature. They had to take enormous risks in unfamiliar environments, and in some cases expansion of agriculture was encouraged by sequences of "good" years, which were, however, not necessarily representative of long-term conditions. A continuing controversy concerns the appropriate bounds on where wheat can be grown profitably. In different centuries and decades, opinions have varied about this, and the interior margin of the wheatbelt has fluctuated considerably over the years.

The costs of learning which enterprises should be most profitable are usually irrevocably sunk, but with a general taxation environment that has sometimes supported farming endeavors, the community at large is usually a significant shareholder in such learning experiences.

1.4.2. Implications of longer-term climatic change

There are some indications that long-term, perhaps sustained, climatic changes are taking place in addition to the inherent variability of rainfall (*see* Section 2). The nature of climatic change is such that for any reasonable presumption about the rate at which people learn about slowly changing phenomena, recognition of the existence of change must necessarily be very slow. Interventions are intended to mitigate the extent of the impact of untoward changes, and to the extent that they are effective or ineffective (a major question taken up later in Sections 3 and 4) then recognition and learning about the new environmental regime can be seriously impaired. Effective government policies may well serve to exacerbate poor decision making with respect to investment and resource allocation in areas that are suffering what amounts to climatic deterioration. Intervention policies must be couched in such a manner as not to shield individual decision makers from emerging reality, whatever the source of change in natural or economic environments may be.

The cause of any future climatic change is perhaps not important for this study, although it should be recognized that increasing atmospheric carbon dioxide concentrations are the most commonly cited likely cause. In the case of carbon dioxide-induced change, prevention is hardly a viable option. Some modification of increasing rates of accumulation may be possible through maximum utilization of nonfossil energy sources, but there are still all sorts of uncertainties about the natural working of the carbon cycle. Adaptation may be possible, but whether it will occur in a planned or unplanned way really depends on wide acceptance of sound predictions of the consequences. This in turn points to obvious needs for realistic and soundly based predictions, whatever the cause of change. As far as carbon dioxide is concerned, there is, on the one hand, the very complex problem of estimating how concentrations might vary, and, on the other hand, what sorts of biological and climatic responses may be produced. There is a need for a continuing assessment of our knowledge of the processes involved and of the accuracy of various techniques of numerical synthesis of both climatic factors and plant physiological functioning.

Equally important is the reliability to be attached to predictions of climatic response, if the predictions are to influence national policies. Many areas of uncertainty need resolution before reliable forecasts are possible of the precise timing of carbon dioxide increases or of their detailed consequences. The same applies to any other causes of change, and to some extent to any causes of variability. Reliable predictions are required of the timing, scale and nature of events.

The implications for agriculture of climatic change depend strongly on the pattern of precipitation change, at least as far as Australia is concerned. This may be more important than patterns of temperature change, although, of course, the two will be interrelated. They will also depend on better assessments than those currently available of any changes in the length of growing seasons and of seasonal distribution of rainfall.

Any suggestion of climatic changes, whatever their cause, also points to the need to consider such factors as possible changes in the radiation environment, in seasonal and diurnal temperature characteristics, and also the likely direction of temperature and rainfall changes in particular. It is possible that climatic change could lead to changes in crop species, but the actual shifts will depend very much on factors such as consumer acceptance as well as on purely climatic considerations such as diurnal and seasonal patterns of change. There is a need to improve methods of isolating the climatic factor in crop yield from other factors such as fertilizers, crop strains, farming techniques, soils and government policies.

Quantitative estimates of the nature, magnitude and timing of change are only just starting to become available, and they are still of questionable reliability. Overall, it does appear possible that net changes could increase total agricultural productive capacity in Australia, perhaps with a 10 to 20% increase in yields. Part of this may come from expansion of viable production areas, which could mean a contraction (or would it be expansion?) of risk zones. If there is expansion of the wheatbelt as a result of climatic change, there may be dangers of overcropping if favorable changes are not sustained. Such expansion may reduce the likelihood of drought conditions prevailing in all parts of the wheatbelt simultaneously, thus effectively spreading the risk; but expansion could also mean greater areas exposed to risk.

An increase in yields might take place over the next 80 to 100 years. However, consideration of related aspects, such as regional distribution, year-to-year variability (which itself may change), and the overall infrastructure, all require more precise specification of the climatic situation than is currently available.

Effects of changes on national production and marketing have to be considered, and this involves Australia's present customers, potential customers and competitors. Population trends and their effects on consumer demands will further modify the trade situation. There are also many other factors to consider in relation to population trends, such as the possibility of general rural depopulation for a variety of reasons. Net international supply/demand situations will determine price trends and variations, and they may in turn be reflected in general commercial activity and perhaps in rural depopulation or population growth. It is difficult, in fact, to isolate the planning and policy issues in Australia, or anywhere else, from a broader consideration of the global impacts, particularly in view of the complex interrelationships of international trade. The implications of changing climatic conditions, whether in the form of year-to-year variability or as sustained longer-term change, must be assessed in terms of maintaining an annual clearance of stocks, which will depend on markets; and in terms of making sufficient supplies available to meet the requirements of regular overseas buyers whose situation may also be changing.

1.4.3. Mechanisms of response to climatic variations

Strategies for handling the problems caused by climatic variability and change are not necessarily difficult to devise, but implementation is not always easy and practical complexities frequently confound theoretical simplicity. Current policies tend to emphasize assistance-giving, perhaps through the provision of generous credit facilities or with regional unemployment schemes. In future, however, changes may become necessary and appropriate as the merits of assistance-giving are questioned. There is some belief that assistance distorts producers' perceptions of their environment. It is possible that intervention by governments, such as relief measures, has been developed and institutionalized beyond the level desirable on the grounds of economic efficiency. As alternatives to assistance-giving, perhaps governments should be promoting approaches such as diversification of land use, better management and less frequent cropping. To mitigate the impact of drought when it occurs, producers should be encouraged to adopt drought-resistance strategies such as keeping reserves, insurance and new crop strains.

The development and application of drought strategies include a variety of activities. While knowledge of risk of occurrence of rainfall deficiencies is sound, there is room for improvement in the dissemination of this information, not just to producers but also to planners and policy makers. There is also a need for better dissemination of knowledge of drought impacts to water users and decision makers at all levels: to farmers, to administrators, to utilities, and to government services in all parts of the political arena. Governments could improve the way in which they can assist decision making by providing better information about seasonal conditions and decision options.

According to Gibbs (1984), there are two streams of activity that should be encouraged to improve the ability of producers to handle drought; and it seems reasonable to suggest that these could also apply to problems caused by variability and change. On the one hand, research activity should seek improvement of knowledge of intensity, duration and extent of deficiencies, variability and change, including a better understanding of impacts on economic and social conditions at all levels. On the other hand, there is room for improvement in the use and application of existing knowledge to areas such as monitoring of drought, variability or change, and the development of appropriate coping strategies.

The sections that follow develop the themes outlined in this introduction, examining climatic variability and change, the impacts of these on Australian wheat production in a variety of scenarios, and the implications of such impacts for the planners and decision makers at all levels.

SECTION 2

Climatic Patterns and Variability in the Australian Wheatbelt

2.1. Introduction

This section sets the climatic background against which the crop response and economic and social impact sections of the Australian case study are developed. A general description of the climate of the Australian wheatbelt has been given in Section 1. The underlying causes of the observed patterns are examined here. The evidence for climatic fluctuations in Australia during the period of instrumental record (practically, because of data limitations, during the twentieth century only) is examined, as it adds further detail to the background against which wheat production must be set. Such evidence illustrates the degree of relatively sustained variability to which it might be argued producers have become accustomed, and it also points to the possible nature of sustained trends to which longer-term adjustments, and for which longer-term planning, might be necessary. Studies of any tendency for dry or wet periods to persist, and so indicate longer-term variation in rainfall patterns, may also aid the ultimate prediction of extremes, leading to reduced impacts on the rural sector.

Several workers have recently looked closely at a variety of atmospheric teleconnections as playing major roles in the production of climatic patterns and climatic variability. Much attention for the Australian region has been focused on the apparent relationships between the Walker circulation–Southern Oscillation/El Niño complex and the occurrence of unusually dry or wet spells. These relationships are discussed in detail in this section.

There has always been a keen interest in the whole question of drought predictability in Australia and indeed elsewhere. Since the possibility of prediction should be of concern to planners, policy makers and decision-makers, this is examined with particular reference to the large-scale climatic controls concerned.

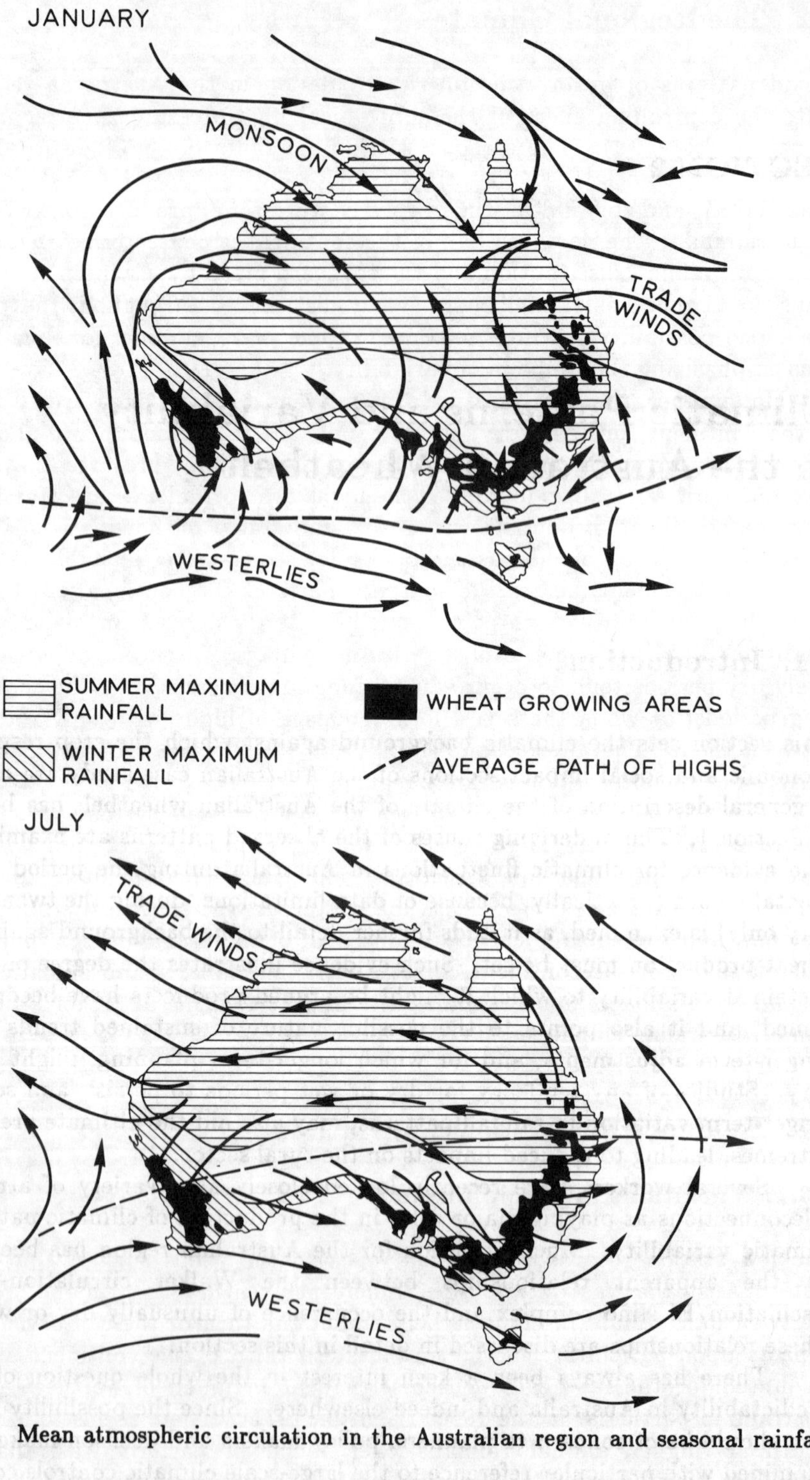

Figure 2.1. **Mean atmospheric circulation in the Australian region and seasonal rainfall zones.**

2.2. The Regional Climate

Broad patterns of mean atmospheric circulation in the Australian region illustrate the controlling effect of the subtropical high-pressure zone on Australian climate (*Figure 2.1*). The seasonal latitudinal shifts of the high-pressure zone, together with associated movements of the low-latitude intertropical convergence zone (ITCZ) and the mid-latitude westerly wind zone, produce marked seasonality in rainfall. The northern half of the continent is characterized by a summer maximum rainfall distribution and the southern areas by a winter maximum. A transitional zone has rainfall more evenly distributed throughout the year, with the added possibility of winter maxima in some years, summer maxima in others, thus emphasizing the problem of variability.

In summer, the high-pressure systems tend to track eastward to the south of the continent, bringing dry, stable conditions to the south and allowing the inflow of warm, moist and unstable air to the low-pressure area that develops over northern Australia. In winter, the highs are normally located further north over the continent, producing warm, dry and stable outflowing winds over the north, but resulting in the passage of regular sequences of unstable, cold frontal low-pressure systems across the southern parts of the continent. Although the controlling subtropical high-pressure cell is clearly defined on maps of average conditions, it really consists of a series of temporary, traveling systems. Relatively dry periods tend to occur when strong highs persist over the continent for long periods, or when there is a long sequence of highs with only short breaks between them. The alternation of highs and troughs exacerbates rainfall variability, since the alternation is rarely regular.

The critical role of the subtropical high-pressure systems in controlling Australian rainfall is highlighted above, but other major circulation features also have a significant influence on Australian climates. These include the intertropical convergence zone, the mid-latitude westerlies, the southeast trades, tropical cyclones, and the subtropical jetstreams. All of these at different times, and in different parts of the continent, affect Australian climates and climatic variability and have impacts upon rainfall patterns (*see* Gentilli, 1971, for a full discussion of Australian climates and climatic controls).

2.3. Evidence for Climatic Variations

2.3.1. Variations in precipitation

The few station records that extend back to the 1800s suggest that much of eastern Australia was wetter in the nineteenth century than in the first half of the twentieth. Shifts in climatic belts about the turn of the century had major impacts upon agriculture in the southeastern third of the continent and partly accounted for the failure of denser settlement in much of northeastern South Australia in particular in the 1890s (Gentilli, 1971). There seems to have been a return to higher annual rainfalls over much of eastern Australia around 1945–1946 (Pittock, 1975). Annual rainfalls have since increased by 10–20%

over a large area of southeast Australia, although it is interesting to note Pittock's estimate that the area of significant (95% level) rainfall change in Australia (1941–74 means compared with 1913–40 means) is about what would be expected from a random data set.

Statistically identifiable changes and persistence effects in mean annual rainfalls have been reported by a number of researchers. Short-term and medium-term changes in rainfall, with a break in the early to mid-1940s, have been identified, for example, by Pittock (1975), Cornish (1977) and Russell (1981).

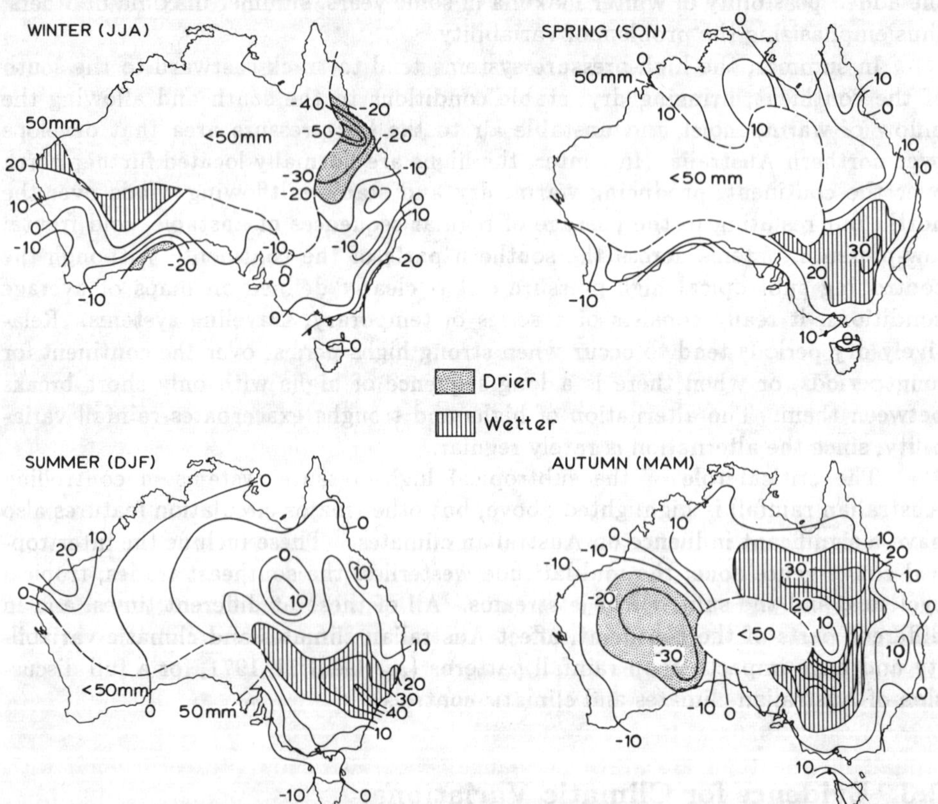

Figure 2.2. Changes in seasonal mean district precipitation in percentage of the 66-year average, between 1913–45 and 1946–78. Areas with district mean seasonal rainfall of less than 50 mm have not been plotted, because high year-to-year variability makes changes too dependent on chance occurrence of isolated heavy falls. Source: Pittock (1983).

Pittock (1983) has updated his earlier analysis and examined the changes in terms of three-month seasons (*Figure 2.2*). Precise estimates of statistical significance have not been calculated because the data are not normally distributed in low-rainfall regions and there may be a high degree of spatial correlation to complicate interpretations of statistical significance. The main features of his recent analysis, as far as the wheatbelt in particular is concerned, are:

(1) A winter decrease in mean rainfall in southwestern Australia and on the western slopes of the Divide in New South Wales.
(2) In spring, most of inland New South Wales and parts of South Australia and Victoria showing an increase in excess of 20%, with a maximum in excess of 30% in central New South Wales (rainfall district 50).
(3) In summer, an increase in excess of 20% in parts of north-central New South Wales.
(4) Large areas in western New South Wales, central-southern Queensland and part of South Australia showing autumn increases of up to 40%; with much of inland Western Australia having a decrease.

These data support the increase in annual rainfall reported in Pittock (1975) and point to its concentration in spring, summer and autumn, which in turn points to a change in pattern of seasonal distribution of rainfall. Examination of this changed seasonal distribution for rainfall district 50, in the New South Wales sector of the wheatbelt, shows percentage increases up to 52% in October (spring) and 50% in February (summer), but a 41% decrease in June (winter) [*Figure 2.3(a)*]. Frequency distributions of February and June rainfalls in the two 33-year periods show that the changes in the means are due to shifts in the whole distribution and not to isolated extreme rainfalls in a few years [*Figure 2.3(b)*].

Russell (1981), on the basis of an analysis for the period 1895–1974, also identified a trend of increasing summer rainfall over much of southeastern Australia, including parts of the wheatbelt (*Figure 2.4*). In addition, his analysis found some discontinuities in summer and winter rainfalls over short and medium periods. Pronounced discontinuities of summer rainfall occurred in 1944–46 and in winter rainfall in 1950 over large areas of Australia. The period 1946 to 1955 was identified as one of above-average summer rainfall (October to March), and 1950 to 1965 was identified as a period of above-average winter rainfall in wheatbelt areas. Such changes were also shown to correspond with changes in wheat yields. Unfortunately, it is not easy to compare results from Pittock and Russell because they used different definitions of seasons.

Cornish's (1977) analysis of changes in seasonal and annual rainfall was confined to New South Wales. His work indicated a significant increase in annual and summer rainfall in central New South Wales since 1945, with annual increases in the wheatbelt of up to 20%, and summer (January to March) increases up to as much as 50%, but generally between 20 and 30%. These results seem to be consistent with those of Pittock and Russell, but direct comparison is difficult, again because of Cornish's use of a different definition of the summer period.

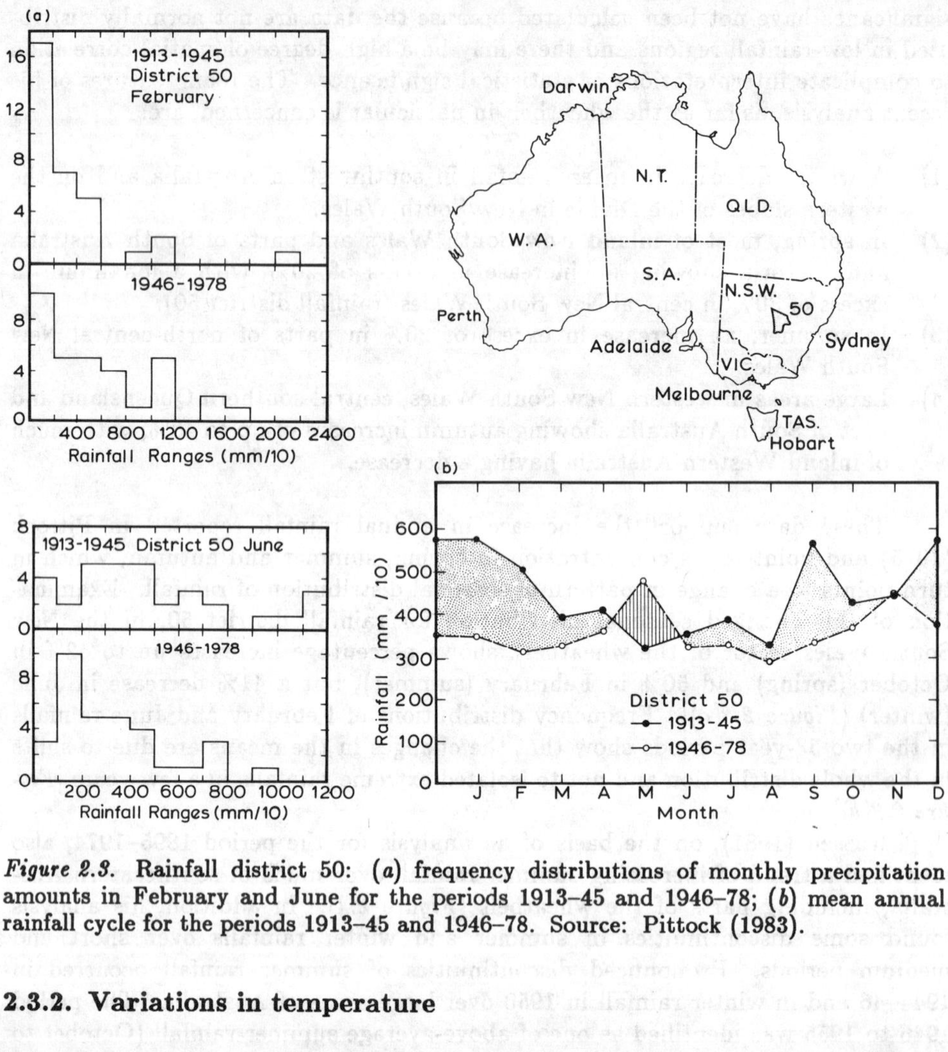

Figure 2.3. Rainfall district 50: (*a*) frequency distributions of monthly precipitation amounts in February and June for the periods 1913–45 and 1946–78; (*b*) mean annual rainfall cycle for the periods 1913–45 and 1946–78. Source: Pittock (1983).

2.3.2. Variations in temperature

Information about possible temperature changes during the period of instrumental record is less readily available, because of the relative paucity of temperature data for long periods. On a hemispheric basis, the increase in annual mean temperature between the 1910s and the 1970s has been about 1.3°C in the 30–50°S zone (Paltridge and Woodruff, 1981) (*Figure 2.5*). Tucker (1980) analyzed yearly average spring temperatures for wheatbelt stations for the period 1950–1979 to determine coefficients of regression with time. The only significant results were positive trends in daily minimum temperatures for four stations in South Australia. Coughlan (1979) found that between 1946 and 1975 most of Australia experienced a rise in annual mean maximum temperatures, with the greatest increases in inland southeastern Australia (*Figure 2.6*). In relation to the wheatbelt it will be noted, however, that considerable variations are apparent, ranging from relatively large increases in the New South Wales sector to decreases in the Western Australia sector.

Figure 2.4. Areas showing significant (Prob. < 0.05) linear trends of summer rainfall (*a*) October to March, (*b*) December to February. Source: Russell (1981).

The middle-latitude southern hemisphere warming reported by Paltridge and Woodruff (1981) was about half of that which some climatic models suggest would occur with a doubling of present atmospheric carbon dioxide concentrations (*see*, for example, Manabe and Wetherald, 1980; Manabe and Stouffer, 1980). The temperature increase itself is more than can be accounted for solely by increasing carbon dioxide during the twentieth century, so other effects, such as volcanic dust, also have to be considered. However, it is worth noting that the observed temperature increase from the 1910s to the 1970s occurred on a time scale similar to that over which the carbon dioxide concentration is expected to increase by perhaps 50% in the coming decade. It seems reasonable to suggest that regional climatic changes associated with the temperature increase observed through this century may be analogous to those that can be expected due to near future increases in carbon dioxide concentrations. On this basis, Pittock (1983) suggests that perhaps we can expect to see a continuation of current regional climatic trends, and that the climatic change that has occurred in Australia during this century provides the best analogy so far as to what might occur with a further increase by about 50% in carbon dioxide concentration.

2.3.3. Possible future climatic changes

Pittock and Salinger (1982) looked at the likely consequence of increasing carbon dioxide on the basis of modeling results, physical reasoning, analogies with the period of maximum Holocene temperatures and year-to-year fluctuations. Their scenarios suggested increased rainfall across most of Australia north of about 30°S, possibly about as far south as 37°S, with a possible decrease in southern Australia, especially the southwest. The difference between 30°S and 37°S as the southern limit of increased rainfall clearly has major implications for the

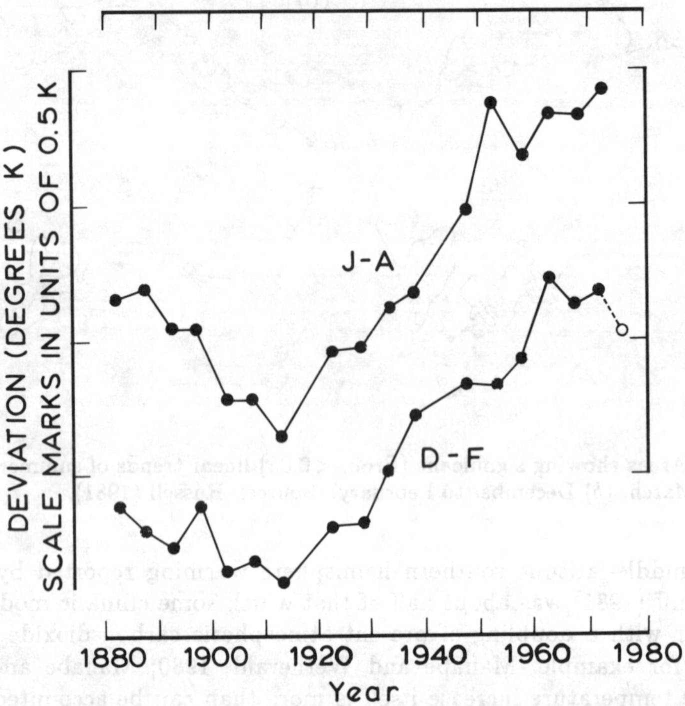

Figure 2.5. Five-year average deviations from the long-term means of spatially mean surface temperature data for the southern hemisphere as a function of time. Source: Paltridge and Woodruff (1981).

wheatbelt. Their work also pointed to a preferential increase in summer rainfall, which is consistent with changes observed so far.

If rainfall changes observed in Australia are related to changing carbon dioxide concentrations producing temperature changes of about 1° C, then a doubling of carbon dioxide concentrations, as suggested in some scenarios, could be expected to be associated with more dramatic precipitation changes.

The physical explanation for the observed patterns of climatic change is generally consistent with a more positive Southern Oscillation index and a poleward movement of the subtropical high-pressure belt, at least over eastern Australia. This would in turn mean a more southerly location for the winter rain-bearing frontal systems in the westerlies, and perhaps a change in the patterns of cyclogenesis and blocking in the Australian region. Pittock (1983) points to the possible significance of maximum precipitation changes over eastern Australia having occurred towards the poleward limit of the southern rainfall regime, and into the spring and autumn months. This may indicate that the critical mechanism is an enhancement of the interaction between mid-latitude synoptic disturbances and the moist subtropical air in these longitudes.

Results imply that very large percentage changes in precipitation (greater than 50%) and in seasonal distribution of precipitation could occur over the next

Figure 2.6. Value of Kendall's rank statistic ($\times$ 100) giving the trend of annual mean maximum temperature for the period 1946–1975. Confidence limits of 95% give a value of approximately +0.25. Kendall's rank statistic $= 4P/n(n-1)-1$, where P is the total number of cases in a series $T_1 \ldots T_n$ in which $T_j > T_i$ for $j > i$. The coefficient may vary from -1 to $+1$ and its expected value in a random series is zero. Source: Coughlan (1979).

few decades, in addition to changes already recorded. While it may be possible to postulate larger warmings (say, greater than 1°C zonal average annual warming) than those so far observed, there is no doubt that caution is necessary in view of still considerable uncertainty about the complex interrelationships in the atmosphere, including a variety of feedback mechanisms. Further, as Coughlan (1979) has pointed out, if we consider two atmospheric factors, A and B, it seems likely that while a change in factor A either positively or negatively will be accompanied by a rainfall variation of the same sign but varying strengths throughout the wheatbelt, a change in factor B may lead to substantial variations in rainfall in opposite directions in western and eastern sectors of the wheatbelt.

Clearly, the results of work by several researchers seem to support changes in the seasonal distribution of rainfall in the wheatbelt, although the detailed picture is complex. On the other hand, temperature changes are less clear, although some twentieth century warming does seem apparent. Detailed relationships between the identified climatic variations and wheat yields remain to be determined, in particular in terms of the implications of such fluctuations for changing risk levels. The overall picture is further complicated by recent work by Doran and McGilchrist (1983) concerned with the detection of climatic change in Australia. Their conclusion is that there is no evidence of continental climatic change in Australia and that, while cyclical patterns of wet and dry periods still occur, they lie well within the bounds expected of randomly behaved phenomena.

2.4. Large-Scale Climatic Controls and Drought Predictability

2.4.1. Rainfall cycles

Attempts to identify drought cycles in Australia date back to the first half of the nineteenth century. By the end of the last century there were claims for drought cycles of two, seven, nine, 10, 11, 12 and 19 years' duration (Jeans, 1972).

Quayle (1938) supported an association of rainfall variation with a sunspot cycle. Vines (1977) and King *et al.* (1974) claimed that Australian wheat yields per unit area were closely related to 22-year and 11-year sunspot cycles. As Nicholls (1978) pointed out, however, King *et al.* (1974) did not indicate what data they had used, and the filtering methods used by Vines (1977) would probably indicate 11-year and 22-year cycles in almost any time series. Nicholls (1978) calculated the power spectrum of the time series of wheat yields in Victoria and Western Australia and found no evidence to support the idea that production is affected by 11-year and 22-year sunspot cycles. Nicholls (1981) further examined relationships between annual rainfalls at 18 Australian stations and variations in sunspot numbers and concluded that there was no evidence of a relationship strong enough to be of any use in long-range weather prediction.

In his major study of Australian droughts, Foley (1957) could identify no regular rhythm in the timing of drought onset, duration or extent. Wonder and Howlett (1977) regarded the incidence of drought and non-drought years in central Australia as random. On the other hand Maher (1967) and Gibbs and Maher (1967) concluded that runs of rainfall both equal to/above median and equal to/below median were almost always longer than would be expected in random series without persistence. Dury (1983) suggested that sequences of precipitation in pastoral Australia can be described by square waves, so that their variation is step-functional. Runs of months or years in which observed values vary about the long-term mean seem to be the exception. Instead, blocks of time with high average values alternate with blocks of low values, pointing to variations of the precipitation record as either random, or as random with persistence. The effect of persistence exaggerates the step-functional patterns of rainfall time series, tends to exacerbate the effects of drought when it sets in, and points to the prospects of developing simple methods of forecasting onset, intensity or duration of drought as being remote. There is no way of predicting the end of a run in progress or the onset of the next. On the basis of such evidence, the best hope for seasonal forecasting may rest with analogue methods.

2.4.2. The Southern Oscillation

Some studies have shown that there may be some promise for forecasting Australian droughts in terms of relationships with the Walker circulation–Southern Oscillation/El Niño complex. During the baseline period selected for this project (1951–80) Australia has suffered major droughts of varying severity in 1951, 1957, 1961, 1965–67 and 1972, and more recently in 1982–1983. El Niño events of varying intensity happened around these periods, except the drought of 1961 – i.e., in 1951, 1953, 1957, 1963, 1965, 1969, 1972, 1976 and 1982.

Streten (1981 and 1983) has shown that anomalously low sea-surface temperatures (SSTs) around Australia, particularly at low latitudes, seem to be associated with dry years, and higher temperatures with wet years. He found that SSTs in the eastern Pacific could be linked to Australian rainfall. When the eastern Pacific is colder than normal, Australia tends to be wet. Analysis indicated that SSTs before the onset of a dry period usually showed a sudden warming of the eastern Pacific during the summer. Such a rise was particularly marked in the drought years of 1957 and 1965, and less marked in 1961.

El Niño events certainly seem to be associated with some of Australia's severest droughts, but there are anomalies, such as the absence of El Niño in 1961 and the good rainfall of 1963, when there was an El Niño. Some of these differences may be explained by varying intensity of El Niño events. In 1965, for example, SSTs off South America were 4°C above the long-term average, while in 1963 the increase was less than 1°C. Further, in 1979 and 1980 changes in winds, sea level and rainfall in the western Pacific resembled those occurring during the development of El Niño, but no El Niño occurred in those years. Yet dry spells set in over eastern Australia at that time, their effects perhaps being exacerbated by the 1982 El Niño. Streten's work points to the possibility of some connection between rainfall and the Southern Oscillation, since its operation is generally considered to be linked to SSTs. In fact, Streten found that although the extreme rainfall years tended to be related to extremes of a Southern Oscillation index, this is not always the case. El Niño itself still cannot be predicted reliably, and there is still a great deal to learn about the dynamics of the Pacific circulation.

There is evidence that atmospheric patterns during droughts seem to force changes in the behavior of the oceans, which are in turn reflected in atmospheric behavior. Patterns of variation in Australian rainfall and temperatures have been related to indices of the large-scale atmospheric circulation, such as indices of the Southern Oscillation (e.g., Troup, 1965; Wright, 1975; McBride and Nicholls, 1983), and the latitude of the subtropical high-pressure belt over eastern Australia (Pittock, 1973) (*Figure 2.7*). These and many other studies for other parts of the southern hemisphere or for the southern hemisphere as a whole seem to establish that hemisphere-wide variations are tied to circulations on a hemispheric, perhaps global, scale. Eigenvector analysis of annual rainfalls in Australia gives patterns for the first and second eigenvectors, which show good correlations with an index of the Southern Oscillation (Troup, 1965) and with an index of the latitude of the subtropical high-pressure belt (Pittock,

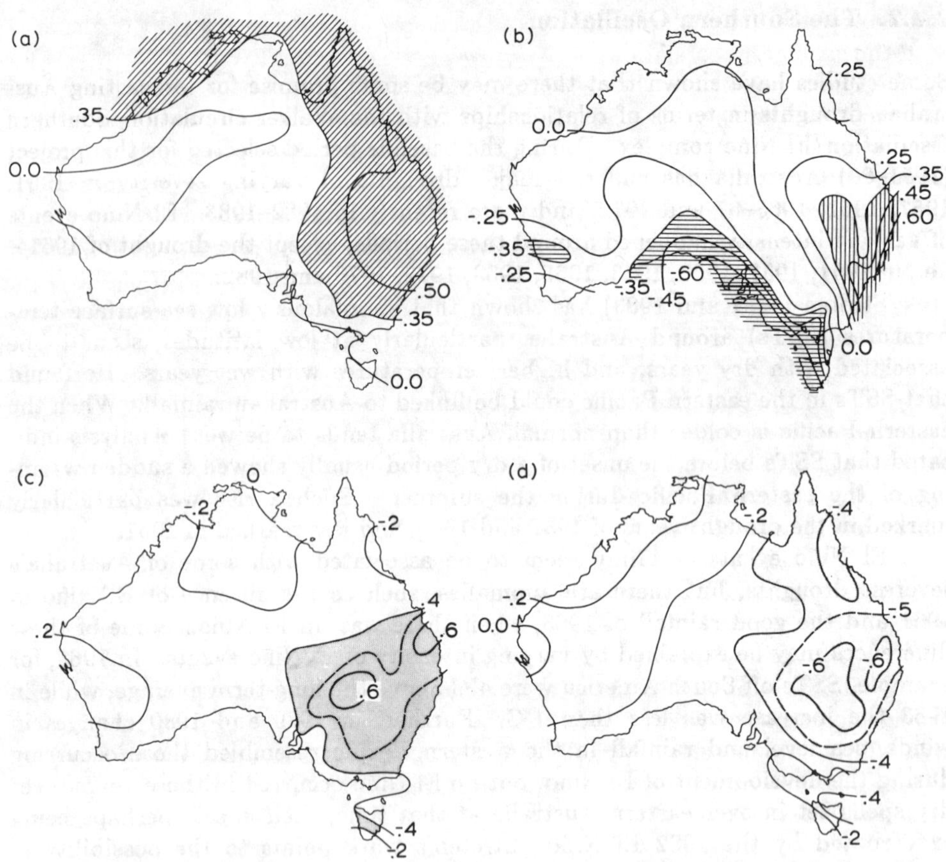

Figure 2.7. (*a*) Isopleths of equal correlation coefficient between the annual mean Southern Oscillation index (Troup, 1965) and district mean annual rainfall for the period 1941–1970. ($R = 0.35$, significant at the 95% level; $R = 0.45$ at the 99% level.) Source: Pittock (1975). (*b*) Isopleths of equal correlation coefficients between the annual mean latitude of the surface high-pressure belt over eastern Australia and district mean annual rainfall for the period 1941–1970. Source: Pittock (1975). (*c*) Correlation between the annual mean value of the Wright (1975) Southern Oscillation index and district mean annual rainfall for the period 1933–1974. Source: McBride and Nicholls (1983). (*d*) Correlation between annual Darwin pressure and district mean annual rainfall for the period 1933–1974. Source: McBride and Nicholls (1983).

1973), respectively (Pittock, 1983), the correlations being +0.60 and −0.69 respectively, both significant at the 99% level.

The seasonal relationships between Southern Oscillation indices and Australian rainfall have been examined by McBride and Nicholls (1983). They found a distinct seasonal cycle in the correlations, with spring (September to November) rainfalls showing the strongest correlations and summer (December to February) the weakest. In some seasons, especially spring, lag correlations were strong enough to suggest the possibility of seasonal rainfall predictions. The strongest correlations do seem to cover the eastern sector of the wheatbelt, and the correlation patterns show some evidence for a westward shift of the area with the strongest correlations.

Nicholls (1980) has shown that for a few locations at least in northern and eastern Australia there is evidence supporting the feasibility of accurate prediction, a few months ahead, of the general character of the weather in a season. Quayle (1920) indicated that during late winter and spring, rainfall in eastern and northern Australia could be predicted some months ahead from observations of atmospheric pressure at Darwin. This work has more recently been verified by Nicholls (1981 and 1983a), Nicholls and Woodcock (1981), and Nicholls *et al.* (1982), and is significant because Darwin pressure is one of the measures commonly used in indices of the Southern Oscillation.

If Darwin pressure averaged over a month or two is above the long-term mean for that period, then less rain than usual can be expected in the next few (1–5) months over eastern and northern Australia. Useful predictions are limited to the July to November period, using Darwin pressure data from May to June onwards. Correlation coefficients between Darwin pressure anomalies (deviations from long-term means) and subsequent rainfall anomalies during the stated period are typically −0.5. Unfortunately, the area for which this predictor seems to be most useful lies largely outside the wheatbelt, and there is still not an accurate long-range prediction method for the west and extreme south of the continent or during summer and autumn.

Nicholls *et al.* (1982) have also demonstrated that Darwin pressures can provide forecasts of whether the tropical Australian wet season in the Darwin area will be late or early. Nicholls (1984) has extended this to show similar prediction possibilities for the whole of the Australian tropics.

Similar results for the long-range prediction of seasonal mean temperatures have also been produced by Nicholls (1983b). Synoptic scale weather fluctuations can make a major contribution to interannual variability of seasonal mean temperatures. In areas where synoptic scale events play a major role in determining interannual variability, accurate long-range prediction is unlikely in the immediate future; but in other areas and in seasons where synoptic events play only minor roles in determining interannual variability, long-range prediction could be feasible. This includes areas where there may be external forcing of the atmosphere by SST anomalies or by variations in surface albedo. Therefore, there would seem to be higher potential for predictability in the tropics, with predictability decreasing southwards. Thus, the prospects do not appear too promising for those parts of Australia containing most of the wheatbelt.

The most recent major drought, of 1982–83, provides some evidence to support some of the interrelationships so far referred to as having possible predictive value. Charts showing atmospheric circulation and SSTs over the southern hemisphere (*Figures 2.8–2.10*) reveal anomalies of pressure, circulation and SSTs associated with the drought. The mean 500 mb circulation for August and November 1982 shows an area over Australia where pressures were significantly above average since April 1982. Consistently high pressures were accompanied by persistent long-wave ridges, which minimized the influence of lows and associated cold fronts over southeastern Australia (Bureau of Meteorology, 1983). This situation also appears on the mean sea-level charts for August and

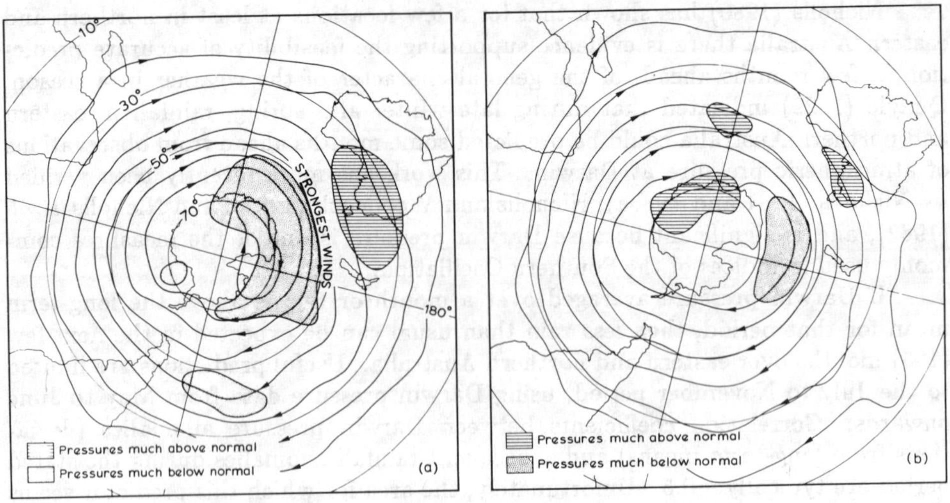

Figure 2.8. **Mean flow patterns at 500 mb (*a*) during August 1982, (*b*) during November 1982. Source: Bureau of Meteorology (1983).**

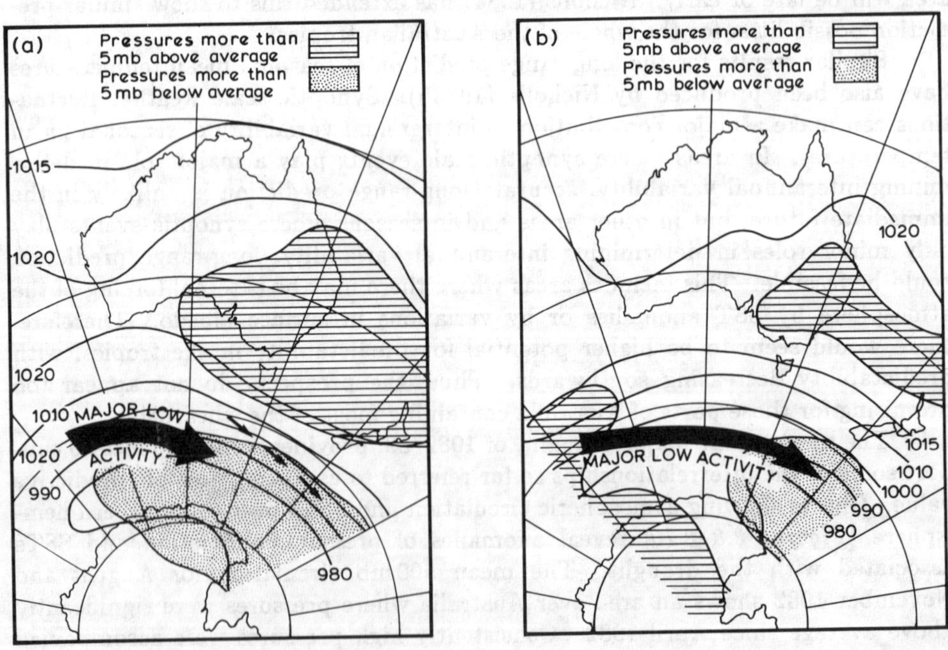

Figure 2.9. Mean sea-level pressures (*a*) for August 1982, (*b*) for November 1982. Source: Bureau of Meteorology (1983).

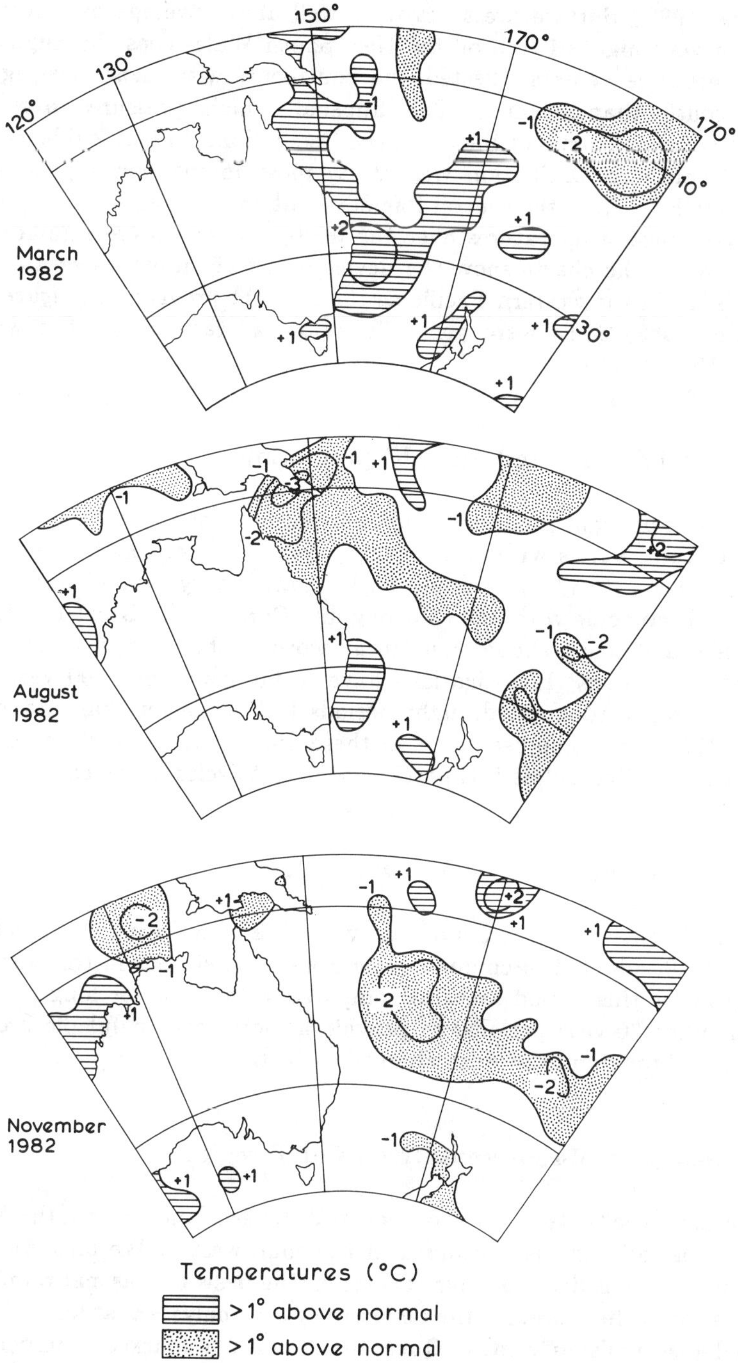

Figure 2.10. Sea-surface temperature anomalies for March, August and November 1982. Source: Bureau of Meteorology (1983).

November 1982. Surface pressures were well above average over Australia and reached a very high +10 mb on the New South Wales coast in August. Only a few low-pressure centers affected the continent, with most passing eastward further south than normal. The unusually high pressures over Australia represented one extreme of the Southern Oscillation. In fact, evidence suggested that the Southern Oscillation was at its most intense since pressure records became available from the central Pacific about 45 years ago.

Sea-surface temperatures over the western Pacific were significantly lower than normal. The charts show the development of an extensive area of below-normal SSTs in the western Pacific as the drought progressed (*Figure 2.10*). By November 1982, SSTs were generally below average across the west Pacific between 15 and 30°S.

2.4.3. Coral density and drought prediction

Some hope of drought prediction has been raised by the recent discovery that the density of coral varies with each year's growth (Isdale, 1984). Denser bands of coral have been found to correlate with known heavy rainfall periods, and less dense bands coincide with known droughts. Coral in the Great Barrier Reef, off the Queensland coast, should contain a record of the weather conditions during the past 1000 years. Investigations have so far gone back 300 years and have already detected apparent drought periods lasting longer than any recorded in the last 100 years. The usefulness of these findings for predictive purposes will, however, be limited unless it is possible to detect cycles in the coral record.

2.5. The Baseline Period, 1951–1980

The focus in this subsection is on analysis of rainfall during the baseline period, 1951 to 1980, adopted in common with other case studies in this volume. However, data for this period are also compared with those for long period records and for other 30-year periods, to provide a more meaningful picture of rainfall variability throughout the Australian wheatbelt.

2.5.1. Selection of representative rainfall regions

As the wheatbelt stretches from central Queensland to South Australia in eastern Australia and is continued in the southwest of Western Australia, it is convenient to consider it as four regions on the basis of seasonal rainfall patterns and location. Therefore, for the purposes of the analyses that follow, the division adopted distinguishes central Queensland (with a marked summer maximum rainfall distribution); southern Queensland and northern New South Wales (summer maximum less marked, rainfall more evenly distributed throughout the year); central and southern New South Wales, Victoria and South Australia (grading from rainfall evenly distributed through the year to winter maximum rainfall); and Western Australia (marked winter maximum rainfall).

For convenience in the following discussion the regions of the wheatbelt will be identified simply as Region 1 (central Queensland), Region 2 (southern Queensland/northern New South Wales), Region 3 (central and southern New South Wales/Victoria/South Australia), and Region 4 (Western Australia).

Rainfall of significance for wheat production in these regions occurs at different times of the year according to the region. Therefore, the analyses have been confined to those parts of the year considered most important, rather than looking at whole-year rainfall. Thus, the inclusive periods analyzed are: for Region 1, December to March; for Region 2, December to April; for Region 3, March to November; and for Region 4, March to October.

2.5.2. Selection of climatic scenarios

Long-term rainfall variations in the appropriate critical periods for selected stations are shown in *Figure 2.11.* All stations show considerable variability through their records, both over the long period and within the baseline period (1951–80). Some measures of this variability are summarized in *Table 2.1.* The graphs and the table also give some indication of how representative the baseline period may be in relation to full available records.

Consistent patterns of relatively dry or relatively wet periods are difficult to identify within or between the regions for the long-term records. For the baseline period, however, it is possible to detect similar variations at different stations within the same regions. This emphasizes the considerable regional variation within the whole wheatbelt.

Extreme dry years

Extreme dry years for the long-term and baseline periods are listed in *Table 2.2.* Some patterns do emerge here. In the summer rainfall Regions 1 and 2, the summer of 1964–65 was clearly one with widespread dry conditions, and for some of the stations was the driest on record. In the winter rainfall Regions 3 and 4, however, the 1964–65 period was nowhere the driest, with the winters of 1957 and 1969 being the most common driest seasons. The recent drought of 1982–83 in eastern Australia produced conditions drier than any during the baseline period and for many areas drier than any on record (*Figure 2.12*) (*see* later in this section).

Back-to-back dry years

Back-to-back dry years are harder to identify for most stations. Some examples are listed in *Table 2.3*, which again points to considerable within-region and between-region variability.

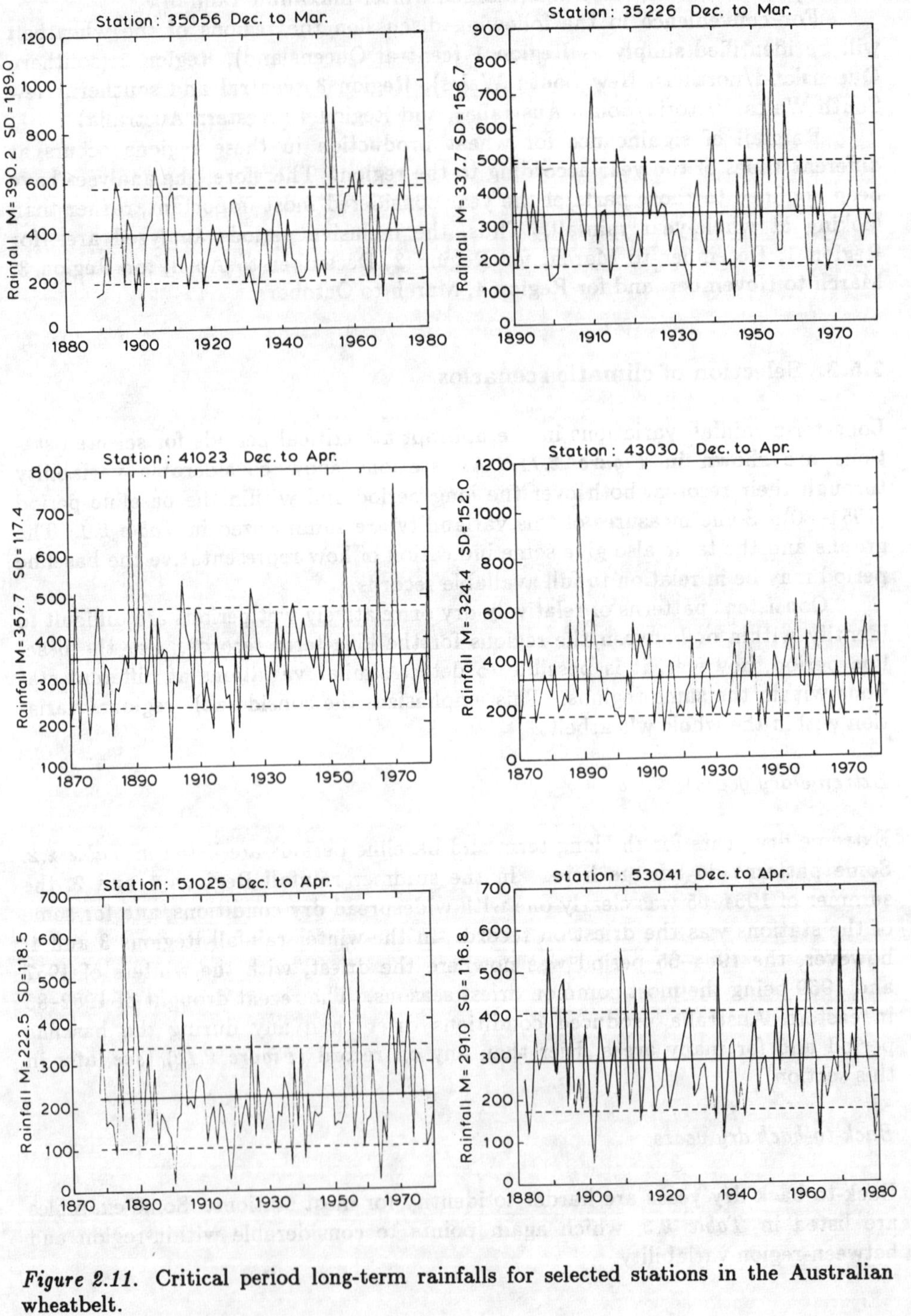

Figure 2.11. **Critical period long-term rainfalls for selected stations in the Australian wheatbelt.**

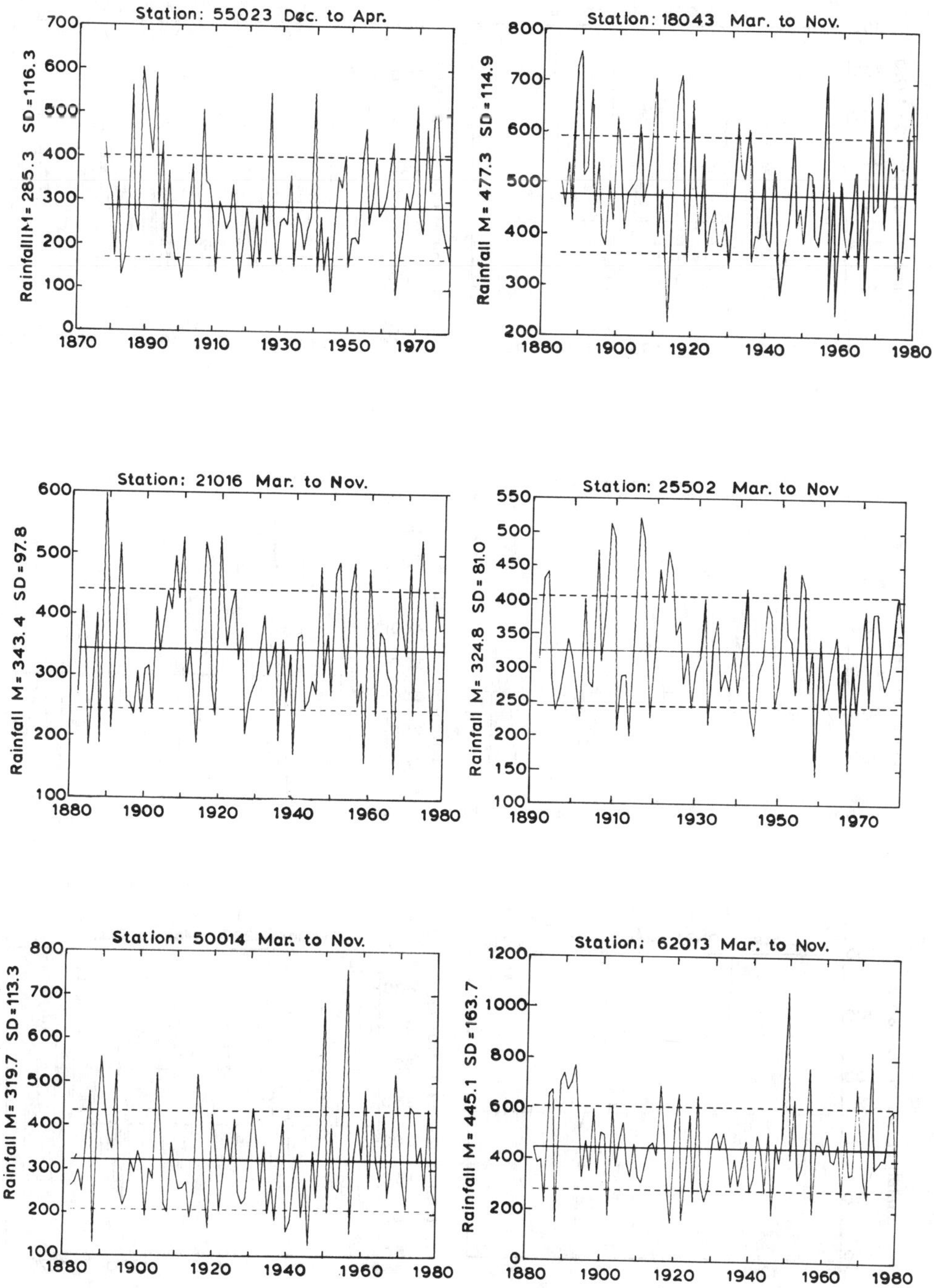

Figure 2.11 continued.

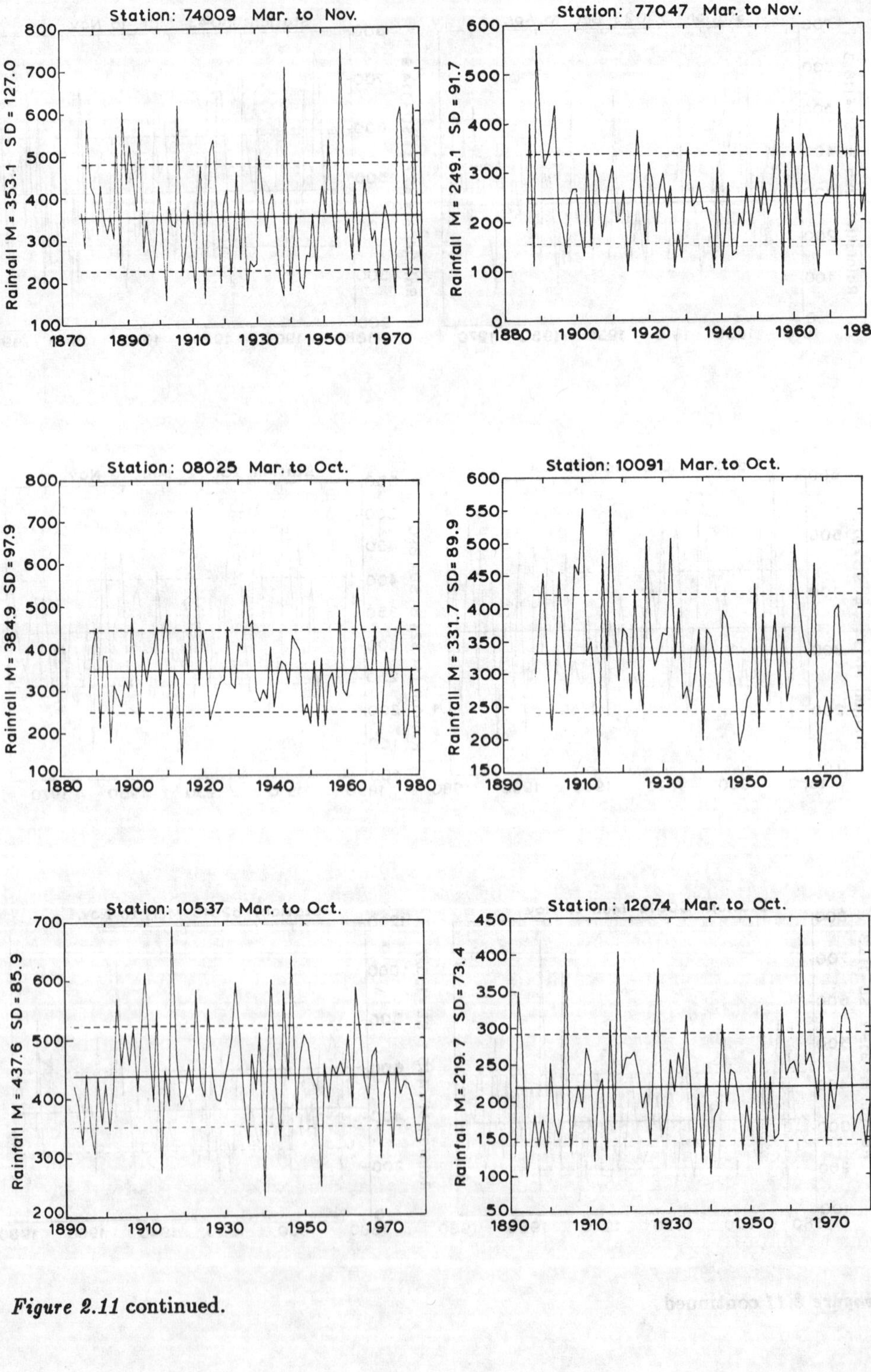

Figure 2.11 continued.

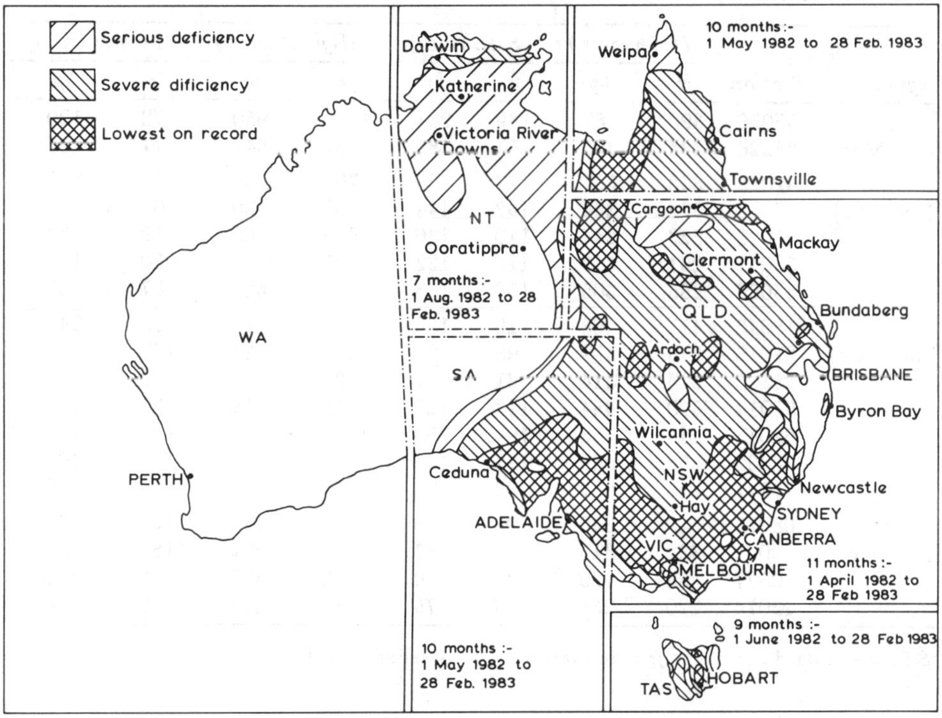

Figure 2.12. Rainfall deficiencies during the 1982–83 drought in eastern Australia. Source: Bureau of Meteorology (1983).

1-in-10 rainfall events

The one-year-in-ten rainfall amounts for the different regions of the wheatbelt for the critical periods have been mapped using all available and suitable station records (*Figures 2.13–2.16*). For each region the one-year-in-ten rainfall amounts have been mapped for four periods: the full record, the baseline period, and the two preceding 30-year periods of 1891–1920 and 1921–1950. This permits identification of area shifts of risk through time and also allows comparison of the baseline period with the full record. Further detail of the shifts through time is given in *Figure 2.17*, which shows for selected stations the changing one-year-in-ten rainfall amounts for successive overlapping 30-year periods, going as far back in the record as possible.

In Region 1 (*Figure 2.13*), for the baseline period, the one-year-in-ten rainfall ranges from about 275 mm in the southeast to 175 mm in the northwest, with a clear east-to-west decrease across the region, and a steeper gradient in the northern part of the region. Considering the 175 mm and 200 mm isohyets, the map points to a westward shift from the earliest to the most recent period, suggesting decreasing risk, as the amounts of rainfall expected once in ten years have increased.

Table 2.1. Critical period rainfall[a] for selected stations in the wheatbelt.

		Mean (mm)		*S.D.* (mm)		*Highest* (mm)		*Lowest* (mm)	
Region	*Station*	*fr*	*bp*	*fr*	*bp*	*fr*	*bp*	*fr*	*bp*
1	35056	390	457	189	201	1023	950	72	120
(Dec–Mar)	35226	334	365	157	181	842	842	77	97
2	41023	358	367	117	126	797	762	116	227
(Dec–Apr)	43030	324	346	152	144	1096	639	100	100
	51025	223	244	119	136	676	568	19	24
	53041	291	313	117	122	601	601	50	105
	55023	285	305	116	113	604	522	87	87
3	18043	477	468	115	123	757	716	228	243
(Mar–Nov)	21016	343	357	98	102	597	524	142	142
	25502	325	314	81	75	522	453	142	142
	50014	320	344	113	123	758	758	125	152
	62013	445	456	164	150	1066	833	147	194
	74009	353	369	127	140	748	748	156	168
	77047	248	262	92	94	558	438	69	69
4	08025	349	340	98	96	736	544	126	174
(Mar–Oct)	10091	332	313	90	87	557	497	151	165
	10537	438	420	86	70	692	587	233	249
	12074	220	233	73	75	439	439	99	113

[a] *S.D.* = standard deviation, *fr* = full record, *bp* = baseline period.

Table 2.2. Extreme dry years for selected stations in the wheatbelt.

		Full record		*Baseline period*	
Region	*Station*	*Year*	*Rainfall* (mm)	*Year*	*Rainfall* (mm)
1	35056	1901–02	72	1968–69	120
(Dec–Mar)	35226	1945–46	77	1964–65	97
2	41023	1901–02	116	1964–65	227
(Dec–Apr)	43030	1964–65	100	1964–65	100
	51025	1901–02	19	1964–65	24
	53041	1901–02	50	1964–65	105
	55023	1964–65	87	1964–65	87
3	18043	1914	228	1959	243
(Mar–Nov)	21016	1967	142	1967	142
	25502	1969	142	1969	142
	50014	1888	125	1957	152
	62013	1888	147	1957	194
	74009	1902	156	1972	168
	77047	1957	69	1957	69
4	08025	1914	126	1969	174
(Mar–Oct)	10091	1914	151	1969	165
	10537	1940	233	1954	249
	12074	1894	99	1952	113

Table 2.3. Examples of back-to-back dry years for selected stations in the wheatbelt.

		Full Period		*Baseline Period*	
Region	*Station*	*Years*	*Rainfall* (mm)	*Years*	*Rainfall* (mm)
1	35056	1932–33/33–34	169/95	1963–64/64–65	259/189
(Dec–Mar)	35226	1944–45/45–46	192/77	1963–64/64–65	210/97
2	41023	1919–20/20–21	268/166	1967–68/68–69	246/267
(Dec–Apr)	43030	1901–02/02–03	136/155	1964–65/65–66	100/261
	51025	1900–01/01–02	128/19	1964–65/65–66	24/169
	53041	1979–80/80–81	108/169	1979–80/80–81	108/169
	55023	1900–01/01–02	165/122	1964–65/65–66	87/173
3	18043	1944/45	289/374	1976/77	318/372
(Mar–Nov)	21016	1927/28	205/256	1958/59	290/159
	25502	1943/44	235/201	1961/62	240/270
	50014	1940/41	156/179	1979/80	250/221
	62013	1928/29	237/292	1971/72	325/255
	74009	1937/38	207/165	1976/77	208/220
	77047	1943/44	172/180	1976/77	183/196
4	08025	1976/77	187/219	1976/77	187/219
(Mar–Oct)	10091	1949/50	189/214	1969/70	165/231
	10537	1896/97	346/307	1978/79	324/326
	12074	1894/95	99/134	1956/57	148/155

Region 2 (*Figure 2.14*) shows a broadly similar pattern of decreasing one-year-in-ten rainfall from around 250 mm in the northeast to about 100 mm in the southeast. Any westward shift of risk is less marked for this region than for Region 1. There are in fact signs of a two-way shift about a north–south axis running through the center of the region, with western areas perhaps tending toward slightly decreased risk and eastern, especially northeastern, areas tending toward increased risk. The 150 mm isohyet has, for example, shifted westward, while the 200 and 250 mm isohyets have drifted eastward, giving less marked risk gradients during the baseline period.

In Region 3 (*Figure 2.15*), for the baseline period, the one-year-in-ten rainfall ranges from around the 350 mm mark in the southeast and in the extreme northwest, to generally below 150 mm in the northern interior parts of the region. Particularly sharp risk gradients are found in the northwest. Variations between the successive 30-year periods are not marked, although there is some indication of an inland shift of higher rainfall amounts, mainly through the western part of the region in New South Wales and in northern Victoria. Greater within-region variation of risk is apparent for this region than for any other parts of the wheatbelt.

The risk gradient in Region 4 (*Figure 2.16*) trends clearly from northeast to southwest across the region. For the baseline period the one-year-in-ten rainfall ranges from around 100 mm in the northeast to about 350 mm for a small area of the southwest. There is evidence of a westward shift of all isohyets during the baseline period, again pointing to increasing risk for much of this part of the wheatbelt during the most recent period.

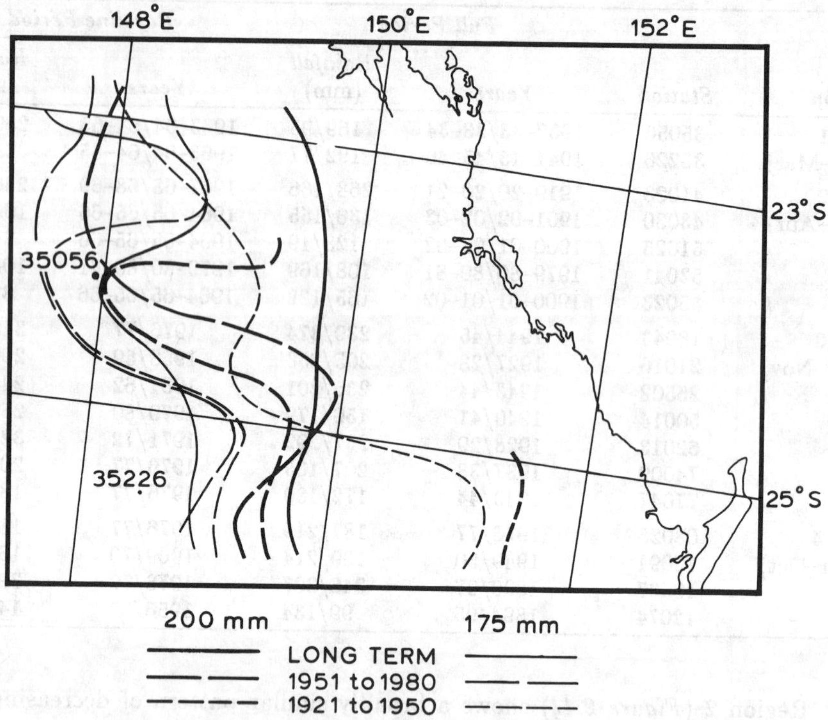

Figure 2.13. Location of the 175 mm and 200 mm critical period one-year-in-ten rainfall isohyets for Region 1.

Spatial patterns of low-rainfall probability

For individual stations, and by implication for whole regions, the probability levels of different amounts of rainfall over periods of a standard length (in this case, 30 years) can vary considerably from one period to the next. Some examples are included in *Figure 2.18*, which shows cumulative frequency distributions for eight stations throughout the wheatbelt.

Further evidence for the level of risk in different parts of the wheatbelt, and for the variation between and within the four regions used in this study, is shown in *Figure 2.19*. This illustrates, for long-term records, the possibilities that critical period rainfalls will be no more than 200 mm. These maps clearly indicate the risk gradients across all of the regions and also point to the central Queensland region as generally having least risk; and to the western parts of northern New South Wales as having the largest area of relatively high risk, mostly between 20% and 50% probability that the region will not receive 200 mm in the critical period.

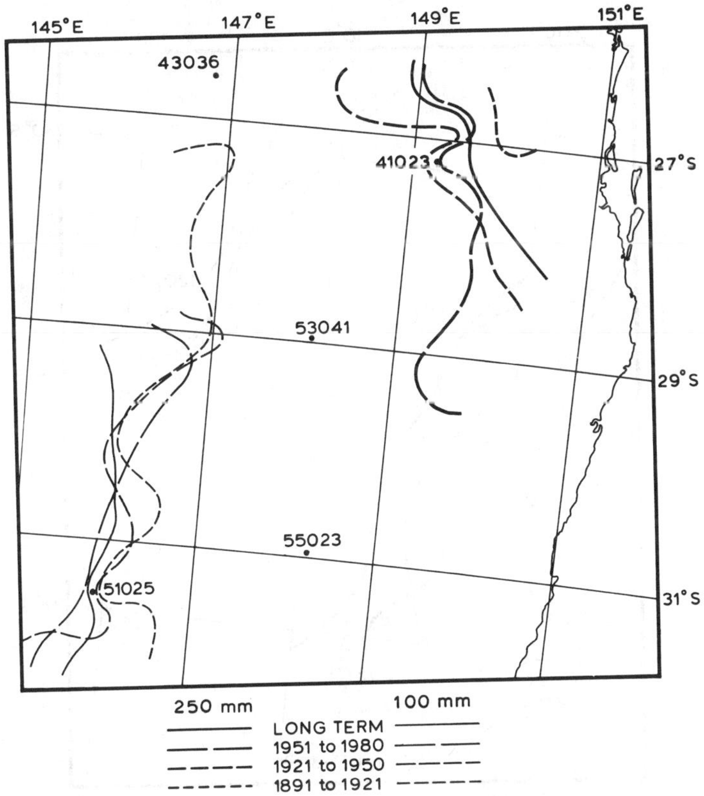

Figure 2.14. Location of the 100 mm and 250 mm critical period one-year-in-ten rainfall isohyets for Region 2.

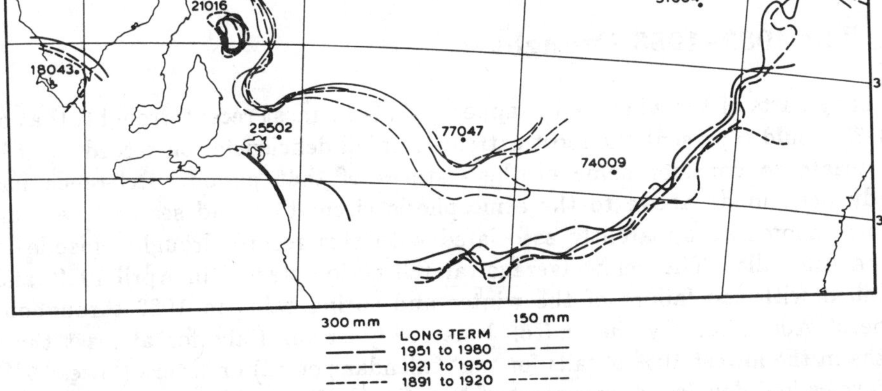

Figure 2.15. Location of the 150 mm and 300 mm critical period one-year-in-ten rainfall isohyets for Region 3.

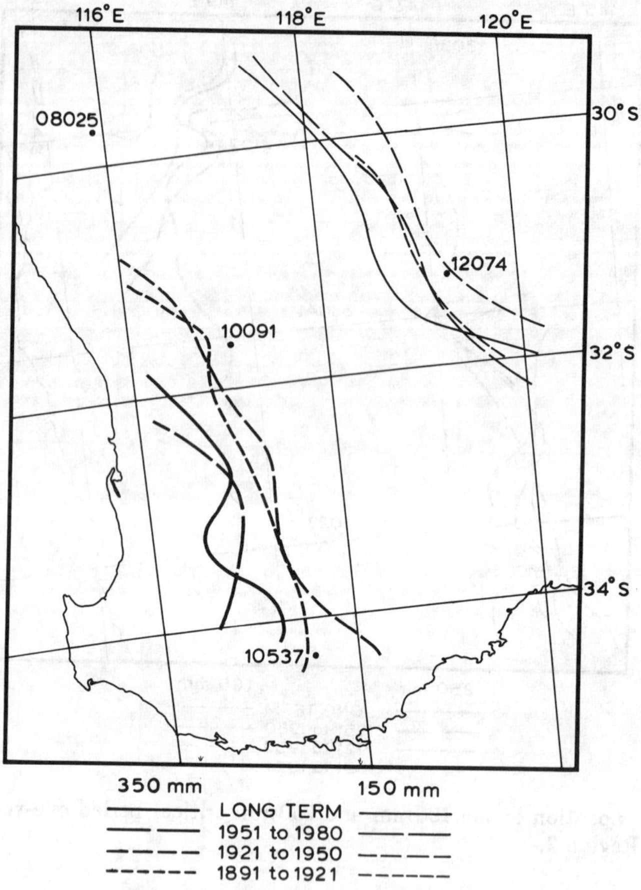

Figure 2.16. Location of the 150 mm and 350 mm critical period one-year-in-ten rainfall isohyets for Region 4.

2.6. The 1982–1983 Drought

For many parts of the wheatbelt it appears that the most recent drought, that of 1982–83, could represent the most extreme rainfall deficiencies on record, so it is appropriate to consider some of the features of that period. Reference has already been made above to the atmospheric circulation and sea-surface temperature anomalies apparently associated with this record drought episode in eastern Australia. The below-average rainfall period started in April 1982, and continued with the failure of the winter and spring rains in 1982 throughout southeast Australia. By the end of June 1982, serious (falls for at least three months in the lowest 10% of falls for that calendar period) or severe (lowest 5%) deficiencies had developed over most of New South Wales, central and southwest Queensland and northern Victoria. By the end of August 1982, deficiencies of 3–5 months' duration covered much of South Australia, nearly all of Victoria and

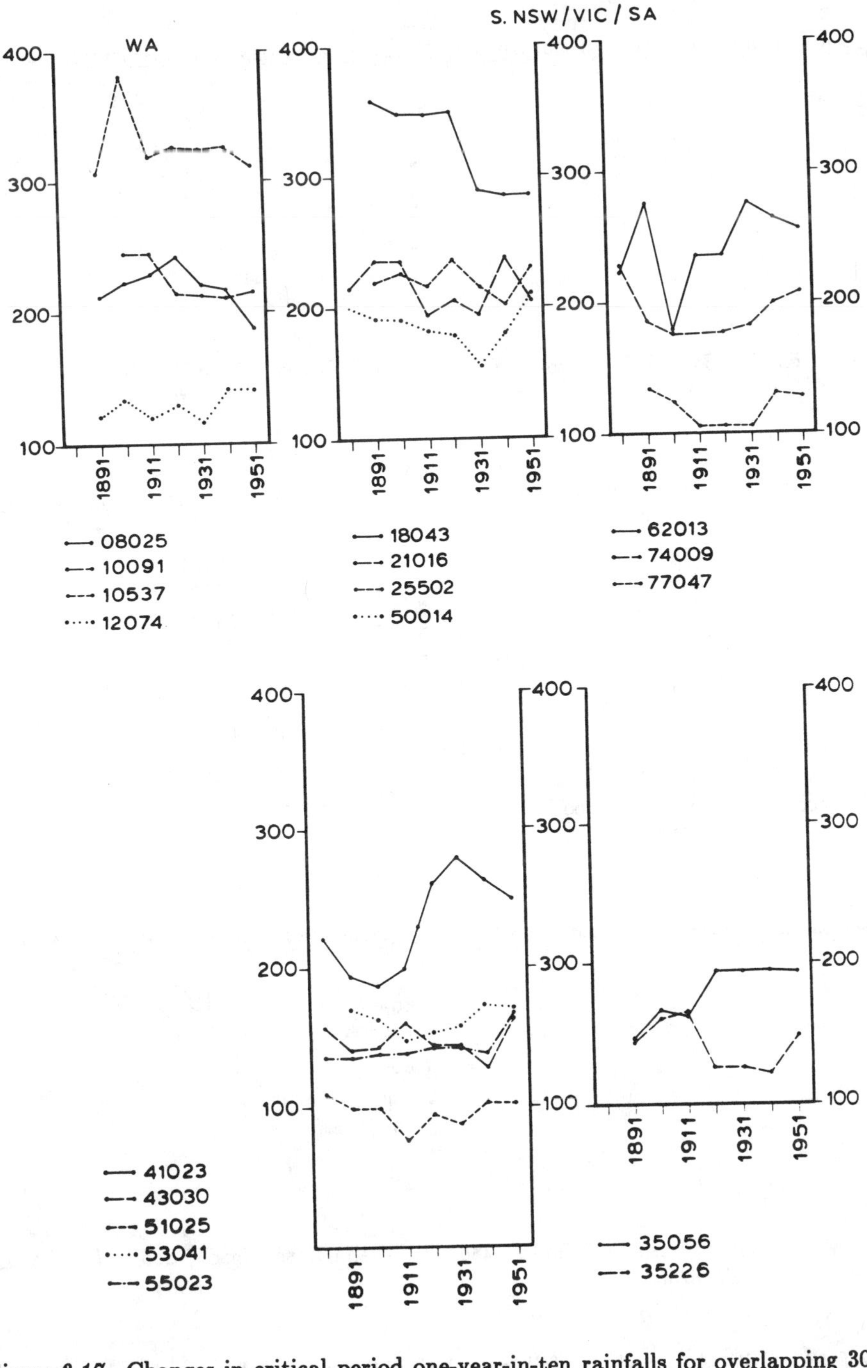

Figure 2.17. Changes in critical period one-year-in-ten rainfalls for overlapping 30-year periods (beginning every 10 years from 1881 or 1891) for selected stations in the Australian wheatbelt.

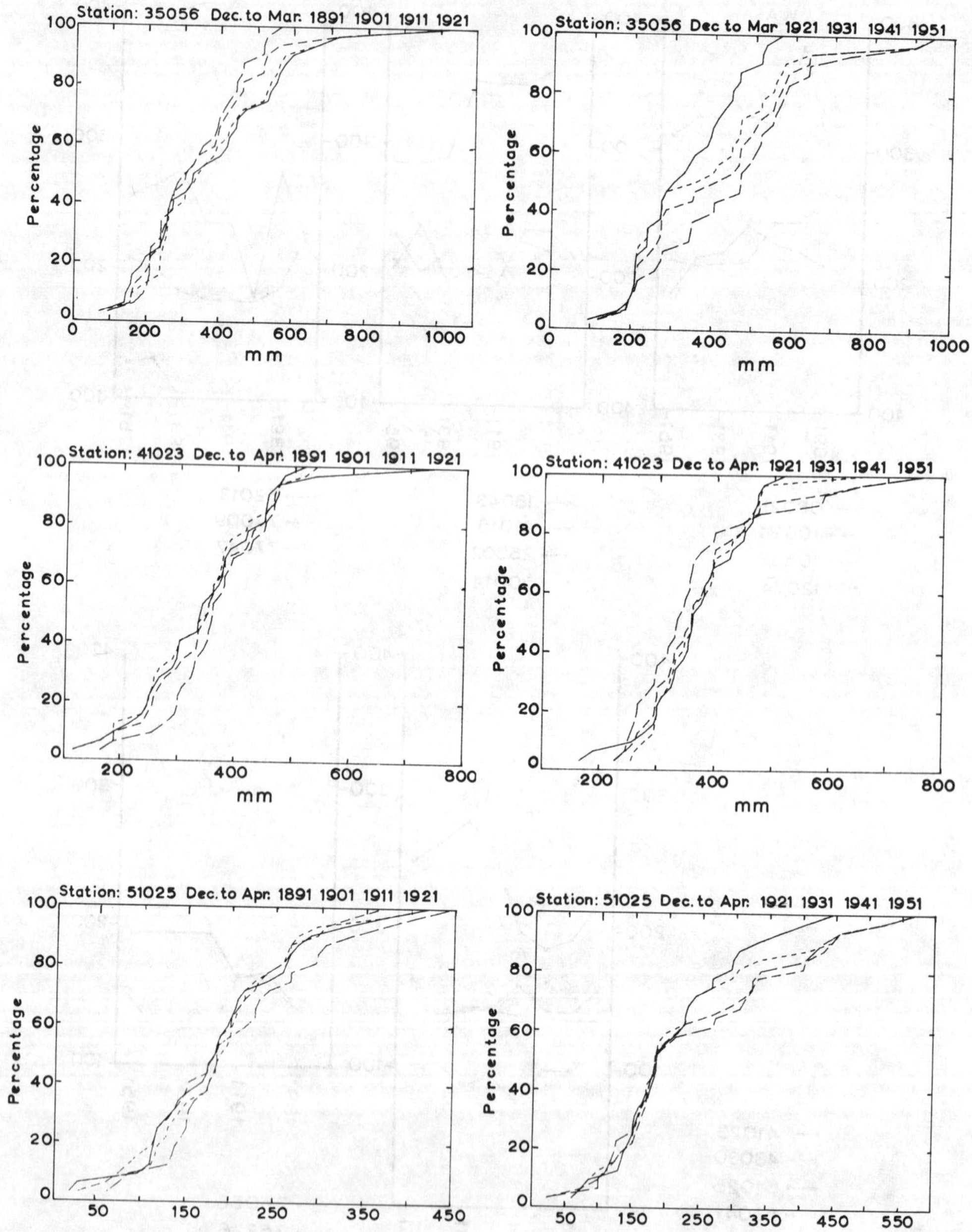

Figure 2.18. Critical period rainfall cumulative frequencies for overlapping 30-year periods (beginning every 10 years from 1891 or 1901) for selected station in the Australian wheatbelt. Key as for *Figure 2.16*.

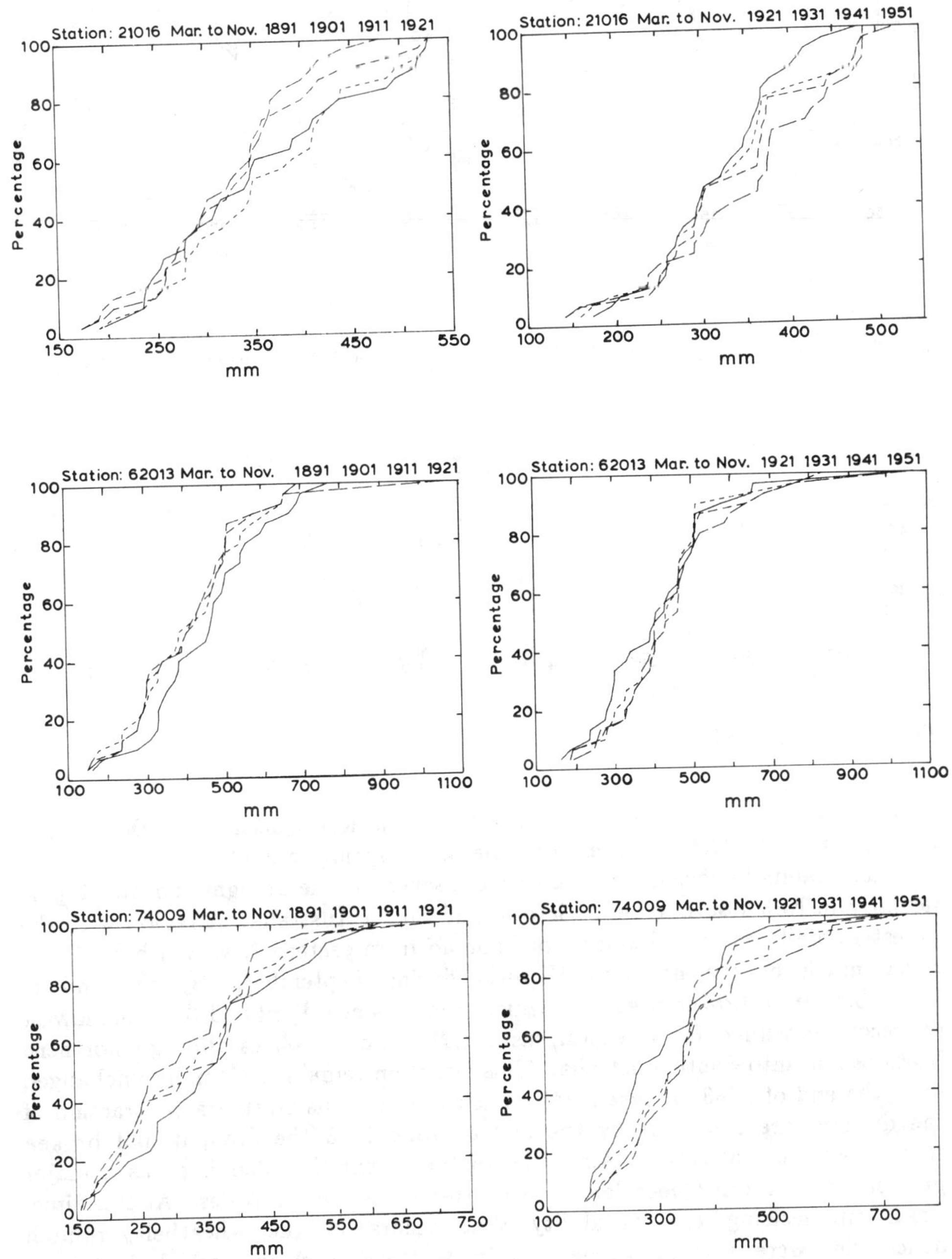

Figure 2.18 continued.

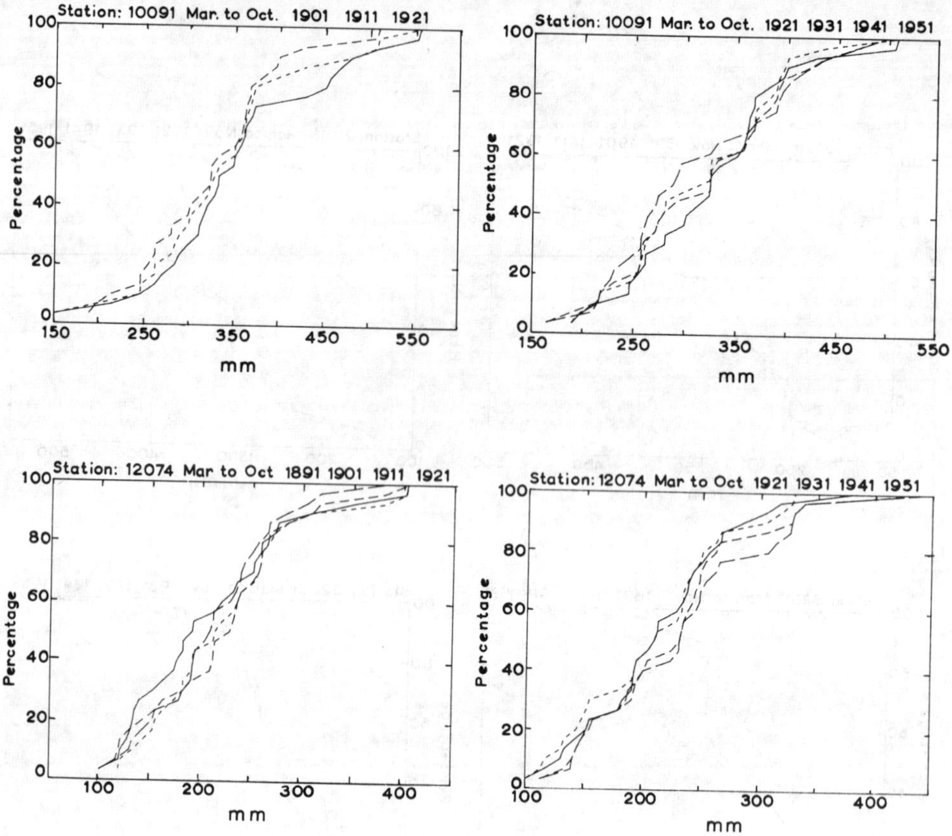

Figure 2.18 continued.

New South Wales, Tasmania, and central and southern Queensland. Deficiencies worsened through October after abnormally low spring rainfalls.

The rainfall deficiencies became more severe as the drought continued, particularly in New South Wales, Victoria and South Australia. A region with the lowest rainfalls on record since April spread from eastern New South Wales to cover much of the northern Victoria during September. By the end of December, an unbroken area over which rainfall since April had been the lowest on record extended from central/eastern New South Wales through northern Victoria and into South Australia. This situation remained virtually unchanged until the end of 1983 (*Figure 2.12*). Then March rains started a contraction of the drought area, so that by the end of June 1983 the drought had broken throughout most of eastern Australia, to the extent that flooding was a major problem in southern Queensland and northern New South Wales. At this time, again emphasizing the variability that occurs in the wheatbelt, rainfall deficiencies were starting to develop in Western Australia, which had been largely unaffected during the eastern drought.

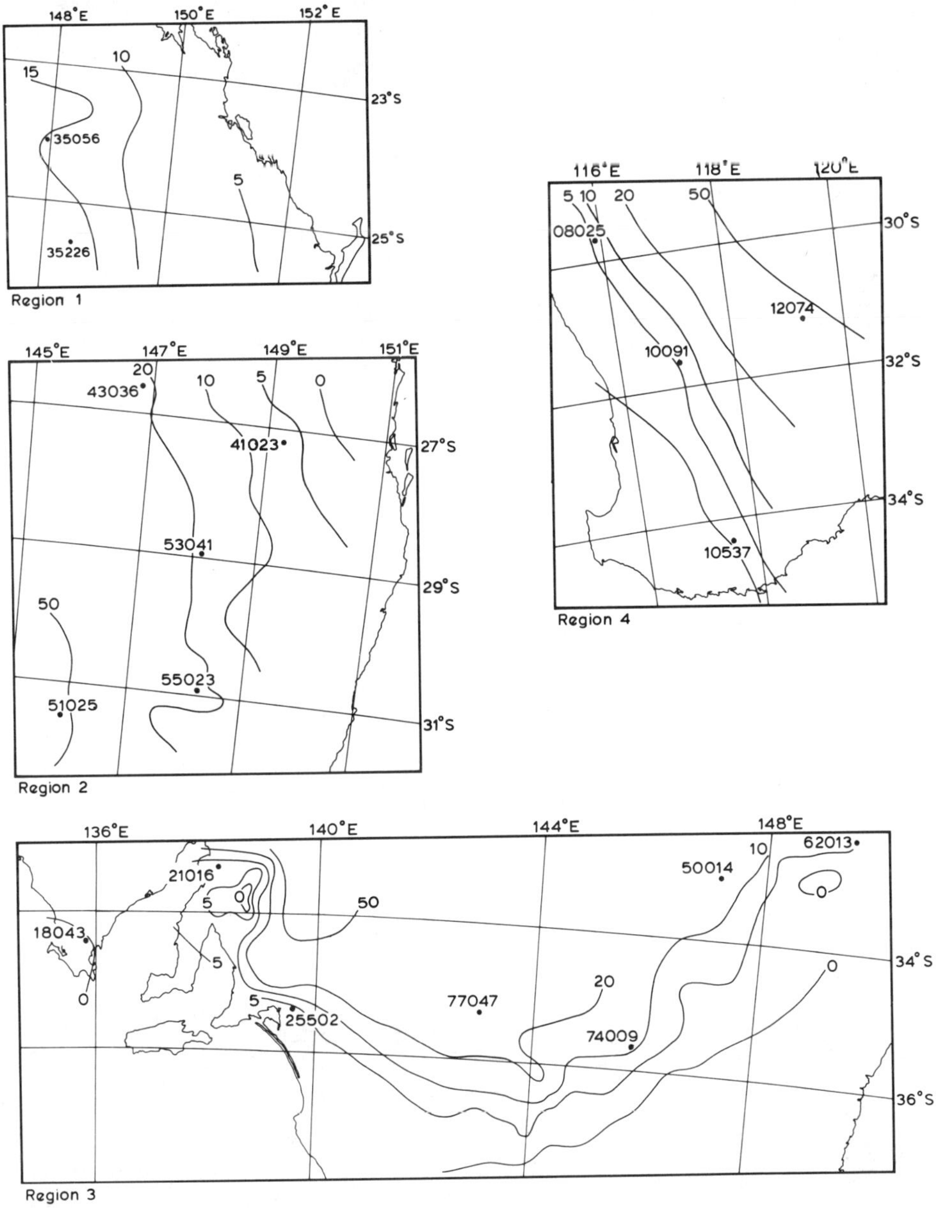

Figure 2.19. Critical period probabilities that rainfall will be no more than 200 mm, based on long-term records.

Some of the implications of this drought, and of earlier drought episodes, on wheat production and agriculture in general, are described in Section 3.

SECTION 3

Socioeconomic Impacts of Climatic Variability

3.1. Introduction

As outlined in Section 1 of this case study, climatic variability is a fundamental fact of life of Australian agriculture. It and its impacts have already been the subject of reviews from several other points of view (*see*, e.g., Williams, 1946; Rutherford, 1950; Everest and Moule, 1952; Anon., 1966; Campbell, 1966; Anon., 1967; Anon., 1968; Lovett, 1973; McQuigg, 1975; Chapman, 1976; Butler and Doesell, 1979; Heathcote and Thom, 1979; Young and Wilson, 1979; Heathcote, 1983). However, the observation of Campbell (1958, p. 3) that "one cannot but be impressed by the extremely limited progress which we, as a people, have made .... in coming to terms with the most insidious and pervasive characteristic of our climate – its variability" is still cogent. The objective in this section is to undertake a broad review that spans agriculture from farm to the macro-economy and spans climate generally rather than one that addresses only drought – the focus of Section 4.

However, within this broad charter, the present review is selective in scope and is confined to economic (as opposed to the important ecological, environmental, meteorological, physiological and sociological) aspects. First, a framework is developed to provide a structure in which past studies might be considered. This features behavioral motivations and probability distributions of financial performance, physical performance and prices. Second, investigations of the impacts of climatic variability on domestic agriculture are reviewed against this framework. These studies are categorized first by level (farm, industry, region and sector) and, within levels, variously by region, industry or type of impact.

3.2. A Framework for Reviewing the Impacts of Climatic Variability on Agriculture

The complexity of the relationships between climatic variations and the people in, and dependent on, the agricultural sector can be tackled by a simplified disaggregation of the major effects. A starting point is the basic behavioral assumption about what motivates the key economic agents in farming – the farmers who make the economic decisions. This objective can be stated as: farmers' satisfaction depends, *inter alia*, on the probability distribution of financial performance. There is mounting evidence (Francisco and Anderson, 1972; Just, 1974; Anderson *et al.*, 1977; Bond and Wonder, 1980) that farmers' attitudes are typically ones of aversion to risk so that features of the probability distribution other than the expected value (e.g., variance, semivariance, lower range, skewness, etc.) are also relevant to decision making. An important personal characteristic that influences attitudes to risk is the wealth or asset base of the farmer. This changes over time, partly as a result of climatic fortune.

The second general relationship is a mix of behavioral and environmental effects that enter the probability distribution of financial performance. This distribution depends on allocative decisions, the probability distribution of yields, the probability distribution of prices, institutions, and government interventions. The probability distributions noted may not, in fact, be independent, e.g., drought-reduced cuts of wool may be associated with reduced strength and quality (see Chislett, 1973) and price. Interventions by government other than through prices may be passive (e.g., the progressive income taxation structure) or active (e.g., various forms of assistance for relief from natural disasters such as drought, flood and tempest).

The third general relationship summarizes the main direct effects of climate at the farm level in that the probability distribution of yields depends on both climate and allocative decisions.

The influence of climate on yields is obvious, even if not precisely understood. The following generalization for the growing of plants in Australia is based on the discussion of Fitzpatrick and Nix (1970). Solar radiation is rarely an important restraining factor. The thermal regime sets the major seasonal growth patterns and constrains growth in most of the country during winter, and the dominating climatic factor in growth is the moisture regime. This is determined by evaporation and precipitation (mainly as rainfall) which is typically meager, strongly seasonal and inevitably variable. The validity of these generalizations is supported by the empirical work of Easter (1975), who related several measures of pastoral production to rainfall and/or to estimated soil moisture levels over different periods.

Many of the allocative decisions of farmers are influenced by climate and are, in turn, often intended to mitigate the effects of climatic variation – consider, e.g., drought feeding of livestock (Hill, 1973), irrigation scheduling, trapping soil moisture by fallowing, choosing between wheat and barley, etc. Moreover, there is typically a strong interaction between controlled decision variables and uncontrolled climatic factors in production, e.g., between rainfall and other climatic factors, and crop variety (Anderson, 1974), stocking rate (Chisholm, 1965; Byrne, 1968; McArthur and Dillon, 1971) or fertilizer (Anderson, 1973; Watson and Anderson, 1977).

The fourth and final general relationship is concerned with what are normally thought of as economic considerations, but which surely also have a climatic component. The probability distribution of prices depends on consumer demand (i.e., incomes, tastes, habits, fashions, etc.), institutions, government intervention and supply – and thus also climate.

Climate may also directly impinge on consumer demand (as noted by Kefford, 1979) and may also act directly on price (e.g., rain on wine grapes in the pre-harvest week). However, the major effects are the aggregate shifts in supply caused by variations in climate. Some of these variations will, of course, be experienced by Australia's trading partners and may be experienced in Australia as shifts in export demand (e.g., USSR crop failures).

The prices in mind are mainly prices for agricultural products. However, prices for farm inputs can also be influenced through general equilibrium effects of widespread climatic events (e.g., price increases for farm materials through increased demand in a good season following a drought). Following Campbell (1958), it has been widely held that most farm investment derives from residual funds available for such purposes after income-unresponsive consumption outlays have been made. If this generalization is true, climatic variability, through its effect on disposable incomes, must have a direct impact on variability of investment in the agricultural sector. There is some evidence (BAE, 1976b, pp. 20–22) that variability is associated with differences in the amount and type of investment undertaken. However, the significance of this evidence is difficult to assess, given the cross-sectional interactions among instabilities of income, enterprise orientations and investment decisions.

Taken together, these relationships depict an agricultural sector directly dependent on climate, and hence farmers are subject to the vagaries of climatic variability. In short, this sector of the economy (like others) is risky and, when the impact of climatic uncertainty is exacerbated by market uncertainties, the environment of agriculture could well be described as turbulent.

Agricultural performance certainly varies with climate. One of the evident roles of the farmer is to buffer crop and animal production against natural climatic variations. This role of stabilizer can be frustrated further by variations arising from the nonphysical environment of agriculture – especially from consumers' changing tastes, reflected in varying prices. For example, at meteorological stations in the eastern pastoral zone, coefficients of variation (i.e., standard deviation/mean) of annual rainfall are about 0.5, while for annual wool production on individual farms they are about 0.3. However, after allowance for variable costs and product prices, coefficients of variation of net farm incomes (and rates of return) of individual wool producers in this zone are about 0.8 (Anderson, 1972).

Each of the relationships has a dual interpretation, depending on whether it is being construed as a planning (*ex ante*) device or as a recording (*ex post*) device. Broadly speaking (see Quiggin and Anderson, 1979), the *ex ante* relationships can be characterized as risk (or uncertainty) and the *ex post* relationships as instability. Of course, these interpretations are connected (by the extent of predictability), and improvements in forecasting imply reductions in some of the costs associated with risk toward the lower costs associated with instability. However, the prospects for long-term forecasting of rainfall are so bleak that it appears that Australian farming will continue to be both unstable *and* risky. Unfortunately for the student of climatic variability, only the effects of instability can be studied with relative ease because most available data are *ex post* observations. Study of *ex ante* effects requires either elicitation of subjective probability distributions or inference about such distributions through econometric methods.

3.3. Farm-Level Impacts

Most of the pertinent farm-level economic research has involved interview surveys in drought-affected and drought-prone regions. It is usually difficult to pin down particular changes in the financial performance of farms as being due to climate. Most droughts have been associated with price movements not entirely dependent on the drought itself.

Several studies (e.g., Cumming, 1966) followed the major drought of eastern Australia in the mid-1960s, which, in the language of this volume, is fairly representative of a worse than 1-in-10 extreme low-precipitation event. Such studies serve to exemplify some of the relationships as *ex post* descriptors. In a study centered on the sheep and cattle industries of southwest Queensland, Haug and Hoy (1970) concluded that substantial negative net farm incomes occurred in drought years. These arose partly from reduced productivity and partly from falls in the size and value of livestock inventories.

Herd and flock performance and size are also highly variable in this region (Rutherford, 1948; Young and Wilson, 1979). For example, over the six-year period 1962/3 to 1967/8, annual herd branding rates (branding to breeders) ranged from 28 to 60%, herd mortality from 5 to 39% and net farm incomes in the beef area ranged from \$72 to \$31 per cattle equivalent. (Note that in this and in the following sections, all money amounts are expressed in early 1982 Australian dollars, which were approximately at parity with US dollars. Note, too, that we have adopted the convention of writing the fiscal year to end-June as 1962/3 and the chronological years 1962 and 1963 as 1962–63.) Generally because of falling herd and flock numbers, property asset values fell, while borrowings rose to meet drought expenditures and post-drought restocking needs (Anderson, 1979). The impacts of drought on livestock farms can be very persistent and, as Rickards and Anderson (1967) have argued, planning (especially financial planning) for recovery must take explicit account of the dynamics of breeding programs.

A similar picture has emerged from other surveys. For example, net farm incomes in the period 1964/5 to 1966/7 relative to those prevailing in the preceding three years fell by about 30% in the wheat–sheep and high-rainfall zones of New South Wales and by about 80% in the pastoral zone of both states (BAE, 1969). It appears that very few properties were financially affected by the drought so seriously that the owners were forced to leave the industry. This contrasts with the rather more drastic consequences of severe drought in the semi-arid areas of some other countries, for example, India (Jodha, 1978; *see* Part V, this volume).

Officers of the Reserve Bank (Anon., 1968) made a detailed sample survey of drought-affected properties in northern New South Wales. Farms were visited after the drought conditions of 1965 and, in view of the persistent adverse seasons, again in 1967. The impacts of drought on farm operations were found to include not only the usually accepted moves to reduce stock numbers and to improve feed supplies, but also a variety of other measures including reduction of development and living expenditures, increased crop-growing under reduced stocking pressures, and increased borrowing and asset run-down. It was revealed that some producers had substantial off-farm reserves available for use in times of low farm income.

Drought relief measures are set on an individual state basis (*see* Section 4). They include concessions on road and rail freight on livestock to and from agistment (rented pasture), restocking movement, forced sales of livestock, fodder, water and grains. Loans to primary producers are also available at concessional rates and repayment conditions for carry-on finance and restocking. Other measures are clearly of importance to affected individuals but, in aggregate, the cost is usually not very high. In 1966/7, drought relief amounted to only 0.6% of all government expenditure (Heathcote, 1967). However, in 1977/8, drought payments amounted to 1.2% of government expenditure.

The Alice Springs district is a semi-arid one, which has been the focus of several BAE farm level studies (BAE, 1968; Reeves and Crellin, 1972; Wonder and Howlett, 1977). Cattle numbers in the district rose to a peak of 353 000 in 1958 under the influence of a run of relatively good seasons, improved communications, improving prospects for beef, and property investments supplementing the existing waterholes and opening up new country. Severe drought over nine years caused numbers to be reduced to 134 000 by June 1966. Numbers had increased to 200 000 by 1971, but the combination of poor markets and the run of subsequent good to excellent seasons created conditions where the Alice Springs herd approached half a million head in 1978.

Pastoral companies, of course, have embodied climatic risk-spreading in their management since the early days of pastoral settlement by gaining ownership of strings of properties in areas unlikely to be influenced by highly correlated climatic events (Anderson, 1970a, b), as well as branching out into investment in other sectors of the economy. Some of the costs and benefits of spatial diversification of grazing properties have been reviewed by Anderson (1971).

Farm business management, involving decision making under price and climatic uncertainty, lags in production, and institutional and legal restraints, is never easy to do well. Management must constantly make adjustments to resource use to cope. During and after droughts there are inevitably recriminations about the virtue of various strategies and tactics that have been adopted by different farmers (Candler, 1958). For instance, with the reduced agricultural maintenance and investment rates in the 1960s and 1970s, the pastoral resource base was run down. This may have predisposed increased sensitivity to subsequent droughts. Whatever is done, there is clearly an element of luck in the success of any planning. Moreover, given the individuality of the mentioned conceptual relationships, it is not prudent to offer general guidelines for drought management on farms – before, during or after a drought (Campbell, 1966; Anderson and Hardaker, 1973). Useful analysis must incorporate the individual nuances of environment and farmer.

3.4. Industry-Level Impacts

The impact of climate at the industry level can be gauged approximately by the extent of variation in the volume of production. A measure of variation in volume, which facilitates comparison across industries of diverse size and nature, is the coefficient of variation based on detrended observations. Motha *et al.* (1975) have estimated such coefficients for the period 1949/50 to 1972/3 and these range from low values of about 0.05 for the wool and whole-milk industries to high values of around 0.37 for the sorghum, cotton and oilseeds industries. Wheat, which experienced an intermediate coefficient of variation of 0.23, is the industry selected for most attention in this section and in this case study.

As for other Australian agricultural industries, the history of the wheat industry has been punctuated by climatic extremes – particularly droughts. Callaghan (1973) documented aspects of 11 major droughts that afflicted this industry in the previous nine decades. He argued that the impact of droughts, while still serious, has become less dramatic because of productivity improvements (especially through mechanization and better varieties), expansion of the industry into more meteorologically dispersed regions and investment in bulk storages. The situation is complicated by recent shifts of production into more marginal wheat growing areas that traditionally have been devoted to grazing sheep for wool production. The shift is primarily in response to farmers' perceptions of changing relative profitability. There is a tendency, however, for variability around trend to have increased for wheat and other cereals (Hazell and Anderson, 1985).

3.4.1. Some impacts on the wheat industry

The effect of the 1965/6 drought on the wheat industry at the national level was not drastic. However, at the state, regional and local levels, the effects of the drought were severe. In New South Wales, the actual area sown to wheat was about 80% of the intended area of 2.3 million ha. The pre-sowing rainfall (averaged cross-sectionally) was only 17 mm a month from March through to May 1956, compared with the mean of 49 mm a month (standard deviation 21 mm) for the 26 years 1951–76. All other states received below-average pre-sowing rainfall in 1965/6, although conditions were considered "dry" rather than "drought".

Yield is critically dependent on weather. In New South Wales, the 1965/6 yield of 0.57 t/ha was the second lowest for that state for the 28-year period to 1977/8 (mean 1.23 t/ha, standard deviation 0.37 t/ha) and can be contrasted with the more recent and more severe disaster of the 1982/3 crop, which yielded only 0.46 t/ha. Yields in other eastern states were only marginally affected in 1965/6. Australian yield was not drastically affected by this drought and actual sowings fell short of intended sowings by only 9%. The national yield series has been virtually trendless since the 1950s and is characterized (over the 30-year period to 1978–79) by a mean of 1.21 t/ha and a coefficient of variation of 0.16. Even at this level of aggregation (for Victorian districts, *see* Ryan, 1974), yields (with median 1.25 t/ha and first decile 0.90 t/ha) are not especially normal.

Rainfall is but one of several climatic factors influencing yields. Hail, wind and temperature, especially frost (Single, 1975), can be even more crucial on occasions. However, rainfall variability does account for much of the overall impact of the many effects of climatic influence on yields. Especially important is rainfall during the growing season (*see*, e.g., Cornish, 1950; Byerlee and Anderson, 1969; Guise, 1969 and Francisco, 1980) although pre-sowing rainfall is also important and can substitute to some extent for early growing season rainfall. At aggregations beyond particular plots, the precision of the relationship between a rainfall index and yield necessarily suffers, although Fisher (1978) found some relationships of acceptable precision using wheat yields for local government areas in several states. Low precision is evident in the state and national regressions reported by Anderson (1979). These were log-linear equations of yield on growing season rainfall so that the slope could be interpreted directly as a (partial) elasticity of production (i.e., ratio of proportional change in output to a small proportional change in the factor of production). This specification for ordinary least-squares estimation is rather restrictive in imposing a constant elasticity of response to rainfall, irrespective of its level. With this qualification in mind, it was notable that, for all these equations, response is inelastic, so that a 1% increase in growing season rainfall induces a yield response raging from 0.3% in Queensland through to 0.8% in New South Wales. A further implication of the constant and inelastic proportional response model is that the marginal product of rainfall diminishes with increasing rainfall. In other words, the gain from a given above-average rainfall does not symmetrically compensate the loss from an equal below-average rainfall. Implications of such elasticities in a context of resource allocation have been noted by Edwards (1978).

3.4.2. Wheat yield variability in New South Wales

The random nature of major climatic variables, such as rainfall, makes farm planning difficult and can lead to wide variations in area sown as well. In examining the magnitude of these climatic impacts on the wheat industry, data from 16 selected local government areas in New South Wales, representing "traditional" and "marginal" parts of the wheatbelt, were employed (Anderson, 1979).

The eastern (traditional) side of the New South Wales wheatbelt generally has a longer history of wheat growing than the western (marginal) side (MacIndoe, 1975). This is due in part to the earlier settlement of the eastern areas and the associated development of transport and other grain marketing facilities, and in part to climatic patterns. Rainfall, in particular, tends to be lower and somewhat more variable in the west.

Since the mid-1950s, the area sown to wheat in New South Wales has increased dramatically and in 1983/4 was 3.85 million ha compared with only 0.76 million ha in 1956/7. A feature of this expansion has been the rapid increase that has occurred in western areas (even to areas administered by the Western Lands Commissioner) that were once considered marginal. Substantial amounts of wheat are now grown in these areas, which demonstrates that the concept of wheat growing area is dynamic if not ephemeral (cf. McLennan,

1963). Expansion of wheat growing into climatically less favorable areas can be presumed to mean lower average yields (Russell, 1973). The question of whether wheat yields are lower in marginal areas than in traditional areas has been the subject of occasional attention (Lewis and Dawson, 1948; Rutherford, 1958; Anon., 1967). Average shire wheat yields for the period 1956/7 to 1975/6 ranged from 1.60 t/ha in Waugoola to 0.90 t/ha in Walgett. Between these extremes, however, several shires have recorded similar average yields. A test of whether the marginal average yields are different from traditional average yields was made. The result was null at the 1% level of significance. Productivity developments in the form of improved technology, better varieties and improved management practices have possibly resulted in more efficient use of available moisture, and this may have enabled farmers to achieve profitable yields in areas where wheat production was previously considered too risky.

Previous researchers (e.g., BAE, 1961; Nix, 1975) have found that Australian yields do not show a definite inverse relationship between level and variability. For the marginal and traditional areas, it was found that yield variances were also not significantly different. There has, of course, been considerable production variability. The vagaries of weather are recognized as influencing the area and yield of wheat, but it is generally true that farmers have very little control over the performance of the crop once it is sown. Correlations over time between area sown and yield (actually realized and also lagged exponentially-smoothed) for nearly all local government areas are less than 0.1. The impact of climate on yields is reflected in year-to-year variability. Therefore, any definition of the agroclimatic limit to wheat production must be somewhat arbitrary. Despite seemingly stable climatic averages, the area in eastern Australia where farmers are prepared to grow wheat has shifted through time in a generally westward direction.

3.4.3. Some effects of climatic variability on wheat industries in other countries

Every country is subjected to climatic extremes, and of particular relevance for the Australian wheat industry are the other large wheat producers and importers [e.g., McKay and Allsopp (1977) on Canada; Committee on Climate and Weather Fluctuations and Agricultural Production (1976) and Thompson (1975) on the USA]. Anderson (1979) reported some yield characteristics for major wheat producers. The major effect apparent among the correlations reported was related to "hemisphere" – the average correlation between countries' yields in the same hemisphere is 0.55 and in different hemispheres is 0.12.

In the USSR, for example, the wheat crop has frequently suffered drastically through unfavorable climatic events. The Soviet wheat crop is particularly susceptible to prolonged cold weather conditions, winter kill, spring drought and wind and hailstorms at the time of harvest (*see*, for example, Part VI, this volume). During the period from 1963 to 1976, several sharp falls in wheat production occurred in the USSR. This analysis is concentrated on two particular falls in production, in 1972 and and 1975, although these were by no means the most drastic. The decline in production in 1972 was due to a reduction in the area sown and a severe drought, especially in the Ukraine. During 1972/3, wheat production in the USSR fell to 86 mt, some 22% below that of 110 mt in 1973/4. The result of this decrease in production, together with a change in policy favoring the grain feeding of livestock, was that, in order to meet its commitments, the USSR had to import much more than anticipated. Imports of wheat and flour totaled 15 mt in 1972/3 compared with a yearly average of 1 mt from 1967/8 to 1971/2.

The effect of the change in Soviet import requirements in 1972/3 had a dramatic impact on world trade and prices. The 15 mt purchase represented over 20% of world wheat trade. However, to disaggregate the effects of the drought in the USSR from changes in Soviet agricultural policy and the policies of other trading nations at that time is a challenging task. The Soviet purchase came at a time when the USA had opted for a policy of reduced stockholding, world fertilizer production had slumped, and increasing oil prices were adding to an inflationary spiral in most countries. The net result of all of these factors was that not only did Australia benefit from the direct sales of around 1 mt to the USSR, but also the total Soviet purchases contributed to raising the world price of wheat. For example, the average f.o.b. export price of ASW wheat rose from $120/t in 1971/2 to $170/t in 1972/3.

Due to drought conditions in 1975, Soviet wheat production slumped to 66 mt, the lowest level for the decade. Soviet policy, however, dictated that only 10 mt of wheat would be imported at the higher prevailing world prices. Thus, although the 1975 drought was more severe than the one in 1972, it had a lesser impact on world wheat trade. The USSR, however, did purchase much larger quantities (15 mt) of coarse grains in 1975–76 compared with the quantities (6 mt) in 1972/3 (International Wheat Council, 1976).

In contemplating the impact of climate on Australian agriculture, there are thus dangers in taking a perspective that is too parochial. A severe drought may bankrupt some growers and cause many others extreme discomfort in both farm and household matters. Drought on the scale of that in 1965/6 may reduce industry returns by, say, $700 million. However, an analogous event in the USSR may well result in a windfall gain to the same Australian growers of, say, $400 m.

The repercussions for the Australian wheat industry of adverse seasonal conditions in North America are unlikely to be large. Yield of USA wheat is relatively stable, and the chances of a poor national harvest are slight. Moreover, in the USA, in particular, there is capacity to hold large quantities of wheat (in excess of 30 mt or over 50% of typical production), and the effects of poor seasonal conditions can thus be mitigated. Although the USA has considerable storage capacity, the desirability of stockholding has been vigorously debated. In general, importing countries benefit and exporting countries lose through a stockholding policy in the major exporting countries. In insuring against unfavorable seasonal conditions, conflict thus arises between gainers and losers in the distribution of benefits and costs associated with the buffering stocks. For example, using an assumption of rather inelastic demand for wheat, Reutlinger (1976) has estimated that, in managing a reserve stock with a total of 20 mt storage capacity, the USA, Canada and Australia together could lose about \$200 m a year, while the developing countries, and Japan and Europe, could gain about \$150 m and \$100 m a year on average, respectively.

3.5. Regional Impacts

The importance of agriculture in many regions of Australia is such that the impacts of climatic variations on local farms and agricultural industries reverberate strongly in other parts of regional economies. A ready example of such a transmitted effect is the variable activity in farm machinery retailing as seasonal fortunes vary (Davis *et al.*, 1979).

A simplified procedure for measuring such impacts is regional input–output analysis in which the effects are conveniently summarized by various multipliers. Only a few Australian investigations are available at this level, but they are probably good guides to the magnitudes of multipliers that will be found in more widespread subsequent studies.

Agricultural-to-regional output multipliers are of the order of 2 (Mandeville and Powell, 1976; Jensen, 1976), so that a drought-induced reduction of, say, \$10 m in regional farm output will lead to a reduction of about \$20 m in total regional output. These output multipliers are not as high as some others (e.g., government services = about 3) because of linkages beyond the given region, especially linkages involving imports from (and thus transfer of some instability to) outside the region (Powell and Mandeville, 1978). However, they are of particular significance because of the relative instability (through climate and export markets) of the agricultural sector. A more recent input–output study of the major drought of the early 1980s in the Darling Downs (largely cropping) area of Queensland featured a more refined output multiplier effect of 1.7, based on a reduction of rural sector output of \$140 m, and a total regional effect of \$240 m (Powell and Saeed, 1984).

Regional income multipliers are typically even higher (about 3) than output multipliers, so it follows that climatic variations can induce substantial impacts on regional incomes and especially on employment (Phillips, 1974). Consider the regional illustration of Powell (1978), wherein agriculture contributes on average 30% of regional output. A minor drought might cause agricultural output to fall by 10% and thus direct regional output to fall by 3% with an associated direct income decline of, say, 4%. With an income multiplier of 3, the full effect on regional income would thus exceed 10%.

Doubtless it is in recognition of the extent of such impacts that governments have sought to intervene in droughts with various regional employment schemes (usually through local government authorities). As Powell (1985) elaborates, however, the flow-on effects are not spread very evenly among the sectors of the economy. Furthermore, some of those sectors have as much difficulty in coping with the situation as do farmers. Thus, if government policy is to provide drought assistance, a proposition debated in the following chapter, there is *prima facie* evidence that assistance to some nonfarm sectors could also be justified in severe droughts. Appropriately directed assistance to particular sectors would appear to involve household support oriented to boosting the economy generally.

3.6. Sectoral and National Impacts

So far the effects of extreme climatic regimes have been considered as being unambiguously severe in affected regions. However, as Gibbs and Maher (1967) and Maher (1973) have indicated in their reviews of spatial patterns of annual rainfalls, it is unlikely that large proportions of Australia simultaneously experience similar extremes of climate, especially rainfall. There is considerable "cancelling out" of good and bad experiences, so that impacts at high levels of geographical aggregation are not too severe. Such effects have been modeled within New South Wales by Francisco (1980), who related regional production to spatially correlated monthly rainfall aggregates. Naturally, when the more intensively farmed and grazed regions (e.g., southeast Australia) experience extremes, these effects are reflected more strongly in national aggregates than are extremes in the more extensively managed regions.

3.6.1. Aggregate impacts of climatic variability

Most of the work addressed to measuring the aggregate economic impact of climatic variation has been concerned with major droughts. For instance, there have been several reviews of the droughts of the 1960s. McIntyre (1973) estimated that gross rural output in the drought years 1965/6 and 1967/8 fell by 11% and 15%, respectively, below the smoothed trend values. Estimation of the corresponding changes in gross value of agricultural production implies or requires assumptions about the responsiveness of prices to changed quantities produced in Australia. Bates (1976) estimated corresponding falls in gross value of 15% and 17.5%, which were of the order of $1000 m. Duloy and Woodland (1967) modeled such a drought and assumed that a fall in output of 20% would be associated with a 10% reduction of gross value (i.e., an aggregate price elasticity of demand of −2).

The primary purpose of that study by Duloy and Woodland (1967) was to estimate the multiplier effects of drought on the national economy (e.g., on gross national product, GNP). Using a simple econometric model of the economy, they found, in contrast to the results discussed at the regional level, that there was only a small multiplier (1.07) from gross value of agricultural production to GNP. As Bates (1976) has argued, this small effect is largely a consequence of the now small (about 5%) contribution of agriculture to GNP, although strengthening linkages of agriculture with the nonfarm sector may have subsequently increased the drought multiplier. However, the effect of a major drought on GNP is still unlikely to be very much different from the 1% to 2% (*see*, e.g., 1982/3 GDP in *Table 1.1*) estimated by Duloy and Woodland (1967) and Mules (1973). These authors also reviewed the consequences of drought for the balance of payments, an issue of more active interest in past decades (Campbell, 1958). The consensus is that there is now little cause for concern in this regard.

To broaden attention to climatic variations other than only severe droughts, the recent history of the main rural aggregates was reviewed by Anderson (1979). For the period 1949/50 to 1977/8, annual movements of farm income, value of agricultural production and farm costs, and indexes of prices received and prices paid were charted and piecewise linear trends fitted. The times at which changes are imposed in the trends were arbitrarily determined merely by inspection of the patterns. All series exhibited considerable variation; and it is not obvious just what role climatic variation has played in each, although it seems reasonable to argue that most of the variation about trend in the output series is ascribable to climatic variations. Certainly the major dips in output series could be associated with recognized droughts.

It was assumed, for simplicity, that climate affects directly only aggregate output with no direct impact on input use, prices paid and costs. There is an indirect effect on prices received through the elasticity of the aggregate demand function for Australian agricultural output. It was assumed, following Powell (1960), that the aggregate price elasticity of demand, e, is -1.5 and is constant. This means that, if output in a given period is a random variable, Q, with trend value Q^* and coefficient of variation $C(Q^*)$, then gross farm product at prices that are influenced only by variation in Q can be expressed as $G = AQ^{1+(1/e)}$, where A is a constant that depends on e. For $e = -1.5$, $G = AQ^{1/3}$ and G can be interpreted as a derived measure of value of agricultural production that abstracts from all other sources of variation in prices except for the aggregate impact of climate.

With these assumptions and the further one that Q is approximately normally distributed, a sampling approach was adopted to measure the variation in the derived variable, G. Using the computed residuals, $C(Q^*)$ was estimated as 0.034 and the corresponding coefficient of variation for G was found to be 0.011. This is an assessment of the "pure climate" component of variation in the gross value of agricultural production. It can be contrasted directly with the corresponding coefficient of variation in actual gross value of agricultural production of 0.029, denominated in the 1978/9 trend value. Thus, in a rough way, it can be claimed that climate is responsible for $(0.011/0.029)100 = 38\%$ of the variation in gross value of agricultural production.

Attempting the same rough assessment for farm income is more difficult because it is necessary to account for the effect of changing farm costs on the mean and coefficient of variation of farm income. Using a scaling factor, based on recent trends in costs and value of production, to inflate the derived coefficient of variation, the comparable ratio of climatic to total variation in farm income is $(0.035/0.090)100 = 39\%$.

3.6.2. Aggregate impacts of long-run change in climate

The impacts discussed have been of rather short-term nature. Typically, the effects of extreme climatic events have become insignificant within a few years of occurrence. In the absence of geographically detailed scenarios of potential changes in climate, it has not been possible here to chart or even to contemplate the possible detailed impact of long-run changes on Australian agriculture. Some possibilities are explored in Pittock *et al.* (1978), including an agricultural case study by Arnold and Galbraith.

Some progress in considering such speculative changes can be made by working at a high level of aggregation and assuming a nonspecific shift in aggregate output. The approach taken is to use a small general equilibrium model of the Australian economy (Stoeckel, 1978). In this way it is possible also to identify the likely balance of payments effects on the exchange rate and thus the possible gains to other sectors (e.g., the import-competing sector and the mining sector).

The particular shift considered here is a 13% permanent reduction in gross rural output using the 1973/4 year as a base. This is similar in magnitude to that of a serious (1960s-style) drought. The fall in output is modeled by assuming that the adverse climatic change increases average costs of production, which is equivalent to negative technical change. However, capital losses of livestock and pastures that would be associated with such a drought are ignored, as are aspects of probability distributions other than means.

A selection of results is presented in Anderson (1979). The precision of the consequent changes is indeterminate because of the nonstatistical estimation and subsequent nonlinear transformation of the model. The analysis could be interpreted only as indicating orders of magnitude and likely directions of the major effects. Although national income was indicated as falling by nearly 1%, values added by the mining sector and the import-competing sector increased by 1.2% and 0.6%, respectively. The reason why these sectors gain is that, with the imposed zero external balance, reduced agricultural exports lead to a devaluation of the Australian dollar. The manufacturing export sector, however, contracts since it comprises mainly food processing and relies on agricultural produce as an input. In the model, percentage changes in value added were found by adding the changes in gross output and real price.

Two further qualifications should be made. In imposing the slump in agricultural productivity, supposedly because of climatic change, it was assumed that there is no related primary effect in any other sector. Second, the presumption about this unspecified climatic change is that it affects only Australia. Should such a change occur, it is likely to be associated with other changes throughout the world, in which case prices for rural exports would increase and tend to counter the output changes in determining Australian incomes and agricultural earnings.

3.7. Conclusions

Some of the numerous impacts of variable climate on Australian agriculture have been reviewed by addressing particular impacts at levels of aggregation ranging from the farm, through the industry and the region to the whole sector. Rainfall-related effects are generally the most important in terms of economic impact. The identified effects were always significant, but their sensitivity and importance tend to wane as the level of aggregation increases. The scenario interpretation of some of these impacts is explored briefly toward the end of Section 4.

Apart from the intrinsic interest of understanding the nature of agricultural production, the importance of perceiving the effects of climatic variability in agriculture is the potential insight into possible market failures and consequent roles for governments. The diversity of governmental interventions that prevails indicates that considerable progress has been made in official recognition of some of the impacts of climatic variability (cf. Campbell, 1958; IAC, 1978). Such assistance, however, might easily distort producers' perceptions of their environment and raise problems of the "moral hazard" type (Freebairn, 1978). For instance, there is a strong tradition of governmental intervention in the rural credit market (BAE, 1977), and some of the interventions have been "justified" on the basis of mitigating untoward climatic experiences. However, the effectiveness of such intervention is very doubtful (Standen, 1978). These and other policy issues are taken up in the following section.

SECTION 4

Implications for Policy Makers and Planners

4.1. Introduction

Although the ability to forecast the arrival and duration of particular extreme climatic events is limited, they will naturally and inevitably recur as inherent features of Australian agriculture. Most farmers and other rural business people are aware of fluctuating seasonal (and market) conditions and make their own contingency plans. Such plans result in marked changes in production and incomes in the rural sector, and some transfer effects to the rest of the economy – *see* Lovett (1973) and Anderson (1979) for an overview of these effects; and Campbell *et al.* (1983), Purtill *et al.* (1983) and Gibbs (1984) for discussion of the most recent drought (1982–83). Most farmers balance the good and bad times, and maintain both their ability to continue production and their household living standards until better seasonal conditions return. However, some do not make adequate provision for the stock, income and cash flow losses caused, for example, by drought, flood and fire. Selected reports of the ravages and hardships of drought are irresistible headline material in the media. Sympathy flows from the public, and governments agree, all too readily, to provide subsidies from public funds for drought assistance.

4.2. Climatic Variability and Governmental Intervention

Policy makers have concentrated interventions by governments on the impacts experienced by farmers. Measures such as income equalization deposits certainly cannot stabilize the effects of variable climate completely. However, they operate at the discretion of individuals, who hold their individual attitudes to risk and perceptions of climate, without distortion of the realities of their turbulent environments. These include climatically sensitive agricultural environments and, increasingly, nonagricultural sectors of the economy. The number of rural producers and their families who derive revenue from sources other than the home farm is increasing (Reithmuller and Spillman, 1978). This compensating tendency is only partial because of the multiplier effects of drought on sectors other than agriculture. A further measure that could be important in buffering the impact of future extreme climatic events on producers is the Rural Adjustment Scheme, with its provisions for carry-on finance and household support (Threlfall, 1977). As noted above (Subsection 3.5), however, self-employed business people who are also severely affected by climatic impacts should be considered on a basis similar to that applicable to farmers.

Several features of the dynamic regulations pertaining to income taxation have been promulgated in recognition of the particular difficulties posed for producers by variability in their environment, including those of climatic origin. Such measures as income equalization deposits (and the earlier drought bonds), and the special provisions concerning forced sale of livestock and carry-forward of losses are of undeniable assistance to many producers (particularly those with substantial tax burdens) in mitigating the financial impacts of climatic variability. Such interventions have been reviewed by the IAC (1978).

Natural disaster relief is generally the responsibility of the states although the Commonwealth Government stands ready to join state governments (currently on a \$3 for \$1 basis beyond stated base levels of expenditure) to assist in financing a range of measures, such as drought, bushfire and other disaster relief. Under complementary Commonwealth and state legislation, the Natural Disaster Relief Arrangements (NDRA) support a broad framework of core measures of assistance and for sharing the subsidy costs. Drought relief measures are set internally on an individual state basis and differ between states. They are too varied to catalogue here; but include concessions on road and rail freight on livestock to and from agistment (rented pasture), restocking movements, forced sales of livestock, fodder, water, and grains. Loans to primary producers are also usually available at concessional rates and conditions for carry-on finance and restocking. Other measures are stock slaughter compensation, provision of centralized water points and various forms of unemployment relief. Such measures are clearly of importance to affected individuals, but may increasingly be distorting, ineffective and costly (Freebairn, 1983).

Availability of drought concessions is usually triggered by official drought declarations made by appropriate agriculture ministers, based on regional advice. The timing of these declarations is of importance to governments and primary producers in different ways. Early declaration can have implications for state treasuries by rapidly depleting financial reserves. Late declaration, while likely (at least in short-drought situations) to deplete state treasuries to a lesser extent, can lose political support in adversely affected regions.

The interactions between livestock condition, pasture status, prices in the market place and the timing of onset and retreat of drought are complex enough factors for property managers to handle. The superimposition of a game of guessing when freight concessions might be introduced does not make the managerial task any easier. Sometimes producers who take action early to avoid stressing livestock or pastures can be disadvantaged relative to those who wait for concessions to be introduced. The recent shift to "objective" declaration, based on Bureau of Meteorology rainfall statistics (interpolated decile ranges) as triggers, is imperfect but commendable.

The state-determined measures for primary producers' relief from floods and bushfires exhibit wide variation. The range covers relief from damage by inland flooding, tidal flooding, bushfire, hail, storm, frost, cyclone, humid conditions, earthquake, and even, for potato growers in one state, heavy rain. In some states, the concessions available are well defined, while in others they are introduced according to need on an *ad hoc* basis. In the eligibility criteria, terms such as "necessitous or distressed circumstances" feature prominently.

The Standing Committee on Agriculture has had as one of its supporting advisory bodies the Commonwealth and States Committee on Drought and Other Natural Disasters. There is also the Natural Disasters Organization with interests much wider than agriculture. The current status of Commonwealth and state policies with respect to the impact of climate on rural producers and industries reflects the difficulties which the Australian Commonwealth–state system has in providing an equitable and predictable system to apply when persons or groups of persons are disadvantaged by the adverse effects of climate. Despite collaborative attempts over several years, Commonwealth and state officials have seemingly been able to engineer only marginal improvements in the process and the consistency with which disaster relief is administered.

Agricultural and pastoral activities in Australia span a diversity of climates and production systems. Assessment of climatic impacts for purposes of intervention thus has necessarily to be made at rather disaggregated geographical levels. In this respect, the state system meets the purpose relatively well because climatic adversity can be locally assessed and drawn to the attention of state and, if necessary, Commonwealth bodies.

The financial situations of producers, and the likely requirements for drought carry-on and restocking credit, are probably rather different now from those that prevailed in the 1960s. Although producers would in some ways now be more vulnerable to financial setbacks in the event of adverse climate affecting their farm businesses, there is a compensating aspect in the increasing importance to farmers of nonfarm income.

Whether governments have tended to intervene excessively is a moot point that is taken up affirmatively below. Some affected individuals have neither the desire nor the means to plan for and to cope with all impacts of climate on them. They will be a continuing target of relief measures, but such intervention should be recognized as being part of a national welfare policy. It is politically unrealistic to expect governments not to intervene in natural disasters. However, apart from "welfare" cases, it can be hypothesized that interventions in agriculture occasioned by climatic variability have been developed and institutionalized beyond levels that are desirable on grounds of economic efficiency.

These rather conservative hypotheses about Australian interventions seem to match those reached in other countries. In the USA, for example, a major study concentrating on the Great Plains documented the increasing involvement of the US government in attempts to alleviate impacts of climate variability. Wilhite *et al.* (1982) demonstrated that governments in the USA respond to drought through crisis management, which often results in the implementation of hastily prepared assessment procedures and ineffective, poorly coordinated and untimely responses. Four requirements for more effective response were identified:

(1) Reliable and timely informational products and dissemination plans.
(2) Improved impact assessment techniques, particularly in the agricultural sector.
(3) A common designation procedure administered by a single interagency committee.
(4) Standby assistance programs that address the specific problems associated with drought.

They argued that all assistance programs should be administered by a single agency or interagency committee and recommended that drought response plans that address each of these critical issues be developed by state and federal governments. The advice seems sound for Australia, as well, especially from the perspective of the following policy-oriented commentary, which analyzes the *ad hoc* policies implemented during the 1982–83 Australian drought.

The broad objectives of drought assistance include welfare support and the maintenance of resources, and government subsidies for drought relief represent substantial transfers of resources to the rural sector. From 1973/4 through 1981/2, the Commonwealth Government provided \$152 m in drought assistance for natural disaster relief (IAC, 1983b), and between \$50 m and \$100 m was provided from state treasuries (Freebairn, 1983). (Here, as in previous sections in this case study, the convention is adopted of writing the fiscal year to end June as 1982/3 and the years 1982 and 1983 as 1982–83.) Additional funds were provided in taxation incentives, unemployment benefits and public works programs. Payments for drought relief in 1982/3 were estimated to have exceeded \$165 m (Standing Committee on Agriculture Working Group, 1983; Campbell *et al.*, 1983; Freebairn, 1983). The 1982/3 subsidy was one of the largest subsidies to the rural sector (other measures are recorded in IAC, 1983a). It represented over 8% of the estimated net value of rural production of about \$2000 m (BAE, 1983c).

Drought assistance provided for the 1982-83 drought illustrates the workings of the NDRA (*see* IAC, 1983b; Standing Committee on Agriculture Working Group, 1983). The level of availability of carry-on loans and the rates of freight subsidies were modified in August 1982 and, throughout the drought period, lack of details on the conditions of eligibility confused many farmers and surely also some administrators. In September 1982, new assistance measures in the form of purchased fodder and interest rate subsidies were announced and, even in June 1983, there was indecision as to the duration of these assistance measures. Other forms of assistance included subsidies for the transport of water, for agistment of livestock, for water bores and, in some states, for the disposal of helpless and unsaleable stock.

In many respects, Freebairn's (1983) critical assessment that is followed here is similar to those by the IAC (1983b) and the Standing Committee on Agriculture Working Group (1983) and supplements the commentaries on policy toward previous droughts by others, including Campbell (1973), McIntyre (1973), Musgrave and Lesueur (1973), Bates (1976), and Anderson (1979).

4.3. Drought Policy, 1982–83

The largest government expenditures on drought aid provided in the 1982–83 drought were the subsidies on purchased fodder and on debt interest. The effects of the subsidies on the allocation of resources and as a means of providing welfare support are now analyzed. As background, the effects of the subsidies on prices, quantities and income are assessed.

4.3.1. The fodder subsidy

Two measures reduced the cost of purchased fodder to feed livestock in drought-declared areas during 1982–83. First, a 50% subsidy was payable on fodder purchased for the maintenance of sheep and cattle to a limit of 80¢ per sheep per month and $8 per cattle beast per month. The subsidy did not apply to fodder produced and used on the farm, or to fodder purchased prior to the drought. Second, a 75% subsidy was paid on the cost of transporting purchased fodder. The principal objectives of the subsidies were to induce farmers to feed more animals rather than have them slaughtered or allow them to die, and thus to speed the recovery of the livestock industries after the drought; the schemes were seen as also providing some income support and meeting humanitarian goals.

The first-round effect of the fodder subsidy was to shift demand of the recipient farmers for fodder; the demand of nonassisted farmers and the supply relationship remained unchanged. Fodder is demanded for normal production as well as for drought maintenance. Results from studies of the intensive livestock industries indicate that feed prices, typically measured by grain prices, have a significant effect on quantity produced and, by implication, fodder demand. Farm management studies of drought feeding strategies by Officer and Dillon (1965) and Anderson and Hardaker (1973) show that the demand for fodder for drought feeding is sensitive to fodder prices and, of course, perceptions about the length and severity of drought (Byerlee and Anderson, 1982). Bain (1973) and BAE (1976a), using econometric models of the aggregate demand for fodder, obtained results that are consistent with the hypothesis that the demand for fodder as an aggregate, and the demands for individual fodders, are price-responsive, and that the different fodders are imperfect substitutes for each other.

The supply function for wheat for fodder can be regarded as perfectly elastic about the export price. In practice, the Australian Wheat Board attempts to meet all feed wheat demands by diverting grain from stocks and export sales. In this respect at least, the Australian situation may be unrepresentative of that faced by most other countries. Domestic feed-wheat sales amount to less than 10% of total wheat sales. Over one-half of coarse grains are exported. On the other hand, hay is not an internationally traded good with few alternative uses to livestock feed and whose supply elasticity is very low in the short run. The low correlation between prices of feed-wheat and other fodders (usually less than 0.4; *see* Freebairn, 1983) supports Bain's (1973) contention that wheat has a specialized market and serves as a relatively high-priced reserve source of feed in drought. While the prices of coarse grains and hay are highly correlated, they are far from perfectly correlated. Prices of hay and, to a lesser extent, coarse grains have been more variable than feed-wheat prices. Based on observed price movements, there is a reasonable degree of independence of the prices of the different fodders, and they can thus be regarded as imperfect substitutes.

With this background, Freebairn's (1983) analysis of the effect of the fodder subsidies is as follows. Wheat prices will remain unchanged, feed sales will increase at the expense of export sales and stocks, and wheat will become relatively more important in total domestic fodder sales. The increased demand for coarse grains will be absorbed mostly as a diversion of export sales and a relatively small price increase. With hay, the principal effect will be a higher price and very little increase in quantity. Based on a crude interpretation of available data, these changes did occur in 1982–83 (Freebairn, 1983).

The effect of the fodder subsidies on the numbers and prices of livestock is now addressed. Farmers were encouraged to retain more sheep and cattle on their farms and to increase their nutritional status. As noted above, however, fodder costs did not fall by the full extent of the subsidy and not all farmers availed themselves of it. If, as is argued by Fisher and Munro (1983), farmers' expectations are, in part, rational expectations, then the subsidies, by encouraging the retention of animals, reduced farmers' expected end-of-drought prices and, hence, the incentive to feed animals.

This and other second-round effects modified, but did not eliminate, the initial effect of the fodder subsidies to increase the number of sheep and cattle retained on farms and to increase the productivity of the national flock and herd.

At the aggregate industry level, there were changes in livestock and livestock product prices resulting from the fodder subsidies. Increased wool production resulted in a lower price than would otherwise have been the case. The price fall was proportionately less than the output increase because of the elastic demand. Meat supply fell in the early part of the drought as animals were diverted from slaughter but, in the latter part of the drought and post-drought periods, the sale of retained animals resulted in increased meat supplies compared with the counter-factual situation of no fodder subsidies. These changes in meat supply had proportionately smaller reverse effects on prices because domestic prices are closely tied to export prices, and the export demands for Australian beef, mutton and lamb are elastic.

Many sheep and cattle producers were disadvantaged by the effects that the purchased fodder subsidies had on livestock prices. Farmers who used their own fodder reserves, either home-produced or previously purchased, those who stocked conservatively, those who allowed their animals to die and those who sold some of their stock did not receive the subsidy. As many as one-half of the drought-affected farmers received no direct benefits from the fodder subsidy in 1982/83.

Taking the aggregate Australian picture and data from the Australian Agriculture and Grazing Industry Survey, Purtill *et al.* (1983) estimated that 62% of farmers were affected by drought and that 30% either had used or intended to use the purchased fodder subsidies. In the case of Victoria, where the drought affected most of the state, the Rural Finance Commission had spent about \$23 m on the fodder subsidy by May 1983. At 40¢ per sheep per month, the subsidy outlay amounted to support for 57.7 million sheep. This implies that the subsidy would cover all animals affected for about one month, yet the drought had been declared over most of the state for at least six months (Freebairn, 1983).

Nonparticipating farmers in drought-free areas were affected by the commodity price changes. The effects were adverse for wool producers, for those selling animals in the post-drought period, and for those selling meat products in the latter part of the drought and the post-drought period. Indirect benefits were given to those selling livestock in the early part of the drought and buying at the end of the drought. Because of high price elasticities in demand for meat, these indirect price effects are unlikely to have been very large.

Intensive livestock producers and dairy farmers were also adversely affected. First, the fodder subsidies resulted in higher costs of purchased fodder. Second, in the case of meats, the lower (higher) prices of the red meats adversely (favorably) affected the final-product demand for pigmeat and chicken. Almost certainly, the output-price effects had less influence on incomes than the fodder-price effects.

In summary, the fodder subsidies resulted in a mixed pattern of gainers and losers. No more than one-half of the drought-affected sheep and cattle producers received the subsidy. Some of the subsidy was gained by fodder producers and carriers, and some went to overseas consumers of livestock products. Adverse price changes resulted in losses to sheep and cattle producers in drought-free areas, farmers in drought areas who used alternative strategies such as feeding their own fodder or conservative stocking rates, and intensive livestock producers. While the subsidies stimulated some farmers to increase their herds, the aggregate industry impact was smaller than what might otherwise have been expected because of the modifying price effects.

4.3.2. The interest rate subsidy

Two forms of subsidy on the cost of borrowed credit were available in 1982–83 to farmers in drought-declared areas: concessional carry-on loans and a debt interest subsidy. Farmers assessed to have both viable prospects over the longer term and to be unable to obtain funds from normal commercial sources were eligible for special carry-on loans. There were differences in the schemes between states with regard to maximum levels of carry-on loans (*see* IAC, 1983b; Standing Committee on Agriculture Working Group, 1983) and in the criteria and procedures used to determine "long-term viability" and "unable to obtain funds from normal commercial sources". In general, few people seem to have been disturbed by the apparent contradiction of these two terms (Freebairn, 1983). The loans had a maximum interest rate of 4% per annum, the possibility of interest and principal repayment holidays, and a maximum duration of seven years. A new initiative provided a subsidy on debt interest above 12% per annum. At the time, the bank overdraft rate for loans under \$100 000 was 14.5%, with higher rates being charged by finance companies. The subsidy was generally available to all farmers. The objectives of these measures included assisting farmers to service their debt, helping to maintain the rural capital stock, and releasing funds from debt servicing and repayment for household consumption.

An assessment of the overall debt position, who received the interest subsidies, and the effects of these subsidies on asset ownership and prices requires some background data on farm balance sheets. At June 1982, the gross institutional indebtedness of the farming sector was \$4353 m (Reserve Bank of Australia, 1982). This represents about 9% of the estimated value of land, vehicles and machinery, and livestock assets on farms (ABS, various dates), and can be compared with "interest-bearing assets of farm businesses estimated to have totaled around \$3300 m at June 1981" (Balderstone *et al.*, 1982, p. 53). In practice, farmers drew on their financial reserves and increased their institutional debt as part of a survival strategy. Data analyzed for the 1982–83 drought by Purtill *et al.* (1983), and for the 1960s' droughts by McIntyre (1973) and Bates (1976), show that the increased debt burden was not as dramatic as the fall in farm receipts and income. Evidence submitted to the Standing Committee on Agriculture Working Group (1983), its own conclusions and Haynes *et al.* (1983) have contended that the commercial financial sector was able to accommodate the increased demand for credit by farmers during the 1982–83 drought. McIntyre (1973) reached similar conclusions for the 1960s' droughts. In short, the overall picture is one of generally high equity levels and financial reserves, a strong base to support borrowing, and success in obtaining institutional credit for carry-on purposes during a drought.

The aggregate picture naturally does not reveal the diversity of asset positions of individual farmers. In their detailed analysis of representative farm survey data for 1978/9, Adams and Minnis (1982) found that 29% of all farmers had a zero gross debt position and 52% had debts of less than \$10 000. Small-scale farmers had a more favorable debt position: of those farmers with assets of less than \$150 000, nearly 38% were debt-free compared with 33% for farmers with assets of \$150 000 to \$300 000 and 17% for farmers with assets exceeding \$300 000. In an analysis of sheep farmers over the period 1957/8 through 1975/6, Cornell and Kerridge (1980, p. 2) found that "net cash incomes of indebted farmers were greater than those of debt-free farms, although the difference was not large". The general pattern of these figures applied to the 1982/3 drought year, although absolute debt levels would have been higher. They indicate that the interest subsidies were of benefit to less than one-half of the farm population and that larger farmers with a more-than-proportionate share of the debt burden, and with at least comparable income flows, were the principal beneficiaries.

It has been argued by some people, including the Standing Committee on Agriculture Working Group (1983), that the provision of additional carry-on finance is justified by perceived failures in the rural credit market. Many regulations have interfered with the operation of the credit market (BAE, 1972 and 1977; Standen, 1978 and 1982; Campbell *et al.*, 1981). Special government institutions provide about one-half of the total rural institutional debt. To a large extent, the supply of this credit is controlled by regulation, much of it comes at concessional interest rates and nonprice rationing criteria are used to allocate the funds. Interest rate regulations affect trading bank lending, particularly via the maximal rate on small (less than $100 000) overdrafts, and bank funds are allocated on criteria in addition to interest rates such as equity, personal rapport with the manager and past deposit history (Valentine, 1973; Standen, 1978; Ockwell and Batterham, 1980). The banks clearly met a portion, and probably a large portion, of the additional farmer requests for carry-on finance in the 1982–83 drought. A similar situation seemed to exist in the 1960s (McIntyre, 1973; Standen, 1978; Standing Committee on Agriculture Working Group, 1983). However, because of nonprice rationing, it is likely that some farmers were not able to obtain all the credit that they had sought at the regulated below-market small overdraft interest rate. But farmers were free to seek funds from other elements of the credit market, including pastoral and finance companies and trade credit. Because of farmers' small share in the Australian credit market, it is arguable that the supply of credit to farmers from these sources was highly elastic at market rates of interest. In this context, most of the claims of inadequate credit for carry-on during the 1982–83 drought refer to an inability to obtain funds at concessional interest rates from traditional lenders, and not an inability to obtain funds from other sources at market rates (Freebairn, 1983).

The concessional carry-on finance was available only to a subset of 1982–83 drought-affected farmers, and the subsidy was discriminatory in several ways. A farmer assessed as being unable to obtain funds from normal commercial sources, and to have viable prospects over the long-term, obtained a carry-on loan at 4% interest. At the same time, a neighbor able to obtain funds from, for example, a trading bank overdraft, paid up to 14.5%. Farmers assessed not to have long-term viable prospects were denied the concessional loan. Inevitably, subjectivity was involved in making the assessments. The eligibility criteria were not directly tied to perceived cases of market failure. Further, the market failure argument does not support the need for the subsidy on the carry-on loans.

Freebairn's (1983) interpretation of the effects of the interest subsidies on market prices and quantities was as follows. Farmers receiving the subsidy had an improved cash flow position and lower credit cost, giving them a greater ability to hold on to their farm assets. In aggregate, the subsidies were reflected in an increased demand for farm resources generally. Part of the extra demand went to inputs with a high elasticity of supply (e.g., general supplies, fuel and labor). For these items, the price changes would have been small. Another part of the extra demand went to commodities such as land, livestock and hay with low elasticities of supply. For these commodities, part of the subsidy led to higher prices. The higher prices applied to all farmers, with sellers reaping windfall capital gains (and buyers, losses) as a result of the interest subsidies.

In summary, the interest subsidies were of direct benefit to only a subset of drought-affected farmers, and the subsidies resulted in price changes affecting both the assisted and the nonassisted farmers.

4.4. Drought Subsidies and Efficiency

The purchased fodder and interest subsidies paid to farmers in the 1982–83 drought were input subsidies available to some, but not all, farmers. They changed the allocation of resources between the rural sector and the rest of the economy, between different activities and farmers within the rural sector, and they influenced farmers' choices of future drought strategies.

At the aggregate rural industry level, the drought subsidies, like any subsidy, enhanced the attraction of keeping resources in, or adding new resources, to the industry. However, there are several arguments to counter the need for such assistance on account of a drought (Freebairn, 1978). While a drought causes a temporary fall in returns to agricultural resources, the fall, of itself, does not necessarily lead to a misallocation of society's resources. Given the prospects of an inevitable return to normal or better seasonal conditions, and the low opportunity value of the resources to other industries, most quasi-fixed resources stayed in the industry. There were transfers within, and not losses to, the industry. A large part of the fodder and interest rate subsidies paid in the 1982–83 drought found its way into higher fodder, land and livestock prices than would have been the case without the subsidies. While the drought may have adversely affected longer-term industry expectations in the minds of some decision makers, and caught some entrepreneurs who had gambled against, or even ignored, the possibility of drought, others with sufficient foresight accepted the opportunity for acquiring additional resources at bargain prices. Based on land data series reported by Powell (1974) and Bruyn (1981), previous droughts had a very small effect on land prices. In the 1982–83 drought, the majority of farmers received no direct drought assistance and they, in conjunction with the commercial capital markets, were able to maintain the rural capital stock and reasonable household consumption levels.

The argument that drought assistance is justified on grounds of resource efficiency, because variations in seasonal conditions are a source of uncertainty and farmers are risk-averse, is doubtful (Quiggin and Anderson, 1979), particularly in the case of the 1982/3 measures (Freebairn, 1983). As argued in more detail by the IAC (1978), the proposition ignores the reality that there is uncertainty from other natural hazards and from market forces. At best, the argument is one for assistance on a broad community basis via a pre-announced contingency arrangement. By contrast, the 1982/3 subsidies were partial input subsidies to a subset of drought-affected farmers, and several of the measures (fodder and debt interest subsidies) were proclaimed as one-off subsidies after the commencement of the drought.

Harris *et al.* (1974) and others have argued that drought assistance to the relatively efficient extensive cropping and grazing industries is worthy of support on second-best or "tariff compensation" grounds to balance the high levels of assistance granted to some other parts of the economy. On the criterion of political palatability, assistance for drought (and other perceived crises) has advantages over tariff reductions and continuing general commodity price support for the efficient rural industries. However, as Lloyd (1975), Warr (1978) and others show, it is difficult to be sure that compensating assistance will improve the allocation of society's resources. This is particularly so with the 1982/3 measures, which were characterized by being partial and *ad hoc* forms of assistance.

A number of resource allocation distortions were caused by the fodder and interest rate subsidies. As input subsidies, they influenced adversely the choice of production techniques. The fodder subsidy favored the feeding of animals relative to alternative strategies of a conservative stocking rate, animal sales and animal deaths. Interest rate subsidies favored production using long-lived capital assets relative to labor and materials. The fodder subsidy resulted in reduced export grain sales and more lower-social-value domestic feed sales. By placing an effective ceiling on the cost of borrowed credit, the debt interest subsidy reduced the incentive for borrowers to find the lowest-cost finance. It encouraged lenders to substitute higher interest charges for less expenditure on loan assessment and management advice. For the reasons discussed above, it is most unlikely that the social marginal value of the additional resources drawn to the agricultural sector as a result of the 1982/3 subsidies exceeded their opportunity returns elsewhere in the economy.

Both the fodder and the interest subsidies provided incentives for farmers to lessen their efforts to cope with future droughts from their own resources. They discouraged the conservation of fodder, use of conservative stocking rates and building up financial reserves, because these strategies received no subsidies. By contrast, the strategies of no fodder conservation, heavy stocking rates and free spending during droughts are relatively more attractive because they permit higher returns during good times, and they attract subsidies that lower costs during droughts. The beneficial effect of the subsidies would be greater with a consistent, pre-announced policy of assistance than with *ad hoc* policies of the 1982/3 type because of uncertainty about the form and magnitude of the subsidies. Even so, the reasonable assumption that some assistance will be forthcoming has the side effect of reducing the preparedness of individuals for inevitable future extreme climatic events such as droughts.

4.5. Drought Subsidies and Welfare Support

One motive for drought assistance has been the provision of an adequate income for people dependent on the rural sector for their livelihood. This represents an extension of the societal goal of providing all its members with at least a minimal level of income. Conceptually, drought should not create a welfare problem. Rational decisions to enter an industry, including one in agriculture, require that adequate returns be generated on average, over the long term, to meet satisfactory income levels, including an allowance for the risks involved. Further, contingency arrangements would be made to accumulate funds from good seasons for use during difficult drought periods (*see also* Section 1). In practice, most farmers cope successfully with their fluctuating incomes, and this was shown to be the case in 1982–83. Inevitably, however, some individuals have insufficient funds for their household consumption needs.

Existing social security arrangements provided an income safety net for many families whose incomes were adversely affected by drought.

Unemployment benefit schemes ensured a basic income for all employees. Unemployment benefits were available also to farmers. However, the need to maintain assets, such as livestock, in working condition and problems with interpretation of the regulations resulted in only an insignificantly small number of farmers drawing unemployment benefits in 1982–83. The rehabilitation and household support components of the Rural Adjustment Scheme were available to provide income support for farmers. In practice, these measures of welfare support were virtually unused by farmers in the 1982–83 drought (BAE, 1983c). As noted by Vincent (1976) and several contributors to the Commission of Inquiry into Poverty (1975), there are many conceptual and practical issues in providing welfare support to self-employed persons, including farmers. The present measures supposedly available to Australian farmers have not been used by many.

The 1982/3 drought assistance measures provided additional funds to the recipient farmers. In the case of the purchased fodder and debt interest subsidies, the amount of support was proportional to the amount of fodder bought and debt incurred, respectively. Less than one-half of livestock farmers received the fodder subsidy. While many of these farmers experienced a severe reduction of income, some were large-scale operators with sizeable financial reserves. Given the debt structure of Australian farmers, the debt interest subsidy was regressive because the large farmers with high incomes were more heavily indebted and received more subsidy than the small-scale farmers.

Some of the recipients of the concessional interest rate carry-on loans were probably worthy of income support on social welfare grounds. But the measure missed at least two groups potentially in need of income support: those farmers assessed not to have long-term viable prospects and those who were able to obtain some credit from normal sources but who still confronted a temporary shortage of cash for household consumption. At the same time, some farmers with above-poverty-line funds for household needs received subsidized finance. In general, neither the selection of farmers nor the level of concessional credit provided under the scheme were based on the criterion of funds required for a minimal level of household consumption.

In summary, the interest rate and fodder subsidies paid in the 1982–83 drought were only loosely, if at all, related to minimal household consumption needs, and they were blunt and inefficient instruments for dispensing welfare support.

4.6. A Rational Drought Policy

Most economists who have considered the matter (e.g., Anderson, 1979; Freebairn, 1983) agree that droughts, of themselves, do not justify specific subsidies for reasons concerned with efficiency of resource use and that welfare support is more directly and effectively achieved by direct payments than through product and input subsidies. Policy analysts must assume that farmers are aware of fluctuating seasonal conditions and their effects, that farmers take a long-term view when making investment decisions and that they have contingency arrangements for coping with shortfalls in income and cash in the event of significant climatic variations. In fact, most Australian farmers do cope successfully with droughts. Those that do not probably should not be farming. General ways in which governments can facilitate private decision making in a world with droughts have been advocated over the years.

A basic prerequisite for rational decision making is information about the probabilities of occurrence and duration of droughts, and about the relative costs and benefits of alternative management strategies (e.g., Campbell, 1973). Analyses of meteorological data by Gibbs and Maher (1967) and others in the Bureau of Meteorology, and the work of many people including Nix (1975), Thatcher and Lloyd (1975), Smith *et al.* (1978), Toft and O'Hanlon (1979), and others on water balance models, provide guidelines for advice on the likelihood of different environmental conditions. More research in these areas may be desirable, especially on how the probabilities of drought conditions are influenced by alternative management strategies. Examples of information about the techniques, costs and benefits of drought mitigation strategies, such as alternative stocking rates, fodder reserves, cash reserves and animal sales, are given by Dillon and Lloyd (1962), Officer and Dillon (1965), Campbell (1966), Anderson and Hardaker (1973), Toft and O'Hanlon (1979) and numerous reports prepared by officers of the State Departments of Agriculture and Conservation. There remains scope for further model development, generalization and extension; for tailoring advice to specific situations; and for updating advice to reflect changes in prices and technology. Because of the public-good nature of many of these types of information, some government intervention is justified.

Governments can facilitate private decision making by adopting a consistently applied (i.e., predictable) and widely known set of policies. They would specify the conditions under which a drought would be declared and the forms and levels of assistance (if any) that would be provided. *Ad hoc* reactions to each new drought by government adds to the uncertainty facing farmers, and probably results in less effective farmer decisions with typically less output on average than would be the case without policy-induced uncertainty (MacLaren, 1980, 1983).

An efficient capital market would ensure that farmers and others were able to balance the variability of income receipts with more stable requirements of funds for operating expenses, business investment and household consumption. With a freely competitive market, such as recommended by Campbell *et al.* (1981) and as has been progressively implemented recently in Australia, there should be no problems in obtaining carry-on loans at market rates of interest during droughts and in obtaining loans with flexible repayment schedules. Regulations that have affected capital markets over recent years provided only a necessary, and not a sufficient, condition for further government intervention during droughts. There may have been an argument for providing additional funds for carry-on loans, but only at market interest rates. Banks permit flexibility in the repayment of some loans, particularly overdraft loans, and these informal arrangements are preferred by many borrowers. However, it is arguable that, at least until recently, regulations have deterred the development of formal and explicit flexible repayment loans, such as those proposed by Baker (1974), the IAC (1978) and Kent and Lloyd (1983). The present less encumbered and more competitive financial markets should simplify future drought financial planning. Observers of credit market performance in the aftermath of the 1982–3 drought are optimistic about farmers' access to adequate credit (Haynes *et al.*, 1983).

There have been a number of proposals for insurance schemes as an additional strategy for use by farmers to combat the effects of drought (and other causes of income fluctuations). Swerling (1959), Campbell and Glau (1970) and others have proposed income insurance schemes. Such schemes involve considerable administrative costs and there are problems of moral hazard and adverse selection. Specific risks are insurable in some states [e.g., compulsory hail insurance for wheat and barley, and optional flood and hail insurance for rice, in Queensland (IAC, 1978)]. A rainfall insurance scheme, such as that suggested by the IAC (1978), avoids the problems of high administrative costs in the narrow context of drought, although it may still prove either undesirable or administratively too expensive to be attractive to most producers (Bardsley *et al.*, 1984; Lloyd and Mauldon, 1985). One reason for the failure of these schemes to be developed by the private sector is the expectation that governments will provide some assistance in the event of drought. Because of the wide coverage of some drought experiences, enormous payouts will occur in particular years. There may be a role for government assistance in establishing rainfall insurance schemes and for underwriting (but not subsidizing) them once they begin operation.

While the minimal income needs of employees are seemingly covered adequately by unemployment benefits, analogous satisfactory atime of arrangements are not presently available for self-employed persons. To date, the Rural Adjustment Scheme has played a minor role in welfare support (Musgrave, 1982). There is, however, continuing scope for further investigation of the proposals by the Commission of Inquiry into Poverty (1975) for annuity schemes, negative income tax schemes and other initiatives as efficient ways of providing predictable cash flows for immediate consumption by farm households.

4.7. Conclusion

Several criticisms were made of the purchased fodder and interest rate subsidies used by the government to assist the rural sector through the 1982–83 drought, and to the freight and agistment subsidies. These were partial and selective input subsidies. Little, if any, justification on efficiency, equity or welfare support grounds can be found for diverting funds to one group of farmers, probably less than one-half of those affected by drought, and not to others. The motivation, presumably, was a blend of bureaucratic expediency and political flag-waving. Worse still, it is likely that such subsidies increase the severity of future drought impacts by lowering the incentive for farmers to adopt their own support strategies. The *ad hoc* nature of the policies added unnecessarily to the uncertainty facing private decision makers. Finally, the subsidies were blunt instruments for providing welfare support, and some in genuine need received no assistance.

An alternative strategy for drought assistance would involve the government playing a facilitating role to private decision making and providing direct income grants for welfare support. Farmers are able to adopt many strategies to cope with extreme climatic episodes including saving and borrowing funds, conservative stocking rates, conserved fodder, livestock sales, delayed fertilizer application, etc. Most, in fact, do cope with fluctuating seasonal conditions without government assistance. The rest should be encouraged likewise to "roll with the punches" instead of seeking featherbed solutions.

Government can assist private decision making by providing information about seasonal conditions and decision options (Gibbs, 1984), as well as price and outlook situations. It can continue to remove distorting regulations in the capital market. If some of the current regulations of the capital market continue, there may be a case for government funding for carry-on loans at market rates of interest and with flexible repayment schedules. Similarly, governments could conceivably underwrite a rainfall insurance scheme at market rates of interest. Income needs of employees are covered adequately by the unemployment benefits scheme. Initiatives to enable the government to provide an income safety net for families dependent on self-employed persons (including farmers but also others, especially in rural communities) for their income should be considered as part of a benign infrastructure to deal with uncertainty and economic misfortunes, including those with their origins in climatic variations.

In terms of the scenarios that are addressed throughout this volume, the 1982–83 drought of eastern Australia can be considered variously as: a convenient recent "bad" year; up to mid-1982, as a 1-in-10 drought event; and up to April 1983, as a "back-to-back" drought event or a consecutive anomaly. Possible effects of a long-term change of climate were dealt with in Subsections 1.4.2. and 3.6.2. The lessons of relevance from these three scenarios appear to be, respectively:

(1) In spite of fairly regular opportunity and practice, Australian policy makers and planners, even in the 1980s, have yet to crystalize clear, defensible, announced policies for dealing with anomalous weather as it affects the rural sector.

(2) The implicit policy of "hastening slowly" in extreme events as frequent as 1-in-10 seems appropriate for an agriculture facing a climatic regime as variable as that enjoyed in Australia – that is, relying primarily on passive instruments already in place (e.g., income equalization deposits and an untrammeled credit market).

(3) The problem with back-to-back events is that their severity sparks media attention and government perceptions of demands for "doing something". More or less well intentioned policies get formulated hastily and rushed into implementation with predictably unfortunate consequences. Governments must resist the temptations that are seemingly too often perceived as imperative. The consolation is that, by definition, such events are experienced only rarely. In the interregnum, sound policies should be developed, debated, announced and confidently taken into account in future planning by all decision makers who stand to be affected by climatic variation.

Acknowledgments

Section 3 draws heavily on Anderson (1979), which was greatly assisted by several officers of the Australian Bureau of Agricultural Economics. Section 4 draws heavily on the work of J.W. Freebairn (especially Freebairn, 1983), and also Quiggin and Anderson (1979) and Anderson (1979), with whom we share a perception – presumably also with the Industries Assistance Commission (IAC), as evidenced by Lloyd and Mauldon (1985) – of the general desirability of keeping governments out of climatic interventions and citizens' bedrooms.

References

ABS (various dates) *Agricultural Sector Financial Statistics*, Catalogue No. 7507.0, various issues.

Adams, P. and Minnis, P. (1982). *The Campbell Inquiry and the Rural Sector*. BAE Occasional Paper No. 62. AGPS, Canberra.

Anderson, J.R. (1970a). Spatial diversification of sheep holdings: a case study in the pastoral zone. *Rev. Marketing and Agric. Econ.*, **38**(3), 137–141.

Anderson, J.R. (1970b). Rainfall correlations in the pastoral zone of eastern Australia. *Aust. Meteor. Mag.*, **18**(2), 94–101.

Anderson, J.R. (1971). Spatial diversification of high risk sheep farms. In J.B. Dent and J.R. Anderson (Eds.) *Systems Analysis in Agricultural Management*. Wiley, Sydney, pp. 239–266.

Anderson, J.R. (1972). Risk and farm size in the pastoral zone. *Aust. J. Agric. Econ.*, **16**(1), 1–10.

Anderson, J.R. (1973). Sparse data, climatic variability, and yield uncertainty in response analysis. *Amer. J. Agric. Econ.*, **55**(1), 77–82.

Anderson, J.R. (1974). Risk efficiency in the interpretation of agricultural production research. *Rev. Marketing and Agric. Econ.*, **42**(3), 131–184.

Anderson, J.R. (1979). Impacts of climatic variability in Australian agriculture: a review. *Rev. Marketing and Agric. Econ.*, **47**(3), 147–178.

Anderson, J.R. (1985). Societal impacts: economic. In D.D. Houghton (Ed.) *Handbook of Applied Meteorology*. Wiley, New York, 1164–1174.

Anderson, J.R. and Hardaker, J.B. (1973). Management decisions and drought. In J.V. Lovett (Ed.) *The Environmental, Economic and Social Significance of Drought*. Angus and Robertson, Sydney, pp. 220–244.

Anderson, J.R., Dillon, J.L. and Hardaker, J.B. (1977). *Agricultural Decision Analysis*. Iowa State University Press, Ames, IA.

Anonymous (1966). Drought. *Current Affairs Bulletin*, **38**(4), 51–64.

Anonymous (1967). *Report of the ANZAAS Symposium on Drought*. Bureau of Meteorology, Melbourne, November.

Anonymous (1968). *Physical and Financial Effects of Drought*. Reserve Bank of Australia, Sydney, February.

BAE (1961). *The Variability of Wheat Yield per Acre from 1949–50 to 1958–59*. Bureau of Agricultural Economics, Canberra.

BAE (1968). *The Northern Territory Beef Cattle Industry: An Economic Survey 1962–63 to 1964–65*. Bureau of Agricultural Economics, Canberra.

BAE (1969). *An Economic Survey of Drought Affected Pastoral Properties New South Wales and Queensland 1964–65 to 1965–66*. Wool Economic Research Report No. 15. Bureau of Agricultural Economics, Canberra.

BAE (1972). *Rural Credit in Australia*. AGPS, Bureau of Agricultural Economics, Canberra.

BAE (1976a). *Agricultural Supply Projections 1979–80*. Occasional Paper No. 37. AGPS, Bureau of Agricultural Economics, Canberra.

BAE (1976b). *Rural Income Fluctuations Submission to the IAC*. Occasional Paper No. 39. AGPS, Bureau of Agricultural Economics, Canberra.

BAE (1977). *A Review of Credit in the Australian Rural Sector*. AGPS, Bureau of Agricultural Economics, Canberra.

BAE (1983a). *Rural Industry in Australia*. Australian Government Publishing Service, Bureau of Agricultural Economics, Canberra, 126 pp.

BAE (1983b). ***Trends in the Australian Rural Sector***, No. 106. Australian Government Publishing Service, Bureau of Agricultural Economics, Canberra.

BAE (1983c). *Q. Rev. Rural Econ.*, **5**(1), and earlier issues. Bureau of Agricultural Economics.

BAE (1985). Statistics. *Q. Rev. Rural Econ.*, **7**(1), 194–213.

Bain, R.A. (1973). Consumption and prices for major feed grains in Australia. *Q. Rev. Agric. Econ.*, **26**(3), 186–197.

Baker, C.B. (1974). An economic alternative to concessional farm interest rates. *Aust. J. Agric. Econ.*, **18**(3), 171–192.

Balderstone, J.S. *et al.* (1982). Agricultural policy: issues and options for the 1980s. ***Working Group Report to the Minister for Primary Industry.*** AGPS, Canberra.

Bardsley, P., Abey, A. and Davenport, S. (1984). The economics of insuring crops against drought. *Aust. J. Agric. Econ.*, **28**(1), 1–14.

Bates, W.R. (1976). National economic effects of drought in Australia. In T.G. Chapman (Ed.) *Drought.* Australian UNESCO Seminar, AGPS, Canberra, pp. 217–242.

Bond, G.E. and Wonder, B.S. (1980). Risk attitudes amongst Australian farmers. *Aust. J. Agric. Econ.*, **24**(1), 16–35.

Bruyn, P.A. (1981). ***Rural Land Values in Victoria.*** Agricultural Economics Working Paper. Department of Agriculture of Victoria, Melbourne.

Bureau of Meteorology (1983). ***Commentary on the Current Drought.*** Department of Science and Technology, Melbourne, 17 pp.

Butler, J.R.G. and Doesell, D.P. (1979). ***Natural Disaster Relief in Australia.*** Australian National University Press, Canberra.

Byerlee, D.R. and Anderson, J.R. (1969). Value of predictors of uncontrolled factors in response functions. *Aust. J. Agric. Econ.*, **13**(2), 28–37.

Byerlee, D.R. and Anderson, J.R. (1982). Risk utility and the value of information in farmer decision making. *Rev. Marketing and Agric. Econ.*, **50**(3), 231–246.

Byrne, P.F. (1968). ***An Evaluation of Different Stocking Rates of Merino Sheep.*** Wool Economic Research Report No. 13. Bureau of Agricultural Economics, Canberra.

Callaghan, A.R. (1973). The impact of drought on Australian wheat production and marketing. In J.V. Lovett (Ed.) *The Environmental, Economic and Social Significance of Drought.* Angus and Robertson, Sydney, pp. 99–120.

Campbell, J.K. *et al.* (1981). ***Australia Financial System: Final Report of the Committee of Inquiry.*** AGPS, Canberra.

Campbell, K.O. (1958). The challenge of production instability in Australian agriculture. *Aust. J. Agric. Econ.*, **2**(1), 3–23.

Campbell, K.O. (1966). Problems of adaptation of pastoral businesses in the arid zone. *Aust. J. Agric. Econ.*, **10**(1), 14–26.

Campbell, K.O. (1973). The future role of agriculture in the Australian economy. In J.V. Lovett (Ed.) *The Environmental, Economic and Social Significance of Drought.* Angus and Robertson, Sydney, pp. 3–14.

Campbell, K.O. and Glau, T.E. (1970). ***An Income Stabilization Scheme for the Wool Industry.*** Report No. 4. Department of Agricultural Economics, University of Sydney.

Campbell, R., Crowley, P. and Demura, P. (1983). Impact of drought on national income and employment. *Q. Rev. Rural Econ.* **5**(3), 254–257.

Candler, W. (1958). On the economics of drought reserves. *Rev. Marketing and Agric. Econ.*, **26**(4), 215–226.

Ceplecha, V.J. (1971). The distribution of the main components of the water balance in Australia. *Aust. Geographer*, **11**, 455–462.

Chapman, T.G. (Ed.) (1976). *Drought.* Australian UNESCO Seminar, AGPS, Canberra.

Chisholm, A.H. (1965). Towards the determination of optimum stocking rates in the high rainfall zone. *Rev. Marketing and Agric. Econ.*, **33**(1), 5–31.

Chislett, G.D.A. (1973). Effects of drought on wool. In J.V. Lovett (Ed.) *The Environmental, Economic and Social Significance of Drought.* Angus and Robertson, Sydney, pp. 121–130.

Commission of Inquiry into Poverty (1975). *Poverty in Australia.* AGPS, Canberra.

Committee on Climate and Weather Fluctuations and Agricultural Production (1976). *Climate and Food: Climate Fluctuation and U.S. Agricultural Production.* National Academy of Sciences, Washington, DC.

Cornell, L.D. and Kerridge, K.W. (1980). *Borrowing and Investment on Australian Sheep Farms: A Review of Some Past Developments.* BAE Occasional Paper No. 54, AGPS, Canberra.

Cornish, E.A. (1950). The influence of rainfall in the yield of wheat in South Australia. *Aust. J. Sci. Res.*, **3**(B), 178–218.

Cornish, P.M. (1977). Changes in seasonal and annual rainfall in New South Wales. *Search*, **8**(2), 38–40.

Coughlan, M.J. (1979). Recent variations in annual-mean maximum temperatures over Australia. *Q. J. Roy. Meteor. Soc.*, **105**(445), 707–719.

Cumming, J.N. (1966). The effects of the 1965–66 drought on sheep numbers and the expected rate of recovery. *Q. Rev. Agric. Econ.*, **19**(4), 169–176.

Davis, D.C., Mullen, J.D. and Bryant, M.J. (1979). Farmer income and expenditure patterns in a wheat-sheep region. *Rev. Marketing and Agric. Econ.*, **47**(1), 27–50.

Dillon, J.L. and Lloyd, A.G. (1962). Inventory analysis of drought reserves for Queensland graziers: some empirical analytics. *Aust. J. Agric. Econ.*, **6**(2), 50–67.

Doran, D.G. and McGilchrist, C.A. (1983). On the detection of climate change. *Paper presented to Hydrology and Water Resources Symposium*, Hobart, November.

Duloy, J.H. and Woodland, A.D. (1967). Drought and the multiplier. *Aust. J. Agric. Econ.*, **11**(1), 82–86.

Dury, G.H. (1983). Step-functional incidence and impact of drought in pastoral Australia. *Aust. Geog. Studies*, **21**(1), 69–91.

Easter, C.D. (1975). Some agronomic aspects of factors underlying production in the New South Wales grazing industry. *Q. Rev. Agric. Econ.*, **28**(3), 177–199.

Edwards, C. (1978). Gambling, insuring, and the production function. *Agric. Econ. Res.*, **30**(2), 25–28.

Everest, S.L. and Moule, G.R. (1952). Studies in the environment of Queensland – 2. The climatic factor in drought. *Queensland J. Agric. Sci.*, **9**(3), 185–299.

Fisher, B.S. (1978). *An Analysis of Storage Policies for the Australian Wheat Industry.* Research Report No. 7. Department of Agricultural Economics, University of Sydney.

Fisher, B.S. and Munro, R.G. (1983). Supply response in the Australian extensive livestock and cropping industries: a study of intentions and expectations. *Aust. J. Agric. Econ.*, **27**(1), 1–11.

Fitzpatrick, E.A. and Nix, H.A. (1970). The climatic factor in Australian grassland and ecology. In R.M. Moore (Ed.) *Australian Grasslands.* Australian National University Press, Canberra, pp. 3–26.

Foley, J.C. (1957). Drought in Australia. *Bulletin*, **43**, Commonwealth Bureau of Meteorology, Melbourne.

Francisco, E.M. (1980). *Regional Rainfall, Crop Yields and Farm Incomes in New South Wales – A Simulation Approach.* A.P.M.A.A. Report No. 9. University of New England, Armidale.

Francisco, E. and Anderson, J.R. (1972). Chance and choice west of the Darling. *Aust. J. Agric. Econ.*, **16**(2), 82–93.

Freebairn, J.W. (1978). Pros and cons of temporary industry assistance. *Aust. J. Agric. Econ.*, **22**(3), 194–205.

Freebairn, J.W. (1983). Drought assistance policy. *Aust. J. Agric. Econ.*, **27**(3), 185–199.

Gentilli, J. (Ed.) (1971). Climates of Australia and New Zealand. *World Survey of Climatology*, Vol. 13. Elsevier, Amsterdam.

Gibbs, W.J. (1968). *Benefits of Meteorological Services to Australia.* World Weather Watch Planning Report No. 27. World Meteorological Organization, Geneva.

Gibbs, W.J. (1984). The great Australian drought: 1982–83. *Disasters*, **8**(2), 89–104.

Gibbs, W.J. and Maher, J.V. (1967). Rainfall deciles as drought indicators. *Bulletin No. 48*, Bureau of Meteorology, Melbourne.

Guise, J.W.B. (1969). Factors associated with variation in the aggregate average yield of New Zealand wheat (1918–1967). *Amer. J. Agric. Econ.*, **51**(4), 866–881.

Harris, S.F. *et al.* (1974). *The Principles of Rural Policy in Australia: A Discussion Paper.* AGPS. Canberra.

Haug, N.G. and Hoy, R.J. (1970). The economic and financial position of graziers in arid south-west Queensland: 1962–63 to 1967–68. *Q. Rev. Agric. Econ.*, **23**(4), 222–243.

Haynes, J., Minnis, P. and Urban, P. (1983). Rural indebtedness and the need for post-drought credit. *Q. Rev. Rural Econ.*, **5**(3), 258–263.

Hazell, P.B.R. and Anderson, J.R. (1985). *Changes in Stability of Australian Cereal Production.* FPRI Working Paper. Washington, DC.

Heathcote, R.L. (1967). The effects of past droughts on the national economy. In Anon., *Report of the ANZAAS Symposium on Drought.* Bureau of Meteorology, Melbourne, November.

Heathcote, R.L. (1983). *Discussion paper for study conference on the impact of climatic change on natural and managed terrestrial ecosystems.* Villach, Austria, September.

Heathcote, R.L. and Thom, B.G. (Eds.) (1979). *Natural Hazards in Australia.* Australian Academy of Science, Canberra.

Hill, M.K. (1973). Farm management for drought mitigation. In J.V. Lovett (Ed.) *The Environmental, Economic and Social Significance of Drought.* Angus and Robertson, Sydney, pp. 195–219.

IAC (1978). *Rural Income Fluctuations.* AGPS, Industries Assistance Commission, Canberra.

IAC (1979). *Rural Income Fluctuations.* Report No. 161, AGPS. Industries Assistance Commission, Canberra, February.

IAC (1983a). *Assistance to Australian Agriculture.* AGPS, Industries Assistance Commission, Canberra.

IAC (1983b). *Rural Adjustment (Interim Report).* Industries Assistance Commission, Canberra.

International Wheat Council (1976). *Review of the World Wheat Situation, 1975–76.* London.

Isdale, P. (1984) Fluorescent bands in massive corals record centuries of coastal rainfall. *Nature*, **310**(5978), 578–9.

Jeans, J.N. (1972). *An Historical Geography of New South Wales to 1901.* Reed Education, Sydney.

Jensen, R.C. (1976). An interindustry study of the Central Queensland economy. *Econ. Record*, **52**(139), 315–28.

Jodha, N.S. (1978). Effectiveness of farmers' adjustments to risk. *Econ. and Pol. Weekly*, **8**(25), A38–A48.

Just, R.E. (1974). An investigation of the importance of risk in farmers' decisions. *Amer. J. Agric. Econ.*, **56**(1), 14–25.

Kefford, J.F. (1979). Impact of climate variability on food processing. In Anon., *The Impact of Climate on Australian Society and Economy.* CSIRO, Melbourne, Appendix A3(6).

Kent, M.A. and Lloyd, A.G. (1983). Coping with income instability: some estimates of probability of default under various loan repayment arrangements. *Paper presented to the Annual Conference to the Australian Agricultural Economics Society*, Brisbane.

King, J.E., Hurst, E., Slater, A.J., Smith, P.A. and Tamkin, B. (1974). Agriculture and sunspots. *Nature*, **252**, 2–3.

Lewis, J.N. and Dawson, B. (1948). Reliability of wheat yields: variations in New South Wales by shires. *Q. Rev. Agric. Econ.*, **1**(2), 12–13.

Lloyd, A.G. and Mauldon, R.G. (1985). Agricultural instability and alternative government policies: the Australian experience. In P. Hazell, C. Pomareda and A. Valdes (Eds.) *Agricultural Risks and Insurance: Issues and Policies.* Johns Hopkins University Press, Baltimore, MD.

Lloyd, P.J. (1975). Tariff compensation: an undesirable policy. *Aust. J. Agric. Econ.*, **19**(3), 146–153.

Lovett, J.V. (Ed.) (1973). *The Environmental, Economic and Social Significance of Drought.* Angus and Robertson, Sydney.

McArthur, I.D. and Dillon, J.L. (1971). Risk, utility and stocking rate. *Aust. J. Agric. Econ.*, **15(1)**, 20–35.

McBride, J.L. and Nicholls, N. (1983). Seasonal relationships between Australian rainfall and the Southern Oscillation. *Mon. Wea. Rev.*, **109**, 1998–2004.

MacIndoe, S.L. (1975). History of production of wheat in Australia. In A.A. Lazenby and E.A. Matheson (Eds.) *Australian Field Crops: Wheat and Other Temperate Cereals* Vol. 1. Angus and Robertson, Sydney, Vol.1, pp. 99–121.

McIntyre, A.J. (1973). Effects of drought in the economy. In J.V. Lovett (Ed.) *The Environmental, Economic and Social Significance of Drought.* Angus and Robertson, Sydney, pp. 181–192.

McKay, G.A. and Allsopp, T. (1977). Climate and climate variability. In Canada Committee on Agrometeorology, *Climatic Variability in Relation to Agricultural Productivity and Practices.* Canada Department of Agriculture, Research Branch, Ottawa.

MacLaren, D. (1980). Agricultural policy uncertainty and the risk averse firm. *European Rev. Agric. Econ.*, **7**(4), 395–411.

MacLaren, D. (1983). The output response of the risk averse firm: some comparative statics for agricultural policy. *J. Agric. Econ.*, **34**(1), 45–56.

McLennan, L.W. (1963). Recent wheat acreage changes in Australian states and likely future movements. *Q. Rev. Agric. Econ.*, **16**(3), 136–144.

McQuigg, J.D. (1975). *Economic Impacts of Weather Variability*. Atmospheric Science Department, University of Missouri, Columbia, MO.

Maher, J.V. (1967). Drought assessment by statistical analysis of rainfall. *Report, ANZAAS Symposium on Drought*. Commonwealth Bureau of Meteorology, Melbourne.

Maher, J.V. (1973). Meteorological aspects of drought. In J.V. Lovett (Ed.) *The Environmental, Economic and Social Significance of Drought*. Angus and Robertson, Sydney, pp. 41–54.

Manabe, S. and Stouffer, R.J. (1980). Sensitivity of a global climate model to an increase of CO_2 concentration in the atmosphere. *J. Atmos. Sci.*, **37**, 99–118.

Manabe, S. and Wetherald, R.T. (1980). On the distribution of climate change resulting from an increase in CO_2 content of the atmosphere. *J. Atmos. Sci.*, **37**, 99–118.

Mandeville, T.D. and Powell, R.A. (1976). *Structural Interrelationships in a New South Wales Wheat Region*. Agric. Econ. Bull. No. 22. Department of Agricultural Economics and Business Management, University of New England, Armidale.

Motha, G., Sheales, T.C. and Saad, M.M. (1975). Fluctuations in Australian rural production and prices – some implications for support policies. *Q. Rev. Agric. Econ.*, **28**(1), 38–48.

Mules, T.J. (1973). *Supply and Cost Equations for Australian Agriculture with Applications to Income Determination*. E.D.P. 3. Department of Economics, University of Adelaide.

Musgrave, W.F. (1982). Rural adjustment. In D.B. Williams (Ed.) *Agriculture in the Australian Economy*, 2nd edition. Sydney University Press, pp. 292–307.

Musgrave, W.F. and Lesueur, J.F. (1973). Large dams and drought. In J.V. Lovett (Ed.) *The Environmental, Economic and Social Significance of Drought*. Angus and Robertson, Sydney, pp. 281–296.

Nicholls, N. (1978). Sunspots and Australian wheat yield. *Search*, **9**(8–9), 319.

Nicholls, N. (1980). Long-range weather forecasting: value, status and prospects. *Rev. Geophys. Space Phys.*, **18**, 771–788.

Nicholls, N. (1981). Sunspot cycles and and Australian rainfall. *Search*, **12**(3–4), 83–85.

Nicholls, N. (1983a). Predictability of the 1982 Australian drought. *Search*, **14**(5–6), 154–155.

Nicholls, N. (1983b). The potential for long-range prediction of seasonal mean temperature in Australia. *Aust. Meteor. Mag.*, **31**(4), 203–207.

Nicholls, N. (1984). A system for predicting the onset of the North Australian wet-season. *J. Climatol.*, **4**(4), 425–435.

Nicholls, N. and Woodcock, F. (1981). Verification of an empirical long-range weather forecasting technique. *Q. J. Roy. Meteor. Soc.*, **107**, 973–976.

Nicholls, N., McBride, J.L. and Ormerod, R.J. (1982). On predicting the onset of the Australian wet season at Darwin. *Mon. Wea. Rev.*, **110**, 14–17.

Nix, H.A. (1975). The Australian climate and its effects on grain yield and quality. In A.A. Lazenby and E.A. Matheson (Eds.) *Australian Field Crops: Wheat and Other Temperate Cereals*, Vol. 1. Angus and Robertson, Sydney, 183–226.

Ockwell, A.P. and Batterham, R.L. (1980). Non-price rationing of agricultural credit by two trading banks. *Aust. J. Agric. Econ.*, **24**(3), 183–195.

Officer, R.R. and Dillon, J.L. (1965). *Calculating the Best-Bet Fodder Reserve*. Professional Farm Management Guidebook No. 1. ABRI, University of New England, Armidale.

Paltridge, G. and Woodruff, S. (1981). Changes in global surface temperature from 1880 to 1977 derived from historical records of sea surface temperatures. *Q. J. Roy. Meteor. Soc.*, **99**, 424–437.

Phillips, J. (1974). The relation between urban and rural populations: empirics and implications. *Rev. Marketing and Agric. Econ.*, **42**(2), 114–122.

Pittock, A.B. (1973). Global meridional interactions in stratosphere and troposphere. *Q. J. Roy. Meteor. Soc.*, **99**, 424–437.

Pittock, A.B. (1975). Climate change and the patterns of variation in Australian rainfall. *Search*, **6**(11–12), 498–504.

Pittock, A.B. (1983). Recent climatic change in Australia: implications for a CO_2-warmed earth. *Climatic Change*, **5**, 321–340.

Pittock, A.B. and Salinger, M.J. (1982). Toward regional scenarios for a CO_2-warmed earth. *Climatic Change*, **4**, 23–40.

Pittock, A.B., Frakes, L.A., Jenssen, D., Peterson, J.A. and Zillman, J.W. (Eds.) (1978). *Climatic Change and Variability: A Southern Perspective*. Cambridge University Press, London.

Powell, R.A. (1960). Production and income uncertainty in the world industry: an aggregate approach. *Aust. J. Agric. Econ.*, **42**(2), 114–122.

Powell, R.A. (1974). *Technological Change in Australian Agriculture 1920–21 to 1969–70*. Ph.D. thesis, University of New England, Armidale.

Powell, R.A. (1978). Stagnating economic growth: its effect on agriculture and regional implications. *Paper presented to the Fourth Advanced Studies Institute in Regional Science*, Siegen, Germany, August.

Powell, R.A. (1985). Estimates of the economic impact of the 1980–83 drought. *Paper presented to the Annual Conference of the Australian Agricultural Economics Society*, Armidale, February.

Powell, R.A. and Mandeville, T.D. (1978). Analysis of sector interactions and stability in a rural region of N.S.W. *Econ. Record*, **54**(146), 239–255.

Powell, R.A. and Saeed, I. (1984). The regional effects of drought: a case study of the Darling Downs, 1980–81. In A.D. Sorensen, R.A. Powell and M. Weinard (Eds.) *Papers of the Eighth Meeting of the Australia and New Zealand Sectors of the Regional Science Association*, Armidale, pp. 18–38.

Purtill, A. *et al.* (1983). A study of the drought. *Supplement to Q. Rev. Rural Econ.*, **5**(1), 3–11.

Quayle, E.T. (1929). Long-range rainfall forecasting from tropial (Darwin) air pressures. *Proc. Roy. Soc. Victoria*, **41**, 160–164.

Quayle, E.T. (1938). Australian rainfall in sunspot cycles. *Bulletin 22*, Commonwealth Bureau of Meteorology, Melbourne.

Quiggin, J.C. and Anderson, J.R. (1979). Stabilization and risk reduction. *Aust. J. Agric. Econ.*, **23**(3), 191–206.

Reeves, G.W. and Crellin, J.R. (1972). Pastoral properties in the Alice Springs district (N.T.) – their economic and financial position: 1965–66 to 1969–70. *Q. Rev. Agric. Econ.*, **25**(1), 27–49.

Reithmuller, P.C. and Spillman, N.D. (1978). Off-farm employment in the grazing industry. *Q. Rev. Rural Econ.*, Introductory Issues, 44–47.

Reserve Bank of Australia (1982). *Bulletin*. Various issues. Sydney.

Reutlinger, S. (1976). A simulation model for evaluating worldwide buffer stocks of wheat. *Amer. J. Agric. Econ.*, **58**(1), 1–12.

Reynolds, R.G., Watson, W.D. and Collins, D.J. (1983). Water resources aspects of drought in Australia. *Water 2000: Consultants Report No. 13.* Department of Resources and Energy, Australian Government Publishing Service, Canberra.

Rickards, P.A. and Anderson, J.R. (1967). The influence of technical productivity and availability of carry-over finance on optimum post-drought recovery. *Farm Management,* **3**(1), 1–10.

Russell, J.S. (1973). Yield trends for different crops in different areas and reflections on the sources of crop yield improvement in the Australian environment. *J. Aust. Inst. Agric. Sci.*, **39**(3), 156–166.

Russell, J.S. (1981). Geographic variation in seasonal rainfall in Australia – an analysis of the 80-year period 1895–1974. *J. Aust. Inst. Agric. Sci.*, **47**(1), 59–66.

Rutherford, J. (1950). The measurement of climatic risk in the Western Division of New South Wales. *Rev. Marketing and Agric. Econ.*, **18**(4), 246–264.

Rutherford, J. (1958). Integration of irrigation and dryland farming in the southern Murray Basin. *Rev. Marketing and Agric. Econ.*, **26**(4), 227–283.

Rutherford, R.S.G. (1948). Fluctuations in the sheep population of New South Wales, 1860–1940: a preliminary study. *Econ. Record,* **24**(2), 56–71.

Ryan, T.J. (1974). A note on Victorian wheat yield distributions. *Aust. J. Agric. Econ.*, **18**(2), 1–10.

Single, W.V. (1975). Frost injury. In A. Lazenby and E.M. Matheson (Eds.) *Australian Field Crops: Wheat and Other Temperate Cereals,* Vol. 1. Angus and Robertson, Sydney, Vol.1, pp. 364–383.

Smith, R.C.G. *et al.* (1978). A systems approach to the adaptation of sunflower to new environments. IV: Yield variability and optimum cropping strategies. *Field Crops Res.*, **1**, 229–242.

Standen, B.J. (1978). Rural credit policy and rural bank lending. *Aust. J. Agric. Econ.*, **22**(1), 41–53.

Standen, B.J. (1982). Credit and agriculture. In D.B. Williams (Ed.) *Agriculture in the Australian Economy*, 2nd edition, Sydney University Press, Sydney.

Standing Committee on Agriculture Working Group (1983). *An Evaluation of Existing Drought Policies Given the Current Drought Experience.* Annex A, mimeo, Canberra.

Stoeckel, A.B. (1978). *A General Equilibrium Study of Mining and Agriculture in the Australian Economy.* Ph.D. thesis, Duke University, Durham, NC.

Stoeckel, A.B. and Miller, G. (1982). Agriculture in the economy. In D.B. Williams (Ed.) *Agriculture in the Australian Economy.* Sydney University Press, 166–185.

Streten, N.A. (1981). Southern hemisphere sea surface temperature variability and apparent associations with Australian rainfall. *J. Geophys. Res.*, **86**, 485–497.

Streten, N.A. (1983). Extreme distributions of Australian annual rainfall in relation to sea surface temperature. *J. Climatol.*, **3**(2), 143–153.

Swerling, B.C. (1959). Income protection for farmers – a possible approach. *J. Pol. Econ.*, **67**(2), 173–186.

Thatcher, L.P. and Lloyd, A.G. (1975). Least-cost feed reserve using drought probabilities derived from a grazing model. *Rev. Marketing and Agric. Econ.*, **43**(3), 113–125.

Thompson, L.M. (1975). Weather variability, climatic change, and grain production. *Science*, **188**(4188), 535–541.

Threlfall, P. (1977). Government reconstruction and adjustment assistance measures in the Australian rural sector. *Q. Rev. Agric. Econ.*, **30**(3), 177–200.

Toft, H.I. and O'Hanlon, P.W. (1979). A dynamic programming model for on-farm decision making in drought. *Rev. Marketing and Agric. Econ.*, **47**(1), 5–16.

Troup, A.J. (1965). The Southern Oscillation. *Q. J. Roy. Meteor. Soc.*, **91**, 490–506.

Tucker, G.B. (1980). On assessing the climatic response to a continuing increase in the carbon dioxide content of the atmosphere. In G.I. Pearman (Ed.) *Carbon Dioxide and Climate: Australian Research.* Australian Academy of Science, Canberra, pp. 21–32.

Valentine, T.J. (1973). The loan supply function of Australian trading banks: an empirical analysis. *Aust. Econ. Papers*, **12**(1), 57–69.

Vincent, D.P. (1976). Economic aspects of farm poverty. *Aust. J. Agric. Econ.*, **20**(2), 103–118.

Vines, R.G. (1977). Possible relationships between rainfall, crop yields and the sunspot cycle. *J. Aust. Inst. Agric. Sci.*, **43**(1), 3–13.

Waring, E.J. (1976). Local and regional effects of drought in Australia. In T.G. Chapman (Ed.) *Drought.* Australia UNESCO Seminar, AGPS, Canberra, pp. 243–256.

Warr, P.G. (1978). The case against tariff compensation. *Aust. J. Agricul. Econ.*, **22**(2), 85–98.

Watson, W.D. and Anderson, J.R. (1977). Spatial vs. time series data for assessing response risk. *Rev. Marketing and Agric. Econ.*, **45**(3), 80–84.

Wilhite, D.A. Rosenberg, N.J. and Glantz, M.H. (1982). *Government Response to Drought in the United States: Lessons from the Mid-1970s.* Climate Dynamics Program, Final Report on NSF Grant No. ARM-8108447, pp. 69.

Williams, J.R. (1946). Some aspects of climatic variability, with special reference to rainfall in New South Wales. *Rev. Marketing and Agric. Econ.*, **14**(7), 239–250.

Wonder, B. and Howlett, R. (1977). The influence of season and property characteristics upon the viability of cattle producers in drought prone areas: a case study of the Alice Springs District. *Q. Rev. Agric. Econ.*, **30**(3), 235–253.

Wright, P.B. (1975). *An Index of the Southern Oscillation.* CRU Research Publication 4. Climatic Research Unit, University of East Anglia, Norwich, 22 pp.

Young, M.D. and Wilson, A.D. (1979). The influence of climatic variability on the pastoral zone. In Anon., *The Impact of Climate on Australian Society and Economy.* CSIRO, Melbourne, Appendix A3(2).

Index

Editors' note: Reference to authors, institutes and places is made only where these are necessary to identify a climatic phenomenon, a method of impact assessment or an analytical model.